The practice of engineering noise control demands a solid understanding of the fundamentals of acoustics, the practical application of current noise control technology and the underlying theoretical concepts. This fully revised and updated fourth edition provides a comprehensive explanation of these key areas clearly, yet without oversimplification.

Written by experts of their field, the practical focus echoes advances in the discipline, reflected in the fourth edition's new material, including:

completely updated coverage of sound transmission loss, mufflers and exhaust stack directivity
a new chapter on practical numerical acoustics
thorough explanation of the latest instruments for measurements and analysis.

Essential reading for advanced students or those already well versed in the art and science of noise control, this distinctive text can be used to solve real world problems encountered by noise and vibration consultants as well as engineers and occupational hygienists.

David A. Bies is now retired having served as a Reader and then Visiting Research Fellow at the University of Adelaide's School of Mechanical Engineering. He is an expert and widely published acoustics physicist who has also worked as a senior consultant in industry.

Colin H. Hansen is Professor and Head of the School of Mechanical Engineering at the University of Adelaide. With a wealth of experience in consultanting, research and teaching in acoustics, he has authored numerous books, journal articles and conference proceedings on the topic.

Engineering Noise Control

Theory and practice

Fourth edition

David A. Bies and Colin H. Hansen

Spon Press
an imprint of Taylor & Francis
LONDON AND NEW YORK

This book is dedicated to Susan, to Carrie, to Kristy and to Laura.

First published 1988 by E & FN Spon, an imprint of Routledge
Second edition 1996
Third edition 2003 by Spon Press
Fourth edition 2009 by Spon Press
2 Park Square, Milton Park, Abingdon, OX14 4RN

Simultaneously published in the USA and Canada by Taylor & Francis
270 Madison Avenue, New York, NY 10016, USA

Spon Press is an imprint of the Taylor & Francis Group, an informa business

© 2009 David A. Bies and Colin H. Hansen

Typeset in Times New Roman by
the authors
Printed and bound in Great Britain by
CPI Antony Rowe, Chippenham, Wiltshire

This publication presents material of a broad scope and applicability. Despite stringent efforts by all concerned in the publishing process, some typographical or editorial errors may occur, and readers are encouraged to bring these to our attention where they represent errors of substance. The publisher and author disclaim any liability, in whole or in part, arising from information contained in this publication. The reader is urged to consult with an appropriate licensed professional prior to taking any action or making any interpretation that is within the realm of a licensed professional practice.

British Library Cataloguing in Publication Data
A catalogue record for this book is available from the British Library

Library of Congress Cataloging-in-Publication Data
  Bies, David A., 1925-
  Engineering noise control: theory and pratice / David A. Bies and Colin H. Hansen.—4th ed.
    p. cm.
  Includes bibliographical references.
  1. Noise control. 2. Machinery—Noise. I. Hansen, Colin H., 1951-II. Title.
  TD892.B54 2009
                                                620.2\'3—dc22 2009002335

ISBN 13: 978-0-415-48706-1 (hbk)
ISBN 13: 978-0-415-48707-8 (pbk)
ISBN 13: 978-0-203-87240-6 (ebk)

ISBN 10: 0-415-48706-4 (hbk)
ISBN 10: 0-415-48707-2 (pbk)
ISBN 10: 0-203-87240-1 (ebk)

CONTENTS

PREFACE

Although this fourth edition follows the same basic style and format as the first, second and third editions, the content has been considerably updated and expanded, yet again. This is partly in response to significant advances in the practice of acoustics and in the associated technology during the six years since the third edition and partly in response to improvements, corrections, suggestions and queries raised by various practitioners and students. The major additions are outlined below. However, there are many other minor additions and corrections that have been made to the text but which are not specifically identified here.

The emphasis of this edition is purely on passive means of noise control and the chapter on active noise control that appeared in the second and third editions has been replaced with a chapter on practical numerical acoustics, where it is shown how free, open source software can be used to solve some difficult acoustics problems, which are too complex for theoretical analysis. The removal of Chapter 12 on active noise control is partly due to lack of space and partly because a more comprehensive and a more useful treatment is available in the book, *Understanding Active Noise Cancellation* by Colin H. Hansen.

Chapter 1 includes updated material on the speed of sound in compliant ducts and the entire section on speed of sound has been rewritten with a more unified treatment of solids, liquids and gases.

Chapter 2 has been updated to include some recent discoveries regarding the mechanism of hearing damage.

Chapter 3 has been considerably updated and expanded to include a discussion of expected measurement precision and errors using the various forms of instrumentation, as well as a discussion of more advanced instrumentation for noise source localisation using near field acoustic holography and beamforming. The discussion on spectrum analysers and recording equipment has been completely rewritten to reflect more modern instrumentation.

In Chapter 4, the section on evaluation of environmental noise has been updated and rewritten.

Additions in Chapter 5 include a better definition of incoming solar radiation for enabling the excess attenuation due to meteorological influences to be determined. Many parts of Section 5.11 on outdoor sound propagation have been rewritten in an attempt to clarify some ambiguities in the third edition. The treatment of a vibrating sphere dipole source has also been considerably expanded.

In Chapter 7, the section on speech intelligibility in auditoria has been considerably expanded and includes some guidance on the design of sound reinforcement systems. In the low frequency analysis of sound fields, cylindrical rooms are now included in addition to rectangular rooms. The section on the measurement of the room constant has been expanded and explained more clearly. In the section on auditoria, a discussion of the optimum reverberation time in classrooms has now been included.

In Chapter 8, the discussion on STC and weighted sound reduction index has been revised. The prediction scheme for estimating the transmission loss of single isotropic panels has been extended to low frequencies in the resonance and stiffness controlled ranges and the Davy method for estimating the Transmission Loss of double panel walls has been completely revised and corrected. The discussion now explains how to calculate the TL of multi-leaf and composite panels. Multi-leaf panels are described as those made up of different layers (or leaves) of the same material connected together in various ways whereas composite panels are described as those made up of two leaves of different materials bonded rigidly together. A procedure to calculate the transmission loss of very narrow slits such as found around doors with weather seals has also been added. A section on the calculation of flanking transmission has now been included with details provided for the calculation of flanking transmission via suspended ceilings. The section on calculating the Insertion Loss of barriers according to ISO9613-2 has been rewritten to more clearly reflect the intention of the standard. In addition, expressions are now provided for calculating the path lengths for sound diffracted around the ends of a barrier.

Chapter 9 has had a number of additions: Transmission Loss calculations (in addition to Insertion Loss calculations) for side branch resonators and expansion chambers; a much more detailed and accurate analysis of Helmholtz resonators, including better estimates for the effective length of the neck; an expanded discussion of higher order mode propagation, with expressions for modal cut-on frequencies of circular section ducts; a number of new models for calculating the Transmission Loss of plenum chambers; and a more detailed treatment of directivity of exhaust stacks.

In Chapter 10, the effect of the mass of the spring on the resonance frequency of isolated systems has been included in addition to the inclusion of a discussion of the surge phenomenon in coil springs. The treatment of vibration absorbers has been revised and expanded to include a discussion of vibration neutralisers, and plots of performance of various configurations are provided. The treatment of two-stage vibration isolation has been expanded and non-dimensional plots provided to allow estimation of the effect of various parameters on the isolation performance.

Chapter 11 remains unchanged and chapter 12 has been replaced with Chapter 13, where the previous content of Chapter 13 now serves as an introduction to a much expanded chapter on practical numerical acoustics written by Dr Carl Howard. This chapter covers the analysis of complex acoustics problems using boundary element analysis, finite element analysis and MatLab. Emphasis is not on the theoretical aspects of these analyses but rather on the practical application of various software packages including a free open source boundary element package.

Appendix A, which in the first edition contained example problems, has been replaced with a simple derivation of the wave equation. A comprehensive selection of example problems tailored especially for the book are now available on the internet for no charge at: http://www.causalsystems.com.

Appendix B has been updated and considerably expanded with many more materials and their properties covered. In Appendix C, the discussion of flow resistance measurement using an impedance tube has been expanded and clarified. Expressions for the acoustic impedance of porous fibreglass and rockwool materials have been extended to include polyester fibrous materials and plastic foams. The impedance expressions towards the end of Appendix C now include a discussion of multi-layered materials.

ACKNOWLEDGMENTS

We would like to thank all of those who took the time to offer constructive criticisms of the first, second and third editions, our graduate students and the many final year mechanical engineering students at the University of Adelaide who have used the first, second and third editions as texts in their engineering acoustics course.

The first author would like to sincerely thank his daughter Carrie for all the support and understanding she has freely given to enable completion of this fourth edition.

The second author would like to express his deep appreciation to his family, particularly his wife Susan and daughters Kristy and Laura for the patience and support which was freely given during the many years of nights and weekends that were needed to complete this edition.

We would also like to thank the following people for their contributions to the book: Con Doolan for proof reading and commenting on Section 5.3.2, dealing with the analysis of sound radiation from a vibrating sphere; John Davy for some insights on his double wall transmission loss theory in Chapter 8; John Davy for providing his theory on the directivity of sound radiation from exhaust ducts prior to its publication; John Davy for his comments and corrections to the section in Chapter 9 on directivity of exhaust stacks; and Athol Day for supplying his exhaust stack directivity data prior to its publication.

We are most grateful for the generous moral support and assistance from our friend, Peter McAllister, without whom it is unlikely that this fourth edition would have been completed.

Finally, the authors are deeply indebted to Carl Howard who wrote the new Chapter 12 on Numerical Acoustics. Although this meant that Chapter 12 on active noise control in the third edition had to be omitted for space reasons, we feel that the new Chapter 12 will be of more practical use to the noise control engineers and consultants who may use this book as a reference. Those interested in active noise control will find a more comprehensive treatment in the book, *Understanding Active Noise Cancellation*, by Colin H. Hansen.

CHAPTER ONE

Fundamentals and Basic Terminology

LEARNING OBJECTIVES

In this chapter the reader is introduced to:

- fundamentals and basic terminology of noise control;
- noise-control strategies for new and existing facilities;
- the most effective noise-control solutions;
- the wave equation;
- plane and spherical waves;
- sound intensity;
- units of measurement;
- the concept of sound level;
- frequency analysis and sound spectra;
- adding and subtracting sound levels;
- three kinds of impedance; and
- flow resistance.

1.1 INTRODUCTION

The recognition of noise as a source of annoyance began in antiquity, but the relationship, sometimes subtle, that may exist between noise and money seems to be a development of more recent times. For example, the manager of a large wind tunnel once told one of the authors that in the evening he liked to hear, from the back porch of his home, the steady hum of his machine 2 km away, for to him the hum meant money. However, to his neighbours it meant only annoyance and he eventually had to do without his evening pleasure.

The conflicts of interest associated with noise that arise from the staging of rock concerts and motor races, or from the operation of airports, are well known. In such cases the relationship between noise and money is not at all subtle. Clearly, as noise may be the desired end or an inconsequential by-product of the desired end for one group, and the bane of another, a need for its control exists. Each group can have what it wants only to the extent that control is possible.

The recognition of noise as a serious health hazard is a development of modern times. With modern industry has come noise-induced deafness; amplified music also takes its toll. While amplified music may give pleasure to many, the excessive noise of much modern industry probably gives pleasure to very few, or none at all. However, the relationship between noise and money still exists and cannot be ignored. If paying people, through compensation payments, to go deaf is little more expensive than implementing industrial noise control, then the incentive definitely exists to do nothing, and hope that decision is not questioned.

A common method of noise control is a barrier or enclosure and in some cases this may be the only practical solution. However, experience has shown that noise control at the design stage is generally accomplished at about one-tenth of the cost of adding a barrier or an enclosure to an existing installation. At the design stage the noise producing mechanism may be selected for least noise and again experience suggests that the quieter process often results in a better machine overall. These unexpected advantages then provide the economic incentive for implementation, and noise control becomes an incidental benefit. Unfortunately, in most industries engineers are seldom in the position of being able to make fundamental design changes to noisy equipment. They must often make do with what they are supplied, and learn to apply effective "add-on" noise-control technology. Such "add-on" measures often prove cumbersome in use and experience has shown that quite often "add-on" controls are quietly sabotaged by employees who experience little benefit and find them an impediment to their work.

In the following text, the chapters have been arranged to follow a natural progression, leading the reader from the basic fundamentals of acoustics through to advanced methods of noise control. However, each chapter has been written to stand alone, so that those with some training in noise control or acoustics can use the text as a ready reference. The emphasis is upon sufficient precision of noise-control design to provide effectiveness at minimum cost, and means of anticipating and avoiding possible noise problems in new facilities.

Simplification has been avoided so as not to obscure the basic physics of a problem and possibly mislead the reader. Where simplifications are necessary, their consequences are brought to the reader's attention. Discussion of complex problems has also not been avoided for the sake of simplicity of presentation. Where the discussion is complex, as with diffraction around buildings or with ground-plane reflection, results of calculations, which are sufficient for engineering estimates, are provided. In many cases, procedures also are provided to enable serious readers to carry out the calculations for themselves. For those who wish to avoid tedious calculations, there is a software package, ENC, available that follows this text very closely. See www.causalsystems.com.

In writing the equations that appear throughout the text, a consistent set of symbols is used: these symbols are defined following their use in each chapter. Where convenient, the equations are expressed in dimensionless form; otherwise SI units are implied unless explicitly stated otherwise.

To apply noise-control technology successfully, it is necessary to have a basic understanding of the physical principles of acoustics and how these may be applied to the reduction of excessive noise. Chapter 1 has been written with the aim of providing the basic principles of acoustics in sufficient detail to enable the reader to understand the applications in the rest of the book.

Chapter 2 is concerned with the ear, as it is the ear and the way that it responds to sound, which generally determines the need for noise control and criteria for acceptable minimum levels. The aim of Chapter 2 is to aid in understanding criteria for acceptability, which are the subject of Chapter 4. Chapter 3 is devoted to instrumentation, data collection and data reduction. In summary, Chapters 1 to 4 have been written with the aim of providing the reader with the means to quantify a noise problem.

Chapter 5 has been written with the aim of providing the reader with the basis for identifying noise sources and estimating noise levels in the surrounding environment, while Chapter 6 provides the means for rank ordering sources in terms of emitted sound power. It is to be noted that the content of Chapters 5 and 6 may be used in either a predictive mode for new proposed facilities or products or in an analytical mode for analysis of existing facilities or products to identify and rank order noise sources.

Chapter 7 concerns sound in enclosed spaces and provides means for designing acoustic treatments and for determining their effectiveness. Chapter 8 includes methods for calculating the sound transmission loss of partitions and the design of enclosures, while Chapter 9 is concerned with the design of dissipative and reactive mufflers.

Chapter 10 is about vibration isolation and control, and also gives attention to the problem of determining when vibration damping will be effective in the control of emitted noise and when it will be ineffective. The reader's attention is drawn to the fact that less vibration does not necessarily mean less noise, especially since vibration damping is generally expensive.

Chapter 11 provides means for the prediction of noise radiated by many common noise sources and is largely empirical, but is generally guided by considerations such as those of Chapter 5.

Chapter 12 is concerned with numerical acoustics and its application to the solution of complex sound radiation problems and interior noise problems.

1.2 NOISE-CONTROL STRATEGIES

Possible strategies for noise control are always more numerous for new facilities and products than for existing facilities and products. Consequently, it is always more cost effective to implement noise control at the design stage than to wait for complaints about a finished facility or product.

In existing facilities, controls may be required in response to specific complaints from within the work place or from the surrounding community, and excessive noise levels may be quantified by suitable measurements. In proposed new facilities, possible complaints must be anticipated, and expected excessive noise levels must be estimated by some procedure. Often it is not possible to eliminate unwanted noise entirely and more often to do so is very expensive; thus minimum acceptable levels of noise must be formulated, and these levels constitute the criteria for acceptability.

Criteria for acceptability are generally established with reference to appropriate regulations for the work place and community. In addition, for community noise it is advisable that at worst, any facility should not increase background (or ambient) noise levels in a community by more than 5 dB(A) over existing levels without the facility, irrespective of what local regulations may allow. Note that this 5 dB(A) increase applies to broadband noise and that clearly distinguishable tones (single frequencies) are less acceptable.

When dealing with community complaints (predicted or observed) it is wise to be conservative; that is, to aim for adequate control for the worst case, noting that community noise levels may vary greatly (±10 dB) about the mean as a result of atmospheric conditions (wind and temperature gradients and turbulence). It is worth careful note that complainants tend to be more conscious of a noise after making a complaint and thus subconsciously tend to listen for it. Thus, even after considerable noise reduction may have been achieved and regulations satisfied, complaints may continue. Clearly, it is better to avoid complaints in the first place and thus yet another argument supporting the assertion of cost effectiveness in the design stage is provided.

In both existing and proposed new facilities and products an important part of the process is to identify noise sources and to rank order them in terms of contributions to excessive noise. When the requirements for noise control have been quantified, and sources identified and ranked, it is possible to consider various options for control and finally to determine the cost effectiveness of the various options. As was mentioned earlier, the cost of enclosing a noise source is generally much greater than modifying the source or process producing the noise. Thus an argument, based upon cost effectiveness, is provided for extending the process of source identification to specific sources on a particular item of equipment and rank ordering these contributions to the limits of practicality.

Community noise level predictions and calculations of the effects of noise control are generally carried out in octave frequency bands. Current models for prediction are not sufficiently accurate to allow finer frequency resolution and less fine frequency resolution does not allow proper account of frequency-dependent effects. Generally, octave band analysis provides a satisfactory compromise between too much and too little detail. Where greater spectrum detail is required, one-third octave band analysis is often sufficient.

If complaints arise from the work place, then regulations should be satisfied, but to minimise hearing damage compensation claims, the goal of any noise-control program should be to reach a level of no more than 80 dB(A). Criteria for other situations in the work place are discussed in Chapter 4. Measurements and calculations are generally carried out in standardised octave or one-third octave bands, but particular care must be given to the identification of any tones that may be present, as these must be treated separately.

More details on noise control measures can be found in the remainder of this text and also in ISO 11690/2 (1996).

Any noise problem may be described in terms of a sound source, a transmission path and a receiver, and noise control may take the form of altering any one or all of these elements. When considered in terms of cost effectiveness and acceptability, experience puts modification of the source well ahead of either modification of the transmission path or the receiver. On the other hand, in existing facilities the last two may be the only feasible options.

1.2.1 Sound Source Modification

Modification of the energy source to reduce the noise generated often provides the best means of noise control. For example, where impacts are involved, as in punch presses, any reduction of the peak impact force (even at the expense of the force acting over a longer time period) will dramatically reduce the noise generated. Generally, when a choice between various mechanical processes is possible to accomplish a given task, the best choice, from the point of view of minimum noise, will be the process that minimises the time rate of change of force or jerk (time rate of change of acceleration). Alternatively, when the process is aerodynamic a similar principle applies; that is, the process that minimises pressure gradients will produce minimum noise. In general, whether a process is mechanical or fluid mechanical, minimum rate of change of force is associated with minimum noise.

Mechanical shock between solids should be minimised; for example, impact noise may be generated by parts falling into metal bins and the height that the parts fall could be reduced by using an adjustable height collector (see Figure 1.1a) or the collector could be lined with conveyor belt material. Alternatively the collector could have rubber flaps installed to break the fall of the parts (see Figure 1.1b).

The control of noise at its source may require maintenance, substitution of materials, substitution of equipment or parts of equipment, specification of quiet equipment, substitution of processes, substitution of mechanical power generation and

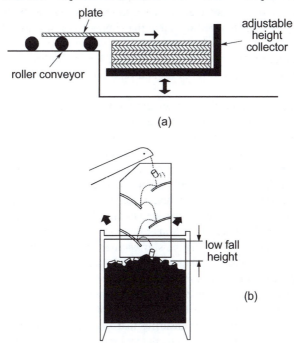

Figure 1.1 Impact noise reduction: (a) variable height collector;
(b) interrupted fall.

transmission equipment, change of work methods, reduction of vibration of large structures such as plates, beams, etc. or reduction of noise resulting from fluid flow.

Maintenance includes balancing moving parts, replacement or adjustment of worn or loose parts, modifying parts to prevent rattles and ringing, lubrication of moving parts and use of properly shaped and sharpened cutting tools.

Substitution of materials includes replacing metal with plastic, a good example being the replacement of steel sprockets in chain drives with sprockets made from flexible polyamide plastics.

Substitution of equipment includes use of electric tools rather than pneumatic tools (e.g. hand tools), use of stepped dies rather than single-operation dies, use of rotating shears rather than square shears, use of hydraulic rather than mechanical presses, use of presses rather than hammers and use of belt conveyors rather than roller conveyors.

Substitution of parts of equipment includes modification of gear teeth, by replacing spur gears with helical gears – generally resulting in 10 dB of noise reduction, replacement of straight edged cutters with spiral cutters (for example, in wood working machines a 10 dB(A) reduction may be achieved), replacement of gear drives with belt drives, replacement of metal gears with plastic gears (beware of additional maintenance problems) and replacement of steel or solid wheels with pneumatic tyres.

Substitution of processes includes using mechanical ejectors rather than pneumatic ejectors, hot rather than cold working, pressing rather than rolling or forging, welding or squeeze rivetting rather than impact rivetting, use of cutting fluid in machining processes, changing from impact action (e.g. hammering a metal bar) to progressive pressure action (e.g. bending a metal bar with pliers), replacement of circular saw blades with damped blades and replacement of mechanical limit stops with micro-switches.

Substitution of mechanical power generation and transmission equipment includes use of electric motors rather than internal combustion engines or gas turbines, or the use of belts or hydraulic power transmissions rather than gear boxes.

Change of work methods includes replacing ball machines with selective demolition in building demolition, replacing pneumatic tools by changing manufacturing methods, such as moulding holes in concrete rather than cutting after production of the concrete component, use of remote control of noisy equipment such as pneumatic tools, separating noisy workers in time, but keeping noisy operations in the same area, separating noisy operations from non-noisy processes. Changing work methods may also involve selecting the slowest machine speed appropriate for a job (selecting large, slow machines rather than smaller, faster ones), minimising the width of tools in contact with the workpiece (2 dB(A) reduction for each halving of tool width) and minimising protruding parts of cutting tools.

Reductions of noise resulting from the resonant vibration of structures (plates, beams, etc.) may be achieved by ensuring that machine rotational speeds do not coincide with resonance frequencies of the supporting structure, and if they do, in some cases it is possible to change the stiffness or mass of the supporting structure to change its resonance frequencies (increasing stiffness increases resonance frequencies and increasing the mass reduces resonance frequencies). In large structures, such as a roof or ceiling, attempts to change low order resonance frequencies by adding mass or stiffness may not be practical.

Another means for reducing sound radiation due to structural vibration involves reducing the acoustic radiation efficiency of the vibrating surface. Examples are the replacement of a solid panel or machine guard with a woven mesh or perforated panel or the use of narrower belt drives. Damping a panel can be effective (see Section 10.7) if it is excited mechanically, but note that if the panel is excited by an acoustic field, damping will have little or no effect upon its sound radiation. Blocking the transmission of vibration along a noise radiating structure by the placement of a heavy mass on the structure close to the original source of the noise can also be effective.

Reduction of noise resulting from fluid flow may involve providing machines with adequate cooling fins so that noisy fans are no longer needed, using centrifugal rather than propeller fans, locating fans in smooth, undisturbed air flow, using fan blades designed using computational fluid dynamics software to minimise turbulence, using large low speed fans rather than smaller faster ones, minimising the velocity of fluid flow and maximising the cross-section of fluid streams. Fluid flow noise reduction may also involve reducing the pressure drop across any one component in a fluid flow system, minimising fluid turbulence where possible (e.g. avoiding obstructions in the flow), choosing quiet pumps in hydraulic systems, choosing quiet nozzles for compressed air systems (see Figure 11.4), isolating pipes carrying the fluid from support structures, using flexible connectors in pipe systems to control energy travelling in the fluid as well as the pipe wall and using flexible fabric sections in low pressure air ducts (near the noise source such as a fan).

Another form of source control is to provide machines with adequate cooling fins so that noisy fans are no longer needed. In hydraulic systems the choice of pumps, and in compressed air systems the choice of nozzles, is important.

Other alternatives include minimising the number of noisy machines running at any one time, relocating noisy equipment to less sensitive areas or if community noise is a problem, avoiding running noisy machines at night.

1.2.2 Control of the Transmission Path

In considering control of the noise path from the source to the receiver some or all of the following treatments need to be considered: barriers (walls), partial enclosures or full equipment enclosures, local enclosures for noisy components on a machine, reactive or dissipative mufflers (the former for low frequency noise or small exhausts, the latter for high frequencies or large diameter exhaust outlets), lined ducts or lined plenum chambers for air-handling systems, vibration isolation of machines from noise-radiating structures, vibration absorbers and dampers, active noise control and the addition of sound-absorbing material to reverberant spaces to reduce reflected noise fields.

1.2.3 Modification of the Receiver

In some cases, when all else fails, it may be necessary to apply noise control to the receiver of the excessive noise. This type of control may involve use of ear-muffs, ear-plugs or other forms of hearing protection; the enclosure of personnel if this is practical; moving personnel further from the noise sources; rotating personnel to reduce noise exposure time; and education and emphasis on public relations for both in-plant and community noise problems.

Clearly, in the context of treatment of the noise receiver, the latter action is all that would be effective for a community noise problem, although sometimes it may be less expensive to purchase complainants' houses, even at prices well above market value.

1.2.4 Existing Facilities

In existing facilities or products, quantification of the noise problem requires identification of the noise source or sources, determination of the transmission paths from the sources to the receivers, rank ordering of the various contributors to the problem and finally, determination of acceptable solutions.

To begin, noise levels must be determined at potentially sensitive locations or at locations from which the complaints arise. For community noise, these measurements may not be straightforward, as such noise may be strongly affected by variable weather conditions and measurements over a representative time period may be required. This is usually done using remote data logging equipment in addition to periodic manual measurements.

The next step is to apply acceptable noise level criteria to each location and thus determine the required noise reductions, generally as a function of octave or one-third octave frequency bands (see Section 1.10). Noise level criteria are usually set by regulations and appropriate standards.

Next, the transmission paths by which the noise reaches the place of complaint are determined. For some cases this step is often obvious. However, cases may occasionally arise when this step may present some difficulty, but it may be very important in helping to identify the source of a complaint.

Having identified the possible transmission paths, the next step is to identify (understand) the noise generation mechanism or mechanisms, as noise control at the source always gives the best solution. Where the problem is one of occupational noise, this task is often straightforward. However, where the problem originates from complaints about a product or from the surrounding community, this task may prove difficult. Often noise sources are either vibrating surfaces or unsteady fluid flow (air, gas or steam). The latter aerodynamic sources are often associated with exhausts. In most cases, it is worthwhile determining the source of the energy that is causing the structure or the aerodynamic source to radiate sound, as control may best start there. For a product, considerable ingenuity may be required to determine the nature and solution to the problem. In existing facilities and products, altering the noise generating mechanism may range from too expensive to acceptable and should always be considered as a means for possible control.

For airborne noise sources, it is important to determine the sound power and directivity of each to determine their relative contributions to the noise problem. The radiated sound power and directivity of sources can be determined by reference to the equipment manufacturer's data, reference to Chapter 11, or by measurement, using methods outlined in Chapters 5 and 6. The sound power should be characterised in octave or one-third octave frequency bands (see Section 1.10) and dominant single frequencies should be identified. Any background noise contaminating the sound power measurements must be taken into account (see Section 1.11.5).

Having identified the noise sources and determined their radiated sound power levels, the next task is to determine the relative contribution of each noise source to the noise level at each location where the measured noise levels are considered to be excessive. For a facility involving just a few noise sources this is a relatively straightforward task. However, for a facility involving tens or hundreds of noise sources, the task of rank ordering can be intimidating, especially when the locations of complaint are in the surrounding community. In the latter case, the effect of the ground terrain and surface, air absorption and the influence of atmospheric conditions must also be taken into account, as well as the decrease in sound level with distance due to the "spreading out" of the sound waves.

Commercial computer software is available to assist with the calculation of the contributions of noise sources to sound levels at sensitive locations in the community or in the work place. Alternatively, one may write one's own software (see Chapter 5). In either case, for an existing facility, measured noise levels can be compared with predicted levels to validate the calculations. Once the computer model is validated, it is then a simple matter to investigate various options for control and their cost effectiveness.

In summary, a noise control program for an existing facility includes:

- undertaking an assessment of the current environment where there appears to be a problem, including the preparation of noise level contours where required;
- establishment of the noise control objectives or criteria to be met;
- identification of noise transmission paths and generation mechanisms;
- rank ordering noise sources contributing to any excessive levels;
- formulating a noise control program and implementation schedule;
- carrying out the program; and
- verifying the achievement of the objectives of the program.

More detail on noise control strategies for existing facilities can be found in ISO 11690/1 (1996).

1.2.5 Facilities in the Design Stage

In new facilities and products, quantification of the noise problem at the design stage may range from simple to difficult. At the design stage the problems are the same as for existing facilities and products; they are identification of the source or sources, determination of the transmission paths of the noise from the sources to the receivers, rank ordering of the various contributors to the problem and finally determination of acceptable solutions. Most importantly, at the design stage the options for noise control are generally many and may include rejection of the proposed design. Consideration of the possible need for noise control in the design stage has the very great advantage that an opportunity is provided to choose a process or processes that may avoid or greatly reduce the need for noise control. Experience suggests that processes chosen because they make less noise, often have the additional advantage of being more efficient.

The first step for new facilities is to determine the noise criteria (see Chapter 4) for sensitive locations, which may typically include areas of the surrounding residential community that will be closest to the planned facility, locations along the boundary of the land owned by the industrial company responsible for the new facility, and within the facility at locations of operators of noisy machinery. Again, care must be taken to be conservative where surrounding communities are concerned so that initial complaints are avoided.

In consideration of possible community noise problems following establishment of acceptable noise criteria at sensitive locations, the next step may be to develop a computer model or to use an existing commercial software package to estimate expected noise levels (in octave frequency bands) at the sensitive locations, based on machinery sound power level and directivity information (the latter may not always be available), and outdoor sound propagation prediction procedures. Previous experience or the local weather bureau can provide expected ranges in atmospheric weather conditions (wind and temperature gradients and turbulence levels) so that a likely range and worst case sound levels can be predicted for each community location. When directivity information is not available, it is generally assumed that the source radiates uniformly in all directions.

If the estimated noise levels at any sensitive location exceed the established criteria, then the equipment contributing most to the excess levels should be targeted for noise control, which could take the form of:

- specifying lower equipment sound power levels, or sound pressure levels at the operator position, to the equipment manufacturer;
- including noise-control fixtures (mufflers, barriers, enclosures, or factory walls with a higher sound transmission loss) in the factory design; or
- rearrangement and careful planning of buildings and equipment within them.

Sufficient noise control should be specified to leave no doubt that the noise criteria will be met at *every* sensitive location. Saving money at this stage is not cost effective. If predicting equipment sound power levels with sufficient accuracy proves difficult, it may be helpful to make measurements on a similar existing facility or product.

More detail on noise control strategies and noise prediction for facilities at the design stage can be found in ISO 11690/3 (1997).

1.2.6 Airborne versus Structure-borne Noise

Very often in existing facilities it is relatively straightforward to track down the original source(s) of the noise, but it can sometimes be difficult to determine how the noise propagates from its source to a receiver. A classic example of this type of problem is associated with noise on board ships. When excessive noise (usually associated with the ship's engines) is experienced in a cabin close to the engine room (or in some cases far from the engine room), or on the deck above the engine room, it is necessary to determine how the noise propagates from the engine. If the problem arises from airborne noise passing through the deck or bulkheads, then a solution may include one or more of the following: enclosing the engine, adding sound-absorbing material to the engine room, increasing the sound transmission loss of the deck or bulkhead by using double wall constructions or replacing the engine exhaust muffler.

On the other hand, if the noise problem is caused by the engine exciting the hull into vibration through its mounts or through other rigid connections between the engine and the hull (for example, bolting the muffler to the engine and hull), then an entirely different approach would be required. In this latter case it would be the mechanically excited deck, hull and bulkhead vibrations which would be responsible for the unwanted noise. The solution would be to vibration isolate the engine (perhaps through a well-constructed floating platform) or any items such as mufflers from the surrounding structure. In some cases, standard engine vibration isolation mounts designed especially for a marine environment can be used.

As both types of control are expensive, it is important to determine conclusively and in advance the sound transmission path. The simplest way to do this is to measure the noise levels in octave frequency bands at a number of locations in the engine room with the engine running, and also at locations in the ship where the noise is excessive. Then the engine should be shut down and a loudspeaker placed in the engine room and driven so that it produces noise levels in the engine room sufficiently high for them to be readily detected at the locations where noise reduction is required.

Usually an octave band filter is used with the speaker so that only noise in the octave band of interest at any one time is generated. This aids both in generating sufficient level and in detection. The noise level data measured throughout the ship with just the loudspeaker operating should be increased by the difference between the engine room levels with the speaker as source and with the engine as source, to give corrected levels for comparison with levels measured with the engine running. In many cases, it will be necessary for the loudspeaker system to produce noise of similar level to that produced by the engine to ensure that measurements made elsewhere on the ship are above the background noise. In some cases, this may be difficult to achieve in practice with loudspeakers. The most suitable noise input to the speaker is a recording of the engine noise, but in some cases a white noise generator may be acceptable. If the corrected noise levels in the spaces of concern with the speaker excited are substantially less than those with the engine running, then it is clear that engine isolation is the first noise control that should be implemented. In this case, the best control that could be expected from engine isolation would be the difference in corrected noise level with the speaker excited and noise level with the engine running.

If the corrected noise levels in the spaces of concern with the speaker excited are similar to those measured with the engine running, then acoustic noise transmission is the likely path, although structure-borne noise may also be important, but at a slightly lower level. In this case, treatment to minimise airborne noise should be undertaken and after treatment, the speaker test should be repeated to determine if the treatment has been effective and to determine if structure-borne noise has subsequently become the problem.

Another example of the importance of determining the noise transmission path is demonstrated in the solution to an intense tonal noise problem in the cockpit of a fighter aircraft, which was thought to be due to a pump, as the frequency of the tone corresponded to a multiple of the pump rotational speed. Much fruitless effort was expended to determine the sound transmission path until it was shown that the source was the broadband aerodynamic noise at the air-conditioning outlet into the cockpit and the reason for the tonal quality was because the cockpit responded modally (see Chapter 7). The frequency of strong cockpit resonance coincided with a multiple of the rotational speed of the pump but was unrelated. In this case the obvious lack of any reasonable transmission path led to an alternative hypothesis and a solution.

1.3 ACOUSTIC FIELD VARIABLES

1.3.1 Variables

Sound is the sensation produced at the ear by very small pressure fluctuations in the air. The fluctuations in the surrounding air constitute a sound field. These pressure fluctuations are usually caused by a solid vibrating surface, but may be generated in other ways; for example, by the turbulent mixing of air masses in a jet exhaust. Saw teeth in high-speed motion ($60 \ \text{ms}^{-1}$) produce a very loud broadband noise of aerodynamic origin, which has nothing to do with vibration of the blade. As the disturbance that produces the sensation of sound may propagate from the source to the ear through any elastic medium, the concept of a sound field will be extended to include structure-borne as well as airborne vibrations. A sound field is described as a perturbation of steady-state variables, which describe a medium through which sound is transmitted.

For a fluid, expressions for the pressure, particle velocity, temperature and density may be written in terms of the steady-state (mean values) and the variable (perturbation) values as follows, where the variables printed in bold type are vector quantities:

Pressure:	$P_{tot} = P + p(\mathbf{r}, t) \ (\text{Pa})$
Velocity:	$\mathbf{U}_{tot} = \mathbf{U} + \mathbf{u}(\mathbf{r}, t) \ (\text{m/s})$
Temperature:	$T_{tot} = T + r(\mathbf{r}, t) \ (^{\circ}\text{C})$
Density:	$\rho_{tot} = \rho + \sigma(\mathbf{r}, t) \ (\text{kg/m}^3)$

Pressure, temperature and density are familiar scalar quantities that do not require discussion. However, explanation is required for the particle velocity $\mathbf{u}(\mathbf{r}, t)$ and the vector equation involving it, identified above by the word "velocity". The notion of particle velocity is based upon the assumption of a continuous rather than a molecular medium. "Particle" refers to a small bit of the assumed continuous medium and not to the molecules of the medium. Thus, even though the actual motion associated with the passage of an acoustic disturbance through the conducting medium, such as air at high frequencies, may be of the order of the molecular motion, the particle velocity describes a macroscopic average motion superimposed upon the inherent Brownian motion of the medium. In the case of a convected medium moving with a mean velocity U, which itself may be a function of the position vector r and time t, the particle velocity $u(r, \underline{t})$ associated with the passage of an acoustic disturbance may be thought of as adding to the mean velocity to give the total velocity. Combustion instability provides a notorious example.

Any variable could be chosen for the description of a sound field, but it is easiest to measure pressure in a fluid and strain, or more generally acceleration, in a solid. Consequently, these are the variables usually considered. These choices have the additional advantage of providing a scalar description of the sound field from which all other variables may be derived. For example, the particle velocity is important for the determination of sound intensity, but it is a vector quantity and requires three measurements as opposed to one for pressure. owever, instrumentation (Microflown) is available that allows the instantaneous measurement of particle velocity along all three cartesian coordinate axes at the same time. In solids, it is generally easiest to measure acceleration, especially in thin panels, although strain might be preferred as the measured variable in some special cases. If non-contact measurement is necessary, then instrumentation known as laser vibrometers are available that can measure vibration velocity along all three cartesian coordinate axes at the same time and also allow scanning of the surface being measured so a complete picture of the surface vibration response can be obtained for any frequency of interest. However, these instruments are quite expensive.

1.3.2 The Acoustic Field

In the previous section, the concept of sound field was introduced and extended to include structure-borne as well as airborne disturbances, with the implicit assumption that a disturbance initiated at a source will propagate with finite speed to a receiver. It is of interest to consider the nature of the disturbance and the speed with which it propagates. To begin, it should be understood that the small perturbations of the acoustic field may always be described as the sum of cyclic disturbances of appropriate frequencies, amplitudes and relative phases. In a fluid, a sound field will be manifested by variations in local pressure of generally very small amplitude with associated variations in density, displacement, particle velocity and temperature. Thus in a fluid, a small compression, perhaps followed by a compensating rarefaction, may propagate away from a source as a sound wave. The associated particle velocity lies parallel to the direction of propagation of the disturbance, the local particle displacement being first in the direction of propagation, then reversing to return the particle to its initial position after passage of the disturbance. A compressional or longitudinal wave has been described.

The viscosity of the fluids of interest in this text is sufficiently small for shear forces to play a very small part in the propagation of acoustic disturbances. A solid surface, vibrating in its plane without any normal component of motion, will produce shear waves in which the local particle displacement is parallel to the exciting surface, but normal to the direction of propagation of the disturbance. However, such motion is always confined to a very narrow region near to the vibrating surface and does not result in energy transport away from the near field region. Alternatively, a compressional wave propagating parallel to a solid surface will give rise to a similar type of disturbance at the fixed boundary, but again the shear wave will be confined to a very thin viscous boundary layer. Temperature variations associated with the passage of an acoustic disturbance through a gas next to a solid boundary, which is characterised by a very much greater thermal capacity, will likewise give rise to a thermal wave propagating into the boundary; but again, as with the shear wave, the thermal wave will be confined to a very thin thermal boundary layer of the same order of size as the viscous boundary layer. Such viscous and thermal effects, generally referred to as the acoustic boundary layer, are usually negligible for energy transport, and are generally neglected, except in the analysis of sound propagation in tubes and porous media, where they provide the energy dissipation mechanisms.

It has been shown that sound propagates in liquids and gases predominantly as longitudinal compressional waves; shear and thermal waves play no significant part. In solids, however, the situation is much more complicated, as shear stresses are readily supported. Not only are longitudinal waves possible, but so are transverse shear and torsional waves. In addition, the types of waves that propagate in solids strongly depend upon boundary conditions. In thin plates for example, bending waves, which are really a mixture of longitudinal and shear waves, predominate, with important consequences for acoustics and noise control. Bending waves are of importance in the consideration of sound radiation from extended surfaces, and the transmission of sound from one space to another through an intervening partition.

1.3.3 Magnitudes

The minimum acoustic pressure audible to the young human ear judged to be in good health, and unsullied by too much exposure to excessively loud music, is approximately 20×10^{-6} Pa, or 2×10^{-10} atmospheres (since one atmosphere equals 101.3×10^3 Pa). The minimum audible level occurs between 3000 and 4000 Hz and is a physical limit; lower sound pressure levels would be swamped by thermal noise due to molecular motion in air.

For the normal human ear, pain is experienced at sound pressures of the order of 60 Pa or 6×10^{-4} atmospheres. Evidently, acoustic pressures ordinarily are quite small fluctuations about the mean.

1.3.4 The Speed of Sound

Sound is conducted to the ear through the surrounding medium, which in general will be air and sometimes water but sound may be conducted by any fluid or solid. In fluids, which readily support compression, sound is transmitted as longitudinal waves and the associated particle motion in the transmitting medium is parallel to the direction of wave propagation. However, as fluids support shear very weakly, waves dependent upon shear are weakly transmitted and often may be neglected. Consequently, longitudinal waves are often called sound waves. For example, the speed of sound waves travelling in plasma has provided information about the interior of the sun. In solids, which can support both compression and shear, energy may be transmitted by all types of waves, but only longitudinal wave propagation is referred to as "sound".

The concept of an "unbounded medium" will be introduced as a convenient and often used idealisation. In practice, the term, unbounded medium, has the meaning that longitudinal wave propagation may be considered sufficiently remote from the influence of any boundaries that such influence may be neglected. The concept of unbounded medium generally is referred as "free field" and this alternative expression will also be used where appropriate in this text.

The propagation speed of sound waves, called the phase speed in any conducting medium (solid or fluid), is dependent upon the stiffness, D, and the density, ρ, of the medium. The stiffness, D, however may be complicated by the boundary conditions of the medium and in some cases it may also be frequency dependent. These matters will be discussed in the following text. In this format the phase speed, c, takes the following simple form:

$$c = \sqrt{D/\rho} \quad (\text{ms}^{-1}) \tag{1.1}$$

The effect of boundaries on the longitudinal wave speed will now be considered but with an important omission for the purpose of simplification. The discussion will not include boundaries between solids, which generally is a seismic wave propagation problem not ordinarily encountered in noise control problems. Only propagation at boundaries between solids and fluids and between fluids will be considered, as they affect longitudinal wave propagation. At boundaries between solids and gases the characteristic impedance mismatch (see Section 1.12) is generally so great that the effect of the gas on wave propagation in the solid may be neglected; in such cases the gas may be considered to be a simple vacuum in terms of its effect on wave propagation in the solid.

In solids, the effect of boundaries is to relieve stresses in the medium at the unsupported boundary faces as a result of expansion at the boundaries normal to the direction of compression. Well removed from boundaries, such expansion is not possible and in a solid medium, the free field is very stiff. On the other hand, for the case of boundaries being very close together, wave propagation may not take place at all and in this case the field within such space, known as evanescent, commonly is assumed to be uniform. It may be noted that the latter conclusion follows from an argument generally applied to an acoustic field in a fluid within rigid walls. Here the latter argument has been applied to an acoustic field within a rigid medium with unconstrained walls.

For longitudinal wave propagation in solids, the stiffness, D, depends upon the ratio of the dimensions of the solid to the wavelength of a propagating longitudinal wave. Let the solid be characterised by three orthogonal dimensions h_i, $i=1,2,3$, which determine its overall size. Let h be the greatest of the three dimensions of the solid, where E is Young's modulus and f is the frequency of a longitudinal wave propagating in the solid. Then the criterion proposed for determining D is that the ratio of the dimension, h, to the half wavelength of the propagating longitudinal wave in the solid is greater than or equal to one. For example, wave propagation may take place along dimension h when the half wavelength of the propagating wave is less than or just equal to the dimension, h. This observation suggests that the following inequality must be satisfied for wave propagation to take place.

$$2hf \geq \sqrt{D/\rho} \qquad (1.2)$$

For the case that only one dimension, h, satisfies the inequality and two dimensions do not then the solid must be treated as a wire or thin rod along dimension, h, on which waves may travel. In this case the stiffness constant, D, is that of a rod, D_r, and takes the following form:

$$D_r = E \qquad (1.3)$$

The latter result constitutes the definition of Young's modulus of elasticity, E.

In the case that two dimensions satisfy the inequality and one dimension does not the solid must be treated as a plate over which waves may travel. In this case, where v is Poisson's ratio (v is approximately 0.3 for steel), the stiffness, $D = D_p$, takes the following form:

$$D_p = E/(1 - v^2) \qquad (1.4)$$

For a material for which Poisson's ratio is equal to 0.3, $D = 1.099E$.

If all three dimensions, h_i, satisfy the criterion then wave travel may take place in all directions in the solid. In this case, the stiffness constant, $D = D_s$, takes the following form:

$$D_s = \frac{E(1 - v)}{(1 + v)(1 - 2v)} \qquad (1.5)$$

For fluids, the stiffness, D_F, is the bulk modulus or the reciprocal of the more familiar compressibility. That is:

$$D_F = -V(\partial V/\partial P)^{-1} = \rho(\partial P/\partial \rho) \qquad \text{(Pa)} \qquad (1.6a,b)$$

In Equation (1.6), V is a unit volume and $\partial V/\partial P$ is the incremental change in volume associated with an incremental change in static pressure P.

The effect of boundaries on the longitudinal wave speed in fluids will now be considered. For fluids (gases and liquids) in pipes at frequencies below the first higher order mode cut-on frequency (see Section 9.8.3.2), where only plane waves propagate, the close proximity of the wall of the pipe to the fluid within may have a very strong effect in decreasing the medium stiffness. The stiffness of a fluid in a pipe, tube or more generally, a conduit, will be written as D_C. The difference between D_F and D_C represents the effect of the pipe wall on the stiffness of the contained fluid. This effect will depend upon the ratio of the mean pipe radius, R, to wall thickness, t, the ratio of the density, ρ_w of the pipe wall to the density ρ of the fluid within it, Poisson's ratio, v, for the pipe wall material, as well as the ratio of the fluid stiffness, D_F, to the Young's modulus, E, of the pipe wall. The expression for the stiffness, D_C, of a fluid in a conduit follows (Pavic, 2006):

$$D_C = \frac{D_F}{1 + \dfrac{D_F}{E}\left(\dfrac{2R}{t} + \dfrac{\rho_w}{\rho}v^2\right)} \tag{1.7}$$

The compliance of a pipe wall will tend to increase the effective compressibility of the fluid and thus decrease the speed of longitudinal wave propagation in pipes. Generally, the effect will be small for gases, but for water in plastic pipes, the effect may be large. In liquids, the effect may range from negligible in heavy-walled, small-diameter pipes to large in large-diameter conduits.

For fluids (gases and liquids), thermal conductivity and viscosity are two other mechanisms, besides chemical processes, by which fluids may interact with boundaries.

Generally, thermal conductivity and viscosity in fluids are very small, and such acoustical effects as may arise from them are only of importance very close to boundaries and in consideration of damping mechanisms. Where a boundary is at the interface between fluids or between fluids and a solid, the effects may be large, but as such effects are often confined to a very thin layer at the boundary, they are commonly neglected.

Variations in pressure are associated with variations in temperature as well as density; thus, heat conduction during the passage of an acoustic wave is important. In gases, for acoustic waves at frequencies well into the ultrasonic frequency range, the associated gradients are so small that pressure fluctuations may be considered to be essentially adiabatic; that is, no sensible heat transfer takes place between adjacent gas particles and, to a very good approximation, the process is reversible. However, at very high frequencies, and in porous media at low frequencies, the compression process tends to be isothermal. In the latter cases heat transfer tends to be complete and in phase with the compression.

For gases, use of Equation (1.1), the equation for adiabatic compression (which gives $D = \gamma P$) and the equation of state for gases, gives the following for the speed of sound:

$$c = \sqrt{\gamma P/\rho} = \sqrt{\gamma RT/M} \quad \text{(m/s)} \tag{1.8a,b}$$

where γ is the ratio of specific heats (1.40 for air), T is the temperature in degrees Kelvin (°K), R is the universal gas constant which has the value 8.314 Jmol^{-1} °K^{-1}, and M is the molecular weight, which for air is 0.029 kg mol^{-1}. Equations (1.1) and (1.8) are derived in many standard texts: for example Morse (1948); Pierce (1981); Kinsler *et al.* (1982).

For gases, the speed of sound depends upon the temperature of the gas through which the acoustic wave propagates. For sound propagating in air at audio frequencies, the process is adiabatic. In this case, for temperature, T, in degrees Celsius (and not greatly different from 20°C), the speed of sound may be calculated to three significant figures using the following approximation.

$$c = 331 + 0.6T \quad (\text{m/s}) \tag{1.9}$$

For calculations in this text, unless otherwise stated, a temperature of 20°C for air will be assumed, resulting in a speed of sound of 343 m/s and an air density of 1.206 kg/m^3 at sea level, thus giving $\rho c = 414$. Some representative speeds of sound are given in Appendix B.

1.3.5 Dispersion

The speed of sound wave propagation as given by Equation (1.1) is quite general and permits the possibility that the stiffness, D, may either be constant or a function of frequency. For the cases considered thus far, it has been assumed that the stiffness, D, is independent of the frequency of the sound wave, with the consequence that all associated wave components of whatever frequency will travel at the same speed and thus the wave will propagates without dispersion, meaning wave travel takes place without changing wave shape.

On the other hand, there are many cases where the stiffness, D, is a function of frequency and in such cases the associated wave speed will also be a function of frequency. A familiar example is that of an ocean wave, the speed of which is dependent upon the ocean depth. As a wave advances into shallow water its higher frequency components travel faster than the lower frequency components, as the speed of each component is proportional to the depth of water relative to its wavelength. The greater the depth of water relative to the component wavelength, the greater the component speed. In deep water, the relative difference in the ratio of water depth to wavelength between low and high frequency components is small. However, as the water becomes shallow near the shore, this difference becomes larger and eventually causes the wave to break. A dramatic example is that of an ocean swell produced by an earthquake deep beneath the ocean far out to sea that becomes the excitement of a tsunami on the beach.

In Chapter 8, bending waves that occur in panels, which are a combination of longitudinal and shear waves are introduced as they play an important role in sound transmission through and from panels. Bending wave speed is dependent upon the frequency of the disturbance and thus is dispersive. Particle motion associated with bending waves is normal to the direction of propagation, whereas for longitudinal waves, it is in the same direction.

In liquids and gases, dispersive propagation is observed above the audio frequency range at ultrasonic frequencies where relaxation effects are encountered. Such effects make longitudinal wave propagation frequency dependent and consequently dispersive. Dispersive sound effects have been used to investigate the chemical kinetics of dissociation in liquids. Although not strictly dispersive, the speed of propagation of longitudinal waves associated with higher order modes in ducts is an example of a case where the effective wave speed along the duct axis is frequency dependent. However, this is because the number of reflections of the wave from the duct wall per unit axial length is frequency dependent, rather than the speed of propagation of the wave through the medium in the duct.

When an acoustic disturbance is produced, some work must be done on the conducting medium to produce the disturbance. Furthermore, as the disturbance propagates, as must be apparent, energy stored in the field is convected with the advancing disturbance. When the wave propagation is non-dispersive, the energy of the disturbance propagates with the speed of sound; that is, with the phase speed of the longitudinal compressional waves. On the other hand, when propagation is dispersive, the frequency components of the disturbance all propagate with different phase speeds; the energy of the disturbance, however, propagates with the group speed. Thus in the case of dispersive propagation, one might imagine a disturbance which changes its shape as it advances, while at the same time maintaining a group identity, and travelling at a group speed different from that of any of its frequency components. The group speed is defined later in Equation (1.33).

1.3.6 Acoustic Potential Function

The hydrodynamic equations, from which the equations governing acoustic phenomena derive, generally are complex and well beyond solution in closed form. Fortunately, acoustic phenomena generally are associated with very small perturbations. Thus in such cases it is possible to greatly simplify the governing equations to obtain the relatively simple linear equations of acoustics. Phenomena, that may be described by relatively simple linear equations, are referred to as linear acoustics and the equations are referred to as linearised.

However, situations may arise in which the simplifications of linear acoustics are inappropriate; the associated phenomena are then referred to as nonlinear. For example, a sound wave incident upon a perforated plate may incur large energy dissipation due to nonlinear effects under special circumstances. Convection of sound through or across a small hole, due either to a superimposed steady flow or to relatively large amplitudes associated with the sound field, may convert the cyclic flow of the sound field into local fluid streaming. Such nonlinear effects take energy from the sound field thus reducing the sound to produce local streaming of the fluid medium, which produces no sound. Similar nonlinear effects also may be associated with acoustic energy dissipation at high sound pressure levels, in excess of 130 dB re 20 μPa.

In general, except for special cases such as those mentioned, which may be dealt with separately, the losses associated with an acoustic field are quite small and consequently the acoustic field may be treated as conservative, meaning that energy dissipation is insignificant and may be neglected. Under such circumstances, it is possible to define a potential function, φ, which, as will be shown in the next section, is a solution to the wave equation (Pierce, 1981) with two very important advantages. The potential function may be either real or complex and most importantly it provides a means for determining both the acoustic pressure and the particle velocity by simple differentiation.

The acoustic potential function, φ, is defined so that its negative gradient provides the particle velocity, u, as follows:

$$u = -\nabla\varphi \tag{1.10}$$

Alternatively, differentiation of the acoustic potential function with respect to time, t, provides the acoustic pressure, which for negligible convection velocity, U, is given by the following equation:

$$p = \rho \partial\varphi/\partial t \tag{1.11}$$

At high sound pressure levels, or in cases where the particle velocity is large (as in the case when intense sound induces streaming through a small hole or many small holes in parallel), Equation (1.11) takes the form given by Equation (1.12) (Morse and Ingard, 1968, p.244), which follows, where the coordinate, x, is along the centre-line (axis) of a hole.

$$p = \rho\left[\partial\varphi/\partial t - \frac{1}{2}(\partial\varphi/\partial x)^2\right] \tag{1.12}$$

In writing Equation (1.12) a third term on the right side of the equation given in the reference has been omitted as it is inversely proportional to the square of the phase speed and thus in the cases considered here it is negligible. Alternatively, if a convection velocity, U, is present and large and the particle velocity, u, is small, Equation (1.11) takes the following form:

$$p = \rho[\partial\varphi/\partial t - U\partial\varphi/\partial x] \tag{1.13}$$

Taking the gradient of Equation (1.11), interchanging the order of differentiation on the RHS of the equation and introducing Equation (1.10) gives Euler's famous equation of motion for a unit volume of fluid acted upon by a pressure gradient:

$$\rho\frac{\partial u}{\partial t} = -\nabla p \tag{1.14}$$

1.4 WAVE EQUATION

In the previous section it was postulated that an acoustic potential function, φ, may be defined which by simple differentiation provides solutions for the wave equation for the particle velocity, $\boldsymbol{u}$, and acoustic pressure, p. The acoustic potential function satisfies the well-known linearised wave equation as follows (Kinsler *et al.*, 1982):

$$\nabla^2\varphi = (1/c^2)\,\partial^2\varphi/\partial t^2 \tag{1.15}$$

Equation (1.15) is the general three-dimensional form of the acoustic wave equation in which the Laplacian operator, ∇^2, is determined by the choice of curvilinear coordinates (Morse and Ingard, 1968, p307-8). For the present purpose it will be sufficient to restrict attention to rectangular and spherical coordinates. However, cylindrical coordinates also are included in the latter reference.

Equation (1.15) also applies if the acoustic pressure variable, p, is used to replace φ in Equation (1.15). However, the wave equation for the acoustic particle velocity is more complicated. Derivations of the wave equation in terms of acoustical particle velocity with and without the presence of a mean flow are given in Chapter 2 of Hansen and Snyder (1997). Other useful books containing derivations of the wave equation are Fahy and Walker (1998) and Fahy (2001). A brief derivation of the wave equation is given in this text in Appendix A

1.4.1 Plane and Spherical Waves

In general, sound wave propagation is quite complicated and not amenable to simple analysis. However, sound wave propagation can often be described in terms of the propagation properties of plane and spherical waves. Plane and spherical waves, in turn, have the convenient property that they can be described in terms of one dimension. Thus, the investigation of plane and spherical waves, although by no means exhaustive, is useful as a means of greatly simplifying and rendering tractable what in general may be a very complicated problem. The investigation of plane and spherical wave propagation is the subject of the following sections.

1.4.2 Plane Wave Propagation

For the case of plane wave propagation, only one spatial dimension, x, the direction of propagation is required to describe the acoustic field. An example of plane wave propagation is sound propagating along the centre line of a rigid wall tube. In this case, Equation (1.15) written in terms of the potential function, φ, reduces to:

$$\partial^2\varphi/\partial x^2 = (1/c^2)\partial^2\varphi/\partial t^2 \tag{1.16}$$

A solution for Equation (1.16), which may be verified by direct substitution, is:

$$\varphi = \mathrm{f}(ct \pm x) \tag{1.17}$$

The function, f, in Equation (1.17) describes a distribution along the x axis at any fixed time, t, as well as the variation with time at any fixed place, x, along the direction of propagation. If the argument $(ct \pm x)$ is fixed and the positive sign is chosen then with increasing time, t, x, must decrease with speed, c. Alternatively, if the argument $(ct \pm x)$ is fixed and the negative sign is chosen then with increasing time, t, x, must increase with speed, c. Consequently, a wave travelling in the positive x direction is represented by taking the negative sign and a wave travelling in the negative x direction is represented by taking the positive sign in the argument of Equation (1.17).

A very important relationship between acoustic pressure and particle velocity will now be determined. A prime sign, $'$, will indicate differentiation of a function by its argument as for example, $df(w)/dw = f'(w)$. Substitution of Equation (1.17) in Equation (1.10) gives Equation (1.18) and substitution in Equation (1.11) gives Equation (1.19) as follows:

$$u = \mp f'(ct \pm x) \tag{1.18}$$

$$p = \rho c f'(ct \pm x) \tag{1.19}$$

Division of Equation (1.19) by Equation (1.18) gives a very important result, the characteristic impedance, ρc, of a plane wave:

$$p/u = \pm \rho c \tag{1.20}$$

In Equation (1.20), the positive sign is taken for positive travelling waves, while the negative sign is taken for negative travelling waves. The characteristic impedance is one of three kinds of impedance used in acoustics. It provides a very useful relationship between acoustic pressure and particle velocity in a plane wave. It also has the property that a duct terminated in its characteristic impedance will respond as an infinite duct as no wave will be reflected at its termination.

Fourier analysis enables the representation of any function, $f(ct \pm x)$, as a sum or integral of harmonic functions. Thus it will be useful for consideration of the wave equation to investigate the special properties of harmonic solutions. Consideration will begin with the following harmonic solution for the acoustic potential function, where k is a constant, which will be investigated, and β is an arbitrary constant representing an arbitrary relative phase:

$$\varphi = A \cos(k(ct \pm x) + \beta) \tag{1.21}$$

In Equation (1.21) as β is arbitrary, for fixed time β may be chosen so that:

$$kct + \beta = 0 \tag{1.22}$$

In this case, Equation (1.17) reduces to the following representation of the spatial distribution:

$$\varphi = A \cos kx = A \cos(2\pi x/\lambda) \tag{1.23a,b}$$

From Equations (1.23) it may be concluded that the unit of length, λ, defined as the wavelength of the propagating wave and the constant, k, defined as the wave number are related as follows:

$$2\pi/\lambda = k \tag{1.24}$$

An example of harmonic (single frequency) plane wave propagation in a tube is illustrated in Figure 1.2. The type of wave generated is longitudinal, as shown in Figure 1.2(a) and the corresponding pressure fluctuations as a function of time are shown in Figure 1.2(b).

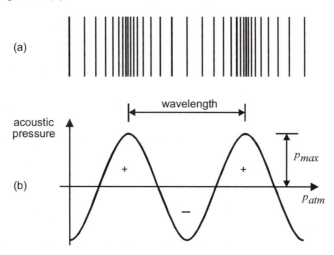

Figure 1.2 Representation of a sound wave: (a) compressions and rarefactions of a sound wave in space at a fixed instance in time; (b) graphical representation of sound pressure variation.

The distribution in space has been considered and now the distribution in time for a fixed point in space will be considered. The arbitrary phase constant, β, of Equation (1.17) will be chosen so that, for fixed position x:

$$\beta \pm kx = 0 \tag{1.25}$$

Equation (1.21) then reduces to the following representation for the temporal distribution:

$$\varphi = A\cos kct = A\cos\frac{2\pi}{T}t \tag{1.26a,b}$$

The unit of time, T is defined as the period of the propagating wave:

$$2\pi/kc = T \tag{1.27}$$

Its reciprocal is the more familiar frequency, f. Since the angular frequency, ω, is quite often used as well, the following relations should be noted:

$$2\pi/T = 2\pi f = \omega \qquad (1.28a,b)$$

and from Equations (1.27) and (1.28b):

$$k = \omega/c \qquad (1.29)$$

and from Equations (1.24), (1.28b), and (1.29):

$$f\lambda = c \qquad (1.30)$$

The relationship between wavelength and frequency is illustrated in Figure 1.3.

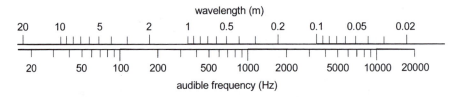

Figure 1.3 Wavelength in air versus frequency under normal conditions.

Note that the wavelength of audible sound varies by a factor of about one thousand. The shortest audible wavelength is 17 mm (corresponding to 2000 Hz) and the longest is 17 m (corresponding to 20 Hz). Letting $A = B/\rho\omega$ in Equation (1.21) and use of Equation (1.29) and either (1.10) or (1.11) gives the following useful expressions for the particle velocity and the acoustic pressure respectively for a plane wave:

$$u = \pm\frac{B}{\rho c}\sin(\omega t \mp kx + \beta) \qquad (1.31)$$

$$p = \pm B\sin(\omega t \mp kx + \beta) \qquad (1.32)$$

The wavenumber, k, may be thought of as a spatial frequency, where k is the analog of frequency, f, and wavelength, λ, is the analog of the period, T. It may be mentioned in passing that the group speed, briefly introduced in Section 1.3.5, has the following form:

$$c_g = d\omega/dk \qquad (1.33)$$

By differentiating Equation (1.29) with respect to wavenumber k, it may be concluded that for non-dispersive wave propagation where the wave speed is independent of frequency, as for longitudinal compressional waves in unbounded media, the phase and group speeds are equal. Thus, in the case of longitudinal waves propagating in unbounded media, the rate of acoustic energy transport is the same as the speed of sound, as earlier stated.

A convenient form of harmonic solution for the wave equation is the complex solution written in either one or the other of following equivalent forms:

$$\varphi = A \mathrm{e}^{j(\omega t \pm kx + \beta)} = A\cos(\omega t \pm kx + \beta) + jA\sin(\omega t \pm kx + \beta) \qquad (1.34\text{a,b})$$

In either form the negative sign represents a wave travelling in the positive x-direction, while the positive sign represents a wave travelling in the negative x-direction.

The real parts of Equations (1.34) are just the solutions given by Equation (1.21). The imaginary parts of Equations (1.34) are also solutions, but in quadrature (90° out of phase) with the former solutions. By convention, the complex notation is defined so that what is measured with an instrument corresponds to the real part; the imaginary part is then inferred from the real part. The complex exponential form of the harmonic solution to the wave equation is used as a mathematical convenience, as it greatly simplifies mathematical manipulations, allows waves with different phases to be added together easily and allows graphical representation of the solution as a rotating vector in the complex plane. Setting $\beta = 0$ and $x = 0$, allows Equation (1.34a,b) to be rewritten in the following simplified useful form:

$$A\mathrm{e}^{j\omega t} = A(\cos\omega t + j\sin\omega t) \qquad (1.35)$$

Equation (1.35) represents harmonic motion that may be represented at any time, t, as a rotating vector of constant magnitude A, and constant angular velocity, ω, as illustrated in Figure1.4. Referring to the figure, the projection of the rotating vector on the abscissa, x-axis, is given by the real term on the RHS of Equation (1.35) and the projection of the rotating vector on the ordinate, y-axis, is given by the imaginary term.

For the special case of single frequency sound, complex notation may be introduced. For example, the acoustic pressure of amplitude, p_0, and the particle velocity of amplitude, $\boldsymbol{u}_0$, may then be written in the following general form where the wavenumber, k, is given by Equation (1.25):

$$p(\boldsymbol{r},t) = p_0(\boldsymbol{r}) \mathrm{e}^{jk(ct + |\boldsymbol{r}| + \theta/k)} = p_0(\boldsymbol{r}) \mathrm{e}^{j(\omega t + \theta_p(\boldsymbol{r}))} = A\mathrm{e}^{j\omega t} \qquad (1.36\text{a-c})$$

and

$$\boldsymbol{u}(\boldsymbol{r},t) = \boldsymbol{u}_0(\boldsymbol{r}) \mathrm{e}^{j(\omega t + \theta_u(\boldsymbol{r}))} = B\mathrm{e}^{j\omega t} \qquad (1.37\text{a,b})$$

where A and B are complex numbers.

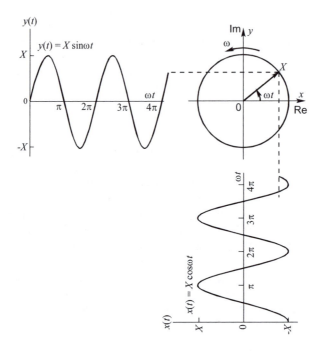

Figure 1.4 Harmonic motions represented as a rotating vector.

In writing Equations (1.36b) and (1.37a) it has been assumed that the origin of vector r is at the source centre and θ_p and θ_u are, respectively, the phases of the pressure and particle velocity and are functions of location r.

Use of the complex form of the solution makes integration and differentiation particularly simple. Also, impedances are conveniently described using this notation. For these reasons, the complex notation will be used throughout this book. However, care must be taken in the use of the complex notation when multiplying one function by another. In the calculation of products of quantities expressed in complex notation it is important to remember that the product implies that only like quantities are multiplied. In general, the real parts of the quantities are multiplied. This is important, for example, in the calculation of intensity associated with single frequency sound fields expressed in complex notation.

1.4.3 Spherical Wave Propagation

A second important case is that of spherical wave propagation; an example is the propagation of sound waves from a small source in free space with no boundaries nearby. In this case, the wave Equation (1.15) may be written in spherical coordinates in terms of a radial term only, since no angular dependence is implied. Thus Equation (1.15) becomes (Morse and Ingard, 1968, p. 309):

$$\frac{1}{r^2}\frac{\partial}{\partial r}\left[r^2\frac{\partial\varphi}{\partial r}\right] = \frac{1}{c^2}\frac{\partial^2\varphi}{\partial t^2} \tag{1.38}$$

As

$$\frac{1}{r^2}\frac{\partial}{\partial r}\left[r^2\frac{\partial\varphi}{\partial r}\right] = \frac{2}{r}\frac{\partial\varphi}{\partial r} + \frac{\partial^2\varphi}{\partial r^2} = \frac{1}{r}\frac{\partial}{\partial r}\left[\varphi + r\frac{\partial\varphi}{\partial r}\right] = \frac{1}{r}\frac{\partial^2(r\varphi)}{\partial r^2} \tag{1.39a,b}$$

the wave equation may be rewritten as:

$$\frac{\partial^2(r\varphi)}{\partial r^2} = \frac{1}{c^2}\frac{\partial^2(r\varphi)}{\partial t^2} \tag{1.40}$$

The di
fference between, and similarity of, Equations (1.16) and (1.40) should be noted.
Evidently, $r\varphi = f(ct \mp r)$ is a solution of Equation (1.40) where the source is located
at the origin.

Thus:

$$\varphi = \frac{f(ct \mp r)}{r} \tag{1.41}$$

The implications of the latter solution will now be investigated.

To proceed, Equations (1.10) and (1.11) are used to write expressions for the
acoustic pressure and particle velocity in terms of the potential function given by
Equation (1.41). The expression for the acoustic pressure is:

$$p = \rho c\frac{f'(ct \mp r)}{r} \tag{1.42}$$

and the expression for the acoustic particle velocity is:

$$u = \frac{f(ct \mp r)}{r^2} \pm \frac{f'(ct \mp r)}{r} \tag{1.43}$$

In Equations (1.41), (1.42) and (1.43) the upper sign describes a spherical wave
that decreases in amplitude as it diverges outward from the origin, where the source
is located. Alternatively, the lower sign describes a converging spherical wave, which
increases in amplitude as it converges towards the origin.

The characteristic impedance of the spherical wave may be computed, as was
done earlier for the plane wave, by dividing Equation (1.42) by Equation (1.43) to
obtain the following expression:

$$\frac{p}{u} = \rho c\frac{rf'(ct \mp r)}{f(ct \mp r) \pm rf'(ct \mp r)} \tag{1.44}$$

If the distance, r, from the origin is very large, the quantity, rf', will be sufficiently large compared to the quantity, f, for the latter to be neglected; in this case, for outward-going waves the characteristic impedance becomes ρc, while for inward-going waves it becomes $-\rho c$. In summary, at large enough distance from the origin of a spherical wave, the curvature of any part of the wave finally becomes negligible, and the characteristic impedance becomes that of a plane wave, as given by Equation (1.21). See the discussion following Equation (1.17) in Section 1.4.2 for a definition of the use of the prime, '.

A moment's reflection, however, immediately raises the question: how large is a large distance? The answer concerns the curvature of the wavefront; a large distance must be where the curvature or radius of the wavefront as measured in wavelengths is large. For example, referring to Equation (1.24), a large distance must be where:

$$kr \gg 1 \tag{1.45}$$

For harmonic waves, the solution given by Equation (1.41) can also be written in the following particular form.

$$\varphi = \frac{f(k(ct \pm r))}{r} = \frac{f(\omega t \pm kr)}{r} = \frac{A}{r} e^{j(\omega t \pm kr)} \tag{1.46a,b,c}$$

Substitution of Equation (1.46c) into Equation (1.11) gives an expression for the acoustic pressure for outwardly travelling waves (corresponding to the negative sign in Equation (1.46c)), which can be written as:

$$p = \frac{j\omega A\rho}{r} e^{j(\omega t - kr)} = \frac{jk\rho cA}{r} e^{j(\omega t - kr)} \tag{1.47a,b}$$

while substitution of Equation (1.46c) into Equation (1.10) gives an expression for the acoustic particle velocity:

$$u = \frac{A}{r^2} e^{j(\omega t - kr)} + \frac{jkA}{r} e^{j(\omega t - kr)} \tag{1.48}$$

Dividing Equation (1.47b) by Equation (1.48) gives the following result which holds for a harmonic wave characterised by a wavenumber k, and also for a narrow band of noise characterised by a narrow range of wavenumbers around k:

$$\frac{p}{u} = \rho c \frac{jkr}{1 + jkr} \tag{1.49}$$

For inward-travelling waves, the signs of k are negative.

Note that Equation (1.49) can also be derived directly by substituting Equation (1.46c) into Equation (1.44).

Consideration of Equation (1.49) now gives explicit meaning to large distance as, according to Equations (1.24) and (1.45), large distance means that the distance measured in wavelengths is large. Note that when Equation (1.45) is satisfied, Equation (1.49) reduces to the positive, outward-travelling form of Equation (1.20), which is a plane wave. For the case of a narrow band of noise, for example an octave band, the wavelength is conveniently taken as the wavelength associated with the band centre frequency.

1.4.4 Wave Summation

It will be shown that any number of harmonic waves of the same frequency travelling in one particular direction combine to produce one wave travelling in the same direction. For example, a wave that is reflected back and forth between two terminations many times may be treated as a single wave travelling in each direction.

Assume that many waves, all of the same frequency, travel together in the same direction. The waves may each be represented by rotating vectors as shown in Figure 1.5. The wave vectors in the figure will all rotate together with the passage of time and thus they will add vectorially as illustrated in the figure for the simple case of two waves separated in phase by β.

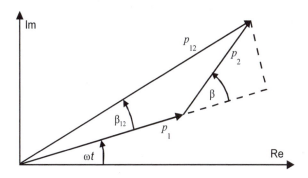

Figure 1.5 Combining two harmonic waves travelling in the same direction.

Consider any two waves travelling in one direction, which may be described as $p_1 = A_1 e^{j\omega t}$ and $p_2 = A_2 e^{j\omega t + \beta}$, where β is the phase difference between the two waves. The cosine rule gives for the magnitude of the sum, A_{12}, and the relative phase, β_{12}, of the combined wave, the following expressions:

$$A_{12}^2 = A_1^2 + A_2^2 + 2A_1 A_2 \cos\beta \tag{1.50}$$

$$\beta_{12} = \tan^{-1} \frac{A_2 \sin\beta}{A_1 + A_2 \cos\beta} \tag{1.51}$$

Equations (1.50) and (1.51) define the vector sum as:

$$p_{12} = A_{12}e^{j(\omega t + \beta_{12})} \qquad (1.52)$$

The process is then repeated, each time adding to the cumulative sum of the previous waves, a new wave not already considered, until the sum of all waves travelling in the same direction has been obtained. It can be observed that the sum will always be like any of its constituent parts; thus it may be concluded that the cumulative sum may be represented by a single wave travelling in the same direction as all of its constituent parts.

1.4.5 Plane Standing Waves

If a loudspeaker emitting a tone is placed at one end of a closed tube, there will be multiple reflections of waves from each end of the tube. As has been shown, all of the reflections in the tube result in two waves, one propagating in each direction. These two waves will also combine, and form a "standing wave". This effect can be illustrated by writing the following expression for sound pressure at any location in the tube as a result of the two waves of amplitudes A and B, respectively, travelling in the two opposite directions:

$$p = A e^{j(\omega t + kx)} + B e^{j(\omega t - kx + \beta)} \qquad (1.53)$$

Equation (1.53) can be rewritten making use of the following identity:

$$0 = -B e^{j(kx + \beta)} + B e^{j(kx + \beta)} \qquad (1.54)$$

Thus:

$$p = (A - B e^{j\beta})e^{j(\omega t + kx)} + 2B e^{j(\omega t + \beta)}\cos kx \qquad (1.55)$$

Consideration of Equation (1.55) shows that it consists of two terms. The first term on the right hand side is a left travelling wave of amplitude $(A - B e^{j\beta})$ and the second term on the right side is a standing wave of amplitude $2B e^{j\beta}$. In the latter case the wave is described by a cosine, which varies in amplitude with time, but remains stationary in space.

1.4.6 Spherical Standing Waves

Standing waves are most easily demonstrated using plane waves, but any harmonic wave motion may produce standing waves. An example of spherical standing waves is provided by the sun, which may be considered as a fluid sphere of radius r in a vacuum. At the outer edge, the acoustic pressure may be assumed to be effectively zero. Using Equation (1.47b), the sum of the outward travelling wave and the reflected inward travelling wave gives the following relation for the acoustic pressure, p, at the surface of the sphere:

$$p = jk\rho c \frac{2Ae^{j\omega t}}{r} \cos kr = 0 \qquad (1.56)$$

where the identity, $e^{-jkr} + e^{jkr} = 2\cos kr$, has been used.

Evidently, the simplest solution for Equation (1.56) is $kr = (2N - 1)/2$ where N is an integer. If it is assumed that there are no losses experienced by the wave travelling through the media making up the sun, the first half of the equation is valid everywhere except at the centre, where $r = 0$ and the solution is singular. Inspection of Equation (1.56) shows that it describes a standing wave. Note that the largest difference between maximum and minimum pressures occur in the standing wave when $p = 0$ at the boundary. However, standing waves (with smaller differences between the maximum and minimum pressures) will be also be generated for conditions where the pressure at the outer boundary is not equal to 0.

1.5 MEAN SQUARE QUANTITIES

In Section 1.3.1 the variables of acoustics were listed and discussed. For the case of fluids, they were shown to be small perturbations in steady-state quantities, such as pressure, density, velocity and temperature. Alternatively, in solids they were shown to be small perturbations in stress and strain variables. In all cases, acoustic fields were shown to be concerned with time-varying quantities with mean values of zero; thus, the variables of acoustics are commonly determined by measurement as mean square or as root mean square quantities.

In some cases, however, one is concerned with the product of two time-varying quantities. For example, sound intensity will be discussed in the next section where it will be shown that the quantity of interest is the product of the two time-varying quantities, acoustic pressure and acoustic particle velocity averaged over time. The time average of the product of two time dependent variables $f(t)$ and $g(t)$ will frequently be referred to in the following text and will be indicated by the following notation: $\langle f(t)g(t) \rangle$. Sometimes the time dependence indicated by (t) will be suppressed to simplify the notation. The time average of the product of $f(t)$ and $g(t)$ is defined as follows:

$$\langle f(t)g(t) \rangle = \langle fg \rangle = \lim_{T \to \infty} \frac{1}{T} \int_0^T f(t)g(t)\, dt \qquad (1.57)$$

When the time dependent variables are the same, the mean square of the variable is obtained. Thus the mean square sound pressure $\langle p^2(r, t) \rangle$ at position r is as follows:

$$\langle p^2(\mathbf{r}, t) \rangle \; = \; \lim_{T \to \infty} \frac{1}{T} \int_0^T p(\mathbf{r}, t)\, p(\mathbf{r}, t)\, \mathrm{d}t \tag{1.58}$$

The brackets, $\langle\ \rangle$, are sometimes used to indicate other averages of the enclosed product; for example, the average of the enclosed quantity over space. Where there may be a possibility of confusion, the averaging variable is added as a subscript; for example, the mean square sound pressure averaged over time and space may also be written as $\langle p^2(\mathbf{r}, t) \rangle_{S,t}$.

Sometimes the amplitude g_A of a single frequency quantity is of interest. In this case, the following useful relation between the amplitude and the root mean square value of a sinusoidally varying single frequency quantity is given by:

$$g_A \; = \; \sqrt{2 \langle g^2(t) \rangle} \tag{1.59}$$

1.6 ENERGY DENSITY

Any propagating sound wave has both potential and kinetic energy associated with it. The total energy (kinetic + potential) present in a unit volume of fluid is referred to as the energy density. Energy density is of interest because it is used as the quantity that is minimised in active noise cancellation systems for reducing noise in enclosed spaces. The kinetic energy per unit volume is given by the standard expression for the kinetic energy of a moving mass divided by the volume occupied by the mass. Thus:

$$\psi_k(t) \; = \; \frac{1}{2}\rho u^2(t) \tag{1.60}$$

The derivation of the potential energy per unit volume is a little more complex and may be found in Fahy (2001) or Fahy and Walker (1998). The resulting expression is:

$$\psi_p(t) \; = \; \frac{p^2(t)}{2\rho c^2} \tag{1.61}$$

The total instantaneous energy density at time, t, can thus be written as:

$$\psi_{tot}(t) \; = \; \psi_k(t) + \psi_p(t) \; = \; \frac{\rho}{2}\left[u^2(t) + \frac{p^2(t)}{(\rho c)^2} \right] \tag{1.62}$$

Note that for a plane wave, the pressure and particle velocity are related by $u(t) = p(t)/\rho c$, and the total time averaged energy density is then:

$$\psi \; = \; \frac{\langle p^2(t) \rangle}{\rho c^2} \tag{1.63}$$

where the brackets, $\langle\ \rangle$, in the equation indicate the time average.

1.7 SOUND INTENSITY

Sound waves propagating through a fluid result in a transmission of energy. The time averaged rate at which the energy is transmitted is the acoustic intensity. This is a vector quantity, as it is associated with the direction in which the energy is being transmitted. This property makes sound intensity particularly useful in many acoustical applications.

The measurement of sound intensity is discussed in Section 3.13 and its use for the determination of sound power is discussed in Section 6.5. Other uses include identifying noise sources on items of equipment, measuring sound transmission loss of building partitions, measuring impedance and sound-absorbing properties of materials and evaluating flanking sound transmission in buildings. Here, discussion is restricted to general principles and definitions, and the introduction of the concepts of instantaneous intensity and time average intensity. The concept of time average intensity applies to all kinds of noise and for simplicity, where the context allows, will be referred to in this text as simply the intensity.

For the special case of sounds characterised by a single frequency or a very narrow-frequency band of noise, where either a unique or at least an approximate phase can be assigned to the particle velocity relative to the pressure fluctuations, the concept of instantaneous intensity allows extension and identification of an active component and a reactive component, which can be defined and given physical meaning. Reactive intensity is observed in the near field of sources (see Section 6.4), near reflecting surfaces and in standing wave fields. The time average of the reactive component of instantaneous intensity is zero, as the reactive component is a measure of the instantaneous stored energy in the field, which does not propagate. However, this extension is not possible for other than the cases stated. For the case of freely propagating sound; for example, in the far field of a source (see Section 6.4), the acoustic pressure and particle velocity are always in phase and the reactive intensity is identically zero in all cases.

1.7.1 Definitions

In the following analysis and throughout this book, vector quantities are represented as bold font. The subscript, 0, is used to represent an amplitude.

Sound intensity is a vector quantity determined as the product of sound pressure and the component of particle velocity in the direction of the intensity vector. It is a measure of the rate at which work is done on a conducting medium by an advancing sound wave and thus the rate of power transmission through a surface normal to the intensity vector. As the process of sound propagation is cyclic, so is the power transmission and consequently an instantaneous and a time-average intensity may be defined. However, in either case, the intensity is the product of pressure and particle velocity. For the case of single frequency sound, represented in complex notation, this has the meaning that intensity is computed as the product of like quantities; for example, both pressure and particle velocity must be real quantities.

The instantaneous sound intensity, $I_i(r, t)$, in an acoustic field at a location given by the field vector, r, is a vector quantity describing the instantaneous acoustic power transmission per unit area in the direction of the vector particle velocity, $u(r, t)$. The general expression for the instantaneous sound intensity is:

$$I_i(r,t) = p(r,t)\, u(r,t) \tag{1.64}$$

A general expression for the sound intensity $I(r)$, is the time average of the instantaneous intensity given by Equation (1.64). Referring to Equation (1.57), let $f(t)$ be replaced with $p(r, t)$ and $g(t)$ be replaced with $u(r, t)$, then the sound intensity may be written as follows:

$$I(r) = \langle p(r,t)\, u(r,t) \rangle = \lim_{T\to\infty} \frac{1}{T} \int_0^T p(r,t)\, u(r,t)\, dt \tag{1.65a,b}$$

Integration with respect to time of Equation (1.14), introducing the unit vector $n = r/r$, taking the gradient in the direction n and introduction of Equation (1.36c) gives the following result:

$$u(r,t) = \frac{nj}{\omega\rho} \frac{\partial p(r,t)}{\partial r} = \frac{n}{\omega\rho}\left[-p_0 \frac{\partial \theta_p}{\partial r} + j\frac{\partial p_0}{\partial r} \right] e^{j(\omega t + \theta_p(r))} \tag{1.66a,b}$$

Substitution of the real parts of Equations (1.36b) and (1.37a) into Equation (1.64) gives the following result for the sound intensity in direction, n:

$$I_n(r,t) = -\frac{n}{\omega\rho}\left[p_0^2 \frac{\partial \theta_p}{\partial r} \cos^2(\omega t + \theta_p) + p_0 \frac{\partial p_0}{\partial r} \cos(\omega t + \theta_p)\sin(\omega t + \theta_p) \right] \tag{1.67}$$

The first term in brackets on the right-hand side of Equation (1.67) is the product of the real part of the acoustic pressure and the in-phase component of the real part of the particle velocity and is defined as the active intensity. The second term on the right-hand side of the equation is the product of the real part of the acoustic pressure and the in-quadrature component of the real part of the particle velocity and is defined as the reactive intensity. The reactive intensity is a measure of the energy stored in the field during each cycle but is not transmitted.

Using well known trigonometric identities (Abramowitz and Stegun, 1965), Equation (1.67) may be rewritten as follows:

$$I_n(r,t) = -\frac{n}{2\omega\rho}\left[p_0^2 \frac{\partial \theta_p}{\partial r}\left(1 + \cos 2(\omega t + \theta_p)\right) + p_0 \frac{\partial p_0}{\partial r}\sin 2(\omega t + \theta_p) \right] \tag{1.68}$$

Equation (1.68) shows that both the active and the reactive components of the instantaneous intensity vary sinusoidally but the active component has a constant part. Taking the time average of Equation (1.68) gives the following expression for the intensity:

$$I(r) = -\frac{n}{2\omega\rho} p_0^2 \frac{\partial\theta_p}{\partial r} \tag{1.69}$$

Equation (1.69) is a measure of the acoustic power transmission in the direction of the intensity vector.

Alternatively substitution of the real parts of Equations (1.36) and (1.37) into Equation (1.64) gives the instantaneous intensity:

$$I_n(r,t) = n p_0 u_0 \cos(\omega t + \theta_p)\cos(\omega t + \theta_u) \tag{1.70}$$

Using well-known trigonometric identities (Abramowitz and Stegun, 1965), Equation (1.70) may be rewritten as follows:

$$I_i(r,t) = \frac{p_0 u_0}{2}\left[\left(1 + \cos 2(\omega t + \theta_p)\right)\cos(\theta_p - \theta_u) + \sin 2(\omega t + \theta_p)\sin(\theta_p - \theta_u)\right] \tag{1.71}$$

Equation (1.71) is an alternative form to Equation (1.68). The first term on the right-hand side of the equation is the active intensity, which has a mean value given by the following equation:

$$I(r) = \frac{p_0 u_0}{2}\cos(\theta_p - \theta_u) = \frac{1}{2}\text{Re}\{AB^*\} \tag{1.72a,b}$$

The second term in Equation (1.71) is the reactive intensity, which has an amplitude given by the following equation (Fahy, 1995):

$$I_r(r) = \frac{p_0 u_0}{2}\sin(\theta_p - \theta_u) = \frac{1}{2}\text{Im}\{AB^*\} \tag{1.73a,b}$$

where the * indicates the complex conjugate (see Equations (1.36) and (1.37)).

1.7.2 Plane Wave and Far Field Intensity

Waves radiating outward, away from any source, tend to become planar. Consequently, the equations derived in this section also apply in the far field of any source. For this purpose, the radius of curvature of an acoustic wave should be greater than about ten times the radiated wavelength.

For a propagating plane wave, the characteristic impedance ρc is a real quantity and thus, according to Equation (1.20), the acoustic pressure and particle velocity are in phase and consequently acoustic power is transmitted. The intensity is a vector quantity but where direction is understood the magnitude is of greater interest and will frequently find use throughout the rest of this book. Consequently, the intensity will be written in scalar form as a magnitude. If Equation (1.20) is used to replace u in

Equation (1.65a) the expression for the plane wave acoustic intensity at location r becomes:

$$I = \langle p^2(r,t) \rangle / \rho c \tag{1.74}$$

In Equation (1.74) the intensity has been written in terms of the mean square pressure.

If Equation (1.20) is used to replace p in the expression for intensity, the following alternative form of the expression for the plane wave acoustic intensity is obtained:

$$I = \rho c \langle u^2(r,t) \rangle \tag{1.75}$$

where again the vector intensity has been written in scalar form as a magnitude. The mean square particle velocity is defined in a similar way as the mean square sound pressure.

1.7.3 Spherical Wave Intensity

If Equations (1.42) and (1.43) are substituted into Equation (1.65a) and use is made of Equation (5.2) (see Section 5.2.1) then Equation (1.74) is obtained, showing that the latter equation also holds for a spherical wave at any distance r from the source. Alternatively, similar reasoning shows that Equation (1.75) is only true of a spherical wave at distances r from the source, which are large (see Section 1.4.3).

To simplify the notation to follow, the r dependence (dependence on location) and time dependence t of the quantities p and u will be assumed, and specific reference to these discrepancies will be omitted.

It is convenient to rewrite Equation (1.49) in terms of its magnitude and phase. Carrying out the indicated algebra gives the following result:

$$\frac{p}{u} = \rho c \, e^{j\beta} \cos \beta \tag{1.76}$$

where $\beta = (\theta_p - \theta_u)$ is the phase angle by which the acoustic pressure leads the particle velocity and is defined as:

$$\beta = \tan^{-1}[1/(kr)] \tag{1.77}$$

Equation (1.71) gives the instantaneous intensity for the case considered here in terms of the pressure amplitude, p_0, and particle velocity amplitude, u_0. Solving Equation (1.76) for the particle velocity in terms of the pressure shows that $u_0 = p_0/(\rho c \cos b)$. Substitution of this expression and Equation (1.77) into Equation (1.71) gives the following expression for the instantaneous intensity of a spherical wave, $I_{si}(r, t)$:

$$I_{si}(\textbf{\textit{r}},t) = \frac{p_0^2}{2\rho c}\left[\left(1 + \cos 2\left(\omega t + \theta_p\right)\right) + \frac{1}{kr}\sin 2\left(\omega t + \theta_p\right)\right] \qquad (1.78)$$

Consideration of Equation (1.78) shows that the time average of the first term on the right-hand side is non-zero and is the same as that of a plane wave given by Equation (1.74), while the time average of the second term is zero and thus the second term is associated with the non-propagating reactive intensity. The second term tends to zero as the distance r from the source to observation point becomes large; that is, the second term is negligible in the far field of the source. On the other hand, the reactive intensity becomes quite large close to the source; this is a near field effect.

Integration over time of Equation (1.78), taking note that the integral of the second term is zero, gives the same expression for the intensity of a spherical wave as was obtained previously for a plane wave (see Equation (1.74)).

1.8 SOUND POWER

As mentioned in the previous section, when sound propagates, transmission of acoustic power is implied. The intensity, as a measure of the energy passing through a unit area of the acoustic medium per unit time, was defined for plane and spherical waves and found to be the same. It will be assumed that the expression given by Equation (1.74) holds in general for sources that radiate more complicated acoustic waves, at least at sufficient distance from the source so that, in general, the power, W, radiated by any acoustic source is:

$$W = \int_S \textbf{\textit{I}}\cdot\textbf{\textit{n}}\ \mathrm{d}S \qquad (1.79)$$

where $\textbf{\textit{n}}$ is the unit vector normal to the surface of area S.

For the cases of the plane wave and spherical waves, the mean square pressure, $\langle p^2 \rangle$, is a function of a single spatial variable in the direction of propagation. The meaning is now extended to include; for example, variations with angular direction, as is the case for sources that radiate more power in some directions than in others. A loudspeaker that radiates most power on axis to the front would be such a source.

According to Equation (1.79), the sound power, W, radiated by a source is defined as the integral of the acoustic intensity over a surface surrounding the source. Most often, a convenient surface is an encompassing sphere or spherical section, but sometimes other surfaces are chosen, as dictated by the circumstances of the particular case considered. For a sound source producing uniformly spherical waves (or radiating equally in all directions), a spherical surface is most convenient, and in this case Equation (1.79) leads to the following expression:

$$W = 4\pi r^2 I \qquad (1.80)$$

where the magnitude of the acoustic intensity, I, is measured at a distance r from the source. In this case the source has been treated as though it radiates uniformly in all directions. Consideration is given to sources which do not radiate uniformly in all directions in Section 5.8.

1.9 UNITS

Pressure is an engineering unit, which is measured relatively easily; however, the ear responds approximately logarithmically to energy input, which is proportional to the square of the sound pressure. The minimum sound pressure that the ear may detect is 20 μPa, while the greatest sound pressure before pain is experienced is 60 Pa. A linear scale based on the square of the sound pressure would require 10^{13} unit divisions to cover the range of human experience; however, the human brain is not organised to encompass such an enormous range in a linear way. The remarkable dynamic range of the ear suggests that some kind of compressed scale should be used. A scale suitable for expressing the square of the sound pressure in units best matched to subjective response is logarithmic rather than linear (see Sections 2.4.3 and 2.4.4).

The logarithmic scale provides a convenient way of comparing the sound pressure of one sound with another. To avoid a scale that is too compressed, a factor of 10 is introduced, giving rise to the decibel. The level of sound pressure, p, is then said to be L_p decibels (dB) greater than or less than a reference sound pressure, p_{ref}, according to the following equation:

$$L_p = 10 \log_{10} \frac{\langle p^2 \rangle}{p_{ref}^2} = 10 \log_{10} \langle p^2 \rangle - 10 \log_{10} p_{ref}^2 \quad \text{(dB)} \qquad \text{(1.81a,b)}$$

For the purpose of absolute level determination, the sound pressure is expressed in terms of a datum pressure corresponding to the lowest sound pressure which the young normal ear can detect. The result is called the sound pressure level, L_p (or SPL), which has the units of decibels (dB). This is the quantity that is measured with a sound level meter.

The sound pressure is a measured root mean square (r.m.s.) value and the reference pressure $p_{ref} = 2 \times 10^{-5}$ N/m² or 20 μPa. When this value for the reference pressure is substituted into Equation (1.81b), the following convenient alternative form is obtained:

$$L_p = 10 \log_{10} \langle p^2 \rangle + 94 \quad \text{(dB)} \qquad \text{(1.82)}$$

In the discipline of underwater acoustics (and any other liquids), the same equations apply, but the reference pressure used is 1 μPa

In Equation (1.82), the acoustic pressure, p, is measured in pascals. Some feeling for the relation between subjective loudness and sound pressure level may be gained by reference to Figure 1.6 and Table 1.1, which show sound pressure levels produced by a range of noise sources.

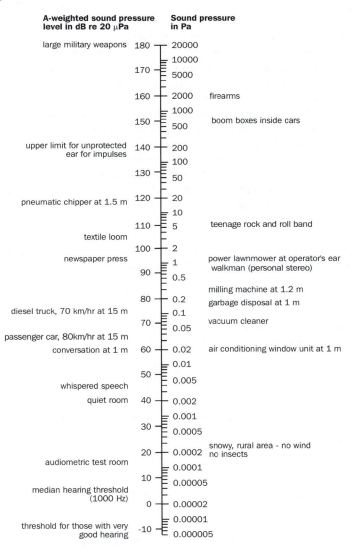

Figure 1.6 Sound pressure levels of some sources.

The sound power level, L_w (or PWL), may also be defined as follows:

$$L_w = 10 \log_{10} \frac{(\text{sound power})}{(\text{reference power})} \quad (\text{dB}) \quad (1.83)$$

Table 1.1 Sound pressure levels of some sources

Sound pressure level (dB re 20 μPa)	Description of sound source	Typical subjective description
140	Moon launch at 100 m; artillery fire, gunner's position	Intolerable
120	Ship's engine room; rock concert, in front and close to speakers	
100	Textile mill; press room with presses running; punch press and wood planers, at operator's position	Very noisy
80	Next to busy highway, shouting	Noisy
60	Department store, restaurant, speech levels	
40	Quiet residential neighbourhood, ambient level	Quiet
20	Recording studio, ambient level	Very quiet
0	Threshold of hearing for normal young people	

The reference power is 10^{-12}W. Again, the following convenient form is obtained when the reference sound power is introduced into Equation (1.83):

$$L_w = 10 \log_{10} W + 120 \quad \text{(dB)} \tag{1.84}$$

In Equation (1.84), the power W is measured in watts.

A sound intensity level, L_I, may be defined as follows:

$$L_I = 10 \log_{10} \frac{\text{(sound intensity)}}{\text{(ref. sound intensity)}} \quad \text{(dB)} \tag{1.85}$$

A convenient reference intensity is 10^{-12} W/m^2, in which case Equation (1.85) takes the following form:

$$L_I = 10 \log_{10} I + 120 \quad \text{(dB)} \tag{1.86}$$

The introduction of the magnitude of Equation (1.74) into Equation (1.86) and use of Equation (1.82) gives the following useful result:

$$L_I = L_p - 10 \log_{10}(\rho c / 400) = L_p + 26 - 10 \log_{10}(\rho c) \quad \text{(dB)} \tag{1.87a,b}$$

Reference to Appendix B allows calculation of the characteristic impedance, ρc. At sea level and 20°C the characteristic impedance is 414 kg m^{-2} s^{-1}, so that for plane and spherical waves, use of Equation (1.87a) gives the following:

$$L_I = L_p - 0.2 \quad \text{(dB)} \tag{1.88}$$

1.10 SPECTRA

A propagating sound wave has been described either as an undefined disturbance as in Equation (1.17), or as a single frequency disturbance as given, for example, by Equation (1.21). Here it is shown how an undefined disturbance may be described conveniently as composed of narrow frequency bands, each characterised by a range of frequencies. There are various such possible representations and all are referred to broadly as spectra.

It is customary to refer to spectral density level when the measurement band is one hertz wide, to one-third octave or octave band level when the measurement band is one-third octave or one octave wide, respectively, and to spectrum level for measurement bands of other widths.

In air, sound is transmitted in the form of a longitudinal wave. To illustrate longitudinal wave generation, as well as to provide a model for the discussion of sound spectra, the example of a vibrating piston at the end of a very long tube filled with air is used, as illustrated in Figure 1.7.

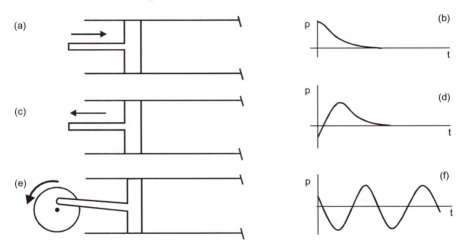

Figure 1.7 Sound generation illustrated. (a) The piston moves right, compressing air as in (b). (c) The piston stops and reverses direction, moving left and decompressing air in front of the piston, as in (d). (e) The piston moves cyclically back and forth, producing alternating compressions and rarefactions, as in (f). In all cases disturbances move to the right with the speed of sound.

Let the piston in Figure 1.7 move forward. Since the air has inertia, only the air immediately next to the face of the piston moves at first; the pressure in the element of air next to the piston increases. The element of air under compression next to the piston will expand forward, displacing the next layer of air and compressing the next elemental volume. A pressure pulse is formed which travels down the tube with the speed of sound, c. Let the piston stop and subsequently move backward; a rarefaction is formed next to the surface of the piston which follows the previously formed compression down the tube. If the piston again moves forward, the process is repeated with the net result being a "wave" of positive and negative pressure transmitted along the tube.

If the piston moves with simple harmonic motion, a sine wave is produced; that is, at any instant the pressure distribution along the tube will have the form of a sine wave, or at any fixed point in the tube the pressure disturbance, displayed as a function of time, will have a sine wave appearance. Such a disturbance is characterised by a single frequency. The motion and corresponding spectrum are illustrated in Figure 1.8(a) and (b).

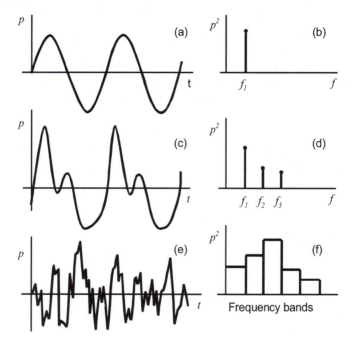

Figure 1.8 Spectral analysis illustrated. (a) Disturbance p varies sinusoidally with time t at a single frequency f_1, as in (b). (c) Disturbance p varies cyclically with time t as a combination of three sinusoidal disturbances of fixed relative amplitudes and phases; the associated spectrum has three single-frequency components f_1, f_2 and f_3, as in (d). (e) Disturbance p varies erratically with time t, with a frequency band spectrum as in (f).

If the piston moves irregularly but cyclically, for example, so that it produces the waveform shown in Figure 1.8(c), the resulting sound field will consist of a combination of sinusoids of several frequencies. The spectral (or frequency) distribution of the energy in this particular sound wave is represented by the frequency spectrum of Figure 1.8(d). As the motion is cyclic, the spectrum consists of a set of discrete frequencies.

Although some sound sources have single-frequency components, most sound sources produce a very disordered and random waveform of pressure versus time, as illustrated in Figure 1.8(e). Such a wave has no periodic component, but by Fourier analysis it may be shown that the resulting waveform may be represented as a collection of waves of all frequencies. For a random type of wave the sound pressure squared in a band of frequencies is plotted as shown, for example, in the frequency spectrum of Figure 1.8(f).

Two special kinds of spectra are commonly referred to as white random noise and pink random noise. White random noise contains equal energy per hertz and thus has a constant spectral density level. Pink random noise contains equal energy per measurement band and thus has an octave or one-third octave band level that is constant with frequency.

1.10.1 Frequency Analysis

Frequency analysis is a process by which a time-varying signal is transformed into its frequency components. It can be used for quantification of a noise problem, as both criteria and proposed controls are frequency dependent. When tonal components are identified by frequency analysis, it may be advantageous to treat these somewhat differently than broad band noise. Frequency analysis serves the important function of determining the effects of control and it may aid, in some cases, in the identification of sources. Frequency analysis equipment and its use is discussed in Chapter 3.

To facilitate comparison of measurements between instruments, frequency analysis bands have been standardised. The International Standards Organisation has agreed upon "preferred" frequency bands for sound measurement and by agreement the octave band is the widest band for frequency analysis. The upper frequency limit of the octave band is approximately twice the lower frequency limit and each band is identified by its geometric mean called the band centre frequency.

When more detailed information about a noise is required, standardised one-third octave band analysis may be used. The preferred frequency bands for octave and one-third octave band analysis are summarised in Table 1.2.

Reference to the table shows that all information is associated with a band number, *BN*, listed in column one on the left. In turn the band number is related to the centre band frequencies, f_C, of either the octaves or the one-third octaves listed in the columns two and three. The respective band limits are listed in columns four and five as the lower and upper frequency limits, f_ℓ and f_u. These observations may be summarised as follows:

$$BN = 10\log_{10}f_C \quad \text{and} \quad f_C = \sqrt{f_\ell f_u} \qquad (1.89\text{a,b})$$

A clever manipulation has been used in the construction of Table 1.2. By small adjustments in the calculated values recorded in the table, it has been possible to arrange the one-third octave centre frequencies so that ten times their logarithms are the band numbers of column one on the left of the table. Consequently, as may be observed the one-third octave centre frequencies repeat every decade in the table. A practitioner will recognise the value of this simplification.

Table 1.2 Preferred frequency bands (Hz)

Band number	Octave band centre frequency	One-third octave band centre frequency	Band limits Lower	Band limits Upper
14		25	22	28
15	31.5	31.5	28	35
16		40	35	44
17		50	44	57
18	63	63	57	71
19		80	71	88
20		100	88	113
21	125	125	113	141
22		160	141	176
23		200	176	225
24	250	250	225	283
25		315	283	353
26		400	353	440
27	500	500	440	565
28		630	565	707
29		800	707	880
30	1,000	1,000	880	1,130
31		1,250	1,130	1,414
32		1,600	1,414	1,760
33	2,000	2 ,000	1,760	2,250
34		2,500	2,250	2,825
35		3,150	2,825	3,530
36	4,000	4,000	3,530	4,400
37		5,000	4,400	5,650
38		6,300	5,650	7,070
39	8,000	8,000	7,070	8,800
40		10,000	8,800	11,300
41		12,500	11,300	14,140
42	16,000	16,000	14,140	17,600
43		20,000	17,600	22,500

In the table, the frequency band limits have been defined so that the bands are approximately equal. The limits are functions of the analysis band number, *BN*, and the ratios of the upper to lower frequencies, and are given by:

$$f_u/f_\ell = 2^{1/N} \qquad N = 1,3 \qquad (1.90)$$

where $N = 1$ for octave bands and $N = 3$ for one-third octave bands.

The information provided thus far allows calculation of the bandwidth, Δf of every band, using the following equation:

$$\Delta f = f_C \frac{2^{1/N} - 1}{2^{1/2N}} \qquad = 0.2316 f_C \text{ for } 1/3 \text{ octave bands}$$
$$= 0.7071 f_C \text{ for octave bands} \qquad\qquad (1.91)$$

It will be found that the above equations give calculated numbers that are always close to those given in the table.

When logarithmic scales are used in plots, as will frequently be done in this book, it will be well to remember the one-third octave band centre frequencies. For example, the centre frequencies of the 1/3 octave bands between 12.5 Hz and 80 Hz inclusive, will lie respectively at 0.1, 0.2, 0.3, 0.4, 0.5, 0.6, 0.7, 0.8 and 0.9 of the distance on the scale between 10 and 100. The latter two numbers in turn will lie at 1.0 and 2.0, respectively, on the same logarithmic scale.

Instruments are available that provide other forms of band analysis (see Section 3.12). However, they do not enjoy the advantage of standardisation so that the comparison of readings taken on such instruments may be difficult. One way to ameliorate the problem is to present such readings as mean levels per unit frequency. Data presented in this way are referred to as spectral density levels as opposed to band levels. In this case the measured level is reduced by ten times the logarithm to the base ten of the bandwidth. For example, referring to Table 1.2, if the 500 Hz octave band which has a bandwidth of 354 Hz were presented in this way, the measured octave band level would be reduced by $10 \log_{10}(354) = 25.5$ dB to give an estimate of the spectral density level at 500 Hz.

The problem is not entirely alleviated, as the effective bandwidth will depend upon the sharpness of the filter cut-off, which is also not standardised. Generally, the bandwidth is taken as lying between the frequencies, on either side of the pass band, at which the signal is down 3 dB from the signal at the centre of the band.

The spectral density level represents the energy level in a band one cycle wide whereas by definition a tone has a bandwidth of zero.

There are two ways of transforming a signal from the time domain to the frequency domain. The first requires the use of band limited digital or analog filters. The second requires the use of Fourier analysis where the time domain signal is transformed using a Fourier series. This is implemented in practice digitally (referred to as the DFT – discrete Fourier transform) using a very efficient algorithm known as the FFT (fast Fourier transform). Digital filtering is discussed in Appendix D.

1.11 COMBINING SOUND PRESSURES

1.11.1 Coherent and Incoherent Sounds

Thus far, the sounds that have been considered have been harmonic, being characterised by single frequencies. Sounds of the same frequency bear fixed phase relationships with each other and as observed in Section 1.4.4, their summation is strongly dependent upon their phase relationship. Such sounds are known as coherent sounds.

Coherent sounds are sounds of fixed relative phase and are quite rare, although sound radiated from different parts of a large tonal source such as an electrical transformer in a substation are an example of coherent sound. Coherent sounds can also be easily generated electronically. When coherent sounds combine they sum vectorially and their relative phase will determine the sum (see Section 1.4.4).

It is more common to encounter sounds that are characterised by varying relative phases. For example, in an orchestra the musical instruments of a section may all play in pitch, but in general their relative phases will be random. The violin section may play beautifully but the phases of the sounds of the individual violins will vary randomly, one from another. Thus the sounds of the violins will be incoherent with one another, and their contributions at an observer will sum incoherently. Incoherent sounds are sounds of random relative phase and they sum as scalar quantities on an energy basis. The mathematical expressions describing the combining of incoherent sounds may be considered as special limiting cases of those describing the combining of coherent sound.

Sound reflected at grazing incidence in a ground plane at large distance from a source will be partially coherent with the direct sound of the source. For a source mounted on a hard ground, the phase of the reflected sound will be opposite to that of the source so that the source and its image will radiate to large distances near the ground plane as a vertically oriented dipole (see Section 5.3). The null plane of a vertically oriented dipole will be coincident with the ground plane and this will limit the usefulness of any signalling device placed near the ground plane. See Section 5.10.2 for discussion of reflection in the ground plane.

1.11.2 Addition of Coherent Sound Pressures

When coherent sounds (which must be tonal and of the same frequency) are to be combined, the phase between the sounds must be included in the calculation.

Let $p = p_1 + p_2$ and $p_i = p_{i0}\cos(\omega t + \beta_i)$, $i = 1, 2$; then:

$$p^2 = p_{10}^2 \cos^2(\omega t + \beta_1) + p_{20}^2 \cos^2(\omega t + \beta_2) + 2p_{10}p_{20}\cos(\omega t + \beta_1)\cos(\omega t + \beta_2)$$

(1.92)

where the subscript, 0, denotes an amplitude.

Use of well known trigonometric identities (Abramowitz and Stegun, 1965) allows Equation (1.92) to be rewritten as follows:

$$p^2 = \frac{1}{2}p_{10}^2\big(1 + \cos 2(\omega t + \beta_1)\big) + \frac{1}{2}p_{20}^2\big(1 + \cos 2(\omega t + \beta_2)\big) + p_{10}p_{20}\big[\cos(2\omega t + \beta_1 + \beta_2) + \cos(\beta_1 - \beta_2)\big]$$

(1.93)

Substitution of Equation (1.93) into Equation (1.58) and carrying out the indicated operations gives the time average total pressure, $\langle p^2 \rangle$. Thus for two sounds of the same frequency, characterised by mean square sound pressures $\langle p_1^2 \rangle$ and $\langle p_2^2 \rangle$ and phase difference $\beta_1 - \beta_2$, the total mean square sound pressure is given by the following equation:

$$\langle p^2 \rangle = \langle p_1^2 \rangle + \langle p_2^2 \rangle + 2\langle p_1 p_2 \rangle \cos(\beta_1 - \beta_2) \tag{1.94}$$

1.11.3 Beating

When two tones of very small frequency difference are presented to the ear, one tone, which varies in amplitude with a frequency modulation equal to the difference in frequency of the two tones, will be heard. When the two tones have exactly the same frequency, the frequency modulation will cease. When the tones are separated by a frequency difference greater than what is called the "critical bandwidth", two tones are heard. When the tones are separated by less than the critical bandwidth, one tone of modulated amplitude is heard where the frequency of modulation is equal to the difference in frequency of the two tones. The latter phenomenon is known as beating. For more on the beating phenomenon, see Section 2.2.6.

Let two tonal sounds of amplitudes A_1 and A_2 and of slightly different frequencies, ω and $\omega + \Delta\omega$ be added together. It will be shown that a third amplitude modulated sound is obtained. The total pressure due to the two tones may be written as:

$$p_1 + p_2 = A_1 \cos \omega t + A_2 \cos(\omega + \Delta\omega)t \tag{1.95}$$

where one tone is described by the first term and the other tone is described by the second term in Equation (1.95).

Assuming that $A_1 \geq A_2$, defining $A = A_1 + A_2$ and $B = A_1 - A_2$, and using well known trigonometric identities, Equation (1.95) may be rewritten as follows:

$$\begin{aligned} p_1 + p_2 &= A \cos(\Delta\omega/2)t \cos(\omega + \Delta\omega/2)t \\ &+ B \cos(\Delta\omega/2 - \pi/2)t \cos(\omega + \Delta\omega/2 - \pi/2)t \end{aligned} \tag{1.96}$$

When $A_1 = A_2$, $B = 0$ and the second term in Equation (1.96) is zero. The first term is a cosine wave of frequency $(\omega + \Delta\omega)$ modulated by a frequency $\Delta\omega/2$. At certain values of time, t, the amplitude of the wave is zero; thus, the wave is described as fully modulated. If B is non-zero as a result of the two waves having different amplitudes, a modulated wave is still obtained, but the depth of the modulation decreases with increasing B and the wave is described as partially modulated. If $\Delta\omega$ is small, the familiar beating phenomenon is obtained (see Figure 1.9). The figure shows a beating phenomenon where the two waves are slightly different in amplitude resulting in partial modulation and incomplete cancellation at the null points.

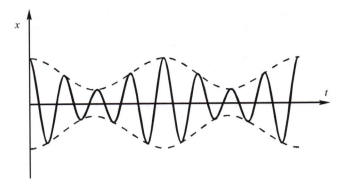

Figure 1.9 Illustration of beating.

It is interesting to note that if the signal in Figure 1.9 were analysed on a very fine resolution spectrum analyser, only two peaks would be seen; one at each of the two interacting frequencies. There would be no peak seen at the beat frequency as there is no energy at that frequency even though we apparently "hear" that frequency.

1.11.4 Addition of Incoherent Sounds (Logarithmic Addition)

When bands of noise are added and the phases are random, the limiting form of Equation (1.96) reduces to the case of addition of two incoherent sounds:

$$\langle p_t^2 \rangle = \langle p_1^2 \rangle + \langle p_2^2 \rangle \qquad (1.97)$$

which may be written in a general form for the addition of N incoherent sounds as:

$$L_{pt} = 10 \log_{10}\left(10^{L_1/10} + 10^{L_2/10} \ldots\ldots\ldots + 10^{L_N/10}\right) \qquad (1.98)$$

Incoherent sounds add together on a linear energy (pressure squared) basis. The simple procedure embodied in Equation (1.98) may easily be performed on a standard calculator. The procedure accounts for the addition of sounds on a linear energy basis and their representation on a logarithmic basis. Note that the division by 10, rather than 20 in the exponent is because the process involves the addition of squared pressures.

Example 1.1

Assume that three sounds of different frequencies (or three incoherent noise sources) are to be combined to obtain a total sound pressure level. Let the three sound pressure levels be (a) 90 dB, (b) 88 dB and (c) 85 dB. The solution is obtained by use of Equation (1.81b).

Solution

For source (a):

$$\langle p_1^2 \rangle = p_{ref}^2 \times 10^{90/10} = p_{ref}^2 \times 10 \times 10^8$$

For source (b):

$$\langle p_2^2 \rangle = p_{ref}^2 \times 6.31 \times 10^8$$

For source (c):

$$\langle p_3^2 \rangle = p_{ref}^2 \times 3.16 \times 10^8$$

The total mean square sound pressure is:

$$\langle p_t^2 \rangle = \langle p_1^2 \rangle + \langle p_2^2 \rangle + \langle p_3^2 \rangle = p_{ref}^2 \times 19.47 \times 10^8$$

The total sound pressure level is:

$$L_{pt} = 10 \log_{10}[\langle p_t^2 \rangle / p_{ref}^2] = 10 \log_{10}[19.47 \times 10^8] = 92.9 \text{ dB}$$

Alternatively, in short form:

$$L_{pt} = 10 \log_{10}\left(10^{90/10} + 10^{88/10} + 10^{85/10}\right) = 92.9 \text{ dB}$$

Some useful properties of the addition of sound levels will be illustrated with two further examples. The following example will show that the addition of two sounds can never result in a sound pressure level more than 3 dB greater than the level of the louder sound.

Example 1.2

Consider the addition of two sounds of sound pressure levels L_1 and L_2 where $L_1 \geq L_2$. Compute the total sound pressure level on the assumption that the sounds are incoherent; for example, that they add on a squared pressure basis:

$$L_{pt} = 10 \log_{10}\left(10^{L_1/10} + 10^{L_2/10}\right)$$

then,

$$L_{pt} = L_1 + 10 \log_{10}\left(1 + 10^{(L_2 - L_1)/10}\right)$$

Since,

$$10^{(L_2 - L_1)/10} \leq 1$$

then,

$$L_{pt} \leq L_1 + 3 \text{ dB}$$

Example 1.3

Consider the addition of N sound levels each with an uncertainty of measured level $\pm \Delta$.

$$L_{pt} = 10 \log_{10}\left(\sum_{i=1}^{N} 10^{(L_i \pm \Delta)/10} \right)$$

Show that the level of the sum is characterised by the same uncertainty:

$$L_{pt} = 10 \log_{10}\left(\sum_{i=1}^{N} 10^{L_i/10} \right) \pm \Delta$$

Evidently the uncertainty in the total is no greater than the uncertainty in the measurement of any of the contributing sounds.

1.11.5 Subtraction of Sound Pressure Levels

Sometimes it is necessary to subtract one noise from another; for example, when background noise must be subtracted from total noise to obtain the sound produced by a machine alone. The method used is similar to that described in the addition of levels and is illustrated here with an example.

Example 1.4

The sound pressure level measured at a particular location in a factory with a noisy machine operating nearby is 92.0 dB(A). When the machine is turned off, the sound pressure level measured at the same location is 88.0 dB(A). What is the sound pressure level due to the machine alone?

Solution

$$L_{pm} = 10 \log_{10}\left(10^{92/10} - 10^{88/10} \right) = 89.8 \text{ dB(A)}$$

For noise-testing purposes, this procedure should be used only when the total sound pressure level exceeds the background noise by 3 dB or more. If the difference is less than 3 dB a valid sound test probably cannot be made. Note that here subtraction is between squared pressures.

1.11.6 Combining Level Reductions

Sometimes it is necessary to determine the effect of the placement or removal of constructions such as barriers and reflectors on the sound pressure level at an observation point. The difference between levels before and after an alteration (placement or removal of a construction) is called the noise reduction, *NR*. If the level decreases after the alteration, the NR is positive; if the level increases, the *NR* is negative. The problem of assessing the effect of an alteration is complex because the number of possible paths along which sound may travel from the source to the observer may increase or decrease.

In assessing the overall effect of any alteration, the combined effect of all possible propagation paths must be considered. Initially, it is supposed that a reference level L_{pR} may be defined at the point of observation as a level which would or does exist due only to straight-line propagation from source to receiver. Noise reduction due to propagation over any other path is then assessed in terms of this reference level. Calculated noise reductions would include spreading due to travel over a longer path, losses due to barriers, reflection losses at reflectors and losses due to source directivity effects (see Section 5.11.3).

For octave band analysis, it will be assumed that the noise arriving at the point of observation by different paths combines incoherently. Thus, the total observed sound level may be determined by adding together logarithmically the contributing levels due to each propagation path.

The problem that will now be addressed is how to combine noise reductions to obtain an overall noise reduction due to an alteration. Either before alteration or after alteration, the sound pressure level at the point of observation due to the ith path may be written in terms of the ith path noise reduction, NR_i, as:

$$L_{pi} = L_{pR} - NR_i \qquad (1.99)$$

In either case, the observed overall noise level due to contributions over n paths, including the direct path, is:

$$L_p = L_{pR} + 10 \log_{10} \sum_{i=1}^{n} 10^{-(NR_i/10)} \qquad (1.100)$$

The effect of an alteration will now be considered, where note is taken that, after alteration, the propagation paths, associated noise reductions and number of paths may differ from those before alteration. Introducing subscripts to indicate cases A (before alteration) and B (after alteration) the overall noise reduction ($NR = L_{pA} - L_{pB}$) due to the alteration is:

$$NR = 10 \log_{10} \sum_{i=1}^{n_A} 10^{-(NR_{Ai}/10)} - 10 \log_{10} \sum_{i=1}^{n_B} 10^{-(NR_{Bi}/10)} \qquad (1.101)$$

Example 1.5

Initially, the sound pressure level at an observation point is due to straight-line propagation and reflection in the ground plane between the source and receiver. The arrangement is altered by introducing a very long barrier, which prevents both initial propagation paths but introduces four new paths (see Section 8.5). Compute the noise reduction due to the introduction of the barrier. In situation A, before alteration, the sound pressure level at the observation point is L_{pA} and propagation loss over the path reflected in the ground plane is 5 dB. In situation B, after alteration, the losses over the four new paths are respectively 4, 6, 7 and 10 dB.

Solution

Using Equation (1.101) gives the following result:

$$NR = 10 \log_{10}\left[10^{-0/10} + 10^{-5/10}\right]$$
$$- 10 \log_{10}\left[10^{-4/10} + 10^{-6/10} + 10^{-7/10} + 10^{-10/10}\right]$$
$$= 1.2 + 0.2 = 1.4 \text{ dB}$$

1.12 IMPEDANCE

In Section 1.4, the specific acoustic impedance for plane and spherical waves was introduced and it was argued that similar expressions relating acoustic pressure and associated particle velocity must exist for waves of any shape. In free space and at large distances from a source, any wave approaches plane wave propagation and the characteristic impedance of the wave always tends to ρc.

Besides the specific acoustic impedance, two other kinds of impedance are commonly used in acoustics. These three kinds of impedance are summarised in Table 1.3 and their uses will be discussed in the sections that follow. All of the definitions suppose that, with the application of a periodic force or pressure at some point in a dynamical system, a periodic velocity of fixed phase relative to the force or pressure will ensue. Note the role of cross-sectional area, S, in the definitions of the impedances shown in the table. In the case of mechanical impedance (radiation impedance) or ratio of force to velocity, the area S is the area of the radiating surface. For the case of acoustic impedance, the area S is the cross-sectional area of the sound-conducting duct.

1.12.1 Mechanical Impedance, Z_m

The mechanical impedance is the ratio of a force to an associated velocity and is commonly used in acoustics to describe the radiation load presented by a medium to a vibrating surface. Radiation impedance, which is a mechanical impedance, will be encountered in Chapters 5 and 6.

Table 1.3 Three impedances used in acoustics

Type	Definition	Dimensions
1. Mechanical impedance	$Z_m = F/u = pS/u$	(M/T)
2. Specific acoustic impedance	$Z_s = p/u$	$(MT^{-1} L^{-2})$
3. Acoustic impedance	$Z_A = p/v = p/uS$	$(MT^{-1} L^{-4})$
where F = sinusoidally time-varying force		(MLT^{-2})
u = sinusoidally time-varying acoustic particle velocity		(LT^{-1})
p = sinusoidally time-varying acoustic pressure		$(MT^{-2} L^{-1}$
v = sinusoidally time-varying acoustic volume velocity		$(L^3 T^{-1})$
S = area		(L^2)

1.12.2 Specific Acoustic Impedance, Z_s

The specific acoustic impedance is the ratio of acoustic pressure to associated particle velocity. It is important in describing the propagation of sound in free space and is continuous at junctions between media. It is important in describing the reflection and transmission of sound at an absorptive lining in a duct or on the wall or ceiling of a room. It is also important in describing reflection of sound in the ground plane. The specific acoustic impedance will find use in Chapters 5, 6 and 9.

1.12.3 Acoustic Impedance, Z_A

The acoustic impedance will find use in Chapter 9 in the discussion of propagation in reactive muffling devices, where the assumption is implicit that the propagating wavelength is long compared to the cross-dimensions of the sound conducting duct. In the latter case, only plane waves propagate and it is then possible to define a volume velocity as the product of duct cross-sectional area, S, and particle velocity. The volume velocity is continuous at junctions in a ducted system as is the related acoustic pressure. Consequently, the acoustic impedance has the useful property that it is continuous at junctions in a ducted system (Kinsler *et al.*, 1982).

1.13 FLOW RESISTANCE

Porous materials are often used for the purpose of absorbing sound. Alternatively, it is the porous nature of many surfaces, such as grass-covered ground, that determines their sound reflecting properties. As discussion will be concerned with ground reflection in Chapters 5 and 8, with sound absorption of porous materials in Chapters 7 and 8 and Appendix C, with attenuation of sound propagating through porous materials in Chapter 8 and Appendix C, and with absorption of sound propagating in ducts lined with porous materials in Chapter 9, it is important to consider the property of porous materials that relates to their use for acoustical purposes.

A solid material that contains many voids is said to be porous. The voids may or may not be interconnected; however, for acoustical purposes it is the interconnected voids that are important; the voids, which are not connected, are generally of little importance. The property of a porous material which determines its usefulness for acoustical purposes, is the resistance of the material to induced flow through it, as a result of a pressure gradient. Flow resistance, an important parameter that is a measure of this property, is defined according to the following simple experiment. A uniform layer of porous material of thickness, ℓ, and area, S, is subjected to an induced mean volume flow, V_0 (m^3/s), through the material and the pressure drop, ΔP, across the layer is measured. Very low pressures and mean volume velocities are assumed (of the order of the particle velocity amplitude of a sound wave having a sound pressure level between 80 and 100 dB). The flow resistance of the material, R_f, is defined as the induced pressure drop across the layer of material divided by the resulting mean volume velocity per unit area of the material:

$$R_f = \Delta P S / V_0 \tag{1.102}$$

The units of flow resistance are the same as for specific acoustic impedance, ρc; thus it is sometimes convenient to specify flow resistance in dimensionless form in terms of numbers of ρc units.

The flow resistance of unit thickness of material is defined as the flow resistivity R_1 which has the units Pa s m^{-2}, often referred to as MKS rayls per metre. Experimental investigation shows that porous materials of generally uniform composition may be characterised by a unique flow resistivity. Thus, for such materials, the flow resistance is proportional to the material thickness, ℓ, as follows:

$$R_f = R_1 \ell \tag{1.103}$$

Note that flow resistance characterises a layer of specified thickness, whereas flow resistivity characterises a bulk material in terms of resistance per unit thickness.

For fiberglass and rockwool fibrous porous materials, which may be characterised by a mean fibre diameter, d, the following relation holds (Bies, 1984):

$$\frac{R_1 \ell}{\rho c} = 27.3 \left(\frac{\rho_m}{\rho_f} \right)^{1.53} \left(\frac{\mu}{d \rho c} \right) \left(\frac{\ell}{d} \right) \tag{1.104}$$

In the above equation, in addition to the quantities already defined, the gas density, ρ, (1.206 kg/m^2 for air at 20°C) the porous material bulk density, ρ_m, and the fibre material density, ρ_f have been introduced. The remaining variables are the speed of sound, c, of the gas and the dynamic gas viscosity, μ (1.84×10^{-5} kg m^{-1} s^{-1} for air at 20°C). The dependence of flow resistance on bulk density, ρ_m, and fibre diameter, d of the porous material is to be noted. Here, the importance of surface area is illustrated by Equation (1.104) from which one may readily conclude that the total surface area is proportional to the inverse of the fibre diameter in a fibrous material. Decrease in fibre diameter results in increase of flow resistivity and increase in sound absorption. A useful fibrous material will have very fine fibres.

Values of flow resistivity for a range of products found in Australia have been measured and published (Bies and Hansen, 1979, 1980). For further discussion of flow resistance, its method of measurement and other properties of porous materials which follow from it, the reader is referred to Appendix C.

CHAPTER TWO

The Human Ear

LEARNING OBJECTIVES

This chapter introduces the reader to:

- the anatomy of the ear;
- the response of the ear to sound;
- relationships between noise exposure and hearing loss;
- an understanding of the needs of the ear;
- loudness measures; and
- masking of some sound by other sound.

2.1 BRIEF DESCRIPTION OF THE EAR

The comfort, safety and effective use of the human ear are the primary considerations motivating interest in the subject matter of this book; consequently, it is appropriate to describe that marvellous mechanism. The discussion will make brief reference to the external and the middle ears and extensive reference to the inner ear where the transduction process from incident sound to neural encoding takes place. This brief description of the ear will make extensive reference to Figure 2.1.

2.1.1 External Ear

The pinna, or external ear, will be familiar to the reader and no further reference will be made to it other than the following few words. As shown by Blauert (1983) the convolutions of the pinna give rise to multiple reflections and resonances within it, which are frequency and direction dependent. These effects and the location of the pinna on the side of the head make the response of the pinna directionally variable to incident sound in the frequency range of 3 kHz and above. For example, a stimulus in the latter frequency range is heard best when incident from the side of the head.

If there were a mechanism tending to maintain levels in the ear within some narrow dynamic range, the variability in response resulting from the directional effects imposed by the pinna would be suppressed and would not be apparent to the listener. However, the information could be made available to the listener as a sense of location of the source. Amplification through undamping provided by the active response of the outer hair cells, as will be discussed in Section 2.2.3, seems to provide just such a mechanism. Indeed, the jangling of keys are interpreted by a single ear in such a way as to infer the direction and general location of the source in space without moving the head.

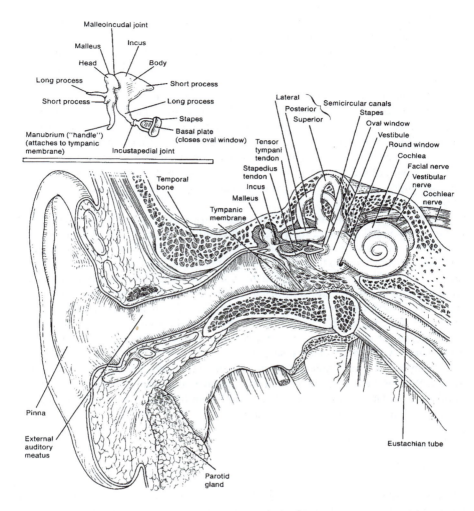

Figure 2.1 A representation of the pinna, middle and inner ear (right ear, face forward). Copyright by R. G. Kessel and R. H. Kardon, *Tissues and Organs: A Text-Atlas of Scanning Electron Microscopy*, W. H. Freeman, all rights reserved.

2.1.2 Middle Ear

Sound enters the ear through the auditory duct, a more or less straight tube between 23 and 30 mm in length, at the end of which is the eardrum, a diaphragm-like structure known as the tympanic membrane. Sound entering the ear causes the eardrum to move in response to acoustic pressure fluctuations within the auditory canal and to transmit motion through a mechanical linkage provided by three tiny bones, called ossicles, to a second membrane at the oval window of the middle ear. Sound is transmitted through the oval window to the inner ear (see Figure 2.1).

The middle ear cavity is kept at atmospheric pressure by occasional opening, during swallowing, of the eustachian tube also shown in Figure 2.1. If an infection causes swelling or mucus to block the eustachian tube, preventing pressure equalisation, the air in the middle ear will be gradually absorbed, causing the air pressure to decrease below atmospheric pressure and the tympanic membrane to implode. The listener then will experience temporary deafness.

Three tiny bones located in the air-filled cavity of the middle ear are identified in Figure 2.1 as the malleus (hammer), incus (anvil) and stapes (stirrup). They provide a mechanical advantage of about 3:1, while the relative sizes of the larger eardrum and smaller oval window result in an overall mechanical advantage of about 15:1. As the average length of the auditory canal is about 25 mm, the canal is resonant at about 4 kHz, giving rise to a further mechanical advantage about this frequency of the order of three.

The bones of the middle ear are equipped with a muscular structure (see stapedius and tensor tympani tendons, Figure 2.1), which allows some control of the motion of the linkage, and thus transmission of sound to the inner ear. For example, a moderately loud buzz introduced into the earphones of a gunner may be used to induce tensing of the muscles of the middle ear to stiffen the ossicle linkage, and to protect the inner ear from the loud percussive noise of firing. On the other hand, some individuals suffer from what is most likely a hereditary disease, which takes the form of calcification of the joints of the middle ear, rendering them stiff and the victim deaf. In this case the cure may, in the extreme case, take the form of removal of part of the ossicles and replacement with a functional prosthesis.

A physician, who counted many miners among his patients, once told one of the authors that for those who had calcification of the middle ear a prosthesis gave them good hearing. The calcification had protected them from noise induced hearing loss for which they also received compensation.

2.1.3 Inner Ear

The oval window at the entrance to the liquid-filled inner ear is connected to a small vestibule terminating in the semicircular canals and cochlea. The semicircular canals are concerned with balance and will be of no further concern here, except to remark that if the semicircular canals become involved, an ear infection can sometimes induce dizziness or uncertainty of balance.

In mammals, the cochlea is a small tightly rolled spiral cavity, as illustrated for humans in Figure 2.1. Internally, the cochlea is divided into an upper gallery (scala vestibuli) and a lower gallery (scala tympani) by an organ called the cochlear duct (scala media), or cochlear partition, which runs the length of the cochlea (see Figure 2.2). The two galleries are connected at the apex or apical end of the cochlea by a small opening called the helicotrema. At the basal end of the cochlea the upper and lower galleries terminate at the oval and round windows respectively. The round window is a membrane-covered opening situated in the same general location of the lower gallery as the oval window in the upper gallery (see Figure 2.1). A schematic representation of the cochlea unrolled is shown in Figure 2.2 and a cross-sectional view is shown in Figure 2.3.

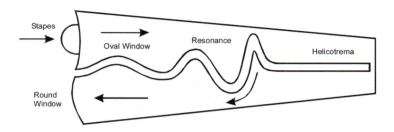

Figure 2.2 Schematic model of the cochlea (unrolled).

In humans, the average length of the cochlea from the basal end to the apical end is about 34 mm. The fluids within the cochlea, perilymph (sodium rich), which fills the upper and lower galleries and endolymph (potassium rich), which fills the cochlear partition, are essentially salt water.

As will be shown the central partition acts as a mechanical shunt between the upper and lower galleries. Any displacement of the central partition, which tends to increase the volume of one gallery, will decrease the volume of the other gallery by exactly the same amount. Consequently, it may be concluded that the fluid velocity fields in the upper and lower galleries are essentially the same but of opposite phase. For later reference these ideas may be summarised as follows. Let subscripts 1 and 2 refer respectively to the upper and lower galleries then the acoustic pressure p and volume velocity v (particle velocity multiplied by gallery cross sectional area) may be written as follows:

$$p = p_1 = -p_2 \qquad\qquad (2.1)$$

$$v = v_1 = -v_2 \qquad\qquad (2.2)$$

2.1.4 Cochlear Duct or Partition

The cochlear duct (see Figure 2.3), which divides the cochlea into upper and lower galleries, is triangular in cross-section, being bounded on its upper side next to the scala vestibuli by Reissner's membrane, and on its lower side next to the scala tympani by the basilar membrane. On its third side it is bounded by the stria vascularis, which is attached to the outer wall of the cochlea. The cochlear duct is anchored at its apical end to a bony ridge on the inner wall of the cochlear duct formed by the core of the cochlea. The auditory nerve is connected to the central partition through the core of the cochlea.

The closed three sides of the cochlear duct form a partition between the upper and lower galleries and hence the alternative name of cochlear partition. It has been

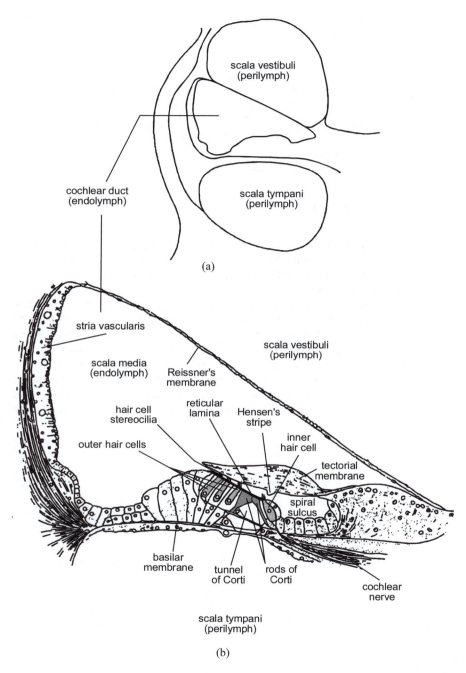

Figure 2.3 (a) Cross-section of the cochlea. (b) Cross-sectional detail of the organ of Corti.

suggested that the potassium rich endolymph of the cochlear duct supplies the nutrients for the cells within the duct, as there are no blood vessels in this organ. Apparently, the absence of blood vessels avoids the background noise which would be associated with flowing blood, because it is within the cochlear duct, in the organ of Corti, that sound is sensed by the ear.

The organ of Corti, shown in Figure 2.3, rests upon the basilar membrane next to the bony ridge on the inner wall of the cochlea and contains the sound-sensing hair cells. The sound-sensing hair cells in turn are connected to the auditory nerve cells which pass through the bony ridge to the organ of Corti. The supporting basilar membrane attached, under tension, at the outer wall of the cochlea to the spiral ligament (see Figure 2.3) provides a resilient support for the organ of Corti.

The cochlear partition, the basilar membrane and upper and lower galleries form a coupled system much like a flexible walled duct discussed in Section 1.3.4. In this system sound transmitted into the cochlea through the oval window proceeds to travel along the cochlear duct as a travelling wave with an amplitude that depends on the flexibility of the cochlear partition, which varies along its length. Depending on its frequency, this travelling wave will build to a maximum amplitude at a particular location along the cochlear duct as shown in Figure 2.2 and after that location, it will decay quite rapidly. This phenomenon is analysed in detail in Section 2.2.1.

Thus, a tonal sound, incident upon the ear results in excitation along the cochlear partition that gradually increases up to a place of maximum response. The tone is sensed in the narrow region of the cochlear partition where the velocity response is a maximum. The ability of the ear to detect the pitch (see Section 2.4.5) of a sound appears to be dependent (among other things that are discussed in Section 2.1.6 below) upon its ability to localise the region of maximum velocity response in the cochlear partition and possibly to detect the large shift in phase of the partition velocity response from leading to lagging the incident sound pressure in the region of maximum response.

The observations thus far may be summarised by stating that any sound incident upon the ear ultimately results in a disturbance of the cochlear partition, beginning at the stapes (basal) end, up to a length determined by the frequency of the incident sound. It is to be noted that all stimulus components of frequencies lower than the maximum response frequency result in some motion at all locations towards the basal end of the cochlear partition where high frequencies are sensed. For example, a heavy base note drives the cochlear partition over its entire length to be heard at the apical end. The model, as described thus far, provides a plausible explanation for the gross observation that with time and exposure to excessive noise, high-frequency sensitivity of the ear is progressively lost more rapidly than is low-frequency sensitivity (see Section 4.2.3).

As is well known, the extent of the subsequent disturbance induced in the fluid of the inner ear will depend upon the frequency of the incident sound. Very low-frequency sounds, for example 50 Hz, will result in motion of the fluid over nearly the entire length of the cochlea. Note that such motion is generally not through the helicotrema except, perhaps, at very low frequencies well below 50 Hz.

High-frequency sounds, for example 6 kHz and above, will result in motion restricted to about the first quarter of the cochlear duct nearest the oval window. The situation for an intermediate audio frequency is illustrated in Figure 2.2. An explanation for these observations is proposed in Section 2.2.1

2.1.5 Hair Cells

The sound-sensing hair cells of the organ of Corti are arranged in rows on either side of a rigid triangular construction formed by the rods of Corti, sometimes called the tunnel of Corti. As shown in Figure 2.3, the hair cells are arranged in a single row of inner hair cells and three rows of outer hair cells. The hair cells are each capped with a tiny hair bundle, hence the name, called hair cell stereocilia, which are of the order of 6 or 7 μm in length.

The stereocilia of the inner hair cells are free standing in the cleft between the tectorial membrane above and the reticular lamina below, on which they are mounted. They are velocity sensors responding to any slight motion of the fluid which surrounds them (Bies, 1999). Referring to Figure 2.3, it may be observed that motion of the basilar membrane upward in the figure results in rotation of the triangular tunnel of Corti and a decrease of the volume of the inner spiral sulcus and an outward motion of the fluid between the tectorial membrane and the reticular lamina. This fluid motion is amplified by the narrow passage produced by Hensen's stripe and suggests its function (see Figure 2.3). The inner hair cells respond maximally at maximum velocity as the tectorial membrane passes through its position of rest.

By contrast with the inner hair cells, the outer hair cells are firmly attached to and sandwiched between the radially stiff tectorial membrane and the radially stiff reticular lamina. The reticular lamina is supported on the rods of Corti, as shown in the Figure 2.3. The outer hair cells are capable of both passive and active response. The capability of active response is referred to as mortility. When active, the cells may extend and contract in length in phase with a stimulus up to about 5 kHz. Since the effective hinge joint of the tectorial membrane is at the level of its attachment to the limbus, while the hinge joint for the basilar membrane is well below at the foot of the inner rods of Corti, any slight contraction or extension of the outer hair cells will result in associated motion of the basilar membrane and rocking of the tunnel of Corti.

Motility of the outer hair cells provides the basis for undamping and active response. In turn, undamping provides the basis for stimulus amplification. Undamping of a vibrating system requires that work must be done on the system so that energy is replaced as it is dissipated during each cycle. The motility of the outer hair cells provides the basis for undamping. A mechanism by which undamping is accomplished in the ear has been described by Mammano and Nobili (1993) and is discussed in Section 2.2.3.

2.1.6 Neural Encoding

As illustrated in Figure 2.3, the cochlear nerve is connected to the hair cells through the inner boney ridge on the core of the cochlea. The cochlear nerve is divided about equally between the afferent system which carries information to the brain and the efferent system which carries instructions from the brain to the ear. The cells of the afferent nervous system, which are connected to the inner hair cells, convert the analog signal provided by the inner hair cells into a digital code, by firing once each cycle in phase, for excitation up to frequencies of about 5 kHz. At frequencies above about 5 kHz the firing is random. As the sound level increases, the number of neurons firing increases, so that the sound level is encoded in the firing rate. Frequency information also is encoded in the firing rate up to about 5 kHz. Pitch and frequency, though related, are not directly related (see Section 2.4.5). At frequencies higher than 5 kHz, pitch is associated with the location of the excitation on the basilar membrane.

In Section 2.1.7 it will be shown that one way of describing the response of the cochlear partition is to model it as a series of independent short segments, each of which has a different resonance frequency. However, as also stated in that section this is a very approximate model due to the basilar membrane response being coupled with the response of the fluid in the cochlear duct and upper and lower galleries. When sound pressure in the upper gallery at a segment of the cochlear partition is negative, the segment is forced upward and a positive voltage is generated by the excited hair cells. The probability of firing of the attached afferent nerves increases. When the sound pressure at the segment is positive, the segment is pushed downward and a negative voltage is generated by the hair cells. The firing of the cells of the afferent nervous system is inhibited. Thus, in-phase firing occurs during the negative part of each cycle of an incident sound.

Neurons attached to the hair cells also exhibit resonant behaviour, firing more often for some frequencies than others. The hair cells are arranged such that the neuron maximum response frequencies correspond to basilar membrane resonance frequencies at the same locations. This has the effect of increasing the sensitivity of the ear.

The digital encoding favoured by the nervous system implies a finite dynamic range of discrete steps, which seems to be less than 170 counts per second in humans. As the dynamic range of audible sound intensity bounded by "just perceptible" at the lower end and "painful" at the higher end is 10^{13}, a logarithmic-type encoding of the received signal is required. In an effort to provide an adequate metric, the decibel system has been adopted through necessity by the physiology of the ear.

Furthermore, it is seen that if the digital signal to the brain is to be encoded at a relatively slow rate of less than 170 Hz, and the intensity of the sound is to be described by the firing rate, then frequency analysis at the ear, rather than the brain, is essential. Thus, the ear decomposes the sound incident upon it into frequency components, and encodes the amplitudes of the components in rates of impulses forming information packets, which are transmitted to the brain for reassembly and interpretation.

In addition to the afferent nervous system, which conducts information from the ear to the brain, there also exists an extensive efferent nervous system of about the same sise, which conducts instructions from the brain to the ear. A further distinction between the inner hair cells lying next to the inner rods of Corti and the outer hair cells lying on the opposite side of the rods of Corti (see Figure 2.3) is to be observed.

Whereas about 90% of the fibres of the afferent nerve connect directly with the inner hair cells, and only the remaining 10% connect with the more numerous outer hair cells, the efferent system connections seem to be about equally distributed between inner and outer hair cells. The efferent system is connected to the outer hair cells and to the afferent nerves of the inner hair cells (Spoendlin, 1975). Apparently, the brain is able to control the state (on or off) of the outer hair cells and similarly the state of the afferent nerves connected to the inner hair cells. As will be shown in Section 2.2.3, control of the state of the outer hair cells enables selective amplification.

It is suggested here that control of amplification may also allow interpretation of directional information imposed on the received acoustic signal by the directional properties of the external ear, particularly at high frequencies (see Section 2.2.3). In support of the latter view is the anatomical evidence that the afferent nerve system and density of hair cells, about 16,000 in number, is greatest in the basilar area of the cochlea nearest to the oval window, where high-frequency sounds are sensed.

The connection of the efferent system to the afferent nerves of the inner hair cells suggests volume control to maintain the count rate within the dynamic range of the brain. In turn this supports the suggestion that the inner hair cells are the sound detectors. The function of the outer hair cells will be discussed in Section 2.2.3.

2.1.7 Linear Array of Uncoupled Oscillators

Voldřich (1978) has investigated the basilar membrane in immediate post mortem preparations of the guinea pig and he has shown that rather than a membrane it is accurately characterised as a series of discrete segments, each of which is associated with a radial fibre under tension. The fibres are sealed in between with a gelatinous substance of negligible shear viscosity to form what is referred to as the basilar membrane. The radial tension of the fibres varies systematically along the entire length of the cochlea from large at the basal end to small at the apical end. As the basilar membrane response is coupled with the fluid response in the cochlear duct, this is consistent with the observation that the location of maximum response of the cochlear partition varies in the audible frequency range from highest frequency at the basal end to lowest frequency at the apical end. This has led researchers in the past to state that the basilar membrane may be modelled as an array of linear, uncoupled oscillators. As the entire system is clearly coupled, this model is an approximate one only, but it serves well as an illustration.

Mammano and Nobili (1993) have considered the response of a segment of the central partition to an acoustic stimulus and have proposed the following differential equation describing the displacement, ξ, of the segment in response to two forces acting upon the segment. One force, F_S, is due to motion of the stapes and the other force, F_B, is due to motion of all other parts of the membrane. In the following equation, m is the segment mass, κ is the segment stiffness, z is a normalised longitudinal coordinate along the duct centre line from $z = 0$ at the stapes (basal end) to $z = 1$ at the helicotrema (apical end) and t is the time coordinate. The damping term, K, has the form given below in Equation (2.5). The equation of motion of a typical segment of the basilar membrane as proposed by Mammano and Nobili is:

$$m\frac{\partial^2 \xi}{\partial t^2} + K\frac{\partial \xi}{\partial t} + \kappa \xi = F_S + F_B \qquad (2.3)$$

The total force $F_S + F_B$ is the acoustic pressure p multiplied by the area, $w\Delta z$, of the segment of the basilar membrane upon which it acts, where w is the width of the membrane and Δz is the thickness of a segment in the direction along the duct centre-line expressed as follows:

$$F_S + F_B = pw\Delta z \qquad (2.4)$$

The damping term K, expressed as an operator, has the following form:

$$K = C + \frac{\partial s(z)}{\partial z}\frac{\partial}{\partial z} \qquad (2.5)$$

In Equation (2.5) the first term on the right hand side, C, is the damping constant due to fluid viscosity and the second term provides an estimate of the contribution of longitudinal viscous shear. The quantity, $s(z)$, is the product of two quantities: the effective area of the basilar membrane cross-section at location, z; and the average shearing viscosity coefficient (≈ 0.015 kg m^{-1} s^{-1}) of a section of the organ of Corti at location, z.

In the formulation of Mammano and Nobili, the first term in Equation (2.5) is a constant when the cochlea responds passively but is variable when the cochlea responds actively. In an active state, the variable C implies that Equation (2.3) is nonlinear. The second term in Equation (2.5) implies longitudinal coupling between adjacent segments and also implies that Equation (2.3) is nonlinear. However, it may readily be shown that the second term is negligible in all cases; thus, the term, K, will be replaced with the variable damping term, C, in subsequent discussion.

Variable damping will be expressed in terms of damping and undamping as explained below (see Section 2.2.3). When K is replaced with C, Equation (2.3) becomes the expression which formally describes the response of a simple one-degree-of-freedom linear oscillator for the case that C is constant or varies slowly. It will be shown that in a passive state, C is constant and in an active state it may be expected that C varies slowly. In the latter case, the cochlear response is quasi-stationary, about which more will be said later.

It is proposed that the cochlear segments of Mammano and Nobili may be identified with a series of tuned mechanical oscillators. It is proposed to identify a segment of the basilar membrane, including each fibre that has been identified by Voldřich (1978) and the associated structure of the central partition, as parts of an oscillator.

Mammano and Nobili avoid discussion of nonlinearity when the cochlear response is active. Instead, they tacitly assume quasi-stationary response and provide a numerical demonstration that varying the damping in their equation of motion gives good results. Here it will be explicitly assumed that slowly varying damping characterises cochlear response, in which case the response is quasi-stationary. Justification for the assumption of quasi-stationary response follows.

Quasi-stationary means that the active response time is long compared with the period of the lowest frequency that is heard as a tone. As the lowest audible frequency is, by convention, assumed to be 20 Hz, it follows that the active response time is longer than 0.05 seconds. As psycho-acoustic response times seem to be of the order of 0.3 to 0.5 seconds, a quasi-stationary solution seems justified. This assumption is consistent with the observation that the efferent and afferent fibres of the auditory nerve are about equal in number and also with the observation that no other means of possible control of the outer hair cells has been identified.

The observation that longitudinal viscous shear forces may be neglected leads to the conclusion that each cochlear partition segment responds independently of any modal coupling between segments. Consequently, the cochlear partition may be modelled approximately as a series of modally independent linear mechanical oscillators that respond to the fluid pressure fields of the upper and lower galleries. Strong fluid coupling between any cochlear segment and all other segments of the cochlea accounts for the famous basilar membrane travelling wave discovered by von Békésy (1960).

The modally independent segments of the cochlea will each exhibit their maximum velocity response at the frequency of undamped resonance for the corresponding mechanical oscillator. Thus a frequency of maximum response (loosely termed resonance), which remains fixed at all sound pressure levels, characterises every segment of the cochlear duct. The frequency of un-damped resonant response will be referred to here as the characteristic frequency or the resonance frequency. The characteristic frequency has the important property that it is independent of the system damping (Tse *et al.*, 1979). The amplitude of response, on the other hand, is inversely proportional to the system damping. Thus, variable damping provides variable amplification but does not affect the characteristic frequency.

2.2 MECHANICAL PROPERTIES OF THE CENTRAL PARTITION

2.2.1 Basilar Membrane Travelling Wave

The pressure fields observed at any segment of the basilar membrane consist not only of contributions due to motion of the stapes and, as shown here, to motion of the round window but, very importantly, to contributions due to the motion of all other parts of the basilar membrane as well. Here, it is proposed that the upper and lower galleries may each be modelled as identical transmission lines coupled along their entire length by the central partition, which acts as a mechanical shunt between the galleries (Bies, 2000).

Introducing the acoustic pressure, p, volume velocity, v (particle velocity multiplied by the gallery cross-sectional area), defined in Equations (2.1) and (2.2) and the acoustical inductance, L_A, per unit length of either gallery, the equation of motion of the fluid in either gallery takes the following form:

$$\frac{\partial p}{\partial z} = L_A \frac{\partial v}{\partial t} \qquad\qquad (2.6)$$

The acoustical inductance is an inertial term and is defined below in Equation (2.10).

Noting that motion of the central partition, which causes a loss of volume of one gallery, causes an equal gain in volume in the other gallery, the following equation of conservation of fluid mass may be written for either gallery:

$$\frac{\partial v}{\partial z} = 2 C_A \frac{\partial p}{\partial t} \tag{2.7}$$

where C_A is the acoustic compliance per unit length of the central partition and is defined below in Equation (2.16).

Equations (2.6) and (2.7) are the well-known transmission line equations due to Heaviside (Nahin, 1990), which may be combined to obtain the well-known wave equation.

$$\frac{\partial^2 p}{\partial z^2} = \frac{1}{c^2} \frac{\partial^2 p}{\partial t^2} \tag{2.8}$$

The phase speed, c, of the travelling wave takes the form:

$$c^2 = \frac{1}{2 C_A L_A} \tag{2.9}$$

The acoustical inductance per unit length, L_A, is given by:

$$L_A = \frac{\rho}{S_g} \tag{2.10}$$

where ρ is the fluid density and S_g is the gallery cross-sectional area.

The acoustical compliance, C_A, per unit length of the central partition is readily obtained by rewriting Equation (2.3). It will be useful for this purpose to introduce the velocity, u, of a segment of the basilar membrane, defined as follows:

$$u = \frac{\partial \xi}{\partial t} \tag{2.11}$$

Sinusoidal time dependence of amplitude, ξ_0, will also be assumed. Thus:

$$\xi = \xi_0 e^{j\omega t} \tag{2.12}$$

Introducing the mechanical compliance, C_M, Equation (2.3) may be rewritten in the following form:

$$-\frac{ju}{C_M \omega} = F_S + F_B \tag{2.13}$$

The mechanical compliance, C_M, is defined as follows:

$$C_M = (\kappa - m\omega^2 + jK\omega)^{-1} \tag{2.14}$$

The acoustical compliance per unit length, C_A, is obtained by multiplying the mechanical compliance by the square of the area of the segment upon which the total force acts and dividing by the length of the segment in the direction of the gallery centre line. The expression for the acoustical compliance per unit length is related to the mechanical compliance as follows:

$$C_A = w^2 \Delta z \, C_M \tag{2.15}$$

Substitution of Equation (2.14) in Equation (2.15) gives the acoustical compliance as follows:

$$C_A = \frac{w^2 \Delta z}{(\kappa - m\omega^2 + jC\omega)} \tag{2.16}$$

Substitution of Equations (2.10) and (2.16) into Equation (2.9) gives the following equation for the phase speed, c, of the travelling wave on the basilar membrane:

$$c = \sqrt{\frac{S_g(\kappa - m\omega^2 + jC\omega)}{2\rho w^2 \Delta z}} \tag{2.17}$$

To continue the discussion it will be advantageous to rewrite Equation (2.17) in terms of the following dimensionless variables, which characterise a mechanical oscillator. The un-damped resonance frequency or characteristic frequency of a mechanical oscillator, ω_N, is related to the oscillator variables, stiffness, κ, and mass, m, as follows (Tse *et al.*, 1979):

$$\omega_N = \sqrt{\kappa/m} \tag{2.18}$$

The frequency ratio, X, defined as the stimulus frequency, ω, divided by the characteristic frequency, ω_N, will prove useful and here is the frequency of maximum response at a particular location along the basilar membrane. That is:

$$X = \omega/\omega_N \tag{2.19}$$

The critical damping ratio, ζ, defined as follows, will play a very important role in the following discussion (see Section 10.2.1, Equation (10.12)):

$$\zeta = \frac{C}{2m\omega_N} = \frac{C}{2\sqrt{\kappa m}} \tag{2.20a,b}$$

It will be convenient to describe the mass, m, of an oscillator as proportional to the mass of fluid in either gallery in the region of an excited segment. The proportionality constant, α, is expected to be of the order of one.

$$m = \alpha \rho S_g \Delta z \tag{2.21}$$

Substitution of Equations (2.18) to (2.21) in Equation (2.17) gives the following equation for the speed of sound, which will provide a convenient basis for understanding the properties of the travelling wave on the basilar membrane.

$$c = \frac{\alpha S_g \omega_N}{w\sqrt{2}} \sqrt{1 - X^2 + j2\zeta X} \tag{2.22}$$

At locations on the basal side of a place of maximum response along the cochlear partition, where frequencies higher than the stimulus frequency are sensed, the partition will be driven below its frequency of maximum response. In this region the partition will be stiffness-controlled and wave propagation will take place. In this case, $X < 1$ and Equation (2.22) takes the following approximate form, which is real, confirming that wave propagation takes place.

$$c = \frac{\alpha S_g \omega_N}{w\sqrt{2}} \tag{2.23}$$

At distances on the apical side of a place of maximum response, the partition will be driven above the corresponding frequency of maximum response, the shunt impedance of the basilar membrane will be mass controlled and wave propagation in this region is not possible. In this case $X \gg 1$ and Equation (2.22) takes the following imaginary form, confirming that no real wave propagates. Any motion will be small and finally negligible, as it will be controlled by fluid inertia.

$$c = \frac{\alpha S_g \omega_N}{w\sqrt{2}} jX \tag{2.24}$$

In the region of the cochlear partition where its stiffness and mass, including the fluid, are in maximum response with the stimulus frequency, the motion will be large, and only controlled by the system damping. In this case $X = 1$ and Equation (2.22) takes the following complex form.

$$c = \frac{\alpha S_g \omega_N}{w\sqrt{2}} \sqrt{\zeta}(1 + j) \tag{2.25}$$

As shown by Equation (2.25), at a place of maximum response on the basilar membrane, the mechanical impedance becomes complex, having real and imaginary parts, which are equal. In this case, the upper and lower galleries are shorted together. At the place of maximum response at low sound pressure levels when the damping ratio, ζ, is small, the basilar membrane wave travels very slowly.

Acoustic energy accumulates at the place of maximum response and is rapidly dissipated doing work transforming the acoustic stimulus into neural impulses for transmission to the brain. At the same time, the wave is rapidly attenuated and conditions for wave travel cease, so that the wave travels no further, as first observed by von Békésy (1960). The model is illustrated in Figure 2.2, where motion is shown as abruptly stopping at about the centre of the central partition.

2.2.2 Energy Transport and Group Speed

In a travelling wave, energy is transported at the group speed. Lighthill (1991) has shown by analysis of Rhode's data that the group speed of a travelling wave on the basilar membrane tends to zero at a place of maximum response. Consequently, each frequency component of any stimulus travels to the place where it is resonant and there it slows down, accumulates and is rapidly dissipated doing work to provide the signal that is transmitted to the brain. The travelling wave is a marvellous mechanism for frequency analysis.

The group speed, c_g, is defined in Equation (1.33) in terms of the component frequency, ω, and the wave number, k. Rewriting Equation (1.33) in terms of the frequency ratio, X, given by Equation (2.19), gives the following expression for the group speed:

$$c_g = \omega_N \frac{\mathrm{d}X}{\mathrm{d}k} \tag{2.26}$$

The wave number is defined in Equation (1.24). Substitution of Equation (2.22) into Equation (1.24) gives an expression relating the wave number, k, to the frequency ratio, X, as follows:

$$k = \frac{w\sqrt{2}}{\alpha S_g} X(1 - X^2 + j2\zeta X)^{-1/2} \tag{2.27}$$

Substitution of Equation (2.27) in Equation (2.26) gives, with the aid of a bit of tedious algebra, the following expression for the group speed:

$$c_g = \frac{\alpha S_g \omega_N}{\sqrt{2}w} \frac{(1 - X^2 + j2\zeta X)^{3/2} (1 - j\zeta X)}{(1 + \zeta^2 X^2)} \tag{2.28}$$

In Equation (2.28), the damping ratio, ζ, appears always multiplied by the frequency ratio, X. This has the physical meaning that the damping ratio is only important near a place of resonant response, where the frequency ratio tends to unity. Furthermore, where the basilar membrane responds passively, the frequency ratio is small and the damping ratio then is constant, having its maximum value of 0.5 (see Section 2.2.4). It may be concluded that in regions removed from places of resonant response, the group speed varies slowly and is approximately constant.

As a stimulus component approaches a place of maximum response and at the same time the frequency ratio tends to 1, the basilar membrane may respond actively, depending upon the level of the stimulus, causing the damping ratio to become small. At the threshold of hearing, the damping ratio will be minimal, of the order of 0.011. However, at sound pressure levels of the order of 100 dB, the basilar membrane response will be passive, in which case the damping ratio will be 0.5, its passive maximum value. See Section 2.2.4 (Bies, 1996).

When a stimulus reaches a place of maximum response, the frequency ratio, $X = 1$, and the group speed is controlled by the damping ratio. The damping ratio in turn is determined by the active response of the place stimulated, which in turn, is determined by the level of the stimulus.

As a stimulus wave travels along the basilar membrane, the high frequency components are sequentially removed. When a stimulus component approaches a place of maximum response the corresponding frequency ratio, X, tends to the value 1 and the group speed of that component becomes solely dependent upon the damping ratio. The numerator of Equation (2.28) then becomes small, dependent upon the value of the damping ratio, ζ, indicating, as observed by Lighthill in Rhode's data, that the group speed tends to zero as the wave approaches a place of maximum response (Lighthill, 1996).

2.2.3 Undamping

It was shown in Section 2.1.6 that a voltage is generated at the outer hair cells in response to motion of the basilar membrane. In the same section it was shown that the outer hair cells may respond in either a passive or an active state, presumably determined by instructions conveyed to the hair cells from the brain by the attached efferent nerves. In an active state the outer hair cells change length in response to an imposed voltage (Brownell *et al.*, 1985) and an elegant mathematical model describing the biophysics of the cochlea, which incorporates this idea, has been proposed (Mammano and Nobili, 1993). In the latter model, the stereocilia of the outer hair cells are firmly embedded in the tectorial membrane and the extension and contraction of the outer hair cells results in a greater motion of the basilar membrane in the direction of the imposed motion. The associated rocking of the tunnel of Corti into and out of the sulcus increases the rate of flow through the cleft in which the inner hair cell cilia are mounted and thus amplifies their motion (velocity) relative to the surrounding fluid. In this model, the outer hair cells act upon the cochlea in a region of resonant response, resulting in amplification of as much as 25 dB at very low sound pressure levels.

The effect of the intervention of the outer hair cells is to un-damp the cochlea in a region of resonant response when the stimulation is of low level. For example, at sound levels corresponding to the threshold of hearing, undamping may be very large, so that the stimulated segment of the cochlea responds like a very lightly damped oscillator. Tuning is very sharp and the stimulus is greatly amplified. At increasing levels of stimulation undamping decreases, apparently to maintain the basilar membrane velocity response at the location of maximum response within a relatively narrow dynamic range. It is suggested here that this property may be the basis for interpretation of the distortions on an incident sound field imposed by the pinna. The latter distortions are associated with direction in the frequency range above 3 kHz. Thus, the direction of jangling keys may be determined with just one ear, without moving the head (see Section 2.1.1).

As shown by Equation (2.18) the frequency of maximum velocity response does not depend upon the system damping. By contrast, as shown by Equation (10.15), the frequency of maximum displacement response of a linear mechanical oscillator is dependent upon the system damping. As shown by the latter equation, when the damping of an oscillator is small the frequency of maximum displacement response approaches that of the un-damped resonance frequency (or frequency of maximum response), but with increasing damping, the frequency of maximum displacement response shifts to lower frequencies, dependent upon the magnitude of the damping ratio (see Equation (10.15)).

The inner hair cells, which are velocity sensors (Bies, 1999), are the cells that convert incident sound into signals, which are transmitted by the afferent system to the brain where they are interpreted as sound. Thus inner hair cells are responsible for conveying most of the amplitude and frequency information to the brain. At low sound pressure levels, in a region of resonant response, the outer hair cells amplify the motion of the inner hair cells, which sense the sound, by undamping the corresponding segments of the cochlea. Thus the outer hair cells play the role of compressing the response of the cochlea (Bacon, 2006) so that our hearing mechanism is characterised by a huge dynamic range of up to 130 dB, which would not have been possible without some form of active compression.

At high sound pressure levels undamping ceases, apparently to protect the ear. In summary, undamping occurs at relatively low sound pressure levels and within a narrow frequency range about the frequency of maximum response at a place of stimulation. At all other places on the cochlea, which do not respond to such an extent to the particular stimulus, and at all levels of stimulation, the cochlear oscillators are heavily damped and quite linear. Only in a narrow range of the place on the cochlea where a stimulus is sensed in resonant response and at low sound pressure levels is the cochlea nonlinear.

From the point of view of the engineer it is quite clear that the kind of nonlinearity, which is proposed here to explain the observed nonlinear response of the cochlea, is opposite to that which is generally observed in mechanical systems. Generally nonlinearity is observed at high levels of stimulation in systems, which are quite linear at low levels of stimulation.

2.2.4 The Half Octave Shift

Hirsh and Bilger (1955) first reported an observation that is widely referred to as the "half octave shift". They investigated the effect upon hearing levels at 1 kHz and

1.4 kHz of six subjects exposed to a 1 kHz tone for intervals of time ranging from 10 seconds to 4 minutes and for a range of sensation levels from 10 dB to 100 dB. Sensation levels are understood to mean levels relative to the threshold of the individual test subject.

Reporting mean results of their investigation, Hirsh and Bilger (1955) found that for one minute duration in each case and for all sensation levels from 10 to 100 dB the threshold shift at 1 kHz was essentially a constant 6 dB but the threshold shift at 1.4 kHz was an increasing monotonic function of sensation level. At a sensation level of 60 dB the shifts at the two frequencies were essentially the same, but at higher sensation levels, the shift at 1.4 kHz was greater than at 1 kHz. Many subsequent investigations have confirmed these results for other frequencies and for other species as well.

A result typical of such investigations, has been used to construct Figure 2.4. In the figure, temporary threshold shift (TTS) in decibels is shown as a function of frequency in kilohertz distributed on the abscissa generally as along the cochlear duct with high frequencies at the basal end on the left and low frequencies at the apical end on the right.

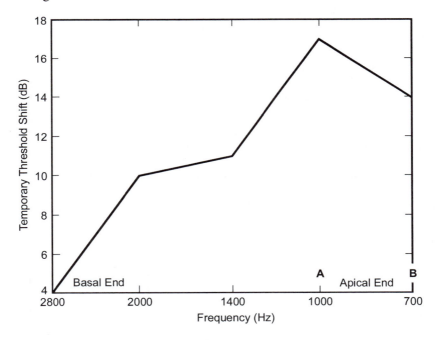

Figure 2.4 Typical half octave shift due to exposure to a loud 700 Hz tone.

In the figure, TTS is shown after exposure to an intense 700 Hz tone. It is observed that an 11 dB shift at 700 Hz is associated with a larger shift of about 17 dB at 1 kHz, one-half octave higher in frequency. Significant loss is observed also at even higher frequencies but essentially no loss is observed at frequencies below 700 Hz.

Crucial to the understanding of the explanation which will be proposed for the half-octave shift is the observation that the outer hair cells are displacement sensors and the inner hair cells, which provide frequency and amplitude information to the brain, are velocity sensors. While the idea that the outer hair cells are displacement sensors is generally accepted, the quantity that the inner hair cells sense is controversial. Bies (1999) has shown that the inner hair cells are velocity sensors as first proposed by Billone (1972) and that the published papers which have claimed otherwise and have created the controversy have in every case made errors, which negate their conclusions.

For explanation of Ward's data shown in Figure 2.4, reference will be made to Figure 2.5. To facilitate the explanation proposed here for the half octave shift, points **A** and **B** have been inserted in Figures 2.4 and 2.5. In Figure 2.4 they indicate the one-half octave above and the stimulus frequencies, respectively, while in Figure 2.5 they indicate locations on the cochlear duct corresponding to the places, respectively, where the one-half octave above and the stimulus frequencies are sensed. In the two figures, the Apical end and the Basal end have been inserted to remind the reader that low frequencies are sensed at the apical end and high frequencies are sensed at the basal end of the cochlear duct.

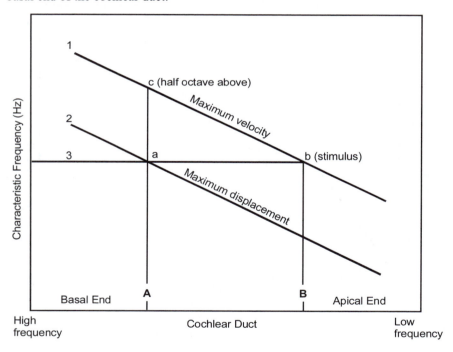

Figure 2.5 Half-octave shift model.

Reference is made now to Figure 2.5 on which the ordinate is the characteristic frequency associated with location on the central partition and the abscissa is the location on the central partition. In the figure, line (1), which remains fixed at all sound pressure levels, represents the locus of characteristic frequency (maximum velocity response) versus location on the duct. The location of line (2), on the other hand, represents the locus of the frequency of maximum displacement response at high sound pressure levels. The location of line (2) depends upon the damping ratio according to Equation (10.15) which, in turn, depends upon the sound pressure level.

Equation (10.15) shows that for the frequency of maximum displacement response to be one-half octave below the frequency of maximum velocity response for the same cochlear segment, the damping ratio must equal 0.5. In the figure, line (2) is shown at high sound pressure levels (> 100 dB) at maximum damping ratio and maximum displacement response. As the sound pressure level decreases below 100 dB, the damping decreases and line (2) shifts toward line (1) until the lines are essentially coincident at very low sound pressure levels.

Consider now Ward's investigation (Figure 2.4) with reference to Figure 2.5. Ward's 700 Hz loud exposure tone is represented by horizontal line (3), corresponding to exposure of the ear to a high sound pressure level for some period of time at the place of maximum displacement response at (a) and at the same time at the place of maximum velocity response at (b). The latter point (b) is independent of damping and independent of the amplitude of the 700 Hz tone, and remains fixed at location B on the cochlear partition. By contrast, the maximum displacement response for the loud 700 Hz tone is at a location on the basilar membrane where 700 Hz is half an octave lower than the characteristic frequency at that location for low level sound. Thus, the maximum displacement response occurs at intersection (a) at location **A** on the cochlear partition, which corresponds to a normal low-level characteristic frequency of about 1000 Hz, which is one-half of an octave above the stimulus frequency of 700 Hz.

The highest threshold shift, when tested with low level sound, is always observed to be one half octave higher than the shift at the frequency of the exposure tone at (b). Considering the active role of the outer hair cells, which are displacement sensors, it is evident that point (a) is now coincident with point (c) and that the greater hearing level shift is due to damage of the outer hair cells when they were excited by the loud tone represented by point (a). This damage may prevent the outer hair cells from performing their undamping action, resulting in an apparent threshold shift at the characteristic frequency for low level sound (half an octave higher than the high level sound used for the original exposure).

The lesser damage to the outer hair cells at frequencies higher than the frequency corresponding to one half octave above the exposure tone, may be attributed to the effect of being driven by the exposure tone at a frequency less than the frequency corresponding to the maximum velocity response. Estimation of the expected displacement response in this region on the basal side of point A on the cochlear duct, at the high damping ratio expected of passive response, is in reasonable agreement with this observation.

Here, a simple explanation has been proposed for the well-known phenomenon referred to as "the half octave shift". The explanation given here was previously reported by Bies (1996).

2.2.5 Frequency Response

It is accepted that the frequency response of the central partition ranges from the highest audible frequency at the basal end to the lowest audible frequency at the apical end and, by convention, it also is accepted that the lowest frequency audible to humans as a tone is 20 Hz and the highest frequency is 20 kHz. In the following discussion, it will be assumed that the highest frequency is sensed at the basal end at the stapes and the lowest frequency is sensed at the apical end at the helicotrema.

To describe the frequency response along the central partition, it will be convenient to introduce the normalised distance, z, which ranges from 0 at the basal end of the basilar membrane to 1 at the apical end. Based upon work of Greenwood (1990), the following equation is proposed to describe the frequency response of the central partition:

$$f(z) = 20146\,e^{-4.835z} - 139.8z \tag{2.29}$$

Substitution of $z = 0.6$ in Equation (2.29) gives the predicted frequency response as 1024 Hz. For $z \leq 0.6$, comparison of the relative magnitudes of the two terms on the right hand side of Equation (2.29) shows that the second term is always less than 8% of the first term and thus may be neglected. In this case Equation (2.29) takes the following form:

$$\log_e f = \log_e 20146 - 4.835\,z \tag{2.30}$$

Equation (2.30) predicts that, for frequencies higher than about 1 kHz, the relationship between frequency response and basilar membrane position will be log-linear, in agreement with observation.

2.2.6 Critical Frequency Band

A variety of psycho acoustic experiments have required for their explanation the introduction of the familiar concept of the band pass filter. In the literature concerned with the ear, the band pass filter is given the name critical frequency band (Moore, 1982). It will be useful to use the latter term in the following discussion in recognition of its special adaptations for use in the ear.

Of the 16000 hair cells in the human ear, about 4000 are the sound sensing inner hair cells, suggesting the possibility of exquisite frequency discrimination at very low sound pressure levels when the basilar membrane is very lightly damped. On the other hand, as will be shown, frequency analysis may be restricted to just 35 critical bands and as has been shown, variable damping plays a critical role in the functioning of the basilar membrane. Further consideration is complicated by the fact that damping may range from very small to large with concomitant variation in frequency response of the segments of the basilar membrane. Clearly, active response plays a critical role in determining the critical bandwidth, but the role played is not well understood.

For the case of 1 kHz and higher frequencies ($z \leq 0.6$), the derivative of Equation (2.30) may be written in the following differential form:

$$\frac{\Delta f}{f(z)} = -4.835 \, \Delta z \tag{2.31}$$

In Equation (2.31), the critical bandwidth may be associated with Δf and the centre band frequency with $f(z)$. Moore (1982) has summarised the work of several authors (Scharf, 1970), who provided experimental determinations of critical bandwidth as a function of frequency. This summary was adapted to construct Figure 2.6.

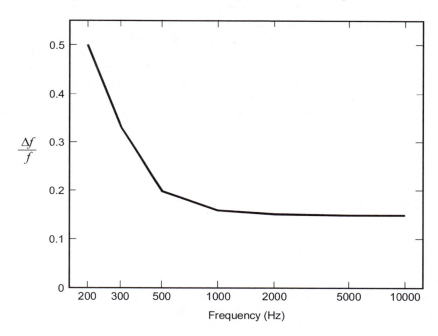

Figure 2.6 Normalised critical bandwidth as a function of centre band frequency based on Figure 3.6 of Moore (1982).

Referring to Figure 2.6, it may be observed that the ratio of critical bandwidth to centre band frequency is constant in the frequency range above 1 kHz. Equation (2.31) shows that in this frequency range each filter extends over a "constant length" of the central partition. A simple calculation using Equation (2.31), with information taken from Figure 2.6 and taking the average length of the basilar membrane as 34 mm (see Section 2.1.3), gives a value of about 1 mm for the "constant length" of the central partition (Moore, 1982). A maximum of 34 filters is suggested by this calculation, which is in very good agreement with experiments suggesting that the cochlear response may be described with about 35 critical bands (Moore, 1982). Each critical band is associated with a segment of the basilar membrane.

In Section 7.3.2, it is shown that the ratio of the resonance frequency, f, of an oscillator to the bandwidth, Δf, measured between the half power point frequencies, is given the name Quality Factor or simply, Q. The quality factor is a measure of the energy dissipated per radian in a cyclic system and thus is a measure of the sharpness of tuning. Consequently, it is reasonable to conclude from Figure 2.6 that the sharpness of tuning of the ear is greatest above about 2000 Hz.

An example of a psycho acoustic experiment in which the critical frequency band plays an important role is the case of masking of a test tone with a second tone or band of noise (see Section 2.4.1). An important application of such investigations is concerned with speech intelligibility in a noisy environment.

The well known phenomenon of beating between two pure tones of slightly different frequencies and the masking of one sound by another are explained in terms of the critical band. The critical bandwidth about a pure tone or a narrow band of noise is defined as the frequency band about the stimulus tone or narrow band of noise within which all sounds are summed together and without which all sounds are summed separately. This consideration suggests that the critical band is associated with a segment of the central partition. For example, energy summing occurs in the net response of a single stimulated segment.

Two pure tones will be heard as separate tones, unless their critical bands overlap, in which case they will be heard as one tone of modulated amplitude. This phenomenon is referred to as beating (see Section 1.11.3). In the case of masking of a test tone with a second tone or a narrow band of noise, only those frequency components of the masker that are in the critical band associated with the test tone will be summed with the test tone. Energy summing of test stimulus and masker components takes place at a place of resonance on the central partition.

2.2.7 Frequency Resolution

As shown in the discussion of spectra (see Section 1.10), noise of broad frequency content can best be analysed in terms of frequency bands, and within any frequency band, however narrow, there are always an infinite number of frequencies, each with an indefinitely small energy content. A tone, on the other hand, is characterised by a single frequency of finite energy content. The question is raised, "What bandwidth is equivalent to a single frequency?" The answer lies with the frequency analysing system of the ear, which is active and about which very little is known. As shown in Section 2.2.6, the frequency analysing system of the ear is based upon a very clever strategy of transporting all components of a sound along the basilar membrane, without dispersion, to the places of resonance where the components are systematically removed from the travelling wave and reported to the brain.

As has been shown, the basilar membrane is composed of about 35 separate segments, which are capable of resonant response and which apparently form the mechanical basis for frequency analysis. Although the number of inner hair cells is of the order of 4000, it appears that the basic mechanical units of frequency analysis are only 35 in number. The recognition of the existence of such discrete frequency bands has led to the definition of the critical band. The postulated critical band has provided the basis for explanation of several well known psycho-acoustic phenomena, which will be discussed in Section 2.4.

The significance of a limited number of basic mechanical units or equivalently critical bands from which the ear constructs a frequency analysis, is that all frequencies within the response range (critical band) of a segment will be summed as a single component frequency. A well known example of such summing, referred to as beating, was discussed in the previous section.

2.3 NOISE INDUCED HEARING LOSS

From the point of view of the noise-control engineer interested in protecting the ear from damage, it is of interest to note what is lost by noise-induced damage to the ear. The outer hair cells are most sensitive to loud noise and may be damaged or destroyed before the inner hair cells suffer comparable damage. Damage to the outer hair cells, which are essential to good hearing, seriously impairs the normal function of the ear.

It is of interest to explore the mechanism of noise-induced hearing loss. It is well accepted that the loss is due to failed hair cells and until recently, it was thought that stereocelia on the hair cells were mechanically damaged so that they were unable to perform their intended function. This is most likely the case where the hearing loss is caused by a single exposure to sudden, very intense sound. However, recent research (Bohne, *et al.*, 2007) has shown that regular exposure to excessive noise causes hearing damage in a different way. It has been shown that rather than mechanical damage, it is chemical damage that causes hearing loss. Regular exposure to excessive noise results in the formation of harmful molecules in the inner ear as a result of stress caused by noise-induced reductions in blood flow in the cochlea. The harmful molecules build up toxic waste products known as free oxygen radicles which injure a wide variety of essential structures in the cochlea causing cell damage and cell death, resulting in eventual widespread cell death and noise induced hearing loss. Once damaged in this way, hair cells cannot repair themselves or grow back and the result is a permanent hearing loss. This type of hearing damage is often accompanied by permanent tinnitus or "ringing in the ears". In fact the onset of temporary tinnitus after a few hours exposure to excessive noise levels such as found in a typical night club is a good indicator that some permanent loss has occurred even though the tinnitus eventually goes away and the loss is not noticeable by the individual.

One good thing about noise induced hearing loss being chemically instead of mechanically based is that one day there is likely to be a pill that can be taken to ameliorate the damage due to regular exposure to excessive noise.

It was suggested in Section 2.1.6 that the outer hair cell control of amplification would allow interpretation of directional information imposed upon an auditory stimulus by the directional properties of the external ear. Apparently, outer hair cell loss may be expected to result in an ear unable to interpret directional information encoded by the pinna on the received acoustic stimulus. A person with outer hair cell loss may have the experience of enjoying seemingly good hearing and yet be unable to understand conversation in a noisy environment. It is to be noted that outer hair cell destruction may be well under way before a significant shift in auditory threshold and other effects such as have been mentioned here are noticed. However, the ability of the ear to hear very low-level sound may be somewhat compromised by outer hair cell damage because, as explained in Section 2.2.4, the outer hair cells are responsible for the "undamping" action that decreases the hearing threshold of the inner hair cells.

A role for the outer hair cells in interpretation of the distortions of the received sound imposed by the pinna is suggested. If the suggestion is true, then in a noisy environment a person with outer hair cell loss will be unable to focus attention on a speaker, and thereby discriminate against the noisy environment. For example, a hearing aid may adequately raise the received level to compensate for the lost sensitivity of the damaged ear, but it cannot restore the function of the outer hair cells and it bypasses the pinna altogether.

With a single microphone hearing aid, all that a person may hear with severe outer hair cell loss in a generally noisy environment will be noise. In such a case a microphone array system may be required, which will allow discrimination against a noisy background and detection of a source in a particular direction. Just such an array was reported by Widrow (2001).

Some people with hearing loss suffer an additional problem, known as recruitment, which is characterised by a very restricted dynamic range of tolerable sound pressure levels between loud enough and too loud. Here it is suggested that severe outer hair cell loss would seem to provide the basis for an explanation for recruitment. For example, it was suggested in Section 2.1.6 that the function of the outer hair cells is to maintain the response of the inner hair cells within a fairly narrow dynamic range. Clearly, if the outer hair cells cannot perform this function, the overall response of the ear will be restricted to the narrow dynamic range of the inner hair cells. For further discussion of recruitment, see Section 2.4.2.

It has been shown that the basilar membrane may be modelled approximately as a series of independent linear oscillators, which are modally independent but are strongly coupled through the fluid in the cochlear. It has been shown also that nonlinearity of response occurs at low to intermediate levels of stimulation in a region about resonant response, through variable damping. It is postulated that the efferent system controls the level of damping, based upon cerebral interpretation of signals from the afferent system and that a time lag of the order of that typical of observed psycho-acoustic integration times, which seem to range between 0.25 and 0.5 seconds, is required for this process (Moore, 1982).

It is postulated here that the ear's response is quasi-stationary, and thus the ear can only respond adequately to quasi-stationary sounds; that is, sounds that do not vary too rapidly in level. It is postulated that the ear will respond inadequately to non-stationary sounds. When the ear responds inadequately to sound of rapidly variable level, it may suffer damage by being tricked into amplifying stimuli that it should be attenuating, and thereby forced to contribute to its own destruction.

In Chapter 4, criteria are presented and their use are discussed for the purpose of the prevention of hearing loss among individuals who may be exposed to excessive noise. The latter criteria, which are widely accepted, make specific recommendations for exposure defined in terms of level and length of time of exposure, which should not be exceeded. The latter criteria are based upon observed hearing loss among workers in noisy industrial environments.

It has been shown that exposure to loud sound, as described above (see Section 4.4), of symphonic musicians often exceeds recommended maximum levels (Jansson and Karlsson, 1983) suggesting, according to the accepted criteria, that symphonic musicians should show evidence of hearing loss due to noise exposure. The hearing of symphonic musicians has been investigated and no evidence of noise induced hearing loss has been observed (Karlsson, *et al.*, 1983).

It is to be noted that symphonic music is generally quasi-stationary, as defined here, whereas industrial noise is certainly not quasi-stationary. The suggestion made here that the ear is capable of coping adequately with quasi-stationary sound, but incapable of coping adequately with sound that is not quasi-stationary, provides a possible explanation for the observation that symphonic music does not produce the hearing loss predicted using accepted criteria. The observations made here would seem to answer the question raised by Brüel (1977) when he asked "Does A-weighting alone provided an adequate measure of noise exposure for hearing conservation purposes?" The evidence presented here seems to suggest not.

2.4 SUBJECTIVE RESPONSE TO SOUND PRESSURE LEVEL

Often it is the subjective response of people to sound, rather than direct physical damage to their hearing, which determines the standard to which proposed noise control must be compared, and which will determine the relative success of the effort. For this reason, the subjective response of people to sound will now be considered, determined as means of large samples of the human population (Moore, 1982). The quantities of concern are loudness and pitch. Sound quality, which is concerned with spectral energy distribution, will not be considered.

2.4.1 Masking

Masking is the phenomenon of one sound interfering with the perception of another sound. For example, the interference of traffic noise with the use of a public telephone on a busy street corner is probably well known to everyone. Examples of masking are shown in Figure 2.7, in which is shown the effect of either a tone or a narrow band of noise upon the threshold of hearing across the entire audible spectrum. The tone or narrow band of noise will be referred to as the masker.

Referring to Figure 2.7, the following may be observed:

1.   The masker is an 800 Hz tone at three sound pressure levels. The masker at 80 dB has increased the level for detection of a 600 Hz tone by 25 dB and the level for detection of a 1,100 Hz tone by 52 dB. The masker is much more effective in masking frequencies higher than itself than in masking frequencies lower than itself.

2.   The masker is a narrow band of noise 90 Hz wide centred at 410 Hz. The narrow band of noise masker is seen to be very much more effective in masking at high frequencies than at low frequencies, consistent with the observation in (a).

As shown in Figure 2.7, high frequencies are more effectively masked than are low frequencies. This effect is well known and is always present, whatever the masker. The analysis presented here suggests the following explanation. The frequency component energies of any stimulus will each be transported essentially without loss at a relatively constant group speed, to a place of resonance. As a component

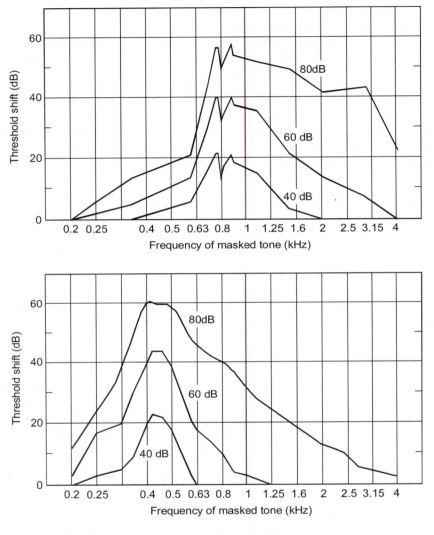

Figure 2.7 Example of masked audible spectra where the masker is either a tone or a narrow band of noise. The masker at three levels is: (a) 800 Hz tone; (b) a narrow band of noise 90 Hz wide centred on 410 Hz.

approaches a place of resonance, the group speed of the component slows down and reaches a minimum at the place of resonance, where the component's energy is dissipated doing work to provide a stimulus, which is transmitted to the brain. Only in the region of resonance of the masker, will the masker and test tone components be summed, giving rise to high threshold levels for detection of the test tone.

Evidently, the higher levels of threshold shift at high frequencies are due to the passage of the masker components through the places of resonance for high frequencies. It is suggested here that the most likely explanation is that the outer hair cells, which act to amplify a test stimulus at low levels, are inhibited by the high levels of excitation resulting from transmission of the masker. Consequently, the threshold level is elevated. By contrast, any residual components of the masker must decay very rapidly so that little or no masker is present on the apical side of the place of masker stimulation. Masking, which is observed at low frequencies, is due to the low frequency response of the filter. In a linear filter, unique relations exist which can provide guidance, but for the ear, the system is active and such relations are unknown.

In all of the curves of Figure 2.7(a), where the masker is a tone, a small dip is to be noted when the variable tone approaches the masking tone. This phenomenon may be interpreted as meaning that the tones are close enough for their critical bands to overlap. In the frequency range of critical bandwidth overlap, one tone of modulated amplitude will be heard. For example, consider two closely spaced but fixed frequencies. As the phases of the two sound disturbances draw together, their amplitudes reinforce each other, and as they subsequently draw apart, until they are of opposite phase, their amplitudes cancel. The combined amplitude thus rises and falls, producing beats (see Section 1.11.3).

Beats are readily identified, provided that the two tones are not greatly different in level; thus the dip is explained in terms of the enhanced detectability due to the phenomenon of beating. In fact, the beat phenomenon provides a very effective way of matching two frequencies. As the frequencies draw together the beating becomes slower and slower until it stops with perfect matching of tones. Pilots of propeller driven twin-engine aircraft use this phenomenon to adjust the two propellers to the same speed.

Reference is made now to Figure 2.8 where the effectiveness, as masker, of a tone and a narrow band of noise is compared. The tone is at 400 Hz and the band of noise is 90 Hz wide centred at 410 Hz. Both maskers are at 80 dB sound pressure level. It is evident that the narrow band of noise is more effective as a masker over most of the audio frequency range, except at frequencies above 1000 Hz where the tone is slightly more effective than a narrow band of noise.

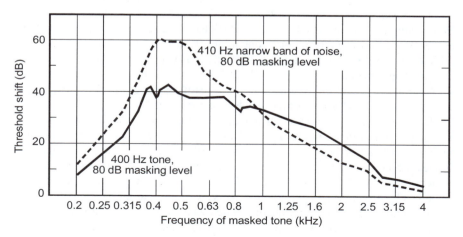

Figure 2.8 Comparison of a tone and a narrow band of noise as maskers.

It is of interest to note that the crossover, where the narrow band of noise becomes less effective as a masker, occurs where the ratio of critical bandwidth to centre band frequency becomes constant and relatively small (see Section 2.2.6 and Figure 2.6). In this range, the band filters are very sharply tuned. That is, the tone is more effective as a masker in the frequency range where the cochlear response is most sharply tuned, suggesting that the band pass filter is narrow enough to reject part of the narrow band of noise masker.

In the foregoing discussion of Figures 2.7 and 2.8, a brief summary has been presented of the effect of the masking of one sound by another. This information is augmented by reference to the work of Kryter (1970). Kryter has reviewed the comprehensive literature which was available to him and based upon his review he has prepared the following summary of his conclusions.

1. Narrowband noise causes greater masking around its frequency than does a pure tone of that frequency. This should be evident, since a larger portion of the basilar membrane is excited by the noise.

2. Narrowband noise is more effective than pure tones in masking frequencies above the band frequency.

3. A noise bandwidth is ultimately reached above which any further increase of bandwidth has no further influence on the masking of a pure tone at its frequency. This implies that the ear recognises certain critical bandwidths associated with the regions of activity on the basilar membrane.

4. The threshold of the masked tone is normally raised to the level of the masking noise only in the critical bandwidth centred on that frequency.

5. A tone, which is a few decibels above the masking noise, seems about as loud as it would sound if the masking noise were not present.

2.4.2 Loudness

The subjective response of a group of normal subjects to variation in sound pressure has been investigated (Stevens, 1957, 1972; Zwicker, 1958; Zwicker and Scharf, 1965). Table 2.1 summarises the results, which have been obtained for a single fixed frequency or a narrow band of noise containing all frequencies within some specified and fixed narrow range of frequencies. The test sound was in the mid audio-frequency range at sound pressures greater than about 2×10^{-3} Pa (40 dB re 20 μPa).

Note that a reduction in sound energy (pressure squared) of 50% results in a reduction of 3 dB and is just perceptible by the normal ear.

The consequence for noise control of the information contained in Table 2.1 is of interest. Given a group of noise sources all producing the same amount of noise, their number would have to be reduced by a factor of 10 to achieve a reduction in apparent loudness of one-half. To decrease the apparent loudness by half again, that is to one-quarter of its original subjectively judged loudness, would require a further reduction of sources by another factor of 10. Alternatively, if we started with one trombone player behind a screen and subsequently added 99 more players, all doing their best, an audience out in front of the screen would conclude that the loudness had increased by a factor of four.

Table 2.1 Subjective effect of changes in sound pressure level

Change in sound level (dB)	Change in power		Change in apparent loudness
	Decrease	Increase	
3	1/2	2	Just perceptible
5	1/3	3	Clearly noticeable
10	1/10	10	Half or twice as loud
20	1/100	100	Much quieter or louder

In contrast to what is shown in Table 2.1, an impaired ear with recruitment (see Section 2.3), in which the apparent dynamic range of the ear is greatly compressed, can readily detect small changes in sound pressure, so Table 2.1 does not apply to a person with recruitment. For example, an increase or decrease in sound power of about 10%, rather than 50% as in the table, could be just perceptible to a person with recruitment.

It has been observed that outer hair cells are more sensitive to excessive noise than are inner hair cells. It has also been observed that exposure to loud noise for an extended period of time will produce effects such as recruitment. These observations suggest that impairment of the outer hair cells is associated with recruitment. With time and rest the ear will recover from the effects of exposure to loud noise if the exposure has not been too extreme. However, with relentless exposure, the damage to the hair cells will be permanent and recruitment may be the lot of their owner.

2.4.3 Comparative Loudness and the Phon

Variation in the level of a single fixed tone or narrow band of frequencies, and a person's response to that variation has been considered. Consideration will now be given to the comparative loudness of two sounds of different frequency content. Reference will be made to two experiments, the results of which are shown in Figure 2.9 as cases (a) and (b).

Referring to Figure 2.9, the experiments have been conducted using many young people with undamaged normal ears. In the experiments, a subject was placed in a free field with sound frontally incident. The subject was presented with a 1 kHz tone used as a reference and alternately with a second sound used as a stimulus. In case (a), the stimulus was a tone and in case (b), the stimulus was an octave band of noise. The subject was asked to adjust the level of the stimulus until it sounded equally loud as the reference tone. After the subject had adjusted a stimulus sound so that subjectively it seemed equally as loud as the 1 kHz tone, the sound pressure of the stimulus was recorded. Maps based on mean lines through the resulting data are shown in Figure 2.9 (a) and (b). It is evident from the figures that the response of the ear as subjectively reported, is both frequency and pressure-amplitude dependent.

The units used to label the equal-loudness contours in the figure are called phons. The lines in the figure are constructed so that all variable sounds of the same number of phons sound equally loud. In each case, the phon scale is chosen so that the number of phons equals the sound pressure level of the reference tone at 1 kHz. For example, according to the figure at 31.5 Hz a stimulus of 40 phons (a tone in Figure 2.9(a) and an octave band of noise in Figure 2.9(b)) sounds equally loud as a 1000 Hz tone of 40 phons, even though the sound pressure levels of the lower-frequency sounds are about 35 dB higher. Humans are quite "deaf" at low frequencies. The bottom line in the figures represents the average threshold of hearing, or minimum audible field (*MAF*).

The phon has been defined so that for a tonal stimulus, every equal loudness contour must pass through the point at 1000 Hz where the sound pressure level is equal to the corresponding phon number. While what has been said is obvious, it is certainly not obvious what the sound pressure level will be at 1000 Hz for an octave-band of noise equal loudness contour of a given phon number. For this consideration attention is drawn to Table 2.2.

Table 2.2 has been constructed by subtracting the recorded sound pressure level shown in (b) from the sound pressure level shown in (a) of Figure 2.9 for all corresponding frequencies and phon numbers. At 1000 Hz the table shows that for the whole range of phon numbers from 20 to 110 phons the sound pressure level of the reference tone is greater than the sound pressure level of the corresponding equal loudness band of noise. Alternatively, and in every case at the 1000 Hz centre band frequency, the octave band of noise of lower sound pressure level sounds equally loud as the 1000 Hz reference tone.

Reference to Table 2.2 shows that the 1000 Hz reference tone ranges above the 1000 Hz octave band from a low of +1 dB at a phon number of 20 to a high of +5 dB at a phon number of 110. In view of the fact mentioned earlier (see Section 2.4.2), that a change of 3 dB in level is just noticeable it must be recognised that precision is very difficult to achieve. Consequently, with respect to the data in Table 2.2, it is suggested here that if phon differences at 4000 Hz levels above 60 phons are ignored as aberrant, all phon differences at frequencies of 1000 Hz to 4000 Hz might be approximated as 4 dB and 8000 Hz might be approximated as 11 dB. Similarly, it is suggested that all phon differences at frequencies of 500 Hz or less might be approximated as +1 dB. This suggestion is guided by the observation made earlier, in consideration of the critical band, and the proposed interpretation put upon the data shown in Figure 2.6.

2.4.4 Relative Loudness and the Sone

In the discussion of the previous section, the comparative loudness of either a tone or an octave band of noise, in both cases of variable centre frequency, compared to a reference tone, was considered and a system of equal loudness contours was established. However, the labelling of the equal loudness contours was arbitrarily chosen so that at 1 kHz the loudness in phons was the same as the sound pressure level of the reference tone at 1 kHz. This labelling provides no information about relative loudness; that is, how much louder is one sound than another.

In this section the relative loudness of two sounds, as judged subjectively, will be considered. Reference to Table 2.1 suggests that an increase in sound pressure level of 10 decibels will result in a subjectively judged increase in loudness of a factor of

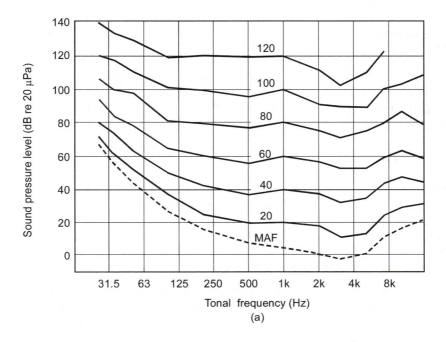

(a)

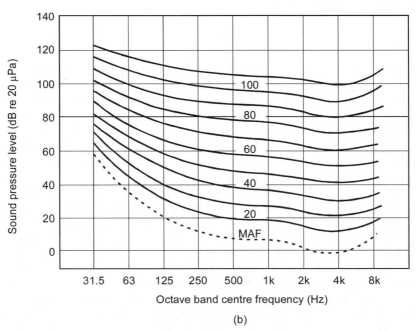

(b)

Figure 2.9 Equal loudness free-field frontal incidence contours in phons: (a) tonal noise; (b) octave band noise. MAF is the mean of the minimum audible field.

Table 2.2 Sound pressure level differences between tones and octave bands of noise of equal loudness based on data in Figure 2.9

Band centre frequency (Hz)	Loudness level differences (phons)					
	20	40	60	80	100	110
31.5	3	2	0	3	3	6
63	3	1	-2	-1	0	3
125	3	0	-2	0	1	2
250	0	-2	-1	-2	0	1
500	1	1	1	-1	2	3
1000	1	4	4	3	5	6
2000	4	6	4	4	3	3
4000	3	4	3	0	-1	-4
8000	14	16	14	11	7	9

2. To take account of the information in the latter table, yet another unit, called the sone, has been introduced. The 40-phon contour of Figure 2.9 has arbitrarily been labelled 1 sone. Then the 50-phon contour of the figure, which, according to Table 2.1, would be judged twice as loud, has been labelled two sones, etc. The relation between the sone, S, and the phon, P, is summarised as follows:

$$S = 2^{(P - 40)/10} \qquad (2.32)$$

At levels of 40 phons and above, the preceding equation fairly well approximates subjective judgment of loudness. However, at levels of 100 phons and higher, the physiological mechanism of the ear begins to saturate, and subjective loudness will increase less rapidly than predicted by Equation (2.32). On the other hand, at levels below 40 phons the subjective perception of increasing loudness will increase more rapidly than predicted by the equation. The definition of the sone is thus a compromise that works best in the mid-level range of ordinary experience, between extremely quiet (40 phons) and extremely loud (100 phons).

In Section 2.2.7, the question was raised "What bandwidth is equivalent to a single frequency?" A possible answer was discussed in terms of the known mechanical properties of the ear but no quantitative answer could be given. Fortunately, it is not necessary to bother with the narrow band filter properties of the ear, which are unknown. The practical solution to the question of how one compares tones with narrow bands of noise is to carry out the implied experiment with a large number of healthy young people, and determine the comparisons empirically.

The experiment has been carried out and an appropriate scheme has been devised for estimating loudness of bands of noise, which may be directly related to loudness of tones (see Moore, 1982 for discussion). The method will be illustrated here for octave bands by making reference to Figure 2.10(a). To begin, sound pressure levels in bands are first determined (see Sections 1.10.1 and 3.2). As Figure 2.10(a) shows, nine octave bands may be considered.

The band loudness index for each of the octave bands is read for each band from Figure 2.10(a) and recorded. For example, according to the figure, a 250 Hz octave band level of 50 dB has an index S_4 of 1.8. The band with the highest index S_{max} is determined, and the loudness in sones is then calculated by adding to it the weighted sum of the indices of the remaining bands. The following equation is used for the calculation, where the weighting B is equal to 0.3 for octave band and 0.15 for one-third octave band analysis, and the prime on the sum is a reminder that the highest-level band is omitted from the sum (Stevens, 1961):

$$L = S_{max} + B \sum_i{}' S_i \quad \text{(sones)} \tag{2.33}$$

When the composite loudness level, L (sones), has been determined, it may be converted back to phons and to the equivalent sound pressure level of a 1 kHz tone. For example, the composite loudness number computed according to Equation (2.33) is used to enter the scale on the left and read across to the scale on the right of Figure 2.10(a). The corresponding sound level in phons is then read from the scale on the right. The latter number, however, is also the sound pressure level for a 1 kHz tone.

Figure 2.10(b) is a more accurate, alternative representation of Figure 2.10(a), which makes it easier to read off the sone value for a given sound pressure level value.

Example 2.1

Given the octave band sound pressure levels shown in the example table in row 1, determine the loudness index for each band, the composite loudness in sones and in phons, and rank order the various bands in order of descending loudness.

Example 2.1 Table

Row		Octave band centre frequencies (Hz)							
	31.5	63	125	250	500	1000	2000	4000	8000
1. Band level (dB re 20 μPa)	57	58	60	65	75	80	75	70	65
2. Band loudness index (sones)	0.8	1.3	2.5	4.6	10	17	14	13	11
3. Ranking	9	8	7	6	5	1	2	3	4
4. Adjustment	0	3	6	9	12	15	18	21	24
5. Ranking level	57	61	66	74	87	95	93	91	89

Solution

1. Enter the band levels in row 1 of the example table, calculated using Figure 2.10(b), read the loudness indices S_i and record them in row 2 of the example table.

2.   Rank the indices as shown in row 3.
3.   Enter the indices of row 2 in Equation (2.33):
$$L = 17 + 0.3 \times 57.2 = 34 \text{ sones}$$
4.   Enter the computed loudness, 34, in the scale on the left and reading across of Figure 2.10(a), read the corresponding loudness on the right in the figure as 91 phons.

Example 2.2

For the purpose of noise control, a rank ordering of loudness may be sufficient. Given such a rank ordering, the effect of concentrated control on the important bands may be determined. A comparison of the cost of control and the effectiveness of loudness reduction may then be possible. In such a case, a short-cut method of rank ordering band levels, which always gives results similar to the more exact method discussed above is illustrated here. Note that reducing the sound level in dB(A) does not necessarily mean that the perceived loudness will be reduced, especially for sound levels exceeding 70 dB(A). Referring to the table of the previous example and given the data of row 1 of the example table, use a short-cut method to rank order the various bands.

Solution

1.   Enter adjustment levels shown in row 4 of the Example 2.1 table.
2.   Add the adjustment levels to the band levels of row 1.
3.   Enter adjusted levels in row 5.
4.   Note that the rank ordering is exactly as shown previously in row 3.

2.4.5 Pitch

The lowest frequency, which can be identified as a tone by a person with normal hearing, is about 20 Hz. At lower frequencies, the individual pressure pulses are heard; the sound is that of a discrete set of events rather than a continuous tone. The highest frequency that a person can hear is very susceptible to factors such as age, health and previous exposure to high noise levels. With acute hearing, the limiting frequency may be as high as 20 kHz, but normally the limit seems to be about 18 kHz.

Pitch is the subjective response to frequency. Low frequencies are identified as "flat" or "low-pitched", while high frequencies are identified as "sharp" or "high-pitched". As few sounds of ordinary experience are of a single frequency (for example, the quality of the sound of a musical instrument is determined by the presence of many frequencies other than the fundamental frequency), it is of interest to consider what determines the pitch of a complex note.

If a sound is characterised by a series of integrally related frequencies (for example, the second lowest is twice the frequency of the lowest, the third lowest is

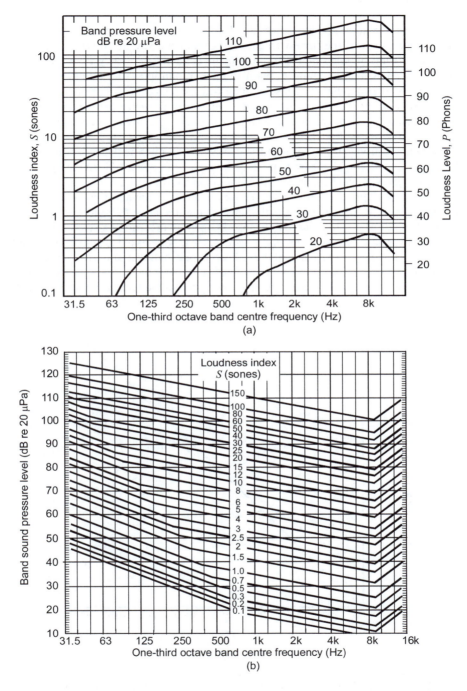

Figure 2.10 Relationship between loudness indices and band sound pressure level (octave or 1/3 octave).

three times the lowest, etc.), then the lowest frequency determines the pitch. Furthermore, even if the lowest frequency is removed, say by filtering, the pitch remains the same; the ear supplies the missing fundamental frequency. However, if not only the fundamental is removed, but also the odd multiples of the fundamental as well, say by filtering, then the sense of pitch will jump an octave. The pitch will now be determined by the lowest frequency, which was formerly the second lowest. Clearly, the presence or absence of the higher frequencies is important in determining the subjective sense of pitch.

Pitch and frequency are not linearly related, and pitch is dependent on sound level. The situation with regard to pitch is illustrated in Figure 2.11. In the figure are shown two lines, A and B. Line A, with a slope of unity, illustrates a linear relationship between sense of pitch and frequency. However, an experimental study produced the empirical curve B, which describes the sense of pitch relative to frequency for tones of 60 dB re 20 µPa. The latter curve was obtained by presenting a reference 1 kHz tone and a variable tone, sequentially, to listeners who were asked to adjust the second tone until its pitch was twice that of the reference tone, half that of the reference tone, etc., until the curve shown could be constructed. The experimenters assigned the name "mel" to the units on the ordinate, such that the pitch of the reference tone was 1000 mels.

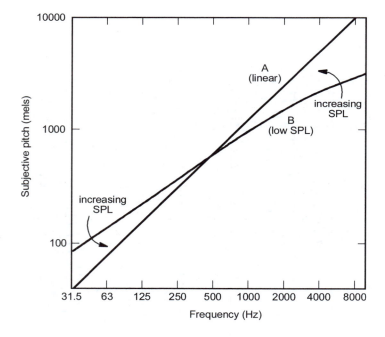

Figure 2.11 Subjective sense of pitch as a function of frequency. Line A is a linear relation observed in the limit at high sound pressure levels. Line B is a nonlinear relation observed at low sound pressure levels.

As mentioned previously, sense of pitch also is related to level. For example, consider a data point on line B as a reference and consider following the steps by which the data point was obtained. When this has been done, tones of level well above 60 dB and frequencies below 500 Hz tend to be judged flat and must be shifted right toward line A, while tones above 500 Hz tend to be judged sharp and must be shifted left toward line A. Referring to Figure 2.11, this observation may be interpreted as meaning that the subjective response (curve B) tends to approach the linear response (curve A) at high sound pressure levels, with the crossover at about 500 Hz as indicated in the figure. It is worthy of note that the system tends to linearity at high sound pressure levels.

CHAPTER THREE

Instrumentation for Noise Measurement and Analysis

LEARNING OBJECTIVES

In this chapter the reader is introduced to:

- acoustic instrumentation;
- condenser, electret and piezo-electric microphones;
- microphone sensitivity, definition and use;
- acoustic instrumentation calibration;
- sound level meters;
- noise dosimeters;
- tape recording of acoustical data;
- sound intensity analysers and particle velocity sensors;
- statistical noise analysers;
- frequency spectrum analysers; and
- instrumentation for sound source localisation.

3.1 MICROPHONES

A wide variety of transduction devices have been demonstrated over the years for converting sound pressure fluctuations to measurable electrical signals, but only two such devices are commonly used for precision measurement (Beranek, 1971, Chapters 3 and 4). As this chapter is concerned with the precision measurement of sound pressure level, the discussion will be restricted to these two types of transducer.

The most commonly used sound pressure transducer for precision measurement is the condenser microphone. To a lesser extent piezoelectric microphones are also used. Both microphones are used because of their very uniform frequency response and their long-term sensitivity stability. The condenser microphone generally gives the most accurate and consistent measure but it is much more expensive to construct than is the piezoelectric microphone. The condenser microphone is available in two forms, which are either externally polarised by application of a bias voltage in the power supply or pre-polarised internally by use of an electret. The externally polarised microphone is sensitive to dust and moisture on its diaphragm, but it is capable of reliable operation at elevated temperatures. The pre-polarised type is not nearly as sensitive to dust and moisture and is the microphone of choice in current instrumentation for accurate measurement of sound. It is also the most commonly used microphone in active noise control systems. Both forms of condenser microphone are relatively insensitive to vibration.

The piezoelectric microphone is less sensitive to dust and moisture but it can be damaged by exposure to elevated temperatures and, in general, it tends to be quite microphonic; that is, a piezoelectric microphone may respond about equally well to vibration and sound whereas a condenser microphone will respond well to sound and effectively not at all to vibration.

3.1.1 Condenser Microphone

A condenser microphone consists of a diaphragm that serves as one electrode of a condenser, and a polarised backing plate, parallel to the diaphragm and separated from it by a very narrow air gap, which serves as the other electrode. The condenser is polarised by means of a bound charge, so that small variations in the air gap due to pressure-induced displacement of the diaphragm result in corresponding variations in the voltage on the condenser.

The bound charge on the backing plate may be provided either by means of an externally supplied bias voltage of the order of 200 V, or by use of an electret, which forms either part of the diaphragm or the backing plate. Details of the electret construction and its use are discussed in the literature (Frederiksen *et al.*, 1979). For the purpose of the present discussion, however, the details of the latter construction are unimportant. In either case, the essential features of a condenser microphone and a sufficient (but simplified) representation of its electrical circuit for the present purpose are provided in Figure 3.1. A more detailed treatment is provided by Brüel and Kjær (1996)

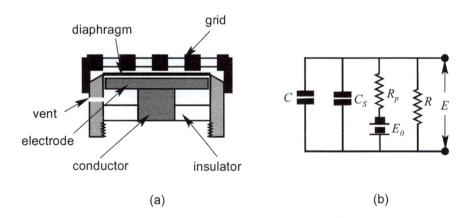

(a)   (b)

Figure 3.1 A schematic representation of a condenser microphone and equivalent electrical circuit.

Referring to Figure 3.1, the bound charge Q may be supplied by a d.c. power supply of voltage E_0 through a very large resistor R_p. Alternatively, the branch containing the d.c. supply and resistor R_p may be thought of as a schematic representation of the electret. The microphone response voltage is detected across the load resistor R. A good signal can be obtained at the input to a high internal impedance detector, even though the motion of the diaphragm is only a small fraction of the wavelength of light.

An equation relating the output voltage of a condenser microphone to the diaphragm displacement will be derived. It is first observed that the capacitance of a condenser is defined as the stored charge on it, Q, divided by the resulting voltage across the capacitance. Using this observation, it can be seen by reference to Figure 3.1, where C is the capacitance of the microphone and C_s is the stray capacitance of the associated circuitry, that for the diaphragm at rest with a d.c. bias voltage of E_0:

$$\frac{Q}{C + C_s} = E_0 \tag{3.1}$$

The microphone capacitance is inversely proportional to the spacing at rest, h, between the diaphragm and the backing electrode. If the microphone diaphragm moves a distance x inward (negative displacement, positive pressure) so that the spacing becomes $h - x$, the microphone capacitance will increase from C to $C + \delta C$ and the voltage across the capacitors will decrease in response to the change in capacitance by an amount E to $E_0 - E$. Thus:

$$E = -\frac{Q}{C + \delta C + C_s} + E_0 \tag{3.2}$$

The microphone capacitance is inversely proportional to the spacing between the diaphragm and the backing electrode, thus:

$$\frac{C + \delta C}{C} = \frac{h}{h - x} \tag{3.3}$$

Equation (3.3) may be rewritten as:

$$\delta C = C\left(\frac{1}{1 - x/h} - 1\right) \tag{3.4}$$

Substitution of Equation (3.4) into Equation (3.2) and use of Equation (3.1) gives the following relation:

$$E = -\frac{Q}{C}\left[\frac{1 - x/h}{1 + (C_s/C)(1 - x/h)} - \frac{1}{1 + C_s/C}\right] \tag{3.5}$$

Equation (3.5) may be rewritten in the following form:

$$E = -\frac{Q}{C}\left[\frac{1 - x/h}{1 + C_s/C}\left(1 + \frac{(x/h)(C_s/C)}{1 + C_s/C} + \dots\right) - \frac{1}{1 + C_s/C}\right] \tag{3.6}$$

The empirical constant K_1 is now introduced and defined as follows:

$$K_1 = \frac{1}{Ch} \tag{3.7}$$

By design, $C_s/C \ll 1$ and $x/h \ll 1$; thus in a well-designed microphone the higher order terms in Equation (3.6) may be omitted and by introducing Equation (3.7), Equation (3.6) takes the following approximate form:

$$E \approx \frac{K_1 Qx}{1 + C_s/C} - \frac{K_1 Qx^2}{h(1 + C_s/C)^2} \approx \frac{K_1 Qx}{1 + C_s/C} \tag{3.8a,b}$$

Equation (3.8b) reflects the fact that a positive pressure, which causes a negative displacement x (inward motion), results in a negative value of the induced voltage E.

Reference to Equation (3.7) shows that the constant K_1 depends upon the spacing at rest between the microphone diaphragm and the backing electrode, and the capacitance of the device, both of which are generally very difficult to determine by design, and consequently the constant must be determined by calibration. For good linear response, the capacitance ratio C_s/C must be kept as small as possible, and similarly, the microphone displacement relative to the condenser diaphragm backing electrode spacing, x/h, must be very small. Thus, in a well-designed microphone, the second term in Equation (3.8a) is negligible.

The vent shown in Figure 3.1 is an essential element of a microphone in that it allows the mean pressure to be equalised on both sides of the microphone diaphragm. Without the presence of this vent, any changes in atmospheric pressure would act to push the diaphragm one way or the other and drastically change the frequency response of the microphone. Unfortunately, the presence of the vent limits the low frequency response of the microphone so that below the cut off frequency caused by the vent, the microphone will be insensitive to acoustic pressure. The cut-off frequency is proportional to the size of the vent; the larger the vent the higher will be the cut-off frequency, which is defined as (Olsen, 2005):

$$f_{co} = \frac{1}{2\pi R_v C_{dc}} \tag{3.9}$$

where R_v is the resistance of the vent and for a vent consisting of a tube of radius, a, and length, ℓ, it is given by (Beranek, 1954):

$$R_v = \frac{8\mu\ell}{\pi a^4} \tag{3.10}$$

where μ is the dynamic viscosity for air and is equal to 1.84×10^{-5} N-s/m^5 at 20°C. The coefficient of viscosity varies with absolute temperature, T, in degrees Kelvin such that:

$$\mu \propto T^{0.7} \tag{3.11}$$

A typical value for R_v is 36×10^9 N-s/m^5. The quantity, C_{dc} in Equation (3.9) is the compliance of the diaphragm cavity system and is the reciprocal of the sum of the reciprocals of the compliance, C_c, of the cavity behind the microphone diaphragm and the compliance, C_d, of the diaphragm as these latter two compliances act together in parallel. Thus:

$$\frac{1}{C_{dc}} = \frac{1}{C_c} + \frac{1}{C_d} = \frac{C_d + C_c}{C_d C_c} \tag{3.12}$$

A typical value for the diaphragm compliance, C_d, is 0.3×10^{-12} m^5/N (Brüel and Kjær, 1996). The compliance of the cavity of volume V_c (m^3) is given by:

$$C_c = \frac{V_c}{\gamma P} \tag{3.13}$$

where $\gamma = 1.4$ is the ratio of specific heats for air and P is the mean atmospheric pressure (Pa). A typical value for C_c is 2.8×10^{-12} (Brüel and Kjær, 1996), which thus defines a typical value for V_c.

3.1.2 Piezoelectric Microphone

A sketch of the essential features of a typical piezoelectric microphone and a schematic representation of its electrical circuit are shown in Figure 3.2. In this case, sound incident upon the diaphragm tends to stress or unstress the piezoelectric element which, in response, induces a bound charge across its capacitance. The effect of the variable charge is like that of a voltage generator, as shown in the circuit.

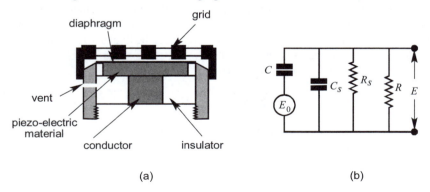

(a)   (b)

Figure 3.2 A schematic representation of a piezoelectric microphone and equivalent electrical circuit.

For a piezoelectric microphone, the equivalent circuit of which is shown in Figure 3.2, the microphone output voltage is given by:

$$E = \frac{Q}{C + C_s} = \frac{Q/C}{1 + C_s/C} \tag{3.14a,b}$$

As $E_0 = Q/C$:

$$E = \frac{E_0}{1 + C_s/C} \tag{3.15}$$

The voltage E_0 generated by the piezoelectric crystal is proportional to its displacement. Thus:

$$E_0 = K_2 x \tag{3.16}$$

Substituting Equation (3.16) into (3.15) gives:

$$E = \frac{K_2 x}{1 + C_s/C} \tag{3.17}$$

3.1.3 Pressure Response

Equations (3.8) and (3.17) show that the response voltage, E for both types of microphone, is essentially linearly related to the displacement x. In either case, an acoustic wave of pressure p incident upon the surface of the microphone diaphragm of area, S, will induce a mean displacement, x, determined by the compliance, K_3, of the diaphragm. Thus, the relation between acoustic pressure and mean displacement is:

$$x = -K_3 S p \tag{3.18}$$

Substitution of Equation (3.18) into either Equation (3.13) or Equation (3.17) results in a relation, formally the same in either case, between the induced microphone voltage and the incident acoustic pressure. Thus, for either microphone:

$$E = -\frac{K S p}{(1 + C_s/C)} \tag{3.19}$$

In Equation (3.19) the constant K must be determined by calibration.

3.1.4 Microphone Sensitivity

It is customary to express the sensitivity of microphones in decibels relative to a reference level. Following accepted practice, the reference voltage, E_{ref}, is set equal to 1V and the reference pressure, p_0, is set equal to 1 Pa. The following equation may then be written for the sensitivity of a microphone:

$$S = 20 \log_{10}[Ep_0/(E_{ref}p)] \tag{3.20}$$

Using Equation (1.78), Equation (3.20) may, in turn, be rewritten in terms of sound pressure level L_p as:

$$S = 20 \log_{10} E - L_p + 94 \quad \text{(dB)} \tag{3.21}$$

Typical microphone sensitivities range between -25 and -60 dB re 1VPa^{-1}. For example, if the incident sound pressure level is 74 dB, then a microphone of -30 dB re 1V Pa^{-1} sensitivity will produce a voltage which is down from 1V by 50 dB. The voltage thus produced is $E = 10^{-50/20} = 3.15$ mV.

3.1.5 Field Effects and Calibration

Reference to Equation (3.19) shows that the output voltage of a microphone is directly proportional to the area of the diaphragm. Thus the smaller a microphone of a given type, the smaller will be its sensitivity. As the example shows, the output voltage may be rather small, especially if very low sound pressure levels are to be measured, and the magnitude of the gain that is possible in practice is limited by the internal noise of the amplification devices. These considerations call for a microphone with a large diaphragm, which will produce a corresponding relatively large output voltage.

In the design of a good microphone one is concerned with uniform frequency response, besides general sensitivity as discussed above, and the demand for high-frequency response recommends against a microphone with a large diaphragm. The problem is that, at high frequencies, the wavelength of sound becomes very small. Thus, for any diaphragm, there will be a frequency at which the diameter of the diaphragm and the wavelength of sound are comparable. When this happens, large diffraction effects begin to take place which will make the microphone response to the incident sound quite irregular, and also sensitive to the direction of incidence. This may be undesirable when the direction of incidence is unknown.

Unfortunately, the problem of diffraction cannot be avoided; in fact diffraction effects begin to be apparent when the wavelength of the measured sound is still as much as 10 times the diameter of the microphone, and this is always well within the expected measurement range of the instrument. At still higher frequencies and shorter wavelengths the response of the microphone becomes quite sensitive to the angle of incidence of the measured sound, as illustrated in Figure 3.3. In the figure the increase or decrease in sound pressure level due to the presence of the microphone, relative to the sound pressure level that would exist in the absence of the microphone, is shown. Thus for a microphone characterised by the diffraction effects shown in the figure, and for a microphone diameter to wavelength ratio of 0.63, a normally incident sound will produce a sound pressure at the microphone which is 8 dB higher than the same sound will produce at grazing incidence.

Essentially, what is affected is the phase as well as the amplitude of the sound pressure distribution over the diaphragm of the microphone. Since the problem of

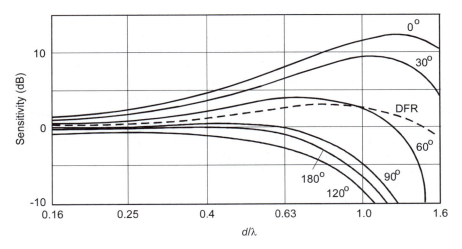

Figure 3.3 Microphone free-field correction: the sound pressure level on the diaphragm relative to the free-field level in the absence of the microphone, as a function of angle of incidence. Angles are measured relative to the normal to the diaphragm. DFR is the diffuse-field response, i.e. sound pressure level on the diaphragm in a diffuse field (random-incidence calibration); adapted from Bruel and Kjaer, 1973.

diffraction cannot be avoided for practical reasons of sensitivity, as has been explained, it is necessary to take account of the expected angle of incidence of sound upon a microphone during use. This has led to the so-called free-field calibration, which is a function of angle of incidence. Alternatively, if sound is assumed to be incident from all directions at once, then the properly weighted average of the free-field calibrations produces the random-incidence calibration. Such a weighted average is shown in Figure 3.3 as the diffuse-field response (DFR).

Yet another possibility exists. If a uniform pressure is imposed on the diaphragm (done in practice electrostatically), then a pressure response can be determined. The latter response is affected by the design of the backing electrode, the size and shape of the backing cavity, and the mass and tension of the diaphragm. It is thus possible to shape the high-frequency pressure response to compensate for the effective increase in pressure at the diaphragm due to diffraction. For example, if the pressure-response roll-off is shaped by design to just compensate for the increase of the free-field pressure at normal incidence, giving a combined response which is fairly flat for normally incident sound, then a free-field microphone is produced. Alternatively, the pressure response roll-off can be adjusted to compensate for the free-field increase at any angle of incidence, or it can be adjusted to compensate for the random incidence response, in which case a random-incidence or diffuse-field microphone is produced.

In summary, three types of calibration are recognised. These are free field, random incidence and pressure sensitivity. Microphones are commonly sold as (1) free field, generally meaning flat response for normal incidence, or (2) as random incidence, meaning generally flat response in a diffuse sound field characterised by sound of equal intensity incident from all directions. Alternatively, some sound level meters provide a filter network to allow simulation of diffuse field response for any particular microphone. In some applications where diffraction is not present (such as when the microphone is mounted flush with the wall of a duct to measure turbulent pressure fluctuations), the pressure-response calibration only is used.

Both free-field and diffuse-field microphones are used for industrial noise measurement. When it is obvious from which direction the noise is coming, then a free-field microphone is used and pointed directly at the source. When it is not obvious where the noise is coming from or when there are noise sources all around the microphone, then best results are obtained by using a diffuse-field microphone and pointing it straight up into the air. In this case most sound will then be incident at 90° to the microphone axis, and as the diffuse-field response is reasonably close to the 90° incidence response for many microphones the measurement error is minimised. For more accurate results, the actual error can be compensated by using the microphone characteristics for a particular microphone, similar to those shown in Figure 3.3.

All microphones must trade sensitivity for frequency response. Frequency response is inversely related to the microphone diaphragm diameter D, while the sensitivity is directly related to the fourth power of the diameter, as may be inferred by reference to Equations (3.19) and (3.21). It can be seen that good high-frequency response is obtained at the expense of a rapid deterioration in sensitivity. The effect is illustrated in Figure 3.4, where the respective sensitivities of three free-field microphones, having diameters of approximately 25, 12 and 6 mm, are shown. Fortunately, in the audio-frequency range, which is of interest here, it has been possible to produce, and make commercially available, microphones which have sufficient frequency response and sensitivity for most purposes; amplifiers are available which can adequately cope with their small output signals. Recent advances in microphone technology have resulted in 12 mm microphones that are 10 dB more sensitive than shown in Figure 3.4. Thus, 25 mm diameter microphones are not used very often any more as 12 mm microphones with similar sensitivity and better frequency response are available.

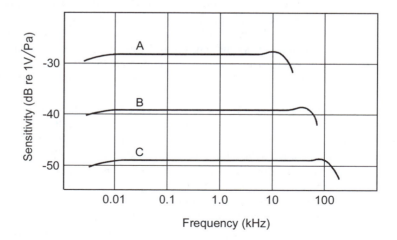

Figure 3.4 Free-field condenser microphone sensitivities. Diaphragm diameters: curve A, 25 mm; curve B, 12 mm; and curve C, 6 mm.

3.1.6 Microphone Accuracy

The accuracy of a microphone is dependent on the frequency of the sound being measured, the quality of the microphone and the accuracy of the calibrator. The accuracy and frequency response of modern digital instrumentation to which the microphone is attached is usually well in excess of the microphone accuracy and thus need not be considered here.

In a calibration laboratory, it is possible to calibrate microphones to an accuracy of 0.05 dB. However, outside of a calibration laboratory, pistonphones held on to the microphone while it is attached to a sound level meter or analyser are used. Microphone calibrators for use in the field or non certified laboratories usually put out a single tonal frequency, although some are available that provide multiple tones for calibration at different frequencies. However, when using a calibrator of this type, it is assumed that the frequency response of the microphone is flat over the entire frequency range. To ensure that the microphone frequency response remains inside the band specified by the manufacturer, it is usually necessary to have the microphone and attached instrumentation calibrated once per year at a certified calibration laboratory.

Microphone calibrators used during field measurements are intended to calibrate not just the microphone but the entire measurement system. Most such calibrators have an accuracy of ± 0.2 dB. There are also some precision pistonphone calibrators that have an accuracy of better than ± 0.1 dB. However, the frequency response of a microphone over the audio frequency range is such that even with a type 1 sound level meter, it is difficult to achieve a measurement accuracy of better than ± 0.5 dB in any frequency band and especially for the overall A-weighted level.

3.2 WEIGHTING NETWORKS

In the previous chapter it was shown that the apparent loudness of a sound (that is, the subjective response of the ear) varies with frequency as well as with sound pressure, and that the variation of loudness with frequency also depends to some extent on the sound pressure. Most sound-measuring instruments have the option to make allowances for this behaviour of the ear by providing electronic "weighting" networks. The various standards organisations recommend the use of three weighting networks (A, B and C), as well as a linear (unweighted) network for use in sound level meters.

The A-weighting circuit was originally designed to approximate the response of the human ear at low sound levels. Similarly, B and C networks were intended to approximate the response of the ear at levels of 55-85 dB and above 85 dB, respectively. The characteristics of these networks are shown in Figure 3.5, with 1/3 octave band values tabulated in Table 3.1. A fourth network, the D-weighting, has been proposed specifically for aircraft noise measurements. However, it has not gained favour and instead, the C-weighting network is currently used for aircraft noise characterisation. For all other noise measurements, the trend appears to be toward exclusive use of the A-weighting network. This is because the A-weighting curve is a good approximation of the ear response to low level sound such as may be typical of environmental noise. Hearing loss as a result of exposure to high level industrial

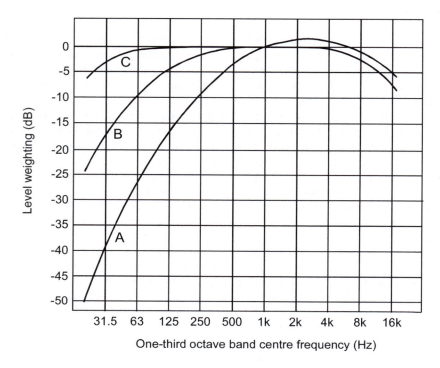

One-third octave band centre frequency (Hz)

Figure 3.5 International standard A-, B- and C-weighting curves for sound level meters.

Table 3.1 A-weighting network corrections (dB)

Frequency (Hz)	A-weighting correction	Frequency (Hz)	A-weighting correction	Frequency (Hz)	A-weighting correction
10	-70.4	160	-13.4	2500	1.3
12.5	-63.4	200	-10.9	3150	1.2
16	-56.7	250	-8.6	4000	1.0
20	-50.5	315	-6.6	5000	0.5
25	-44.7	400	-4.8	6300	-0.1
31.5	-39.4	500	-3.2	8000	-1.1
40	-34.6	630	-1.9	10000	-2.5
50	-30.2	800	-0.8	12500	-4.3
63	-26.2	1000	0.0	16000	-6.6
80	-22.5	1250	0.6	20000	-9.3
100	-19.1	1600	1.0		
125	-16.1	2000	1.2		

noise also seems to be a function of the A-weighted sound level, even though the apparent loudness of high level noise is closer to the C-weighting curve.

When use of a weighting network proves desirable, Figure 3.5 shows the correction which must be added to a linear reading to obtain the weighted reading for a particular frequency. For convenience, corrections for the A-weighting network at 1/3 octave band centre frequencies, are listed in Table 3.1. For example, if the linear reading at 125 Hz were 90 dB re 20 μPa, then the A-weighted reading would be 74 dB(A).

Example 3.1

Given the sound spectrum shown in line 1 of the table below, find the overall unweighted (linear) sound level in decibels and the A-weighted sound level in dB(A).

Example 3.1 Table

Octave band centre frequency (Hz)	31.5	63	125	250	500	1000	2000	4000	8000
Linear level (dB)	95	95	90	85	80	81	75	70	65
A-wt correction	-39.4	-26.2	-16.1	-8.6	-3.2	0	1.2	1.0	-1.1
A-wt level		55.6	68.6	73.9	76.4	76.8	81.0	76.2	71.0	63.9

Solution

The linear level is calculated using:

$$L_{pt} = 10\log_{10}\sum_{i=1}^{9} 10^{L_{pi}/10} = 99.0 \text{ dB}$$

where L_{pi} are the levels shown in line 1 of the table. The A-weighted overall level is found by adding the A-weighting corrections (see Table 3.1) to line 1 to obtain line 3 and then adding the levels in line 3 using the above expression to give:

$$L_{pt} = 84.9 \text{ dB(A)}$$

3.3 SOUND LEVEL METERS

The most common and convenient instrument for measuring the sound pressure level of a sound field is a sound level meter. Currently available sound level meters are all digital with dynamic ranges up to 100 dB. They consist of an analog to digital converter connected to a digital signal processor board that converts the time varying signal to whatever is specified by the user through the GUI (graphical user interface). Sound level meters range from the very simple to the very complex. At the very simple end there may be no user adjustments other than an on-off switch and they usually have a digital readout showing a single number A-weighted sound level that updates every second or so. Slightly better sound level meters have a sensitivity adjustment that allows it to be calibrated and they display the A-weighted sound level averaged over a specified time. Top of the range sound level meters can display octave and 1/3 octave band spectra, as well as L_{eq}, L_{Aeq} and statistical quantities such as L_{10} (sound pressure level exceeded 10% of the time) and L_{90}. They also have A- and C-weighted internal filter networks and an all-pass (linear) arrangement without weighting, as well as some adjustment that allows the user to select the measurement range so that the instrument can be optimised to measure low level or high level noises. These sound level meters are particularly useful for assessing noise for control purposes, especially if the noise level is non-steady or fluctuates with time. In this latter case, it is essential that the sound level meter is able to measure L_{eq}, and L_{Aeq}.

L_{eq}, the equivalent continuous sound pressure level, is defined in terms of the time-varying sound pressure level, $L(t)$, as follows:

$$L_{eq} = 10 \log_{10} \left[\frac{1}{T} \int_0^T 10^{L(t)/10} \, dt \right] \tag{3.22}$$

which is essentially the time averaged sound pressure squared converted to decibels.

The quantity L_{Aeq} is the equivalent continuous A-weighted noise level, which characterises fluctuating noise as an equivalent steady-state level. The A-weighted equivalent continuous noise level is found by replacing the unweighted sound pressure level in Equation (3.22) with the A-weighted sound pressure level and is defined as:

$$L_{Aeq} = 10 \log_{10} \left[\frac{1}{T} \int_o^T 10^{(L_{pA}(t)/10)} \, dt \right] \quad dB(A) \tag{3.23}$$

where T is a long period of time and $L_{pA}(t)$ is the A-weighted sound pressure level at time t. If the quantity T (hours) outside of the integral (but not the integration limit) in Equation (3.23) is replaced with 8 (hours), then L_{Aeq} will become $L_{Aeq,8h}$.

The dynamic range (difference between the maximum and minimum measurable sound levels) of a sound level meter as well as its electronic noise floor are two quantities used to distinguish better equipment. These quantities are also characteristic of the microphone chosen for the task. When high sound levels are to be measured (such as above 140 dB) a low sensitivity microphone should be chosen and when low sound levels are to be measured (such as below 30 dB), a high sensitivity microphone should be used so that the noise floor of the sound level meter or the microphone is not a problem. Most microphones have a dynamic range of 100 to 120 dB.

Most sound level meters have a choice of two meter response times. One, labelled "fast", has a time constant of 100 milliseconds and is designed to approximate the response of the ear. The other, labelled "slow", has a time constant of 1 second, and while not simulating the response of the ear, it is was useful in older sound level meters for determining mean levels when the measured sound fluctuated continuously and violently during the course of measurement when the sound level meter did not have the ability to calculate L_{eq}. Some sound level meters have a third response referred to as "impulse" response (with a standard time constant of 35 milliseconds) for the measurement of impulsive-type noises such as drop forges. Many sound level meters with an impulse response characteristic also have a capability to measure peak noise levels using a standard time constant of 50 µ seconds. Measurement standards and statutory regulations generally specify which meter response should be used for particular measurements.

3.4 CLASSES OF SOUND LEVEL METER

The various classes of sound level meter are discussed in IEC 61672/1 – 2002. Earlier standards published by other organisations are out of date and do not reflect current sound level meter technology. The standard divides sound level meters into two classes: class 1 or class 2, in order of decreasing accuracy. These classes are described as follows:

- **Class 1:**   precision sound level meter, intended for laboratory use or for field use where accurate measurements are required;
- **Class 2:**   general-purpose sound level meter, intended for general field use and for recording noise level data for later frequency analysis.

3.5 SOUND LEVEL METER CALIBRATION

The battery voltage should be checked regularly and the sound level meter and microphone calibration should be checked and adjusted according to the operating instructions. It is recommended that the instrument be overhauled and sent to a calibration facility for full calibration every two years.

Strictly, calibration is an end to end check of an instrument over its entire useful frequency range, and is generally a time-consuming procedure involving specialised equipment dedicated to that purpose. Reference checks at one or more selected frequencies, quite often loosely called calibration, may be used in the field and are strongly recommended. Facilities for making such reference calibration checks differ between manufacturers and types of instruments, but calibration usually involves placing a noise generating device on the microphone that generates a known sound pressure level (often 94 dB or 124 dB) at one specific frequency as described in Section 3.5.2.

3.5.1 Electrical Calibration

An externally or internally generated electrical signal of known amplitude and frequency is injected into the microphone amplifier circuit. If the output meter deflection is incorrect, slight compensation can be made to the amplification by adjustment of a preset control. Although this calibration can check the amplifier, and the weighting networks and filters, the microphone sensitivity is not checked. Thus, it is important to supplement this form of calibration with regular acoustic calibration.

3.5.2 Acoustic Calibration

A tonal acoustic signal of known sound pressure level is applied to the microphone and the meter reading is compared with the reference level. Any error outside the quoted calibrator tolerance may be adjusted by the preset gain control (if available). Most devices use 1000 Hz as the calibration frequency, as the same result is obtained with the A-weighting network switched in or out. Calibrators operating at other frequencies require all weighting networks to be switched out during the calibration process. The noise generating device is usually called a "pistonphone" and care should be taken to ensure that there is a good seal between the microphone housing and calibrator cavity (usually achieved with an "O-ring" on the calibrator part that slips over the microphone).

During calibration, the sensitivity adjustment on the sound level meter is adjusted to make it read a value equal to the sound pressure level generated by the pistonphone calibrator. Calibration is accurate to ± 0.5 dB for most instruments intended for use in the field (general purpose sound level meters) and ± 0.2 dB for precision instruments for laboratory use. Such calibration is limited to one or a few discrete frequencies, so that the procedure can only serve as a spot check at these frequencies. Large errors may indicate damage to either the sound level meter or the calibrator, and in such cases both should be returned to the manufacturer for checking.

3.5.3 Measurement Accuracy

The accuracy of measurements taken with a sound level meter depends on the calibration accuracy as well as the flatness of the frequency response of the sound level meter electronics and microphone. So it is important to examine the calibration chart supplied with the instrument to determine the accuracy over various frequency ranges. In practice, an accuracy of ± 0.5 dB is usually achievable with a type 1 sound level meter and repeatability should be ± 0.1 dB provided that the noise source does not vary and measurements are taken indoors in a stable environment with no air movement or temperature variation. Thus when checking to determine whether an item of equipment meets its specified noise level, which could be sound power level or sound pressure level at a specified relative location, it is unrealistic to expect to be able to measure with a better accuracy than ± 0.5 dB with a type 1 sound level meter and and ± 1dB with a type 2 sound level meter. In many cases of outdoor or on-site sound pressure level measurements, it is difficult to justify reporting results with any more precision than 1 dB.

3.6 NOISE MEASUREMENTS USING SOUND LEVEL METERS

Good data are the result of careful attention to detail and the elimination of possible sources of measurement error. It should be noted that the following considerations are quite general and may apply whether or not measurements are made with a sound level meter. The main sources of error are listed below.

3.6.1 Microphone Mishandling

If the microphone or its stand is placed upon a surface which is vibrating, then the microphone may produce spurious signals which will be read on the meter as sound pressure levels. Care must be taken to keep the microphone away from sources of vibration. Exposure to extreme noise levels (see manufacturer's data), or shocks due to dropping the microphone may cause a serious change in calibration.

3.6.2 Sound Level Meter Amplifier Mishandling

The amplifier of the sound level meter must be protected from magnetic fields, intense sound and vibration, otherwise it may itself act as a microphone and give an erroneous output.

As some older-style sound level meters have both an input and output amplifier, it is possible to overload the input amplifier and still obtain a meter deflection on scale. This condition produces erroneous readings and is easily undetected unless the amplifiers have visual overload indicators. A common situation where overloading of the input amplifier can occur is when octave or one-third octave band sound levels are measured, and the band level is small when compared with the overall level. The manufacturer's instructions should be carefully followed to avoid this error. However, most recently constructed sound level meters do not have separately adjustable amplifiers, but rather a choice of dynamic range for the instrument. In this case, it is important to make sure that the instrument range is selected to best cover the range in noise levels to be measured.

3.6.3 Microphone and Sound Level Meter Response Characteristics

The manufacturer's data should be consulted when a noise measurement is to be made, to ensure that both the pressure amplitude (minimum as well as maximum) and frequency of the signal are within the specified ranges of the microphone and SLM.

3.6.4 Background Noise

Background noise is the factor most likely to affect sound pressure level readings. To determine whether or not background noise may interfere with measurements, the source to be measured should be turned off (if possible). If the background noise is less than 10 dB below the total level when the source is turned on, a correction must be made to each reading as described and summarised in Section 1.11.5. If the difference between the total noise level with the sound source in operation and the background noise level in any octave or 1/3 octave measurement band is less than 3 dB, then meaningful acoustic measurements in that band probably cannot be made.

3.6.5 Wind Noise

Even light winds blowing past a microphone produce spurious readings, particularly at frequencies below 200 Hz. These effects are best reduced by fitting a porous acoustic foam windscreen over the microphone. Air movement within some industrial plants can be enough to warrant taking such a precaution. However, most commercially available foam microphone wind screens are inadequate when wind speeds exceed 20 km h^{-1}. If it is necessary to take noise measurements when the wind is stronger than this and one can be sure that noise due to wind rustling in nearby vegetation or other objects is not going to be a problem, then an alternative screen is needed. This may take the form of a cubic enclosure of dimensions approximately 1 metre × 1 metre × 1 metre, made of shade cloth and open at the top with the microphone in the centre of the cube. Whenever foam windscreens are used, note should be taken of microphone manufacturer's advice regarding the slight effect on the microphone sensitivity at high frequencies. A useful study of the effect of wind and windscreens on environmental noise measurements was reported by Hessler *et al.* (2008). Some instrument manufacturers provide instruments that compensate their output to account for the attenuation of their windscreens (which is frequency dependent and greater at high frequencies than low frequencies).

3.6.6 Temperature

The temperature-sensitive elements in the measurement system are the batteries and the microphone. As the temperature increases, the life of the batteries decreases. Some microphones are particularly insensitive to temperature. The best grade of condenser microphone typically retains its calibration within 0.5 dB over a range from -40°C to 150°C. The piezoelectric microphone is also quite stable over a range of -40°C to 60°C. It may be damaged, however, at higher temperatures.

3.6.7 Humidity and Dust

Condenser microphones in particular are sensitive to both humidity and dust on the diaphragm, both of which cause a high level of self-noise in the instrument. When not in use the equipment should be stored in a dry place, and the microphone protected with a desiccator cap. A better plan might be to remove and store the microphone in a tight enclosure containing a desiccant. Alternatively, condenser microphones that use an electret, referred to commercially as pre-polarised, do not suffer these disadvantages and no special precautions are necessary for their care and use, except that they should not be subjected to rainfall or immersed in water, or have their diaphragms scratched or contaminated with dirt.

3.6.8 Reflections from Nearby Surfaces

Nearby objects, which may affect the sound radiation of a source, must often remain undisturbed for the purposes of the measurement; for example, while assessing operator noise levels. At other times their presence may be highly undesirable, such as when assessing the true acoustic power output of a particular machine.

Such possible effects may be expected when an object's dimensions are comparable with or larger than the wavelength of the sound being measured. The possible effect on the sound field may be judged by reference to Table 5.1 and the accompanying discussion of Section 5.9.1 of Chapter 5.

When it is desirable to avoid the effect on sound measurements of nearby objects which cannot be removed, the objects should be covered with sound-absorptive material.

3.7 TIME-VARYING SOUND

Most industrial noise is not steady (although it may be continuous), but rather fluctuates significantly in level over a short period of time. Traffic noise, which is the predominant source of environmental noise in urban areas, is similarly characterised.

Time-varying sound is usually described statistically, either in terms of L_{eq}, the equivalent continuous sound pressure level for a given time period, or in terms of L_x, the sound pressure level that is exceeded for $x\%$ of the time. Both L_{eq} and L_x are generally expressed in dB(A) units. The two most commonly used exceedence levels are L_{10} and L_{90}. The quantity L_{10}, the level exceeded for 10% of the time, is a measure of the higher level (and hence more intrusive) components of noise, whereas L_{90} is a measure of the background or residual level (also referred to as ambient level). For an older-style sound level meter with an analog meter display and without the capability to display statistical quantities, both L_{10} and L_{90} may be estimated, with the meter set to A-weighting and "fast" response. The quantity L_{10} corresponds approximately to the average of the maximum pointer deflections, and L_{90} to the average of the minima. Both quantities may be weighted (see Sections 3.2 and 4.3.6) or unweighted.

In cases where the fluctuating component of a noise level is relatively small, that is $L_{10} - L_{90}$ does not exceed 10 dB, L_{eq} corresponds approximately to the average of the maximum pointer deflections with the sound level meter set to a "slow" response.

3.8 NOISE LEVEL MEASUREMENT

The type of noise measurement made will depend upon the purpose of the measurement. For example, is it desired to establish dB(A) levels to determine whether allowable criteria have been exceeded? Or are octave band levels needed for comparison with model predictions or for determining machine noise reduction requirements? Or are narrow band measurements needed for the purpose of identifying noise sources?

The type of noise measurement made will also depend upon the type of noise source; steady state and impulsive noises require different types of measurement. These measurements and the appropriate instrumentation are summarised in Table 3.2 for various noise sources.

3.9 DATA LOGGERS

These instruments (sometimes referred to as statistical analysers) measure the distribution of fluctuating noise with time for the purpose of community noise and hearing damage risk assessment. In addition to providing energy averaged noise levels (L_{eq}), they readily provide statistical quantities such as L_{10}, L_{50} and L_{90}, which are respectively the levels exceeded 10%, 50% and 90% of the time and, provided that they are well protected or hidden to avoid theft or vandalism, can often be left in the field unattended, gathering data for weeks at a time.

3.10 PERSONAL SOUND EXPOSURE METER

These devices (sometimes referred to as dosimeters) measure the cumulative noise dose in terms of A-weighted sound energy over a period of a normal working day for the purpose of assessing hearing damage risk. The instruments perform the necessary integration to determine L_{Aeq} according to Equation (3.23) (see Sections 4.3.3, 4.3.4 and 4.3.6). The L_{Aeq} measurement can be converted to the eight-hour equivalent, $L_{Aeq,8h}$, using the analysis of Chapter 4 and in many cases this is also done by the instrument. In addition, most instruments provide a measure of exposure level, $E_{A,T}$ as well. The technical specifications of dosimeters are the same as for type II sound level meters and are provided in the standard, IEC61672/1 (2002).

Personal sound exposure meters are designed to be carried by individuals who are exposed to varying levels of high-intensity noise; thus they are small in size and simple to operate, with a microphone that can clip on to the lapel of a coat or shirt. Personal sound exposure meters are available to comply with European, Australian and American regulations. The meters complying with Australian and European regulations have the following characteristics: an eight-hour exposure to a level of 90 dB(A) constitutes a 100% daily noise dose; and for each 3 dB increase above 90 dB(A) the noise dose is doubled for the same exposure time. Some older personal sound exposure meters inhibit their energy integration below a level of 80 dB(A) and these should not be used as they are obsolete. In the USA a 5 dB weighting rather than the 3 dB weighting mentioned is used to determine exposure, and energy integration is inhibited below 80 dB(A). Some personal sound exposure meters also allow a 6 dB weighting (see discussion, Section 4.3). The upper limit for some personal sound exposure meters is 130 dB(A) but 140 dB(A) is more desirable, especially when the user is exposed to impact noise.

Table 3.2 Noise types and their measurement

	Characteristics	Type of source	Type of measurement	Type of instrument	Remarks
0 5 10 15 20 s / small variation / 10dB	Constant continuous noise	Pumps, electric motors, gearboxes, conveyors	Direct reading of A-weighted value	Sound level meter	Octave or 1/3 octave analysis if noise is excessive.
intermittent noise / background noise / 10dB	Constant but intermittent noise	Air compressor, automatic machinery during a work cycle	dB value and exposure time or L_{Aeq}	Sound level meter Integrating sound level meter	Octave or 1/3 octave analysis if noise is excessive.
large fluctuations / 10dB	Periodically fluctuating noise	Mass production, surface grinding	dB value, L_{Aeq} or noise dose	Sound level meter Integrating sound level meter	Octave or 1/3 octave analysis if noise is excessive.
large irregular fluctuations / 10dB	Fluctuating non-periodic noise	Manual work, grinding, welding, component assembly	L_{Aeq} or noise dose Statistical analysis	Noise dosimeter Integrating sound level meter	Long term measurement usually required.
similar impulses / 10dB / 10dB	Repeated impulses	Automatic press, pneumatic drill, rivetting	L_{Aeq} or noise dose and "Impulse" noise level Check "peak" value	Impulse sound level meter or SLM with "peak" hold	Difficult to assess. More harmful to hearing than it sounds.
isolated impulse / 10dB / 10dB	Single impulse	Hammer blow, material handling, punch press	L_{Aeq} and "peak" value	Impulse sound level meter or SLM with "peak" hold	Difficult to assess. Very harmful to hearing.

Noise characteristics classified according to the way they vary with time. Constant noise remains within 5 dB for a long time. Constant noise which starts and stops is called intermittent. Fluctuating noise varies significantly but has a constant long term average ($L_{Aeq,T}$). Impulse noise lasts for less than one second.

Personal sound exposure meters have replaced the majority of equipment used for the assessment of hazard to hearing in industry, as they provide a simple, unified and accurate measure of noise exposure.

In some countries the term "A-weighted equivalent level", L_{Aeq}, rather than noise dose is used in legislation and regulations dealing with worker noise exposure. The maximum allowed time of exposure is then specified so that the L_{Aeq} determined during a normal working day (eight hours) does not exceed 85 dB(A). When L_{Aeq} is used to define the sound level averaged over an eight hour period, it is defined as $L_{Aeq,8h}$.

Many personal sound exposure meters are capable of providing the user with information in the form of L_{Aeq} or any statistical parameter (such as L_{10} or L_{90}) as well as the peak level over any predetermined increment of time (such as 1 second) and thus provide a time history of the noise exposure. The accuracy of this instrumentation is usually ± 0.5 dB.

3.11 RECORDING OF NOISE

The recording of noise for subsequent analysis is sometimes very convenient, especially when undertaking fieldwork. Use of old analog recorders can introduce serious errors. However, digital recorders are very accurate and are used almost universally. Until recently many people used DAT (digital audio tape) recorders to record industrial or environmental sound for later analysis. The accuracy that can be expected with this technology is of the order of +1, -0.5 dB. The upper frequency limit of DAT recorders is dependent upon the internal clock rate and is generally 10 to 20 kHz and their dynamic range is about 70 dB.

More recently tapes have been made redundant and recordings are made using relatively low cost solid state storage devices such as SD cards or compact flash cards. Recorders that record on to these devices are known as DARs (digital audio recorders). The accuracy of these devices is variable and users are advised to check the equipment specifications prior to use. Low cost devices often have a +1 dB, -3 dB frequency response and mid range devices can have a frequency response from 20 Hz to 20 kHz within ± 0.2 dB and a dynamic range of 80 dB. Most of these devices have high pass filters that limit their low frequency capability to 20 Hz. Some (very few) devices can record frequencies down to to D.C. but these devices are in the category of special purpose sound and vibration instrumentation and are quite expensive. Recordings can also be made using a data acquisition card and a notebook computer. Available data acquisition cards vary considerably in cost and performance in terms of frequency response and dynamic range. It is also important to make sure that data acquisition cards used with a computer include low pass anti-aliasing filtering (see Appendix D) to prevent aliasing of high-frequency data into the relatively low audio range.

To make full use of the dynamic range of any recording device it is necessary to record with as high an input signal level as possible without overloading the recorder. It is important to check for overloading using either a meter inbuilt into the recorder or to check an initial recording to make sure that no part of the signal is within 10%

of the maximum allowable input voltage. For impact or impulse type noise (see Sections 4.4.2 and 4.4.3) an impulse sound level meter should be used. The input level to the recorder should then be adjusted accordingly.

When recording a signal for subsequent analysis, it is recommended that calibration tones be introduced at the microphone while keeping careful track of all gains and subsequent gain changes in the recording system. This will allow determination of recorded levels with minimum likelihood of significant error. Alternatively, the signal level could be logged at the time of recording and then the playback levels could be recovered on playback, but generally with less precision.

It is good practice to record at the beginning of a recording the following information: microphone position, unweighted sound pressure level measured with a sound level meter, calibration level, attenuator settings and a description of the test configuration and weather conditions. A redundancy of recorded information can be very valuable when something goes wrong unexpectedly during a recording session which cannot conveniently be repeated.

3.12 SPECTRUM ANALYSERS

When it is desirable to have a detailed description of noise levels as a function of frequency, narrowband spectrum analysers are used. They are especially useful for the analysis of signals containing pure tone components. Often, the frequency of a pure tone gives an indication of its origin, and such information indicates areas for the initiation of noise control. Spectrum analysers sample and digitise the input signal and then use digital filters or the Discrete Fast Fourier Transform (DFFT) algorithm to convert the signal from the time domain to the frequency domain. The sampling rate, which is generally 2.5 times the upper frequency limit of the analyser, sets the upper analysis limit.

Digital filters are used to obtain precision octave and one third octave band data as the DFFT analysis usually does not result in sufficient accuracy to satisfy the relevant standards. However, when narrower band analysis is necessary, DFFT analysis is used.

Analysers generally use either 1024 or 2048 sample points in the time domain to give a 400-line or 800-line spectrum in the frequency domain. Within the range of the analyser, the upper bound of the frequency analysis may be selected, but the smaller the upper frequency selected, the longer will be the required time to acquire the 1024 (or 2048) samples. Each record of 1024 (or 2048) samples is processed while the next record is being acquired.

The analyser real-time frequency limit (or real-time bandwidth) is that frequency at which the processing time for a record is just equal to the time taken to acquire the next record. At higher frequencies some of the data must be discarded between records. Thus the analyser real-time frequency limit is important when transient events are analysed, as high-frequency data may be omitted. Commercially available analysers generally have real-time frequency limits varying from 0.5 kHz to 20 kHz for single-channel operation and half this for two-channel operation. Upper frequency limits range from 20 kHz to 100 kHz.

Most analysers have very powerful options for presentation of the frequency (or time) domain data, including a "zoom" capability where a small part of the spectrum may be analysed in detail. A "waterfall" display capability may also be included, which allows the changes in the frequency spectrum to be observed as a function of time. Various averaging options are also offered such that a number of frequency spectra may be averaged to give a more representative result. Multi-channel analysers (with two-channel being the most common) offer the further capability of investigation of the interrelationships between different input signals.

Another important facility possessed by some analysers is an external clock input which allows the analyser to sample at an externally determined rate. This facility is useful for monitoring rotating machinery noise and vibration where speed variations exist over short periods of time. The external clock signal may be derived from a marker on the rotating shaft (using an optical tachometer). The result of using this external clock is that the rotating machinery harmonics always appear at the same frequency and are not "washed out" in the averaging process. Aspects of frequency analysis are discussed in more detail in Appendix D.

Spectrum analysers are also available as plug in cards for personal computers or notebook computers. Sometimes these cards have no anti-aliasing filters (see Appendix D) and great care is required in their use due to the "folding back" (Randall, 1987) to lower frequencies of signals having a frequency in excess of the digital sampling rate of the analyser. This can be a serious problem in many measurements and analog anti-aliasing filters should be used at all times. Note that switched capacitor filters are not suitable for anti-aliasing filters as the high frequency carrier signal will "fold-back" into the signal to be sampled.

3.13 INTENSITY METERS

Both the two-channel real-time and parallel filter analysers mentioned above offer the possibility of making a direct measurement of the acoustic intensity of propagating sound waves using two carefully matched microphones and band limited filters. The filter bandwidth is generally not in excess of one octave.

The measurement of sound intensity requires the simultaneous determination of sound pressure and particle velocity. The determination of sound pressure is straightforward, but the determination of particle velocity presents some difficulties; thus, there are two principal techniques for the determination of particle velocity and consequently the measurement of sound intensity. Either the acoustic pressure and particle velocity are measured directly (p–u method) or the acoustic pressure is measured simultaneously at two closely spaced points and the mean sound pressure is calculated by averaging the two measurements and the particle velocity is calculated by taking the difference between the two measurements and dividing the result by the microphone spacing and some constants (p–p method). In either case the pressure is multiplied by the particle velocity to produce the instantaneous intensity and the time average intensity (see Section 1.7).

Errors inherent in intensity measurements and limitations of instrumentation are discussed in Sections 3.13.1.1, 3.13.2.1 and by Fahy (1995).

3.13.1 Sound Intensity by the p–u Method

In the p–u method the acoustic particle velocity is measured directly. One method involves using two parallel ultrasonic beams travelling from source to receiver in opposite directions. Any particle movement in the direction of the beams will cause a phase difference in the received signals at the two receivers. The phase difference is related to the acoustic particle velocity in the space between the two receivers and may be used to calculate an estimate of the particle velocity up to a frequency of 6 kHz.

The ultrasound technique is not used very much any more as it has been overtaken by a much more effective technology known as the "Microflown" particle velocity sensor (de Bree *et al.*, 1996; Druyvesteyn and de Bree, 2000) which uses a measure of the temperature difference between two resistive sensors spaced 40 μm apart to estimate the acoustic particle velocity. The temperature difference between the two sensors is caused by the transfer of heat from one sensor to the other by convection as a result of the acoustic particle motion. This in turn leads to a variation in resistance of the sensor, which can be detected electronically. To get a temperature difference which is sufficiently high to be detected, the sensors are heated with a d.c. current to about 500°K. The sensor consists of two cantilevers of silicon nitride (dimensions 800×40×1 μm) with an electrically conducting platinum pattern, used as both the sensor and heater, placed on them. The base of the sensor is silicon, which allows it to be manufactured using the same wafer technology as used to make integrated circuit chips. Up to 1000 sensors can be manufactured in a single wafer. The sensitivity of the sensors (and signal to noise ratio) can be increased by packaging of the sensor in such a way that the packaging increases the particle velocity near the sensor.

The spacing of 40 μm between the two resistive sensors making up the particle velocity sensor is an optimal compromise. Smaller spacing reduces the heat loss to the surroundings so that more of the heat from one sensor is convected to the other, making the device more sensitive. On the other hand, as the sensors come closer together, conductive heat flow between the two of them becomes an important source of heat loss and thus measurement error.

It has been shown (Jacobsen and Liu, 2005) that measurement of sound intensity using the "Microflown" sensor to determine the particle velocity directly is much more accurate than the indirect method involving the measurement of the pressure difference between two closely spaced microphones, which is described in the following section. Jacobsen and Liu (2005) also showed that the "Microflown" sensor was more accurate than microphones for acoustic holography which involves using the measurement of acoustic pressure OR particle velocity in a plane front of a noise source to predict the acoustic pressure AND particle velocity in another plane.

Particle velocity sensors such as the "Microflown" are more useful than microphones for measuring particle velocity as they are directional, which makes them less susceptible to background noise, and they are more sensitive. When used close to a reasonably stiff noise emitting structure their superiority over microphones is even more apparent as the pressure associated with any background noise is approximately doubled at the surface, whereas the particle velocity associated with the background noise will be close to zero where it is reflected. It is possible to purchase a small sound intensity probe (12 mm diameter) that includes a microphone and a "microflown" and this can be used to directly measure sound intensity.

The active and reactive intensities for a harmonic sound field are given by Equations (1.72) and (1.73) in Chapter 1, respectively.

3.13.1.1 Accuracy of the p–u Method

With sound intensity measurements, there are systematic errors, random errors and calibration uncertainty. The calibration uncertainty for a typical p-u probe is approximately ± 0.5 dB. However, the systematic error arises from the inaccuracy in the model that compensates for the phase mismatch between the velocity sensor and the pressure sensor in the probe. This model is derived by exposing the p-u probe to a sound field where the relative amplitudes and phases between the acoustic particle velocity and pressure are well known (Jacobsen and de Bree, 2008).

Although the phase error is not as important for a p-u probe as it is for a p-p probe (as the p-p probe also relies on the phase between the two microphones to estimate the particle velocity), it is still the major cause of uncertainty in the intensity measurement, especially for reactive fields where the reactive intensity amplitude is larger or comparable with the active intensity. Reactive intensity fields occur close to noise radiating structures. For such fields, the errors in sound intensity measurement using a p-u probe increase as the phase between the acoustic pressure and particle velocity increases. The Microflown handbook gives the systematic error for intensity measurements using a p-u probe at a particular frequency as:

$$\text{Error (dB)} = 10\log_{10}\left[1 + \left|\frac{\hat{I} - I}{I}\right|\right] = 10\log_{10}(1 + \beta_e \tan\beta_f) \qquad (3.24)$$

where I is the actual active intensity at the p-u probe location, $\hat{I}$ is the intensity measured by the probe β_e (in radians) is the phase calibration error which for a typical p-u probe is of the order of $2.5°$ or 0.044 radians and β_f is the phase between the acoustic pressure and particle velocity in the sound field. It can be seen that this systematic error will exceed 1 dB if the phase between the acoustic pressure and particle velocity exceeds $80°$. In practice, if the sound field reactivity is too high, the probe can be moved further from the noise source and the reactivity will decrease. Also as reflected sound and sound from sources other than the one being measured do not usually increase the reactivity of the sound field, they will not influence the accuracy of the intensity measurement made with the p-u probe, in contrast to their significant influence on intensity measurements made with the p-p probe (see section 3.13.2.1). Note that very reactive fields are unlikely to occur except at low frequencies, whereas the phase error in the p-p probe affects the accuracy of the intensity measurement over the entire audio frequency range. Thus the p-u probe usually gives a more accurate value of sound intensity at high frequencies and in fact is useful up to 20 kHz whereas the limit of commercially available p-p probes is 10 kHz.

Estimations of the magnitude of random errors are very difficult to make and no estimates have been reported in the literature.

3.13.2 Sound Intensity by the p–p Method

In the p–p method, the determinations of acoustic pressure and acoustic particle velocity are both made using a pair of high quality condenser microphones. The microphones are generally mounted side by side or facing one another and separated by a fixed distance (6 mm to 50 mm) depending upon the frequency range to be investigated. A signal proportional to the particle velocity at a point midway between the two microphones and along the line joining their acoustic centres is obtained using the finite difference in measured pressures to approximate the pressure gradient while the mean is taken as the pressure at the midpoint.

The useful frequency range of p–p intensity meters is largely determined by the selected spacing between the microphones used for the measurement of pressure gradient, from which the particle velocity may be determined by integration. The spacing must be sufficiently small to be much less than a wavelength at the upper frequency bound so that the measured pressure difference approximates the pressure gradient. On the other hand, the spacing must be sufficiently large for the phase difference in the measured pressures to be determined at the lower frequency bound with sufficient precision to determine the pressure gradient with sufficient accuracy. Clearly, the microphone spacing must be a compromise and a range of spacings is usually offered to the user.

The assumed positive sense of the determined intensity is in the direction of the centre line from microphone 1 to microphone 2. For convenience where appropriate in the following discussion the positive direction of intensity will be indicated by unit vector n and this is in the direction from microphone 1 to microphone 2.

Taking the gradient of Equation (1.10) in the direction of unit vector n, and using Equation (1.11) gives the equation of motion relating the pressure gradient to the particle acceleration. That is:

$$n\frac{\partial p}{\partial n} = -\rho\frac{\partial u_n}{\partial t} \tag{3.25}$$

where u_n is the component in direction n of particle velocity u, p and u are both functions of the vector location r and time t, and ρ is the density of the acoustic medium. The normal component of particle velocity, u_n, is obtained by integration of Equation (3.25) where the assumption is implicit that the particle velocity is zero at time $t = -\infty$:

$$u_n(t) = -\frac{n}{\rho}\int_{-\infty}^{t}\frac{\partial p(\tau)}{\partial n}\,d\tau \tag{3.26}$$

The integrand of Equation (3.26) is approximated using the finite difference between the pressure signals p_1 and p_2 from microphones 1 and 2 respectively and Δ is the separation distance between them:

$$u_n(t) = -\frac{n}{\rho\Delta} \int_{-\infty}^{t} \left[p_2(\tau) - p_1(\tau) \right] d\tau \tag{3.27}$$

The pressure midway between the two microphones is approximated as the mean:

$$p(t) = \frac{1}{2} \left[p_1(t) + p_2(t) \right] \tag{3.28}$$

Thus the instantaneous intensity in direction n at time t is approximated as:

$$I_i(t) = -\frac{n}{2\rho\Delta} \left[p_1(t) + p_2(t) \right] \int_{-\infty}^{t} \left[p_1(\tau) - p_2(\tau) \right] d\tau \tag{3.29}$$

For stationary sound fields the instantaneous intensity can be obtained from the product of the signal from one microphone and the integrated signal from a second microphone in close proximity to the first (Fahy, 1995):

$$I_i(t) = \frac{n}{\rho\Delta} p_2(t) \int_{-\infty}^{t} p_1(\tau) d\tau \tag{3.30}$$

The time average of Equation (3.30) gives the following expression for the time average intensity in direction n (where n is the unit vector).

$$I = \frac{n}{\rho\Delta} \lim_{T\to\infty} \frac{1}{T} \int_0^T \left[p_2(t) \int_{-\infty}^{t} p_1(\tau) d\tau \right] dt \tag{3.31}$$

Commercial instruments with digital filtering (one third octave or octave) are available to implement Equation (3.31). As an example, consider two harmonic pressure signals from two closely spaced microphones:

$$p_i(t) = p_{0i} e^{j(\omega t + \theta_i)} \quad i = 1, 2 \tag{3.32}$$

Substitution of the real components of these quantities in Equation (3.30) gives for the instantaneous intensity the following result:

$$I_i(t) = \frac{n}{4\rho\Delta\omega} \left[p_{01}^2 \sin(2\omega t + 2\theta_1) - p_{02}^2 \sin(2\omega t + 2\theta_2) + 2p_{01}p_{02}\sin(\theta_1 - \theta_2) \right] \tag{3.33}$$

Taking the time average of Equation (3.33) gives the following expression for the (active) intensity:

$$I = \frac{n p_{01} p_{02}}{2\rho\omega\Delta} \sin(\theta_1 - \theta_2) \tag{3.34}$$

If the argument of the sine is a small quantity then Equation (3.34) becomes approximately:

$$I_n = \frac{n P_1 P_2}{2 \rho \omega \Delta} (\theta_1 - \theta_2) , \qquad \theta_1 - \theta_2 \ll 1 \tag{3.35}$$

This equation also follows directly from Equation (1.69), where the finite difference approximation is used to replace $\partial \theta_p / \partial r$ with $(\theta_1 - \theta_2)/\Delta$ and p_0^2 is approximated by $p_{01} p_{02}$. The first two terms of the right-hand side of Equation (3.33) describe the reactive part of the intensity. If the phase angles θ_1 and θ_2 are not greatly different, for example, the sound pressures p_{01} and p_{02} are measured at points that are closely spaced compared to a wavelength, the magnitude of the reactive component of the intensity is approximately:

$$I_r = \frac{1}{4 \omega \rho \Delta} \left[p_{01}^2 - p_{02}^2 \right] \tag{3.36}$$

Equation (3.36) also follows directly from the second (reactive) term of Equation (1.68) where p_0 is replaced by $(p_{01} + p_{02})/2$ and $\partial p / \partial r$ is replaced with the finite difference approximation $(p_{01} - p_{02})/\Delta$. Measurement of the intensity in a harmonic stationary sound field can be made with only one microphone, a phase meter and a stable reference signal if the microphone can be located sequentially at two suitably spaced points. Indeed, the 3-D sound intensity vector field can be measured by automatically traversing, stepwise, a single microphone over an area of interest. Use of a single microphone for intensity measurements eliminates problems associated with microphone, amplifier and integrator phase mismatch as well as enormously reducing diffraction problems encountered during the measurements.

In general the determination of the total instantaneous intensity vector, $I_i(t)$ requires the simultaneous determination of three orthogonal components of particle velocity. Current instrumentation is available to do this using a single p-u probe or single p-p probe. Note that with a p-p probe, the finite difference approximation used to obtain the acoustic particle velocity has problems at both low and high frequencies which means that at least three different microphone spacings are needed to cover the audio frequency range. At low frequencies, the instantaneous pressure signals at the two microphones are very close in amplitude and a point is reached where the precision in the microphone phase matching is insufficient to accurately resolve the difference. At high frequencies the assumption that the pressure varies linearly between the two microphones is no longer valid. The p-u probe suffers from neither of these problems as it does not use an approximation to the sound pressure gradient to determine the acoustic particle velocity.

3.13.2.1 Accuracy of the p–p Method

The accuracy of the p-p method is affected by both systematic and random errors. The systematic error stems from the amplitude sensitivity difference and phase mismatch between the microphones and is a result of the approximations inherent in the finite difference estimation of particle velocity from pressure measurements at two closely spaced microphones.

The error due to phase mismatch can be expressed in terms of the difference, δ_{pI}, between sound pressure and intensity levels measured in the sound field being evaluated and the difference, δ_{pIO}, between sound pressure and intensity levels measured by the instrumentation in a specially controlled uniform pressure field in which the phase at each of the microphone locations is the same and for which the intensity is zero. This latter quantity is a measure of the accuracy of the phase matching between the two microphones making up the sound intensity probe and the higher its value, the higher is the quality of the instrumentation (that is, the better is the microphone phase matching). The quantity, δ_{pI}, is known as the "Pressure-Intensity Index" for a particular sound field and δ_{pIO}, is known as the "Residual Pressure-Intensity Index" which is a property of the instrumentation and should be as large as possible. If noise is coming from sources other than the one being measured or there are reflecting surfaces in the vicinity of the noise source, the Pressure Intensity Index will increase and so will the error in the intensity measurement. Other sources or reflected sound do not affect the accuracy of intensity measurements taken with a p-u probe, but the p-u probe is more sensitive than the p-p probe in reactive sound fields (such as the near field of a source - Jacobsen, 2008).

The normalised systematic error in intensity due to microphone phase mismatch is a function of the actual phase difference, β_f between the two microphone locations in the sound field and the phase mismatch error, β_s. The normalised error is given by Fahy (1995) as:

$$e_\beta(I) = \beta_s / \beta_f \qquad (3.37)$$

The Pressure-Intensity Index may be written in terms of this error as:

$$\delta_{pIO} - \delta_{pI} = 10\log_{10}|1 + (1/e_\beta(I))| = 10\log_{10}|1 + (\beta_f / \beta_s)| \qquad (3.38a,b)$$

A normalised error of Equation (3.37) equal to 0.25 corresponds to a sound intensity error of approximately 1 dB (= $10\log_{10}(1 +0.25)$), and a difference, $\delta_{pIO} - \delta_{pI} = 7$ dB. A normalised error equal to 0.12 corresponds to a sound intensity error of approximately 0.5 dB and a difference, $\delta_{pIO} - \delta_{pI} = 10$ dB.

The pressure intensity index will be large (leading to relatively large errors in intensity estimates) in near fields and reverberant fields and this can extend over the entire audio frequency range.

The phase mismatch between the microphones in the p-p probe is related to the Residual Pressure-Intensity Index (which is often supplied by the p-p probe suppliers) by:

$$\beta_s = kd10^{-\delta_{pIO}/10} \qquad (3.39)$$

Phase mismatch also distorts the directional sensitivity of the p-p probe so that the null in response of the probe (often used to locate noise sources) is changed from the $90°$ direction to that given by (Fahy, 1995):

$$\beta_m = \cos^{-1}(\beta_s / kd) \tag{3.40}$$

where k is the wavenumber at the frequency of interest and d is the spacing between the two microphones in the p-p probe.

In state of the art instrumentation, microphones are available that have a phase mismatch of less than $0.05°$. In cases where the phase mismatch is larger than this, the instrumentation sometimes employs phase mismatch compensation in the signal processing path.

The error due to amplitude mismatch is zero for perfectly phase matched microphones. However, for imperfectly phase matched microphones, the error is quite complicated to quantify and depends on the characteristics of the sound field being measured.

Fahy (1995) shows that random errors in intensity measurements add to the uncertainty due to systematic errors and in most sound fields where the signals received by the two microphones are random and have a coherence close to unity, the normalised random error is given by $(BT)^{-1/2}$, corresponding to an intensity error of:

$$e_r(I) = 10\log_{10}\left(1 + (BT)^{-1/2}\right) \tag{3.41}$$

where B is the bandwidth of the measurement in Hz and T is the effective averaging time, which may be less than the measurement time unless real-time processing is performed. The coherence of the two microphone signals will be less than unity in high frequency diffuse fields or where the microphone signals are contaminated by electrical noise, unsteady flow or cable vibration. In this case the random error will be greater than indicated by Equation (3.41).

Finally, there are errors due to instrument calibration that add to the random errors. This is approximately ± 0.2 dB for one particular manufacturer but the reader is advised to consult calibration charts that are supplied with the instrumentation.

3.13.3 Frequency Decomposition of the Intensity

In the measurement and active control of acoustic power transmission, it is often necessary to decompose the intensity signal into its frequency components. This may be done either directly or indirectly.

3.13.3.1 Direct Frequency Decomposition

For a p–u probe, the frequency distribution of the mean intensity may be obtained by passing the two output signals (p and u) through identical bandpass filters prior to performing the time averaging.

With a p–p probe, the frequency distribution may be determined by passing the two signals through appropriate identical bandpass filters, either before or after performing the sum, difference and integration operations of Equation (3.29) and then time averaging the resulting outputs.

3.13.3.2 Indirect Frequency Decomposition

Determination of the intensity using the indirect frequency decomposition method is based upon Fourier analysis of the two probe signals (either the p–u signals or the p–p signals). Fahy (1995, pp. 95-97) shows that for a p–u probe, the intensity as a function of frequency is given by the single-sided cross-spectrum (or cross-spectral density) G_{pu} between the two signals:

$$I(\omega) = \text{Re}\{G_{pu}(\omega)\} \tag{3.42}$$

$$I_r(\omega) = -\text{Im}\{G_{pu}(\omega)\} \tag{3.43}$$

As before, $I(\omega)$ represents the real (or active) time averaged intensity at frequency ω and $I_r(\omega)$ represents the amplitude of the reactive component.

The cross-spectrum of two signals is defined as the product of the complex instantaneous spectrum of one signal with the complex conjugate of the complex instantaneous spectrum of the other signal (see Appendix D). Thus, if $G_p(\omega)$ and $G_u(\omega)$ represent the complex single-sided spectra of the pressure and velocity signals respectively, then the associated cross-spectrum is given by:

$$G_{pu}(\omega) = G_p^*(\omega) G_u(\omega) \tag{3.44}$$

where the * represents the complex conjugate (see Appendix D).

Note that for random noise $G_{pu}(\omega)$ is a cross-spectral density function in which case the expressions on the left side of Equations (3.42) and (3.43) represent intensity per hertz. For single frequency signals and harmonics, $G_{pu}(\omega)$ is the cross-spectrum obtained by multiplying the cross-spectral density by the bandwidth of each FFT filter (or the frequency resolution).

For the case of the p–p probe, Fahy (1995) shows that the mean active intensity I and amplitude I_r of the reactive intensity in direction n at frequency ω are:

$$I(\omega) = -\frac{1}{\rho \omega \Delta} \text{Im}\{G_{p1p2}(\omega)\} \tag{3.45}$$

$$I_r(\omega) = -\text{Im}\{G_{pu}(\omega)\} = \frac{1}{2\rho\omega\Delta}\left[G_{p1p1}(\omega) - G_{p2p2}(\omega)\right] \tag{3.46}$$

where G_{p1p2} is the cross-spectrum of the two pressure signals and G_{p1p1} and G_{p2p2} represent the auto spectral densities (see Appendix D).

For the case of a stationary harmonic sound field, it is possible to determine the sound intensity by using a single microphone and the indirect frequency decomposition method just described by taking the cross-spectrum between the microphone signal and a stable reference signal (referred to as A) of the same frequency, for two locations p_1 and p_2 of the microphone. Thus, the effective cross-spectrum G_{p1p2} for use in the preceding equations can be calculated as follows:

$$G_{p1p2}(\omega) = \frac{G_{p1}^{*}(\omega)\, G_{A}(\omega)\, G_{p2}^{*}(\omega)\, G_{p2}(\omega)}{G_{p2}(\omega)^{*}\, G_{A}(\omega)} = \frac{G_{p1A}(\omega)\, G_{p2p2}(\omega)}{G_{p2A}(\omega)} \qquad (3.47a,b)$$

Remember that the results for intensity must be multiplied by the frequency resolution of the cross-spectral density to find the single-frequency intensity. Note however that most spectrum analysers measure a cross-spectrum as well as a cross-spectral density, and in the cross-spectrum the frequency multiplication has already been done.

Equations (3.45) to (3.47) are easily implemented using a standard commercially available FFT spectrum analyser. Both the two-channel real-time and parallel filter analysers are commercially available and offer the possibility of making a direct measurement of the acoustic intensity of propagating sound waves in a sound field using two carefully matched microphones and band limited filters. The filter bandwidth is generally not in excess of one octave. If the microphones are not well matched, measurements are still possible (Fahy, 1995).

3.14 ENERGY DENSITY SENSORS

As mentioned in Chapter 1, energy density is a convenient quantity to minimise when actively controlling enclosed sound fields. However, the measurement of energy density is not as straightforward as the measurement of sound pressure, although it is simpler than the measurement of sound intensity. Both sound pressure and acoustical particle velocity must be sensed but, unlike the case of acoustic intensity measurement, a measurement of the phase between the acoustic pressure and particle velocity is not needed to enable an estimation of the energy density to be made.

It is possible to construct an energy density sensor from low cost (temperature-compensated) electret microphones (Cazzolato, 1999; Parkins, 1998), as illustrated in Figure 3.6. However, the microphones used as the sensing elements in the energy density sensor need to be omni-directional over the frequency range of interest. Cazzolato showed that for the frequency range between 60 Hz and 600 Hz (a typical range used in active noise control systems), a microphone spacing of 50 mm was optimal. The arrangement proposed by Cazzolato (1999) involved the use of four microphones positioned as shown in Figure 3.6, which allowed the measurement of the particle velocity in three orthogonal directions (using Equation (3.27)) as well as the average pressure in the vicinity of the centre of the device. The circuitry is illustrated in Figure 3.7. Alternatively, a 3-Dimensional version of the Microflown particle velocity probe described in Section 3.13.1 may be used to obtain the particle velocity in the three orthogonal directions.

Once the sound pressure and acoustic particle velocities in the three orthogonal directions (1, 2 and 3) are found, Equation (1.62) can be rewritten to calculate the 3-dimensional energy density as:

$$\psi_{tot}(t) = \psi_{k}(t) + \psi_{p}(t) = \frac{\rho}{2}\left[u_{1}^{2}(t) + u_{2}^{2}(t) + u_{3}^{2}(t) + \frac{p^{2}(t)}{(\rho c)^{2}}\right] \qquad (3.48)$$

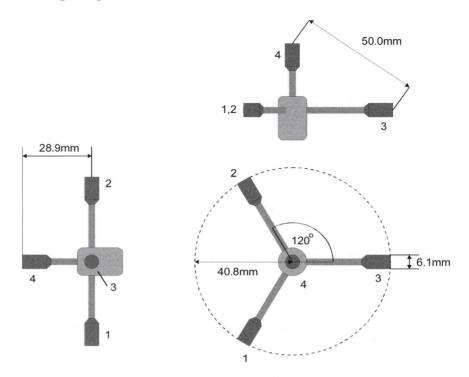

Figure 3.6 3-axis, 3-dimensional energy density sensor with four microphones.

3.15 SOUND SOURCE LOCALISATION

The identification of the actual location of sources of sound on vibrating structures or surfaces in air flows (aerodynamic sound) is of great interest to any noise control engineer. There is a range of instrumentation available for this purpose and practically all of it is quite expensive.

There are four main techniques for locating sound sources: nearfield acoustic holography (NAH), statistically optimised nearfield acoustic holography (SONAH), beamforming (BF) and direct acoustic intensity measurement. Instrumentation using each of these techniques is discussed below. The physical principles under-pinning each technique is discussed and a limited amount of mathematical analysis will be presented to illustrate the fundamental principles of each technique. More detail is provided in the references.

It is worth noting that none of the techniques to be discussed can distinguish between a sound wave radiated directly from a source and a wave reflected from an object or surface. Thus reflecting surfaces can often be identified as noise sources. In addition to determine whether the sound is coming from in front or behind the array, either the array must be non-planar or made up of microphones in two parallel planes with more complex signal processing as well.

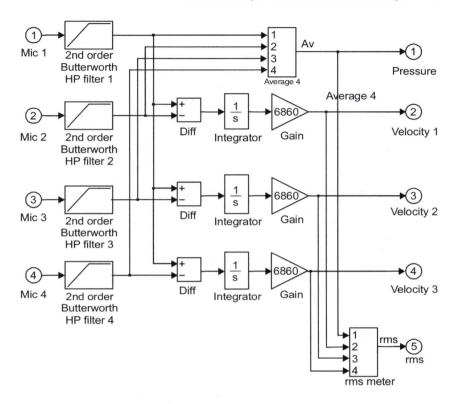

Figure 3.7 Circuit used to process signals from an energy density sensor to produce an r.m.s signal proportional to energy density as well as individual pressure and velocity signals (the latter being particularly useful for active noise control applications).

3.15.1 Nearfield Acoustic Holography (NAH)

Acoustic holography involves the measurement of the amplitude and phase of a sound field at many locations on a plane at some distance from a sound source, but in its near field so that evanescent waves contribute significantly to the microphone signals. The measurements are than used to predict the complex acoustic pressure and particle velocity on a plane that approximates the surface of the source (the prediction plane). Multiplying these two quantities together gives the sound intensity as a function of location in the prediction plane, which allows direct identification of the relative strength of the acoustic radiation from various areas of a vibrating structure. The theoretical analysis that underpins this technique is described in detail in a book devoted just to this topic (Williams, 1999) and so it will only be summarised here.

Acoustic holography may involve only sound pressure measurements in the measurement plane (using phase matched microphones) and these are used to predict both the acoustic pressure and particle velocity in the plane of interest. The product of these two predicted quantities is then used to determine the acoustic intensity on the plane of interest which is usually adjacent to the sound radiating structure. Alternatively, only acoustic particle velocity may be measured in the measurement plane (using the "Microflown") and then these measurements may be used to predict the acoustic pressure as well as the acoustic particle velocity in the measurement plane. Finally, both acoustic pressure and acoustic particle velocity may be measured in the measurement plane. Then the acoustic particle velocity measurement is used to predict the acoustic particle velocity in the plane of interest and the acoustic pressure measurement is used to predict the acoustic pressure in the plane of interest.

Jacobsen and Liu (2005) showed that measurement of both acoustic pressure and particle velocity gave the best results followed by measurement of only particle velocity with the measurement of only acoustic pressure being a poor third. The reason that using like quantities to predict like quantities gives the best results is intuitively obvious, but the reasons that the particle velocity measurement gives much better results than the pressure measurement need some explanation. The reasons are summarised below:

- Particle velocity decays more rapidly than sound pressure towards the edges of the measurement region and has a larger dynamic range. This means that spatial windowing, (similar to the time domain windowing discussed in Appendix D, but in this case in the spatial domain), which is a necessary part of the application of the technique to finite size structures, does not have such an influence when particle velocity measurements are used.
- Predicting particle velocity from acoustic pressure measurements results in amplification of high spatial frequencies which increases the inherent numerical instability of the prediction. On the other hand, predicting acoustic pressure from acoustic particle velocity results in reduction of the amplitudes of high spatial frequencies.
- Phase mismatch between transducers in the measurement array has a much greater effect on the prediction of acoustic particle velocity from acoustic pressure measurements than vice versa.

The practical implementation of acoustic holography is complex and expensive, involving many sensors whose relative phase calibration must be accurately known, and whose relative positions must be known to a high level of accuracy. However, commercially available instrumentation exists, which does most of the analysis transparently to the user, so it is relatively straightforward to use.

3.15.1.1 Summary of the Underlying Theory

Planar near field acoustic holography involves the measurement of the amplitudes and relative phases of either or both of the acoustic pressure and acoustic particle velocity at a large number of locations on a plane located in the nearfield of the radiating structure on which it is desired to locate and quantify noise sources. The measurement plane must be in the near field of the structure and as close as practical to it, but no closer than the microphone spacing. As the measurement surface is a plane, the noise source map will be projected on to a plane that best approximates the structural surface. If the structural surface is curved, it is possible to use a curved sensor array but the analysis becomes much more complicated. The resolution (or accuracy of noise source location) of this technique is equal to the spacing of the measurement array from the noise radiating surface at low frequencies and at frequencies above the frequency where this distance is equal to half a wavelength, the resolution is half a wavelength. The following analysis is for a measurement array of infinite extent with microphones sufficiently close together that the measured sound pressure on the measurement plane is effectively continuous. The effect of deviating from this ideal situation by using a finite size array and a finite number of sensors will be discussed following the analysis.

For the analysis, it is assumed that the plane encompassing the radiating structure is in the x-y plane at a z coordinate, $z_p \leq 0$ and the measurement plane is located at $z_m > 0$ and is parallel to the plane at z_p. The wavenumber transform corresponding to complex pressure measurements made on the measurement plane is a transform from the spatial domain to the wavenumber domain and may be written as:

$$K(k_x, k_y) = \int_{-\infty}^{\infty} \int_{-\infty}^{\infty} p(x, y, z_m) e^{j(k_x x + k_y y)} \, dx \, dy \qquad (3.49)$$

where k_x and k_y are the wavenumber components in the x and y directions and their values cover the wavenumber range from $-\infty$ to $+\infty$. Of course, in practice the finite size of the array and the physical spacing of the microphones will limit the wavenumber spectrum and hence the frequency range that can be covered. The acoustic pressure is given by the inverse transform as:

$$p(x, y, z_m) = \frac{1}{(2\pi)^2} \int_{-\infty}^{\infty} \int_{-\infty}^{\infty} K(k_x, k_y) e^{-j(k_x x + k_y y)} \, dk_x \, dk_y \qquad (3.50)$$

From the preceding two equations, it can be seen that the wave number spectrum, $K(k_x, k_y)$ for any given value of k_x and k_y may be interpreted as the amplitude of a plane wave provided that:

$$k_x^2 + k_y^2 \leq k^2 \qquad (3.51)$$

The plane wave propagates in the (k_x, k_y, k_z) direction and k_z must satisfy:

$$k_x^2 + k_y^2 + k_z^2 = k^2 = (\omega/c)^2 \qquad (3.52)$$

If Equation (3.51) is not satisfied, then the particular component of the wavenumber spectrum represents an evanescent wave whose amplitude decays with distance from the sound source.

The wavenumber transform in the prediction plane, $z = z_p$ can be calculated by multiplying Equation (3.50) with an exponential propagator defined as:

$$G(z_p, z_m, k_x, k_y) \ = \ e^{-jk_z(z_p - z_m)} \tag{3.53}$$

After the multiplication of the wavenumber transform by G in Equation (3.53), the inverse transform (see Equation (3.55)) is taken to get the sound pressure in the prediction plane.

The acoustic particle velocity in the prediction plane is calculated from the wavenumber transform of Equation (3.49) by multiplying it by the *pu* propagator defined below and then taking the inverse transform, as for the pressure.

$$G_{pu}(z_p, z_m, k_x, k_y) \ = \ \frac{k_z}{\rho c k} e^{-jk_z(z_p - z_m)} \tag{3.54}$$

where:

$$k_z \ = \ \begin{cases} \sqrt{k^2 - k_x^2 - k_y^2} & \text{for } k_x^2 + k_y^2 \le k^2 \\[2ex] -j\sqrt{k_x^2 + k_y^2 - k^2} & \text{for } k_x^2 + k_y^2 > k^2 \end{cases} \tag{3.55}$$

It is also possible to complete the entire process using particle velocity measurements instead of pressure measurements. In this case, the particle velocity is substituted for the acoustic pressure in Equation (3.49) to give the particle velocity wave number transform and then Equation (3.50) gives the particle velocity on the LHS instead of the acoustic pressure. The propagator in equation (3.53) is then multiplied with the particle velocity wavenumber transform and then the inverse transform is taken as in Equation (3.50) to give the particle velocity on the prediction plane. The acoustic pressure on the prediction plane can be calculated by multiplying the particle velocity wavenumber transform by the following predictor prior to taking the inverse transform.

$$G_{up}(z_p, z_m, k_x, k_y) \ = \ \frac{\rho c k}{k_z} e^{-jk_z(z_p - z_m)} \tag{3.56}$$

Implementation of the above procedure in practice requires the simplifications of a finite size measurement plane (which must be slightly larger than the radiating structure being analysed) and a finite spacing between measurement transducers. The adverse effect of the finite size measurement plane is minimised by multiplying the wavenumber transform by a spatial window that tapers towards the edge of the array so less and less weighting is placed on the measurements as one moves from the centre to the edge of the array. To prevent wrap around errors, the array used in the wavenumber transform is larger than the measurement array and all points outside the actual measurement array are set equal to zero.

The effect of the finite spacing between sensors in the measurement array is to limit the ability of the array to sample high spatial frequency components that exist if the sound field varies strongly as a function of location. To avoid these high frequency components, the array must be removed some distance from the noise radiating structure but not so far that the evanescent modes are so low in intensity that they cannot be measured.

The other effect of finite microphone spacing is to limit the upper frequency for the measurement as a result of the spatial sampling resolution. In theory this spacing should be less than half a wavelength but in practice, good results are obtained for spacings less than about one quarter of a wavelength.

The lower frequency ability of the measurement is limited by the array size, which should usually be at least a wavelength, but in some special cases, if the sound pressure at the edges of the array has dropped off sufficiently, an array size of 1/3 of a wavelength can be used.

To be able to calculate the finite Fast Fourier Transforms, the measurement grid for NAH must be uniform; that is, all sensors must be uniformly spaced.

3.15.2 Statistically Optimised Nearfield Acoustic Holography (SONAH)

The SONAH method (Hald *et al.*, 2007) is a form of nearfield acoustic holography in which the FFT calculation is replaced with a least squares matrix inversion. The advantage of this method is that the measurement array does not need to be as large as the measurement source and the measurement sensors need not be regularly spaced. In addition, the sound field can be calculated on a surface that matches the contour of the noise radiating surface. However, the computer power needed for the inversion of large matrices can be quite large. The noise source location resolution is similar to that for NAH.

The requirements for maximum microphone spacing are similar to NAH and typical maximum operating frequencies of commercially available SONAH systems range from 1 kHz to 6 kHz and the typical dynamic range (difference in level between strongest and weakest sound sources that can be detected) is 15 to 20 dB. There seems to be no lower limiting frequency specified by equipment manufacturers.

The array size affects the lower frequency limit of the measurement but the requirements for SONAH are much less stringent than those for NAH. It is possible to undertake measurements down to frequencies for which the array size is 1/8 of a wavelength.

The analysis begins by representing the sound field as a set of plane evanescent wave functions defined as:

$$\Phi_k(x,y,z) = e^{-j(k_x x + k_y y + k_z(z_p - z_m))} \tag{3.57}$$

where k_z is defined by Equation (3.55). In practice, the sound field at any location, (x, y, z) is represented by N plane wave functions, chosen to cover the wave number spectrum of interest, so that:

$$p(x,y,z) \approx \sum_{n=1}^{N} a_n \Phi_{kn}(x,y,z) \tag{3.58}$$

The coefficients, a_n are determined by using the pressure measurements over L locations, $(x_\ell, y_\ell, z_\ell, \ell = 1,...., L)$, in the measurement array so that:

$$p(x_\ell, y_\ell, z_\ell) \approx \sum_{n=1}^{N} a_n \Phi_{kn}(x_\ell, y_\ell, z_\ell) \qquad \ell = 1,......, L \qquad (3.59)$$

The plane wave functions corresponding to all the measurement points, L, can be expressed in matrix form as:

$$A_m = \begin{bmatrix} \Phi_{k1}(x_1,y_1,z_1) & \Phi_{k2}(x_1,y_1,z_1) & \cdots & \Phi_{kN}(x_1,y_1,z_1) \\ \Phi_{k1}(x_2,y_2,z_2) & \Phi_{k2}(x_2,y_2,z_2) & \cdots & \Phi_{kN}(x_2,y_2,z_2) \\ \vdots & & & \vdots \\ \Phi_{k1}(x_L,y_L,z_L) & \Phi_{k2}(x_L,y_L,z_L) & \cdots & \Phi_{kN}(x_L,y_L,z_L) \end{bmatrix} \qquad (3.60)$$

Then the pressure at the measurement locations may be written as:

$$P_m = A_m a \qquad (3.61)$$

where:

$$P_m = \begin{bmatrix} p(x_1,y_1,z_1) \\ p(x_2,y_2,z_2) \\ \vdots \\ p(x_L,y_L,z_L) \end{bmatrix} \quad \text{and} \quad a = \begin{bmatrix} a_1 \\ a_2 \\ \vdots \\ a_N \end{bmatrix} \quad \text{and} \quad \varphi(x,y,z) = \begin{bmatrix} \Phi_{k_1}(x,y,z) \\ \Phi_{k_2}(x,y,z) \\ \vdots \\ \Phi_{k_N}(x,y,z) \end{bmatrix} \qquad (3.62\text{a,b,c})$$

Equation (3.58) may then be used with Equation (3.62) to write an expression for the pressure at any other point not in the measurement array (and usually on the surface of the noise source being examined) as:

$$p(x,y,z) = a^T \varphi(x,y,z) \qquad (3.63)$$

The matrix a is found by inverting Equation (3.61) which includes the measured data. As the matrix is non-square a pseudo inverse is obtained so we obtain:

$$a = \left(A_m^H A_m + \alpha I\right)^{-1} A_m^H P_m \qquad (3.64)$$

where I is the identity matrix and α is the regularisation parameter, which is usually selected with the following equation to give a value close to the optimum:

$$\alpha = \left(1 + \frac{1}{2(kd)^2}\right) \times 10^{SNR/10} \qquad (3.65)$$

where *SNR* is the signal-to-noise ratio for the measured data and $d = z_m - z_p$ is the distance between the sound source and the measurement plane.

The particle velocity is obtained using the same procedure and same equations as above, except that Equation (3.57) is replaced with:

$$\Phi_k(x,y,z) = \frac{k_z}{\rho c k} \, e^{-j(k_x x + k_y y + k_z(z_p - z_m))} \tag{3.66}$$

This measurement technique is accurate in terms of quantifying the sound intensity as a function of location on the noise emitting structure and it has the same resolution and frequency range as NAH. No particle velocity measurement is needed.

3.15.3 Helmholtz Equation Least Squares Method (HELS)

The HELS method (Wu, 2000; Isakov and Wu, 2002) is very similar to the SONAH method discussed above, except instead of using plane wave functions to describe the sound field as in Equation (3.57), the HELS method uses spherical wave functions. The analysis is quite a bit more complicated than it is for the SONAH method and as this method does not produce any better results than the SONAH method, it will not be discussed any further.

3.15.4 Beamforming

Beamforming measures the amplitude and phase of the sound pressure over a planar or spherical or linear array of many microphones and this is used to maximise the total summed output of the array for sound coming from a specified direction, while minimising the response due to sound coming from different directions. By inserting an adjustable delay in the electronic signal path from each microphone, it is possible to "steer" the array so that the direction of maximum response can be varied, In this way the relative intensity of sound coming from different directions can be determined and the data analysed to produce a map of the relative importance of different parts of a sound source to the total far field sound pressure level. This principle underpins the operation of a device known as an acoustic camera. Unlike NAH and SONAH, beamforming operates in the far field of the sound source and it is more accurate at higher frequencies.

Typical frequency ranges and recommended distances between the array and the radiating noise source for various commercially available array types are listed in Table 3.3. The dynamic range of a beamforming measurement varies from about 6 dB for ring arrays to up to 15 dB for spiral arrays. However, spiral arrays have a disadvantage of poor depth of field, so that it is more difficult to focus the array on the sound source, especially if the sound source is non-planar. Poorly focussed arrays used on more than one sound source existing at different distances from the array but in a similar direction can result in sources cancelling one another so that they disappear from the beamforming image altogether.

Table 3.3 Beamforming array properties

Array name	Array size	Number of mics.	Distance from source (m)	Frequency range (Hz)	Backward attenuation[a]
Star	3×2m arms	36	3–300	100–7,000	-21 dB
Ring	0.75 m dia	48	0.7–3	400–20,000	0 dB
Ring	0.35 m dia	32	0.35–1.5	400–20,000	0 dB
Ring	1.4 m dia	72	2.5–20	250–20,000	0 dB
Cube	0.35m across	32	0.3–1.5	1,000–10,000	-20 dB
Sphere	0.35 m dia	48	0.3–1.5	1,000–10,000	-20 dB

[a]This is how much a wave is attenuated if it arrives from behind the array.

When a spherical array is used inside an irregular enclosure such as a car passenger compartment, it is necessary to use a CAD model of the interior of the enclosure and accurately position the beamforming array within it. Then the focus plane of the array can be adjusted in software for each direction to which the array is steered.

Beamformers have the disadvantage of poor spatial resolution of noise source locations, especially at low frequencies. For a beamforming array of largest dimension, D, and distance from the source, L, the resolution (or accuracy with which a source can be located) is given by:

$$Res = 1.22 \frac{L}{D} \lambda \qquad (3.67)$$

For acceptable results, the array should be sufficiently far from the source that it does not subtend an angle greater than 30° in order to cover the entire source. In general, the distance of the array from the sound source should be at least the same as the array diameter, but no greater if at all possible. A big advantage of the beamforming technique is that it can image distant sources as well as moving sources. It is also possible to get quantitative measures of the sound power radiated by the source (Hald, 2005). One disadvantage of beamforming compared to NAH is that it is not possible to distinguish between sound radiated directly by the source and sound reflected from the source (as the measurements are made in the far field of the source). Also, for planar arrays, one cannot distinguish between sound coming from in front of or behind the array.

Beamforming can give erroneous results in some situations. For example, if the array is not focussed at the source distance, the source location will not be clear and sharp - it can look quite fuzzy. If two sources are at different distances from the array, it is possible that neither will be identified.

Beamforming array design is also important as there is a trade off between depth of focus of the array and its dynamic range. The spiral array has the greatest dynamic range (up to 15 dB) but a very small depth of focus whereas the ring array only has a dynamic range of 6 dB but a large depth of focus, allowing the array to focus on noise sources at differing distances and not requiring such precision in the estimate of the distance of the noise source from the array. The dynamic range is greatest for broadband noise sources and least for low frequency and tonal sources.

3.15.4.1 Summary of the Underlying Theory

Beamforming theory is complicated so only a brief summary will be presented here. For more details, the reader is referred to Christensen and Hald (2004) and Johnson and Dudgeon (1993). There are two types of beamforming: infinite- focus distance and finite-focus distance. For the former, plane waves are assumed and for the latter, spherical waves are assumed to originate from the focal point of the array.

In essence, infinite-focus beamforming in the context of interest here is the process of summing the signals from an array of microphones and applying different delays to the signals from each microphone so that sound coming from a particular direction causes a maximum summed microphone response and sound coming from other directions causes no response. Of course in practice, sound from any direction will still cause some response but the principle of operation is that these responses will be well below the main response due to sound coming from the direction of interest.

It is also possible to scale the beamformer output so that a quantitative measure of the active sound intensity at the surface of the noise radiator can be made (Hald, 2005).

Consider a planar array made up of L microphones at locations $(x_\ell, y_\ell, \ell = 1, ..., L)$ in the x - y plane. If the measured pressure signals, p_ℓ are individually delayed and then summed, the output of the array is:

$$p(\boldsymbol{n}, t) = \sum_{\ell=1}^{L} w_\ell p_\ell(t - \Delta_\ell(\boldsymbol{n})) \tag{3.68}$$

where w_ℓ is the weighting coefficient applied to pressure signal p_ℓ, and its function is to reduce the importance of the signals coming from the array edges, which in turn reduces the amplitudes of side lobes in the array response. Side lobes are peaks in the array response in directions other than the design direction and serve to reduce the dynamic range of the beamformer. The quantity $\boldsymbol{n}$ in Equation (3.68) is the unit vector in the direction of maximum sensitivity of the array and the time delays Δ_ℓ are chosen to maximise the array sensitivity in direction, $\boldsymbol{n}$. This is done by delaying the signals associated with a plane wave arriving from direction, $\boldsymbol{n}$, so that they are aligned in time before being summed. The time delay, Δ_ℓ, is the dot product of the unit vector, $\boldsymbol{n}$, and the vector, $\boldsymbol{r}_\ell = (x_\ell, y_\ell)$ divided by the speed of sound, c. That is:

$$\Delta_\ell = \frac{\boldsymbol{n} \cdot \boldsymbol{r}_\ell}{c} \tag{3.69}$$

If the analysis is done in the frequency domain, the beamformer output at angular frequency, ω, is:

$$P(\boldsymbol{n}, \omega) \;=\; \sum_{\ell=1}^{L} w_{\ell} P_{\ell}(\omega) e^{-j\omega \Delta_{\ell}(\boldsymbol{n})} \;=\; \sum_{\ell=1}^{I} w_{\ell} P_{\ell}(\omega) e^{-j\boldsymbol{k}\cdot\boldsymbol{r}_{\ell}} \qquad (3.70\mathrm{a,b})$$

where $\boldsymbol{k} = -k\boldsymbol{n}$ is the wave number vector of a plane wave incident from the direction, $\boldsymbol{n}$, which is the direction in which the array is focussed. More detailed analysis of various aspects affecting the beamformer performance are discussed by Christensen and Hald (2004) and Johnson and Dudgeon (1993).

Finite-focus beamforming using a spherical wave assumption to locate the direction of a source and its strength at a particular distance from the array (array focal point) follows a similar but slightly more complex analysis than outlined above for infinite-focus beamforming. For the array to focus on a point source at a finite distance, the various microphone delays should align in time, the signals of a spherical wave radiated from the focus point. Equation (3.70a) still applies but the delay, Δ_{ℓ}, is defined as:

$$\Delta_{\ell} \;=\; \frac{|\boldsymbol{r}| - |\boldsymbol{r} - \boldsymbol{r}_i|}{c} \qquad (3.71)$$

where $\boldsymbol{r}$ is the vector location of the source from an origin point in the same plane as the array, $\boldsymbol{r}_{\ell}$ is the vector location of microphone, ℓ, in the array with respect to the same origin and $|\boldsymbol{r} - \boldsymbol{r}_{\ell}|$ is the scalar distance of microphone ℓ, from the source location.

More complex beamforming analyses applicable to aero-acoustic problems, where the array is close to the source and there is a mean flow involved, are discussed by Brooks and Humphreys (2006).

3.15.5 Direct Sound Intensity measurement

The use of a stethoscope, which is essentially a microphone, to manually scan close to the surface of an item of equipment to locate noise sources is well known. However, in the presence of significant levels of background noise, this method is no longer effective. Manually scanning a particle velocity sensor such as the "Microflown" to measure the normal acoustic particle velocity over a surface has a number of advantages: the particle velocity signal is larger than the pressure signal close to a source; background noise reflected from the surface being scanned produces close to zero particle velocity at the surface whereas the acoustic pressure is approximately doubled; and the particle velocity sensor is directional (in contrast to the omni-directional nature of a microphone), thus further reducing the influence of background noise. There is also equipment available that transforms the particle velocity signal to an audible signal by feeding it into a head set. This crude method of source location in the presence of high levels of background noise seems to be very effective (de Bree and Druyvesteyn, 2005; de Vries and de Bree, 2008), even though in theory the relative sound power radiated by the various locations on a surface or structure can only be determined if the near field sound intensity is measured.

If more accurate measurements are needed then a small intensity probe made up of a miniature microphone and a "Microflown" transducer can be used to scan the surface over which the noise source identification and quantification is required. The scan should be as close as possible to the surface and as far as is practicable, it should follow the surface contour. It is quite feasible for this scan to be done manually by taking intensity measurements adjacent to a large number of points on the noise radiating surface and having the intensity probe stationary for each measurement. A less accurate measurement is to manually scan the intensity probe over an imaginary surface adjacent to the noise radiating surface.

One big advantage of the direct measurement of sound intensity adjacent to a noise radiating surface is the relatively large dynamic range (difference in dB between maximum and minimum measurable intensities) that can be achieved. A dynamic range between 30 and 60 dB is common compared to 20 dB for NAH and SONAH using microphones and 40 dB for NAH and SONAH using particle velocity sensors. The dynamic range of beamforming measurements varies from 6 to 15 dB (Heilmann *et al.*, 2008).

Another advantage of the direct intensity measurement method is the wide bandwidth possible (20 Hz to 20 kHz) compared to 200Hz to 2kHz for planar holography and 2kHz to 10kHz for beamforming arrays.

However, direct measurement of sound intensity close to a surface does have some problems due the dominance of the reactive sound field in that region. This means that any errors in the phase matching between pressure and particle velocity sensors can have a relatively large impact on the accuracy of the intensity measurement. For this reason, intensity measurements can only be made accurately if the phase difference between the acoustic pressure and particle velocity is less than $85°$, which corresponds to a reactivity of 10 dB. The reactivity for a harmonic sound field is defined as:

$$Re = 10\log_{10}(I/I_r) \tag{3.72}$$

where I and I_r are defined in Equations (1.72) and (1.73) respectively. If the reactivity is too high, the intensity probe must be moved further from the noise radiating structure.

CHAPTER FOUR

Criteria

LEARNING OBJECTIVES

In this chapter the reader is introduced to:

- various measures used to quantify occupational and environmental noise;
- hearing loss associated with age and exposure to noise;
- hearing damage risk criteria, requirements for speech recognition and alternative interpretations of existing data;
- hearing damage risk criteria and trading rules;
- speech interference criteria for broadband noise and intense tones;
- psychological effects of noise as a cause of stress and effects on work efficiency;
- Noise Rating (*NR*), Noise Criteria (*NC*), Room Criteria (*RC*), Balanced Noise Criteria (NCB) and Room Noise Criteria (RNC) for ambient level specification; and
- environmental noise criteria.

4.1 INTRODUCTION

An important part of any noise-control program is the establishment of appropriate criteria for the determination of an acceptable solution to the noise problem. Thus, where the total elimination of noise is impossible, appropriate criteria provide a guide for determining how much noise is acceptable. At the same time, criteria provide the means for estimating how much reduction is required. The required reduction in turn provides the means for determining the feasibility of alternative proposals for control, and finally the means for estimating the cost of meeting the relevant criteria.

For industry, noise criteria ensure the following:

- that hearing damage risk to personnel is acceptably small;
- that reduction in work efficiency due to a high noise level is acceptably small;
- that, where necessary, speech is possible; and
- that noise at plant boundaries is sufficiently small for noise levels in the surrounding community to be acceptable.

Noise criteria are important for the design of assembly halls, classrooms, auditoria and all types of facilities in which people congregate and seek to communicate, or simply seek rest and escape from excessive noise. Criteria are also essential for specifying acceptable environmental noise limits. In this chapter, criteria and the basis for their formulation are discussed. It is useful to first define the various noise measures that are used in standards and regulations to define acceptable noise limits.

4.1.1 Noise Measures

4.1.1.1 A-weighted Equivalent Continuous Noise Level, L_{Aeq}

The un-weighted continuous noise level was defined in Chapter 3, Equation (3.22). The A-weighted Equivalent Continuous Noise Level has a similar definition except that the noise signal is A-weighted before it is averaged. After A-weighting, the pressure squared is averaged and this is often referred to as energy averaging. The A-weighted Equivalent Continuous Noise Level is used as a descriptor of both occupational and environmental noise and for an average over time, T, it may be written in terms of the instantaneous A-weighted sound pressure level, $L_A(t)$ as:

$$L_{Aeq,T} = 10 \log_{10} \left[\frac{1}{T} \int_0^T 10^{L_A(t)/10} \, dt \right] \tag{4.1}$$

For occupational noise, the most common descriptor is $L_{Aeq,8h}$, which implies a normalisation to 8 hours, even though the contributing noises may be experienced for more or less than 8 hours. Thus, for sound experienced over T hours:

$$L_{Aeq,8h} = 10 \log_{10} \left[\frac{1}{8} \int_0^T 10^{L_A(t)/10} \, dt \right] \tag{4.2}$$

If the sound pressure level is measured using a sound level meter at m different locations where an employee may spend some time, then Equation (4.2) becomes:

$$L_{Aeq,8h} = 10 \log_{10} \frac{1}{8} \left[t_1 10^{L_{A1}/10} + t_2 10^{L_{A2}/10} + \dots\dots\ t_m 10^{L_{Am}/10} \right] \tag{4.3}$$

where L_{Ai} are the measured equivalent A-weighted sound pressure levels and t_i are the times in hours which an employee spends at the m locations. Note that the sum of t_1............t_m does not have to equal 8 hours.

4.1.1.2 A-weighted Sound Exposure

Industrial sound exposure may be quantified using the A-weighted Sound Exposure, $E_{A,T}$, defined as the time integral of the squared, instantaneous A-weighted sound pressure, $p_A^2(t)$ (Pa²) over a particular time period, $T = t_2 - t_1$ (hours). The units are pascal-squared-hours (Pa².h) and the defining equation is:

$$E_{A,T} = \int_{t_1}^{t_2} p_A^2(t) \, dt \tag{4.4}$$

Using Equations (4.1) and (4.4), the relationship between the A-weighted Sound Exposure and the A-weighted Equivalent Continuous Noise Level, $L_{Aeq,T}$, can be shown to be:

$$E_{A,T} = 4T \times 10^{(L_{Aeq,T} - 100)/10} \qquad (4.5)$$

4.1.1.3 A-weighted Sound Exposure Level, L_{AE} or SEL

The A-weighted Sound Exposure Level is defined as:

$$L_{AE} = 10\log_{10}\left[\int_{t_1}^{t_2} \frac{p_A^2(t)}{p_{ref}^2}\,\mathrm{d}t\right] = 10\log_{10}\left[\frac{E_{A,T} \times 3600}{p_{ref}^2}\right] \qquad (4.6a,b)$$

where the times t_1, t_2 and $\mathrm{d}t$ are in seconds (not hours as for A-weighted sound exposure in Equation (4.4)) and $T = t_2 - t_1$.

A "C-weighted" Sound Exposure Level is defined by substituting the C-weighted noise level for the A-weighted level in Equation (4.6). These two exposure level quantities are sometimes used for assessment of general environmental noise, but mainly for the assessment of transient environmental noise, such as traffic noise, aircraft noise and train noise. When the event is a transient, the time interval, t_1 - t_2 must include the 10 dB down points as shown in Figure 4.1.

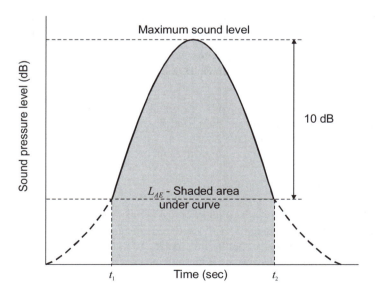

Figure 4.1 Sound exposure of a single event.

An equivalent Continuous Noise Level for a nominal 8-hour working day may be calculated from $E_{A,8h}$ or L_{AE} using:

$$L_{Aeq,8h} = 10\log_{10}\left(\frac{E_{A,8h}}{3.2 \times 10^{-9}}\right) = 10\log_{10}\left(\frac{1}{28,800}\, 10^{L_{AE,8h}/10}\right) \qquad (4.7a,b)$$

4.1.1.4 Day–Night Average Sound Level, L_{dn} or DNL

The Day–Night Average Sound Level is used sometimes to quantify traffic noise and some standards regarding the intrusion of traffic noise into the community are written in terms of this quantity, L_{dn}, which is defined as:

$$L_{dn} = 10\log_{10}\frac{1}{24}\left[\int_{22:00}^{07:00} 10 \times 10^{L_A(t)/10}\,dt + \int_{07:00}^{22:00} 10^{L_A(t)/10}\,dt\right] \quad \text{dB} \qquad (4.8)$$

For traffic noise, the Day–Night Average Sound Level for a particular vehicle class is related to the Sound Exposure Level by:

$$L_{dn} = L_{AE} + 10\log_{10}(N_{day} + N_{eve} + 10 \times N_{night}) - 49.4 \quad \text{dB(A)} \qquad (4.9)$$

where, L_{AE} is the A-weighted Sound Exposure Level for a single vehicle pass-by, N_{day}, N_{eve} and N_{night} are the numbers of vehicles in the particular class that pass by in the daytime (0700 to 1900 hours), evening (1900 to 2200 hours) and nighttime (2200 to 0700 hours), respectively and the normalisation constant, 49.4, is $10\log_{10}$ of the number of seconds in a day. To calculate the L_{dn} for all vehicles, the above equation is used for each class and the results added together logarithmically (see Section 1.11.4).

4.1.1.5 Community Noise Equivalent Level, L_{den} or CNEL

The Community Noise Equivalent Level is used sometimes to quantify industrial noise and traffic noise in the community and some regulations are written in terms of this quantity, L_{den}, which is defined as:

$$L_{den} = 10\log_{10}\frac{1}{24}\left[\int_{22:00}^{07:00} 10 \times 10^{L_A(t)/10}\,dt + \int_{07:00}^{19:00} 10^{L_A(t)/10}\,dt\right.$$

$$\left. + \int_{19:00}^{22:00} 3 \times 10^{L_A(t)/10}\,dt\right] \quad \text{dB} \qquad (4.10)$$

For traffic noise, the Community Noise Equivalent Level for a particular vehicle class is related to the Sound Exposure Level by:

$$L_{den} = L_{AE} + 10\log_{10}\left(N_{day} + 3N_{eve} + 10N_{night}\right) - 49.4 \quad \text{(dB)} \quad (4.11)$$

where, L_{AE} is the A-weighted Sound Exposure Level for a single vehicle pass-by, the constant, $49.4 = 10\log_{10}$ (number of seconds in a day) and N_{day}, N_{eve} and N_{night} are the numbers of vehicles in the particular class that pass by in the daytime (0700 to 1900 hours), evening (1900 to 2200 hours) and nighttime (2200 to 0700 hours), respectively. To calculate the L_{den} for all vehicles, the above equation is used for each class and the results added together logarithmically (see Section 1.11.4).

4.1.1.6 Effective Perceived Noise Level, L_{PNE}

This descriptor is used solely for evaluating aircraft noise. It is derived from the Perceived Noise Level, L_{PN}, which was introduced some time ago by Kryter (1959). It is a very complex quantity to calculate and is a measure of the annoyance of aircraft noise. It takes into account the effect of pure tones (such as engine whines) and the duration of each event.

The calculation procedure begins with a recording of the sound pressure level vs time curve, which is divided into 0.5 second intervals over the period that the aircraft noise exceeds background noise. Each 0.5 second interval (referred to as the kth interval) is then analysed to give the noise level in that interval in 24 1/3 octave bands from 50 Hz to 10 kHz. The noy value for each 1/3 octave band is calculated using published tables (Edge and Cawthorne, 1976) or curves (Raney and Cawthorne, 1998). The total noisiness (in noys) corresponding to each time interval is then calculated from the 24 individual 1/3 octave band noy levels using:

$$n_t = n_{max} + 0.15\left(\sum_{i=1}^{24} n_i - n_{max}\right) \quad \text{(noy)} \quad (4.12)$$

where n_{max} is the maximum 1/3 octave band noy value for the time interval under consideration. The perceived noise level for each time interval is then calculated using:

$$L_{PN} = 40 + 33.22\log_{10} n_t \quad \text{(dB)} \quad (4.13)$$

The next step is to calculate the tone-corrected perceived noise level (L_{PNT}) for each time interval. This correction varies between 0 dB and 6.7 dB and it is added to the L_{PN} value. It applies whenever the level in any one band exceeds the levels in the two adjacent 1/3 octave bands. If two or more frequency bands produce a tone correction, only the largest correction is used. The calculation of the actual tone-correction is complex and is described in detail in the literature (Edge and Cawthorne, 1976). The maximum tone corrected perceived noise level over all time intervals is denoted L_{PNTmax}. Then next step in calculating L_{PNE} is to calculate the duration correction, D, which is usually negative and is given by Raney and Cawthorne (1998), corrected here, as:

$$D = 10\log_{10}\left(\sum_{k=i}^{i+2d} 10^{L_{PNT(k)}/10} \right) - 13 - L_{PNT\text{max}}$$ (4.14)

where $k = i$ is the time interval for which L_{PNT} first exceeds $L_{PNT\text{max}}$ and d is the length of time in seconds that L_{PNT} exceeds $L_{PNT\text{max}}$.

Finally the effective perceived noise level is calculated using:

$$L_{PNE} = L_{PNT\text{max}} + D$$ (4.15)

In a recent report (Yoshioka, 2000), it was stated that a good approximate and simple method to estimate L_{PNE} was to measure the maximum A-weighted sound level, $L_{A\text{max}}$, over the duration of the aircraft noise event (which lasts for approximately 20 seconds in most cases) and add 13 dB to obtain L_{PNE}.

4.1.1.7 Other Descriptors

There are a number of other descriptors used in the various standards, such as "long time average A-weighted sound pressure level" or "long-term time average rating level", but these are all derived from the quantities mentioned in the preceding paragraphs and defined in the standards that specify them, so they will not be discussed further here.

4.2 HEARING LOSS

Hearing loss is generally determined using pure tone audiometry in the frequency range from about 100 Hz to 8 kHz, and is defined as the differences in sound pressure levels of a series of tones that are judged to be just audible compared with reference sound pressure levels for the same series of tones. It is customary to refer to hearing level which is the level at which the sound is just audible relative to the reference level when referring to hearing loss. However, the practice will be adopted here of always using the term hearing loss rather than the alternative term hearing level.

4.2.1 Threshold Shift

In Chapter 2 the sensitivity of the ear to tones of various frequencies was shown to be quite non-uniform. Equal loudness contours, measured in phons, were described as well as the minimum audible field or threshold of hearing. The latter contours, and in particular the minimum audible field levels, were determined by the responses of a great many healthy young people, males and females in their 20s, who sat facing the source in a free field. When the subject had made the required judgment; that is, that two sounds were equally loud or the sound was just audible, the subject vacated the testing area, and the measured level of the sound in the absence of the subject was determined and assigned to the sound under test. In other words, the assigned sound pressure levels were the free-field levels, unaffected by diffraction effects due to the presence of the auditor.

In Chapter 3 the problem of characterising the sensitivity of a microphone was discussed. It was shown that diffraction effects, as well as the angle of incidence, very strongly affect the apparent sensitivity. Clearly, as the human head is much larger than any commercial microphone, the ear as a microphone is very sensitive to the effects mentioned. In fact, as mentioned in Chapter 2, the ear and brain, in close collaboration, make use of such effects to gain source location information from a received signal. Thus, it is apparent that the sound pressure level at the entrance to the ear may be very different from the level of the freely propagating sound field in the absence of the auditor.

The threshold of audibility has been chosen as a convenient measure of the state of health of the auditory system. However, the provision of a free field for testing purposes is not always practical. Additionally, such a testing arrangement does not offer a convenient means for testing one ear at a time. A practical and much more convenient method of test is offered by use of earphones. Such use forms the basis of pure tone audiometry. The assumption is then implicit that the threshold level determined as the mean of the responses of a great many healthy young people, males and females in their 20s, corresponds to the minimum audible field mentioned earlier. The latter interpretation will be put upon published data for hearing loss based upon pure tone audiometric testing. Thus, where the hearing sensitivity of a subject may be 20 dB less than the established threshold reference level, the practice is adopted in this chapter of representing such hearing loss as a 20 dB rise in the free-field sound pressure level which would be just audible to the latter subject. This method of presentation is contrary to conventional practice, but it better serves the purpose of illustrating the effect of hearing loss upon speech perception.

4.2.2 Presbyacusis

It is possible to investigate the hearing sensitivity of populations of people who have been screened to eliminate the effects of disease and excessive noise. Hearing deterioration with age is observed in screened populations and is called presbyacusis. It is characterised by increasing loss with increasing frequency and increasing rate of loss with age. Men tend to lose hearing sensitivity more rapidly than women. There is evidence to show that hearing deterioration with age may also be race specific (Driscoll and Royster, 1984). Following the convention proposed in the preceding section, the effect of presbyacusis is illustrated in Figure 4.2 as a rise in the mean threshold of hearing level. For comparison, the range of quiet speech sounds is also indicated in the figure. As the fricative parts of speech lie generally at the right and lower portion of the speech range, it is evident that old folks may not laugh as readily at the jokes, not because of a jaded sense of humour, but rather because they missed the punch lines.

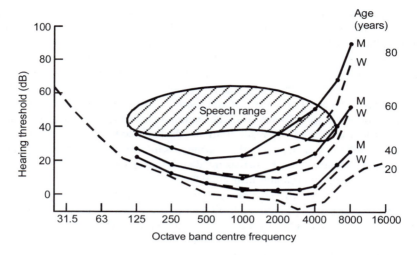

Figure 4.2 Threshold shift due to presbyacusis: M = men; W = women. Speech sounds: male, normal voice, at 1 m. Data from Beranek (1971), Chapter 17.

4.2.3 Hearing Damage

Hearing loss may be not only the result of advancing age but also the result of exposure to excessive noise. Loss caused by exposure to excessive noise usually occurs first in the frequency range from about 4000 Hz to 6000 Hz, which is the range of greatest sensitivity of the human ear. Following the proposed method of presentation, the plight of women habitually exposed to excessive noise in a jute mill is illustrated in Figure 4.3. The dismal effect upon their ability to understand speech is clearly illustrated. It is equally clear that, for these ladies, gossip in public places cannot possibly be a private matter.

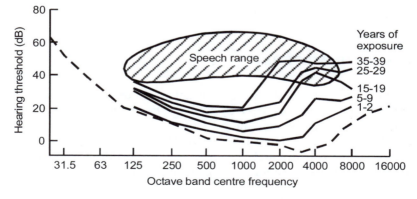

Figure 4.3 Threshold shift due to excessive noise exposure. Speech sounds: male, normal voice, at 1 m.

Exposure to excessive noise for a short period of time may produce a temporary loss of hearing sensitivity. If this happens, the subject may notice a slight dulling in hearing at the end of the exposure. This effect is often accompanied by a ringing in the ears, known as tinnitus, which persists after the noise exposure. This temporary loss of hearing sensitivity is known as temporary threshold shift (TTS) or auditory fatigue. Removal from the noise generally leads to more or less complete recovery if the exposure has not been too severe. If the noise exposure is severe or is repeated sufficiently often before recovery from the temporary effect is complete, a permanent noise-induced hearing loss may result. Initially, the loss occurs in the frequency range from about 4000 to 6000 Hz, but as the hearing loss increases it spreads to both lower and higher frequencies. With increasing deterioration of hearing sensitivity the maximum loss generally remains near 4000 Hz.

The first handicap due to noise-induced hearing loss to be noticed by the subject is usually some loss of hearing for high-pitched sounds such as squeaks in machinery, bells, musical notes, etc. This is followed by a diminution in the ability to understand speech; voices sound muffled, and this is worse in difficult listening conditions. The person with noise-induced hearing loss complains that everyone mumbles. High-frequency consonant sounds of low intensity are missed, whereas vowels of low frequency and higher intensity are still heard. As consonants carry much of the information in speech, there is little reduction in volume but the context is lost. However, by the time the loss is noticed subjectively as a difficulty in understanding speech, the condition is far advanced. Fortunately, present-day hearing aids, which contain spectral shaping circuitry, can do much to alleviate this problem, although the problem of understanding speech in a noisy environment such as a party will still exist. Recently, a hearing aid, which makes use of a directional antenna worn as a band around one's neck as described in Chapter 2, has become available to alleviate this problem (Widrow, 2001).

Data have been presented in the USA (Royster *et al.*, 1980) which show that hearing loss due to excessive noise exposure may be both race and sex specific. The study showed that, for the same exposure, white males suffered the greatest loss, with black males, white females and black females following with progressively less loss, in that order. The males tend to have greatest loss at high frequencies, whereas the females tend to have more uniform loss at all frequencies.

4.3 HEARING DAMAGE RISK

The meaning of "damage risk" needs clarification in order to set acceptable noise levels to which an employee may be exposed. The task of protecting everyone from any change in hearing threshold in the entire audio-frequency range is virtually impossible and some compromise is necessary. The accepted compromise is that the aim of damage risk criteria must be to protect most individuals in exposed groups of employees against loss of hearing for everyday speech. Consequently the discussion begins with the minimum requirements for speech recognition and proceeds with a review of what has been and may be observed.

The discussion will continue with a review of the collective experience upon which a data base of hearing level versus noise exposure has been constructed. It will conclude with a brief review of efforts to determine a definition of exposure, which accounts for both the effects of level and duration of excessive noise and to mathematically model the data base (ISO 1999, 1990) in terms of exposure so defined. The purpose of such mathematical modelling is to allow formulation of criteria for acceptability of variable level noise, which is not covered in the data base. Criteria are designed to ensure exposed people retain the minimum requirements for speech recognition.

4.3.1 Requirements for Speech Recognition

For good speech recognition the frequency range from 500 to 2000 Hz is crucial; thus, criteria designed to protect hearing against loss of ability to recognise speech are concerned with protection for this frequency range. In the United States, loss for speech recognition purposes is assumed to be directly related to the arithmetic average of hearing loss in decibels at 500, 1000 and 2000 Hz. For compensation purposes, the 3000 Hz loss is included in the average. In Australia, a weighted mean of loss in the frequency range from 500 to 6000 Hz is used as the criterion. An arithmetic average of 25 dB loss defines the boundary between just adequate and inadequate hearing sensitivity for the purpose of speech recognition. For practical purposes, a hearing loss of 25 dB will allow speech to be just understood satisfactorily, while a loss of 92 dB is regarded as total hearing loss. If a person suffers a hearing loss between 25 dB and 92 dB, that person's hearing is said to be impaired, where the degree of impairment is determined as a percentage at the rate of 1.5 percentage points for each decibel loss above 25 dB.

As an indication of the amount of compensation awarded, in the USA it is paid in terms of weeks of salary where the number of weeks salary to be paid is calculated as the % loss multiplied by 1.5 for a person with a dependent family and 1.33 for a person with no dependents. If the loss is different in different ears, then an average loss must be obtained by multiplying the % loss in the better ear by 5, adding this to the poorer ear and then dividing by 6.

4.3.2 Quantifying Hearing Damage Risk

In a population of people who have been exposed to excessive noise and who have consequently suffered observable loss of hearing, it is possible to carry out retrospective studies to determine quantitative relationships between noise exposure and hearing threshold shift. Two such studies have been conducted (Burns and Robinson, 1970 and Passchier-Vermeer, 1968, 1977) and these are referenced in the ISO1999, 1990. The standard states that neither of the latter studies form part of its data base. The International Standards Organisation document makes no reference to any other studies, including those used to generate its own data base.

The standard provides equations for reconstructing its data base and these form the basis for noise regulations around the world. It is important to note that studies to determine the quantitive relationship between noise exposure and hearing threshold shift are only feasible in situations where noise levels are effectively "steady state", and it has only been possible to estimate these noise levels and duration of exposure retrospectively. The International Standards Organisation data base shows that permanent threshold shift is dependent upon both the level of the sound and the duration of the exposure, but it cannot provide information concerning the effects of variable level sound during the course of exposure.

A retrospective study of time-varying level exposure is needed from a regulatory point of view but is not available. From this point of view, a knowledge of the relationship between hearing loss and a time-varying level of exposure over an extended period of time is required to establish the relationships between noise level, duration of exposure and permanent threshold shift. A knowledge of this relationship is required to establish trading rules between length of exposure and level of exposure. Since such information is not available, certain arbitrary assumptions have been made which are not uniformly accepted, and therein lies the basis for contention.

It is reasonable to assume that aging makes some contribution to loss of hearing in people exposed to excessive noise (see Section 4.2.2). Consequently, in assessing the effects of excessive noise, it is common practice to compare a noise-exposed person with an unexposed population of the same sex and age when making a determination of loss due to noise exposure. As there is no way of directly determining the effect of noise alone it is necessary in making any such comparison that some assumption be made as to how noise exposure and aging collectively contribute to the observed hearing loss in people exposed to excessive noise over an extended time.

An obvious solution, from the point of view of compensation for loss of hearing, is to suppose that the effects of age and noise exposure are additive on a decibel basis. In this case, the contribution due to noise alone is computed as the decibel difference between the measured threshold shift and the shift expected due to aging. However, implicit in any assumption that might be made is some implied mechanism. For example, the proposed simple addition of decibel levels implies a multiplication of analog effects. That is, it implies that damage due to noise and age are characterised by different mechanisms or damage to different parts of the hearing system. From the point of view of determining the relationship between hearing loss due to noise and noise exposure it is important to identify the appropriate mechanisms. For example, if summation of analog effects (which would imply similar damage mechanisms for age and noise or damage to the same parts of the hearing system), rather than multiplication is the more appropriate mechanism, then the addition of age and noise-induced effects might be more appropriately represented as the addition of the antilogarithms of hearing threshold shifts due to age and noise and a very different interpretation of existing data is then possible (Kraak, 1981; Bies and Hansen, 1990).

At present there exists no generally accepted physical hearing loss model to provide guidance on how the effects of age and noise should be combined. This is of importance, as it is the relationship between hearing loss due to age and noise exposure that must be quantified to establish acceptable exposure levels. In particular, it is necessary to establish what constitutes exposure, as it is the exposure that must be quantified.

4.3.3 International Standards Organisation Formulation

The International Standard, ISO-1999 (1990) and the American Standard, ANSI S3.44 - 1996 (R2006) provide the following empirical equation for the purpose of calculating the hearing threshold level, H', associated with age and noise of a noise-exposed population:

$$H' = H + N - HN/120 \tag{4.16}$$

H is the hearing threshold level associated with age and N is the actual or potential noise-induced permanent threshold shift, where the values of H, H' and N vary and are specific to the same fractiles of the population.

Only the quantities H and H' can be measured in noise-exposed and non-noise-exposed populations respectively. The quantity N cannot be measured independently and thus is defined by Equation (4.16). It may be calculated using the empirical procedures provided by the Standard.

The values to be used in Equation (4.16) are functions of frequency, the duration of exposure, Θ (number of years), and the equivalent continuous A-weighted sound pressure level for a nominal eight-hour day, $L_{Aeq,8h}$, averaged over the duration of exposure, Θ. For exposure times between 10 and 40 years the median (or 50% fractile) potential noise induced permanent threshold shift values, N_{50} (meaning that 50% of the population will suffer a hearing loss equal to or in excess of this) are given by the following equation:

$$N_{50} = (u + v \log_{10}\Theta)(L_{Aeq,8h} - L_0)^2 \tag{4.17}$$

If $L_{Aeq,8h} < L_0$, then $L_{Aeq,8h}$ is set equal to L_0 to evaluate Equation (4.17). This equation defines the long-term relationship between noise exposure and hearing loss, where the empirical constants u, v and L are listed in Table 4.1.

Table 4.1 Values of the coefficients u, v and L used to determine the NIPTS for the median value of the population, $N_{0,50}$

Frequency (Hz)	u	v	L_0 (dB)
500	-0.033	0.110	93
1000	-0.02	0.07	89
2000	-0.045	0.066	80
3000	+0.012	0.037	77
4000	+ 0.025	0.025	75
6000	+ 0.019	0.024	77

For exposure times less than 10 years:

$$N_{50} = \frac{\log_{10}(\Theta + 1)}{\log_{10}(11)} N_{50:\Theta=10} \tag{4.18}$$

For other fractiles, Q, the threshold shift is given by:

$$N_Q = N_{50} + kd_u ; \qquad 5 < Q < 50$$

$$N_Q = N_{50} - kd_L ; \qquad 50 < Q < 95$$

(4.19a,b)

The constant k is a function of the fractile, Q, and is given in Table 4.2. The parameters d_u and d_L can be calculated as follows:

$$d_u = (X_u + Y_u \log_{10}\Theta)(L_{Aeq8} - L_0)^2 \tag{4.20}$$

$$d_L = (X_L + Y_L \log_{10}\Theta)(L_{Aeq8} - L_0)^2 \tag{4.21}$$

If $L_{Aeq,8h} < L_0$, then $L_{Aeq,8h}$ is set equal to L_0 for the purposes of evaluating Equations (4.20) and (4.21). The constants, X_u Y_u X_L and Y_L are listed in Table 4.3.

Table 4.2 Values of the multiplier k or each fractile Q

Q		k
0.05	0.95	1.645
0.10	0.90	1.282
0.15	0.85	1.036
0.20	0.80	0.842
0.25	0.75	0.675
0.30	0.70	0.524
0.35	0.65	0.385
0.40	0.60	0.253
0.45	0.55	0.126
	0.50	0.0

The threshold shift, H_{50}, for the 50% fractile due to age alone is given in the standard as:

$$H_{50} = a (Y - 18)^2 \tag{4.22}$$

For other fractiles, Q, the threshold shift is given by the following equations.

$$H_Q = H_{50} + kS_u ; \qquad 5 < Q < 50$$

$$H_Q = H_{50} - kS_L ; \qquad 50 < Q < 95$$

(4.23)

where k is given in Table 4.2, Q is the percentage of population that will suffer the loss H_Q, and where:

$$S_L = b_L + 0.356 \, H_{50} \qquad\qquad (4.24)$$

$$S_u = b_u + 0.445 \, H_{50} \qquad\qquad (4.25)$$

Table 4.3 Constants for use in calculating N_Q fractiles

Frequency (Hz)	X_u	Y_u	X_L	Y_L
500	0.044	0.016	0.033	0.002
1000	0.022	0.016	0.020	0.000
2000	0.031	-0.002	0.016	0.000
3000	0.007	0.016	0.029	-0.010
4000	0.005	0.009	0.016	-0.002
6000	0.013	0.008	0.028	-0.007

Values of a, b_u and b_L differ for men and women and are listed in Table 4.4 as a function of octave band centre frequency.

Table 4.4 Values of the parameters b_u b_L and a used to determine respectively the upper and lower parts of the statistical distribution H_Q

Frequency (Hz)	Value of b_u		Value of b_L		Value of a	
	Males	Females	Males	Females	Males	Females
125	7.23	6.67	5.78	5.34	0.0030	0.0030
250	6.67	6.12	5.34	4.89	0.0030	0.0030
500	6.12	6.12	4.89	4.89	0.0035	0.0035
1000	6.12	6.12	4.89	4.89	0.0040	0.0040
1500	6.67	6.67	5.34	5.34	0.0055	0.0050
2000	7.23	6.67	5.78	5.34	0.0070	0.0060
3000	7.78	7.23	6.23	5.78	0.0115	0.0075
4000	8.34	7.78	6.67	6.23	0.0160	0.0090
6000	9.45	8.90	7.56	7.12	0.0180	0.0120
8000	10.56	10.56	8.45	8.45	0.0220	0.0150

4.3.4 Alternative Formulations

The authors have demonstrated that an alternative interpretation of the International Standard ISO 1999 data base is possible, and that the interpretation put upon it by the standard is not unique (Bies and Hansen, 1990). Alternatively, very extensive work carried out in Dresden, Germany, over a period of about two decades between the mid-1960s and mid-1980s has provided yet a third interpretation of the existing data base. These latter two formulations lead to the conclusion that for the purpose of determining hearing loss, noise exposure should be determined as an integral of the root mean square (r.m.s.) pressure with time rather than the accepted integral of mean square pressure. This in turn leads to a 6 dB trading rule rather than the 3 dB trading rule that is widely accepted. Trading rules are discussed below in Section 4.3.6.

Recently, it has been shown that neither the formulation of Bies and Hansen nor the standard, ISO 1999, accounts for post exposure loss observed in war veterans (Macrae, 1991). Similarly it may be shown that the formulation of the Dresden group (Kraak *et al.*, 1977, Kraak, 1981) does not account for the observed loss. However, the formulation of Bies and Hansen (1990) as well as that of the Dresden group may be amended to successfully account for post-exposure loss (Bies, 1994).

4.3.4.1 Bies and Hansen Formulation

Bies and Hansen (1990) introduce sensitivity associated with age, STA and with noise, STN (as amended by Bies (1994)) and they propose that the effects of age and noise may be additive on a hearing sensitivity basis. They postulate the following relationship describing hearing loss, H', with increasing age and exposure to noise, which may be contrasted with the ISO 1999 formulation embodied in Equation (4.16):

$$H' = 10 \log_{10}(STA + STN) \qquad (4.26)$$

Additivity of effects on a sensitivity basis rather than on a logarithmic basis (which implies multiplication of effects) is proposed.

Hearing sensitivity associated with age is defined as follows:

$$STA = 10^{H/10} \qquad (4.27)$$

In the above equations, H is the observed hearing loss in a population unexposed to excessive noise, called presbyacusis, and is due to aging alone. It may be calculated by using Equation (4.22).

Bies and Hansen (1990) proposed an empirically determined expression for the sensitivity to noise, STN. Their expression, modified according to Bies (1994), accounts for both loss at the time of cessation of exposure to excessive noise, $STN(Y_{ns})$ where Y_{ns} (years) is the age when exposure to excessive noise stopped and to post-exposure loss, M_c, after exposure to excessive noise has stopped. The former term, $STN(Y_{ns})$, accounts for loss up to cessation of exposure at Y_{ns} years, while the latter term, M_c accounts for continuing hearing loss after exposure to excessive noise ceases.

Loss at the cessation of exposure is a function of the length of exposure, $\Theta = Y - 18$ (years) and the A-weighted sound pressure of the excessive noise, p_A. Here, Y is the age of the population and following the international standard ISO 1999 it is assumed that exposure to excessive noise begins at age 18 years. The quantity STN is defined as zero when Θ is zero. Use of Equations (4.16), (4.26) and (4.27) gives the following expression for $STN(Y_{ns})$ in terms of N given by Equation (4.17) or (4.18) and H given by Equation (4.22):

$$STN(Y_{ns}) = 10^{H/10}\left(10^{(N - 0.0083HN)/10} - 1\right) \qquad (4.28)$$

Hearing sensitivity, *STN*, associated with noise exposure is then:

$$STN = STN(Y_{ns}) + M_c(Y_{ns}, Y)$$
$$Y > Y_{ns} \qquad (4.29)$$

The post-exposure term, M_c, has been determined empirically for one frequency (Bies, 1994) and may be expressed in terms of the age of the population, Y, and the age when exposure to excessive noise, Y_{ns}, stopped. The proposed post-exposure correction is based upon data provided by Macrae (1991) and is limited to loss at 4 kHz as no information is available for other frequencies:

$$M_c = 0.0208\, Y_{ns}\,(Y - Y_{ns}) \qquad (4.30)$$

For the case of the reconstructed data base of the International Standard, the quantity, M_c is assumed to be zero, because the standard provides no post-exposure information.

Implicit in this formulation is the assumption that the A-weighted sound pressure, p_A is determined in terms of the equivalent A-weighted sound pressure level, L'_{Aeq} as follows:

$$p_A = 10^{L'_{Aeq}/20} \qquad (4.31)$$

where

$$L'_{Aeq} = 20\log_{10}\left(\frac{1}{T}\int_0^T \left[p_A^2(t)\right]^{1/2}\,dt\right) \qquad (4.32)$$

which may be contrasted with the traditional Equation (4.1). Equation (4.32) implies that an equivalent noise level may be calculated by integrating acoustic pressures rather than pressures squared as implied by Equation (4.1). This leads to a 6 dB trading rule for exposure time versus exposure level (see Section 4.3.6).

4.3.4.2 Dresden Group Formulation

The Dresden group investigated the relationship between noise exposure and hearing loss using retrospective studies of noise exposed persons, temporary threshold shift investigations, and animal experiments. Their major result supported by all three types of investigation describing the long-term effect of noise on the average hearing loss of an exposed population is summarised for the 4 kHz frequency below. The relationship describes exposure to all kinds of industrial and other noise including interrupted, fluctuating, and impulsive noise with peak sound pressure levels up to 135 dB re 20 μPa. At higher levels, the observed loss seems to be dependent upon pressure squared or energy input.

An A-weighted linear noise dose, B_s is defined in terms of the total time, T_s of exposure to noise in seconds as follows:

$$B_s = \int_0^{T_s} \left[p_A^2(t) \right]^{1/2} dt \qquad (4.33)$$

An age-related noise dose, B_a in terms of the age of the person, T_{sa} in seconds is defined as follows:

$$B_a = 0.025 \, (T_{sa} - T_s) \qquad (4.34)$$

The permanent threshold shift, H', is given by the following equation:

$$H' = k_f \log_{10} \left[\frac{B_s + B_a}{B_0} \right] \qquad (4.35)$$

The quantities of Equation (4.35), not already defined, are k_f, a constant specific for each audiometric frequency, with the value of 50 for 4 kHz and B_0, a critical noise dose used as a reference with the value of 2×10^7 Pa s.

Consideration of Equation (4.35) shows that if the term associated with noise exposure, B_s is very much larger than the term associated with age, B_a, then with cessation of exposure to noise, no further threshold shift should be observed until the term associated with age also becomes large. However, as pointed out above, Macrae (1991) has provided data showing the threshold shift continues and as suggested above, the expression given by Equation (4.35) may be corrected by the simple device of adding Mc, given by Equation (4.30).

4.3.5 Observed Hearing Loss

In Figure 4.4, observed median loss in hearing at 4000 Hz is plotted as a function of the percentage risk of incurring that loss for a specified length of exposure at a specified sound pressure level. The presentation is based upon published data (Beranek, 1971; Burns and Robinson, 1970). Length of exposure is expressed in years, where it is assumed that a person would be exposed to the stated level for about 1900 hours during each year. In the figure, the curve labelled 80 dB(A) represents a lower bound for hearing loss that can be attributed to noise exposure; presumably all lower exposure levels would lie on the same curve, because this loss is attributed to age and other causes.

Referring to Figure 4.4, it is evident that hearing deterioration is very rapid during the first 10 years and progressively more so as the exposure level rises above

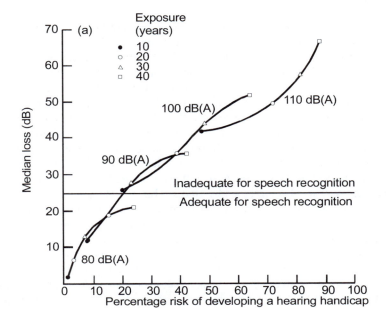

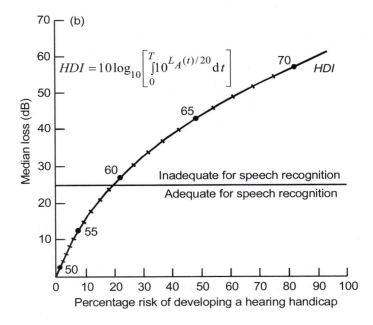

Figure 4.4 Hearing damage as a function of exposure. The percentage risk of developing a hearing handicap (averaged over 500 Hz, 1000 Hz and 2000 Hz) and the median loss (at 4000 Hz) incurred with exposure are shown (a) as function of mean sound pressure level in the workplace (dB(A)) and exposure time (years); and (b) as a function of hearing deterioration index *HDI*. The quantity $L_A(t)$ is the mean exposure level (dB(A)) over time dt, and T is the exposure time.

80 dB(A). The data for percentage risk of developing a hearing handicap refer to loss averaged arithmetically over 500 Hz, 1000 Hz and 2000 Hz, whereas the data for median loss (meaning that 50% of the population has this or greater loss) refer to loss measured at 4000 Hz. This choice of representation is consistent with the observation that noise induced hearing loss always occurs first and proceeds most rapidly at 4000 Hz to 6000 Hz, and is progressively less at both lower and higher frequencies.

Inspection of Figure 4.4(a) shows that the 10-year exposure point for any given level always lies close to the 30-year exposure point for the level 10 dB lower. For example, the point corresponding to 30 years of exposure at a level of 80 dB(A) lies close to the point corresponding to 10 years at a level of 90 dB(A). This observation may be summarised by the statement that 30 years are traded for ten years each time the sound pressure level is increased by 10 dB. This observation in turn suggests the metric proposed in Figure 4.4(b), which fairly well summarises the data shown in Figure 4.4(a) and indicates that hearing loss is a function of the product of acoustic pressure and time, not pressure squared and time. Thus a hearing deterioration index, *HDI* is proposed, based upon sound pressure, not sound energy, which is the cumulative integral of the r.m.s. sound pressure with time. Figure 4.4(b) shows that to avoid hearing impairment in 80% of the population, a strategy should be adopted that avoids acquiring a hearing deterioration index greater than 59 during a lifetime.

4.3.6 Some Alternative Interpretations

In Europe and Australia, the assumption is implicit in regulations formulated to protect people exposed to excessive noise, that hearing loss is a function of the integral of pressure squared with time as given by Equation (4.1). In the United States the same assumption is generally accepted, but a compromise has been adopted in formulating regulations where it has been assumed that recovery, during periods of non-exposure, takes place and reduces the effects of exposure. The situation may be summarised by rewriting Equation (4.1) more generally as in the following equation.

$$L'_{Aeq,8h} = \frac{10}{n} \log_{10} \left\{ \frac{1}{8} \int_0^T \left[10^{L_A(t)/10} \right]^n dt \right\} \tag{4.36}$$

The prime has been used to distinguish the quantity from the traditional, energy averaged $L_{Aeq,8h}$, defined in Equation (4.1).

Various trading rules governing the equivalence of increased noise level versus decreased exposure time are used in regulations concerning allowed noise exposure. For example, in Europe and Australia it is assumed that, for a fixed sound exposure, the noise level may be increased by 3 dB(A) for each halving of the duration of exposure, while in United States industry, the increase is 5 dB(A) and in the United States Military the increase is 3 dB(A).

Values of n in Equation (4.36), corresponding to trading rules of 3 or 5 dB(A), are approximately 1 and 3/5 respectively. If the observation that hearing loss due to noise exposure is a function of the integral of r.m.s pressure with time, then $n = \frac{1}{2}$ and the trading rule is approximately 6 dB(A). The relationship between n and the trading rule is:

$$n = 3.01/L \tag{4.37}$$

where L is the decibel trading level which corresponds to a change in exposure by a factor of two for a constant exposure time. Note that a trading rule of 3 results in n being slightly larger than 1, but it is close enough to 1, so that for this case, it is often assumed that it is sufficiently accurate to set $L'_{Aeq,8h} = L_{Aeq,8h}$.

Introduction of a constant base level criterion, L_B, which $L'_{Aeq,8h}$ should not exceed and use of Equation (4.37) allows Equation (4.36) to be rewritten in the following form:

$$L'_{Aeq,8h} = \frac{L}{0.301} \log_{10} \left\{ \frac{1}{8} \int_0^T \left[10^{0.301(L_A(t) - L_B)/L} \right] 10^{0.301 L_B/L} \, dt \right\} \tag{4.38}$$

Equation (4.38) may, in turn be written as follows:

$$L'_{Aeq,8h} = \frac{L}{0.301} \log_{10} \left\{ \frac{1}{8} \int_0^T \left[10^{0.301(L_A(t) - L_B)/L} \right] dt \right\} + L_B \tag{4.39}$$

or,

$$L'_{Aeq,8h} = \frac{L}{0.301} \log_{10} \left\{ \frac{1}{8} \int_0^T \left[2^{(L_A(t) - L_B)/L} \right] dt \right\} + L_B \tag{4.40}$$

Note that if discrete exposure levels were being determined with a sound level meter as described above, then the integral would be replaced with a sum over the number of discrete events measured for a particular person during a working day. For example, for a number of events, m, for which the ith event is characterised by an A-weighted sound level of $L_{A\,i}$, Equation (4.40) could be written as follows.

$$L'_{Aeq,8h} = \frac{L}{0.301} \log_{10} \left\{ \frac{1}{8} \sum_{i=1}^m \left[2^{(L_{Ai} - L_B)/L} \right] \times t_i \right\} + L_B \tag{4.41}$$

When $L'_{Aeq,8h} = L_B$ reference to Equation (4.40) shows that the argument of the logarithm on the right-hand side of the equation must be one. Consequently, if an employee is subjected to higher levels than L_B, then to satisfy the criterion, the length of time, T, must be reduced to less than eight hours. Setting the argument equal to one, $L_A(t) = L_B = L'_{Aeq,8h}$ and evaluating the integral using the mean value theorem, the maximum allowed exposure time to an equivalent noise level, $L'_{Aeq,8h}$ is:

$$T_a = 8 \times 2^{-(L'_{Aeq,8h} - L_B)/L} \tag{4.42}$$

If the number of hours of exposure is different to 8, then to find the actual allowed exposure time to the given noise environment, denoted $L'_{Aeq,T}$, the "8" in Equation (4.42) is replaced by the actual number of hours of exposure, T.

The daily noise dose (*DND*), or "noise exposure", is defined as equal to 8 hours divided by the allowed exposure time, T_a with L_B set equal to 90. That is:

$$DND = 2^{(L'_{Aeq,8h} - 90)/L} \qquad (4.43)$$

In most developed countries (with the exception of USA industry), the equal energy trading rule is used with an allowable 8-hour exposure of 85 dB(A), which implies that in Equation (4.39), $L = 3$ and $L_B = 85$. In industry in the USA, $L = 5$ and $L_B = 90$, but for levels above 85 dB(A) a hearing conservation program must be implemented and those exposed must be given hearing protection. Interestingly, the US Military uses $L = 3$ and $L_B = 85$. Additionally in industry and in the military in the USA, noise levels less than 80 dB(A) are excluded from Equation (4.39). No levels are excluded for calculating noise dose (or noise exposure) according to Australian and European regulations, but as levels less than 80 dB(A) contribute such a small amount to a person's exposure, this difference is not significant in practice.

Example 4.1

An Australian timber mill employee cuts timber to length with a cut-off saw. While the saw idles it produces a level of 85 dB(A) and when it cuts timber it produces a level of 96 dB(A) at the work position.
1. If the saw runs continuously and actually only cuts for 10% of the time that it is switched on, compute the A-weighted, 8-hour equivalent noise level.
2. How much must the exposure be reduced in hours to comply with the 85 dB(A) criterion?

Solution:

1. Making use of Equation (4.2) (or Equation (4.38) with $L = 3$, in which case $L_{Aeq,8h} \approx L'_{Aeq,8h}$), the following can be written:

$$L_{Aeq,8h} = 10 \log_{10}\left[\frac{1}{8}\left(7.2 \times 10^{85/10} + 0.8 \times 10^{96/10}\right)\right] = 88.3 \text{ dB(A)}$$

2. Let T_a be the allowed exposure time. Then:

$$L_{Aeq,8h} = 85.0 \text{ dB(A)} = 10 \log_{10}\left[\frac{T_a}{8}\left(0.9 \times 10^{85/10} + 0.1 \times 10^{96/10}\right)\right]$$

Solving this equation gives $T_a = 3.7$ hours. The required reduction = 8 - 3.7 = 4.3 hours. Alternatively, use Equation (4.42) and let $L = 3$, $L_B = 85$:

$$T_a = 8 \times 2^{-(88.34 - 85.0)/3} = 8/2^{1.11} = 3.7 \text{ hours}$$

Alternatively, for an American worker, $L = 5$ and use of Equation (4.41) gives $L'_{Aeq,8h} = 87.2$. Equation (4.42) with $L= 5$ and $L_B = 90$ gives for the allowed exposure time T_a:

$$T_a = 8 \times 2^{-(87.2 - 90.0)/5} = 8 \times 2^{0.56} = 11.8 \text{ hours}$$

4.4 HEARING DAMAGE RISK CRITERIA

The noise level below which damage to hearing from habitual exposure to noise should not occur in a specified proportion of normal ears is known as the hearing damage risk criterion (DRC). It should be noted that hearing damage is a cumulative result of level as well as duration, and any criterion must take both level and duration of exposure into account. Note that it is not just the workplace that is responsible for excessive noise. Many people engage in leisure activities that are damaging to hearing, such as going to night clubs with loud music, shooting or jet ski riding. Also listening to loud music through headphones can be very damaging, especially for children.

4.4.1 Continuous Noise

A continuous eight-hour exposure each day to ordinary broadband noise of a level of 90 dB(A) results in a hearing loss of greater than 25 dB (averaged over 0, 5, 1 and 2 kHz) for approximately 25% of people exposed for 30 years or more. This percentage is approximate only, as it is rare to get agreement between various surveys that are supposedly measuring the same quantity. This is still a substantial level of hearing damage risk. On the other hand, a criterion of 80 dB(A) for an eight-hour daily exposure would constitute a negligible hearing damage risk for speech. Therefore, to minimise hearing loss, it is desirable to aim for a level of 80 dB(A) or less in any plant design. Limits higher than 80 dB(A) must be compromises between the cost of noise control, and the risk of hearing damage and consequent compensation claims.

Although an exposure to 80 dB(A) for eight hours per day would ensure negligible hearing loss for speech due to noise exposure, a lower level would be required to ensure negligible hearing loss at all audible frequencies. One viewpoint is that 97% of the population should be protected from any measurable noise-induced permanent shift in hearing threshold at all frequencies, even after 40 years of exposure for eight hours per day for 250 days of the year. If we assume that, for about 10% of each eight-hour working day, a worker is out of the area of maximum noise (owing to visits to other areas) and, further, that he or she is exposed to noise levels which are over 5 dB lower during the remaining sixteen hours of the day, then studies worldwide show that for 97% protection at *all* frequencies, the noise level must not exceed 75 dB(A). If a worker is exposed to continuous noise for 24 hours per day, the level must not exceed 70 dB(A).

Another viewpoint is that it is only necessary to protect people from hearing damage for speech, and that to aim for the above levels is unnecessarily conservative and economically unrealistic. In 1974, having reviewed the published data, the Committee of American Conferences of Governmental Industrial Hygienists determined that an exposure level of 85 dB(A) during a working life would result in 90% of people suffering a hearing loss of less than 25 dB when averaged over the frequencies, 0.5, 1 and 2 kHz.

Current standards in most countries now recommend 85 dB(A) as an acceptable level for eight hours of exposure, although most people agree that a level of 80 dB(A) is more desirable from the point of view of minimising hearing damage.

4.4.2 Impulse Noise

Impulse noise is defined as a short-duration sound characterised by a shock front pressure waveform (i.e. virtually instantaneous rise), usually created by a sudden release of energy; for example, as encountered with explosives or gun blasts (Rice, 1974; Rice and Martin, 1973). Such a characteristic impulse pressure waveform is often referred to as a Friedlander wave, and is illustrated in Figure 4.5(a). This single-impulse waveform is typically generated in free-field environments, where sound-reflecting surfaces that create reverberation are absent. With gunfire, mechanically generated noise is also present in addition to the shock pulse, and in this case the waveform envelope can take the form illustrated in Figure 4.5(b).

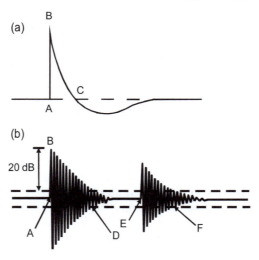

Figure 4.5 Idealised waveforms of impulse noises. Peak level = pressure difference AB; rise time = time difference AB; A duration = time difference AC; B duration = time difference AD (+ EF when a reflection is present).

The durations of impulsive noises may vary from microseconds up to 50 ms, although in confined spaces reverberation characteristics may cause the duration to extend considerably longer. In all cases, however, the characteristic shock front is present. For the purpose of assessing hearing damage risk, a "B duration" time has been defined as the time required for the peak level to drop 20 dB, as illustrated in Figure 4.5.

In general, people are not habitually exposed to impulsive noises. In fact, only people exposed to explosions such as quarry blasting or gunfire are exposed to impulse noises (as opposed to impact noises). Estimates of the number of pulses likely to be received on any one occasion vary between 10 and 100, although up to 1000 impulses may sometimes be encountered.

4.4.3 Impact Noise

Impact noises are normally produced by non-explosive means, such as metal-to-metal impacts in industrial plant processes. In such cases the characteristic shock front is not always present, and due to the reverberant industrial environments in which they are heard, the durations are often longer than those usually associated with impulse noise. The background noise present in such situations, coupled with the regularity with which impacts may occur, often causes the impacts to give the impression of running into one another. People in industry are often habitually exposed to such noises, and the number of impacts heard during an eight-hour shift usually runs into thousands.

Figure 4.6 shows one researcher's (Rice, 1974) recommended impulse and impact upper bound criteria for daily exposure, over a wide range of peak pressure levels, as a function of the product of the B duration of each impulse (or impact) and the number of impulses (or impacts). The criterion of Figure 4.6 is arranged to be equivalent to a continuous exposure to 90 dB(A) for an eight-hour period, and this point is marked on the chart. It is interesting to note that if, instead of using the equal energy concept (3 dB(A) allowable increase in noise level for each halving of the exposure time) as is current Australian and European practice, a 5 dB(A) per halving of exposure is used (as is current US practice), the criteria for impulse and impact noise would essentially become one criterion. If the person exposed is wearing ear-muffs, the US military allows 15 dB to be added to the impulse criteria (MIL-STD-1474D, 1991).

The equivalent noise dose (or noise exposure) corresponding to a particular *B* duration multiplied by the number of impacts or impulses, and a corresponding peak pressure level may be calculated using Figure 4.6. The first step is to calculate the product of the impulse or impact *B* duration and number of impulses (impacts). This value is entered on the abscissa of Figure 4.6 and a vertical line drawn until it intersects the appropriate curve. For impulse noise of less than 1000 impulses per exposure the upper curve is used, while for impact noise the lower curve is used. From the point of intersection of the appropriate curve and the vertical line, a horizontal line is drawn to intersect the ordinate at the value of peak sound level corresponding to a noise dose of unity.

Alternatively, the peak level of an individual impact is entered on the ordinate and a horizontal line drawn until it intersects the lower curve. A vertical line is drawn downwards from the point of intersection. Where the vertical line intersects the abscissa indicates the product of B duration and number of impacts that will correspond to a noise dose of unity. The noise dose is halved for each 3 dB that the measured peak level is exceeded by the peak level corresponding to a noise dose of unity. The noise dose is also halved if the number of impacts is halved.

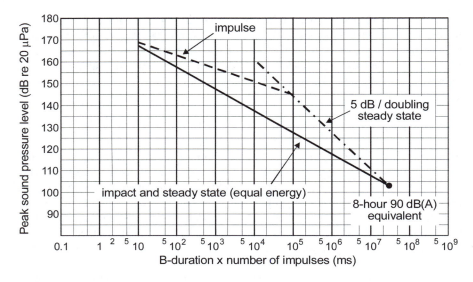

Figure 4.6 A comparison of impulse and impact damage risk criteria with steady-state criteria (after Rice, 1974).

Due to the difficulty in measuring the duration of the impact noise, the preceding procedure has not seen much use. In fact, the ISO standard, ISO1999 (1990), states, "The prediction method presented is based primarily on data collected with essentially broadband steady non-tonal noise. The application of the data base to tonal or impulsive/impact noise represents the best available extrapolation." This implies that impact and impulse noise should be treated in a similar way to continuous noise in terms of exposure and that the best way of assessing its effect is to use an integrating sound level meter that has a sufficiently short time constant (50 µs) to record the energy contained in short impacts and impulses and include this energy with the continuous noise energy in determining the overall energy averaged L_{Aeq}. However, there is still a feeling in the audiometry community (von Gierke *et al.*, 1982) that impulse and impact noise should be assessed using a C-weighted energy average or L_{Ceq} and this should be combined with the A-weighted level, L_{Aeq}, for continuous noise on a logarithmic basis to determine the overall level, which is then used in place of the continuous L_{Aeq} in the assessment of damage risk. Others feel that the impact noise should not be weighted at all, but this results in undue emphasis being placed on very low frequency sound such as that produced by slamming a car door and is not recommended as a valid approach. With modern digital instrumentation, the 50 µs time constant is achieved if the instrument has a frequency response to 20 kHz and a corresponding sampling rate of 48 kHz. Instruments with a higher frequency response may measure a higher peak for the same impact noise.

It is generally accepted in most standards that no-one should be exposed to a peak noise level that exceeds 140 dB, and for children the limit should be 120 dB. Some specify that this should be measured with a C-weighting network implemented on the measuring instrument and some specify that no weighting should be used. The latter specification is rather arbitrary as the measured level then depends on the upper and lower frequency range of the measuring instrument. The time constant of the instrument measuring the peak noise level should be 50 μs.

4.5 IMPLEMENTING A HEARING CONSERVATION PROGRAM

To protect workers in noisy industries from the harmful effects of excessive noise, it is necessary to implement a well-organised hearing conservation program. The key components of such a program include:

- regular noise surveys of the work environment which includes:
    - making a preliminary general survey to determine the extent of any problems and to provide information for planning the detailed survey;
    - determining the sound power and directivity (or sound pressure at the operator locations) of noisy equipment;
    - identification, characterisation and ranking of noise sources;
    - identification of high noise level areas and their contribution to worker exposures;
    - determination of individual worker exposures to noise using noise measurements and dosimeters (ISO9612-1997, ANSI S12.19-1996 (R2006));
    - prediction of the risk of hearing loss for individual or collective groups of workers using ISO1999 (1990); and
    - identifying hearing conservation requirements.
- regular audiometric testing of exposed workers to evaluate the program effectiveness and to monitor temporary threshold shift (TTS) at the end of the work shift as well as permanent threshold shift (measured by testing after a quiet period) (see ANSI S3.6-1996, ANSI S3.1-1991, ISO 8253-part 1-1989, ISO 8253-part 2-1992, ISO 8253-part 3-1996, IEC 60645-part 1-2001, part 2-1993, part 3-2007, part 4-1994 with the following notes:
    - elimination of temporary threshold shift will eliminate permanent threshold shift that will eventually occur as a result of sufficient incidences of TTS;
    - anyone with a permanent threshold shift in addition to the shift they had at the beginning of their employment should be moved to a quieter area and if necessary be given different work assignments;
- installation and regular monitoring of the effectiveness of noise control equipment and fixtures;
- consideration of noise in the specification of new equipment;
- consideration of administrative controls involving the reorganisation of the workplace to minimise the number of exposed individuals and the duration of their exposure;
- education of workers;
- regular evaluation of the overall program effectiveness, including noting the reduction in temporary threshold shift in workers during audiometric testing;

- careful record keeping including noise data and audiometric test results, noise control systems purchased, instrumentation details and calibration histories, program costs and periodic critical analysis reports of the overall program; and
- appropriate use of the information to:
  - ▸ inform workers of their exposure pattern and level;
  - ▸ act as a record for the employer;
  - ▸ identify operators whose exposure is above the legal limits;
  - ▸ identify areas of high noise level;
  - ▸ identify machines or processes contributing to excessive noise levels;
  - ▸ indicate areas in which control is necessary;
  - ▸ indicate areas where hearing protection must be worn prior to engineering noise controls have been implemented;
  - ▸ indicate areas where hearing protection must be worn even after engineering noise controls being implemented; and
  - ▸ identification of the most appropriate locations for new machines and processes.

To be successful, a hearing conservation program requires:
- well defined goals and objectives;
- competent program management;
- commitment from management at the top of the organisation;
- commitment from the workers involved;
- adequate financial resources;
- access to appropriate technical expertise;
- good communication and monitoring systems;
- a philosophy of continuous improvement.

4.6 SPEECH INTERFERENCE CRITERIA

In this section the interfering effect of noise upon oral communication is considered. Table 4.5 lists some of the significant frequency ranges that are of importance for these considerations. For comparison, the frequency range of a small transistor radio speaker is typically 200 to 5000 Hz.

4.6.1 Broadband Background Noise

Maintenance of adequate speech communication is often an important aspect of the problem of dealing with occupational noise. The degree of intelligibility of speech is dependent upon the level of background noise in relation to the level of spoken words. Additionally, the speech level of a talker will depend upon the talker's subjective response to the level of the background noise. Both effects can be quantified, as illustrated in Figure 4.7.

To enter Figure 4.7, the Speech Interference Level (SIL) is computed as the arithmetic average of the background sound pressure levels in the four octave bands,

Table 4.5 Significant frequency ranges for speech communication

	Approximate frequency range (Hz)
Range of hearing	16 to 20,000
Speech range	200 to 6,000
Speech intelligibility (containing the frequencies most necessary for understanding speech)	500 to 4000
Speech privacy range (containing speech sounds which intrude most objectionably into adjacent areas)	250 to 2500
Male voice (peak frequency of energy output)	350
Female voice (peak frequency of energy output)	700

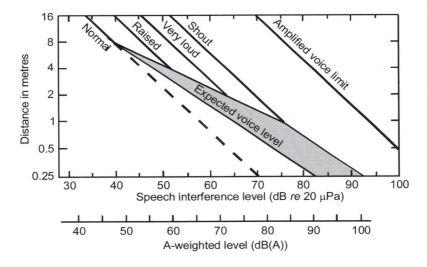

Figure 4.7 Rating noise with respect to speech interference.

500 Hz, 1000 Hz, 2000 Hz and 4000 Hz. Alternatively, the figure may be entered using the A-weighted scale, but with less precision. Having entered the figure, the voice level required for just adequate speech communication may then be determined for the various distances indicated on the abscissa. The area below each curve represents the combinations of talker/listener separation and background noise level for which reliable face-to-face communication is possible. For example, for a speech interference level of 60 dB and distances between the talker and the listener less than 1.5 m, a normal speech level would seem to be adequate, while at 2 m, a raised voice level would be necessary for just adequate speech recognition. As a further example, at a speech Interference level of 80 dB, shouting would seem to be necessary at 0.5 m for just adequate speech recognition.

The figure also shows the voice level that the talker would automatically use (expected voice level) as a result of the background noise level. The range of expected voice level represents the expected range in a talker's subjective response to the background noise. If the talker is wearing an ear protection device such as ear plugs or earmuffs, the expected voice level will decrease by 4 dB.

For face-to-face communication with "average" male voices, the background noise levels shown by the curves in Figure 4.7 represent upper limits for just acceptable speech communication, i.e. 95% sentence intelligibility, or alternatively 60% word-out-of-context recognition. For female voices the speech interference level, or alternatively the A-weighted level shown on the abscissa, should be decreased by 5 dB, i.e. the scales should be shifted to the right by 5 dB. The figure assumes no reflecting surfaces to reflect speech sounds. Where reflecting surfaces exist the scale on the abscissa should be shifted to the right by 5 dB. Where the noise fluctuates greatly in level, the scale on the abscissa may be shifted to the left by 5 dB.

For industrial situations where speech and telephone communication are important, such as in foremen's offices, control rooms, etc. an accepted criterion for background noise level is 70 dB(A).

4.6.2 Intense Tones

Intense tones may mask sounds associated with speech. The masking effect of a tone is greatest on sounds of frequency higher than the tone; thus low-frequency tones are more effective than high-frequency tones in masking speech sounds. However, tones in the speech range, which generally lies between 200 and 6000 Hz, are the most effective of all in interfering with speech recognition. Furthermore, as the frequencies 500-5000 Hz are the most important for speech intelligibility, tones in this range are most damaging to good communication. However, if masking is required, then a tone of about 500 Hz, rich in harmonics, is most effective. For more on the subject of masking refer to Section 2.4.1.

4.7 PSYCHOLOGICAL EFFECTS OF NOISE

4.7.1 Noise as a Cause of Stress

Noise causes stress: the onset of loud noise can produce effects such as fear, and changes in pulse rate, respiration rate, blood pressure, metabolism, acuity of vision, skin electrical resistance, etc. However, most of these effects seem to disappear rapidly and the subject returns to normal, even if the noise continues, although there is evidence to show that prolonged exposure to excessive loud noise will result in permanently elevated blood pressure. Excessive environmental noise has been shown to accelerate mental health problems in those predisposed to mental health problems. Continuous noise levels exceeding 30 dB(A) or single event levels exceeding 45 dB(A) disturb sleep. This also can lead to elevated blood pressure levels.

4.7.2 Effect on Behaviour and Work Efficiency

Behavioural responses to noise are usually explained in terms of arousal theory: there is an optimum level of arousal for efficient performance; below this level behaviour is sluggish and above it, tense and jittery. It seems reasonable to suppose, therefore, that noise improves performance when arousal is too low for the task, and impairs it when arousal is optimal or already too high. The complex task, multiple task or high repetition rate task is performed optimally under relatively quiet conditions, but performance is likely to be impaired under noisy conditions. Quiet conditions, on the other hand, are sub-optimal for the simple task, and performance is improved by the addition of noise. The important variable is the kind of task being performed, and not the kind of noise present. To generalise, performance in doing complex tasks is likely to be impaired in the presence of noise and for simple tasks it is likely to be improved.

However, various studies have shown that if the noise level is far in excess of that required for the optimum arousal level for a particular task, workers become irritable as well as less efficient. This irritability usually continues for some time after the noise has stopped.

4.8 AMBIENT NOISE LEVEL SPECIFICATION

The use of a room or space for a particular purpose may, in general, impose a requirement for specification of the maximum tolerable background noise; for example, one would expect quiet in a church but not in an airport departure lounge. All have in common that a single number specification is possible.

The simplest way of specifying the maximum tolerable background noise is to specify the maximum acceptable A-weighted level. As the A-weighted level simulates the response of the ear at low levels, and has been found to correlate well with subjective response to noise, such specification is often sufficient. Table 4.6 gives some examples of maximum acceptable A-weighted sound levels and reverberation times in the 500 to 1000 Hz octave bands for unoccupied spaces. A full detailed list is published in Australian Standard AS 2107-1987.

The values shown in Table 4.6 are for continuous background noise levels within spaces, as opposed to specific or intermittent noise levels. In the table the upper limit of the range of values shown is the maximum acceptable level and the lower limit is the desirable level. Recommended noise levels for vessels and offshore mobile platforms are listed for various spaces in AS2254-1988.

For ambient noise level specification, a number of quantities are used. For indoor noise and situations where noise control is necessary, noise weighting curves are often used (see Section 4.8.1). For quantifying occupational noise and for environmental noise regulations, $L_{Aeq,8h}$ or L_{Aeq} respectively (see Section 4.1.1.1) are commonly used. Broner and Leventhall (1983) conclude that the A-weighted measure is also acceptable for very low frequency noise (20-90 Hz). For impulsive environmental noise, characterised by very short duration impulses, the "C-weighted" sound exposure level (L_{CE}, defined in Section 4.1.1.3) is used. For transient environmental noise such as aircraft flyovers, the "A-weighted" sound exposure level, L_{AE} (see Section 4.1.1.3), is used. For traffic noise, L_{10} (see Section 3.7) or L_{dn} (see Section 4.1.1.4) are used. L_{dn} is also used for specifying acceptable noise levels in houses.

Table 4.6 Recommended ambient sound levels for different areas of occupancy in buildings (space furnished but unoccupied)

Types of occupancy/activity	Recommended sound level (dB (A))	Recommended reverberation time at 500 to 1000 Hz (sec)
Lecture rooms, assembly halls, conference venues	30	0.6 for 300 m³ to 1.4 for 50 000 m³ varying with $\log_{10}$ (room volume)
Audio-visual areas	40	0.6-0.8
Churches (250 or less people)	30	–
Computer rooms (teaching)	40	0.4-0.6
Computer rooms (working)	45	0.4-0.6
Conference rooms, seminar rooms, tutorial rooms	30	0.6-0.7
Corridors and lobbies	45	–
Drama studios	30	10% to 20% higher than lecture rooms above
Libraries (reading)	40	10% to 20% higher than lecture rooms above
Libraries (stack area)	45	10% to 20% higher than lecture rooms above
Music studios and concert halls	30	0.8 for 400 m³ to 2.2 for 50 000 m³, varying linearly with log (room volume)
Professional and admin. offices	35	0.6-0.8
Design offices, drafting offices	40	0.4
Private offices	35	0.6-0.8
Reception areas	40	–
Airport terminals	45	–
Restaurants	40	–
Hotel bar	45	–
Private house (sleeping)	25	–
Private house (recreation)	30	–

4.8.1 Noise Weighting Curves

Although the specification of an A-weighted level is easy and convenient it gives no indication of which frequency components may be the source of non-compliance. For most acoustic design purposes it is more useful to make use of a weighting curve, which defines a spectrum of band levels in terms of a single number. Five currently used sets of single-number weighting curves are shown in Figures 4.8-4.12. These figures provide, respectively, noise rating (*NR*), noise criteria (*NC*), room criteria (*RC*), balanced noise criteria (*NCB*) and room noise criteria (*RNC*) weighting curves.

4.8.1.1 NR Curves

Noise Rating (*NR*) curves have been adopted by the International Standards Organisation and are intended for general use, particularly for rating environmental and industrial noise levels. They are also used in many cases by machinery manufacturers to specify machinery noise levels.

The Noise Rating, *NR*, of any noise characterised in octave band levels may be calculated algebraically. The *NR* of a noise is equal to the highest octave band noise rating (NR_B) which is defined as:

$$NR_B = \frac{L_{p_B} - A_B}{B_B} \qquad (4.44)$$

where A_B and B_B are as listed in Table 4.7. However, the family of curves generated using Equation (4.44) and shown in Figure 4.8 is in more common use than the equation. By convention, the *NR* value of a noise is expressed as an integer.

Table 4.7 Constants used to calculate *NR* curve numbers for octave bands between 31.5 and 8000 Hz

Octave band centre frequency (Hz)	A_B	B_B
31.5	55.4	0.681
63	35.5	0.790
125	22.0	0.870
250	12.0	0.930
500	4.8	0.974
1000	0.0	1.000
2000	-3.5	1.015
4000	-6.1	1.025
8000	-8.0	1.030

To determine the *NR* rating of a noise, measured octave band sound pressure levels are plotted on Figure 4.8 and the rating is determined by the highest weighting curve which just envelopes the data. If the highest level falls between two curves, linear interpolation to the nearest integer value is used. Note that it is also possible to use 1/3 octave band data on 1/3 octave band NR curves, which are obtained by moving the octave band curves down by $10\log10(3) = 4.77$ dB.

Specification of an *NR* number means that in no frequency band shall the octave band sound pressure in the specified space exceed the specified curve (tangent

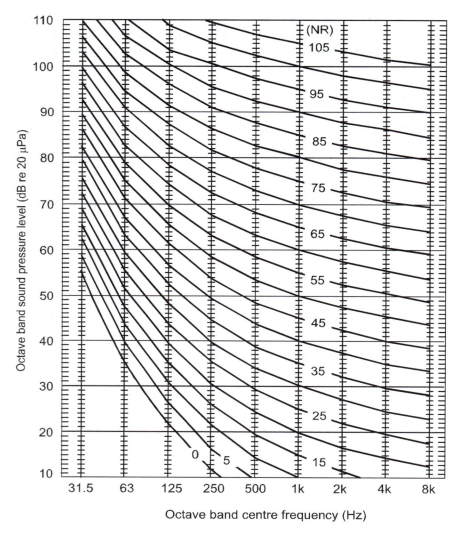

Figure 4.8 Noise rating curves (*NR*).

method). In practice, a tolerance of ±2 dB is implied. In specifications, an allowance of 2 dB above the curve is usually acceptable in any one octave band, provided that the levels in the two adjacent bands are only 1 dB below the criterion curve.

Example 4.2

Find the *NR* number for the sound spectrum of Example 3.1.

Solution

Plot the unweighted sound spectrum on a set of NR curves. The highest curve that envelopes the data is $NR = 81$ (interpolated between the NR80 and NR85 curves).

4.8.1.2 NC Curves (Figure 4.9)

Noise criteria curves (Figure 4.9) were developed in response to the need for specification of just acceptable noise in occupied spaces with all systems running. They are still used extensively in the building services industry even though they have

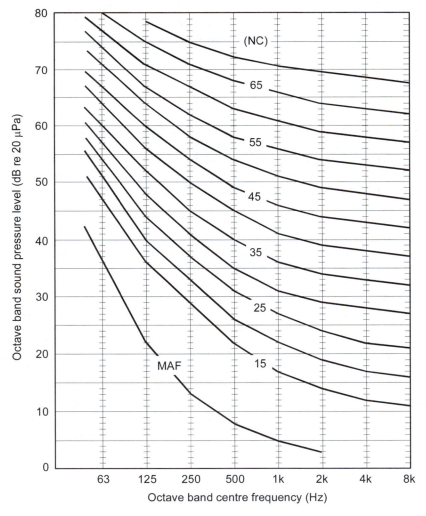

Figure 4.9 Noise criteria curves (*NC*). "MAF" = minimum audible field".

been largely superseded by NCB criteria (see Section 4.8.1.4). Noise criteria curves are not defined in the 31.5 Hz octave band and thus do not account for very low frequency rumble noises. They are also regarded as too permissive in the 2000 Hz and higher octave bands and do not correlate well with subjective response to air-conditioning noise. This has resulted in them now being considered generally unsuitable for rating interior noise. The *NC* rating of a noise is determined in a similar way to the *NR* rating, except that Figure 4.9 is used instead of Figure 4.8. The *NC* rating, which is an integer value, corresponds to the curve that just envelopes the spectrum. No part of the spectrum may exceed the *NC* curve that describes it. Note that linear interpolation is used to generate curves corresponding to integer *NC* numbers between the 5 dB intervals shown in Figure 4.9. The simplicity of the procedure for determining an *NC* rating is the main reason these curves are still in use today. The more complex procedures for determining an *RC* or *NCB* rating (see below) have prevented these latter (and more appropriate) ratings from being universally accepted.

4.8.1.3 RC Curves

Room criteria (*RC*) curves have been developed to replace Noise Criteria curves for rating only air conditioning noise in unoccupied spaces. The *RC* curves include 16 Hz and 31.5 Hz octave band levels (see Figure 4.10), although few sound level meters with external octave band filters include the 16 Hz octave band. Interest in the 31.5 Hz and 16 Hz bands stems from the fact that a level of the order of 70 dB or greater may result in noise-induced vibrations that are just feelable, especially in lightweight structures. Such vibration can also give rise to objectionable rattle and buzz in windows, doors and cabinets, etc.

For spectrum shapes that are ordinarily encountered, the level in the 16 Hz band can be estimated from the difference in levels between the unweighted reading and the 31.5 Hz octave band level. A difference of +4 dB or more is evidence of a level in the 16 Hz band equal to or greater than the level in the 31.5 Hz band.

The *RC* number is the average of the 500 Hz, 1000 Hz and 2000 Hz octave band sound levels, expressed to the nearest integer. If any octave band level below 500Hz exceeds the *RC* curve by more than 5 dB, the noise is denoted "rumbly" (e.g. RC 29(R)). If any octave band level above 500Hz exceeds the RC curve by more than 3 dB, the noise is denoted, "hissy". (e.g. RC 29(H)). If neither of the above occurs, the noise is denoted "neutral" (e.g. RC 29(N)). If the sound pressure levels in any band between and including 16 Hz to 63 Hz lie in the cross hatched regions in Figure 4.10, perceptible vibration can occur in the walls and ceiling and rattles can occur in furniture. In this case, the noise is identified with "RV" (e.g. RC 29(RV)). The level, L_B of the octave band of centre frequency f, corresponding to a particular *RC* curve is given by:

$$L_B = RC + \frac{5}{0.3}\log_{10}\left(\frac{1000}{f}\right) \tag{4.45}$$

RC curves represent the shape of the least objectionable noise spectrum. For this reason they are only defined up to an *RC* number of 50. By contrast, *NR* and *NC* curves are intended for specifying an upper bound of acceptability for background noise, and do not necessarily represent the least objectionable noise spectrum. Thus, in certain cases, such as in open plan offices, where it may be advantageous to introduce background noise to ensure speech privacy, noise with a spectrum shape of an *NR* or NC curve is unsatisfactory, being both too rumbly and too hissy. However, *RC* curves are suitable for specifying the introduction of acoustic "perfume", and noise with a spectrum of that shape has been found to be the least objectionable.

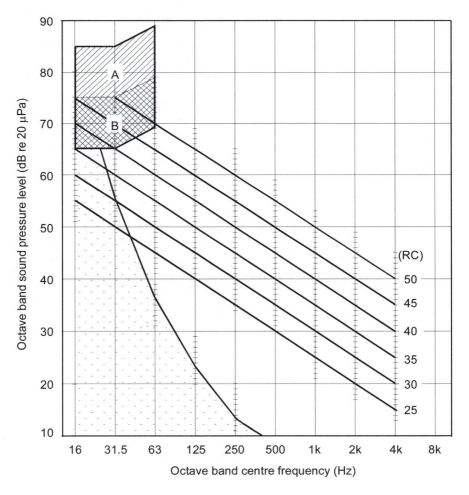

Figure 4.10 Room Criterion (RC) curves. Regions A and B: noise-induced feelable vibration in lightweight structures and induced rattles in light fixtures, doors, windows, etc. Region A, high probability; region B, moderate probability.

Acceptable RC ratings for background sound in rooms as a result of air conditioning noise are listed in Table 4.8

Table 4.8 Acceptable air conditioning noise levels in various types of occupied space. Note that the spectrum shape of the noise should not deviate from an RC curve by more than 3 dB and should contain no easily distinguishable tonal components. Adapted from ASHRAE (2007)

Room type	Acceptable RC(N)	Room type	Acceptable RC(N)
Residence	25-35	performing arts spaces	25
Hotel meeting rooms	25-35	Music practice rooms	30-35
Hotel suites	25-35	Laboratories	40-50
Other hotel areas	35-45	Churches	25-35
Offices and conf. rooms	25-35	Schools, lecture rooms	25-30
Building corridors	40-45	Libraries	30-40
Hospitals			
Private rooms	25-35	Indoor stadiums, gyms	40-50
Wards	30-40	Stadium with speech ampl.	45-55
Operating rooms	25-35	Courtrooms (no mics.)	25-35
Public areas	30-40	Courtrooms (speech ampl.)	30-40

4.8.1.4 NCB Curves

Balanced Noise Criteria (*NCB*) curves are shown in Figure 4.11. They are used to specify acceptable background noise levels in occupied spaces and include air-conditioning noise and any other ambient noise. They apply to occupied spaces with all systems running and are intended to replace the older *NC* curves. More detailed information on *NCB* curves may be found in American National Standard ANSI S12.2-1995 (R1999), "Criteria for Evaluating Room Noise". The designation number of an *NCB* curve is equal to the Speech Interference Level (SIL) of a noise with the same octave band levels as the *NCB* curve. The SIL of a noise is the arithmetic average of the 500 Hz, 1 kHz, 2 kHz and 4 kHz octave band decibel levels, calculated to the nearest integer. To determine whether the background noise is "rumbly", the octave band sound levels of the measured noise are plotted on a chart containing a set of *NCB* curves. If any values in the 500 Hz octave band or lower exceed by more than 3 dB the curve corresponding to the *NCB* rating of the noise, then the noise is labelled "rumbly". To determine if the noise is "hissy", the *NCB* curve which is the best fit of the octave band sound levels between 125 Hz and 500 Hz is determined. If any of the octave band sound levels between 1000 Hz and 8000 Hz inclusive exceed this *NCB* curve, then the noise is rated as "hissy".

4.8.1.5 RNC Curves (Figure 4.12)

The *RC* and *NCB* curves have a number of limitations that can lead to undesirable results. The *RC* curves set criteria that are below the threshold of hearing to protect

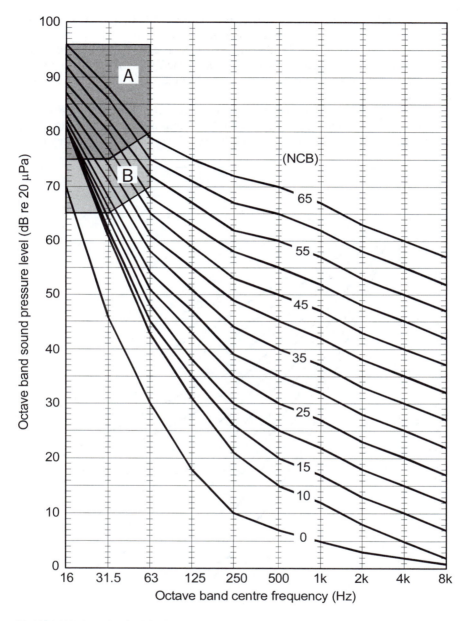

Figure 4.11 Balanced Noise Criterion curves (*NCB*). Regions A and B: noise-induced feelable vibration in lightweight structures, induced audible rattle in light fixtures, doors, windows, etc.; region A, high probability; region B, moderate probability.

against large turbulence fluctuations that generate high levels of low frequency noise for which the level can vary by up to 10 dB in synchronisation with fan surging.

However, the *RC* curves could unnecessarily penalise a well designed HVAC system such as may be used in a concert hall, requiring 10 dB or more of unnecessary noise attenuation at low frequencies. On the other hand, NCB curves are intended for well designed HVAC systems. They do not sufficiently penalise poorly designed systems that may be energy efficient but are characterised by high levels of turbulence induced low frequency noise.

Schomer (2000) proposed a new set of curves (room noise criteria or *RNC* - see Figure 4.12) that are intended to address the limitations associated with the *RC* and *NCB* curves. The intention is for noise criteria to be above the threshold of hearing for

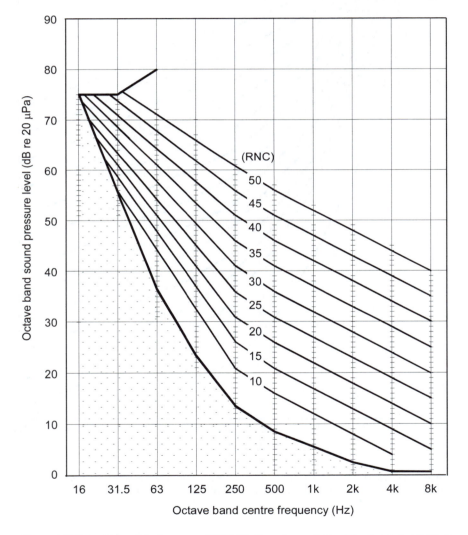

Figure 4.12 Room Noise Criterion curves (*RNC*). The lowest curve is the approximate threshold of hearing for octave band noise.

well-behaved systems, while at the same time preventing a turbulence-producing, fan-surging HVAC system from being labelled acceptable.

It is unlikely that the RNC curves will receive general acceptance because of the complexity of the rating process. Essentially, the RNC rating of a sound is obtained using the tangency method in much the same way as obtaining an NR number. That is, the RNC rating is the integer value of the highest RNC curve that intersects the plotted spectra. The complexity arises in the determination of what values to plot. These are the measured octave band values with correction terms added in the 31.5 Hz, 63 Hz and 125 Hz bands. It is the determination of the correction terms that is complex. For broadband noise radiated from an air conditioning duct without the presence of excessive turbulence or surging, the correction terms are zero. In cases where there is excessive turbulence, the correction at 31.5 Hz can be as high as a 4 dB increase and if surging is present it can be as high as 12 dB. The correction at 125 Hz is usually zero, except in the case of surging it may be up to 2 dB. A straight line is drawn between the plotted corrected value at 31 Hz and the corrected value at 125 Hz to obtain the corrected value at 63 Hz. The 16 Hz value is not plotted for cases involving excessive turbulence or surging. Details on the calculation of the correction to be added in the 31.5 Hz and 125 Hz octave bands are provided by Schomer (2000).

Briefly, the correction is calculated by taking a large number of octave band sound pressure level measurements, L_i, from 16 Hz to 8 kHz, over some reasonable time interval (for example, 20 seconds), using "fast" time weighting and sampling intervals between 50 and 100 ms. The mean sound pressure level, L_m, is calculated by taking the arithmetic mean of all the measured dB levels in each octave band. The energy averaged, L_{eq}, is also calculated for each octave band using:

$$L_{eq} = 10\log_{10}\left[\frac{1}{N}\sum_{i=1}^{N}10^{L_i/10}\right] \tag{4.46}$$

The corrections Δ_{31} and Δ_{125} to be added to the 31.5 Hz and 125 Hz octave band measurements of L_{eq}, is calculated for each of the two bands using:

$$\Delta = 10\log_{10}\left[\frac{1}{N}\sum_{i=1}^{N}10^{(L_i-L_m)/\delta}\right] + L_m - L_{eq} \tag{4.47}$$

The quantity, δ, is set equal to 5 for calculations in the 31.5 Hz band and equal to 8 for calculations in the 125 Hz octave band and N is the total number of measurements taken in each band. Note that for the 31.5 Hz correction term calculation, data for the 16 Hz, 31.5 Hz and 63 Hz bands are all included in each of the three terms in Equation (4.47). Before inclusion in Equation (4.47), the 16 Hz data has 14 dB subtracted from each measurement and the 63 Hz data has 14 dB added to each measurement. The noise is considered well behaved if the correction, Δ, for the 31.5 Hz octave band is less than 0.1 dB. In this case, the actual measured octave band data are plotted on the set of RNC curves from 16 Hz to 8 KHz, with no correction applied to any band.

Note that RNC values are reported, for example, as: RNC41-(63 Hz). In this example the highest RNC curve intersected was the RNC 41 curve and this occurred at 63 Hz.

4.8.2 Comparison of Noise Weighting Curves with dB(A) Specifications

For the majority of occupied spaces, advisory limits can be placed on maximum permissible background noise levels, but recommended levels will vary slightly depending upon the source of information.

As mentioned earlier, when an attempt was made to use *NC* rating curves as a guide for deliberate spectral shaping of background noise, the result was unsatisfactory. The internationally accepted *NR* curves do not provide an improvement in this respect. For example, if the *NR* weighting curves 15 to 50 are taken as background noise spectra, then A-weighted according to Table 3.1, and their overall equivalent A-weighted levels determined as described in Section 3.2, it is found that in all cases the low-frequency bands dominate the overall level. As A-weighting accords with subjective response, one would intuitively expect trouble with a spectrum that emphasises a frequency range which ordinarily contributes little to an A-weighted level. In this respect the *RC* weighting curves are much more satisfactory. When A-weighted it is found that neither the high- nor the low-frequency extremes dominate; rather, the mid-frequency range contributes most to the computed equivalent A-weighted level.

Since the various noise rating schemes are widely used and much published literature has been written in terms of one or another of the several specifications, Table 4.9 has been prepared to allow comparisons between them. In preparing Table 4.8, the dB(A) levels equivalent to an *NR, NC, NCB*, or *RNC* level were calculated by considering levels only in the 500 Hz, 1000 Hz and 2000 Hz bands, and assuming a spectrum shape specified by the appropriate *NR* or *NC* curve. On the other hand, the entries for the *RC* data were calculated using levels in all octave bands from 16 Hz to 4000 Hz and assuming the spectrum shape specified by the appropriate *RC* curve.

It is expected that the reader will understand the difficulties involved in making such comparisons. The table is intended as a guide, to be used with caution.

Table 4.9 Comparison of ambient level criteria

dB(A)	Specification			Comment
	NR	*NC, NCB* and *RNC*	*RC*	
25-30	20	20	20	Very quiet
30-35	25	25	25	
35-40	30	30	30	Quiet
40-45	35	35	35	
45-50	40	40	40	Moderately noisy
50-55	45	45	45	
55-60	50	50	50	Noisy
60-65	55	55	–	
65-70	60	60	–	Very noisy

Judgment is often necessary in specifying a noise rating for a particular application. Consideration must be given to any unusual aspects, such as people's attitudes to noise, local customs and need for economy. It has been found that there are differences in tolerance of noise from one climate to another. People in those countries in which windows are customarily open for most of the year seem to be more tolerant of noise, by 5-10 dB(A), than people in countries in which windows are customarily tightly closed for most of the year.

4.8.3 Speech Privacy

When designing an office building, it is important to ensure that offices have speech privacy so that conversations taking place in an office cannot be heard in adjacent offices or corridors. Generally, the higher the background noise levels from air conditioning and other mechanical equipment, the less one has to worry about speech privacy and the more flimsy can be the office partitions, as will become clear in the following discussion. One can deduce that speech privacy is likely to be a problem in a building with no air conditioning or forced ventilation systems.

To avoid speech privacy problems between adjacent offices, it is important that the following guidelines are followed:

1. Use partitions or separation walls with adequate sound insulation (see Table 4.10). Sound insulation is the reduction in sound pressure level between two adjacent rooms, one of which contains a sound source. It is related to the transmission loss of a panel as discussed in Chapter 8. Use Equation (8.17), to calculate the required partition average transmission loss (defined in Table 4.10 caption) for the required noise reduction given in Table 4.10.

2. Ensure that there are no air gaps between the partitions and the permanent walls and floor.

3. Ensure that the partitions extend all the way to the ceiling, roof or the underside of the next floor; it is common practice for the partitions to stop at the level of the suspended ceiling to make installation of ducting less expensive; however, this has a bad effect on speech privacy as sound travels through the suspended ceiling along the ceiling backing space and back through the suspended ceiling into other offices or into the corridor. Alternatively, ceiling tiles with a high transmission loss (*TL*, see chapter 8) as well as a high absorption coefficient could be used.

4. Ensure that acoustic ceiling tiles with absorption coefficients of at least 0.1 to 0.4 are used for the suspended ceiling to reduce the reverberant sound level in the office spaces, and thus to increase the overall noise reduction of sound transmitted from one office to another.

When speech privacy is essential, there are two alternative approaches that may be used. The first is to increase the sound insulation of the walls (for example, by using double stud instead of single stud walls so that the same stud does not contact both leaves). The second approach is to add acoustic "perfume" to the corridors and offices adjacent to those where privacy is important. This "perfume" could be introduced using a random-noise generator, appropriate filter, amplifier and speakers, with the speakers mounted above the suspended ceiling. The filter would be adjusted to produce an overall noise spectrum (existing plus introduced noise), which followed the shape of one of the *RC* curves shown in Figure 4.10, although it is not usually desirable to exceed an *RC* of 30 for a private office (see AS 2107-1987 and Table 4.6).

Table 4.10 Speech privacy noise insulation requirement

Sound as heard by occupant	Average sound insulation[a] plus ambient noise(dB(A))
Intelligible	70
Ranging between intelligible and unintelligible	75-80
Audible but not obtrusive (unintelligible)	80-90
Inaudible	90

[a] Average sound insulation is the arithmetically averaged sound transmission loss over the 1/3 octave bands from 100 Hz to 3150 Hz.

4.9 ENVIRONMENTAL NOISE LEVEL CRITERIA

A comprehensive document (Berglund *et al.*, 1995, 1999), which addresses many environmental noise issues, has been published by Stockholm University and the World Health Organisation. It is recommended as an excellent source for detailed information. Here, the discussion is limited to an overview of the assessment of environmental noise impacts.

Acceptable environmental (or community) noise levels are almost universally specified in terms of A-weighted sound pressure levels. Standards exist (ANSI S12.9 parts 1 to 5, 2003-2007; ASTM E1686-03, 1996; ISO 1996, parts 1 to 2, 2003-2007; AS1055, parts 1-3, 1997) that specify how to measure and assess environmental noise. The ideas in these standards are summarised below.

A comprehensive review of the effect of vehicle noise regulations on road traffic noise, changes in vehicle emissions over the past 30 years and recommendations for consideration in the drafting of future traffic noise regulations has been provided by Sandberg (2001).

4.9.1 A-weighting Criteria

To minimise annoyance to neighbouring residents or to occupiers of adjacent industrial or commercial premises, it is necessary to limit intrusive noise. The choice of limits is generally determined by noise level criteria at the property line or plant boundary. Criteria may be defined in noise-control legislation or regulations. Typical specifications (details may vary) might be written in terms of specified A-weighted equivalent noise levels, L_{Aeq} dB(A), and might state:

Noise emitted from non-domestic premises is excessive if the noise level at the measurement place (usually the noise sensitive location nearest to the noise emitter) for a period during which noise is emitted from the premises:

1. exceeds by more than 5 dB(A) the background noise level at that place (usually defined as the L_{90} level – see Section 3.7); and
2. exceeds the maximum permissible noise level prescribed for that time of the day and the area in which the premises are situated.

Some regulations also specify that the noise measured at the nearest noise sensitive area must not exceed existing background levels by more than 5 dB in any octave band. In some cases A-weighted statistical measures such as L_{90} are used to specify existing background noise levels.

Permissible plant-boundary noise levels generally are dependent upon the type of area in which the industrial premises are located, and the time of day. Premises located in predominantly residential areas and operating continuously (24 hours per day) face restrictive boundary noise level limits, as it is generally accepted that people are 10-15 dB(A) more sensitive to intrusive noise between 10 pm and 7 am.

Note that applicable standards such as ISO 1996, parts 1 and 2 (2003 and 2007), ANSI S12.9, parts 1-5, (2003, 2008, 2008, 2005 and 2007), ASTM E1686-03, 1996 and AS 1055, parts 1, 2 and 3 (1997) are guides to the assessment of environmental noise measurement and annoyance, and do not have any legal force. They are prepared from information gained primarily from studies of noise generated in industrial, commercial and residential locations. In general they (1) are intended as a guide for establishing noise levels that are acceptable in the majority of residential areas; and (2) are a means for establishing the likelihood of complaints of noise nuisance at specific locations. The general method of assessment involves comparison of noise levels measured in dB(A), with acceptable levels.

For steady noise the measured level is the average of the meter reading on a standard sound level meter with the A-weighting network switched in. For fluctuating or cyclic noise, the equivalent continuous A-weighted noise level must be determined. This can be done using a statistical noise analyser or an integrating/averaging sound level meter. If these instruments are not available, some standards detail alternative means of obtaining approximate values with a standard sound level meter or sound exposure meter.

In the absence of measured data, typical expected background noise levels (L_{90}) for various environments are summarised in Table 4.11. A base level of 40 dB(A) is used and corrections are made, based upon the character of the neighbourhood in which the noise is measured, and the time of day. These corrections, adapted from AS 1055 (1997), are listed in Table 4.11. When the measured noise level exceeds the relevant adjusted background noise level, a guide to probable complaints is provided (see Table 4.12).

Example 4.3

A vintage musical instrument collector finds playing his steam calliope a relaxing exercise. He lives in a generally suburban area with infrequent traffic. When he plays,

Table 4.11 Adjustments to base level of 40 dB(A) (adapted partly from Australian Standard AS1055)

	Adjustment (dB(A))
Character of the sound	
Tones or impulsive noise readily detectable	-5
Tones or impulsive noise just detectable	-2
Time of day	
Evening (6 pm to 10 pm)	-5
Night time (10 pm to 7 am)	-10
Neighbourhood	
Rural and outer suburban areas with negligible traffic	0
General suburban areas with infrequent traffic	+5
General suburban areas with medium density traffic or suburban areas with some commerce or industry	+10
Areas with dense traffic or some commerce or industry	+15
City or commercial areas or residences bordering industrial areas or very dense traffic	+20
Predominantly industrial areas or extremely dense traffic	+25

Table 4.12 Estimated public reaction to noise when the adjusted measured noise level exceeds the acceptable noise level (see Table 4.10 for adjustments to base level 40 dB(A))

Amount in dB(A) by which adjusted measured noise level exceeds the acceptable noise level	Public reaction	Expression of public reactions in a residential situation
0-5	Marginal	From no observed reaction to sporadic complaints
5-10	Little	From sporadic complaints to widespread complaints
10-15	Medium	From sporadic complaints to threats of community action
15-20	Strong	From widespread complaints to threats of community action
20-25	Very strong	From threats of community action to vigorous community action
25 and over	Extreme	Immediate direct community and personal action

the resulting sound level due to his instrument ranges to about 55 dB(A) at the nearby residences. If he finds himself insomnious at 3 am, should he play his calliope as a sedative to enable a return to sleep?

Solution

Begin with the base level of 40 dB(A) and subtract 5 dB(A) to account for the tonal nature of the sound. Next add the adjustments indicated by Table 4.11 for time of day and location. The following corrected criterion is obtained:

$$35 - 10 + 5 = 30 \text{ dB(A)}$$

The amount by which the expected level exceeds the corrected criterion at the nearby residences is as follows:

$$55 - 30 = 25 \text{ dB(A)}$$

Comparison of this level with the levels shown in Table 4.12 suggests that strong public reaction to his playing may be expected. He had best forget playing as a cure for his insomnia!

4.10 ENVIRONMENTAL NOISE SURVEYS

To document existing environmental noise, one or more ambient sound surveys must be undertaken and if a new facility is being planned, the calculated emissions of the facility must be compared with existing noise levels to assess the potential noise impact. Existing noise regulations must be met, but experience has shown that noise problems may arise, despite compliance with all applicable regulations.

When undertaking a noise survey to establish ambient sound levels, it is important to exclude transient events and noise sources, such as insect noise, which may not provide any masking of noise from an industrial facility. Similarly, high ambient noise levels resulting from weather conditions favourable to sound propagation must be recognised and corresponding maximum noise levels determined. If the facility is to operate 24 hours a day, the daily noise variation with time must be understood using continuous monitoring data collected in all four seasons. For installations that shut down at night, measurements can be limited to daytime and one nighttime period.

Existing standards (ASTM E1686-03. ASTM E1779-96a(2004), ISO1996/1&2 (2003, 2007) should always be followed for general guidance.

4.10.1 Measurement Locations

Residential areas closest to the noise source are usually chosen as measurement locations, but occasionally it is necessary to take measurements at other premises such as offices and churches. If the receiver locations are above the planned site or at the other side of a large body of water, measurements may be needed as far as 2 km away from the noise source. Sometimes it is possible to take measurements on residential properties; at other times, it may be necessary to use a laneway. Whatever locations are used, it is important that they are as far from the main access road as the residences themselves and definitely not near the side of the road, as the resulting percentile levels and equivalent sound levels will not represent actual levels at the front or rear of a residence, except for perhaps the L_{90} and perhaps higher percentile levels as well in areas characterised by sparse traffic. Noise emission codes and zoning regulations specify allowable levels at the property line of the noise generator, so it is often necessary to take measurements at a number of property boundary locations also.

If a major roadway is near the planned site, it is useful to measure at locations that are at least two convenient distances from the centre of the roadway, so that the noise propagation pattern of such a major source can be superimposed onto area maps around the planned facility.

4.10.2 Duration of the Measurement Survey

Current practice commonly involves both continuous unattended monitoring over a 24-hour period, and periodic 10 to 15 minute attended sampling. A minimum of 40 hours, or at least two nighttime intervals, is necessary to adequately determine and document a noise environment. It is rarely sufficient to use only one of the two monitoring techniques. However if both are used concurrently and the origins of noise events are recorded during the attended sampling periods, the two techniques provide complementary information that can be combined to obtain a good understanding of the ambient noise environment. It is recommended that at least one continuous monitor (preferably 3 or 4) be used at the most critical locations and that regular attended sampling be undertaken at a number of locations, including the critical locations.

If a regulatory authority assesses the impact of a new facility on the basis of the "minimum" ambient level (over a specified averaging time), it is important to sample over a number of days to determine what the daily average minimum may be. In some cases, the minimum may be different at different times of the year. Environmental noise environments that are dominated by high density daily traffic, are generally very repeatable from day to day (with typical standard deviations of less than 1 dB for the same time of day). However, in the early morning hours, weather conditions become more important than traffic volume and minimum levels vary in accordance with sound propagation conditions. Conversely, measurements in suburban and rural environments, not dominated by traffic, are not very repeatable and can be characterised by standard deviations higher than 5 dB. These data are often affected by wind and temperature gradients as well as by the wind generating noise indirectly, such as by causing leaves in trees to rustle.

4.10.3 Measurement Parameters

The statistical measure, L_{10}, (see Section 3.7) is primarily used for assessing traffic noise; L_{50} is a useful measure of the audibility of noise from a planned facility; and L_{90}

is used to classify and characterise residential area environments. In many cases, L_1 is used as a measure of the peak noise level, but it often underestimates the true peak by as much as 20 dB. In many cases, the energy weighted equivalent noise level, L_{eq} is used as a measure. For regulatory purposes, the above levels are all usually expressed in overall dB(A). However, to gain a good understanding of the environmental noise character and the important contributors to it, it is often useful to express the statistical measures mentioned above in 1/1 or 1/3-octave bands.

When environmental noise is measured, the following items should be included in the measurement report:

- reference to the appropriate noise regulation document (regulation usually);
- date and time of measurement;
- details of measurement locations;
- weather conditions (wind speed and direction, relative humidity, temperature and recent precipitation);
- operating conditions of sound source (e.g. % load etc.), description of the noise source and its condition, and any noticeable characteristics such as tones, modulation or vibration;
- instrumentation used and types of measurements recorded (e.g. spectra, L_{10}, L_{Aeq});
- levels of noise due to other sources;
- measured data or results of any calculations pertaining to the noise source being measured;
- any calculation procedures used for processing the measurements;
- results and interpretation; and
- any other information required by the noise regulation document.

4.10.4 Noise Impact

To calculate the overall noise impact of an industry on the surrounding community, the number of people exposed to various noise levels is used to arrive at a single noise exposure index called the Total Weighted Population (or TWP). This quantity is calculated using day–night average sound levels (L_{dn}), weighting factors to weight higher levels as more important and the number of people exposed to each level as follows:

$$TWP = \sum_i W_i P_i \tag{4.48}$$

where P_i is the number of people associated with the ith weighting factor which, in turn, is related to a particular L_{dn} level as defined in Table 4.13.

The relative impact of one particular noise environment may be compared with another by comparing the Noise Impact Index for each environment, defined as:

$$NII = \frac{TWP}{\sum_i P_i} \tag{4.49}$$

Table 4.13 Annoyance weighting factors corresponding to values of L_{dn}

Range of L_{dn} (dB)	W_i
35–40	0.01
40–45	0.02
45–50	0.05
50–55	0.09
55–60	0.18
60–65	0.32
65–70	0.54
70–75	0.83
75–80	1.20
80–85	1.70
85–90	2.31

CHAPTER FIVE

Sound Sources and Outdoor Sound Propagation

LEARNING OBJECTIVES

In this chapter the reader is introduced to:

- the simple source, source of spherical waves and fundamental building block for acoustical analysis;
- the dipole source, directional properties and modes of generation;
- the quadrupole source, its various forms, directional properties and some modes of generation;
- line sources and their uses for modelling;
- the piston in an infinite wall, far-field and near-field properties, radiation load, and uses for modelling;
- the incoherent plane radiator;
- directional properties of sources and the concept of directivity;
- effects of reflection;
- reflection and transmission at a plane two media interface; and
- sound propagation outdoors, ground reflection, atmospheric effects, methods of prediction.

5.1 INTRODUCTION

Sources of sound generation are generally quite complicated and often beyond the capability of ordinary analysis if a detailed description is required. Fortunately, a detailed description of the noise-generation mechanism is often not necessary for noise-control purposes and in such cases very gross simplifications will suffice. For example, a source of sound of characteristic dimensions much less than the wavelengths it generates can often be approximated as a simple point source of zero dimension. In this case the properties of the idealised point source of sound will provide a sufficient description of the original sound source.

Alternatively, familiarity with the properties of common idealised sources may provide the means for identification of the noise source. For example, in a duct, noise which increases in level with the sixth power of the flow speed can readily be identified as due to a dipole source, and most probably originating at some obstruction to the flow in the duct. For these reasons, as well as others to be mentioned, it is worthwhile considering the properties of some idealised sound sources (Dowling and Ffowcs-Williams, 1982).

Consideration of sound sources out-of-doors presents problems associated with sound propagation over the ground. The problem is often further complicated by the source size, which almost invariably is large. Treatment of this complex problem is considered here and is based upon many references given in the text.

5.2 SIMPLE SOURCE

A simple source is a source of sound that radiates uniformly in all directions and which is very much smaller than the wavelength of the radiated sound. The source could be a small speaker mounted in one wall of a very small rigid box, but as long as the wavelength that it generates is large compared to any of the dimensions of the box, it will qualify as a simple source. That is, the small speaker-box source will generate an acoustic field like that which would be generated by a small pulsating sphere. Thus, the class of simple sources may be thought of as those sources that have the properties of a small pulsating sphere, the diameter of which is very much less than any wavelength it generates. Radiation of sound waves from such a source is illustrated in Figure 5.1.

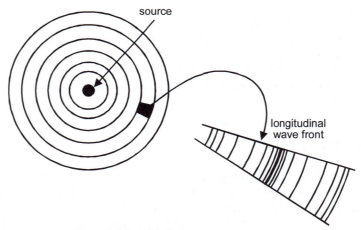

Figure 5.1 Sound radiation from a simple source.

5.2.1 Pulsating Sphere

In Section 1.4.3, an outward-travelling spherical wave solution to the wave equation was considered. Clearly, as the particle velocity of an outward-travelling spherical wave can be made to match the surface velocity of a pulsating sphere, the radiated field of the pulsating sphere can be determined from the spherical wave acoustic potential function given by Equation (1.41).

According to Section 1.7.2, the sound intensity is obtained as the long-time average of the product of the acoustic pressure and particle velocity. Use of Equations (1.42), (1.43) and (1.65b) gives the following equation for the intensity:

$$I = \lim_{T \to \infty} \frac{\rho c}{T} \int_0^T \frac{\mathrm{ff}' + r\mathrm{f}'\mathrm{f}'}{r^3} \, \mathrm{d}t \tag{5.1}$$

However, integration over an encompassing sphere of surface area $4\pi r^2$ shows that if the radiated power W is to be constant, independent of radius r, then:

$$\lim_{T \to \infty} \frac{1}{T} \int_0^T \mathrm{ff}' \, \mathrm{d}t = 0 \tag{5.2}$$

Thus, the intensity of the diverging spherical wave is:

$$I = \rho c \frac{\langle \mathrm{f}'\mathrm{f}' \rangle}{r^2} \tag{5.3}$$

where

$$\langle \mathrm{f}'\mathrm{f}' \rangle = \lim_{T \to \infty} \frac{1}{T} \int_0^T \mathrm{f}'\mathrm{f}' \, \mathrm{d}t \tag{5.4}$$

The symbol $\langle \ \rangle$ indicates a time-averaged quantity (see Section 1.5).

In the form in which the above equations have been written, the functional dependence of the potential function and its derivative has been unspecified to allow for the possibility that a narrow band of noise, rather than a tone, is considered. In the case of a narrow band of noise, for example a one-third octave band, corresponding values of frequency and wavelength may be estimated by using the centre frequency of the narrow band (see Section 1.10.1).

The next step is to satisfy the boundary condition that the surface velocity of the pulsating small spherical source matches the acoustic wave particle velocity at the surface. For the purpose of providing a result useful for narrow frequency bands of noise, the root mean square particle velocity at radius r is considered. The following result is then obtained with the aid of Equations (1.43) and (1.58):

$$\langle u^2 \rangle = \lim_{T \to \infty} \frac{1}{Tr^4} \int_0^T [\mathrm{ff} + 2r\mathrm{ff}' + r^2\mathrm{f}'\mathrm{f}'] \, \mathrm{d}t \tag{5.5}$$

Using Equations (5.2) and (5.5) the following equation may be written for a spherical source having a radius $r = a$:

$$\langle u^2 \rangle = \frac{\langle \mathrm{ff} \rangle}{a^4} + \frac{\langle \mathrm{f}'\mathrm{f}' \rangle}{a^2} \tag{5.6}$$

The r.m.s. volume flux Q_{rms} of fluid produced at the surface of the pulsating spherical source may be defined as follows:

$$Q_{rms} = 4\pi a^2 \sqrt{\langle u^2 \rangle} \qquad (5.7)$$

Substitution of Equation (5.6) in Equation (5.7) gives:

$$Q_{rms} = 4\pi \sqrt{\langle ff \rangle + \langle f'f' \rangle a^2} \qquad (5.8)$$

The particle velocity is now matched to the surface velocity of the pulsating spherical source.

The assumption is made that the following expression, which holds exactly for a pure tone or single frequency, also holds for a narrow band of noise:

$$\langle ff \rangle k^2 = \langle f'f' \rangle \qquad (5.9)$$

The constant k is the wavenumber, equal to ω/c, where for a narrow band of noise the centre frequency in radians per second is used. Substitution of Equation (5.9) into Equation (5.8) gives:

$$Q_{rms} = 4\pi \frac{\sqrt{\langle f'f' \rangle (1 + k^2 a^2)}}{k} \qquad (5.10)$$

Finally, substituting Equation (5.10) into Equation (5.3) gives the intensity of a simple source in terms of the mean flux of the source:

$$I_M = \frac{Q_{rms}^2 k^2 \rho c}{(4\pi r)^2 (1 + k^2 a^2)} \qquad (5.11)$$

Note that for a very small source ($a \ll \lambda$), the term, $k^2 a^2$ may be ignored. As the simple source is also known as a monopole, the subscript "M" has been added to the intensity I. The power radiated by a simple or monopole source is given by the integral of the intensity over the surface of any convenient encompassing sphere of radius r; that is, multiplication of Equation (5.11) by the expression for the surface area of a sphere gives, for the radiated sound power of a monopole:

$$W_M = \frac{Q_{rms}^2 k^2 \rho c}{4\pi (1 + k^2 a^2)} \qquad (5.12)$$

Equation (1.74) may be used to rewrite the expression for the intensity given by Equation (5.11) to give Equation (5.13).

$$\langle p^2 \rangle = \frac{(Q_{rms} k \rho c)^2}{(4\pi r)^2 (1 + k^2 a^2)} \qquad (5.13)$$

Using Equations (5.12) and (5.13), the following may be written for the sound pressure at distance r from the source in terms of the source sound power:

$$\langle p^2 \rangle = \frac{W_M \rho c}{4\pi r^2} \tag{5.14}$$

The r.m.s. flux Q_{rms} must be limited by the source volume to less than $2\sqrt{2}\omega\pi a^3/3$, where a is the radius of the source. The mean square acoustic pressure at distance r from the source is obtained by substituting the expression for Q_{rms} and Equation (1.29) into Equation (5.13). Thus:

$$\langle p^2 \rangle < \left[\frac{\rho c^2 \sqrt{2}}{6} \frac{a}{r} \frac{k^2 a^2}{\sqrt{1 + k^2 a^2}} \right]^2 \tag{5.15}$$

Thus the sound that can be produced by such a source is limited. For example, let the source radius, a, be 0.1λ and the distance r from the source to an observation point be 10λ. Then, $ka = 0.2\pi$ and $a/r = 0.01$ and the sound pressure level at r is less than 135 dB re 20 µPa. The limitation on maximum possible sound pressure arises because the maximum pulsation amplitude of a small source is physically limited to about half of the source radius.

5.2.2 Fluid Mechanical Monopole Source

An example of a fluid mechanical monopole source is the inlet or the exhaust of a reciprocating pump or the exhaust of a reciprocating engine. In either case, the source dimension is small compared to the wavelength of radiated sound. Referring to Equation (5.12) the rms volume flux Q_{rms} may be assumed proportional to the inlet or exhaust cross-sectional area, S multiplied by the local mean flow speed U; that is, $Q_{rms} \propto SU$. The wavenumber, k, which equals the angular frequency ω divided by the speed of sound, c, must likewise be proportional to the local mean flow speed, U, divided by a characteristic length. The characteristic length squared in turn may be considered as proportional to the inlet or exhaust cross-sectional area; thus, $k^2 \propto U^2/Sc^2$. Substitution of the latter expressions in Equation (5.12) and introduction of the Mach number $M = U/c$ gives the following result:

$$W_M \propto (SpU^4/c) \propto (SpU^3) M \tag{5.16a,b}$$

Equation (5.16a) shows that the sound power of a fluid mechanical monopole source may be expected to increase with the fourth power of the stream speed while Equation (5.16b) shows that the sound power is proportional to the quantity SpU^3, which is a measure of the convected kinetic energy or the stream power, and the Mach number M, which for values less than one, may be interpreted as a measure of the source efficiency.

5.3 DIPOLE SOURCE

If two simple sources of equal strength are placed close together and arranged to be always of exactly opposite phase, then as one produces a net outflow the other produces an exactly opposite inflow. Thus, at some distance from the sources there is no net fluid flux. However, being of opposite sign there is a net thrust exerted on the fluid from the positive to the negative source, which fluctuates in direction. It is the time rate of change of this force on the fluid that is important for noise production. Such an arrangement becomes a dipole source in the limit as the separation distance between the sources is made indefinitely small; that is, very much less than the radiated wavelength. A familiar example of a dipole source is one of the prongs of a tuning fork. A less obvious example is the fluctuating flow over a spoiler introduced into an air stream.

5.3.1 Pulsating Doublet or Dipole (Far-field Approximation)

Consider two simple sources of opposite phase placed at $y = h$ and $-h$, and determine the resulting acoustic potential function at the general point (x, y), as shown in Figure 5.2. The potential φ at (x, y) in the far field of the dipole is the sum of the potentials due to sources 1 and 2:

$$\varphi = \varphi_1 + \varphi_2 \qquad (5.17)$$

The form of φ_1 and φ_2 is that given by Equation (1.46), with r replaced respectively with r_1 and r_2.

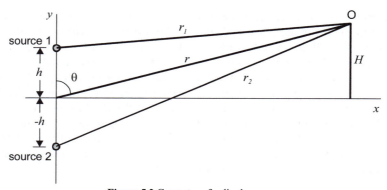

Figure 5.2 Geometry of a dipole source.

The distances r_1 and r_2 may be rewritten in terms of the distance r measured from the origin to the observation point O. That is:

$$r_1 = r\sqrt{1 + (h/r)^2 - 2(h/r)\cos\theta} \qquad (5.18)$$

Thus,

$$r_1 \approx r - h\cos\theta \tag{5.19}$$

and similarly,

$$r_2 \approx r + h\cos\theta \tag{5.20}$$

In writing Equations (5.19) and (5.20) it has been assumed that:

$$(h/r) \ll 1 \tag{5.21}$$

Substitution of Equations (5.19) and (5.20) into Equation (5.17) and use of Equation (1.46) results in the following equation, to first order in small quantities:

$$\varphi = \frac{f(ct - r_1)}{r_1} - \frac{f(ct - r_2)}{r_2} = \frac{f(ct - r + h\cos\theta)}{r} - \frac{f(ct - r - h\cos\theta)}{r} \tag{5.22a,b}$$

Using a Taylor series, i.e.

$$f(x + \Delta) = f(x) + \frac{f'(x)}{1!}\Delta + \frac{f''(x)}{2!}\Delta^2 + \dots \tag{5.23}$$

to expand Equation (5.22), again to first order, the following equation is obtained:

$$\varphi = \frac{1}{r}\Big[f(ct - r) + f'(ct - r)h\cos\theta + \dots \\ - f(ct - r) + f'(ct - r)h\cos\theta - \dots \Big] \tag{5.24}$$

Equation (5.24) reduces to the following expression for the dipole source potential function, valid for distances greater than about one wavelength:

$$\varphi = 2f'(ct - r)(h/r)\cos\theta \tag{5.25}$$

Equations (1.10) and (1.11) of Section 1.3.6 and Equation (5.25) are used to obtain expressions for the acoustic pressure p and particle velocity u:

$$p = \rho\frac{\partial\varphi}{\partial t} = 2\rho c\frac{h}{r}f''(ct - r)\cos\theta \tag{5.26a,b}$$

and

$$u = -\frac{\partial\varphi}{\partial r} = \frac{2h}{r^2}\Big[f'(ct - r) + rf''(ct - r)\Big]\cos\theta \tag{5.27a,b}$$

Division of Equation (5.26b) by Equation (5.27b) produces an equation that is formally the same as obtained for spherical waves (demonstrated using Equation 5.9 and $\langle f'f'\rangle k^2 = \langle f''f''\rangle$, $f'(ct - r) = jkAe^{jk(ct-r)}$ and $f''(ct - r) = -k^2Ae^{jk(ct - r)}$) and given by Equation (1.44)). Thus, for narrow bands of noise or single frequencies, Equation (1.49) also holds for dipoles. At large distances from the source, the characteristic impedance for dipole waves becomes simply ρc.

Using exactly the same procedures and argument used in Section 5.2.1, the dipole field intensity I_D may be shown to be:

$$I_D = \rho c \frac{h^2 Q_{rms}^2 k^4}{(2\pi r)^2 (1 + k^2 a^2)} \cos^2 \theta \qquad (5.28)$$

where Q_{rms} is the rms strength of each monopole source making up the dipole and a is the radius of each of these sources. Finally, integration of the intensity over an encompassing sphere of radius, r, gives for the radiated power of a dipole:

$$W_D = \rho c \frac{k^4 h^2 Q_{rms}^2}{3\pi (1 + k^2 a^2)} \qquad (5.29)$$

The mean square sound pressure at distance r and angular direction θ, measured from the dipole axis, is obtained by substituting into Equation (5.26b), the relations, $f''(ct - r) = -k^2 f(ct - r) = jk f'(ct - r)$, and Equation (5.10) to give:

$$\langle p^2 \rangle = \left[\frac{h Q_{rms} k^2 \rho c \cos \theta}{2\pi r} \right]^2 \frac{1}{(1 + k^2 a^2)} = 3 W_D \frac{\rho c}{4\pi r^2} \cos^2 \theta \qquad (5.30a,b)$$

The characteristic figure-of-eight radiation pattern of the pressure field of the dipole source is illustrated in Figure 5.3. In the plane normal to the axis of the dipole the sound pressure is nil, while along the axis of the dipole the sound pressure is maximum and of opposite phase in opposite directions.

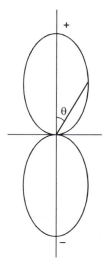

Figure 5.3 The radiation pattern of a dipole source. The axis of the source is vertical. The figure indicates the relative intensity of the sound pressure in the direction θ. In particular, a null in intensity is observed at $\theta = \pi/2$.

It is instructive to compare the radiation efficiencies of the dipole and monopole sources. Using Equations (5.12) and (5.29), their ratio can be determined as:

$$\frac{W_D}{W_M} = \frac{4}{3}(kh)^2 \tag{5.31}$$

As the quantity kh, which is a measure of the sources' separation distance-to-wavelength ratio, is small, it can be seen that the dipole is much less efficient as a radiator of sound than either of its constituent simple sources taken separately. This is because the presence of the second source, of opposite phase and in close proximity, tends to unload the first source so that it produces less sound. The role of an active sound-cancelling device is suggested.

In Equations (5.28) and (5.29) the rms source strength Q_{rms} is multiplied by the sources' separation distance h (see Figure 5.2). In a mathematical sense, the product hQ_{rms} may be set equal to the rms dipole strength D_{rms}, which remains constant while h and Q_{rms} are allowed to become indefinitely small and large respectively.

5.3.2 Pulsating Doublet or Dipole (Near-field)

It is of interest to consider the near field sound pressure in the vicinity of a dipole as it will be useful in the discussion of the oscillating sphere in the next section. Beginning with Equation (1.48), and the definition for flux, Q, from a single pulsating sphere of radius, a as:

$$Q = Q_0 e^{j\omega t} = 4\pi a^2 u = 4\pi a^2 u_0 e^{j\omega t} \tag{5.32a,b,c}$$

the following can be written for the source flux amplitude in terms of the constant, A, of Equation (1.48).

$$Q_0 = 4\pi a^2 \left[\frac{A}{a^2} + \frac{jka}{a} \right] e^{-jka} = 4\pi A (1 + jka) e^{-jka} \tag{5.33a,b}$$

Thus:

$$A = \frac{Q_0}{4\pi} \left[\frac{1}{1 + jka} \right] e^{jka} \tag{5.34}$$

Defining a Greens Function between a source point, r_0 and a measurement point, r, as:

$$G(r, r_0, \omega) = \frac{1}{4\pi |r - r_0|} e^{-jk|r - r_0|} \tag{5.35}$$

and using Equations (1.47b) and (5.34), the following expression (which is accurate in both the near and far fields) may be written for the acoustic pressure at any time, t, radiated by the dipole to some location r:

$$p(r, \theta, t) = G(r_1, h, \omega) - G(r_2, -h, \omega)$$

$$= \frac{jkpcQ_0}{(1 + jka)} \left[G(r - h\cos\theta, h, \omega) - G(r + h\cos\theta, -h, \omega) \right] e^{j(\omega t + ka)} \qquad (5.36a,b)$$

where h is half the distance between the two sources making up the dipole, which is centred at the origin.

If the distance between the two sources making up the dipole is small compared to a wavelength, Equation (5.36) may be written using Taylor's series (see Equation (5.23)) as:

$$p(r, \theta, t) = -\frac{jkpcQ_0 2h\cos\theta}{(1 + jka)} \left[\frac{\partial}{\partial r} G(r, r_0, \omega) \right]_{\{r_0 = 0\}} e^{j(\omega t + ka)} \qquad (5.37)$$

where $r = |r - r_0|$.

Using Equation (5.35), Equation (5.37) may be written as:

$$p(r, \theta, t) = -\frac{k^2 pcQ_0 h\cos\theta}{2\pi r(1 + jka)} \left[1 - \frac{j}{kr} \right] e^{j(\omega t - k(r - a))} = \frac{B\cos\theta}{kr} \left[1 - \frac{j}{kr} \right] e^{j(\omega t - k(r - a))} \quad (5.38a,b)$$

where:

$$B = \frac{pchQ_0 k^3}{2\pi(1 + jka)} \qquad (5.39)$$

Using Equations (1.10) and (1.11) allows the following to be written for the particle velocity:

$$u_r = \frac{B\cos\theta}{kr pc} \left[1 - \frac{2}{(kr)^2} - \frac{2j}{kr} \right] e^{j(\omega t - k(r - a))} \qquad (5.40)$$

Inspection of Equations (5.38) and (5.40) shows that, for small values of kr (that is, close to the source), the acoustic pressure and particle velocity are virtually in quadrature (90° out of phase), suggesting a source that radiates very poorly.

In the far field, where kr is large, the mean square pressure can be derived from Equation (5.38a) as:

$$\langle p^2 \rangle = \left[\frac{hQ_{rms} k^2 pc \cos\theta}{2\pi r} \right]^2 \frac{1}{(1 + k^2 a^2)} \qquad (5.41)$$

Which is identical to Equation (5.30a).

5.3.3 Oscillating Sphere

An oscillating sphere acts in a similar way to a dipole in terms of its sound radiation. It is interesting to consider Equation (5.27b) and note that it is of the form, $u = u_0 \cos\theta \, e^{j\omega t}$. The latter equation, however, is just the expression for the normal component of velocity over the surface of a sphere, which vibrates with velocity amplitude U along the dipole axis. For example, in Figure 5.2 the pair of sources might be replaced with a small sphere of radius a at the origin, which is made to vibrate up and down parallel to the y axis. The net force F exerted on the surrounding medium by the sphere vibrating along the y axis (see Figure 5.2) is determined by integrating the y component of the force due to the acoustic pressure over the surface of the sphere. However, to proceed, expressions for the acoustic pressure and particle velocity in the near field of the source are required.

Slight rearrangement of the expression given by Dowling and Ffowcs-Williams (1983) results in the following for the sound pressure radiated by an oscillating sphere:

$$p(r, \theta, t) = -\frac{j4\pi k\rho c u_0 a^3 \cos\theta}{2(1 + jka) - k^2 a^2} \left[\frac{\partial}{\partial r} G(r, r_0, \omega) \right]_{\{r_0 = 0\}} e^{j(\omega t - k(r - a))} \tag{5.42}$$

which can be rewritten as:

$$p(r, \theta, t) = -\frac{k^2 \rho c u_0 a^3 \cos\theta}{r[2(1 + jka) - k^2 a^2]} \left[1 - \frac{j}{kr} \right] e^{j(\omega t - k(r - a))} \tag{5.43}$$

For a compact sphere ($ka \ll 1$), Equation (5.43) may be written as:

$$p(r, \theta, t) = -\frac{k^2 \rho c u_0 a^3 \cos\theta}{2r(1 + jka)} \left[1 - \frac{j}{kr} \right] e^{j(\omega t - k(r - a))} \tag{5.44}$$

Comparing Equations (5.38) and (5.44) for a compact oscillating sphere, which is much smaller than a wavelength, shows that a compact oscillating sphere, of radius, a, and oscillating with a velocity amplitude of u_0, radiates the same sound field as a dipole made up of two simple sources separated by $2h$, and each having a flux amplitude given by:

$$Q_0 = \pi a^3 u_0 / h \tag{5.45}$$

If the oscillating sphere is not compact, the analysis becomes much more complicated and this is beyond the scope of this text.

Using Equations (1.10), (1.11) (5.44) and (5.45) allows the following to be written for the particle velocity radiated by an oscillating sphere:

$$u_r = \frac{B \cos\theta}{kr\rho c} \left[1 - \frac{2}{(kr)^2} - \frac{2j}{kr} \right] e^{j(\omega t - k(r - a))} \tag{5.46}$$

It can be seen that this equation is identical to Equation (5.40) for a dipole. However, in this case,

$$B = \frac{\rho c a^3 u_0 k^3}{2(1 + jka)}; \quad |B| = \frac{\rho c a^3 u_0 k^3}{2\sqrt{(1 + k^2 a^2)}} \qquad (5.47a,b)$$

According to Equation (1.65b) of Section 1.7.1, the intensity is determined as the long-time average of the product of the real parts of the acoustic pressure given by Equation (5.38), and the particle velocity given by Equation (5.40). In the far field (that is, more than a wavelength from the source), the expression for the intensity is:

$$I_D = \frac{|B|^2 \cos^2 \theta}{2\rho c (kr)^2} \qquad (5.48)$$

The power radiated by the small vibrating sphere may be determined as the long-time average of the real part of the force acting to move the sphere multiplied by the real part of the velocity of the sphere. However, the power radiated must equal the power flow to the far field; thus, an alternative means of determining the power flow due to vibration of the small sphere is to integrate the expression for the far-field intensity over the surface of the sphere. Using either method, the expression obtained for the power radiated is:

$$W_D = \frac{2\pi |B|^2}{3\rho c k^2} \qquad (5.49)$$

The amplitude of the force acting on the small vibrating sphere and along one axis may be determined by integrating the component of the acoustic pressure given by Equation (5.44) over the surface of the sphere. Note that the expression for the acoustic pressure has the form $p = p_0 \cos \theta$. The force along the principal axis of the dipole (see Figure 5.2) is:

$$F = 2\pi a^2 \int_0^\pi p_0 \cos^2 \theta \sin \theta \, d\theta = \frac{4}{3}\pi a^2 p_0 \qquad (5.50)$$

Making use of Equation (5.44) evaluated at a, the dummy variable p_0 may be replaced and the following expression obtained:

$$F = -\frac{4\pi a B}{3k}\left[1 - \frac{j}{ka}\right]e^{j\omega t} \qquad (5.51)$$

The above equation for B may be solved using $F = |F|e^{j\omega t}$ giving the following expression:

$$|B| = \frac{3k^2}{4\pi\sqrt{(1 + k^2 a^2)}}|F| \qquad (5.52)$$

Substitution of Equation (5.52) into (5.49) gives the following expression for the sound power radiated as a result of the force, F, acting on a small vibrating rigid body:

$$W_D = \frac{3\pi f^2 |F|^2}{2\rho c^3 (1 + k^2 a^2)} \tag{5.53}$$

As the analysis assumed a compact source (wavelength of radiated sound is large compared to the radiating object), the $k^2 a^2$ term in the denominator may be neglected. It is not necessary that the compact source be spherical. It may be of any shape and an example is the tooth of a circular saw turning at high speed, which may be successfully modelled as such a source (Bies, 1992). In this case, the quantity, a would be a characteristic source dimension. For a rectangular-shaped block, it is not clear whether the "characteristic dimension" is half the length in the direction of flow or a different dimension. Equation (5.53) has significance for noise generation and control as it shows that the radiated power is only dependent on the force exerted on the object.

For an oscillating sphere of radius, a, moving with a velocity amplitude of u_0, Equations (5.47) and (5.52) may be combined to give the following expression for the force term in Equation (5.53):

$$|F| = 2\rho\omega\pi a^3 u_0 / 3 \tag{5.54}$$

5.3.4 Fluid Mechanical Dipole Source

As mentioned earlier, turbulent flow over a small obstruction in an air stream provides a good example of a fluid mechanical dipole source of sound. Consideration of dipole noise generation in a fluid stream suggests the following approximations and assumptions. Referring to Equation (5.29), the source flux Q may be assumed to be proportional to the mean stream speed U and the square of the characteristic dipole source dimension h. The characteristic source dimension h squared in turn may be considered as proportional to the stream cross-sectional area, S, thus $S \propto h^2$ and $Q \propto SU$. The wavenumber k, which equals the angular frequency, ω, divided by the speed of sound, c, must be proportional to the local mean flow speed, U, divided by the characteristic dipole length, h. Thus $k \propto U/hc$. Substitution of the latter expressions in Equation (5.29) and introduction of the Mach number $M = U/c$ gives the following:

$$W_D \propto S\rho U^6 / c^3 \propto S\rho U^3 M^3 \tag{5.55a,b}$$

Equation (5.55a) shows that the sound power of a dipole source in a fluid stream tends to increase with the sixth power of the fluid stream speed. However, the idling noise of a saw tooth on a moving saw is dipole in nature but, due to the effect of the blade on the radiation impedance, the sound power of the saw is proportional to the fifth power of the saw speed, not the sixth power as predicted by Equation (5.55). See Chapter 6 for more on the effect of the source radiation impedance upon its radiated sound power.

Equation (5.55b) shows that the radiated sound power is proportional to the stream power, $S\rho U^3$ and the cube of the Mach number, M. As in the case of the monopole source, the Mach number term may be interpreted as an efficiency factor for values of Mach number less than one. This interpretation shows that the dipole source is much less efficient in converting stream power to noise than is the monopole source. However, the efficiency of the dipole source increases much more rapidly with flow speed than does the monopole source.

5.4 QUADRUPOLE SOURCE (FAR-FIELD APPROXIMATION)

It was shown earlier that a dipole could be thought of as a sphere vibrating along an axis coincident with the dipole. Similarly, any multipole source may be thought of as a sphere vibrating in appropriate modal response. For these cases, exact solutions in spheroidal coordinates are possible (Morse, 1948), which describe the entire acoustic field. For the quadrupole source considered here, the analysis is restricted to the far field (see Section 6.4), where simpler approximate expressions apply.

If two dipole sources are combined, then there will be no net flux of fluid, and no net unbalanced force, but in general there will remain a net shear force on the fluid. In the extreme, the net shear force may reduce to a local stress on the fluid. Again it is the time rate of change of shear force or stress that is important in producing sound. As fluids can be expected to support such forces poorly, it will come as no surprise that quadrupole sources are relatively poor radiators of sound. However, they do play a dominant part in the mixing region when a fluid jet is introduced into a quiescent atmosphere.

Referring to Figure 5.4, where the coordinates are indicated, it may be shown in the same way as was done for the dipole that the potential function of the generalised quadrupole has the following form:

$$\varphi = \frac{4hL}{r} f''(ct - r) H(\theta, \psi, \alpha) \tag{5.56}$$

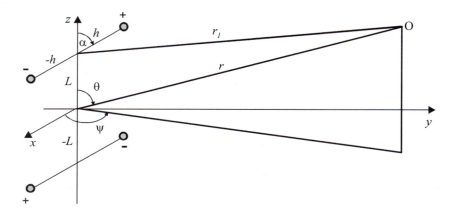

Figure 5.4 Geometry of a quadrupole source.

where

$$H(\theta, \psi, \alpha) = \cos\theta\,(\sin\theta\,\sin\psi\,\sin\alpha + \cos\theta\,\cos\alpha) \tag{5.57}$$

Expressions for the acoustic pressure, p, and particle velocity, u, in the far field immediately follow from Equation (5.56), using Equations (1.10) and (1.11):

$$p = \frac{4hL\rho c}{r} f'''(ct - r)\,H(\theta, \psi, \alpha) \tag{5.58}$$

$$u = \frac{4hL}{r^2}\left[f''(ct - r) + rf'''(ct - r)\right]H(\theta, \psi, \alpha) \tag{5.59}$$

An expression for the near field acoustic pressure may be derived following the procedure outlined by Morse and Ingard (1968, p313) and remembering the positive time dependence used in this text vs the negative time dependence used by Morse and Ingard. The result for the near field pressure is:

$$p = -4j\frac{k\rho cQ_0Lh\cos\theta}{\sqrt{1+k^2a^2}}\,[\frac{\partial}{\partial r}G(\mathbf{r},\mathbf{r}_0,\omega)]_{r_0=0}[\sin\psi\cos\theta + \sin\psi\sin\theta + \cos\psi]e^{j\omega t} \tag{5.60}$$

where $r = |\mathbf{r} - \mathbf{r}_0|$.

Expressions for the intensity, I_q, radiated power, W_q, and alternative expressions for the mean square acoustic pressure $\langle p^2\rangle$ are, respectively:

$$I_q = \rho c\left[\frac{k^3hLQ_{rms}}{\pi r}\right]^2\frac{[H(\theta, \psi, \alpha)]^2}{(1 + k^2a^2)} \tag{5.61}$$

$$W_q = \rho c\frac{(2k^3hLQ_{rms})^2}{5\pi(1 + k^2a^2)}\left(\cos^2\alpha + \frac{1}{3}\sin^2\alpha\right) \tag{5.62}$$

$$\langle p^2\rangle = \left[\rho c\frac{k^3hLQ_{rms}}{\pi r}\right]^2\frac{[H(\theta, \psi, \alpha)]^2}{(1 + k^2a^2)} \tag{5.63}$$

or

$$\langle p^2\rangle = \rho c\frac{5W_q[H(\theta, \psi, \alpha)]^2}{4\pi r^2\left(\cos^2\alpha + \frac{1}{3}\sin^2\alpha\right)} \tag{5.64}$$

5.4.1 Lateral Quadrupole

If $\alpha = \pi/2$ in Equations (5.57) to (5.64), a lateral quadrupole is obtained. This kind of quadrupole characterises the jet-mixing process. In this case the radiated power and mean square acoustic pressure are:

$$W_{lat} = \rho c \frac{[2k^3 h L Q_{rms}]^2}{15\pi(1 + k^2 a^2)} \tag{5.65}$$

$$\langle p_{lat}^2 \rangle = \left[\rho c \frac{k^3 h L Q_{rms} \sin2\theta \sin\psi}{2\pi r \sqrt{(1 + k^2 a^2)}} \right]^2 \tag{5.66}$$

or

$$\langle p_{lat}^2 \rangle = \rho c \frac{15 W_{lat} \sin^2 2\theta \sin^2\psi}{16\pi r^2} \tag{5.67}$$

5.4.2 Longitudinal Quadrupole

If $\alpha = 0$ in Equations (5.57) to (5.64), a longitudinal (or axial) quadrupole is obtained. In this case the radiated power is:

$$W_{long} = \rho c \frac{[2k^3 h L Q_{rms}]^2}{5\pi(1 + k^2 a^2)} \tag{5.68}$$

and the radiated mean square acoustic pressure is:

$$\langle p_{long}^2 \rangle = 5\rho c \frac{W_{long} \cos^4\theta}{4\pi r^2} \tag{5.69}$$

5.4.3 Fluid Mechanical Quadrupole Source

To investigate the parametric dependence of the radiated sound power on the parameters characteristic of a jet, the following assumptions and approximations will be introduced in Equation (5.64). Referring to the latter equation, the source flux Q may be assumed to be proportional to the mean stream speed U and the stream cross-sectional area S; thus, $Q \propto US$, while the stream cross-sectional area, S, in turn will be assumed proportional to the square of either characteristic source dimension, L or h, which in turn will be assumed equal. The wavenumber k, which equals the angular frequency ω divided by the speed of sound c, must be proportional to the local mean flow speed U divided by a characteristic length. The characteristic length squared, in turn, may be considered proportional to the square of either source dimension, L or h. These ideas are summarised as follows:

$$k \propto U/(ch), \quad Q \propto Uh^2, \quad h \approx L \tag{5.70a - c}$$

Substitution of the latter expressions in either Equation (5.65) or (5.68) and introduction of the Mach number $M = U/c$ gives the following results:

$$W \propto S\rho U^8 / c^5 \propto S\rho U^3 M^5 \tag{5.71a,b}$$

Equation (5.71a) shows that the sound power of a quadrupole source in a fluid stream tends to increase with the eighth power of the fluid stream speed. This is Lighthill's famous prediction for jet noise (Lighthill, 1952). Equation (5.71b) shows that the radiated sound power is proportional to the stream power, $S\rho U^3$ and the fifth power of the Mach number M. As in the case of the monopole source, the Mach number term may be interpreted as an efficiency factor for values of Mach number less than one. This interpretation shows that the quadrupole source is much less efficient in converting stream power to noise than the dipole or the monopole source. On the other hand, the efficiency of noise generation of the quadrupole increases much more rapidly with flow speed than the dipole or the monopole. Consequently, what might have been a negligible source of noise may suddenly become a very important source with only a relatively small increase in flow speed.

5.5 LINE SOURCE

5.5.1 Infinite Line Source

Consider a row of incoherent sources in free space away from reflecting surfaces, each radiating a sound of power W and spaced a distance b apart from one another, as shown in Figure 5.5. Since the sources are assumed to be incoherent, that is of random phases, the mean square sound pressure of each will add to give the total mean square sound pressure that will be measured at an observation point at normal distance r_0

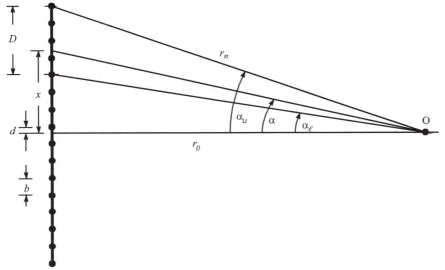

Figure 5.5 Geometry for a line source.

from the row of sources. Let the nearest source be numbered zero, with subsequent sources above and below, as shown in the figure, numbered with positive or negative integer values, n.

Use of Equation (5.14) gives for the mean square sound pressure at the observation point:

$$\langle p^2 \rangle = \rho c \frac{W}{4\pi} \sum_{n=-\infty}^{\infty} [r_0^2 + n^2 b^2]^{-1} \tag{5.72}$$

where

$$r_n^2 = r_0^2 + (nb + d)^2 \tag{5.73}$$

Referring to Figure 5.5, the distance d introduced in Equation (5.72) is less than, or at most equal to, $0.5b$ in magnitude, and takes on a positive value if the zeroeth source lies above the normal, r_0, to the line of sources and a negative value if it lies below. For distance r_0 greater than $3b$, Equation (5.73) reduces to Equation (5.74) below, with an error less than 1 dB. Equation (5.75) is an alternative form of Equation (5.74) from which Equation (5.76) follows without any increase in error:

$$\langle p^2 \rangle = \rho c \frac{W\pi}{4b^2} \left| \frac{b}{\pi r_0} \coth\left[\frac{\pi r_0}{b}\right] \right| \tag{5.74}$$

or,

$$\langle p^2 \rangle \approx \rho c \frac{W}{4br_0}, \qquad r_0 \geq b/2 \text{ (for an error} < 0.5 \text{ dB)} \tag{5.75}$$

According to Equation (5.75), the corresponding sound pressure level at the observation point is:

$$L_p = L_w - 6 - 10\log_{10} r_0 - 10\log_{10} b + 10\log_{10}(\rho c/400) \tag{5.76}$$

where L_w is the sound power level (dB re 10^{-12} W) of each source. At 20°C, where ρc = 414 (SI units), there would be an error of approximately 0.1 dB if the last term in Equation (5.76) is omitted. If r_0 is less than b/π, then the observer distance is close enough to the nearest point source for the sound pressure to be calculated with an error of less than 1 dB by considering this source exclusively. Thus, in this case:

$$\langle p^2 \rangle = \rho c \frac{W}{4\pi r_0^2} \tag{5.77}$$

In the foregoing discussion, the assumption was implicit that because the sources were incoherent they tended not to interfere with each other, so that the calculation could proceed on the assumption that each source radiated independently of the other and the effect at any point of observation of all the sources was obtained as a simple sum of their energy contributions. However, in general the problem can be more complicated, as sources that are coherent do tend to affect one another when placed in close proximity; that is, the energy that a source radiates will depend upon the radiation impedance presented to it and the radiation impedance will be affected by the presence of nearby coherent sources.

Use will again be made of Figure 5.5, where in this case the sources are assumed to be coherent. When a line of coherent point sources is considered, the sound pressure contributions from all sources add at any point of observation. In this case, the phase distribution of the sources along the line, as well as the phase shifts introduced in propagation from the sources to the point of observation, must be taken into account in determining the sum of the contributions. In general, the problem can become quite complicated. However, for the present purpose it will be assumed that all sources radiate in phase and a great simplification is then possible.

When the point of observation r_0 satisfies the condition that $r_0/\lambda \gg 1$, experience suggests that a line of coherent sources radiates like a continuous source of cylindrical waves. Consequently, if the sources are coherent and all radiate in phase with one another, then the mean square sound pressure at an observation point at normal distance r_0 may be calculated assuming that radiation is essentially in the form of cylindrical waves of surface area per unit length $2\pi r_0$. The mean square sound pressure at r_0 is then:

$$\langle p^2 \rangle = \rho c \frac{W}{2\pi b r_0} \tag{5.78}$$

The corresponding sound pressure level is given by the following expression:

$$L_p = L_w - 8 - 10\log_{10} r_0 - 10\log_{10} b + 10\log_{10}(\rho c/400) \tag{5.79}$$

It is to be noted that, according to Equations (5.76) and (5.79), the sound pressure level decreases at the rate of 3 dB for each doubling of distance from a line of coherent or incoherent sources, and the coherent sources produce 2 dB less sound pressure level for the same radiated sound power.

If the distance b between sources is allowed to become indefinitely small, while keeping the quotient W/b constant, then Equations (5.76) and (5.79) may still be used, but in this case L_w becomes the source sound power level per unit length of source and the term $10\log_{10} b$ is omitted.

Traffic and long pipes radiating broadband noise are generally modelled as incoherent line sources. Coherent line sources are relatively uncommon, but long pipes radiating single-frequency noise are generally modelled as such.

5.5.2 Finite Line Source

If the number of sources, each spaced b apart from one another is finite and they are arranged in a line of length, D, as illustrated in Figure 5.5, then the summation of Equation (5.65) may be expressed approximately as an integral. In this case the sound power per unit length becomes W/b and Equation (5.77) (for incoherent sources) takes the following form (Rathe, 1969):

$$\langle p^2 \rangle = \int_{x_1}^{x_2} \frac{W}{b} \frac{\rho c}{4 \pi r^2} \, \mathrm{d}x \tag{5.80}$$

Reference to Figure 5.5 allows the following relationships to be determined:

$$x_1 = r_0 \tan \alpha_\ell, \quad x_2 = r_0 \tan \alpha_u, \quad r = r_0 / \cos \alpha,$$
$$x = r_0 \tan \alpha, \quad \text{and} \quad \mathrm{d}x = r_0 \, \mathrm{d}\alpha \sec^2 \alpha$$

Substitution of these quantities into Equation (5.80) gives the following expression for the mean square sound pressure at the observer (for an incoherent line source):

$$\langle p^2 \rangle = \rho c \int_{\alpha_\ell}^{\alpha_u} \frac{W}{b} \frac{\cos^2 \alpha}{4 \pi r_0^2} \, r_0 \sec^2 \alpha \, \mathrm{d}\alpha \tag{5.81}$$

or

$$\langle p^2 \rangle = \rho c \frac{W}{4 \pi r_0 b} [\alpha_u - \alpha_\ell] \tag{5.82}$$

where α_u and α_ℓ (radians) are the upper and lower bounds of the angle subtended by the source at the observation point. Equation (5.82) is valid provided that (1) the number of sources is equal to or greater than three; and (2) the observation distance r_0 is equal to or greater than $(b/2) \cos \alpha_\ell$.

For an infinite number of incoherent point sources within a finite length D, where W_t is the total power radiated by the source, Equation (5.82) becomes:

$$\langle p^2 \rangle = \rho c \frac{W_t}{4 \pi r_0 D} [\alpha_u - \alpha_\ell] \tag{5.83}$$

For a coherent source of finite length, the sound pressure level will be the same as for a finite length incoherent source for large distances from the source compared with the source length ($r_0 > D/2$). Close to a coherent source of finite length D ($r_0 < D/10$ from the source), the sound pressure level (normal to the centre of the source) will be the same as for an infinite coherent source. In between, the sound pressure level will be between the two levels calculated using Equations (5.78) and (5.83) and interpolation can be used.

The preceding analysis is based on the assumption that the line source is radiating into free space away from any reflecting surfaces. If reflecting surfaces are present, then they must be accounted for as described in Sections 5.9 and 5.10.

5.6 PISTON IN AN INFINITE BAFFLE

The classical circular piston source in an infinite rigid baffle has been investigated by many authors; for example, Kinsler *et al.* (1982), Pierce (1981) and Meyer and

Neumann (1972). Other authors have contributed as well and will be mentioned during the course of the following discussion, which takes a different approach to previously published analyses.

5.6.1 Far Field

The circular piston in an infinite rigid baffle, illustrated in Figure 5.6, is of interest because it has relatively simple geometry; it serves conveniently as an introduction to the behaviour of all radiating surfaces; and it can be approximated by a speaker in a wall. The requirement that the baffle be infinite means that edges are far enough removed for diffraction effects originating there to be ignored. Alternatively, the edges of the baffle might be covered with a sound-absorbent material with the same effect. The presence of the baffle implies that all of the sound power radiated by the piston is radiated into the hemispherical half-space bounded by the plane of the baffle. The piston, generally restricted to a circle of radius a, is assumed to vibrate with uniform normal velocity of amplitude U and frequency ω (rad/s).

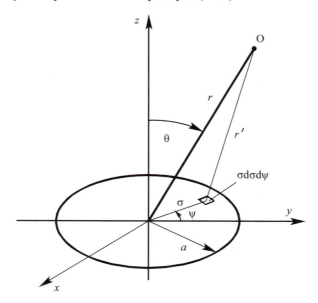

Figure 5.6 A piston source in an infinite baffle. The piston of radius a lies in the x – y plane and vibrates vertically parallel to the z-axis with velocity amplitude U.

A solution for a piston in an infinite baffle will be constructed as an integral of tiny, simple point sources, each of which has a radius of a_i. The analysis begins with recognising that ka is much less than unity and using Equation (5.34) to give:

$$A = Q_0/4\pi \tag{5.84}$$

Substituting Equation (5.84) into Equation (1.48), letting $r = a_i$, and assuming that $ka_i << 1$ gives:

$$u = \frac{(1 + jka_i)}{a_i^2}\frac{Q_0}{4\pi}e^{j(\omega t - ka_i)} \approx \frac{Q_0}{4\pi a_i^2}e^{j(\omega t - ka_i)} \qquad (5.85a,b)$$

As shown by Equation (5.85b), the quantity Q_0 may be interpreted as the strength amplitude or volume velocity amplitude of a source. With this interpretation, substitution of Equation (5.84) into Equation (1.47b) gives an expression for the sound pressure in the acoustic field in terms of the source strength:

$$p = \frac{j\rho c k Q_0}{4\pi r}e^{j(\omega t - kr)} \qquad (5.86)$$

Suppose that the piston in an infinite baffle is composed of a uniform distribution of an infinite number of simple sources such as given by Equation (5.86) and each source has an incremental surface area of $\sigma d\sigma d\psi$ (see Fig. 5.6). The amplitude of the strength of the source corresponding to the incremental area is $Q_0 = 2U\sigma d\sigma d\psi$ where U is the piston velocity amplitude. The factor of 2 is introduced because radiation is only occurring into half space due to the presence of the rigid wall and piston. The effect of the wall can be modelled with an image source so the actual source and its image act together to form a simple source of strength, $Q_0 = 2U\sigma d\sigma d\psi$. It is supposed that radiation from all parts of the surface of the piston are in phase and thus all of the sources are coherent. Thus, the incremental contribution of any source at a point of observation at distance r' is:

$$dp = \frac{j\rho c k U}{2\pi r'}\sigma\, d\sigma\, d\psi\, e^{j(\omega t - kr')} \qquad (5.87)$$

Reference to Figure 5.6 shows that the distance, r', is given by the following expression in terms of the centre-line distance, r.

$$r' = [r^2 + \sigma^2 - 2r\sigma\sin\theta\cos\psi]^{1/2} \qquad (5.88)$$

For the case of the far field where $r \gg a \geq \sigma$, Equation (5.88) becomes approximately:

$$r' \approx r - \sigma\sin\theta\cos\psi \qquad (5.89)$$

Equation (5.89) may be substituted into Equation (5.87), where the further simplification may be made that the denominator is approximated sufficiently by the first term on the right-hand side of Equation (5.89). However, the second term must be retained in the exponent, which reflects the fact that small variations in relative phase of the pressure contributions arriving at the observation point have a very significant effect upon the sum.

Summing contributions given by Equation (5.87) from all points over the surface of the piston to obtain the total pressure p at the far-field observation point O, gives the following integral expression:

$$p = \frac{j\rho ck}{2\pi r} U e^{j(\omega t - kr)} \int_0^a \sigma \, d\sigma \int_0^{2\pi} e^{jk\sigma\sin\theta\cos\psi} \, d\psi \qquad (5.90)$$

Integration of Equation (5.90) begins by noting that the second integral (divided by 2π) is a Bessel Function of the first kind of order zero and argument $k\sigma \sin\theta$ (McLachlan, 1941). Thus, the equation becomes:

$$p = \frac{j\rho ck}{r} U e^{j(\omega t - kr)} \int_0^a \sigma J_0(k\sigma\sin\theta) \, d\sigma \qquad (5.91)$$

Integration of Equation (5.91) gives the following useful result:

$$p = \frac{j\rho ck}{2\pi r} F(w) e^{j(\omega t - kr)} \qquad (5.92)$$

where

$$F(w) = U\pi a^2 [J_0(w) + J_2(w)] = U\pi a^2 \, 2\frac{J_1(w)}{w} \qquad (5.93a,b)$$

$$w = ka \sin\theta \qquad (5.94)$$

and

$$k = \omega/c = 2\pi f/c \qquad (5.95a,b)$$

The quantities $J_0(w)$, $J_1(w)$ and $J_2(w)$ are Bessel functions of the first kind of order 0, 1 and 2 respectively, and argument w. The real part of the pressure given by Equation (5.92) may be written as follows:

$$p = -\frac{\rho\omega}{2\pi r} F(w) \sin(\omega t - kr) \qquad (5.96)$$

The acoustic intensity in the far field is related to the acoustic pressure by Equation (1.74), thus:

$$I = \frac{\rho c k^2}{8\pi^2 r^2} F^2(w) \qquad (5.97)$$

Consideration of Equations (5.92), (5.96) and (5.97) shows that as they are all functions of $F(w)$, and in some directions, θ, there may be nulls in the sound field, according to Equation (5.94). For example, reference to Figure 5.7, where $F(w)/U\pi a^2$ is plotted as a function of w, shows that for w equal to 3.83, 7.0, etc. the function F is zero. On the other hand, for values of w less than 3.83 the function, F, becomes large. Generally, the sound tends to beam on-axis and in appropriate cases to exhibit side lobes.

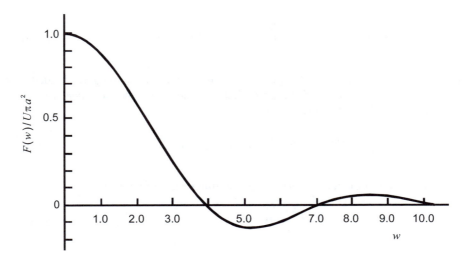

Figure 5.7 The far field directivity function for a piston source in an infinite baffle. The piston has a radius of a and vibrates with a velocity amplitude U. The abscissa is $w = ka \sin \theta$, where k is the wave number and θ is the angle of observation relative to the piston axis (see Figure 5.5).

In the far field, the piston in an infinite baffle is quite directional, and a directivity index, as discussed in the Section 5.8, can be defined. In the near field, by contrast, a directivity index may make little sense because of the presence of local maxima and minima in the pressure field.

5.6.2 Near Field On-axis

In general the near field of the piston source is quite irregular, being characterised by a complicated pattern of local pressure maxima and minima (Meyer and Neumann, 1972). Consequently, no attempt will be made here to calculate the near field except along the axis of symmetry where the integration is simple and the result is instructive. Referring to Figure 5.6, let the point of observation lie at distance r on the axis of symmetry, z. In this case, set $\theta = 0$ in Equation (5.88) and differentiate giving the following result:

$$r'\mathrm{d}r' = \sigma \mathrm{d}\sigma \tag{5.98}$$

Substitution of Equation (5.98) in Equation (5.87) and expressing the pressure on-axis at a distance, r, from the piston of radius, a, as an integral of the pressure contributed by all points on the surface of the piston, gives the following result:

$$p = j\frac{\rho c k U}{2\pi} \int_{0}^{2\pi} \mathrm{d}\psi \int_{r}^{\sqrt{r^2 + a^2}} e^{j(\omega t - kr')} \, \mathrm{d}r' \tag{5.99}$$

Carrying out the integration of Equation (5.99) gives the following expression for the pressure on-axis in the near field of the piston:

$$p = -\rho c U \left[e^{j(\omega t - k\sqrt{r^2 + a^2})} - e^{j(\omega t - kr)} \right] \tag{5.100}$$

Introducing the following identity:

$$e^{-j\alpha} - e^{-j\beta} = -j2 \sin \frac{\alpha - \beta}{2} e^{-j(\alpha + \beta)/2} \tag{5.101}$$

allows Equation (5.100) to be rewritten as follows:

$$p = j2\rho c U e^{j\omega t} \sin\left[\frac{k}{2}\left(\sqrt{r^2 + a^2} - r\right)\right] e^{-j\frac{k}{2}(\sqrt{r^2 + a^2} + r)} \tag{5.102}$$

Equation (5.102) shows that the pressure amplitude has zeroes (nulls) on-axis for values of axial distance r which satisfy the following condition, where n is the number of nulls, counting from the furthest toward the surface of the piston:

$$r = \frac{(a/\lambda)^2 - n^2}{2n/\lambda} \tag{5.103}$$

Equation (5.103) shows that the number of nulls, n, on-axis is bounded by the size of the ratio of piston radius to wavelength, a/λ, because only positive values of distance r are admissible. The equation also shows that there can only be nulls on-axis when $a/\lambda > 1$.

Letting r be large, allows the argument of the sine function of Equation (5.102) to be expressed in series form as in the following equation.

$$\frac{k}{2}\left(\sqrt{r^2 + a^2} - r\right) = \frac{kr}{2}\left[1 + \frac{1}{2}\left(\frac{a}{r}\right)^2 + ... - 1\right] \approx \frac{ka^2}{4r} \tag{5.104a,b}$$

Replacing the sine function in Equation (5.102) with its argument given by Equation (5.104b) gives the following equation, valid for large distances, r:

$$p = j\rho c \frac{Uka^2}{2r} e^{j\left(\omega t - \frac{k}{2}\left(\sqrt{r^2 + a^2} + r\right)\right)} \tag{5.105}$$

In the far field of a source, the sound pressure decreases inversely with distance r; thus, Equation (5.105) describes the sound field in the far field of the piston source.

The sound field in front of the piston which is characterised by the presence of maxima and minima is called the geometric near field. The bound between the far field and the geometric near field cannot be defined precisely, but may be taken as lying beyond the first null on-axis furthest from the surface of the piston (see Equation (5.103)). Thus setting $n = 1$ in Equation (5.103) readily leads to the following condition for a point of observation to lie in the far field (see Section 6.4):

$$r \gg a^2/\lambda \tag{5.106}$$

5.6.3 Radiation Load of the Near Field

As mentioned earlier, the near field of a piston source can be quite complicated, but it is possible to compute rather simply the load that the acoustic field imposes on the vibrating piston, and this information is often quite useful. For example, the net time-varying force divided by the corresponding time-varying piston velocity can be computed and this quantity, known as the radiation impedance, is useful, as will be shown in Section 6.7.

Integration of Equation (5.87) over the surface of the piston gives the pressure p_s acting on surface element $\mathrm{d}S$ and further integration of this pressure over the surface gives the force on the piston. Referring to Figure 5.8 for definition of the coordinates, the following integral expression for the force on the piston is obtained:

$$F_R = \iint_S p_s \, \mathrm{d}S' = \frac{j\rho c k U e^{j\omega t}}{2\pi} \int_0^a r \, \mathrm{d}r \int_0^{2\pi} \mathrm{d}\varphi \int_0^a \sigma \, \mathrm{d}\sigma \int_0^{2\pi} \frac{e^{-jkh}}{h} \, \mathrm{d}\psi \tag{5.107}$$

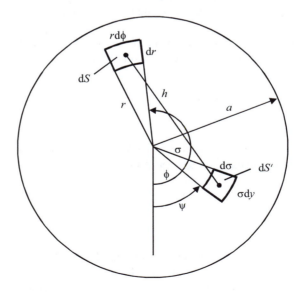

Figure 5.8 Piston coordinates and arrangement for calculating the radiation load.

Carrying out the integration and dividing by the piston velocity $Ue^{j\omega t}$ gives the following expression for the radiation impedance:

$$Z_R = \frac{F_R}{Ue^{j\omega t}} = \rho c \pi a^2 [R_R(2ka) + jX_R(2ka)] \qquad (5.108a,b)$$

The radiation impedance is a complex quantity with a real part, $\rho c \pi a^2 R_R$, and a mass reactive part, $\rho c \pi a^2 X_R$, where $z = 2ka$ and both R_R and X_R are defined by the following equations:

$$R_R(z) = \frac{z^2}{2 \times 4} - \frac{z^4}{2 \times 4^2 \times 6} + \frac{z^6}{2 \times 4^2 \times 6^2 \times 8} - \cdots \qquad (5.109)$$

$$X_R(z) = \frac{4}{\pi} \left[\frac{z}{3} - \frac{z^3}{3^2 \times 5} + \frac{z^5}{3^2 \times 5^2 \times 7} - \cdots \right] \qquad (5.110)$$

The real part, R_R may be used to calculate the radiated power, given the piston velocity; that is:

$$W = \frac{1}{2} \text{Re}\{Z_R\} U^2 = \frac{1}{2} \rho c R_R \pi a^2 U^2 \qquad (5.111a,b)$$

Values for the real term R_R and mass reactance term X_R are plotted in Figure 5.9 for the piston mounted in an infinite baffle (Equations (5.109) and (5.110)) and for comparison the analytically more complex case of a piston mounted in the end of a tube (Levine and Schwinger, 1948). Reference to the figure shows that the mass reactance term in both cases passes through a maximum, then tends to zero with increasing frequency, while the real term approaches and then remains approximately unity with increasing frequency. The figure shows that the effect of the baffle at low frequencies is to double the real term and increase the mass reactance by $\sqrt{2}$ while at frequencies above $ka = 5$, the effect of the baffle is negligible.

Figure 5.9 also shows that, in the limit of high frequencies, the radiation impedance of the piston becomes the same as that of a plane wave, and although the pressure distribution over the surface of the piston is by no means uniform, the power radiated is the same as would be radiated by a section of a uniform plane wave of area πa^2 (Meyer and Neumann, 1972).

The behaviour of the radiation impedance shown in Figure 5.9 is typical of all sound radiators. For example, the open end of a duct will have a similar radiation impedance, with the result that at low frequencies a sound wave propagating within the duct will be reflected back into the duct as though it were reflected from the mass reactance at the free end. Resonance in a musical instrument is achieved in this way. However, at high frequencies, where the impedance becomes essentially that of a freely travelling plane wave, there will be little or no reflection at the open end.

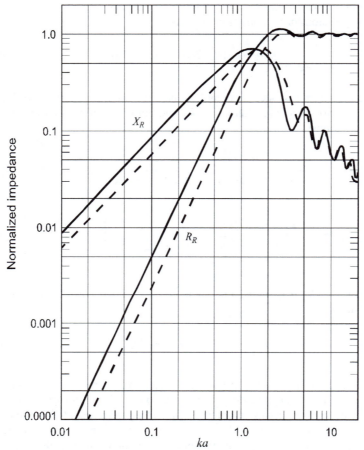

Figure 5.9 Radiation impedance functions for a piston source mounted either in an infinite baffle (solid line) or in the end of a long tube (dashed line). The piston has a radius of a and the impedance has the form: $Z_R = \rho c \pi a^2 [R_R(2ka) + jX_R(2ka)]$.

5.7 INCOHERENT PLANE RADIATOR

5.7.1 Single Wall

The piston in an infinite baffle, which was considered in the previous section, may be considered as an example of a plane distribution of coherent radiators; that is, all parts of the piston vibrate with the same amplitude and phase. By contrast, the wall of a building housing noisy machinery may be modelled as a distribution of elementary incoherent radiators. Openings such as windows or open doors in walls of buildings may be similarly modelled. Consequently it will be of interest to consider the contrasting case of a distribution of elementary incoherent radiators. In all such cases, the contribution at the observer due to reflection in the ground plane must be taken into account separately (see Sections 5.9 and 5.10).

Radiation from a plane distribution of incoherent radiators, as illustrated in Figure 5.10, has been considered by Hohenwarter (1991), who introduced a cosine weighting on the radiation directivity of each elementary source. Note that arbitrary weighting of the directivity of an elementary source is not admissible, because the resulting equation will not necessarily satisfy the wave equation. However, weighting of the energy radiated is admissible as a hypothesis, as the power flow described by the mean square sound pressure is not constrained to satisfy the wave equation. Of course, an analytical solution justifying the proposed hypothesis must satisfy the wave equation, but for the purpose here, the assumption is implicit, following Hohenwarter (1991), that a solution in agreement with observation ensures that an analytical solution exists. The analysis is thus justified on empirical grounds.

Use of the cosine-weighted directivity for each elementary source both greatly simplifies the integration, which is otherwise difficult, and provides an answer in better agreement with observation than an analysis based on unweighted sources. For example, the weighted-source model predicts a finite sound level as one approaches such a distribution (wall of a building), whereas the unweighted model predicts a sound level increasing without bound.

Applying Hohenwarter's cosine weighting function implies that the sound pressure squared at some observer location, O, is obtained by multiplying the squared pressure calculated by assuming that the source radiates uniformly in all directions with the directivity factor, D_i, given by the following equation:

$$D_i = r(x^2 + z^2 + r^2)^{-1/2} = r/r_i \qquad (5.112\text{a,b})$$

The quantities x, z, r and r_i are dependent on the observer location and are defined in Figure 5.10 where a plane incoherent source, centred at distance d along the x-axis from the origin, is represented as L wide and H high in the x - z plane. A general field

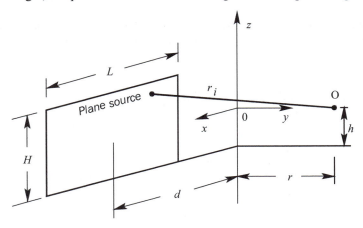

Figure 5.10 Rectangular noise source of length L and height H showing general field observation point, O, and associated coordinates.

observation point in the y - z plane is shown at distance r from the origin and the plane of the wall. The plane source is assumed to be mounted in a large baffle much greater than a wavelength in dimension so that sound is effectively excluded from the rear side.

Using Equation (5.14), including the cosine weighted directivity factor D_i, and replacing the "4" with a "2" to account for radiation only into half space, the mean square sound pressure at the observer location O due to elementary source i located at a distance r_i is:

$$\langle p_i^2 \rangle = \frac{W_i \rho c D_i}{2 \pi r_i^2} \tag{5.113}$$

The sound power W_i radiated by an elementary source i is defined as:

$$W_i = \frac{W \, dx \, dz}{HL} \tag{5.114}$$

where W is the total power radiated by the plane source. Substituting Equations (5.114) and (5.112) for W_i and D_i respectively, into Equation (5.113) gives:

$$\langle p_i^2 \rangle = \frac{\rho c W}{2 \pi HL} \frac{r}{r_i^3} dx \, dz = \frac{\rho c W}{2 \pi HL} [x^2 + z^2 + r^2]^{-3/2} r \, dx \, dz \tag{5.115a,b}$$

Summing over the area of the wall gives the following integral expression for the total mean square sound pressure at the observation point:

$$\langle p^2 \rangle = \frac{\rho c W}{2 \pi HL} \int_{-h}^{H-h} dz \int_{d-L/2}^{d+L/2} [x^2 + z^2 + r^2]^{-3/2} r \, dx \tag{5.116}$$

Integration of the above expression gives the following result where the dimensionless variables that have been introduced are; $\alpha = H/L$, $\beta = h/L$, $\gamma = r/L$ and $\delta = d/L$.

$$\langle p^2 \rangle = \frac{\rho c W}{2 \pi HL} \left[\tan^{-1} \frac{(\alpha - \beta)(\delta + 1/2)}{\gamma \sqrt{(\alpha - \beta)^2 + (\delta + 1/2)^2 + \gamma^2}} \right.$$

$$+ \tan^{-1} \frac{\beta(\delta + 1/2)}{\gamma \sqrt{\beta^2 + (\delta + 1/2)^2 + \gamma^2}} - \tan^{-1} \frac{(\alpha - \beta)(\delta - 1/2)}{\gamma \sqrt{(\alpha - \beta)^2 + (\delta - 1/2)^2 + \gamma^2}} \tag{5.117}$$

$$\left. - \tan^{-1} \frac{\beta(\delta - 1/2)}{\gamma \sqrt{\beta^2 + (\delta - 1/2)^2 + \gamma^2}} \right]$$

As there are no special assumptions that restrict the use of the above equation, it may be used to predict the mean square sound pressure at any point in the field away from the surface of the plane source. For further discussion, see Hohenwarter (1991).

The mean square sound pressure on the axis of symmetry, $\delta = 0$, $\beta = 0.5\alpha$, is also the maximum mean square sound pressure in a plane parallel to the plane of the source, determined by making γ a constant value. Equation (5.117) takes the following form on the axis of symmetry:

$$\langle p^2 \rangle = \frac{2\rho c W}{\pi HL} \tan^{-1}\left[\frac{HL}{2r\sqrt{H^2 + L^2 + 4r^2}}\right] \tag{5.118}$$

At large distances from the source, so that r is very much larger than the dimensions H and L, Equation (5.118) reduces to the following approximate form:

$$\langle p^2 \rangle = \frac{\rho c W}{2\pi r^2} \tag{5.119}$$

Equation (5.119) shows that at large distances from the source, the mean square sound pressure is solely a function of the total radiated sound power and the distance from the plane of the source to the observer, r, and not of any of the other variables. The form of Equation (5.119) is the same as that of a simple monopole source radiating into half space (see Equation (5.14)) so that r may be taken as the distance from the centre of the source to the observation point and the source may be treated like a point source.

Close to the wall, as r tends to zero, Equation (5.118) leads to the following asymptotic form, which is consistent with the expectation based upon the assumed model:

$$\langle p^2 \rangle = \frac{\rho c W}{HL} \tag{5.120}$$

Equations (5.118) and (5.119) have been used to construct Figure 5.11, which shows the difference one might expect in sound pressure level at a particular distance, r, from the source when comparing the field levels due to a point and plane source of the same power output radiating into a hemispherical half-space. Close to the source the sound pressure level due to the point source will be greater, whereas at large distances, where the ratio $r/\sqrt{HL}$ is small, the sound pressure levels due to the two sources will be similar and will decrease at the rate of 6 dB per doubling of distance from the source.

The side of a building which houses noisy machinery, or an open window, is often modelled as an incoherent plane radiator. Although this is not strictly correct, experimental data show that acceptable results are obtained when one-third octave or wider frequency bands are used for the analysis. The sound power radiated by a wall can be calculated in one of two ways. The first method uses calculated values for the interior sound pressure level and measured wall noise reduction properties as

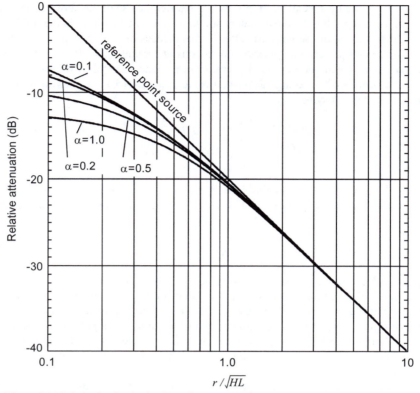

Figure 5.11 Relative levels of point (Equation 5.111) and plane (Equation 5.110) sources on the axis of symmetry. The plane source is of length L and height H and the observation point is at normal distance r from the centre of the plane source. If $H < L$, then $\alpha = H/L$, otherwise if $H > L$, then $\alpha = L/H$.

described in Section 8.2.2. The second method uses measured wall vibration levels, as described in Section 6.7.

5.7.2 Several Walls of a Building or Enclosure

The sound pressure level at a specified distance, r, from a building due to radiation by the walls and roof is calculated, in practice, by assuming that the sound energy is uniformly distributed over the area of an encompassing hemispherical surface of radius, r. Thus:

$$L_p = L_{wt} - 10 \log_{10} S + 10 \log_{10} \frac{\rho c}{400} \qquad (5.121)$$

where S is the surface area of the hemisphere of radius r, and L_{wt} is the total radiated sound power level, calculated on the basis of the total sound power radiated by all of the exterior walls and the roof of the building. The origin of the last term in the equation can be explained by reference to Equation (1.87). It is usually less than 0.2 dB and is often ignored. The sound propagation effects discussed in Section 5.11 would also have to be included in practice. The effect of differences in power radiated by the walls and roof is taken into account by adding a correction to the sound pressure level, L_p, radiated in each direction. For a rectangular building where five directions are of concern, the correction is the difference between L_{wi} and L_{wt} - 7 where L_{wi} is the sound power level radiated by the *i*th surface. This leads to the introduction of the concept of source directivity, which is discussed in the next section. Sound radiation from buildings or machine enclosures is discussed in Section 8.4.2.

5.8 DIRECTIVITY

The near field of most sources is characterised by local maxima and minima in sound pressure (see Sections 5.6.2 and 6.4 for discussion) and consequently the near field cannot be characterised in any unique way as solely a function of direction. However, in the far field the sound pressure will decrease with spreading at the rate of 6 dB per doubling of distance and in this field, a directivity index may be defined that describes the field in a unique way as a function solely of direction.

The simple point source radiates uniformly in all directions. In general, however, the radiation of sound from any source is usually directional, being greater in some directions than in others. In the far field (see Section 6.4), the directional properties of a sound source may be quantified by the introduction of a directivity factor describing the angular dependence of the sound intensity. For example, if the intensity I of Equation (1.79) is dependent upon direction, then the mean intensity, $\langle I \rangle$, averaged over an encompassing spherical surface is introduced and, according to Equation (1.80):

$$\langle I \rangle = \frac{W}{4\pi r^2} \tag{5.122}$$

The directivity factor, D_θ, is defined in terms of the intensity I_θ in direction *(θ, ψ)* and the mean intensity:

$$D_\theta = \frac{I_\theta}{\langle I \rangle} \tag{5.123}$$

The directivity index is defined as:

$$DI = 10\log_{10} D_\theta \tag{5.124}$$

Alternatively, making use of Equations (5.122) and (5.123):

$$DI = 10\log_{10} I_\theta - 10\log_{10} W + 10\log_{10} 4\pi r^2 \tag{5.125}$$

In general, the directivity index is determined by measurement of the intensity I_θ at distance r and angular orientation (θ, ψ) from the source centre. Alternatively, the sound pressure level, L_p, may be measured instead of intensity where Equations (1.74), (1.82) and (1.84) have been used to rewrite Equation (5.125) in the following useful form:

$$DI = L_p - L_w + 20\log_{10}r - 10\log_{10}(\rho c/400) + 11 \qquad \text{(dB)} \qquad (5.126)$$

5.9 REFLECTION EFFECTS

The presence of a reflecting surface near to a source will affect the sound radiated and the apparent directional properties of the source. Similarly, the presence of a reflecting surface near to a receiver will affect the sound received by the receiver. In general, a reflecting surface will affect not only the directional properties of a source but also the total power radiated by the source (Bies, 1961). As the problem can be complicated, the simplifying assumption is often made, and will be made here, that the source is of constant power output; thus only the case of constant power sources will be considered in the following sections. Other source types are discussed in Section 6.2.

5.9.1 Simple Source Near a Reflecting Surface

The concept of directivity may be used to describe the radiation from a simple source in the proximity of one or more bounding planes when it may be assumed that:

- the distance between the source and the reflecting plane is small compared with the distance from the source to the observation point;
- the distance between the source and the reflecting plane is less than or of the order of one tenth of the wavelength of sound radiated; and
- the sound power of the source may be assumed to be constant and unaffected by the presence of the reflecting plane.

For example, a simple source on the ground plane or next to a wall will radiate into the resulting half-space. As the sound power of a simple source may be assumed to be constant then this case may be represented by modifying Equation (5.14) and using Equation (1.74) to give:

$$W = I\frac{4\pi r^2}{D} = \langle p^2\rangle\frac{4\pi r^2}{\rho c D} \qquad (5.127a,b)$$

The intensity, I, is independent of angle in the restricted region of propagation, and the directivity factor, D, takes the value listed in Table 5.1. For example, the value of D for the case of a simple source next to a reflecting wall is 2, showing that all of the sound power is radiated into the half-space defined by the wall.

Table 5.1 Directivity factors for a simple source near reflecting surfaces (a constant power source is assumed – see text)

Situation	Directivity factor, D	Directivity index, DI (dB)
Free space	1	0
Centred in a large flat surface	2	3
Centred at the edge formed by the junction of two large flat surfaces	4	6
At the corner formed by the junction of three large flat surfaces	8	9

If the distance, a, between the source and a reflecting surface is large in comparison, or comparable with the distance, b, between the source and observation point and large in comparison with a wavelength, then the effective sound pressure is calculated by adding the separate intensity contributions due to the source and its image in the reflecting plane, with the image position taken as far behind the reflecting surface as the source is in front, and with the image power equal to the source power (see Figure 5.12).

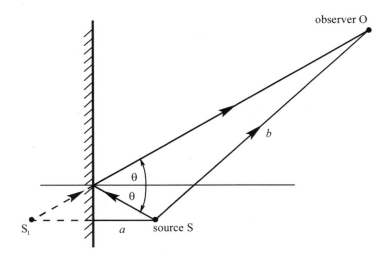

Figure 5.12 Geometry illustrating reflection from a plane rigid surface.

5.9.2 Observer Near a Reflecting Surface

When the observer is located close, of the order of one-tenth of a wavelength or less, to a reflecting surface the sound pressure of the source image (or reflected wave) is added to that at the observer due to the direct sound wave, i.e. pressure doubling occurs with an apparent increase in sound pressure level of 6 dB. On the other hand, when the observer is located further than a tenth wavelength from a reflecting surface then the path difference between the direct and reflected waves is usually sufficiently large for most practical sound sources radiating non-tonal noise for the two waves to combine with random phase; the squared pressures, $\langle p^2 \rangle$, will add with an apparent increase in sound pressure level of 3 dB.

If the noise is tonal in nature then sound pressures of the direct and reflected waves must be added taking into account the phase shift on reflection and relative phase shift due to the differences in lengths of the two propagation paths.

5.9.3 Observer and Source Both Close to a Reflecting Surface

For the case where the source and observer are located close to the same reflecting plane, and their distances h_s and h_r from the plane are small compared to the distance L between them, the path difference between the direct and reflected waves is relatively small. In this case, if there is no turbulence or there are no temperature gradients in the medium in which the wave is propagating and the real part of the characteristic impedance of the reflecting plane, Z_s, satisfies the relation $\mathrm{Re}\{Z_s\}/\rho c > L/(h_s + h_r)$, the addition is coherent; that is, addition of pressures, not intensities. Coherent addition of source and image results in a 6 dB rather than a 3 dB increase in sound pressure level over that for a free field. In this case only, all directivity factors shown in Table 5.1 should be multiplied by two. For the case where the impedance of the reflecting plane does not satisfy the preceding relationship, reference should be made to Section 5.11.11.

5.10 REFLECTION AND TRANSMISSION AT A PLANE / TWO MEDIA INTERFACE

When sound is transmitted over a porous surface such as the surface of the earth, sound energy is absorbed and the direction of sound wave propagation near the surface is affected. The effect is most accurately modelled assuming the reflection of a spherical wave at a plane reflecting surface. Unfortunately, the associated calculations are very complicated. However, the resulting expression for the spherical wave amplitude reflection coefficient may be expressed in terms of the much simpler plane wave amplitude reflection coefficient (see Equation (5.146). Furthermore, the spherical wave amplitude reflection coefficient reduces to the plane wave amplitude reflection coefficient at great distance from the source, as the spherical wave tends to a plane wave. Consequently, it will be convenient to consider first the reflection of a plane wave at a plane interface between two media.

In the literature one of three assumptions is commonly made, often without comment, when considering reflection of sound at an interface between two media, for example, at the surface of the earth. Either it is assumed that the second medium is locally reactive, so that the response of any point on the surface is independent of the response at any other point in the second medium; or it is assumed that the surface of the second medium is modally reactive, where the response of any point on the surface is dependent on the response of all other points on the surface of the second medium. Alternatively, in media in which the sound wave attenuates as it propagates, the response at any point on the surface will depend only on the response of nearby points, depending on the extent of attenuation, and not all other points. In this latter case the surface will be referred to as extensively reactive or as a case of extended reaction. A criterion given by Equation (5.160) for determining how a porous surface, for example the earth, should be treated is discussed in Section 5.10.3.

5.10.1 Porous Earth

When one of the media, such as the earth, is described as porous and the other medium is a gas, such as air, which penetrates the pores of the porous medium, then the term "porous" has the special meaning that sound is transmitted through the pores and not the structure, which is generally far less resilient than the gas in the pores. In such a case, the acoustic properties associated with a porous medium are determined by the combined properties of a rigid gas-filled structure, which may be replaced with a fictitious gas of prescribed properties for the purposes of analysis, as described in Appendix C.

For the case of the earth, which is well modelled as a porous medium (here indicated by subscript m), the characteristic impedance, Z_m and propagation constant k_m (both complex), may be calculated from a knowledge of the earth surface flow resistivity, R_1 in MKS rayls/m as described in Appendix C. Values of flow resistivity, R_1, for various ground surfaces are given in Table 5.2. In Appendix C it is shown that both the characteristic impedance and the propagation constant may be expressed as functions of the dimensionless scaling parameter $\rho f / R_1$, where ρ is the density of the gas in the pores and f is the frequency of the sound considered.

In general, a wavenumber (or propagation constant) may be complex where the real part is associated with the wave speed and the imaginary part is associated with the rate of sound propagation loss. When propagation loss is negligible, the wavenumber takes the form given by Equation (1.24). Alternatively, when sound propagation loss is not negligible, as in the case of propagation in a porous medium, the wavenumber k_m takes the form $k_m = \omega / c_m - j\alpha_m$, where c_m is the wave speed in the porous medium and α_m is the propagation loss factor (see Appendix C).

5.10.2 Plane Wave Reflection and Transmission

The reflection and transmission of a plane sound wave at a plane interface between two media will be considered. As illustrated in Figure 5.13 the interface is assumed to be flat and the incident, reflected and transmitted waves are assumed to be plane. The plane interface is assumed to lie along the abscissa at $y = 0$ and the angles of incidence and reflection, θ, and transmission, ψ, are measured from the normal to the plane of the interface.

Table 5.2 Flow resistivity values for typical ground surfaces (Embleton *et al.*, 1983)

Ground surface	Flow resistivity, R_1 (MKS rayls/m (Pa s m^{-2}))
Dry snow, new fallen 0.1 m over 0.4 m older snow	$10^4 - 3 \times 10^4$
Sugar snow	$2.5 \times 10^4 - 5 \times 10^4$
In forest; pine or hemlock	$2 \times 10^4 - 8 \times 10^4$
Grass; rough pasture, airport	$1.5 \times 10^5 - 3 \times 10^5$
Roadside dirt, small rocks up to 0.1 m mesh size	$3 \times 10^5 - 8 \times 10^5$
Sandy silt, hard packed by vehicles	$8 \times 10^5 - 2.5 \times 10^6$
Thick layer of limestone chips, 0.01 to 0.025 m mesh	$1.5 \times 10^6 - 4 \times 10^6$
Old dirt roadway, fine stones (0.05 m mesh), interstices filled	$2 \times 10^6 - 4 \times 10^6$
Earth, exposed and rain packed	$4 \times 10^6 - 8 \times 10^6$
Quarry dust, fine and hard packed by vehicles	$5 \times 10^6 - 2 \times 10^7$
Asphalt, sealed by dust and light use	3×10^7

The coordinates r_I, r_R, and r_T indicate directions of wave travel and progress of the incident, reflected and transmitted waves, respectively. Medium 1 lies above and medium 2 lies below the x-axis. The latter media extend away from the interface an infinite distance and have characteristic impedances, Z_1 and Z_2, and propagation constants (complex wavenumbers) k_1 and k_2 any or all of which may be complex. For the infinitely extending media, the characteristic impedances are equal to the normal impedances, Z_{N1} and Z_{N2}, respectively, at the interface.

Referring to Figure 5.13 the component propagation constants are defined as follows:

$$k_{1x} = k_1 \sin\theta, \quad k_{1y} = k_1 \cos\theta \qquad (5.128\text{a,b})$$

$$k_{2x} = k_2 \sin\psi, \quad k_{2y} = k_2 \cos\psi \qquad (5.129\text{a,b})$$

The time dependent term $e^{j\omega t}$ may be suppressed, allowing the sound pressure of the propagating incident wave of amplitude A_I to be written as follows:

$$p_I = A_I e^{-jk_1 r_I} \qquad (5.130)$$

Reference to Figure 5.13 shows that the y component of the incident and transmitted waves travels in the negative direction and on reflection in the positive direction whereas the x component travels in the positive direction in all cases. Taking note of these observations, multiplying $k_1 r_I$ by 1, where $1 = \cos^2\theta + \sin^2\theta$, and using Equations (5.128a,b) allows Equation (5.130) to be rewritten as follows:

$$p_I = A_I e^{-j(k_{1x} x - k_{1y} y)} \qquad (5.131)$$

As may readily be shown, the angle of reflection must be equal to the angle of incidence. Thus, expressions for the sound pressure of the propagating, reflected and transmitted waves may be written as follows:

$$p_R = A_R e^{-j(k_{1x}x + k_{1y}y)} \tag{5.132}$$

and

$$p_T = A_T e^{-j(k_{2x}x - k_{2y}y)} \tag{5.133}$$

Continuity of pressure at the interface requires that at $y = 0$:

$$p_I + p_R = p_T \tag{5.134}$$

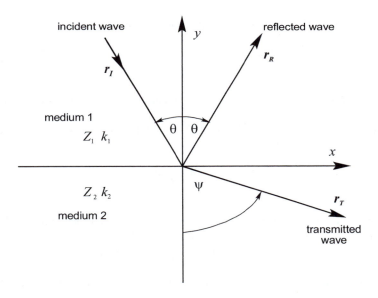

Figure 5.13 Geometry illustrating reflection and transmission at the plane interface between two acoustic media.

Substitution of Equations (5.131), (5.132), and (5.133) into Equation (5.134) gives the following result:

$$A_I e^{-jk_{1x}x} + A_R e^{-jk_{1x}x} = A_T e^{-jk_{2x}x} \tag{5.135}$$

Equation (5.135) must be true for all x; thus it must be true for $x = 0$ which leads to the following conclusion:

$$A_I + A_R = A_T \tag{5.136}$$

Substitution of Equation (5.136) in Equation (5.135) gives the following result:

$$(A_I + A_R)(e^{-jk_{1x}x} - e^{-jk_{2x}x}) = 0 \tag{5.137}$$

The amplitudes A_I and A_R may be positive and non-zero; thus, the second term of Equation (5.135) must be zero for all values of x. It may be concluded that:

$$k_{1x} = k_{2x} \tag{5.138}$$

Using Equations (5.128a) and (5.129b) and introducing the index of refraction $n = c_2/c_1$, Equation (5.138) becomes Snell's law of refraction:

$$\frac{k_1}{k_2} = n = \frac{\sin\psi}{\sin\theta} \tag{5.139a,b}$$

Continuity of particle velocity at the interface requires that at $y = 0$:

$$\frac{p_I - p_R}{Z_{N1}}\cos\theta = \frac{p_T}{Z_{N2}}\cos\psi \tag{5.140}$$

where Z_{N1} and Z_{N2} are the specific normal impedances at the surfaces of media 1 and 2 respectively.

Substitution of Equations (5.131), (5.132) and (5.133) in Equation (5.140) leads directly to the following result:

$$A_I\left[\frac{\cos\theta}{Z_{N1}} - \frac{\cos\psi}{Z_{N2}}\right] = A_R\left[\frac{\cos\theta}{Z_{N1}} + \frac{\cos\psi}{Z_{N2}}\right] \tag{5.141}$$

For later reference when considering reflection from the surface of the earth and without limiting the generality of the equations it will be convenient to consider medium 1 as air of infinite extent having a normal impedance of Z_{N1} (at the earth interface) equal to its characteristic impedance ρc, and propagation constant $k_1 = k = \omega/c$. Similarly the porous earth will be considered as extending infinitely in depth and having a normal impedance of Z_{N2} (at the earth interface) equal to its characteristic impedance, Z_m, and propagation constant, $k_2 = k_m$. On making the indicated substitutions, the complex amplitude reflection coefficient for plane waves, $A_R/A_I = R_p$, may be written as follows:

$$R_p = \frac{Z_{N2}\cos\theta - Z_{N1}\cos\psi}{Z_{N2}\cos\theta + Z_{N1}\cos\psi} = \frac{Z_m\cos\theta - \rho c\cos\psi}{Z_m\cos\theta + \rho c\cos\psi} \tag{5.142a,b}$$

Equation (5.142a) is valid for media of any extent, while Equation (5.142b) only applies to infinitely extending media or media that extend for a sufficient distance that waves reflected from any termination back towards the interface have negligible amplitude on arrival at the interface.

Equation (5.139) may be used to show that:

$$\cos\psi = \sqrt{1 - \left(\frac{k}{k_m}\right)^2 \sin^2\theta} \qquad (5.143)$$

Reference to Equation (5.143) shows that when $k_m = k_2 \gg k_1 = k$, the angle, ψ tends to zero and Equation (5.140b) reduces to the following form:

$$R_p = \frac{Z_{N2}\cos\theta - Z_{N1}}{Z_{N2}\cos\theta + Z_{N1}} = \frac{Z_m\cos\theta - \rho c}{Z_m\cos\theta + \rho c} \qquad (5.144a,b)$$

which is the equation for a locally reactive surface.

The ratio of the amplitude of the transmitted wave to the amplitude of the incident wave may readily be determined by use of Equations (5.136) and (5.142). However, also of interest is the sound power transmission coefficient, τ_p, which is a measure of the energy incident at the interface which is transmitted into the second medium; that is, $\rho c |p_T|^2 / (|Z_m| \ |p_I|^2)$. Multiplication of the left and the right hand sides, respectively, of Equations (5.140) and (5.134), use of Equations (5.139) and (5.142) gives the following expression for the power transmission coefficient for real Z_m:

$$\tau_p = \frac{(1 - |R_p|^2)\cos\theta}{\sqrt{1 - n^2\sin^2\theta}} \qquad (5.145)$$

5.10.3 Spherical Wave Reflection at a Plane Interface

The problem of determining the complex amplitude reflection coefficient for a spherical wave incident upon a plane interface between two media, which is produced by a point source above the interface has been considered by Rudnick (1947) and more recently by Attenborough *et al.* (1980). The results of the later work will be reviewed here. In the following discussion the air above is considered as medium 1 and the porous earth as medium 2. The air above is characterised by air density ρ, propagation constant $k = \omega/c$, and characteristic impedance ρc while the porous earth is characterised by density ρ_m, propagation constant k_m, and characteristic impedance Z_m. In general, the listed variables of the two media may be either real or complex, but in the case of the earth and the air above, only the variables associated with the earth will be considered complex. Where the earth may be characterised by an effective flow resistivity, R_1 (see Table 5.2), the complex quantities, ρ_m, k_m, and Z_m may be calculated by reference to Appendix C.

The expression obtained for the complex amplitude reflection coefficient, R_s, of a spherical wave incident upon a reflecting surface (Attenborough *et al.*, 1980) is as follows:

$$R_s = R_p + B\,G(w)(1 - R_p)$$ (5.146)

In Equation (5.146), R_p is the plane wave complex amplitude reflection coefficient given by either Equation (5.142) or (5.144) as appropriate. For the general case that the reflecting interface is extensively reactive, B is defined as follows:

$$B = \frac{B_1 B_2}{B_3 B_4 B_5}$$ (5.147)

where

$$B_1 = \left[\cos\theta + \frac{\rho c}{Z_m}\left(1 - \frac{k^2}{k_m^2}\sin^2\theta\right)^{1/2}\right]\left[1 - \frac{k^2}{k_m^2}\right]^{1/2}$$ (5.148)

$$B_2 = \left[\left(1 - \frac{1}{\rho_m^2}\right)^{1/2} + \frac{\rho c}{Z_m}\left(1 - \frac{k^2}{k_m^2}\right)^{1/2}\cos\theta + \left(1 - \left(\frac{\rho c}{Z_m}\right)^2\right)\sin\theta\right]^{1/2}$$ (5.149)

$$B_3 = \cos\theta + \frac{\rho c}{Z_m}\left(1 - \frac{k^2}{k_m^2}\right)^{1/2}\left[1 - \frac{1}{\rho_m^2}\right]^{-1/2}$$ (5.150)

$$B_4 = \left[1 - \frac{k^2}{k_m^2}\sin\theta\right]^{1/2}$$ (5.151)

$$B_5 = \left[1 - \frac{1}{\rho_m^2}\right]^{3/2}[2\sin\theta]^{1/2}\left[1 - \left(\frac{\rho c}{Z_m}\right)^2\right]^{1/2}$$ (5.152)

The argument, w, of $G(w)$ in Equation (5.146), is referred to as the numerical distance and is calculated using the following equation, where r_1 and r_2 are defined in Figure 5.14:

$$w = \frac{1}{2}(1 - j)[2k_1(r_1 + r_2)]^{1/2}\frac{B_3}{B_6^{1/2}}$$ (5.153)

In Equation (5.153), B_3 is defined above by Equation (5.150) and B_6 is defined as follows:

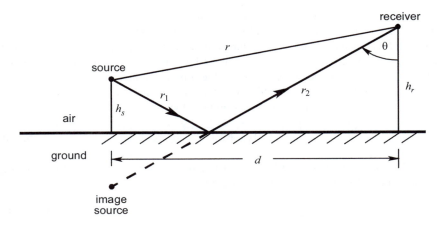

Figure 5.14 Geometry illustrating reflection and transmission above a reflecting plane.

$$B_6 = 1 + \left[\frac{\rho c}{Z_m} \left(1 - \frac{k^2}{k_m^2} \right)^{1/2} \cos\theta + \left(1 - \frac{\rho c}{Z_m} \right)^{1/2} \sin\theta \right] \left[1 - \frac{1}{\rho_m^2} \right]^{-1/2} \qquad (5.154)$$

The term $G(w)$ in Equation (5.146) is defined as follows:

$$G(w) = 1 - j\sqrt{\pi} w g(w) \qquad (5.155)$$

where

$$g(w) = e^{-w^2} \operatorname{erfc}(jw) \qquad (5.156)$$

and "erfc( )" is the error function (Abramowitz and Stegun, 1965).
   For small w ($|w| < 3$):

$$g(w) = e^{-w^2} - \left[\frac{2jw}{\pi^{1/2}} \right] \sum_{n=0}^{\infty} \frac{(-2w^2)^n}{1 \times 3 \times \dots \times (2n+1)} \qquad (5.157)$$

For values of w where the real part is greater than 3 or the imaginary part is greater than 2 and either is less than 6:

$$g(w) = -jw \left| \frac{0.4613135}{w^2 - 0.1901635} + \frac{0.09999216}{w^2 - 1.7844927} + \frac{0.002883894}{w^2 - 5.5253437} \right| \qquad (5.158)$$

For real or imaginary parts of w greater than 6:

$$g(w) = -jw \left[\frac{0.5124242}{w^2 - 0.275255} + \frac{0.05176536}{w^2 - 2.724745} \right] \qquad (5.159)$$

As shown by Rudnick (1947) the numerical distance, w, becomes very large at large distances from the source and the function $G(w)$ tends to zero. Reference to Equation (5.146) shows that as $G(w)$ tends to zero, the complex amplitude reflection coefficient for spherical waves becomes the reflection coefficient for plane waves. On the other hand, w approaches zero close to the source and then $G(w)$ approaches one.

Use of Equation (5.143) and reference to Appendix C gives the following criterion for the porous surface to be essentially locally reactive:

$$\rho f < 10^{-3} R_1 \tag{5.160}$$

When Equation (5.160) is satisfied and the porous surface is essentially locally reactive, the following simplifications are possible:

$$B_1 = B_3 = \cos\theta + \frac{\rho c}{Z_m} \tag{5.161a,b}$$

$$B_2 = (1 + \sin\theta)^{1/2} \tag{5.162}$$

$$B_4 = 1 \tag{5.163}$$

$$B_5 = (2\sin\theta)^{1/2} \tag{5.164}$$

$$B_6 = B_2^2 \tag{5.165}$$

Equation (5.153) becomes:

$$w = \frac{1}{2}(1 - j)[2k(r_1 + r_2)]^{1/2}\left(\cos\theta + \frac{\rho c}{Z_m}\right)(1 + \sin\theta)^{-1/2} \tag{5.166}$$

5.10.4 Effects of Turbulence

Turbulence in the acoustic medium containing the direct and reflected waves has a significant effect on the effective surface spherical wave amplitude reflection coefficient. This effect will now be discussed with particular reference to sound propagation outdoors over the ground. Experience suggests that local turbulence near the ground is especially important because it introduces variability of phase between the reflected sound and the direct sound from the source to the receiver. Variability in phase between the direct and reflected sound determines whether the two sounds, the direct and the reflected sound, should be considered as adding coherently or incoherently. Coherent reflection requires minimal variability of phase and can result in constructive or destructive interference, in which case the variation in level can be very large while incoherent reflection, associated with a large variability of phase, can result in at most a 3 dB variation in observed level.

Solar-driven local air currents near the ground, which result as the ground heats up during the day relative to the cooler air above, will cause local convection and turbulence near the ground of the kind of concern here. Sound of wavelength of the order of or less than the turbulence scale will be observed to warble strongly only a short distance away from the source when observed across a paved parking lot. The model proposed here suggests that coherent reflection should be observed more often at night than during the day.

The effect of turbulence on sound propagation over an acoustically smooth surface has been investigated by Clifford and Lataitis (1983) and by Raspet and Wu (1995). The presence or absence of turbulence may be included by a generalisation of their results to give the following general expression for the reflection term, Γ, which will be used later in Equation (5.195):

$$
\Gamma = \frac{r^2 - (1 - T)(r^2 - (r_1 + r_2)^2)}{(r_1 + r_2)^2} |R_s|^2
$$

$$
+ \frac{2Tr}{r_1 + r_2} \left(\cos[k(r_1 + r_2 - r)] \operatorname{Re}\{R_s\} - \sin[k(r_1 + r_2 - r)] \operatorname{Im}\{R_s\} \right) \tag{5.167}
$$

In the above equation, R_s is the spherical wave complex amplitude reflection coefficient given by Equation (5.146) and the distances, r, r_1 and r_2 are shown in Figure 5.14.

The exact solution (Clifford and Lataitis, 1983; and Raspet and Wu, 1995) for the term T which appears in the above equation is very complicated. However, simplifications are possible that lead to the following approximate expression:

$$
T = e^{-4a\pi^{5/2}\langle n_1^2\rangle 10^3 \Phi} \tag{5.168}
$$

where

$$
\Phi = 0.001 \frac{dL_0}{\lambda^2} \tag{5.169}
$$

In Equation (5.169), d is the horizontal distance between the source and receiver (see Figure 5.14), L_0 is the scale of the local turbulence and λ is the wavelength of the sound under consideration. A value of L_0 of about 1 to 1.2 m is suggested if this quantity is unknown or cannot be measured conveniently. When Φ is greater than 1, incoherent reflection can be expected and when Φ is less than 0.1, coherent reflection can be expected. If reflection is incoherent then the spherical wave amplitude reflection coefficient given by Equation (5.146) reduces to the simpler plane wave amplitude reflection coefficient given by Equation (5.142) or (5.144).

In Equation (5.168), α has the value 0.5 when $d/k \gg L_0^2$ and the value 1 when $d/k \ll L_0^2$. For values of d/k and L_0 which do not satisfy either of the preceding conditions, use of the exact formulation for T may be used (Raspet and Wu, 1995). However, the exact solution is extremely complicated on the one hand and on the other it is reasonable to expect that α will be bounded by its extreme values, 0.5 and 1.

Consequently, as the far field is generally satisfied by the value $\alpha = 0.5$, the latter value is used here to evaluate the exponent of the right-hand side of Equation (5.168).

The term $\langle n_1^2 \rangle$ is the mean square of the fluctuations in the speed of sound relative to its mean value in the absence of turbulence. A value of 5×10^{-6} is suggested (Raspet and Wu, 1995). Evaluation of the exponent allows Equation (5.168) to be rewritten in terms of the figure of merit, Φ, introduced in Equation (5.169) as follows:

$$T = e^{-0.17\Phi} \tag{5.170}$$

As the figure of merit, Φ, approaches 0, indicating little or no turbulence, the term T approaches 1, resulting in coherent ground reflection and a large variation in attenuation as a function of distance and frequency. On the other hand, as the figure of merit becomes large, indicating a large turbulence effect, the term T approaches 0, and the direct and ground reflected waves are combined incoherently, resulting in an expected sound pressure level increase of 3 dB or less.

5.11 SOUND PROPAGATION OUTDOORS, GENERAL CONCEPTS

5.11.1 Methodology

The problems arising from sound propagation outdoors may range from relatively simple to very complex, depending upon the nature of the source and distribution of the affected surrounding areas. If the source is composed of many individual component sources, as would often be the case with an industrial plant, and the surrounding area is extensively affected, then the use of a computer to carry out the analysis associated with level prediction is essential and a number of schemes (Tonin, 1985) have been designed for the purpose. These schemes generally rely heavily upon empirical information determined from field surveys but gradually empiricism is being replaced with well-established analysis based upon extensive research. At present the most successful schemes rely upon a mixture of both theory and empiricism. In all cases the method of forecasting proceeds as follows:

1. Determine power levels L_w of all sources.
2. For a given environment calculate the individual components of excess attenuation A_{Ei} (see Section 5.11.6) for all sources $i = 1$ to N.
3. Compute the resulting sound pressure levels at selected points in the environment for each of the individual sources.
4. Compute the predicted sound pressure levels produced by all of the individual sources at the selected points in the environment by converting the sound pressure level due to each source to pressure squared, adding the squared pressure contributions and converting the total to sound pressure level (see Section 1.11.4).

While it is generally recognised that the various components of attenuation may be inter-related and not simply additive, investigations have not proceeded as yet to the extent that it is possible to quantitatively express all of the possible inter-relations in one encompassing algorithm. Rather, attenuation of sound propagating out-of-doors in addition to that normally expected is approximated as a linear sum of effects, A_{Ei}, when carrying out the computations of step 3 above.

Some inter-relationships between attenuation factors have been explained. For example, in the case of barrier attenuation, the effects of ground reflection on both sides of a barrier have been investigated (Thomasson, 1978). Wind and temperature effects on barrier attenuation have also been investigated (DeJong and Stusnik, 1976). The thickness of barriers is also a variable (Foss, 1979).

5.11.2 Limits to Accuracy of Prediction

The question of accuracy of prediction is always of importance and experience suggests that ±5 dB may be expected (Marsh, 1976). Alternatively an accuracy of ±2 dB has been claimed but in the latter cases measured data were used to refine the prediction scheme (Marsh, 1982; Delany, 1972). The question of accuracy is complicated by the fact that it is difficult to obtain good source power level data. In the absence of manufacturer supplied data, the procedures described in Chapter 6 may be used for source sound power determination.

Another problem that makes accurate prediction of total sound pressure levels at community locations difficult is the variability of weather conditions. In practice, the sound pressure level corresponding to the least favourable weather condition is calculated and cited as a worst case. In addition, the average sound pressure level (see Equation (3.22)), L_{eq}, at various hours during the night and day is usually calculated assuming the most commonly occurring weather conditions for those times. Overall night-time and day-time L_{eq}'s are then calculated on a squared pressure (or energy) average basis.

Using Equation (1.98) (see Section 1.11.4), it is readily shown that the addition of squared sound pressures has the mitigating effect that if a ±3 dB error in sound level prediction for any source is to be expected, the error for 100 such sources combined randomly will be no larger (Tonin, 1985). For example, using logarithmic addition the sum of two levels is at most 3 dB greater than the greater level separately or in the case of several sources, it is at most the greatest level plus ten times the logarithm of the number of sources. As the number of sources is not in doubt the error remains unchanged by the addition.

5.11.3 Outdoor Sound Propagation Prediction Schemes

It will be assumed that the sound existing at an observation point is the sum of the component sounds contributed by many sources. For example, where N sound sources may be required to adequately model an industrial site, then at any observation point, N component sounds will contribute to the sound that is observed. The discussion begins with the method of calculating a component sound at an observation point.

A generalised expression will be written for a single component source that relates the sound pressure level, L_p, at an observation point to the sound power level, L_w, of the source. The sound pressure level at an observation point may be written as the sum of the sound power level of the source; a geometric spreading factor, K, which is dependent upon the type of source and accounts for geometrical spreading as the sound propagates away from the source; a directivity index, DI_M, which accounts for directional properties of the source, including influences of reflections other than those in the ground plane; and an excess attenuation factor, A_E. The excess attenuation factor in turn is the sum of terms including ground reflection, atmospheric effects, etc. as will be explained in Section 5.11.6. In the following analysis of outdoor sound propagation, it is assumed that $\rho c \approx 400$ (SI units), so that $10 \log_{10}(\rho c / 400) = 0$.

The expression relating sound pressure level, L_p, and sound power level, L_w, for a single source is:

$$L_p = L_w - K + DI_M - A_E \tag{5.171}$$

In general, there will be many sources contributing to sound at an observation point. Incoherent sound addition will be assumed, in which case the sound pressure level at the observation point due to N sources may be computed as the sum of contributions as in the following equation.

$$L_p = 10 \log_{10} \sum_{i=1}^{N} 10^{L_{pi}/10} \tag{5.172}$$

where L_{pi} is the sound pressure level due to the ith source.

Enclosed sound sources may be included by using Equations (8.85) and (8.86) of Chapter 8 to calculate the sound pressure level, L_{p1}, averaged over the external enclosure surface of area S_E, where L_w of Equation (8.85) is the sound power level of the enclosed sources. The quantity L_w needed for Equation (5.171) is the sound power level radiated by the enclosure walls, which is equal to $(L_{p1} + 10 \log_{10} S_E)$, and the quantity DI_M is the directivity index of the enclosure in the direction of the receiver.

In the following sections, the calculation of each of the terms in Equation (5.171) will be considered and reference will be made to two outdoor sound propagation prediction schemes that are currently in common use (ISO 9613-2 (1996) and Manning, 1981, known generally as "CONCAWE").

5.11.4 Geometrical Spreading, K

The spreading factor K (referred to as K_1 in CONCAWE) will first be considered. Referring to Equation (5.14) for a monopole source, putting the equation in the form of Equation (5.171) and setting the terms DI_M and A_E equal to zero, it is readily determined that:

$$K = 10 \log_{10} 4\pi + 20 \log_{10} r \tag{5.173}$$

where r is the distance from the source to the point of observation. In the latter case the source is above the ground and radiates spherically in all directions including downward. Alternatively, if the source is on the ground or in a wall and thus radiating hemispherically into half-space then:

$$K = 10 \log_{10} 2\pi + 20 \log_{10} r \qquad (5.174)$$

Similarly, referring to Equation (5.82) it is readily determined that for a line source:

$$K = 10 \log_{10}(4\pi r D/\alpha) \qquad (5.175)$$

In the above equation, D is the length of the line source, r is the distance from the source centre to the point of observation and α is the angle subtended by the source at the point of observation (equal to $\alpha_u - \alpha_\ell$), as shown in Figure 5.5.

In the case of a wall, which may be modelled as a plane incoherent radiator, reference to Equation (5.117) shows that K has the following form:

$$K = 10 \log_{10} 2\pi + 20 \log_{10} r - 10 \log_{10} F(\alpha, \beta, \gamma, \delta) \qquad (5.176)$$

The dimensionless variables of Equation (5.176) are $\alpha = H/L$, $\beta = h/L$, $\gamma = r/L$ and of $\delta = d/L$. In the case of the near field, where $(\gamma/10) < \alpha, \beta, \delta$, the function F takes the following form:

$$
\begin{aligned}
F = \frac{\gamma^2}{\alpha} &\left[\tan^{-1} \frac{(\alpha - \beta)(\delta + 1/2)}{\gamma\sqrt{(\alpha - \beta)^2 + (\delta + 1/2)^2 + \gamma^2}} \right. \\
&+ \tan^{-1} \frac{\beta(\delta + 1/2)}{\gamma\sqrt{\beta^2 + (\delta + 1/2)^2 + \gamma^2}} - \tan^{-1} \frac{(\alpha - \beta)(\delta - 1/2)}{\gamma\sqrt{(\alpha - \beta)^2 + (\delta - 1/2)^2 + \gamma^2}} \\
&\left. - \tan^{-1} \frac{\beta(\delta - 1/2)}{\gamma\sqrt{\beta^2 + (\delta - 1/2)^2 + \gamma^2}} \right]
\end{aligned}
\qquad (5.177)
$$

Alternatively, in the case of the far field, where $\gamma/10 > \alpha, \beta, \delta$, the function $F = 1$ and the corresponding term in Equation (5.176) is zero.

5.11.5 Directivity Index, DI_M

Most sound sources are directional, radiating more sound in some directions than in others. Alternatively, it was shown in Section 5.9.1 that a simple point source, which radiates uniformly in all directions in free space, becomes directional when placed near to a reflecting surface, for example, a wall, floor or the ground. Consequently, a directivity index DI_M, has been introduced in Equation (5.171) to account for the effect of variation in sound intensity with orientation relative to the noise source, but specifically excluding the effect of reflection in the ground plane, which is included as A_g in the excess attenuation factor, A_E, and will be considered separately in the following section.

5.11.6 Excess Attenuation Factor, A_E

The excess attenuation factor, A_E, *is* defined as the sum in decibels (dB) of five separate terms as follows:

$$A_E = A_a + A_{bhp} + A_f + A_g + A_m \tag{5.178}$$

The terms of Equation (5.178) are A_a, the attenuation due to air absorption (K_2 in Manning, 1981); A_{bhp}, the attenuation due to regular barriers, houses and process equipment (K_5, K_6, and K_7 in CONCAWE); A_f, the attenuation due to forests (if present); A_g, the attenuation (which may be a gain rather than a loss) due to reflection in the ground plane (K_3 in CONCAWE); and A_m, the attenuation due to meteorological effects such as wind and temperature gradients (K_4 in CONCAWE, which also may be either a gain or a loss). Each of these terms will be discussed in the following sections, although barrier effects are discussed in greater detail in Section 8.5.

Two early approximate schemes will be mentioned. In the simplest approximation only point sources in an infinitely hard ground plane are considered in which case in Equation (5.171), $DI_M = 0$. The excess attenuation factor, A_E, and the geometrical spreading factor, K, of Equation (5.171) are combined. A value of 4 dB per doubling of distance is commonly assigned to the combined quantity which effectively assigns to A_E a value of -2 dB per doubling of distance from the source.

A more complex early model was developed by the Oil Companies Materials Association (OCMA) (Chambers, 1972). In this scheme only the simplest type of source, the point source in the ground plane is considered. In Equation (5.171), K is replaced using Equation (5.174) and the result is combined with the air absorption term, A_a (see Equation (5.178)) in a single expression, K_1. In this model, the directivity term, $DI_M = 0$, and the remaining terms of Equation (5.178) are combined, with $A_f = 0$, into a second empirically determined term K_2. With these definitions of terms, Equation (5.171) takes the following form:

$$L_p = L_w - K_1 - K_2 \tag{5.179}$$

The terms K_1 and K_2 are defined as follows:

$$K_1 = 10 \log_{10} 2\pi + 20 \log_{10} r + A_a \tag{5.180}$$

and

$$K_2 = A_{bhp} + A_g + A_m \tag{5.181}$$

The term K_1 may be determined from Figure 5.15 and K_2 from Figures 5.16 and 5.17. The term, K_1, includes the effects of geometric spreading from a point source on an infinitely hard ground plane and atmospheric attenuation based in part upon experiment and in part upon the theoretical sound absorption in air at a temperature of 16°C and 70% relative humidity.

The term K_2, accounts for shielding by barriers, hills and plant buildings in an average sense at large distance for minimal and significant shielding respectively. These data are based upon noise measurements conducted in and around two petrochemical plants and are really only applicable to installations similar to those on which the data are based. Computer programs for estimating environmental noise levels, based upon this scheme have been developed (Sutton, 1976; Jenkins and Johnson, 1976), but they are no longer in general use.

In either of the cases that have been cited, all of the component sources contributing to the sound at an observation point in the field are determined and combined following the general procedure outlined in Section 5.11.3 using Equations (5.171) and (5.172).

5.11.7 Air Absorption, A_a

An extensive review of literature on sound propagation in the atmosphere is provided in CONCAWE (Manning, 1981). The author recommends the method of Sutherland *et al.* (1974) as outlined by Gill (1980a) as being the best available scheme for calculating air absorption and quotes an accuracy within ±10% from 0°C to 40°C.

Air absorption, A_a, is dependent upon temperature and relative humidity. Calculated values of absorption rate, m (Sutherland and Bass, 1979; Sutherland, 1975), averaged over an octave band, are listed in Table 5.3 for the frequencies shown for representative values of temperature and relative humidity. For propagation over distance X (in kilometres), the absorption A_a is:

$$A_a = mX \qquad (5.182)$$

ISO 9613-1(1993) contains detailed tables of m for single frequencies over a wide range of temperatures and relative humidities and these values should be used for temperatures and humidities not covered in Table 5.3.

5.11.8 Shielding by Barriers, Houses and Process Equipment/Industrial Buildings, A_{bhp}

The total attenuation due to these effects is denoted A_{bhp}. It is made up of the arithmetic sum of the attenuations due to large barriers (A_b), attenuation due to houses (A_h) and attenuation due to process equipment and industrial buildings (A_p), where:

$$A_{bhp} = A_b + A_h + A_p. \qquad (5.183)$$

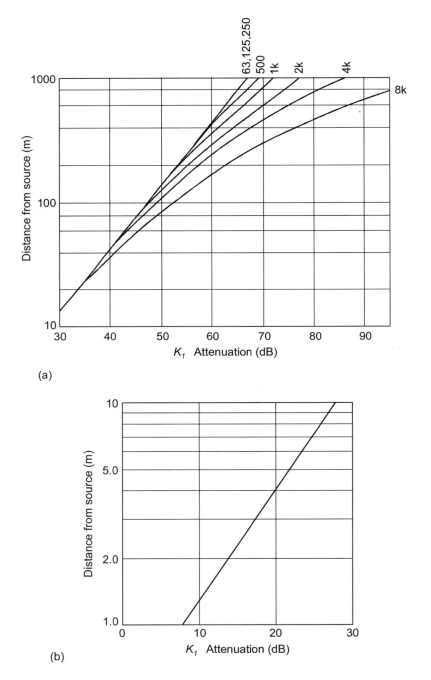

Figure 5.15 OCMA algorithm for determination of distance and air absorption attenuation K_1.

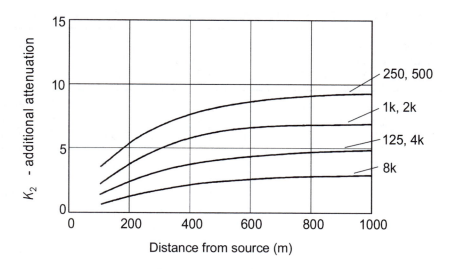

Figure 5.16 OCMA algorithm for determination of the ground attenuation parameter, K_2, for the case of minimal shielding.

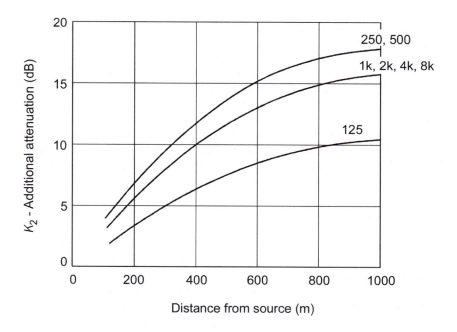

Figure 5.17 OCMA algorithm for determination of the ground attenuation parameter, K_2, for the case of significant shielding.

Table 5.3 Attenuation due to atmospheric absorption (calculated from Sutherland and Bass, 1979)

Relative humidity (%)	Temperature (°C)	63 Hz	125 Hz	250 Hz	500 Hz	1 kHz	2 kHz	4 kHz
				m (dB per 1000 m)				
25	15	0.2	0.6	1.3	2.4	5.9	19.3	66.9
	20	0.2	0.6	1.5	2.6	5.4	15.5	53.7
	25	0.2	0.6	1.6	3.1	5.6	13.5	43.6
	30	0.1	0.5	1.7	3.7	6.5	13.0	37.0
50	15	0.1	0.4	1.2	2.4	4.3	10.3	33.2
	20	0.1	0.4	1.2	2.8	5.0	10.0	28.1
	25	0.1	0.3	1.2	3.2	6.2	10.8	25.6
	30	0.1	0.3	1.1	3.4	7.4	12.8	25.4
75	15	0.1	0.3	1.0	2.4	4.5	8.7	23.7
	20	0.1	0.3	0.9	2.7	5.5	9.6	22.0
	25	0.1	0.2	0.9	2.8	6.5	11.5	22.4
	30	0.1	0.2	0.8	2.7	7.4	14.2	24.0

In most cases only one of the quantities on the RHS of Equation (5.183) will need to be considered.

The attenuation, A_b, due to large barriers and large extended buildings may be calculated using the method of Makaewa (1985) but modified to account for wind and temperature gradients (DeJong and Stusnik, 1976) as described in Section 8.5. The modification is based upon a ray tracing technique and was developed for wind gradients but as the effects of a temperature gradient are similar the modification may also be used to estimate temperature gradient effects, as described in Section 8.5.2.

To include the barrier attenuation, A_b, in Equation (5.183), while retaining the ground effect term, A_g, in Equation (5.178), it is necessary to calculate A_b using Equation (1.101), with $A_b = NR$. This requires the calculation of the noise reduction due to each diffraction path over the barrier, the calculation of the noise reductions of the two paths existing before the installation of the barrier (direct and ground reflected paths) and the combination of these noise reductions using Equation (1.101). The individual terms, NR_{Bi}, used in Equation (1.101) are equal to the A_{bi} terms (in dB) defined in Equation (8.104) plus the arithmetic addition of the reflection loss (in dB) associated with any ground reflections involved in the *i*th path being considered. It is interesting to note that the first term in Equation (1.101), which is concerned with the two paths in the absence of the barrier, is equivalent to the excess attenuation due to the ground with no barrier in place, multiplied by -1. The multiplication by -1 is because we need to subtract out the ground effect calculated with no barrier, as ground effects are automatically included in the second term in Equation (1.101) by including the reflection loss in any paths involving a ground reflection. The calculation procedure is well illustrated in Example 8.7 of Chapter 8.

Meteorological effects influence the barrier calculations as discussed in Section 8.5, so if a barrier is included and meteorological effects are included in the barrier calculations, then they are not also included as a separate term in Equation (5.178). Barrier calculations are discussed in more detail in Section 8.5.

The excess attenuation due to housing (A_h) may be calculated using (ISO 9613-2 1996):

$$A_h = 0.1 B r_b - 10 \log_{10}[1 - (P/100)] \qquad \text{(dB)} \qquad (5.184)$$

where B is the density of buildings along the path (total plan area of buildings divided by the total ground area (including that occupied by the houses), r_b is the distance that the curved sound ray travels through the houses and P is the percentage ($\leq 90\%$) of the length of housing facades relative to the total length of a road or railway in the vicinity. The second term in Equation (5.184) is only used if there are well defined rows of houses perpendicular to the direction of sound propagation. The second term may also not exceed the calculated insertion loss for a continuous barrier the same height as the building facades (see Section 8.5). The quantity r_b is calculated in the same way as $r_f (= r_1 + r_2)$ in Figure 5.18 for travel through foliage, except that the foliage is replaced by houses. It may include both r_1 and r_2. Note that if A_h of Equation (5.183) is non zero, the ground attenuation through the built up area is set equal to zero, unless the ground attenuation with the buildings removed is greater than the first term in Equation (5.184) for A_h. In that case, the ground attenuation for the ground without buildings is substituted for the first term in Equation (5.184).

In large industrial facilities, significant additional attenuation can arise from shielding due to other equipment blocking the line of sight from the source to the receiver. Usually it is best to measure this, but if this cannot be done, it may be estimated using the values in line 1 of Table 5.4 as described in ISO 9613-2 (1996). The distance, r_1, to be used for the calculation is only that part of the curved sound ray (close to the source) that travels through the process equipment and the maximum attenuation that is expected is 10 dB. The distance is equivalent to distance r_1 in Figure 5.18 with the foliage replaced by process equipment. However, in many cases it is not considered worthwhile taking A_p into account, resulting in receiver noise level predictions that may be slightly conservative.

5.11.9 Attenuation due to Forests and Dense Foliage, A_f

If a heavily wooded area (for example, a forest) exists along the propagation path, the excess attenuation may be estimated using the following relationship, which was derived empirically from many measurements (Hoover, 1961) and which is used in the CONCAWE procedure:

$$A_f = 0.01 \, r_f f^{1/3} \qquad (5.185)$$

In the equation, f (Hz) is the frequency of the propagating sound and r_f (m) is the distance of travel through the forest.

For a more recent investigation of sound propagation in a pine forest see Huisman and Attenborough (1991). A number of other investigations of the effectiveness of vegetation along the edges of highways have shown the following results:

- a single row of trees along the highway or near houses results in negligible attenuation of the noise;

- a continuous strip of oleander or equivalent shrubs, at least 2.5 m high and 4.5 to 6 m wide, planted along the edge of a highway shoulder, provides noise attenuation of 1-3 dB(A) at distances of up to 15 m from the rear edge of vegetation;

- a strip of trees, 60 m wide can attenuate traffic noise by up to 10 dB(A); and

- vegetation is not generally considered an effective traffic noise barrier, although it does have an effect in attenuating noise at frequencies above 2 kHz. However, the psychological effect of vegetation as a barrier between a noise source and an observer should not be overlooked – in many cases if the noise source is not visible, it is less noticeable and thus less annoying, even if the level is not significantly changed.

ISO 9613-2 (1996) gives the attenuation values in Table 5.4 for sound propagation through dense foliage. For distances less than 20 m, the values given are absolute dB. For distances between 20 and 200 m, the values given are dB/m and for distances greater than 200 m, the value for 200 m is used.

Table 5.4 Octave band attenuation, A_p due to process equipment and industrial buildings (line 1) and A_f due to dense foliage (lines 2 and 3) (after ISO 9613-2, 1996)

		Octave band centre frequency (Hz)						
	63	125	250	500	1000	2000	4000	8000
A_p (dB/m)	0	.015	.025	.025	.02	.02	.015	.015
A_f (dB) for 10 m $\leq r_f \leq$ 20 m	0	0	1	1	1	1	2	3
A_f (dB/m) for 20 m $\leq r_f \leq$ 200 m	0.02	0.03	0.04	0.05	0.06	0.08	0.09	0.12

The distance of travel through the foliage is not equal to the extent of the foliage between the source and receiver. It depends on the height of the source and receiver and the radius of curvature of the propagating wave as a result of wind and temperature gradients (see Section 5.11.12). ISO 9613-2 (1996) recommends that a radius of 5 km be used for downwind propagation, but a more accurate estimate may be obtained by following the procedure in Section 5.11.12. The distance $r_f = r_1 + r_2$, where r_1 and r_2 are defined in Figure 5.18.

Figure 5.18 Path lengths for sound propagation through foliage.

5.11.10 Ground Effects

Progressively more complicated procedures for taking account of ground effects will be considered in this section. The discussion will begin with the simplest procedures based upon experimental data and will conclude with the most complicated procedure, which requires extensive information about weather conditions, the local geography and the flow resistance of the earth. The procedure outlined in ISO 9613-2 (1996) will also be considered.

5.11.10.1 CONCAWE Method

Experimental data taken in the vicinity of three typical processing plants in Europe have been used to construct the ground effect term, A_g, shown in Figure 5.19. The

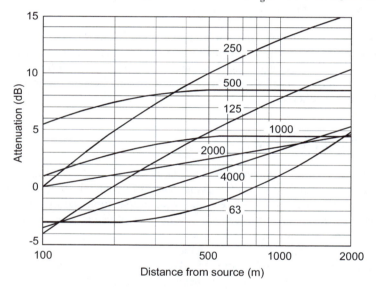

Figure 5.19 CONCAWE ground attenuation curves (after manning, 1981)

ground correction term was separated from the total measured attenuation by subtracting the geometrical and air absorption terms using only data measured during neutral meteorological conditions (zero wind and temperature gradients). The ground effect term shown in the figure applies strictly only to ground similar to that considered in the study, which was flat and undulating land, typical of rural and residential areas (Manning, 1981). However, it is difficult to justify application of these data in a general sense. If Figure 5.19 is used to calculate the ground effect, then for source heights greater than 2 m an additional correction, K_h must be added to A_g. The correction, K_h, is given by:

$$K_h = \begin{cases} (A_g + A_m + 3) \times (\gamma - 1) \text{ dB} & \text{if } (A_g + A_m) > -3 \text{ dB} \\ 0 \text{ dB} & \text{if } (A_g + A_m) \leq -3 \text{ dB} \end{cases} \qquad (5.186)$$

where:

$$\gamma = 1.08 - 0.478(90 - \theta) + 0.068(90 - \theta)^2 - 0.0029(90 - \theta)^3$$
$$\text{and } \gamma_{max} = 1 \qquad (5.187)$$

Note that K_h is always negative, which means that it acts to reduce the excess attenuation. The angle, θ, is in degrees and is defined in Figures 5.13 and 5.14.

5.11.10.2 Simple Method (Hard or Soft Ground)

Alternatively, as a rough first approximation, it is often assumed that the effect of hard ground such as concrete, water or asphalt is to increase noise levels by 3 dB and that soft ground (such as grass, snow, etc.) has no effect. Thus, $A_g = -3$ or 0 dB depending on the ground surface. In this case the assumption is implicit that reflection in the ground plane is incoherent.

5.11.10.3 Plane Wave Method

A better approximation is obtained by calculating the ground amplitude reflection coefficient, which can be done with varying degrees of sophistication and accuracy. The simplest procedure is to assume plane wave propagation and that the effect of turbulence is sufficient that the waves combine incoherently; that is, on an energy basis such that squared pressures add.

The plane wave reflection loss in decibels is given by $A_R = -20 \log_{10}|R_p|$, where R_p has been calculated using Equation (5.144), and A_R is plotted in Figure 5.20 for various values of the dimensionless parameter $\rho f / R_1$. Here, f is the tonal frequency, or the centre frequency of the measurement band. Alternatively, if Equation (5.160) is not satisfied (that is, local reaction cannot be assumed for the ground surface), then R_p should be calculated using Equation (5.142). Figure 5.20 is used to determine the decrease in energy, A_R (dB) of the reflected sound wave on reflection from the ground. The excess attenuation A_g is then calculated (see Equation (1.111)) as follows:

$$A_g = -10 \log_{10}\left[1 + |R_p|^2\right] = -10 \log_{10}\left[1 + 10^{-A_R/10}\right]. \qquad (5.188a,b)$$

The quantity A_g will vary between 0 and -3 dB, depending on the value of A_R (which is always positive).

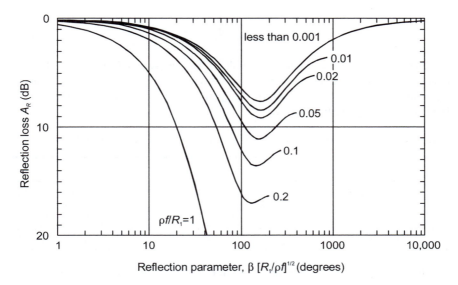

Figure 5.20 Reflection loss as a function of reflection angle β (where β is measured from the horizontal and $\beta = 90 - \theta$), surface flow resistivity R_1, air density ρ and frequency f. Curves are truncated when β reaches $90°$ or the reflection loss exceeds 20 dB. The surface is assumed locally reactive.

5.11.10.4 ISO 9613-2 (1996) Method

The method recommended in ISO9613-2 (1996) for calculating the ground effect assumes the worst case of sound propagating downwind from the source to the receiver and hence the implication is that meteorological effects due to adverse wind and temperature gradients are already included and do not need to be taken into account separately. The ISO method is moderately complex and yields results of moderate accuracy. In this method, the space between the source and receiver is divided into three zones, source, middle and receiver zones. The source zone extends a distance of $30h_s$ from the source towards the receiver and the receiver zone extends $30h_r$ from the receiver towards the source. The middle zone includes the remainder of the path between the source and receiver and will not exist if the source / receiver separation (projected on to the ground plane) is less than $d = 30h_s + 30h_r$, where h_s and h_r are defined in Figure 5.14. The acoustic properties of each zone are quantified using the parameter, G. This parameter has a value of 0.0 for hard ground, a value of 1.0 for soft (or porous) ground and for a mixture of hard and soft ground it is equal to the fraction of ground that is soft. It is assumed in the ISO9613-2 standard that for downwind propagation, most of the ground effect is produced by the ground in the vicinity of the source and receiver so the middle part does not contribute much to the overall value of A_g.

The total excess attenuation due to the ground is the sum of the excess attenuations for each of the three zones. That is:

$$A_g = A_s + A_m + A_r \tag{5.189}$$

Values for each of the three quantities on the right-hand side of Equation (5.189) may be calculated using Table 5.5. Note that if the source receiver separation distance is much larger than their heights above the ground, then $d \approx r$. See Figure 5.14 for a definition of the quantities, d and r.

Table 5.5 Octave band ground attenuation contributions, A_s, A_r and A_m (after ISO 9613-2, 1996)

	Octave band centre frequency (Hz)							
	63	125	250	500	1000	2000	4000	8000
A_s (dB)	-1.5	$-1.5+G_s a_s$	$-1.5+G_s b_s$	$-1.5+G_s c_s$	$-1.5+G_s d_s$	$-1.5(1-G_s)$	$-1.5(1-G_s)$	$-1.5(1-G_s)$
A_r (dB)	-1.5	$-1.5+G_r a_r$	$-1.5+G_r b_r$	$-1.5+G_r c_r$	$-1.5+G_r d_r$	$-1.5(1-G_r)$	$-1.5(1-G_r)$	$-1.5(1-G_r)$
A_m (dB)	$-3q$	$-3q(1-G_m)$	$-3q(1-G_m)$	$-3q(1-G_m)$	$-3q(1-G_m)$	$-3q(1-G_m)$	$-3q(1-G_m)$	$-3q(1-G_m)$

In Table 5.5, G_s, G_r and G_m are the values of G corresponding to the source zone, the receiver zone and the middle zone, respectively. The quantity, A_m is zero for source / receiver separations of less than $r = 30h_s + 30h_r$, and for greater separation distances it is calculated using the last line in Table 5.5 with:

$$q = 1 - \frac{30(h_s + h_r)}{d} \tag{5.190}$$

The coefficients, a_s, b_s, c_s and d_s and the coefficients a_r, b_r, c_r and d_r of Table 5.5 may be calculated using Equations (5.191) to (5.194). In each equation, $h_{s,r}$ is replaced with h_s for calculations of A_s and by h_r for calculations of A_r:

$$a_s, a_r = 1.5 + 3.0e^{-0.12(h_{s,r}-5)^2}(1 - e^{-d/50}) + 5.7e^{-0.09h_{s,r}^2}\left(1 - e^{-2.8 \times 10^{-6} \times d^2}\right) \tag{5.191}$$

$$b_s, b_r = 1.5 + 8.6e^{-0.09h_{s,r}^2}(1 - e^{-d/50}) \tag{5.192}$$

$$c_s, c_r = 1.5 + 14.0e^{-0.46h_{s,r}^2}(1 - e^{-d/50}) \tag{5.193}$$

$$d_s, d_r = 1.5 + 5.0e^{-0.9h_{s,r}^2}(1 - e^{-d/50}) \tag{5.194}$$

5.11.10.5 Detailed, Accurate and Complex Method

The most complex (and perhaps the most accurate) means of determining the ground effect is to assume spherical wave reflection, including the effects of turbulence, to calculate the corresponding reflection term, Γ, as defined by Equation (5.167). The excess attenuation, A_g, may then be calculated using the following equation:

$$A_g = -10 \log_{10}[1 + \Gamma]$$ (5.195)

The excess attenuation, A_g, averaged over a one-third octave or octave band, is calculated by combining logarithmically (see Section 1.11.4) the attenuations calculated at several different frequencies (at least 10) equally spaced throughout the band. This band-averaged attenuation is probably more useful for noise predictions in practice, as atmospheric turbulence and undulating ground will result in considerable fluctuations in the single frequency values.

Calculating the ground effect with the complex procedure just described requires software written specifically for the task. Such software is available and is included in the software package, ENC (www.causalsystems.com).

Example 5.1

Calculate the excess attenuation due to ground effects for the 1000 Hz octave band, given the values for discrete frequencies in the band shown in the following table.

Example 5.1 Table

Frequency (Hz)	A_{gi} (dB)
710	5
781	7
852	9
923	11
994	17
1065	11
1136	9
1207	7
1278	5
1349	4
1420	3

Solution

The band-averaged attenuation is obtained by combining the tabulated values logarithmically, as follows:

$$A_g = -10 \log_{10}\left[\frac{1}{n}\sum_{i=1}^{n} 10^{-A_{gi}/10}\right] = 6.7 \text{ dB}$$

The preceding method for calculating the ground effect has been verified experimentally over very short distances (Embleton *et al.*, 1983).

5.11.11 Image Inversion and Increased Attenuation at Large Distance

Equation (5.144) shows that if the ground impedance, Z_m, is very large (i.e. the ground is very hard), the reflection coefficient will approach unity and, provided that the path difference between the reflected and direct waves is small, the sounds arriving over the two paths will add together to give a 6 dB increase, over the free field value, in sound pressure level at the receiver (see Section 5.9 for discussion).

Equation (5.144) also shows that in the limit as the angle of incidence, θ, approaches $\pi/2$, the reflection coefficient approaches -1, resulting in the reflected wave being of equal amplitude and 180° out of phase with the incident wave. Thus, at grazing incidence or great distances, the source and its ground plane virtual image (which can be considered as being the origin of the reflected wave) coalesce as a dipole. It is to be noted that if the figure of merit Φ is greater than 0.1 (see Equation (5.169)) the reflection will not be coherent and the effect reported here will not be observed. Referring to Figure 5.14, the following expression may be written for large distance, r, such that $r \approx r_1 + r_2$:

$$\cos\theta = \frac{h_s + h_r}{r_1 + r_2} \approx \frac{h_s + h_r}{r} \qquad (5.196)$$

Substitution in Equation (5.30a) and simplification gives:

$$\langle p^2 \rangle = \left[\frac{h Q_{rms} k^2 \rho c (h_s + h_r)}{2\pi r^2 \sqrt{1 + k^2 a^2}} \right]^2 \qquad (5.197)$$

where h is defined in Figure 5.2, a represents the source size (radius of a spherical source), and h_s and h_r are defined in Figure 5.14.

Equation (5.197) shows that, at large distances from the source such that $Z_s \cos\theta \ll \rho c$ and $\cos\theta \approx \rho c (h_s + h_r)/r$, and provided that the direct and reflected waves combine coherently ($\Phi < 0.1$), the sound pressure level decreases as the inverse fourth power and not as the inverse square of the distance r, and thus decreases at the rate of 12 dB for each doubling of distance from the source, not 6 dB as is the case when the direct and reflected waves combine incoherently ($\Phi > 1.0$). Note that Φ is defined in Equation (5.169).

It may be concluded that, near to the sound source, coherent analysis will give good results, but far from the source the actual attenuation due to the ground effect will be somewhere between two values. The first value is calculated using Equations (5.146) and (5.195) and assuming coherent reflection (no turbulence). The second is calculated assuming incoherent reflection (with turbulence) and that sound waves arriving at the receiver by different paths add incoherently. As the distance from the source or frequency increases, the incoherent model (Equation (5.188)) will become more appropriate.

5.11.12 Meteorological Effects

The two principal meteorological variables are wind gradient and vertical temperature gradient. When the temperature increases with height and the temperature gradient is thus positive, the condition is termed an inversion. When the temperature decreases with height and the gradient is thus negative, the condition is termed a lapse. It has been established (Rudnick, 1957) that curvature of sound propagation is mainly dependent upon the vertical gradient of the speed of sound, which can be caused either by a wind gradient or by a temperature gradient, or by both. It has also been shown (Piercy *et al.*, 1977) that refraction due to either a vertical wind gradient or a vertical temperature gradient produces equivalent acoustic effects, which are essentially additive.

A positive temperature gradient (temperature inversion) near the ground will result in sound waves that normally travel upwards being diffracted downwards towards the ground, resulting, in turn, in increased noise levels on the ground. Alternatively, a negative temperature gradient (temperature lapse) will result in reduced noise levels on the ground.

As wind speed generally increases with altitude, wind blowing towards the observer from the source will refract sound waves downwards, resulting in increased noise levels. Conversely, wind blowing from the observer to the source will refract sound waves upward, resulting in reduced noise levels. Here a procedure is given for calculating the sonic gradient and the radius of curvature of the refracted wave.

The wind profile at low altitudes is determined by the ground surface roughness and may be expressed in the following form:

$$U(h) = U_0 h^{\xi} \tag{5.198}$$

In Equation (5.198), $U(h)$ is the wind speed at height h and U_0 is a constant (Manning, 1981). The quantity, ξ, is effectively a constant for a given ground surface and may be derived from data provided by Sutton (1953). Values of ξ for various ground surfaces are listed in Table 5.6. The height h is normally taken as 10 m, although a height of 5 m is sometimes used.

Differentiation of Equation (5.198) gives the following expression for the expected gradient at height, h.

$$\frac{\mathrm{d}U}{\mathrm{d}h} = \xi \frac{U(h)}{h} \tag{5.199}$$

The wind gradient is a vector. Hence U is the velocity component of the wind in the direction from the source to the receiver and measured at height h above the ground. The velocity component, U, is positive when the wind is blowing in the direction from the source to the receiver and negative for the opposite direction. In Section 1.3.4 it

Table 5.6 Values of the empirical constant ξ.

Type of ground surface	ξ
Very smooth (mud flats, ice)	0.08
Snow over short grass	0.11
Swampy plain	0.12
Sea	0.12
Lawn grass, 1 cm high	0.13
Desert	0.14
Snow cover	0.16
Thin grass, 10 cm high	0.19
Air field	0.21
Thick grass, 10 cm high	0.24
Country side with hedges	0.29
Thin grass, 50 cm high	0.36
Beet field	0.42
Thick grass, 50 cm high	0.43
Grain field	0.52

is shown that the speed of sound is related to the atmospheric temperature according to Equation (1.9). The latter expression may be rewritten more conveniently for the present purpose as follows:

$$c = c_0\sqrt{T/273} \tag{5.200}$$

where T is the temperature in °Kelvin and c_0 is the speed of sound at sea level and 0°C (331 m/s). It will be assumed that the vertical temperature profile is linear; that is, the vertical temperature gradient is constant. The vertical sound speed gradient is found by differentiating Equation (5.198) with respect to h to give:

$$\left[\frac{\partial c}{\partial h}\right]_T = 10.0\,\frac{dT}{dh}\left[T_0 + 273\right]^{-1/2} \tag{5.201}$$

In Equation (5.201), dT/dh is the vertical temperature gradient, °C /m, and T_0 is the ambient temperature in °C at 1 m height.

It will be assumed that the gradient due to wind, given by Equation (5.199), gives rise to an effective speed of sound gradient of equal magnitude and it will be assumed that this wind contribution to the sound speed gradient adds to the temperature gradient contribution given by Equation (5.201). The total vertical gradient, dc/dh, then becomes the following:

$$\frac{dc}{dh} = \frac{dU}{dh} + \left[\frac{\partial c}{\partial h}\right]_T \tag{5.202}$$

A sound wave travelling nominally parallel to the ground will have a curved path with radius of curvature, R, given by the following equation (De Jong and Stusnik, 1976). When R is positive the sound rays are curved downward and when R is negative the sound rays are curved upward:

$$R = c_0 \left[\frac{dc}{dh} \right]^{-1}$$

(5.203)

The radius of curvature is used together with the barrier analysis in Section 8.5.2 to determine the effect of wind and temperature gradients on barrier attenuation.

Note that wind and temperature gradients have a negligible effect on sound levels within a few tens of metres from the sound source.

When predicting outdoor sound, it is usual to include downwind or temperature inversion meteorological conditions, corresponding to CONCAWE category 5 (see Section 5.11.12.3). The downwind condition implies that the wind direction makes an angle of less than 45 ° to the line joining the source to the receiver.

5.11.12.1 Attenuation in the Shadow Zone (Negative Sonic Gradient)

A shadow zone is defined as that region where direct sound cannot penetrate due to upwards diffraction. Of course, a small amount of sound will always transmit to the shadow zone as a result of scattering from objects so that one might expect an increase in attenuation in the shadow zone to be up to 30 dB. To create a shadow zone, a negative sonic gradient must exist. If this is the case, then the distance, x, between the source (height, h_s) and receiver (height, h_r) beyond which the receiver will be in the shadow zone is:

$$x = \sqrt{-2R} \left[\sqrt{h_s} + \sqrt{h_r} \right]$$

(5.204)

Note that in the presence of no wind, the shadow zone around a source will only exist when there is no temperature inversion close to the ground and will be symmetrical around the source. In the presence of wind, the distance to the shadow zone will vary with direction from the source, as the sonic gradient in a given direction is dependent on the component of the wind velocity in that direction. Thus, it is possible for a shadow zone to exist in the upwind direction and not exist in the downwind direction.

The angle subtended from a line drawn from the source towards the oncoming wind beyond which there will be no shadow zone is called the critical angle β_c and is given by:

$$\beta_c = \cos^{-1} \left[\frac{(\partial c / \partial h)_T}{dU / dh} \right]$$

(5.205)

This is illustrated in Figure 5.21. The actual excess attenuation due to the shadow zone increases as the receiver moves away from the source further into the zone. It is also dependent on the difference between the angle β, between the receiver / source line

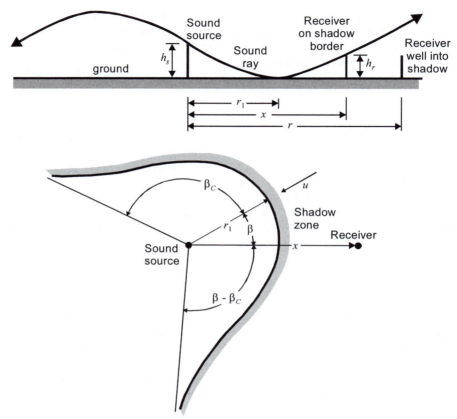

Figure 5.21 Illustration of the shadow zone and the limiting angle β_c, beyond which the shadow zone does not exist. β is the angle subtended between the wind direction and the source / receiver line.

and the line from the source towards the oncoming wind (see Figure 5.21) and the critical angle β_c. The actual excess attenuation as a function of β and β_c may be determined using Figures 5.22 and 5.23 for wind speeds of 5 to 30 km/hr, for octave bands from 250 Hz to 4000 Hz, for ground cover heights less than 0.3 m, for sound source heights of 3 to 5 m and a receiver height of 2 m (Wiener and Keast, 1959). Note that for wind speeds greater than 30 km/hr, noise due to wind blowing over obstacles and rustling leaves in trees usually dominates other environmental noise, unless the latter is particularly severe.

The attenuation for a particular value of β is calculated using Figure 5.22. This is only the excess attenuation due to a negative vertical sonic gradient (upwind propagation and / or no temperature inversion close to the ground). The attenuation cannot exceed the value given by Figure 5.23, which is a function of $\beta_c - \beta$.

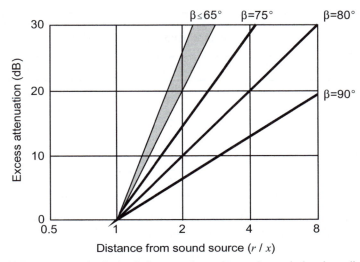

Figure 5.22 Excess attenuation in the shadow zone due to the negative vertical sonic gradient for various angles, β, between the wind direction and the line joining the source and receiver.

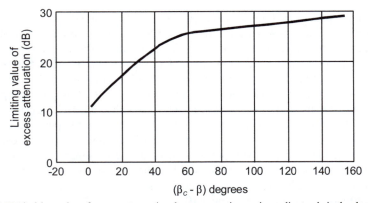

Figure 5.23 Limiting value of excess attenuation due to a negative sonic gradient only in the shadow zone.

5.11.12.2 Meteorological Attenuation Calculated according to Tonin (1985)

For the case of open terrain, review of available data (Parkin and Scholes, 1965; Piercy *et al.*, 1977) has allowed construction of Table 5.7. Excess attenuation of ground-borne aircraft noise measured under a variety of weather conditions has been classified according to the observed vertical sound speed gradient. Attenuation at two distances is recorded in the table.

Attenuations for other distances and sonic gradients may be interpolated from the table with the understanding that no increase is to be expected in the values shown for distances greater than 616 m and for vertical gradients greater than 0.15 s^{-1}.

Table 5.7 Excess atten. A_m (dB) due to wind and temp. gradients at two distances, 110 m and 616 m

Total vertical gradient (s^{-1})	Octave centre band frequencies (Hz)									
	31.5	63	125	250	500	1000	2000	4000	8000	16000
110 metres										
+0.075	-2	-2	-0.5	3	-2	-5	-2	-2	-2	-2
-0.075	1	1	2.5	0	2	6	10	6	6	6
616 metres										
+0.075	-5	-5	-2	0	-9	-9	-6	-7	-7	-7
-0.075	5	5	6	4	7	7	7	6	6	6

Atmospheric turbulence also results in fluctuations of the sound received by the observer, and the effect is usually greater during the day than at night. This effect is not included in this procedure.

5.11.12.3 Meteorological Attenuation Calculated according to CONCAWE

An alternative procedure for calculating the excess attenuation due to meteorological effects, which includes atmospheric turbulence effects, has been proposed by Manning (1981). In this procedure, meteorological effects have been graded into six categories based upon a combined vertical gradient (Manning, 1981). In Table 5.8, levels of incoming solar radiation are defined for use in Table 5.9. In Table 5.9, the temperature gradient is coded in terms of Pasquill stability category A - G. Category A represents a strong lapse condition (large temperature decrease with height). Categories E, F and G on the other hand, represent a weak, moderate and strong temperature inversion respectively with the strong inversion being that which may be observed early on a clear morning. Thus category G represents very stable atmospheric conditions while category A represents very unstable conditions. The wind speed in this table is a non-vector quantity and is included for the effect it has on the temperature gradient. Wind gradient effects are included in Table 5.10.

Table 5.8 Day-time incoming solar radiation (full cloud cover is 8 octas, half cloud cover is 4 octas, etc.)

Altitude of sun	Cloud cover (octas)	Incoming solar radiation
$< 25°$	$0 - 7$	slight
$25° - 45°$	< 4	moderate
$> 45°$	$5 - 7$	moderate
$> 45°$	< 4	strong

The Pasquill stability category is combined with the magnitude of the wind vector using Table 5.10 to determine one of the six meteorological categories (CAT) for which attenuations were experimentally determined from surveys of three European process plants (Manning, 1981). Attenuations are shown in Figures 5.24(a) - (g) for the octave bands from 63 Hz to 4 kHz. The wind speed used in Tables 5.9 and 5.10 is measured at ground level and is the proportion of the wind vector pointing from the source to the receiver.

Table 5.9 CONCAWE determination of Pasquill stability category from meteorological information (full cloud cover is 8 octas, half cloud cover is 4 octas, etc.). After Marsh (1982)

Wind speed[a] m/s	Day-time incoming solar radiation				1 hour before sunset or after sunrise	Night-time cloud cover (octas)		
	slight	moderate	strong	Over-cast		0-3	4-7	8
< 2	A	A-B	B	C	D	F or G[b]	F	D
2.0-2.5	A-B	B	C	C	D	F	E	D
3.0-4.5	B	B-C	C	C	D	E	D	D
5.0-6.0	C	C-D	D	D	D	D	D	D
> 6.0	D	D	D	D	D	D	D	D

[a] Wind speed is measured to the nearest 0.5 m/s and is measured close to the ground.

[b] Category G is restricted to night-time with less than 1 octa of cloud and a wind speed less than 0.5 m/s.

Table 5.10 CONCAWE determination of meteorological category (negative values represent the receiver upwind of the source and positive values are for the receiver downwind of the source). After Marsh (1982)

Meteorological category	Pasquill stability category		
	A, B	*C, D, E* wind speed v (m/s)	*F, G*
1	$v < -3.0$	−	−
2	$-3.0 < v < -0.5$	$v < -3.0$	−
3	$-0.5 < v < +0.5$	$-3.0 < v < -0.5$	$v < -3.0$
4[a]	$+0.5 < v < +3.0$	$-0.5 < v < +0.5$	$-3.0 < v < -0.5$
5	$v > +3.0$	$+0.5 < v < +3.0$	$-0.5 < v < +0.5$
6	−	$v > 3.0$	$+0.5 < v < +3.0$

[a] Category with assumed zero meteorological influence.

If the weather conditions are not accurately known, then it is generally assumed that meteorological effects will result in a sound level variation about the predicted level, as shown in Table 5.11.

5.11.12.4 Meteorological Attenuation Calculated according to ISO 9613-2 (1996)

ISO 9613-2 procedures for calculating ground effects and shielding effects are based on an assumption of downwind propagation from the sound source to the receiver.

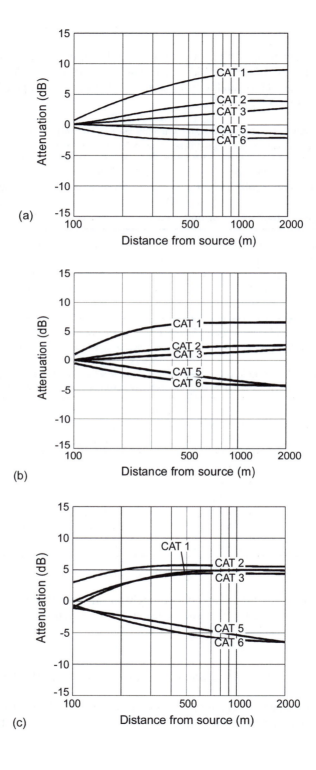

(a)

(b)

(c)

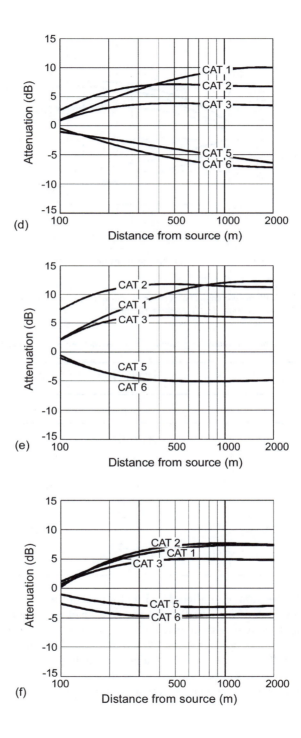

(d)

Distance from source (m)

(e)

Distance from source (m)

(f)

Distance from source (m)

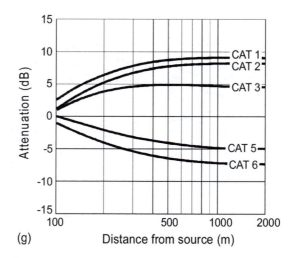

(g)   Distance from source (m)

Figure 5.24 CONCAWE meteorological curves for various octave bands. (a) 63 Hz, (b) 125 Hz, (c) 250 Hz , (d) 500 Hz, (e) 1000 Hz, (f) 2000 Hz, (g) 4000 Hz.

Table 5.11 Variability in sound level predictions due to meteorological influences, A_m (dB)

Octave band centre frequency (Hz)	Distance from source (m)			
	100	200	500	1000
		A_m (dB)		
63	+ 1	+4, -2	+7, -2	+8, -2
125	+1	+4, -2	+6, -4	+7, -4
250	+3, -1	+5, -3	+6, -5	+7, -6
500	+3, -1	+6, -3	+7, -5	+9, -7
1000	+7, -1	+11, -3	+12, -5	+12, -5
2000	+2, -3	+5, -4	+7, -5	+7, -5
4000	+2, -1	+6, -4	+8, -6	+9,-7
8000	+2, -1	+6, -4	+8, -6	+9, -7

Thus the only correction term offered by ISO for meteorological effects is a term to reduce the A-weighted calculated sound pressure level for long-time averages of several months to a year. Essentially, the correction is to allow for the fact that downwind propagation does not occur 100% of the time. The corrections are only to be added to overall A-weighted level calculations and they are not to be included for locations closer to the source than ten times the sum of the source and receiver heights. The correction to account for downwind propagation not occurring 100% of the time is:

$$A_m = A_0[1 - 10(h_s + h_r)/d]$$ (5.206)

where h_s and h_r are the source and receiver heights respectively and d is the horizontal distance between the source and receiver. The quantity, A_0 depends on local meteorological statistics and varies between 0 and 5 dB with values over 2 dB very rare. The standard offers no other procedure for calculating A_0. The value of A_m calculated using this procedure is intended only as a correction to the A-weighted sound level, which is why it contains no frequency dependent terms. However, in the absence of a procedure for calculating A_0, the equation is not very useful.

5.11.13 Combined Excess Attenuation Model

Because it relies on combining different sound rays at the receiver, the method discussed in Section 5.11.10.5 is often referred to as the ray tracing method. However, it should be pointed out that the ground attenuation calculated using this method cannot really be added to that due to barriers and meteorological effects as all three effects interact. The presence of a barrier increases the number of ground reflections and changes their location. The presence of wind and temperature gradients or layers of air of different temperature or wind strength cause sound rays to bend and result in multiple ground reflections (sometimes called "bounces" as the ray bounces up and is then refracted down). Thus current thinking is to use a complex model based on spherical reflection that takes into account all of the different possible ray paths of sound travelling from the source to the receiver (Raspet *et al.*, 1995; Li *et al.*, 1998; Li, 1993, 1994; and Plovsing, 1999). In this way, the effects of atmospheric absorption, ground impedance, meteorological effects (turbulence as well as wind and temperature gradients) and barriers are all included in the one single calculation. However, this approach is extremely complex and is not yet sufficiently developed to appear in commercial software. Thus, at present, acceptable accuracy is obtained by adding the various separate excess attenuation effects as discussed in the previous sections, although ground and barrier effects should be combined as discussed in these sections.

5.11.14 Accuracy of Outdoor Sound Predictions

Although weather conditions can cause large variations in environmental sound levels, the accuracy of prediction is much better than the expected variations if the meteorological conditions are properly accounted for as discussed in the previous sections. The expected accuracy of the overall A-weighted predictions if the procedures discussed in Section 5.11 are followed is ± 3 dB. However, these figures are a guide only as to what the measurement uncertainty really is. In practice, it is important to take into account all contributors to the prediction of environmental noise arriving at a receiver as the result of operation of one or more sound sources. These include uncertainties associated with the: source sound power output; ground surface model; weather conditions (temperature inversions, turbulence and wind); vehicles and other temporary obstacles between the source and receiver; and calculations of attenuation due to any permanent obstacles between source and receiver; background noise levels. There are also uncertainties associated with the measurement of environmental noise and these are discussed in detail by Craven and Kerry (2001).

CHAPTER SIX

Sound Power, Its Use and Measurement

LEARNING OBJECTIVES

In this chapter the reader is introduced to:

- radiation impedance of a source and what influences it;
- the relation between sound power and sound pressure;
- radiation field, near field and far field of a sound source;
- sound power determination using sound intensity measurements, surface vibration measurements and sound pressure measurements, both in the laboratory and in the field; and
- some uses of sound power information.

6.1 INTRODUCTION

In this chapter it will be shown how the quantities, sound pressure, sound intensity and sound power are related. Sound pressure is the quantity most directly related to the response of people and things to airborne sound and this is the quantity that is most often to be controlled. However, sound pressure level is not always the most convenient descriptor for a noise source, since it will depend upon distance from the source and the environment in which the sound and measurement position are located. A better descriptor is usually the sound power level of the source (see Section 1.8).

In many cases, sound power level information is sufficient and no additional information is required for the specification of sound pressure levels. In other cases, information about the directional properties of the source may also be required. For example, the source may radiate much more effectively in some directions than in others. However, given both sound power level and directivity index (see Section 5.8), the sound pressure level at any position relative to the source may then be determined. In fact, a moment's reflection will show that specification of sound pressure level could be quite insufficient for a source with complicated directivity; thus the specification of sound power level for sources is preferred for many noise-control problems.

The sound field generated by a source can be defined in terms of the source properties and the environment surrounding the source. The sound field, in turn, produces a radiation load on the source that affects both: (1) the sound power generated by the source; and (2) the relationship between sound power and sound pressure. In this chapter, the discussion will be restricted to the latter two effects, and will include the experimental determination of source sound power using sound pressure or sound intensity measurements and the use of sound power information for the determination of sound pressure levels in the free field.

When a source radiates into free space one is interested in directivity as well as sound power level. However, if a source radiates into a reverberant field, directivity information is lost, and sound power level is all that one need know about the source. This matter will be considered in detail later in this chapter. Thus, for the present, it will be sufficient to point out that sound power level information is useful for the following purposes:

- it allows comparison of the noise-producing properties of different machines;
- it allows verification that the noise produced by a particular machine meets specifications for noise-control purposes; and
- it provides a means for predicting expected noise levels in reverberant spaces and in the free field when directivity information is also known.

6.2 RADIATION IMPEDANCE

When the sound power level of a source is specified, unless otherwise stated, the assumption is implicit that the radiation impedance presented to the source is the same as it would be in a free unbounded space, commonly referred to as "free field". The radiation impedance is analogous to the load impedance presented to a generator in the more familiar case of electrical circuit theory. In the latter case, as is well known, the power delivered by a generator to a load depends upon both the load impedance and the generator internal impedance. For the case of acoustic sources, the internal impedance is seldom known, although for vibrating and radiating structures the assumption is commonly made that they are constant volume-velocity (or infinite internal impedance) sources. The meaning of this is that the motion of the vibrating surface is assumed to be unaffected by the acoustic radiation load, and this is probably a good approximation in many cases.

Most aerodynamic sources are, however, not well represented as constant volume-velocity sources. In these cases, the problem becomes quite complicated, although such sources are often approximated as constant pressure sources, which means that the acoustic pressure at the source, rather than the source volume-velocity, is unaffected by the acoustic radiation load.

The radiation impedance presented to a source in a confined space or in the presence of nearby reflectors will seldom be the same as that presented to the same source by a free field. For example, if a sound source is placed in a highly reverberant room, as will be discussed later, the radiation impedance will strongly depend on both position and frequency, but in an indeterminate way. However, if the source is moved about to various positions, the average radiation impedance will tend to that of a free field. Thus, in using sound power level as a descriptor, it is tacitly assumed that, on average, the source will be presented with free field, and on average this assumption seems justified.

Some special situations in which the free-field assumption is not valid, however, are worth mentioning. If an omnidirectional constant volume-velocity source is placed next to a large reflecting surface such as a wall, then the source will radiate twice the power that it would radiate in free space (note that a constant pressure source would radiate half the power). This may be understood by considering that the pressure everywhere would be doubled, and thus the intensity would be four times what it would be in free space. As the surface area of integration is reduced by a factor of two from a sphere to a hemisphere encompassing the source, the integral of the intensity, or power radiated, is double what it would be in free space. Thus the sound power radiated by an omnidirectional constant volume-velocity source will be least in the free field, 3 dB greater when the source is placed next to a reflecting surface, 6 dB greater when the source is placed at the junction of two surfaces, and 9 dB greater when the source is placed in the corner at the junction of three surfaces. The corresponding intensity of sound at some distance r from the surfaces will be 6, 12 and 18 dB respectively greater than in free field.

A restriction in application of these ideas must be mentioned. For a source to be close to a reflecting surface it must be closer than one-tenth of a wavelength of the emitted sound. At distances greater than one-tenth of a wavelength the effect upon radiated sound power rapidly diminishes until, for a band of noise at a distance of one-half wavelength, the effect is negligible. For tones, the effect persists to somewhat greater distances than a half wavelength.

If the source is sufficiently far away (half wavelength for octave or one-third octave band sources and two wavelengths for pure tone sources) that its sound power is not at all affected by the reflecting surface, then the sound pressure level (and directivity D) at any location can be estimated by adding the sound field of an image source to that of the source with no reflector present, as discussed in Section 5.9.

Making reference to the discussion of reflection effects in Chapter 5, Section 5.9, and noting that for the cases considered in Table 5.1 and for a constant volume-velocity source located closer than a quarter wavelength to the reflecting surface(s), the following equation may be written for the intensity, I:

$$I = W_0 D^2 / 4\pi r^2 = \langle p^2 \rangle / \rho c \tag{6.1a,b}$$

where W of Equation (5.122) has been replaced with $W_0 D$, W_0 is the power radiated in the free field, and D takes the values shown in Table 5.1.

The internal impedance of a practical source may not be large and it will never be infinite. The case of infinite impedance (i.e. a constant volume-velocity source) may be taken as defining the upper bound on radiated power. For example, a loudspeaker, which, unless backed by a small, airtight enclosure, is not in general a constant volume-velocity source, placed in a corner in a room will produce more sound than when placed in the free field, but the increase in intensity at some reference point on axis will be less than 18 dB.

6.3 RELATION BETWEEN SOUND POWER AND SOUND PRESSURE

The sound pressure level produced by a source may be calculated in terms of the specified sound power level and directivity. Use of Equation (5.126), and the observation that in the SI system of units the characteristic impedance ρc is approximately 400, allows the following approximate equation to be written relating sound pressure level, sound power level and directivity factor:

$$L_p = L_w + 10\log_{10}D - 10\log_{10}S \tag{6.2}$$

where

$$S = 4\pi r^2 \tag{6.3}$$

Equation (6.2) relates the sound pressure level, L_p, at a point to the sound power level, L_w, of a source, its directivity factor, D, dependent upon direction from the source, and the distance, r, from the source to the measurement point. The equation holds as long as the measurement point is in the far field of the source. The far field will be discussed in Section 6.4.

If sound power is not constant and the alternative case of a constant volume-velocity source, radiating equally well in all directions and located within a quarter of a wavelength of a reflecting surface, is considered then, as discussed earlier, W in Equation (6.2) is replaced with W_0D, where W_0 is the sound power that the source would radiate in the free field. In this case, the relation between sound pressure level L_p and sound power level L_{w0} becomes:

$$L_p = L_{w0} + 20\log_{10}D - 10\log_{10}S \tag{6.4}$$

Example 6.1

A swimming pool pump has a sound power level of 60 dB re 10^{-12} W when resting on the ground in the open. It is to be placed next to the wall of a building, a minimum of 2 m from a neighbour's property line. If the pump ordinarily radiates equally well in all directions (omnidirectional), what sound pressure level do you expect at the nearest point on the neighbour's property line?

Solution

It is implicitly assumed that the source is well within one-quarter of a wavelength of the reflecting wall and ground to calculate an upper bound for the expected sound pressure level.

As an upper bound, a constant volume-velocity source is assumed. Previous discussion has shown that, for a source of this type, the radiated sound power level L_{w0} in the presence of no reflecting planes is 3 dB less than it is in the presence of one reflecting plane. Thus, the free field sound power in the absence of any reflecting plane is:

$$L_{w0} = 60 - 3 = 57 \text{ dB}$$

Use is made of Equation (6.4) to write:

$$L_p = 57 + 20\log_{10}4 - 10\log_{10}(16\pi)$$
$$L_p = 52 \text{ dB re } 20\mu\text{Pa}$$

Alternatively, it may be assumed that the sound power is not affected by the reflecting surfaces (either by assuming a constant-power source or by assuming that the source is more than a quarter of a wavelength from the reflecting surfaces) and use is made of Equation (6.2) to calculate a lower bound as follows:

$$L_p = 60 + 10\log_{10}4 - 10\log_{10}(16\pi)$$
$$L_p = 49 \text{ dB re } 20\mu Pa$$

It is concluded that a sound pressure level between 49 and 52 dB may be expected at the neighbour's property line. Note that if a constant pressure aerodynamic source were assumed and its acoustic centre was within a quarter of a wavelength from the reflecting surfaces, the lower bound would be 46 dB. However, in this case, the assumption of a constant pressure source is not justified.

6.4 RADIATION FIELD OF A SOUND SOURCE

The sound field radiated by a source in a free field may be divided into three regions: the hydrodynamic near field, the geometric (or Fresnel) near field, and the far field.

In general, the hydrodynamic near field is considered to be that region immediately adjacent to the vibrating surface of the source, extending outward a distance much less than one wavelength. This region is characterised by fluid motion that is not directly associated with sound propagation. For example, local differences in phase of the displacement of adjacent parts of a vibrating surface will result in fluid motion tangential to the surface, if the acoustic wavelength is long compared with their separation distance. The acoustic pressure will be out of phase with local particle velocity. As sound propagation to the far field is associated with the in-phase components of pressure and particle velocity, it follows that measurements of the acoustic pressure amplitude in the near field give an inaccurate indication of the sound power radiated by the source.

The sound field adjacent to the hydrodynamic near field is known as the geometric near field. In this region, interference between contributing waves from various parts of the source lead to interference effects and sound pressure levels that do not necessarily decrease monotonically at the rate of 6 dB for each doubling of the distance from the source; rather, relative maxima and minima are to be expected. This effect is greater for pure tones than it is for bands of noise. However, in the geometric near field, the particle velocity and pressure of the contributing waves from the various parts of the source are in phase, as for waves in the far field, although the pressure and particle velocity of the resulting combined waves in the geometric near field may not be in phase.

The possibility of determining the sound power radiated by an extended source from pressure measurements made in the geometric near field has been investigated for the case of the baffle mounted piston considered in Section 5.6, using a computer simulation (Bies and Bridges, 1993). It was shown that a simple sound level meter can provide an accurate determination of the radiated sound power even in the geometric near field of the source, where close investigation of the field with an intensity meter would suggest that such determination might not be possible. In fact, in the case investigated, the intensity meter would appear to provide no special advantage.

While the reported study was not exhaustive, it is probably indicative and thus it may be concluded that radiated sound power can be calculated from a sufficient number of sound pressure measurements made in the geometric near field. The problem becomes the determination of a sufficient number of measurements, for which no general rule seems known as of this writing. Consequently, a great many measurements may be necessary to determine when further measurements would appear to provide no improvement in the estimate of radiated sound power.

The region of the sound field extending beyond the geometric near field to infinity is known as the far field, where sound pressure levels decrease monotonically at the rate of 6 dB for each doubling of the distance from the source (for exceptions see Section 5.11.11). In the far field, the source directivity is well defined. The far field is characterised by the satisfaction of three criteria, written as follows (Bies, 1976):

$$r \gg \lambda/(2\pi), \qquad r \gg \ell, \qquad r \gg \pi\ell^2/(2\lambda) \qquad\qquad (6.5\text{a-c})$$

where r is the distance from the source to the measurement position, λ is the wavelength of radiated sound and ℓ is the characteristic source dimension. The "much greater than" criterion in the above three expressions refers to a factor of three or more. More generally, defining $\gamma = 2r/\ell$, and $\kappa = \pi\ell/\lambda$, the above criteria reduce to:

$$\gamma \gg 1/\kappa, \qquad \gamma \gg 2, \qquad \gamma \gg \kappa \qquad\qquad (6.6\text{a-c})$$

These criteria are used to construct Figure 6.1.

The criterion given by Equations (6.5c) or (6.6c) defines the bound between the geometric near field and the far field, as shown in Figure 6.1. It should be pointed out, however, that while satisfaction of the inequality given by the equations is sufficient to ensure that one is in the far field, it may not always be a necessary condition. For example, a very large pulsating sphere has only a far field.

6.4.1 Free-field Simulation in an Anechoic Room

One way to produce a free field for the study of sound radiation is to construct a room that absorbs all sound waves which impinge upon the walls. Such a room is known as an anechoic room, and a sound source placed in the room will produce a sound field similar in all respects to the sound field that would be produced in a boundary free space, except that its extent is limited by the room boundaries.

Figure 6.1 has special significance for the use of anechoic rooms for the purpose of simulating free field. For example, if the characteristic length of a test source is

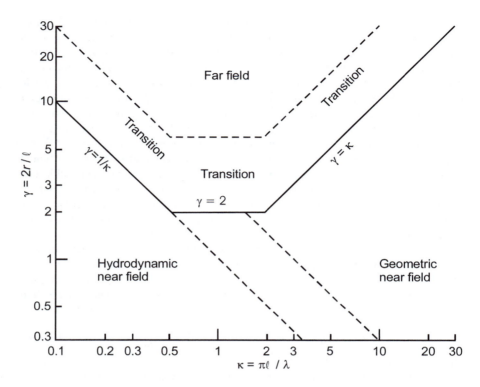

Figure 6.1 Radiation field of a source.

0.2 m and the maximum distance from the source that can be accommodated within an anechoic room is 2 m, then the far field of the source is limited to values of γ ($= 2r/\ell$) less than or equal to 20. When this bound is entered in Figure 6.1, a triangular region is defined in which values of κ and γ may be chosen to ensure that far field is achieved. Similarly, an upper bound, dependent upon the absorptive properties of the walls at high frequencies, may be placed upon κ.

Inspection of Figure 6.1 shows that the region of the far field and even of the geometric near field may be quite limited in a given anechoic room. Indeed, for some sources, it may not even be possible to achieve far-field conditions in a given room. In these instances estimates of sound power levels will have to be made from sound pressure or intensity measurements in the geometric near field. However, any directivity information inferred from measurements made in the geometric near field must be treated with caution, especially if the source is characterised by one or more tones.

6.4.2 Sound Field Produced in an Enclosure

If the sound source is located in a room or enclosure which does not have highly absorptive walls, the sound field radiated by the source will have superimposed upon it additional sound waves caused by reflection of the original sound waves from the room surfaces. The sound field generated by reflection from the room surfaces is called the reverberant field, and at a sufficient distance from the sound source the reverberant field may dominate the direct field generated by the source. In a reverberant field, many reflected wavetrains are usually present and the average sound pressure reaches a level (in the region where the reverberant field dominates) that is essentially independent of distance from the source. The reverberant field is called a diffuse field if a great many reflected wavetrains cross from all possible directions and the sound energy density is very nearly uniform throughout the field. A room in which this is the case is known as a reverberant room. Such a room has boundaries that are acoustically very hard, resulting in a reverberant field that dominates the whole room except for a small region near the source. More will be said about reverberant spaces in the following sections, and in Chapter 7.

6.5 DETERMINATION OF SOUND POWER USING INTENSITY MEASUREMENTS

Recommended practices for the direct measurement of sound intensity are described in various standards (ANSI S12.12-1992 (R2007), ISO 9614/1-1993, ISO 9614/2-1996 and ISO 9614/3-2002). The measurement of sound intensity provides a means for directly determining the magnitude and direction of the acoustic energy flow at any location in space. Measuring and averaging the acoustic intensity over an imaginary surface surrounding a machine allows determination of the total acoustic power radiated by the machine.

To measure sound power, a test surface of area, S, which entirely encloses the sound source, is set up. The sound field is then sampled using an intensity probe scanned over a single test surface, or where a pure tone sound source is measured, a single microphone may be scanned over two test surfaces as described following Equation (3.41). By either method, the sound intensity, I_n, which is the time-averaged acoustic intensity component normal to the test surface averaged over the area of the test surface, is determined. Once the sound intensity has been determined over the enclosing test surface, either by using a number of point measurements or by slowly scanning the intensity probe and averaging the results on an energy (p^2) basis, the sound power may be determined using:

$$W = I_n S \qquad (6.7)$$

Details concerning the use of intensity meters and various procedures for measuring acoustic intensity are discussed in Section 3.13.

Theoretically, sound power measurements made using sound intensity can be conducted in the near field of a machine, in the presence of reflecting surfaces and near other noisy machinery. Any energy from these sources flowing into the test surface at one location will flow out at another location provided that there is a negligible amount of acoustic absorption enclosed by the test surface. Because sound intensity is a vector quantity, when it is averaged over a test surface only the net outflow of energy through that surface is measured. However, if the reactive field (see Section 1.7.1) associated with reflecting surfaces, or the near field of a sound source, is 10 dB or more greater than the active field, or if the contribution of other nearby sound sources is 10 dB or more greater than the sound pressure level of the source under investigation, then in practice, reliable sound intensity measurements cannot be made with currently available precision instrumentation using two phase matched microphones with an accurately measured separation distance. This is because in this situation the acoustic pressures measured by the two microphones are relatively large but the difference between the levels at the two microphones is relatively small. Sound intensity measurements and the corresponding precision that may be expected are discussed in Section 3.13.

The availability of the intensity meter and some experience with its use has shown that, in general, the radiation of sound from a source is much more complicated than might be supposed. It is not uncommon for a vibrating surface to exhibit areas of sound absorption as well as radiation. Thus a map of the acoustic power radiation can be quite complicated, even for a relatively simple source. Consequently, the number of measurements required to determine the field may be quite large, and herein lies the difficulty in the use of the sound intensity meter. When using the intensity meter to determine the net power transmission away from a source, one must be sure to make a sufficient number of measurements on the test surface enclosing the source to adequately describe the resulting sound field. Stated differently, the sound intensity meter may provide too much detail when a few naively conducted pressure measurements may provide an adequate answer.

It is suggested that measurements using an intensity meter be made in the free or semi-free field where possible. Alternatively, intensity measurements can be made close to a radiating surface in a reverberant field (ANSI S12.12-1992 (R2007)). Errors inherent in acoustic intensity measurements and limitations of instrumentation are discussed in Sections 3.12.2.1 and 3.13.2.1 and by Fahy (1995). Guidelines for intensity measurement are provided in ISO 9614/1-1993, ISO 9614/2-1996 and ISO 9614/3-2002.

6.6 DETERMINATION OF SOUND POWER USING CONVENTIONAL PRESSURE MEASUREMENTS

There are a number of accepted methods for the determination of sound power based upon sound pressure measurements made in the vicinity of a source. The choice of method is dependent to a large degree on the precision required, the mobility of the source, the presence of other noise sources if the source to be tested cannot be moved, and the expected field location of the source with respect to reflecting surfaces such as floors and walls.

Each method discussed in the paragraphs to follow is based upon pressure measurements. The discussion begins with the most accurate method and ends with the least accurate; the latter method is used when the source to be tested cannot be moved and other immobile noise sources are in the near vicinity.

6.6.1 Measurement in Free or Semi-free Field

The determination of the sound power radiated by a machine in the free field, using pressure measurements alone, requires that any reverberant sound be negligible. This condition is usually realised only in an anechoic room. Performance specifications of a suitable anechoic room are included in ISO 3745-2003, which also describes internationally accepted test procedures.

The determination of the sound power radiated by a machine in the presence of one or more plane-reflecting surfaces requires that any reverberant sound (i.e. sound returned to the machine from other than the plane reflecting surfaces considered in the measurement) is negligible. In this case the machine, and its one or more acoustical images, may be thought of as the source of sound whose sound power is measured. The standards mentioned earlier describe appropriate measurement arrangements in anechoic rooms for precision measurements. ISO 3744-1994 describes similar but less precise measurement in the open or in large rooms, which are not necessarily anechoic.

The sound power of a machine is determined by the integration of the intensity over a hypothetical spherical surface surrounding it. The centre of the sphere should be the acoustic centre of the machine, and a good approximation to this is generally the geometrical centre of the machine. The sphere should be chosen such that its surface is in the radiation far field of the source (where the sound intensity level is related directly to the sound pressure level and where sound power levels and directivity information can be obtained), or at least in the geometric near field where sound power levels (Bies and Bridges, 1993) but not reliable directivity information can be obtained, but not in the hydrodynamic near field.

If measurements are made in an anechoic room then, according to the standards, the surface of the sphere should be at least one-quarter of a wavelength of sound away from the anechoic room walls and should have a radius of at least twice the major machine dimensions, but not less than 0.6 m. A more accurate means for determining the required radius is to use Figure 6.1. For example, if the major machine dimension is 2 m and the frequency of interest is 125 Hz ($\ell = 2.75$ m) then, from Figure 6.1, the geometric near field will lie in the range from about 1.3 m to about 2.3 m from the source and the far field will lie at distances greater than about 6.5 m. It can be seen that for most machines anechoic rooms of very large size are implied if measurements are to be made in the far field of the source.

The excessively large size of an appropriate anechoic room strongly recommends consideration of the anechoic space above a reflecting plane, as would be provided out-of-doors with the machine mounted on the ground away from any other significant reflectors. Alternatively, such a space can be provided indoors, in a semi-anechoic room. Carrying the consideration further, the mounting arrangement of the machine in use might suggest that its radiated sound power be determined in the presence of a reflecting floor and possibly one or two walls.

The integration of the acoustic intensity over the encompassing spherical surface is achieved by determining time-average squared sound pressures at a discrete set of measurement points arranged to uniformly sample the integration surface. The number of measurement points, N, and their recommended coordinates are summarised in Table 6.1. The first 12 measurement locations of the table are illustrated in Figure 6.2.

Table 6.1 Free and semi-free field measurement locations (ISO 3745-2003)

Numbers	Coordinates in terms of unit radius		
	x	y	z
1	0.36	0.0	0.93
2	0.58	0.58	0.58
3	0.93	0.36	0.0
4	-0.36	0.0	0.93
5	-0.58	0.58	0.58
6	-0.93	0.36	0.0
7	0.0	0.93	0.36
8	0.0	-0.93	0.36
9	0.58	-0.58	0.58
10	-0.58	-0.58	0.58
11	0.93	-0.36	0.0
12	-0.93	-0.36	0.0
13	0.58	0.58	-0.58
14	-0.58	0.58	-0.58
15	0.58	-0.58	-0.58
16	-0.58	-0.58	-0.58
17	0.0	0.93	-0.36
18	0.0	-0.93	-0.36
19	0.36	0.0	-0.93
20	-0.36	0.0	-0.93

The sound power is calculated using the following equations, where the constant C is listed in Table 6.2 for various source mounting configurations.

$$L_w = L_p + 20\log_{10} r + C \quad \text{(dB re } 10^{-12}\,\text{W)} \tag{6.8}$$

$$L_p = 10\log_{10}\left[\frac{1}{N}\sum_1^N 10^{(L_{pi}/10)}\right] \quad \text{(dB re 20 } \mu\text{Pa)} \tag{6.9}$$

If the measurement surface is in the far field, the directivity index corresponding to measurement location i can be determined as follows:

$$DI = L_{pi} - L_p \quad \text{(dB)} \tag{6.10}$$

Example 6.2

Sound pressure levels are measured at 12 points on a hemispherical surface surrounding a machine placed on a reflecting plane, as indicated in Table 6.2 and

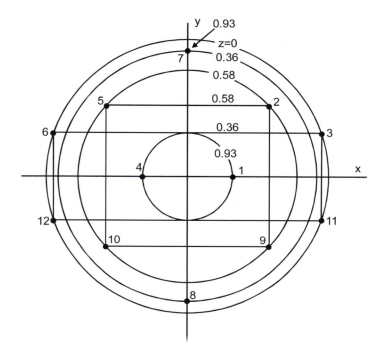

Figure 6.2 The first 12 locations for sound power measurements projected on to the x - y Cartesian coordinate plane. The z coordinate (vertical) is indicated parametrically and is associated with the corresponding circle. The coordinates are normalised with respect to the radius of the spherical surface of integration (taken as unity). Points 1-12 are used for measurements on a hemispherical surface around a noise source on a hard floor. Points 13-20 are added for a free space measurement on a spherical surface with no reflecting plane, and they correspond to points 1, 2, 4, 5, 7, 8, 9 and 10 respectively, except they are on the other side of the sphere.

Table 6.2 Values of the constant C used in Equation (6.8)

Source location	Constant C (dB)	Measurement position numbers (see Fig. 6.2)
1. Near junction of three planes defined by $x = y = z = 0$	2	1, 2, 3
2. Near junction of two planes defined by $y = z = 0$	5	1-6 inclusive
3. Near plane defined by $z = 0$	8	1-12 inclusive
4. Free field	11	1-20 inclusive

Figure 6.2. The Example 6.2 table gives the levels measured in the 500 Hz octave band, in dB re 20 μPa. Find L_w, the sound power level radiated by the machine if the radius of the test hemisphere is 3 m.

Example 6.2 Table

Measurement position number	L_{pi} (dB)	$(p_i/p_{ref})^2$
2	60	1.00×10^6
5	60	1.00×10^6
9	65	3.16×10^6
10	55	0.32×10^6
3	61 (58)	(0.63×10^6)
6	60 (57)	(0.50×10^6)
11	61 (58)	(0.63×10^6)
12	62 (59)	(0.79×10^6)
7	55	0.32×10^6
8	55	0.32×10^6
4	64	2.51×10^6
1	61	1.26×10^6
		Sum = 12.44×10^6

Solution

In the array of points in the table, eight have equal areas (S_1) and four have half areas ($S_1/2$). The half areas are associated with points that lie directly on the reflecting plane surface so that an associated area exists only on one side of the point. The points associated with half areas are 3, 6, 11 and 12, which lie in the plane of the reflecting surface. For these sectors the sound pressure level is reduced by 3 dB (the same as reducing $(p_i/p_{ref})^2$ by a factor of two) and considered to have the same areas as the other sectors (S_1).

Using Equation (6.9), the following is obtained:

$$L_p = 10\log_{10}\left[\frac{1}{12}\sum_{i=1}^{12} 10^{(L_{pi}/10)}\right]$$

Using Equation (1.81a) this becomes:

$$L_p = 10\log_{10}\left[\frac{1}{12}\sum_{i=1}^{12} (p_i/p_{ref})^2\right]$$

From the Example 6.2 table:

$$L_p = 10\log_{10}(12.44 \times 10^6/12) = 10\log_{10}(1.04 \times 10^6)$$

Thus:

$$L_p = 60 \text{ dB re } 20 \text{ } \mu\text{Pa}$$

Alternatively, the sound pressure levels in the table can be averaged by combining them as described in Section 1.11.4 and subtracting from the result, $10\log_{10}N$ (where N is the number of measurements). This procedure requires that all measurements be associated with areas of the same size. Thus, the levels in parentheses in the table for points 3, 6, 11 and 12 are used. Following the procedure outlined in Section 1.11.4, a value of 70.9 dB is obtained.

As there are 12 measurement points, all associated with equal areas, the quantity $10\log_{10}12$ must be subtracted from the result to obtain the required mean value. Thus:

$$L_p = 70.9 - 10\log_{10}12 = 60\ \text{dB re } 20\ \mu\text{Pa}$$

The sound power level may be calculated using Equation (6.8). Substituting values for L_p, r and C into Equation (6.8) gives:

$$L_w = 60.2 + 20\log_{10}3 + 8$$

or

$$L_w = 60 + 9.5 + 8.0 = 78\ \text{dB re } 10^{-12}\ \text{W}$$

That is, the sound power level, L_w, of the machine is 78 dB re 10^{-12} W.

6.6.2 Measurement in a Diffuse Field

To provide a diffuse sound field, a test room is required that has adequate volume and suitable shape, and whose boundaries over the frequency range of interest can be considered acoustically hard. The volume of the room should be large enough so that the number of normal acoustic modes (see Section 7.3.1) in the octave or one-third octave frequency band of interest is enough to provide a satisfactory state of sound diffusion. Various standards state that at least 20 acoustic modes are required in the lowest frequency band used. This implies a minimum room volume of 1.3 λ^3, if measuring in octave bands, and 4.6 λ^3, if measuring in one-third octave bands, where λ is the wavelength of sound corresponding to the lowest band centre frequency.

If the machine emits sound containing discrete frequencies or narrow band components, it is necessary to use a rotating sound diffuser (Ebbing, 1971; Lubman, 1974; Bies and Hansen, 1979). When a rotating diffuser is used, the lowest discrete frequency that can be reliably measured is given by the relation:

$$f_q = 2000(T_{60}/V)^{1/2} \quad \text{(Hz)} \tag{6.11}$$

In Equation (6.11), T_{60} is the time (in seconds) required for a one-third octave band of noise (with centre frequency f_q) to decay by 60 dB after the source is shut off (see Sections 7.5.1 and 7.5.2), and V is the room volume (m³). The rotating sound diffuser should have a swept volume equal to or greater than the cube of the wavelength of sound at the lowest frequency of interest. Procedures to be followed for the measurement of tonal noise in reverberant test chambers are described in ISO 3741-1999.

The shape of the test room for both broadband and tonal measurements should be such that the ratio of any two dimensions is not equal to, or close to, an integer. Ratios of height, width and length of 2:3:5 and 1:1.260:1.588 have been found to give good results.

The absorption coefficient (see Sections 7.7 and 7.8) of the test room should be less than 0.06 in all test frequency bands, although when noise containing pure tone components is measured it is advantageous to increase the absorption coefficient in the low frequencies to 0.15 (to increase the modal overlap – see Sections 7.3.2 and 7.3.3, and thus reduce the sound level variation as a function of frequency). For test purposes, the machine should be placed a distance of at least one-quarter wavelength away from all surfaces not associated with the machine.

There are two methods for determining the sound power of a machine in a reverberant room; the substitution method (ISO 3743/2-1994), and the absolute method (ISO 3741-1999). Both methods require the determination of the space-average sound pressure level produced in the room by the machine. This may be measured using a microphone travelling on a linear traverse across the room, or a circular traverse on a rotating boom. The traverse length should be at least 1.5 λ or 3 m, whichever is greatest. Alternatively, the microphone may be sequentially moved from point to point over a set of discrete measurement positions. In each case an average sound pressure level is determined. The microphone should at all times be kept at least one half of a wavelength away from reflecting surfaces, and out of any region in which the direct field from the sound source dominates the reverberant sound field.

The number of discrete microphone positions required is at least six for broadband noise when a rotating sound diffuser is used. The microphone positions should be at least one-half wavelength of sound apart. When a continuous microphone traverse is used, the equivalent number of discrete microphone positions is equal to twice the traverse length divided by the wavelength of sound corresponding to the centre frequency of the measurement band.

If no rotating diffuser is used, or if the machine radiates discrete frequency components, more microphone positions are required to obtain the same measurement accuracy (± 0.5 dB). The measurement errors associated with various frequency bandwidths and numbers of microphone positions have been well documented (Beranek, 1971).

6.6.2.1 Substitution Method (ISO 3743/2-1994)

The sound source to be tested is placed in the room and the space-average sound pressure level, L_p, is determined for each frequency band. The sound source is then replaced with another sound source of known sound power output, L_{wR} (the reference sound source). The space-average sound pressure level, L_{pR}, produced by the reference source in the room is determined. The sound power level, L_w, for the test source is then calculated using the following relation:

$$L_w = L_{wR} + (L_p - L_{pR})$$

(6.12)

Commonly used reference sound sources include the ILG source (Beranek, 1971), the Bruel and Kjaer type 4204 and the Campanella RSS source.

6.6.2.2 Absolute Method (ISO 3741-1999)

For this method, the sound-absorbing properties of the room are determined in each measurement band from measurements of the frequency band reverberation times, T_{60}, of the room (see Sections 7.5.1 and 7.5.2). The steady-state space-average sound pressure level, L_p, produced by the noise source is also determined for each frequency band, as described earlier. The sound power level L_w produced by the source is then calculated in each frequency band using the following equation (Beranek, 1971):

$$L_w = L_p + 10\log_{10} V - 10\log_{10} T_{60} + 10\log_{10}(1 + S\lambda/8V)$$
$$- 13.9 \quad (\text{dB re } 10^{-12}\,\text{W})$$

(6.13)

where the constant "13.9 dB" has been calculated for a pressure of one atmosphere and a temperature of 20°C, using Equations (7.42) and (7.52) as a basis (with $(1 - \overline{\alpha}) \approx 1$, where $\overline{\alpha}$ is the average Sabine absorption coefficient of the room surfaces - see Section 7.4.2). In Equation (6.13), V is the volume of the reverberant room, S is the total area of all reflecting surfaces in the room, and λ is the wavelength of sound at the band centre frequency. The fourth term on the right of the equation is not derived from Equation (7.52) and represents a correction ("Waterhouse correction") to account for the measurement bias resulting from the space averaged sound pressure level, L_p excluding measurements of the sound field closer than $\lambda/4$ to any room surface (Waterhouse, 1955). All the other terms in the equation can be derived directly from Equations (7.42) and (7.52), where the contribution of the direct field is considered negligible and $S\overline{\alpha} \approx S\overline{\alpha}/(1 - \overline{\alpha})$.

6.6.3 Field Measurement

If the machine is mounted in an environment such that the conditions of free or semi-free field can be met, then sound power measurements can be made using the appropriate method of Section 6.6.1. However, when these conditions are not satisfied and it is inconvenient or impossible to move the machine to be tested, less precise sound power measurements can be made with the machine on site, using one of the methods described in this section.

Most rooms in which machines are installed are neither well damped nor highly reverberant; the sound field in such rooms is said to be semi-reverberant. For the determination of sound power in a semi-reverberant room, no specific assumptions are made concerning the room, except that the room should be large enough so that measurements can be made in the far field of the source and not too close to the room boundaries. The microphone should at all times be at least one-half of a wavelength away from any reflecting surfaces or room boundaries not associated with the machine. The machine should be mounted in its normal position, which will typically include the hard floor but may also include mounting in the corner of a room or at the junction between the floor and a wall. Unusually long or narrow rooms will generally degrade the results obtained and are best avoided if possible. Where the conditions mentioned can be satisfied there are three alternative measurement procedures, as described in the following three sections.

6.6.3.1 Semi-reverberant Field Measurements by Method One (ISO 3747-2000)

To compute the sound power level, the total room absorption (see Section 7.6.1) of the test room must be determined. For this purpose a reference sound source is used (for example, the ILG source mentioned previously). The reference sound source is placed on the floor away from the walls or any other reflecting surfaces, and a hypothetical hemispherical test surface surrounding the reference source is chosen. The radius r of the test surface should be large enough for the test surface to be in the far field of the reference source. Reference to Figure 6.1 shows that this condition is easily satisfied for the ILG reference source, which has a characteristic dimension, ℓ, of the order of 0.1 m. For example, at 500 Hz the distance, r, should be greater than 0.3 m. For best results the test surface should lie in the region about the test source, where its direct field and the reverberant field are about equal (see Section 7.4.4).

Measurements on the surface of the test hemisphere in each octave or one-third octave band (see Table 6.2) allow determination of the reference source average sound pressure level, L_{pR}, due to the combination of direct and reverberant sound fields, using Equation (6.9). At the radius, r, of the test hemisphere, the sound pressure level, L_{p2}, due to the direct sound field of the reference source only is calculated using the reference source sound power levels, L_{wR}, and the following equation obtained by setting $DI = 3$ dB in Equation (5.126):

$$L_{p2} = L_{wR} - 20\log_{10}r - 8 + 10\log_{10}\frac{\rho c}{400} \quad \text{(dB re 20 } \mu\text{Pa)} \quad (6.14)$$

The expression for determining the reciprocal room constant factor $4/R$ (see Section 7.4.4) of the test room is obtained by using the expression relating the sound pressure in a room to the sound power of a sound source derived in Chapter 7, Equation (7.43), as follows:

$$L_{wR} = L_{pR} - 10\log_{10}\left[\frac{D}{4\pi r^2} + \frac{4}{R}\right] - 10\log_{10}\frac{\rho c}{400} \quad \text{(dB re 10}^{-12}\text{ W)} \quad (6.15)$$

For a measurement over a hemispherical surface of area S_H above a hard floor, $D/(4\pi r^2) = 1/S_H$, or $S_H = 2\pi r^2$, as $D = 2$ (see Table 5.1).

Assuming that the reference source radiates the same sound power into the room as it does into free field, the values of L_{wR} in Equations (6.14) and (6.15) are equal. Noting that $20\log_{10}r + 8$ in Equation (6.14) may be replaced with $10\log_{10}S_H$ the following equation is obtained:

$$L_{p2} + 10\log_{10}S_H = L_{pR} - 10\log_{10}\left[\frac{1}{S_H} + \frac{4}{R}\right] \qquad (6.16)$$

or

$$L_{pR} - L_{p2} = 10\log_{10}\left[1 + \frac{4S_H}{R}\right] \qquad (6.17)$$

Rearranging gives an expression for the room constant factor, $4/R$, as follows:

$$4/R = [10^{(L_{pR} - L_{p2})/10} - 1]/S_H \qquad (\text{m}^{-2}) \qquad (6.18)$$

The room constant R will be discussed in Chapter 7, but for the present purpose it may be taken as the total Sabine absorption, $S\bar{\alpha}$, in the room, measured in units of area. Alternative methods for determining the room constant R are given in Section 7.6.

The noise source under test is now operated in the room and the mean square sound pressure level, L_p, over a test surface of radius r and centre at the acoustical centre of the noise source (see previous section) is determined. The radius of the test hemisphere should be chosen large enough for the sound pressure level measurements to be made at least in the geometric near field of the source but preferably in the far field. The sound power level may be computed using Equation (6.15) by replacing L_{wR} with L_w and L_{pR} with L_p.

Table 5.1 provides a guide for the choice of directivity factor D in Equation (6.15). For example, if the noise source is mounted on a hard reflecting plane surface then the test surface should be a hemisphere, and D in Equation (6.15) should take the value 2.

6.6.3.2 Semi-reverberant Field Measurements by Method Two (ISO 3743/1-1994)

If the machine to be tested is located on a hard floor at least one-half of a wavelength away from any other reflecting surfaces, and in addition it can be moved, the measurement of L_w is simplified. In this case, the substitution method described earlier can be employed. The average sound pressure level L_p is determined over a test hemisphere surrounding the machine. The machine is replaced by the reference sound source and the average sound pressure level L_{pR} is determined over the same test hemisphere. The sound power output of the machine is then calculated using the following expression:

$$L_w = L_{wR} + (L_p - L_{pR}) \qquad (\text{dB}) \qquad (6.19)$$

where L_w is the sound power level of the machine and L_{wR} is the sound power level of the reference source. Note that the test hemisphere should be located such that its centre is the acoustical centre of the machine, and its surface is in the far field, or at least the geometric near field, of both the machine and the reference sound source. Any reflecting surfaces present should not be included within the test hemisphere.

6.6.3.3 Semi-reverberant Field Measurements by Method Three

Alternatively, instead of using a reference sound source, the sound power level of any source may be determined by taking sound pressure measurements on two separate test surfaces having different radii. The discussion of Section 6.6.1 may be used as a guide in choosing the test surfaces and measurement locations. In all cases, the centres of the test surfaces are at the acoustic centre of the noise source. The test surface areas are given by $4\pi r^2/D$, where D is given in Table 5.1 and r is the test surface radius. The radii of the test surfaces are such that they are in the far field, or at least in the geometric near field of the source.

The measurement procedure assumes that the background noise levels produced by other machines at the microphone measurement positions make a negligible contribution to the measurements associated with the test machine. This implies that the sound pressure levels at the measurement positions produced by the machine under test are at least 10 dB above the background noise level. If this is not the case, then the sound pressure level data must be corrected for the presence of background noise. All measurements must be repeated with the test machine turned off and the background levels measured. The background levels must be logarithmically subtracted from the test measurement levels, using the method and corresponding limitations outlined in Section 1.11.5.

Let L_{p1} be the average sound pressure level measured over the smaller test surface of area S_1 and L_{p2} be the average sound pressure level measured over the larger test surface of area S_2. L_{p1} and L_{p2} are calculated using Equation (6.9). Since both sets of measurements should give the same result for the sound power, Equation (6.15) gives the following expression relating the measured quantities.

$$L_{p1} - 10 \log_{10} \left[\frac{D}{4\pi r_1^2} + \frac{4}{R} \right] = L_{p2} - 10 \log_{10} \left[\frac{D}{4\pi r_2^2} + \frac{4}{R} \right] \quad \text{(dB)} \qquad (6.20)$$

The unknown quantity, R, is called the room constant and it is evaluated by manipulating the preceding equation and substituting $1/S_1$ for $D/4\pi r_1^2$ and $1/S_2$ for $D/4\pi r_2^2$. Taking antilogs of Equation (6.20) gives:

$$10^{(L_{p1} - L_{p2})/10} = \frac{[1/S_1 + 4/R]}{[1/S_2 + 4/R]} \qquad (6.21)$$

Rearranging gives:

$$\frac{4}{R} = \frac{(1/S_1) - (1/S_2)\,[\,10^{(L_{p1} - L_{p2})/10}\,]}{10^{(L_{p1} - L_{p2})/10} - 1} \tag{6.22}$$

or

$$\frac{1}{S_2} + \frac{4}{R} = \frac{(1/S_1 - 1/S_2)}{(10^{(L_{p1} - L_{p2})/10} - 1)} \tag{6.23}$$

Substituting Equation (6.23) into (6.15) gives the following equation (Diehl, 1977):

$$L_w = L_{p2} - 10\log_{10}[S_1^{-1} - S_2^{-1}]$$
$$+ 10\log_{10}[\,10^{(L_{p1} - L_{p2})/10} - 1\,] - 10\log_{10}\frac{\rho c}{400} \tag{6.24}$$

Example 6.3

A machine is located in a semi-reverberant shop area at the junction of a concrete floor and brick wall. The average sound pressure level in the 1000 Hz octave band over the test surface (a quarter sphere) is 82 dB at a radius of 2 m and 80 dB at a radius of 5 m. Determine the sound power level in the 1000 Hz octave band for the machine, assuming that the sound pressure level measurements were made in the far field of the source.

Solution

$$L = L_{p1} - L_{p2} = 2 \text{ dB}$$

$$S_1^{-1} = 1/\pi r_1^2 = 0.0796 \qquad S_2^{-1} = 1/\pi r_2^2 = 0.0127$$

$$S_1^{-1} - S_2^{-1} = 0.0668$$

Therefore, using Equation (6.24):

$$L_w = 80 - 10\log_{10}(0.0668) + 10\log_{10}(10^{0.2} - 1) - 0.15$$
$$= 80 + 11.75 - 2.33 - 0.15 = 89 \text{ dB re } 10^{-12} \text{ W}$$

6.6.3.4 Near-field Measurements (ISO 3746-1995)

The previous methods outlined for the on-site measurement of sound power radiated by a machine have all required that the room in which the machine is situated be large enough for the measurements to be made in the far field of the source. The previous methods have also assumed that the background noise levels produced by other machines have either made a negligible contribution to the measurements or that a correction can be made for their contribution; that is, it has been assumed that the machine under test has produced a sound pressure level at least 3 dB higher than the background noise level at each microphone measurement position.

In some cases the above assumptions may not be valid. For example, the room in which the machine is situated may be too small for a far field to exist. Alternatively, if the room is of adequate size the background noise levels produced by other machines in the room may be too high to allow valid measurements to be made at a sufficient distance from the test machine to be in its far field. For cases such as these an alternative, but less accurate, procedure may be used to estimate the sound power radiated by the machine. This procedure relies on sound pressure level measurements made close to the machine surface (Diehl, 1977) at a number of points on a hypothetical test surface surrounding the machine (which is usually mounted on a reflective surface such as a hard floor).

The test surface usually conforms approximately to the outer casing of the machine so that its area is easy to calculate (in many cases a parallelepiped may be used), and it is sufficiently close to the machine for the measurements not to be affected too much by nearby reflecting surfaces or background noise. If background noise is a problem it must be accounted for by measuring the sound pressure levels with the machine under test turned on and then with it turned off. The level with it turned off is then subtracted from the level with the machine on, as illustrated in Example 1.4 in Chapter 1.

The effect of nearby reflecting surfaces can be minimised by placing sound-absorbing material on them (e.g. 50 mm thick glass-fibre blanket). The standard and generally accepted distance between the test surface and machine surface is 1 m, but may need to be less in some cases, at the expense of reduced accuracy in the estimation of the radiated sound power.

The average sound pressure level L_p over the test surface is found by measuring the sound pressure level L_{pi} at a number of equally spaced, discrete points over the surface, and then using Equation (6.9). The number of measurement positions, N, is determined by the irregularity of the acoustic field and size of the machine, and should be sufficient to take any irregularities into account. Suitable measurement locations are discussed in ISO 3746-1995 and involve between 5 and 16 locations for a typical machine. Once determined, the value of L_p is used to determine the sound power level of the machine using the following equation:

$$L_w = L_p + 10 \log_{10} S - \Delta_1 - \Delta_2 \qquad (6.25)$$

In Equation (6.25) (estimated to the nearest 1 dB), S is the area of the test surface and Δ_1 and Δ_2 are correction terms.

In the near field of a machine, sound propagation will not necessarily be normal to the arbitrarily chosen measurement surface. As the integration process implied by Equation (6.25) implicitly assumes propagation normal to the measurement surface, the correction factor Δ_2 is introduced to account for possible tangential sound propagation. Values of Δ_2 are given in Table 6.3 as a function of the ratio of the area of the measurement surface S divided by the area of the smallest parallelepiped, S_m, which just encloses the source.

Table 6.3 Correction factor due to near-field effects (Jonasson and Eslon, 1981)

Ratio of test surface area to machine surface area S/S_m	Near-field correction factor Δ_2 (dB)
1- 1.1	3
1.1- 1.4	2
1.4 - 2.5	1
2.5 - ∞	0

The correction factor Δ_1 has been suggested to account for the absorption characteristics of the test room. Values of Δ_1 typical of production rooms are given in Table 6.4 for various ratios of the test room volume V to area S of the test surface.

Table 6.4 Value of room effect correction factor Δ_1 (Pobol, 1976)

Characteristics of production or test room	Ratio of room volume to test surface area V/S (m)			
Usual production room without highly reflective surfaces	20-50	50-90	90-3000	Over 3000
Room with highly reflective surfaces, with no sound-absorbing treatment	50-100	100-200	200-600	Over 600
Δ_1 (dB)	3	2	1	0

In Table 6.4, the distance from the test surface to the machine surface is approximately 1 m. Note that in deriving this equation, ρc has been assumed to be equal to 400.

Both test surfaces should completely surround the machine and correspond roughly to the shape of the machine. The smaller test surface is usually displaced about 1 m from the machine surface. The two test surfaces should be sufficiently far apart that the average sound pressure level over one surface differs from that measured over the other surface by at least 1 dB. Values of Δ_1 as a function of the area ratio of the two test surfaces and the difference in average sound pressure level measured over each surface may be estimated using Figure 6.3, or calculated as described in the text to follow. Neglecting for now the correction term, Δ_2 Equation (6.25), substituting S_1 for S and L_{p1} for L_p in Equation (6.25) and setting the RHS of Equation (6.24) equal to the RHS of Equation (6.25), the following result is obtained:

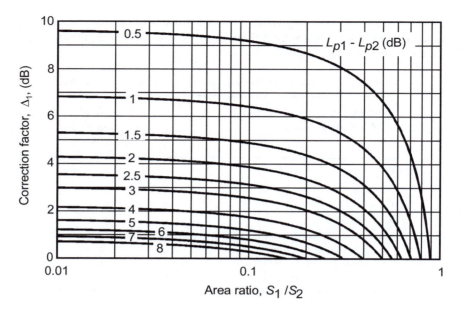

Figure 6.3 Graph for estimating the correction factor, Δ_1.

$$
\begin{aligned}
L_w &= L_{p1} + 10 \log_{10} S - \Delta_1 \\
&= L_{p2} - 10 \log_{10} [S_1^{-1} - S_2^{-1}] + 10 \log_{10} [10^{(L_{p1} - L_{p2})/10} - 1]
\end{aligned}
\tag{6.26}
$$

Rearranging Equation (6.26) gives:

$$
\Delta_1 = L_{p1} - L_{p2} - 10 \log_{10} [10^{(L_{p1} - L_{p2})/10} - 1] + 10 \log_{10} [1 - S_1/S_2]
\tag{6.27}
$$

In Equation (6.27) the subscripts 1 and 2 refer respectively to the measurement surfaces near and remote from the machine. Alternatively, Δ_1 may be estimated using only one measurement surface, the mean acoustic absorption of the room surfaces, and the following procedure. Again, neglecting Δ_2 allows Equation (6.15) (with $D/4\pi r^2 = 1/S_1$) and Equation (6.25) to be set equal. Thus:

$$
L_p + 10 \log_{10} S_1 - \Delta_1 \approx L_p - 10 \log_{10} \left[\frac{1}{S_1} + \frac{4}{R} \right]
\tag{6.28}
$$

where it has been assumed that $\rho c \approx 400$.

Rearranging Equation (6.28) and use of Equation (7.44) (see Section 7.4.4) gives the following expression for the correction term Δ_1:

$$
\Delta_1 = 10 \log_{10} \left[1 + \frac{4S(1 - \overline{\alpha})}{(S_R \overline{\alpha})} \right]
\tag{6.29}
$$

where S is the area of the measurement surface, S_R is the total area of all the room surfaces, and $\bar{\alpha}$ is the mean acoustic Sabine absorption coefficient for the room surfaces (see Chapter 7 for further discussion and measurement methods). Representative values of $\bar{\alpha}$ are included in Table 6.5.

Table 6.5 Approximate value of the mean acoustic absorption coefficient, $\bar{\alpha}$

Mean acoustic absorption coefficient, $\bar{\alpha}$	Description of room
0.05	Nearly empty room with smooth hard walls made of concrete, bricks, plaster or tile
0.1	Partly empty room, room with smooth walls
0.15	Room with furniture, rectangular machinery room, rectangular industrial room
0.2	Irregularly shaped room with furniture, irregularly shaped machinery room or industrial room
0.25	Room with upholstered furniture, machinery or industrial room with a small amount of acoustical material, e.g. partially absorptive ceiling, on ceiling or walls
0.35	Room with acoustical materials on both ceiling and walls
0.5	Room with large amounts of acoustical materials on ceiling and walls

In making near-field measurements, a possible source of error is associated with the microphone response. To avoid such error the microphone response should be as close as possible to ±1 dB over the maximum angle between the line joining the microphone to the furthest point on the machine surface and the normal to the measurement surface. This angle is approximately 60° for a source dimension of 3 m and a measurement surface 1 m from the source. Thus, for a 12 mm microphone, the error due to microphone directional response should be negligible up to frequencies of about 3.15 kHz, as shown by reference to Figure 3.3.

6.7 DETERMINATION OF SOUND POWER USING SURFACE VIBRATION MEASUREMENTS

The sound power radiated by a machine surface can be estimated from a determination of the mean square vibration velocity averaged over the surface (Takatsubo *et al.*, 1983). The measurements are usually made using an accelerometer, an integrating circuit, a bandpass filter and a sound level meter or vibration meter. Generally, five to ten measurements distributed at random, but not too close to the edges of the radiating surface, are sufficient. Referring to Equation (5.111b) for a plane piston noise source a similar expression can be derived by replacing $(U^2/2)$ with the surface mean square velocity, $\langle v^2 \rangle_{S,t}$, replacing the piston surface area, πa^2 with the surface area, S, and renaming the real part of the function, R_R, as the surface radiation efficiency, σ. With these changes, Equation (5.111b) becomes:

$$W = \langle v^2 \rangle_{S,t} S \rho c \sigma \tag{6.30}$$

Taking logarithms to the base 10 of both sides of Equation (6.30) and using octave or one third octave band measurement of the space and time averaged mean square surface velocity, $\langle v^2 \rangle_{S,t}$, the level of sound power radiated in each of the 1/3 octave or octave bands can be computed using the following equation:

$$L_w = 10 \log_{10} \langle v^2 \rangle_{S,t} + 10 \log_{10} S + 10 \log_{10} \sigma + 146 \quad (\text{dB re } 10^{-12}\,\text{W}) \tag{6.31}$$

where S is the surface area. The quantity σ is a measure of the efficiency of radiation, called the radiation efficiency. Since it can be larger than unity, it is sometimes called, "radiation ratio".

A means for estimating the radiation efficiency for flat panels at frequencies above the first panel resonance frequency (see Equation (8.23)) is given in Figure 6.4.

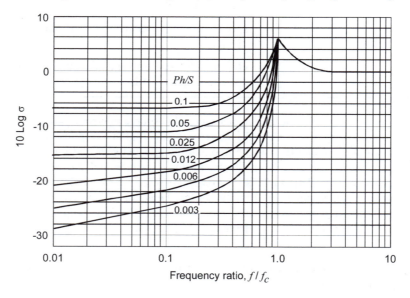

Figure 6.4 The radiation efficiency of steel and aluminum flat panels. The quantity P is the panel perimeter (m), S is the panel surface area (m^2) (one side), and h is the panel thickness (m). The quantities f and f_c are the centre frequency of the octave or 1/3 octave frequency band of interest and the panel critical frequency, respectively. The quantity σ is the panel radiation efficiency.

The curves shown in Figure 6.4 are slightly different to those in the second edition of this book and are considered more accurate. Even more accurate results can be obtained by using the equations in Chapter 7, Section 7.8.2. Reference to Figure 6.4 shows that the driving frequency, relative to a critical frequency determined by the thickness and material characteristics of the panel, is the important parameter that determines the radiation behaviour of the panel. The critical frequency for steel and aluminum panels is given approximately by the following equation:

$$f_c = 12.7/h \tag{6.32}$$

where h is the panel thickness in metres. The critical frequency is the frequency of coincidence discussed in Section 8.2.1. Radiation efficiencies for other structures (e.g. I-beams, pipes, etc.) are available in the published literature (Beranek, 1971; Wallace, 1972; Lyon, 1975; Jeyapalan and Richards, 1979; Anderton and Halliwell, 1980; Richards, 1980; Jeyapalan and Halliwell, 1981; Sablik, 1985).

If the radiation efficiency σ cannot readily be determined, then a less precise option is offered by A-weighting the measured mean square surface velocity. The success of this scheme depends upon the observation that for high frequencies the radiation efficiency is unity, and only for low frequencies is it less than unity and uncertain. The A-weighting process minimises the importance of the low frequencies. Thus the use of A-weighted vibration measurements and setting $\sigma = 1$ in Equation (6.31) generally allows identification of dominant noise sources on many types of machines and an estimate of the radiated sound power.

If vibration measurements are made with an accelerometer and an integrating circuit is not available, then the velocity may be estimated using the following approximation:

$$\langle v^2 \rangle_{S,t} = \langle a^2 \rangle_{S,t} / (2\pi f)^2 \tag{6.33}$$

In the preceding equation, $\langle a^2 \rangle_{S,t}$ is the mean square acceleration averaged in time and space and f is the band centre frequency. If the levels are not A-weighted, the maximum error resulting from the use of this expression is 3 dB for octave bands. On the other hand, if A-weighted levels are determined from unweighted frequency band levels, using Figure 3.5 or Table 3.1, the error could be as large as 10 dB. If a filter circuit is used to determine A-weighted octave band levels then the error would be reduced to at most 3 dB. Use of a filter circuit is strongly recommended.

6.8 SOME USES OF SOUND POWER INFORMATION

The location of a machine with respect to a large reflecting surface such as a floor or wall of a room may affect its sound power output. Thus the interpretation of sound power information must include consideration of the mounting position of the machine (e.g. on a hard floor or wall, at the junction of a wall and floor or in the corner of a room). Similarly, such information should be contained in any sound power specifications for new machines.

The sound power output of a machine may be used to estimate sound pressure levels generated at various locations; for example, in a particular room. In this way, the contribution made by a particular machine to the overall sound level at a particular position in a room may be determined. Means for estimating sound pressure levels from sound power level information are required, dependent upon the situation, and these will now be considered.

6.8.1 The Far Free Field

When the sound pressure level L_p is to be estimated in the far field of the source (see Figure 6.1), the contribution of the reverberant sound field to the overall sound field will be assumed to be negligible for two particular cases. The first is if the machine is mounted in the open, away from any buildings. The second is if the machine in a large room, which has had its boundaries (not associated with the mounting of the machine) treated with acoustically absorptive material, and where the position at which L_p is to be estimated is not closer than one-half wavelength to any room boundaries. In this situation, directivity information about the source is useful.

An expression for estimating L_p at a distance r from a source for the preceding two cases is obtained by rearrangement of Equation (5.126). The directivity index DI that appears in Equation (5.126) is given in Table 5.1. If no directivity information exists, then the average sound pressure level to be expected at a distance r from the source can be estimated by assuming a uniform sound radiation field (see Section 5.11.11 of Chapter 5 for further discussion).

Example 6.4

A subsonic jet has the directivity pattern shown in the following table. Calculate the sound pressure levels at 1 m from a small jet of sound power 100 dB re 10^{-12} W.

Solution

Use Equation (5.171) with K replaced by Equation (5.173), set the excess attenuation term, $A_E = 0$, set $r = 1$ m and write:

$$L_p = 100 - 11 + DI_M$$

Thus:

$$L_p = 89 + DI_M$$

The last column in the table is constructed by adding 89 dB to the numbers shown in the second column.

Example 6.4 Table

Angle relative to direction of jet axial velocity (°)	Directivity index DI (dB)	Predicted sound pressure level at 1 m (dB re 20 μPa)
0	0.0	89
15	3.0	92
30	5.0	94
45	2.5	91.5
60	- 1.0	88
75	- 4.0	85
90	- 6.0	83
105	- 7.5	81.5
120	- 8.0	81
150	- 9.0	80
180	- 10.0	79

6.8.2 The Near Free Field

The near field of a sound source (geometric or hydrodynamic) is generally quite complicated, and cannot be described by a simple directivity index. Thus, estimates of the sound pressure level at fixed points near the surface are based on the simplifying assumption that the sound source has a uniform directivity pattern. This is often necessary as, for example, the machine operator's position is usually in the near field. Rough estimates of the sound pressure level at points on a hypothetical surface of area, S, conforming to the shape of the surface of the machine, and at a specified short distance from the machine surface, can be made using the Equation (6.25). Referring to the latter equation, if the contribution due to the reverberant field can be considered negligible, then $\Delta_1 = 0$.

CHAPTER SEVEN

Sound In Enclosed Spaces

LEARNING OBJECTIVES

In this chapter the reader is introduced to:

- wall-cavity modal coupling, when it is important and when it can be ignored;
- the simplifying assumption of locally reactive walls;
- three kinds of rooms: Sabine rooms, flat rooms and long rooms;
- Sabine and statistical absorption coefficients;
- low-frequency modal description of room response;
- high-frequency statistical description of room response;
- transient response of Sabine rooms and reverberation decay;
- reverberation time calculations;
- the room constant and its determination;
- porous and panel sound absorbers;
- applications of sound absorption; and
- basic auditorium design.

7.1 INTRODUCTION

Sound in an enclosed space is strongly affected by the reflective properties of the enclosing surfaces and to the extent that the enclosing surfaces are reflective, the shape of the enclosure also affects the sound field. When an enclosed space is bounded by generally reflective surfaces, multiple reflections will occur, and a reverberant field will be established in addition to the direct field from the source. Thus, at any point in such an enclosure, the overall sound pressure level is a function of the energy contained in the direct and reverberant fields.

In general, the energy distribution and variation with frequency of a sound field in an enclosure with reflective walls is difficult to determine with precision. Fortunately, average quantities are often sufficient and procedures have been developed for determining these quantities. Accepted procedures divide the problem of describing a sound field in a reverberant space into low- and high-frequency ranges, loosely determined by the ratio of a characteristic dimension of the enclosure to the wavelength of the sound considered. For example, the low-frequency range might be characterised by a ratio of less than 10 while the high-frequency range might be characterised by a ratio of greater than 10; however, precision will be given to the meaning of these concepts in the discussion in the following sections.

7.1.1 Wall-interior Modal Coupling

A complication that arises in consideration of sound in an enclosure is that coupling between modes in the enclosed space (cavity modes) and modes in the enclosure boundaries (wall modes) generally cannot be ignored. For example, in the low-frequency range of the relatively lightweight structures that characterise aircraft and automobiles, coupled wall-cavity modes may be dominant. In such cases, the sound field in the enclosed space cannot be considered in isolation; coupling between the modes in the wall and the cavity must be considered and the wall modes thus become equally as important as the cavity modes in determining the acoustic field in the enclosure (Pan and Bies, 1990a, 1990b, 1990c). Generally, for the lightweight enclosure cases cited above, the low-frequency range extends over most of the audio-frequency range.

It has become traditional to begin a discussion of room acoustics with the assumption that the walls of the enclosure are locally reactive (Kuttruff, 1979) or effectively infinitely stiff; the alternative that the walls may be bulk reactive, meaning that wall-cavity mode coupling is important, seems never to have been considered for the case of sound in rooms. However, it has been shown that when a sound field is diffuse, meaning that the sound energy travels in all directions within the enclosed space with equal probability, the modal response of a bounding surface will be similar to that of a surface which is locally reactive (Pan and Bies, 1988), thus in the frequency range in which the sound field may be assumed to be diffuse, the assumption of locally reactive walls gives acceptable results. However, in the low-frequency range, where the sound field will not be diffuse, cavity-wall modal coupling can be expected to play a part in the response of the room. Modal coupling will affect the resulting steady-state sound field levels as well as the room reverberation time (Pan and Bies, 1990a, 1990b, 1990c). Such a case is considered in Section 7.8.

In the high-frequency range, the concept of a locally reactive boundary is of great importance, as it serves to uncouple the cavity and wall modes and greatly simplify the analysis (Morse, 1939). Locally reactive means that the response to an imposed force at a point is determined by local properties of the surface at the point of application of the force and is independent of forces applied at other points on the surface. That is, the modal response of the boundary plays no part in the modal response of the enclosed cavity.

7.1.2 Sabine Rooms

When the reflective surfaces of an enclosure are not too distant from one another and none of the dimensions is so large that air absorption becomes important, the sound energy density of a reverberant field will tend to uniformity throughout the enclosure. Generally, reflective surfaces will not be too distant, as intended here, if no enclosure dimension exceeds any other dimension by more than a factor of about three. As the distance from the sound source increases in this type of enclosure, the relative contribution of the reverberant field to the overall sound field will increase until it dominates the direct field (Beranek, 1971, Chapter 9; Smith, 1971, Chapter 3). This kind of enclosed space, in which a generally uniform reverberant energy density field, characterised by a mean sound pressure and standard deviation (see Section 7.4), tends to be established, has been studied extensively because it characterises rooms used for assembly and general living and will be the principal topic of this chapter. For convenience, this type of enclosed space will be referred to as a Sabine enclosure named after the man who initiated investigation of the acoustical properties of such rooms (Sabine, 1993).

All enclosures exhibit low- and high-frequency response and generally all such response is of interest. However, only the high-frequency sound field in an enclosure exhibits those properties that are amenable to the Sabine-type analysis; the concepts of the Sabine room are thus strictly associated only with the high-frequency response.

The number of acoustic resonances in an enclosure increases very rapidly as the frequency of excitation increases. Consequently, in the high-frequency range, the possible resonances become so numerous that they cannot be distinguished from one another. Thus, one observes a generally uniform sound field in the regions of the reverberant field not in the vicinity of the source. In this frequency range, the resulting sound field is essentially diffuse and may be described in statistical terms or in terms of average properties.

In the discussion of high-frequency response in Sabine type rooms, the acoustic power transmission into the reverberant sound field has traditionally been treated as a continuum, injected from some source and continually removed by absorption at the boundaries. The sound field is then described in terms of a simple differential equation. The concept of Sabine absorption is introduced and a relatively simple method for its measurement is obtained. This development, which will be referred to as the classical description, is introduced in Section 7.5.1.

In Section 7.5.2 an alternative analysis, based upon a modal description of the sound field, is introduced. It is shown that with appropriate assumptions, the formulations of Norris Eyring and Millington Sette are obtained. Recently it has been shown that, with other assumptions, the analysis leads to the conclusion that the Sabine equation is exact, provided that edge diffraction of the absorbing material is taken into account by appropriately increasing the effective area of absorbing material (Bies, 1995).

7.1.3 Flat and Long Rooms

Enclosed spaces are occasionally encountered in which some of the bounding surfaces may be relatively remote or highly absorptive, and such spaces are also of importance. For example, lateral surfaces may be considered remote when the ratio of enclosure width-to-height or width-to-length exceeds a value of about three. Among such possibilities are flat rooms, characteristic of many industrial sites in which the side walls are remote or simply open, and long rooms such as corridors or tunnels. These two types of enclosure, which have been recognised and have received attention in the technical literature, are discussed in Section 7.9.

7.2 LOW FREQUENCIES

In the low-frequency range, an enclosure sound field is dominated by standing waves at certain characteristic frequencies. Large spatial variations in the reverberant field are observed if the enclosure is excited with pure tone sound, and the sound field in the enclosure is said to be dominated by resonant or modal response.

When a source of sound in an enclosure is turned on, the resulting sound waves spread out in all directions from the source. When the advancing sound waves reach the walls of the enclosure they are reflected, generally with a small loss of energy, eventually resulting in waves travelling around the enclosure in all directions. If each path that a wave takes is traced around the enclosure, there will be certain paths of travel that repeat upon themselves to form normal modes of vibration, and at certain frequencies, waves travelling around such paths will arrive back at any point along the path in phase. Amplification of the wave disturbance will result and the normal mode will be resonant. When the frequency of the source equals one of the resonance frequencies of a normal mode, resonance occurs and the interior space of the enclosure responds strongly, being only limited by the absorption present in the enclosure.

A normal mode has been associated with paths of travel that repeat upon themselves. Evidently, waves may travel in either direction along such paths so that, in general, normal modes are characterised by waves travelling in opposite directions along any repeating path. As waves travelling along the same path but in opposite directions produce standing waves, a normal mode may be characterised as a system of standing waves, which in turn is characterised by nodes and anti-nodes. Where the oppositely travelling waves arrive, for example in pressure anti-phase, pressure cancellation will occur, resulting in a pressure minimum called a node. Similarly, where the oppositely travelling waves arrive in pressure phase, pressure amplification will occur, resulting in a pressure maximum called an anti-node.

In an enclosure at low frequencies, the number of resonance frequencies within a specified frequency range will be small. Thus, at low frequencies, the response of a room as a function of frequency and location will be quite irregular; that is, the spatial distribution in the reverberant field will be characterised by pressure nodes and anti-nodes.

7.2.1 Rectangular rooms

If the source in the rectangular room illustrated in Figure 7.1 is arranged to produce a single frequency, which is slowly increased, the sound level at any location (other than at a node in the room for that frequency) will at first rapidly increase, momentarily reach a maximum at resonance, then rapidly decrease as the driving frequency approaches and then exceeds a resonance frequency of the room. The process repeats with each room resonance. The measured frequency response of a 180 m^3 rectangular reverberation room is shown in Figure 7.2 for illustration. The sound pressure was measured in a corner of the room (where there are no pressure nodes) while the frequency of the source (placed at an opposite corner) was very slowly swept upwards.

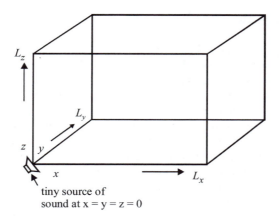

Figure 7.1 Rectangular enclosure.

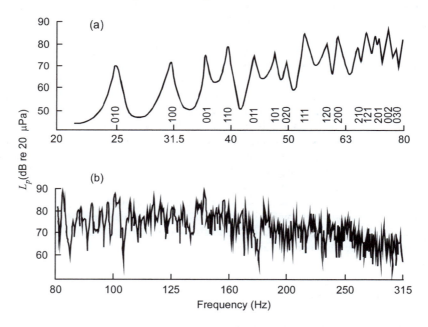

Figure 7.2 Measured frequency response of a 180 m³ rectangular room. (a) In this frequency range room resonances are identified by mode numbers. (b) In this frequency range peaks in the room response cannot be associated with room resonances identified by mode numbers.

Consideration of a rectangular room provides a convenient model for understanding modal response and the placement of sound absorbents for sound control. The mathematical description of the modal response of the rectangular room illustrated in Figure 7.1 takes on a particularly simple form; thus it will be advantageous to use the rectangular room as a model for the following discussion of modal response.

It is emphasised that modal response is by no means peculiar to rectangular or even regular-shaped rooms. Modal response characterises enclosures of all shapes. Splayed, irregular or odd numbers of walls will not prevent resonances and accompanying pressure nodes and antinodes in an enclosure constructed of reasonably reflective walls; nor will such peculiar construction necessarily result in a more uniform distribution in frequency of the resonances of an enclosure than would occur in a rectangular room of appropriate dimensions (see Section 7.3.1). However, it is simpler to calculate the resonance frequencies and mode shapes for rectangular rooms.

For sound in a rectangular enclosure, a standing wave solution for the acoustic potential function takes the following simple form (see Section 1.3.6):

$$\varphi = X(x)\, Y(y)\, Z(z)\, e^{j\omega t} \tag{7.1}$$

Substitution of Equation (7.1) into the wave equation (Equation (1.15)), use of $k^2 = \omega^2/c^2$, and rearrangement gives:

$$\frac{X''}{X} + \frac{Y''}{Y} + \frac{Z''}{Z} = -k^2 \tag{7.2}$$

Each term of Equation (7.2) on the left-hand side is a function of a different independent variable, whereas the right-hand side of the equation is a constant. It may be concluded that each term on the left must also equal a constant; that is, Equation (7.2) takes the form:

$$k_x^2 + k_y^2 + k_z^2 = k^2 \tag{7.3}$$

This implies the following:

$$X'' + k_x^2 X = 0 \tag{7.4}$$

$$Y'' + k_y^2 Y = 0 \tag{7.5}$$

$$Z'' + k^2 Z = 0 \tag{7.6}$$

Solutions to Equations (7.4), (7.5) and (7.6) are as follows:

$$X = A_x e^{jk_x x} + B_x e^{-jk_x x} \tag{7.7}$$

$$Y = A_y e^{jk_y y} + B_y e^{-jk_y y} \tag{7.8}$$

$$Z = A_z e^{jk_z z} + B_z e^{-jk_z z} \tag{7.9}$$

Boundary conditions will determine the values of the constants. For example, if it is assumed that the walls are essentially rigid so that the normal particle velocity, u_x, at the walls is zero, then, using Equations (1.10), (7.1) and (7.7), the following is obtained:

$$u_x = -\partial\varphi/\partial x \qquad (7.10)$$

and

$$-jk_x YZe^{j\omega t}[A_x e^{jk_x x} - B_x e^{-jk_x x}]_{x=0,L_x} = 0 \qquad (7.11)$$

Since

$$jk_x YZe^{j\omega t} \neq 0 \qquad (7.12)$$

then

$$[A_x e^{jk_x x} - B_x e^{-jk_x x}]_{x=0,L_x} = 0 \qquad (7.13)$$

First consider the boundary condition at $x = 0$. This condition leads to the conclusion that $A_x = B_x$. Similarly, it may be shown that:

$$A_i = B_i; \qquad i = x,y,z \qquad (7.14)$$

Next consider the boundary condition at $x = L_x$. This second condition leads to the following equation:

$$e^{jk_x L_x} - e^{-jk_x L_x} = 2j\sin(k_x L_x) = 0 \qquad (7.15)$$

Similar expressions follow for the boundary conditions at $y = L_y$ and $z = L_z$. From these considerations it may be concluded that the k_i are defined as follows:

$$k_i = n_i \frac{\pi}{L_i} \qquad n_i = 0, \pm 1, \pm 2, \ldots; \qquad i = x,y,z \qquad (7.16)$$

Substitution of Equation (7.16) into Equation (7.3) and use of $k^2 = \omega^2/c^2$ leads to the following useful result:

$$f_n = \frac{c}{2}\sqrt{\left[\frac{n_x}{L_x}\right]^2 + \left[\frac{n_y}{L_y}\right]^2 + \left[\frac{n_z}{L_z}\right]^2} \qquad \text{(Hz)} \qquad (7.17)$$

In this equation the subscript n on the frequency variable f indicates that the particular solutions or "eigen" frequencies of the equation are functions of the particular mode numbers n_x, n_y, and n_z.

Following Section 1.3.6 and using Equation (1.11), the following expression for the acoustic pressure is obtained:

$$p = \rho\frac{\partial\varphi}{\partial t} = j\omega\rho X(x)Y(y)Z(z)e^{j\omega t} \qquad (7.18a,b)$$

Substitution of Equations (7.14) and (7.16) into Equations (7.7), (7.8) and (7.9) and in turn substituting these altered equations into Equation (7.18) gives the following expression for the acoustic pressure in a rectangular room with rigid walls:

$$p = \overline{p} \cos\left[\frac{\pi n_x x}{L_x}\right] \cos\left[\frac{\pi n_y y}{L_y}\right] \cos\left[\frac{\pi n_z z}{L_z}\right] e^{j\omega t} \qquad (7.19)$$

In Equations (7.17) and (7.19), the mode numbers n_x, n_y and n_z have been introduced. These numbers take on all positive integer values including zero. There are three types of normal modes of vibration in a rectangular room, which have their analogs in enclosures of other shapes. They may readily be understood as follows:

1. axial modes for which only one mode number is not zero;
2. tangential modes for which one mode number is zero; and
3. oblique modes for which no mode number is zero.

These modes and their significance for noise control will now be discussed.

Axial modes correspond to wave travel back and forth parallel to an axis of the room. For example, the $(n_x, 0, 0)$ mode in the rectangular room of Figure 7.1 corresponds to a wave travelling back and forth parallel to the x axis. Such a system of waves forms a standing wave having n_x nodal planes normal to the x axis and parallel to the end walls. This may be verified by using Equation (7.19). The significance for noise control is that only sound absorption on the walls normal to the axis of sound propagation, where the sound is multiply reflected, will significantly affect the energy stored in the mode. Sound-absorptive treatment on any of the other walls would have only a small effect on an axial mode. The significance for sound coupling is that a speaker placed in the nodal plane of any mode will couple at best very poorly to that mode. Thus, the best place to drive an axial mode is to place the sound source at the end wall where the axial wave is multiply reflected; that is, at a pressure anti-node.

Tangential modes correspond to waves travelling essentially parallel to two opposite walls of an enclosure while successively reflecting from the other four walls. For example, the $(n_x, n_y, 0)$ mode of the rectangular enclosure of Figure 7.1 corresponds to a wave travelling around the room parallel to the floor and ceiling. In this case the wave impinges on all four vertical walls and absorptive material on any of these walls would be most effective in attenuating this mode. Note that absorptive material on the floor or ceiling would be less effective.

Oblique modes correspond to wave travel oblique to all room surfaces. For example the (n_x, n_y, n_z) mode in the rectangular room of Figure 7.1 would successively impinge on all six walls of the enclosure. Consequently, absorptive treatment on the floor, ceiling or any wall would be equally effective in attenuating an oblique mode.

For the placement of a speaker to drive a room, it is of interest to note that every mode of vibration has a pressure anti-node at the corners of a room. This may be verified by using Equation (7.19). A corner is a good place to drive a rectangular room when it is desirable to introduce sound. It is also a good location to place absorbents to attenuate sound and to sample the sound field for the purpose of determining room frequency response.

In Figure 7.2, the first 15 room resonant modes have been identified using Equation (7.17). Reference to the figure shows that of the first 15 lowest order modes, seven are axial modes, six are tangential modes and two are oblique modes. Reference to the figure also shows that as the frequency increases, the resonances become too numerous to identify individually and in this range, the number of axial and tangential modes will become negligible compared to the number of oblique modes. It may be useful to note that the frequency at which this occurs is about 80 Hz in the reverberation room described in Figure 7.2 and this corresponds to a room volume of about 2.25 cubic wavelengths. As the latter description is non-dimensional, it is probably general; however, a more precise boundary between low- and high-frequency behaviour will be given in the following section.

In a rectangular room, for every mode of vibration for which one of the mode numbers is odd, the sound pressure is zero at the centre of the room, as shown by consideration of Equation (7.19); that is, when one of the mode numbers is odd the corresponding term in Equation (7.19) is zero at the centre of the corresponding coordinate (room dimension). Consequently, the centre of the room is a very poor place to couple, either with a speaker or an absorber, into the modes of the room. Consideration of all the possible combinations of odd and even in a group of three mode numbers shows that only one-eighth of the modes of a rectangular room will not have nodes at the centre of the room. At the centre of the junction of two walls, only one-quarter of the modes of a rectangular room will not have nodes, and at the centre of any wall only half of the modes will not have nodes.

7.2.2 Cylindrical Rooms

The analysis of cylindrical rooms follows the same procedure as for rectangular rooms except that the cylindrical coordinate system is used instead of the cartesian system. The result of this analysis is the following expression for the resonance frequencies of the modes in a cylindrical room.

$$f(n_z, m, n) = \frac{c}{2}\sqrt{\left(\frac{n_z}{\ell}\right)^2 + \left(\frac{\psi_{m,n}}{a}\right)^2} \qquad (7.20)$$

where n_z is the number (varying from 0 to ∞) of nodal planes normal to the axis of the cylinder, ℓ is the length of the cylinder and a is its radius. The characteristic values ψ_{mn} are functions of the mode numbers m, n, where m is the number of diametral pressure nodes and n is the number of circumferential pressure nodes. Values of ψ_{mn} for the first few modes are given in Table 7.1.

7.3 BOUND BETWEEN LOW-FREQUENCY AND HIGH-FREQUENCY BEHAVIOUR

Referring to Figure 7.2, where the frequency response of a rectangular enclosure is shown, it can be observed that the number of peaks in response increases rapidly with increasing frequency. At low frequencies, the peaks in response are well separated and can be readily identified with resonant modes of the room. However, at high frequencies, so many modes may be driven in strong response at once that they tend

Table 7.1 Values of $\psi_{m,n}$

$m\backslash n$	0	1	2	3	4
0	0.0000	1.2197	2.2331	3.2383	4.2411
1	0.5861	1.6971	2.7172	3.7261	4.7312
2	0.9722	2.1346	3.1734	4.1923	5.2036
3	1.3373	2.5513	3.6115	4.6428	5.6623
4	1.6926	2.9547	4.0368	5.0815	6.1103

to interfere, so that at high frequencies individual peaks in response cannot be associated uniquely with individual resonances. In this range statistical analysis is appropriate.

Clearly, a need exists for a frequency bound that defines the cross-over from the low-frequency range, where modal analysis is appropriate, to the high-frequency range where statistical analysis is appropriate. Reference to Figure 7.2 provides no clear indication of a possible bound, as a continuum of gradual change is observed. However, analysis does provide a bound, but to understand the determination of the bound, called here the cross-over frequency, three separate concepts are required; modal density, modal damping and modal overlap. These concepts will be introduced in the following three sections and then used to define the cross-over frequency.

7.3.1 Modal Density

The approximate number of modes, N, which may be excited in the frequency range from zero up to f Hz, is given by the following expression for a rectangular room (Morse and Bolt, 1944):

$$N = \frac{4\pi f^3 V}{3c^3} + \frac{\pi f^2 S}{4c^2} + \frac{fL}{8c} \qquad (7.21)$$

In Equation (7.21), c is the speed of sound, V is the room volume, S is the room total surface area and L is the total perimeter of the room, which is the sum of lengths of all edges. It has been shown (Morse and Ingard, 1968) that Equation (7.21) has wider application than for rectangular rooms; to a good approximation it describes the number of modes in rooms of any shape, with the approximation improving as the irregularity of the room shape increases. It should be remembered that Equation (7.21) is an approximation only and the actual number of modes fluctuates above and below the prediction of this equation as the frequency gradually increases or decreases.

For the purpose of estimating the number of modes that, on average, may be excited in a narrow frequency band, the derivative of Equation (7.21), called the modal density, is useful. The expression for the modal density is as follows:

$$\frac{dN}{df} = \frac{4\pi f^2 V}{c^3} + \frac{\pi f S}{2c^2} + \frac{L}{8c} \qquad (7.22)$$

which also applies approximately to rooms of any shape, including cylindrical rooms.

Consideration of Equation (7.22) shows that, at low frequencies, the number of modes per unit frequency that may be excited will be very small but, as the modal density increases quadratically with increasing frequency, at high frequencies the number of modes excited will become very large. Thus, at low frequencies, one can expect large spatial fluctuations in sound pressure level, as observed in Figure 7.2 when a room is excited with a narrow band of noise but, at high frequencies, the fluctuations become small and the reverberant field approximates uniformity throughout the room. The number of oblique modes in a room of any shape is described approximately by the cubic term of Equation (7.21), although the linear and quadratic terms also contribute a little to the number of oblique modes (Morse and Ingard, 1968, pp. 586), with these latter contributions becoming steadily less important as the frequency increases. Similarly the number of tangential modes is dominated by the quadratic term with the linear term also contributing. The number of axial modes is actually 4 times the linear term in equation (7.21) (Morse and Ingard, 1968), but this latter term has also been modified by negative contributions from oblique and tangential modes. Thus, it is evident that at high frequencies the number of oblique modes will far exceed the number of tangential and axial modes and to a good approximation at high frequencies the latter two mode types may be ignored.

7.3.2 Modal Damping and Bandwidth

Referring to Figure 7.2, it may be observed that the recorded frequency response peaks in the low-frequency range have finite widths, which may be associated with the response of the room that was investigated. A bandwidth, Δf, may be defined and associated with each mode, being the frequency range about resonance over which the sound pressure squared is greater than or equal to half the same quantity at resonance. The lower and upper frequencies bounding a resonance and defined in this way are commonly known as the half-power points. The corresponding response at the half-power points is down 3 dB from the peak response. Referring to the figure, the corresponding bandwidths are easily determined where individual resonances may be identified.

The bandwidth, Δf, is dependent upon the damping of the mode; the greater the modal damping, the larger will be the bandwidth. For acoustical spaces such as sound in rooms the modal damping is commonly expressed in terms of the damping factor (similar to the critical damping ratio), which is a viscous based quantity and proportional to particle velocity, whereas for structures, modal damping is commonly expressed in terms of a modal loss factor, η, which is a hysteretic based quantity and proportional to displacement. Alternatively, damping in structures may be viscously based as well and may be expressed in terms of the critical damping ratio, ζ, commonly used to describe damping in mechanical systems. These quantities may be related to each other and to the energy-based quality factor, Q, of the resonant mode, or the logarithmic decrement, δ, by the following relations:

$$\Delta f/f = 1/Q = \eta = \frac{2\zeta}{\sqrt{1 - \zeta^2}} = \delta/\pi \qquad (7.23a,b,c,d)$$

The quality factor Q is discussed in Section 9.7.2.2, the critical damping ratio, ζ, is discussed in Section 10.2.1 and the logarithmic decrement, δ, is discussed in Section 10.8. Here the modal loss factor, η, is presented as an energy-based quantity by its relation to the quality factor Q. The loss factor, η, is sometimes used in acoustics as a viscous based damping quantity. More usually, it has meaning as a structural loss factor based upon a hysteretic damping effect in a structural member. For a solid material, it is defined in terms of a complex modulus of elasticity $E' = E(1 + j\eta)$ where E is Young's modulus of elasticity. This use of the loss factor is discussed in Section 10.2.1. As may be observed by reference to Equation (7.23), when the modal loss factor, η, is small, which is true for most practical cases, the implication is that the critical damping ratio is also small and $\eta = 2\zeta$.

At low frequencies, individual modal bandwidths can be identified and measured directly. At high frequencies, where individual modes cannot be identified, the average bandwidth may be calculated from a measurement of the decay time (see Section 7.5.1) using the following equation (Beranek, 1971):

$$\Delta f = 2.20 / T_{60} \tag{7.24}$$

7.3.3 Modal Overlap

Modal overlap, M, is calculated as the product of the average bandwidth given by either Equation (7.23) or (7.24) and the modal density given by Equation (7.22). The expression for modal overlap is:

$$M = \Delta f \, \mathrm{d}N/\mathrm{d}f \tag{7.25}$$

The modal overlap is a measure of the extent to which the resonances of a reverberant field cover the range of all possible frequencies within a specified frequency range. The concept is illustrated for a hypothetical case of a low modal overlap of 0.6 in Figure 7.3. In the figure, three resonant modes, their respective bandwidths and the frequency range of the specified frequency band are indicated.

7.3.4 Cross-over Frequency

There are two criteria commonly used for determining the cross-over frequency. The criterion that is chosen will depend upon whether room excitation with bands of noise or with pure tones is of interest.

If room excitation with one-third octave, or wider bands of noise is to be considered, then the criterion for statistical (high-frequency) analysis is that there should be a minimum of between 3 and 6 modes resonant in the frequency band. The exact number required is dependent upon the modal damping and the desired accuracy of the results. More modes are necessary for low values of modal damping or if high accuracy is required. If room excitation with a pure tone or a very narrow band of noise is of concern, then the criterion for reliable statistical analysis is that the modal overlap should be greater than or equal to 3.

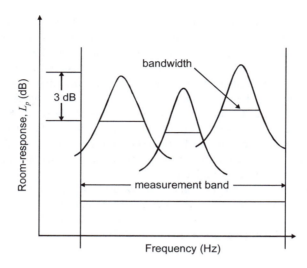

Figure 7.3 Three modes in a specified frequency range with a modal overlap of 0.6.

7.4 HIGH FREQUENCIES, STATISTICAL ANALYSIS

At high frequencies, the sound field in a reverberant space may be thought of as composed of plane waves travelling in all directions with equal probability. If the mean square pressure amplitude of any travelling plane wave is on average the same, independent of direction, then the sound field is said to be diffuse. In a reverberant space, the sound field is diffuse when the modal overlap is three or greater, in which case the sound field steady-state frequency response is essentially a random phenomenon. For excitation with bands of noise, the parameters describing the field are essentially predictable from the room reverberation time (Schroeder, 1969).

The concept of diffuse field implies that the net power transmission along any coordinate axis is negligibly small; that is, any power transmission is essentially the same in any direction. The reverberant sound field may be considered as of constant mean energy density throughout the room. However, the concept does not imply a sound field where the sound pressure level is the same throughout. Even in a perfectly diffuse sound field, the sound pressure level will fluctuate over time at any given location in the room and the long-time-averaged sound pressure level will also fluctuate from point to point within the room. The amount of fluctuation is dependent upon the product of the measurement bandwidth B (in Hz) and the reverberation time, T_{60}, averaged over the band. The expected standard deviation, which describes the spatial fluctuations in a diffuse sound field, is given by the following approximate expression (Lubman, 1969):

$$\sigma \approx 5.57 \, (1 \, + \, 0.238 \, B \, T_{60})^{-1/2} \quad \text{(dB)} \tag{7.26}$$

7.4.1 Effective Intensity in a Diffuse Field

In a diffuse (reverberant) field, sound propagation in all directions is equally likely and consequently the intensity at any point in the field is zero. However, an effective intensity associated with power transmission in a specified direction can be defined. An expression for the effective intensity will now be derived in terms of the reverberant field sound pressure.

Consider sound energy in a reverberant field propagating along a narrow column of circular cross-section, as illustrated in Figure 7.4. Let the column just encompass a small spherical region in the field. The ratio of the volume of the spherical region to the cylindrical section of column that just encompasses the spherical region is:

$$\frac{4\,\pi\,r^3}{3} \cdot \frac{1}{2\,\pi\,r^3} = \frac{2}{3} \tag{7.27}$$

Equation (7.27) shows that the spherical region occupies two-thirds of the volume of the encompassing cylinder.

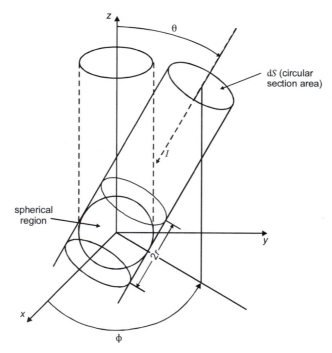

Figure 7.4 Geometrical arrangement for determining effective intensity in a diffuse field.

Referring to Figure 7.4, consider the convergence on the spherical region of sound from all directions. Let the intensity of any incident sound beam of cross-sectional area, dS, be I. The time for the beam to travel through the spherical region (length of the encompassing cylinder) is $2r/c$; thus the incremental contribution per unit area to the energy E in the spherical region due to any beam is:

$$\Delta E = \frac{2}{3} I \frac{2r}{c} \tag{7.28}$$

The total energy is obtained by integrating the incremental energy contribution per unit area of sphere over the area of the sphere. The incremental area of sphere for use in integration is:

$$dS = r^2 \sin\theta \, d\theta \, d\varphi \tag{7.29}$$

Thus:

$$E = \frac{4I}{3c} \int_0^{2\pi} d\varphi \int_0^{\pi} r^3 \sin\theta \, d\theta = \frac{16 I\pi r^3}{3c} \tag{7.30}$$

Let the time averaged acoustic energy density be ψ at the centre of the region under consideration; then the total energy in the central spherical region is:

$$E = \frac{4\pi r^3 \psi}{3} \tag{7.31}$$

Combining Equations (7.30) and (7.31) gives, for the effective intensity, I, in any direction in terms of the time averaged energy density, ψ:

$$I = \psi c / 4 \tag{7.32}$$

To obtain an expression for the energy density, one observes that the length of time required for a plane wave to travel unit distance is just the reciprocal of the speed of sound multiplied by unit distance. Use of this observation and the expression for the intensity of a plane wave given by Equation (1.74) provides the following expression for the time averaged energy density, which also holds for 2-D and 3-D sound fields:

$$\psi = \langle p^2 \rangle / (\rho c^2) \tag{7.33}$$

Substitution of Equation (7.32) in Equation (7.33) gives the following expression for the effective intensity in one direction a diffuse field:

$$I = \langle p^2 \rangle / (4\rho c) \tag{7.34}$$

7.4.2 Energy Absorption at Boundaries

Consider a diffuse sound field in an enclosure and suppose that a fraction of the incident energy is absorbed upon reflection at the enclosure boundaries. Let the average fraction of incident energy absorbed be $\bar{\alpha}$, called the Sabine absorption coefficient. Implicit in the use to which the Sabine absorption will be put is the assumption that this absorption coefficient is strictly a property that may be associated with the absorptive material. Whatever the material, this assumption is not strictly true; it is a useful approximation that makes tractable an otherwise very difficult problem. The concept of absorption coefficient follows from the assumption that the walls of an enclosure may be considered to be locally reactive and thus characterised by an impedance, which is a unique property exhibited by the wall at its surface and is independent of interaction between the incident sound and the wall anywhere else. The assumption is then explicit that the wall response to the incident sound depends solely on local properties and is independent of the response at other points on the surface.

The locally reactive assumption has proven very useful for architectural purposes, but is apparently of very little use in predicting interior noise in aircraft and vehicles of various types. In the latter cases, the modes of the enclosed space couple with modes in the walls, and energy stored in vibrating walls contributes very significantly to the resulting sound field. In such cases, the locally reactive concept is not even approximately true, and neither is the concept of Sabine absorption that follows from it.

7.4.3 Air Absorption

In addition to energy absorption on reflection, some energy is absorbed during propagation between reflections. Generally, propagation loss due to air absorption is negligible, but at high frequencies above 500 Hz, especially in large enclosures, it makes a significant contribution to the overall loss.

Air absorption may be taken into account as follows. As shown in Section 7.5.3, the mean distance, Λ, travelled by a plane wave in an arbitrarily shaped enclosure between reflections, is called the mean free path and is given by the following equation:

$$\Lambda = \frac{4V}{S} \tag{7.35}$$

where V is the room volume and S is the room surface area (Kuttruff, 1979). It will now be assumed that the fraction of propagating sound energy lost due to air absorption between reflections is linearly related to the mean free path. If the fraction lost is not greater than 0.4, then the error introduced by this approximation is less than 10% (0.5 dB).

At this point many authors write $4m'V/S = \alpha_a$ for the contribution due to air absorption, and they provide tables of values of the constant m' as a function of temperature and relative humidity. Here, use will be made of values for air absorption m already given in Table 5.3 for sound propagating outdoors. In a distance of one mean free path the attenuation of sound is:

$$\frac{4mV}{S}10^{-3} = -10\log_{10}e^{-4m'V/S} \tag{7.36}$$

Let

$$4m'V/S = \alpha_a \tag{7.37}$$

Thus:

$$\alpha_a = 4mV/(S10^4 \log_{10}e) = 9.21 \times 10^{-4} mV/S \tag{7.38}$$

Using the above relation, the total mean absorption coefficient, including air absorption may be written as:

$$\overline{\alpha} = \overline{\alpha}_{wcf} + 9.21 \times 10^{-4} mV/S \tag{7.39}$$

Equations (7.32) and (7.34) may be used to write for the power, W_a, or rate of energy absorbed:

$$W_a = \psi Sc\overline{\alpha}/4 = \langle p^2 \rangle S\overline{\alpha}/(4\rho c) \tag{7.40a,b}$$

where $\overline{\alpha}$ is defined by Equation (7.39).

7.4.4 Steady-state Response

At any point in a room, the sound field is a combination of the direct field radiated by the source and the reverberant field. Thus the total sound energy measured at a point in a room is the sum of the sound energy due to the direct field and that due to the reverberant field.

Using Equation (5.14) and introducing the directivity factor, D_θ (see Section 5.8), the sound pressure squared due to the direct field at a point in the room at a distance r and in a direction (θ, φ) from the source may be written as:

$$\langle p^2 \rangle_D = W\rho c D_\theta / 4\pi r^2 \tag{7.41}$$

The quantity D_θ is the directivity factor of the source in direction (θ, φ), ρ is the density of air (kg/m³), c is the speed of sound in air (m/s) and W is the sound power radiated by the source (W). In writing Equation (7.41) it is assumed that the source is sufficiently small or r is sufficiently large for the measurement point to be in the far field of the source.

Consider that the direct field must be once reflected to enter the reverberant field. The fraction of energy incident at the walls, which is reflected into the reverberant field, is $(1 - \overline{\alpha})$. Using Equations (7.40b) and (7.41) and setting the power absorbed equal to the power introduced, W, the sound pressure squared due to the reverberant field may be written as:

$$\langle p^2 \rangle_R = 4W\rho c(1 - \overline{\alpha})/(S\overline{\alpha}) \tag{7.42}$$

The sound pressure level at any point due to the combined effect of the direct and reverberant sound fields is obtained by adding together Equations (7.41) and (7.42).

Thus, using Equations (1.81), (1.84) and (1.87):

$$L_p = L_w + 10 \log_{10} \left[\frac{D_\theta}{4\pi r^2} + \frac{4}{R} \right] + 10 \log_{10} \left[\frac{\rho c}{400} \right] \tag{7.43}$$

At 20°C, where $\rho c = 414$ (SI units), there would be an error of approximately 0.1 dB if the last term in Equation (7.43) is omitted. In common industrial spaces, which have lateral dimensions much greater than their height, Equation (7.43) under predicts reverberant field noise levels (see Section 7.9) close to the noise source and over predicts levels far away from the source (Hodgson, 1994a). The prediction of sound levels in these types of space are discussed in Section 7.9.

Equation (7.43) has been written in terms of the room constant R where the room constant is:

$$R = \frac{S\overline{\alpha}}{1 - \overline{\alpha}} \tag{7.44}$$

7.5 TRANSIENT RESPONSE

If sound is introduced into a room, the reverberant field level will increase until the rate of sound energy introduction is just equal to the rate of sound energy absorption. If the sound source is abruptly shut off, the reverberant field will decay at a rate determined by the rate of sound energy absorption. The time required for the reverberant field to decay by 60 dB, called the reverberation time, is the single most important parameter characterising a room for its acoustical properties. For example, a long reverberation time may make the understanding of speech difficult but may be desirable for organ recitals.

As the reverberation time is directly related to the energy dissipation in a room, its measurement provides a means for the determination of the energy absorption properties of a room. Knowledge of the energy absorption properties of a room in turn allows estimation of the resulting sound pressure level in the reverberant field when sound of a given power level is introduced. The energy absorption properties of materials placed in a reverberation chamber may be determined by measurement of the associated reverberation times of the chamber, with and without the material under test in the room. The Sabine absorption coefficient, which is assumed to be a property of the material under test, is determined in this way and standards (ASTM C423-08a; ISO 354-2003; AS1045-1988) are available that provide guidance for conducting these tests.

In the following sections, two methods will be used to characterise the transient response of a room. The classical description, in which the sound field is described statistically, will be presented first and a second method, in which the sound field is described in terms of modal decay, will then be presented. The second method provides a description in better agreement with experiment than does the classical approach (Bies, 1995).

7.5.1 Classical Description

At high frequencies the reverberant field may be described in terms of a simple differential equation, which represents a gross simplification of the physical process but nonetheless gives generally useful results.

Using Equation (7.40a) and the observation that the rate of change of the energy stored in a reverberant field equals the rate of supply, W_0, less the rate of energy absorbed, W_a, gives the following result:

$$W = V \partial \psi / \partial t = W_0 - \psi S c \bar{\alpha}/4 \tag{7.45}$$

Introducing the dummy variable:

$$X = [4 W_0 / S c \bar{\alpha}] - \psi \tag{7.46}$$

and using Equation (7.46) to rewrite Equation (7.45), the following result is obtained:

$$\frac{1}{X} \frac{dX}{dt} = -\frac{S c \bar{\alpha}}{4 V} \tag{7.47}$$

Integration of the above equation gives:

$$X = X_0 e^{-S c \bar{\alpha} t / 4 V} \tag{7.48}$$

where X_0 is the initial value.

Two cases will be considered. Suppose that initially, at time zero, the sound field is nil and a source of power, W_0, is suddenly turned on. The initial conditions are time $t = 0$ and sound pressure $\langle p^2_0 \rangle = 0$. Use of Equation (7.33) and substitution of Equation (7.46) into Equation (7.48) gives the following expression for the resulting reverberant field at any later time t.

$$\langle p^2 \rangle = \frac{4 W_0 \rho c}{S \bar{\alpha}} \left(1 - e^{-S c \bar{\alpha} t / 4 V} \right) \tag{7.49}$$

Alternatively, consider that a steady-state sound field has been established when the source of sound is suddenly shut off. In this case, the initial conditions are time $t = 0$, sound power $W_0 = 0$, and sound pressure $\langle p^2 \rangle = \langle p_0^2 \rangle$. Again, use of Equation (7.33) and substitution of Equation (7.46) into Equation (7.48) gives, for the decaying reverberant field at some later time t:

$$\langle p^2 \rangle = \langle p_0^2 \rangle e^{-S c \bar{\alpha} t / 4 V} \tag{7.50}$$

Taking logarithms to the base ten of both sides of Equation (7.50) gives the following:

$$L_{p0} - L_p = 1.086 \, S c \bar{\alpha} t / V \tag{7.51}$$

Equation (7.51) shows that the sound pressure level decays linearly with time and at a rate proportional to the Sabine absorption $S\bar{\alpha}$. It provides the basis for the measurement and definition of the Sabine absorption coefficient $\bar{\alpha}$.

Sabine introduced the reverberation time, T_{60} (seconds), as the time required for the sound energy density level to decay by 60 dB from its initial value. He showed that the reverberation time, T_{60}, was related to the room volume, V, the total wall area including floor and ceiling, S, the speed of sound, c, and an absorption coefficient, $\bar{\alpha}$, which was characteristic of the room and generally a property of the bounding surfaces. Sabine's reverberation time equation, which follows from Equations (7.50) and (7.51) with $L_{p0} - L_p = 60$, may be written as follows:

$$T_{60} = \frac{55.25\,V}{Sc\bar{\alpha}} \tag{7.52}$$

It is interesting to note that the effective Sabine absorption coefficient used to calculate reverberation times in spaces such as typical concert halls or factories is not the same as that measured in a reverberation room (Hodgson, 1994b; Kuttruff, 1994), which often leads to inaccuracies in predicted reverberation times. For this reason it is prudent to follow the advice given in the next section.

7.5.2 Modal Description

The discussion thus far suggests that the reverberant field within a room may be thought of as composed of the excited resonant modes of the room. This is still true even in the high-frequency range where the modes may be so numerous and close together that they tend to interfere and cannot be identified separately. In fact, if any enclosure is driven at a frequency slightly off-resonance and the source is abruptly shut off, the frequency of the decaying field will be observed to shift to that of the driven resonant mode as it decays (Morse, 1948). In general, the reflection coefficient, β (the fraction of incident energy that is reflected) characterising any surface is a function of the angle of incidence. It is related to the corresponding absorption coefficient, α (the fraction of incident energy that is absorbed) as $\alpha + \beta = 1$. Note that the energy reflection coefficient referred to here is the modulus squared of the amplitude reflection coefficient discussed in Chapter 5, Section 5.10.

When a sound field decays, all of the excited modes decay at their natural frequencies (Morse, 1948). This implies that the frequency content of the decaying field may be slightly different to that of the steady-state field. Thus, the decay of the sound field is modal decay (Larson, 1978). In the frequency range in which the field is diffuse, it is reasonable to assume that the energy of the decaying field is distributed among the excited modes about evenly within a measurement band of frequencies. In a reverberant field in which the decaying sound field is diffuse, it is also necessary to assume that scattering of sound energy continually takes place between modes so that even though the various modes decay at different rates, scattering ensures that they all contain about the same amount of energy on average during decay. Effectively, in a Sabine room, all modes within a measurement band will decay, on average, at the same rate, because energy is continually scattered from the more slowly decaying modes into the more rapidly decaying modes.

Let $\langle p^2(t) \rangle$ be the mean square band sound pressure level at time t in a decaying field and $\langle p_k^2(0) \rangle$ be the mean square sound pressure level of mode k at time $t = 0$. The decaying field may be expressed in terms of modal mean square pressures, $\langle p_k^2(0) \rangle$, mean energy reflection coefficients, β_k, and modal mean free paths, Λ_k, as follows:

$$\langle p^2(t) \rangle = \sum_{k=1}^{N} \langle p_k^2(0) \rangle \beta_k^{ct/\Lambda_k} \tag{7.53}$$

where

$$\beta_k = \prod_{i=1}^{n} \left[\beta_{ki} \right]^{S_i/S_k} \tag{7.54}$$

In the above equations N is the number of modes within a measurement band. The quantities, β_k, are the energy reflection coefficients and S_i are the areas of the corresponding reflecting surfaces encountered by a wave travelling around a modal circuit associated with mode k and reflection from surface i (Morse and Bolt, 1944). The S_k are the sums of the areas of the S_i reflecting surfaces encountered in one modal circuit of mode, k.

The modal mean free path, Λ_k, is the mean distance between reflections of a sound wave travelling around a closed modal circuit and for a rectangular room is given by the following equation (Larson, 1978):

$$\Lambda_k = \frac{2f_k}{c} \left[\frac{n_x}{L_x^2} + \frac{n_y}{L_y^2} + \frac{n_z}{L_z^2} \right]^{-1} \tag{7.55}$$

The quantities, β_k, represent the energy reflection coefficients encountered during a modal circuit and the symbol, $\displaystyle\prod_{i=1}^{n}$ represents the product of the n reflection coefficients where n is either a multiple of the number of reflections in one modal circuit or a large number. The quantity f_k is the resonance frequency given by Equation (7.17) for mode k of a rectangular enclosure, which has the modal indices n_x, n_y, n_z.

The assumption will be made that the energy in each mode is on average the same, so that in Equation (7.53), $p_k(0)$ may be replaced with $p_0/\sqrt{N}$, where p_0 is the measured initial sound pressure in the room when the source is shut off. Equation (7.53) may be rewritten as follows:

$$\langle p^2(t) \rangle = \langle p_0^2 \rangle \frac{1}{N} \sum_{k=1}^{N} e^{(ct/\Lambda_k)\log_e(1-a_k)} \tag{7.56}$$

A mathematical simplification is now made. In the above expression the modal mean free path length is replaced with the mean of all of the modal mean free paths, $4V/S$, and the modal mean absorption coefficient a_k is replaced with the area weighted mean statistical absorption coefficient, $\bar{a}_{st}$, for the room (see Section 7.7 and Appendix C). The quantity V is the total volume and S is the total wall, ceiling and floor area of the room. In exactly the same way as Equation (7.52) was derived from Equation (7.50), the well known reverberation time equation of Norris-Eyring may be derived from Equation (7.56) as follows:

$$T_{60} = -\frac{55.25\, V}{S c \log_e (1 - \overline{\alpha}_{st})} \tag{7.57}$$

This equation is often preferred to the Sabine equation by many who work in the field of architectural acoustics, as some authors claim that it gives results that are closer to measured data (Neubauer, 2001). However, Beranek and Hidaka (1998) obtained good agreement between measured and predicted reverberation times in concert halls using the Sabine relation. Of course, if sound absorption coefficients measured in a reverberation chamber are to be used to predict reverberation times, then the Sabine equation must be used as the Norris-Eyring equation is only valid if statistical absorption coefficients are used (see Appendix C).

   Note that air absorption must be included in α_{st} in a similar way as it is included in $\overline{\alpha}$ (Equation (7.39)). It is worth careful note that Equation (7.57) is a predictive scheme based upon a number of assumptions that cannot be proven, and consequently inversion of the equation to determine the statistical absorption coefficient $\overline{\alpha}_{st}$ is not recommended.

   With a further simplification, the famous equation of Sabine is obtained. When $\overline{\alpha}_{st} < 0.4$ an error of less than 0.5 dB is made by setting $\overline{\alpha}_{st} \approx - \log_e (1 - \overline{\alpha}_{st})$ in Equation (7.57), and then by replacing $\overline{\alpha}_{st}$ with $\overline{\alpha}$, Equation (7.52) is obtained.

   Alternatively, if in Equation (7.56), the quantity, $(1 - \alpha_k)$, is replaced with the modal energy reflection coefficients β_k and these in turn are replaced with a mean value, called the mean statistical reflection coefficient $\overline{\beta}_{st}$, the following equation of Millington and Sette is obtained.

$$T_{60} = -\frac{55.25\, V}{S c \log_e \overline{\beta}_{st}} \tag{7.58}$$

The quantity, $\overline{\beta}_{st}$, may be calculated using Equation (7.54) but with changes in the meaning of the symbols. β_k is replaced with β_{st}, which is now to be interpreted as the area weighted geometric mean of the random incidence energy reflection coefficients, β_i, for all of the room surfaces; that is:

$$\overline{\beta}_{st} = \left[\prod_{i=1}^{n} \beta_i^{S_i/S} \right] \tag{7.59}$$

The quantity β_i is related to the statistical absorption coefficient α_i for surface i of area S_i by $\beta_i = 1 - \alpha_i$. It is of interest to note that although taken literally, Equation (7.59) would suggest that an open window having no reflection would absorb all of the incident energy and there would be no reverberant field, the interpretation presented here suggests that an open window must be considered as only a part of the wall in which it is placed and the case of total absorption will never occur. Alternatively, reference to Equation (7.53) shows that if any term β_i is zero, it simply does not appear in the sum and thus will not appear in Equation (7.58) which follows from it.

7.5.3 Empirical Description

For calculating reverberation times in rooms for which the distribution of absorption was non-uniform (such as rooms with large amounts of absorption on the ceiling and floor and little on the walls), Fitzroy (1959) proposed the following empirical equation:

$$
T_{60} = \frac{0.16\,V}{S^2}\left[\frac{-S_x}{\log_e(1 - \overline{\alpha}_{xst})} + \frac{-S_y}{\log_e(1 - \overline{\alpha}_{yst})} + \frac{-S_z}{\log_e(1 - \overline{\alpha}_{zst})}\right]
\tag{7.60}
$$

where V is the room volume (m^3), S_x, S_y and S_z are the total areas of two opposite parallel room surfaces (m^2), $\overline{\alpha}_{xst}$, $\overline{\alpha}_{yst}$ and $\overline{\alpha}_{zst}$ are the average statistical absorption coefficients of a pair of opposite room surfaces (see Equation (7.79)) and S is the total room surface area.

Neubauer (2001) presented a modified Fitzroy equation which he called the Fitzroy-Kuttruff equation and which gave more reliable results than the original Fitzroy equation. In fact, this equation has been shown to be even more accurate than the Norris-Eyring equation for architectural spaces with non-uniform sound absorption. The Fitzroy-Kuttruff equation is as follows.

$$
T_{60} = \left(\frac{0.32\,V}{S^2}\right)\left(\frac{L_z(L_x + L_y)}{\overline{\alpha}_w} + \frac{L_x L_y}{\overline{\alpha}_{cf}}\right)
\tag{7.61}
$$

where L_x, L_y and L_z are the room dimensions (m) and:

$$
\overline{\alpha}_w = -\log_e(1 - \overline{\alpha}_{st}) + \left[\frac{\beta_w(\beta_w - \overline{\beta}_{st})S_w^2}{(\overline{\beta}_{st}\,S)^2}\right]
\tag{7.62}
$$

$$
\overline{\alpha}_{cf} = -\log_e(1 - \overline{\alpha}_{st}) + \left[\frac{\beta_{cf}(\beta_{cf} - \overline{\beta}_{st})S_{cf}^2}{(\overline{\beta}_{st}\,S)^2}\right]
\tag{7.63}
$$

where $\overline{\alpha}_{st}$ is the arithmetic mean over the six room surfaces of the surface averaged statistical absorption coefficient, $\beta = (1 - \alpha)$ is the energy reflection coefficient, the subscript, w refers to the walls and the subscript cf refers to the floor and ceiling.

Equations (7.52), (7.57), (7.58), (7.60) and (7.61) for reverberation time are all based on the assumption that the room dimensions satisfy the conditions for Sabine rooms (see Section 7.1.2) and that the absorption is reasonably well distributed over the room surfaces. However, in practice this is not often the case and for rooms that do not meet this criterion, Kuttruff (1994) has proposed that Equation (7.52) be used except that $\overline{\alpha}$ should be replaced with α defined as follows:

$$\alpha = -\log_e(1 - \overline{\alpha}_{st})\Big[1 + 0.5\gamma^2 \log_e(1 - \overline{\alpha}_{st})\Big] + \frac{\sum\limits_{i=1}^{n} \beta_i(\beta_i - 1 + \overline{\alpha}_{st})S_i^2}{S^2(1 - \overline{\alpha}_{st})^2} \qquad (7.64)$$

In Equation (7.64), n is the number of room surfaces (or part room surfaces if whole surfaces are subdivided), $\overline{\alpha}_{st}$ is the statistical absorption coefficient, area averaged over all room surfaces (see Section 7.7.5, Equation (7.79) and β_i is the statistical energy reflection coefficient of surface , i of area S_i. The first term in Equation (7.64) accounts for room dimensions that exceed the Sabine room criterion. The quantity γ^2 is the variance of the distribution of path lengths between reflections divided by the square of the mean free path length. It has a value of about 0.4, provided that the room shape is not extreme. The second term in Equation (7.64) accounts for non-uniform placement of sound absorption.

Neubauer (2000) provided an alternative modified Fitzroy equation for flat and long rooms as:

$$T_{60} = \left[\frac{-0.126 S_x}{\log_e(1 - \overline{\alpha}_{stx})P_x} - \frac{0.126 S_y}{\log_e(1 - \overline{\alpha}_{sty})P_y} - \frac{0.126 S_z}{\log_e(1 - \overline{\alpha}_{stz})P_z}\right]^{1/2} \qquad (7.65)$$

where P_x and P_y are the total perimeters for each of the two pairs of opposite walls and P_z is the total perimeter of the floor and ceiling. Similar definitions apply for S_x, S_y and S_z and also for $\overline{\alpha}_x$, $\overline{\alpha}_y$ and $\overline{\alpha}_z$. Note that for a cubic room, Equation (7.65) may be used with the exponent, ½ replaced by ⅓.

7.5.4 Mean Free Path

When air absorption was considered in Section 7.4.3, the mean free path was introduced as the mean distance travelled by a sound wave between reflections, and frequent reference has been made to this quantity in subsequent sections. Many ways have been demonstrated in the literature for determining the mean free path and two will be presented in this section.

The classical description of a reverberant space, based upon the solution of a simple differential equation presented in Section 7.5.1, leads directly to the concept of mean free path. Let the mean free path be Λ, then in a length of time equal to Λ/c all of the sound energy in the reverberant space will be once reflected and reduced by an amount (one reflection), $e^{-\overline{\alpha}}$. If the energy stored in volume V was initially $V\langle p^2_0\rangle$ ρc^2 and at the end of time Λ/c it is $V\langle p^2\rangle \rho c^2$, then according to Equation (7.50):

$$V\langle p^2\rangle/\rho c^2 = V\langle p_o^2\rangle/\rho c^2 \, e^{-\overline{\alpha}} = V\langle p_o^2\rangle/\rho c^2 \, e^{-S\overline{a}\Lambda/4V} \qquad (7.66a,b)$$

Consideration of Equation (7.66) shows that the mean free path, Λ, is given by Equation (7.35).

Alternatively, a modal approach to the determination of the mean free path may be employed, using modal indices, n_x, n_x, and n_x, respectively. Consideration in this case will be restricted to rectangular enclosures for convenience. For this purpose the following quantities are defined:

$$X = n_x c / 2 f L_x, \quad Y = n_y c / 2 f L_y \text{ and } \quad Z = n_z c / 2 f L_z \tag{7.67}$$

Substitution of Equations (7.67) into Equation (7.17) gives the following result:

$$1 = X^2 + Y^2 + Z^2 \tag{7.68}$$

Letting $a_1 = L_y \times L_z$, $a_2 = L_z \times L_x$, $a_3 = L_x \times L_y$ and $V = L_x \times L_y \times L_z$, multiplying the numerator of the reciprocal of Equation (7.55) by V and the denominator by $L_x \times L_y \times L_z$ and use of Equation (7.67) gives the following result:

$$V / \Lambda_i = a_1 X + a_2 Y + a_3 Z \tag{7.69}$$

An average value for the quantity V/Λ_i may be determined by summing over all possible values of Λ_i. When the modal density is large, it may be assumed that sound is incident from all directions and it is then possible to replace the sum with an integral. Introducing the following spherical coordinates:

$$X = \sin \varphi \cos \theta \quad Y = \sin \varphi \sin \theta \quad Z = \cos \varphi \tag{7.70}$$

substituting Equation (7.70) into Equation (7.69) and forming the integral, the following result is obtained:

$$V / \Lambda = \frac{2}{\pi} \int_0^{\pi/2} d\varphi \int_0^{\pi/2} (a_1 \sin^2 \varphi \cos \theta + a_2 \sin^2 \varphi \sin \theta + a_3 \cos \varphi \sin \varphi) \, d\theta \tag{7.71}$$

Carrying out the indicated integration gives for the mean free path, Λ, the result given previously by Equation (7.35).

7.6 MEASUREMENT OF THE ROOM CONSTANT

Measurements of the room constant, R, given by Equation (7.44), or the related Sabine absorption, $S\bar{\alpha}$, may be made using either a reference sound source or by measuring the reverberation time of the room in the frequency bands of interest. These methods are described in the following sections. Alternatively, yet another method is offered in Chapter 6 (Section 6.6.3.1).

7.6.1 Reference Sound Source Method

The reference sound source is placed at a number of positions chosen at random in the room to be investigated, and sound pressure levels are measured at a number of positions in the room for each source position. In each case, the measurement positions are chosen to be remote from the source, where the reverberant field of the room dominates the direct field of the source. The number of measurement positions for each source position and the total number of source positions used are usually dependent upon the irregularity of the measurements obtained. Generally, four or five source positions with four or five measurement positions for each source position are sufficient, giving a total number of measurements between 16 and 25. The room constant R for the room is then calculated using Equation (7.43) rearranged as follows:

$$R = 4 \times 10^{(L_w - L_p)/10} \tag{7.72}$$

In writing Equation (7.72), the direct field of the source has been neglected, following the measurement procedure proposed above and it has been assumed that $\rho c = 400$. In Equation (7.72), L_p is the average of all the sound pressure level measurements, and is calculated using the following equation:

$$L_p = 10 \log_{10}\left[\frac{1}{N}\sum_{i=1}^{N} 10^{(L_{pi}/10)}\right] \quad (\text{dB re } 20 \text{ } \mu\text{Pa}) \tag{7.73}$$

The quantity L_w is the sound power level (dB re 10^{-12} W) of the reference sound source, and N is the total number of measurements.

7.6.2 Reverberation Time Method

The second method is based upon a measurement of the room reverberation time. When measuring reverberation time in a room, the source of sound is usually a speaker driven by a random noise generator in series with a bandpass filter. When the sound is turned off, the room rate of decay can be measured simply by using a sound level meter attached to a level recorder as illustrated in Figure 7.5. Alternatively, there are many acoustic instruments such as spectrum analysers that can calculate the reverberation time internally for all 1/3 octave bands simultaneously. In this case, the "graphic level recorder" box, "band pass filter" box and the "sound level meter" box are replaced with a "sound analyser" box. However, it is important to ensure that the signal level in each band is at least 45 dB above any background noise. If it is less, the reverberation time results will be less accurate.

The reverberation time, T_{60}, in each frequency band is determined as the reciprocal sound pressure level decay rate obtained using the level recorder or the spectrum analyser. According to Equation (7.51), the recorded level in decibels should decay linearly with time. The slope, generally measured as the best straight line fit to the recorded decay between 5 dB and 35 dB down from the initial steady-state level,

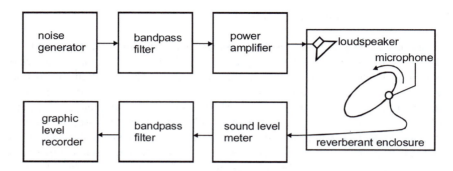

Figure 7.5 Equipment for reverberation time (T_{60}) measurement.

is used to determine the decay rate. The time for the sound to decay from -5 dB to -35 dB from the level at the time the sound source was switched off is estimated from the slope of the straight line fitted to the trace of sound pressure level vs time. The reverberation time, T_{60}, is found by multiplying the result by 2.

It is usual to employ several different microphone positions in the room, and to determine the reverberation times at each position. The value of T_{60} used in the subsequent calculations is then the average of all the values obtained for a given frequency band. Once found, the reverberation time, T_{60}, is used in Equation (7.52), rearranged as follows, to calculate the room absorption. Thus:

$$S\bar{\alpha} = (55.25\,V)/(c\,T_{60}) \qquad (\mathrm{m}^2)$$
(7.74)

At this point a word of caution is in order. When processing the data, determine average decay rates not decay times! Thus, what is required is:

$$\frac{1}{T_{60}} = \frac{1}{N}\sum_{i=1}^{N}\frac{1}{T_{60_i}}$$
(7.75)

When observing reverberation decay curves (average sound pressure level versus time) it will be noted that for almost any room, two different slopes will be apparent. The steeper slope occurs for the initial 7 to 10 dB of decay, the exact number of dB being dependent upon the physical characteristics of the room and contents. When this initial slope is extrapolated to a decay level of 60 dB, the corresponding time is referred to as the early decay time (EDT). The slope of the remainder of the decay curve, when extrapolated to 60 dB corresponds to what is commonly referred to as the reverberation time (RT). The ratio of EDT to RT as well as the absolute values of the two quantities are widely used in the design of architectural spaces, and are discussed in Section 7.11. For more information see Beranek (1962); Mackenzie (1979); Cremer and Muller (1982); and Egan (1987).

7.7 POROUS SOUND ABSORBERS

7.7.1 Measurement of Absorption Coefficients

Sabine absorption coefficients for materials are generally measured in a laboratory using a reverberant test chamber. Procedures and test chamber specifications are described in various standards (ISO 354-2003, ASTM C423-08a and AS 1045- 1988). The material to be tested is placed in a reverberant room and the reverberation time, T_{60}, is measured. The test material is removed and the reverberation time, T'_{60}, of the room containing no test material is measured next. Provided that the absorption of the reverberation room in the absence of the test material is dominated by the absorption of the walls, floor and ceiling, the reverberation times are related to the test material absorption, $S\bar{\alpha}$, by the following equation (derived directly from Equation (7.52)):

$$S\bar{\alpha} = \frac{55.3 V}{c}\left[\frac{1}{T_{60}} - \frac{(S' - S)}{S'T'_{60}}\right] \quad (m^2)$$

(7.76a)

The quantity, S', is the total area of all surfaces in the room including the area covered by the material under test. Equation (7.76a) is written with the implicit assumption that the surface area, S, of the test material is large enough to measurably affect the reverberation time, but not so large as to seriously affect the diffusivity of the sound field, which is basic to the measurement procedure. The standards recommend that S should be between 10 and 12 m^2 with a length-to-breadth ratio between 0.7 and 1.0.

In many cases, the absorption of a reverberation room is dominated by things other than the room walls, such as loudspeakers at low frequencies, stationary and rotating diffuser surfaces at low and mid frequencies and air absorption at high frequencies. For this reason, the contribution of the room to the total absorption is often considered to be the same with and without the presence of the sample. In this case, the additional absorption due to the sample may be written as:

$$S\bar{\alpha} = \frac{55.3 V}{c}\left[\frac{1}{T_{60}} - \frac{1}{T'_{60}}\right] \quad (m^2)$$

(7.76b)

Equation (7.76b) is what appears in most current standards, even though its accuracy is questionable.

The measured value of the Sabine absorption coefficient is dependent upon the sample size, sample distribution and the properties of the room in which it is measured. Because standards specify the room characteristics and sample size and distribution for measurement, similar results can be expected for the same material measured in different laboratories (although even under these conditions significant variations have been reported). However, these laboratory-measured values are used to calculate reverberation times and reverberant sound pressure levels in auditoria and factories that have quite different characteristics, which implies that in these cases, values of reverberation time, T_{60}, and reverberant field sound pressure level, L_p, calculated from measured Sabine absorption coefficients are approximate only.

Statistical absorption coefficients may be estimated from impedance tube measurements, as discussed in Appendix C. A list of Sabine absorption coefficients selected from the literature is included in Table 7.2 for various materials. The approximate nature of the available data makes it desirable to use manufacturer's data or to take measurements where possible.

Table 7.2 Sabine absorption coefficients for some commonly used materials

Material	Octave band centre frequency (Hz)						
	63	125	250	500	1000	2000	4000
Concert hall seats							
Unoccupied – heavily up-holstered seats (Beranek and Hidaka, 1998)		0.65	0.76	0.81	0.84	0.84	0.81
Unoccupied – medium up-holstered seats		0.54	0.62	0.68	0.70	0.68	0.66
Unoccupied – light up-holstered seats		0.36	0.47	0.57	0.62	0.62	0.60
Unoccupied – very light up-holstered seats		0.35	0.40	0.41	0.38	0.33	0.27
Unoccupied – average well-up-holstered seating areas	0.28	0.44	0.60	0.77	0.84	0.82	0.70
Unoccupied – leather-covered upholstered seating areas		0.40	0.50	0.58	0.61	0.58	0.50
Unoccupied – metal or wood seats	0.15	0.19	0.22	0.39	0.38	0.30	
Unoccupied – concert hall, no seats halls lined with thin wood or other materials <3 cm thick	0.16	0.13	0.10	0.09	0.08	0.08	
Halls lined with heavy materials	0.12	0.10	0.08	0.08	0.08	0.08	
100% occupied audience (orchestra and chorus areas) – upholstered seats	0.34	0.52	0.68	0.85	0.97	0.93	0.85
Audience, per person seated $S\overline{\alpha}$ (m²)	0.23	0.37	0.44	0.45	0.45	0.45	
Audience, per person standing $S\overline{\alpha}$ (m²)	0.15	0.37	0.43	0.44	0.44	0.43	
Wooden pews – 100% occupied		0.57	0.61	0.75	0.86	0.91	0.86
Wooden chairs – 100% occupied		0.60	0.74	0.88	0.96	0.93	0.85
Wooden chairs – 75% occupied		0.46	0.56	0.65	0.75	0.72	0.65

Table 7.2 (continued)

Material	Octave band centre frequency (Hz)					
	125	250	500	1000	2000	4000
Acoustic material						
Fibre-glass or rockwool blanket						
16 kg/m³, 25 mm thick	0.12	0.28	0.55	0.71	0.74	0.83
16 kg/m³, 50 mm thick	0.17	0.45	0.80	0.89	0.97	0.94
16 kg/m³, 75 mm thick	0.30	0.69	0.94	1.0	1.0	1.0
16 kg/m³, 100 mm thick	0.43	0.86	1.0	1.0	1.0	1.0
24 kg/m³, 25 mm thick	0.11	0.32	0.56	0.77	0.89	0.91
24 kg/m³, 50 mm thick	0.27	0.54	0.94	1.0	1.0	1.0
24 kg/m³, 75 mm thick	0.28	0.79	1.0	1.0	1.0	1.0
24 kg/m³, 100 mm thick	0.46	1.0	1.0	1.0	1.0	1.0
48 kg/m³, 50 mm thick	0.3	0.8	1.0	1.0	1.0	1.0
48 kg/m³, 75 mm thick	0.43	0.97	1.0	1.0	1.0	1.0
48 kg/m³, 100 mm thick	0.65	1.0	1.0	1.0	1.0	1.0
60 kg/m³, 25 mm thick	0.18	0.24	0.68	0.85	1.0	100
60 kg/m³, 50 mm thick	0.25	0.83	1.0	1.0	1.0	1.0
Polyurethane foam,						
27 kg/m³ 15 mm thick	0.08	0.22	0.55	0.70	0.85	0.75
Floors						
Wood platform with large						
space beneath	0.40	0.30	0.20	0.17	0.15	0.10
Wood floor on joists	0.15	0.11	0.10	0.07	0.06	0.07
Concrete or terrazzo	0.01	0.01	0.01	0.02	0.02	0.02
Concrete block painted	0.01	0.05	0.06	0.07	0.09	0.08
Linoleum, asphalt, rubber						
or cork tile on concrete	0.02	0.03	0.03	0.03	0.03	0.02
Varnished wood joist floor	0.15	0.12	0.10	0.07	0.06	0.07
Carpet, heavy, on concrete	0.02	0.06	0.14	0.37	0.60	0.65
Carpet, heavy, on 1.35 kg/ m²						
hair felt or foam rubber	0.08	0.24	0.57	0.69	0.71	0.73
Carpet, 5 mm thick, on						
hard floor	0.02	0.03	0.05	0.10	0.30	0.50
Carpet, 6 mm thick, on						
underlay	0.03	0.09	0.20	0.54	0.70	0.72
Cork floor tiles (3-4 inch						
thick) – glued down	0.08	0.02	0.08	0.19	0.21	0.22
Glazed tile/marble	0.01	0.01	0.01	0.01	0.02	0.02
Walls						
Acoustic plaster, 10 mm						
thick sprayed on solid wall	0.08	0.15	0.30	0.50	0.60	0.70

Table 7.2 (continued)

Material	Octave band centre frequency (Hz)					
	125	250	500	1000	2000	4000
Hard surfaces (brick walls, plaster, hard floors, etc.)	0.02	0.02	0.03	0.03	0.04	0.05
Gypsum board on 50 × 100 mm studs	0.29	0.10	0.05	0.04	0.07	0.09
Plaster, gypsum or lime, smooth finish; on brick,	0.013	0.015	0.02	0.03	0.04	0.05
on concrete block,	0.012	0.09	0.07	0.05	0.05	0.04
on lath	0.014	0.10	0.06	0.04	0.04	0.03
Solid timber door	0.14	0.10	0.06	0.08	0.10	0.10
Ceilings						
13 mm mineral tile direct fixed to floor slab	0.10	0.25	0.70	0.85	0.70	0.60
13 mm mineral tile suspended 500 mm below ceiling	0.75	0.70	0.65	0.85	0.85	0.90
Curtains						
Light velour, 338 g/m^2, hung flat on wall	0.03	0.04	0.11	0.17	0.24	0.35
hung in folds on wall	0.05	0.15	0.35	0.40	0.50	0.50
Medium velour, 475 g/m^2 draped to half area	0.07	0.31	0.49	0.75	0.70	0.60
Heavy velour, 610 g/m^2 draped to half area	0.14	0.35	0.55	0.72	0.70	0.65
Glass						
Glass, heavy plate	0.18	0.06	0.04	0.03	0.02	0.02
Ordinary window	0.35	0.25	0.18	0.12	0.07	0.04
Other						
Stage openings	0.30	0.40	0.50	0.60	0.60	0.50
Water (surface of pool)	0.01	0.01	0.01	0.015	0.02	0.03
Orchestra with instruments on podium, 1.5 m^2 per person	0.27	0.53	0.67	0.93	0.87	0.80

7.7.2 Noise Reduction Coefficient (NRC)

Sometimes it is useful to use a single number to describe the absorption characteristics of a material. This is particularly useful when a comparison of the relative benefit of a number of different materials has to be made quickly. For this purpose, the frequency-averaged Noise Reduction Coefficient (*NRC*) has been introduced. It is defined as:

$$NRC = \frac{\left(\overline{\alpha}_{250} + \overline{\alpha}_{500} + \overline{\alpha}_{1000} + \overline{\alpha}_{2000}\right)}{4} \tag{7.77}$$

7.7.3 Porous Liners

Where manufacturer's data are not available the statistical absorption coefficient for a porous blanket of thickness, ℓ, backed by a cavity of depth, L, may be calculated as outlined in Appendix C. Some results of such calculations are shown in Figure 7.6. Implicit in the calculations is the assumption that sound propagation within the porous material is predominantly normal to the surface. This condition is sufficiently well satisfied if the porous material is fairly dense. Alternatively, the porous material could be contained between solid partitions arranged to prevent lateral propagation.

The calculated statistical absorption coefficient is optimum when the total flow resistance, R_f, through the material is between $2\rho c$ and $5\rho c$ (see Appendix C for a discussion of flow resistance). This is shown in Figure 7.6, where it can be seen that a porous liner as little as one-tenth of a wavelength thick will give a statistical absorption coefficient of about 0.92. This is close to the maximum that is theoretically possible.

The performance of a porous blanket material can be improved by mounting it so that an air gap between the material and the hard backing wall is provided. Calculations (see Appendix C) show that the optimum air gap depth is equal to the thickness of the porous liner, and that the absorption will be the same as that which would characterise a porous layer equal in thickness to the air gap and liner. To achieve this optimum it would be necessary to place partitions in the air gap to prevent lateral sound propagation in the gap parallel to the wall.

In practice, it is often necessary to protect a porous liner from contamination by moisture, dust, oil, chemicals, etc. which would render it useless. For this reason the material is often wrapped in a thin, limp, impervious blanket made of polyurethane, polyester, aluminum or PVC. The effect of the introduction of a wrapping material is generally to improve the low-frequency absorption at the expense of the high-frequency absorption. However if the wrapping material is sufficiently thin (for example, 20 μm thick polyurethane) then the effect is not measurable. The effect can be calculated as shown in Appendix C.

7.7.4 Porous Liners with Perforated Panel Facings

When mechanical protection is needed for a porous liner, it may be covered using perforated wood, plastic or metal panels. If the open area provided by the perforations is greater than about 20%, the expected absorption is entirely controlled by the properties of the porous liner, and the panel has no effect.

Alternatively, if the facing panel has an open area of less than 20%, then the frequency at which absorption is maximum may be calculated by treating the facing as an array of Helmholtz resonators in a manner similar to that used in Section 9.7.2.2.

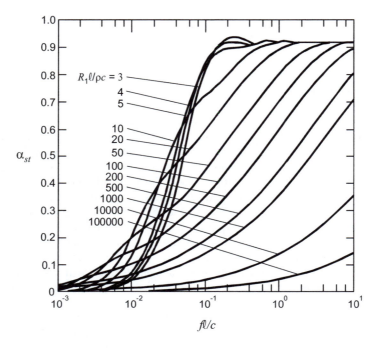

Figure 7.6 The calculated statistical absorption coefficient for a rigidly backed porous material in a reverberant field, as a function of a frequency parameter, for various indicated values of flow resistance. The quantity R_1 is the flow resistivity of the material (MKS rayls/m), ρ is the density of air (kg/m³), c is the speed of sound in air (m/s), f is the frequency (Hz) of the incident sound, and ℓ is the porous material thickness (m).

This procedure leads to the following approximate equation:

$$f_{max} = \frac{c}{2\pi}\left[\frac{P/100}{L[t + 0.85d(1 - 0.22\,d/q)]}\right]^{1/2} \tag{7.78}$$

In Equation (7.78), P is the percentage open area of the panel, L is the depth of the backing cavity including the porous material layer, t is the panel thickness, d is the diameter of the perforations, and q is the spacing between hole centres. If the porous material fills the entire cavity so that the thickness of the porous material is also L, then the speed of sound c should be replaced with $0.85c$ to account for isothermal rather than adiabatic propagation of sound in the porous material at low frequencies. The condition $fL/c < 0.1$ must be satisfied for Equation (7.78) to give results with less than 15% error. Measured data for a panel of 10% open area are presented in Figure 7.7 (see Davern, 1977, for additional data).

   If a perforated facing is used over a porous liner wrapped in an impervious membrane, care must be taken to ensure that the facing and liner are not in contact, otherwise the sound absorbing properties of the combination will be severely

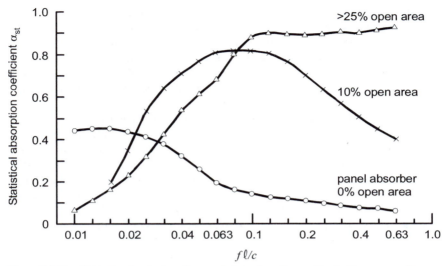

Figure 7.7 The effect of perforations on the sound absorption of a panel backed by a porous liner. The panel surface weight is 2.5 kg/m² and its thickness is 3 mm. The porous liner is 50 mm thick and about 5 ρc flow resistance (see Appendix C).

degraded. Non-contact is usually achieved in practice by placing a 12 mm square wire-mesh spacer between the liner and the perforated sheet.

7.7.5 Sound Absorption Coefficients of Materials in Combination

When a surface is characterised by several different materials having different values of absorption coefficient, it is necessary to determine an average Sabine absorption. For example, for a room characterised by q different materials, the average absorption coefficient will be the area-weighted average absorption of the q different materials calculated using the following equation.

$$\bar{\alpha} = \frac{\sum_{i=1}^{q} S_i \bar{\alpha}_i}{\sum_{i=1}^{q} S_i} \tag{7.79}$$

The quantities in Equation (7.79) are the areas S_i and the Sabine absorption coefficients $\bar{\alpha}_i$ of the q materials.

7.8 PANEL SOUND ABSORBERS

As Figure 7.6 clearly shows, porous liners are not very effective in achieving low-frequency absorption. When low-frequency absorption is required, panel absorbers are generally used. These consist of flexible panels mounted over an air space; for example, on battens (or studs) attached to a solid wall. Alternatively, such panels may be mounted on a suspended ceiling. In any case, to be effective the panels must couple with and be driven by the sound field. Acoustic energy is then dissipated by flexure of the panel. Additionally, if the backing air space is filled with a porous material, energy is also dissipated in the porous material. Maximum absorption occurs at the first resonance of the coupled panel-cavity system.

It is customary to assign a Sabine absorption coefficient to a resonant panel absorber, although the basis for such assignment is clearly violated by the mode of response of the panel absorber; that is, the absorption is not dependent upon local properties of the panel but is dependent upon the response of the panel as a whole. Furthermore, as the panel absorber depends upon strong coupling with the sound field to be effective, the energy dissipated is very much dependent on the sound field and thus on the rest of the room in which the panel absorber is used. This latter fact makes prediction of the absorptive properties of panel absorbers difficult. A typical example of absorption coefficients for a panel absorber is shown in Figure 7.7.

Two methods will be described for estimating the Sabine absorption of panel absorbers. One is empirical and is based upon data measured in auditoria and concert halls and must be used with caution while the other is based upon analysis, but requires considerable experimental investigation to determine all of the required parameters.

7.8.1 Empirical Method

An empirical prediction scheme (Hardwood Plywood Manufacturers' Association, 1962) for flexible panel absorbers that has been found useful in auditoria and concert halls will be outlined. The essence of the prediction scheme is contained in Figures 7.8 and 7.9. First of all, the type of Sabine absorption curve desired is selected from curves A to J in Figure 7.8. The solid curves are for configurations involving a blanket (25 mm thick and flow resistance between $2\rho c$ and $5\rho c$) in the air gap behind the panel, while the dashed curves are for no blanket. Next, the frequency f_0, which is the fundamental panel resonance frequency and the frequency at which maximum absorption is required, is determined and Figure 7.9 is entered for the chosen curve (A to J) and the desired frequency f_0. The required air gap (mm) behind the panel and the required panel surface density (kg/m^2) are read directly from the figure. The resonance frequency used in the preceding procedure is calculated using:

$$f_0 = \frac{1}{2\pi}\sqrt{\frac{\rho c^2}{mL}} \qquad (7.80)$$

which does not take into account the panel rigidity or geometry. A more accurate equation for a plywood panel is (Sendra, 1999):

$$f_0 = \frac{1}{2\pi}\sqrt{\frac{\rho c^2}{mL + 0.6L\sqrt{ab}}} \qquad (7.81)$$

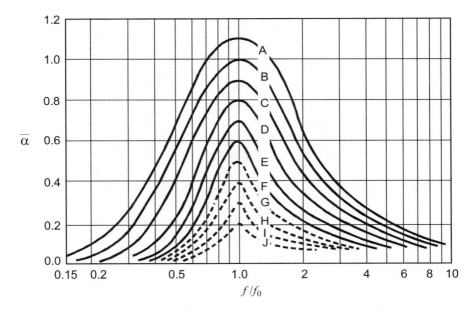

Figure 7.8 Sabine absorption coefficients for resonant plywood panels. The panel configurations corresponding to the curves labelled A-J may be identified using Figure 7.9. Dashed curves (G-J) represent configurations with no absorptive material in the cavity behind the panel. Configurations A-F require a sound-absorbing blanket between the panel and backing wall. The blanket must not contact the panel and should be between 10 and 50 mm thick and consist of glass or mineral fibre with a flow resistance between 1,000 and 2,000 MKS rayls. Panel supports should be at least 0.4 m apart.

where m is the mass per unit area of the panel (kg/m^2), L is the depth of the backing cavity and a, b are the panel dimensions. Thus, it is recommended that before using Figure 7.9, the frequency of maximum desired absorption be multiplied by the ratio:

$$\sqrt{\frac{m}{m + 0.6\sqrt{ab}}} \qquad (7.82)$$

7.8.2 Analytical Method

The Sabine absorption coefficient of n resonant panels of total surface area S and individual surface area A_p may be calculated by using the following formulae. It is to be noted that in this case the absorption coefficient is explicitly a function of the properties of the room as well as the properties of the panel, and consequently fairly good results can be expected. On the other hand, the price paid for good results will be quite a few measurements to determine the properties of both room and panels. The term "panel" includes any backing cavity whether filled with porous material or not, as the case may be. The Sabine absorption coefficient is given by (Pan and Bies, 1990c):

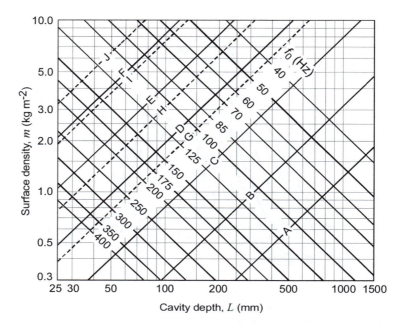

Figure 7.9 Design curves for resonant plywood panels, to be used in conjunction with Figure 7.8. The quantity f_0 is the frequency at which maximum sound absorption is required.

$$\bar{\alpha} = \frac{4V\pi f}{Sc}\left[\left|\left(\eta_A + \eta_{pA}\left(1 + \frac{n_p}{n_A}\right) + \eta_{pT}\right)\right.\right.$$
$$\left.\left. - \left[\left(\eta_A + \eta_{pA}\left(\frac{n_p}{n_A} - 1\right) - \eta_{pT}\right)^2 + 4\frac{n_p}{n_A}\eta_{pA}^2\right]^{1/2}\right| - 2\eta_A\right]$$

$$(7.83)$$

In Equation (7.83), f is the band centre frequency, c is the speed of sound, V is the room volume, $\eta_A = 2.20/fT_{60A}$ is the room loss factor with the panel absent, $\eta_{pT} = 2.20/fT_{60p}$ is the mounted panel loss factor in free space, $\eta_{pA} = \rho c\sigma/2\pi fm$ is the panel coupling loss factor, $n_p = \sqrt{3}\,A_p/(c_L h)$ is the panel modal density which must be multiplied by the number of panels if more than one panel is used, $n_A = dN/df$ is the modal density of the room given by Equation (7.22), S is the total surface area of all panels and A_p is the surface area of one panel. The quantities, T_{60A} and T_{60p}, are respectively, the 60 dB decay times of the room (without panels) and the 60 dB decay times of the panels (in a semi-anechoic space). The quantity, c_L, is the longitudinal wave speed in the panel, h is the panel thickness, ρ, is the density of air, and m is the mass per unit surface area of the panel. The equations used for calculating the panel radiation efficiency, σ, in the second edition of this text were based on those provided by Lyon (1975). They are not used here as they give inaccurate results in many cases.

For frequencies above twice the resonance frequency of the lowest order mode (see Equation (8.23)), the panel radiation efficiency σ may be calculated using the following equations (Maidanik, 1962, corrected as in Price and Crocker, 1970). Above twice the fundamental resonance frequency of the panel and below the critical frequency:

$$\sigma = \left(\frac{2c^2}{f_c^2 A_p} g_1(\xi_c) + \frac{Pc}{f_c A_p} g_2(\xi_c) \right) \gamma \; ; \qquad 2f_{1,1} < f < 0.99 f_c \tag{7.84}$$

where,

$$g_1(\xi_c) = \begin{cases} \dfrac{4}{\pi^4} \dfrac{(1 - 2\xi_c^2)}{\xi_c (1 - \xi_c^2)^{1/2}} ; & f < f_c/2 \\[4mm] 0 ; & f > f_c/2 \end{cases} \tag{7.85a}$$

$$g_2(\xi_c) = \frac{1}{4\pi^2} \left[\frac{(1 - \xi_c^2) \log_e \left(\dfrac{1 + \xi_c}{1 - \xi_c} \right) + 2\xi_c}{(1 - \xi_c^2)^{3/2}} \right] \tag{7.85b}$$

$$\xi_c = \left(\frac{f}{f_c} \right)^{1/2} \tag{7.86}$$

For simply supported panels, γ takes the value of 1 while for clamped edge panels γ takes the value 2. All other conditions lie between these extremes.

For panels supported at the edge and on intermediate battens (or studs), the perimeter, P, in Equation (7.84) is the overall length of the panel perimeter plus twice the length of all the studs. The area, A_p, is the area of the entire panel.

Close to the critical frequency, f_c, the radiation efficiency for a panel of overall dimensions $L_x \times L_y$ is:

$$\sigma \approx \sqrt{\frac{L_x f_c}{c}} + \sqrt{\frac{L_y f_c}{c}} ; \qquad 0.99 f_c < f < 1.01 f_c \tag{7.87}$$

Above the critical frequency, the radiation efficiency is:

$$\sigma = \left(1 - \frac{f_c}{f} \right)^{-1/2} ; \qquad f > 1.01 f_c \tag{7.88}$$

Below the first resonance frequency of the panel, $f_{1,1}$, defined by Equation (8.23) for simply supported panels, the radiation efficiency is (Beranek, 1988):

$$\sigma = \frac{4 A_p}{c^2} f^2 \qquad\qquad (7.89)$$

Note that for square, clamped-edge panels, the fundamental resonance frequency is 1.83 times that calculated using Equation (8.23). For panels with aspect ratios of 1.5, 2, 3, 6, 8 and 10 the factors are 1.89, 1.99, 2.11, 2.23, 2.25 and 2.26 respectively.

Between the lowest order modal resonance and twice that frequency, the radiation efficiency is found by interpolating linearly (on a log σ vs log f plot).

The panel critical frequency, f_c, is defined as follows:

$$f_c = 0.551 \frac{c^2}{c_L h} \qquad\qquad (7.90)$$

In the above equations the quantities P and A_p are the panel perimeter and area respectively. The panel is assumed to be isotropic of uniform thickness h and characterised by longitudinal wave speed, c_L. For steel and aluminum c_L takes the value of about 5150 m/s while for wood, the value lies between 3800 and 4500 m/s. Values of c_L for other materials are given in Appendix B.

7.9 FLAT AND LONG ROOMS

Many enclosures are encountered in practice which have dimensions that are not conducive to the establishment of a reverberant sound field of the kind that has been the topic of discussion thus far and was first investigated by Sabine. Such other types of enclosure (flat and long rooms) are considered briefly in this section and their investigation is based upon work of Kuttruff (1985, 1989).

Reflections at the boundaries of either flat rooms or long rooms produces a reverberant field in addition to the direct field of the source but, whereas in the Sabine-type rooms discussed earlier, the reverberant field could be considered as of constant mean energy density (level) throughout the room, in the case of the non-Sabine-type rooms considered here, the reverberant field will always decay away from the source; there will be no constant mean level reverberant field. However, as in the case of Sabine-type rooms, it will be useful to separately identify the direct and reverberant fields, because the methods of their control will differ. For example, where the direct field is dominant, the addition of sound absorption will be of little value.

Examples of enclosures of the type to be considered here, called flat rooms, are often encountered in factories in which the height, though it may be large, is much smaller than any of the lateral dimensions of the room. Open plan offices provide other familiar examples. For analytical purposes, such enclosures may be considered as contained between the floor and a parallel ceiling but of infinite extent and essentially unconstrained in the horizontal directions except close to the lateral walls of the enclosure. In the latter case, use of the method of images is recommended but is not discussed here (Elfert, 1988).

Examples of long rooms are provided again by factories in which only one horizontal dimension, the length, may be very much greater than either the height or width of the room. Other examples are provided by corridors and tunnels. Enclosed roadways, which are open above, may be thought of as corridors with completely absorptive ceilings and thus also may be treated as long rooms. As with flat rooms, the vertical dimension may be very large; that is, many wavelengths long. The horizontal dimension normal to the long dimension of the room may also be very large. The room cross-section is assumed to be constant and sufficiently large in terms of wavelengths that, as in the case of the Sabine rooms considered earlier, the sound field may be analysed using geometrical analysis.

Reflection at a surface may be quite complicated; thus to proceed, the problem of describing reflection will be simplified to one or the other of two extremes; that is, specular or diffuse reflection. Specular reflection is also referred to in the literature as geometrical reflection. Consideration of specular reflection may proceed by the method of images. In this case the effect of reflection at a flat surface may be simulated by replacement of the bounding surface with the mirror image of the source. Multiple reflections result in multiple mirror image sources symmetrically placed. Diffuse reflection occurs at rough surfaces where an incident wave is scattered in all directions. In the cases considered here, it will be assumed that the intensity, $I(\theta)$, of scattered sound follows Lambert's rule taken from optics. In this case:

$$I(\theta) \propto \cos\theta \qquad (7.91)$$

where θ is the angle subtended by the scattered ray relative to the normal to the surface.

Diffuse reflection at a surface is wavelength dependent; an observation that follows from the consideration that surface roughness is characterised by some size distribution. If the wavelength is large compared to the characteristic dimensions of the roughness, the reflection will be essentially specular, as the roughness will impose only negligible phase variation on the reflected wave at the surface. Alternatively, if the wavelength is small compared to the smallest size of the roughness dimensions, then the reflection, though it may be complicated, must again be specular. In the range where the wavelength is comparable to the surface roughness, the reflection will be diffuse.

In the discussion to follow, specular reflection will be mentioned as a reference case and also as an introduction to the more complicated diffuse reflection cases to follow. However, the discussion will be concerned principally with diffuse reflection based upon the following observation. The floor of a furnished open plan office or the ceiling of a factory with extensive fittings such as piping and conduits may be thought of as a rough surface. Here the simple assumption will be made that sound scattering objects may be considered as part of the surface on which they rest so that the surface with its scatterers may be replaced with an effective diffusely reflecting surface. The use of this concept considerably reduces the complexity of the problem and makes tractable what may be an otherwise intractable problem. However, simplification is bought at the price of some empiricism in determining effective energy reflection coefficients for such surfaces and predictions can only provide estimates of average room sound levels. Limited published experimental data suggests that measurement may exceed prediction by at most 4 dB with proper choice of reflection coefficients (Kuttruff, 1985).

The discussions of the various room configurations in the section to follow are based on theoretical work undertaken by Kuttruff (1985). As an alternative for estimating sound levels and reverberation times in non-Sabine rooms, there are various ray tracing software packages available that work by following the path of packets of sound rays that emanate from the source in all possible directions and eventually arrive at the receiver (specified as a finite volume) after various numbers of reflections from various surfaces. The principles underlying this technique are discussed by many authors including Krokstad *et al.* (1968), Naylor (1993), Lam (1996), Bork (2000), Keränen *et al.* (2003), Xiangyang *et al.* (2003) There are also empirical models based on experimental data that allow the prediction of sound pressure levels in typical workshops as a function of distance from a source of known sound power output (Heerema and Hodgson, 1999; Hodgson, 2003).

7.9.1 Flat Room with Specularly Reflecting Floor and Ceiling

The flat room with specularly reflecting floor and ceiling will be encountered rarely in practice, because the concept really only applies to empty space between two relatively smooth reflecting surfaces. For example, a completely unfurnished open plan office might be described as a room of this type. The primary reason for its consideration is that it serves as a convenient reference for comparison with rooms that are furnished and with rooms which have diffuse reflecting surfaces. It also serves as a convenient starting point for the introduction of the concepts used later.

A source of sound placed between two infinite plane parallel reflecting surfaces will give rise to an infinite series of image sources located along a line through the source, which is normal to the two surfaces. If the source is located at the origin and the receiver is located at $r = r_0$ and each is located midway between the two reflecting surfaces, then the line of image sources will take the form illustrated in Figure 5.5 where, referring to the figure, $d = 0$ and the separation distance b between adjacent image sources in the figure is now the distance, a, between the reflecting planes, i.e. the height of the room. The effective distance from the nth image to the receiving point will be represented by r_n where the index n represents the number of reflections required to produce the image.

The source is assumed to emit a band of noise so that the source and all of its images may be considered as incoherent. In this case, summation at the point of observation may be carried out on an energy basis; sound pressures squared may be added without consideration of phase. It will be assumed that the surfaces below and above have uniform energy reflection coefficients β_1 and β_2 respectively, which are independent of angle of incidence, and that the sound power of the source is W. The mean square sound pressure observed at the receiving point, r, consists of the direct field, given by Equation (5.14) and shown as the first term on the right-hand side, and the reverberant field, given by the summation, where i is the image order, in the following expression:

$$\langle p^2(r)\rangle = \frac{W\rho c}{4\pi}\left[\frac{1}{r^2} + \sum_{i=1}^{\infty}\left(\frac{1/\beta_1 + 1/\beta_2}{r_{2i-1}^2} + \frac{2}{r_{2i}^2}\right)(\beta_1\beta_2)^i\right] \tag{7.92}$$

For $r_n = r_{2i}$ or r_{2i-1}, that is, $n = 2i$ or $2i - 1$:

$$r_n^2 = r^2 + (n\,a)^2 \tag{7.93}$$

Two limiting cases are of interest. If the distance between the receiver and the source is large so that $r \gg a$, then Equation (7.92) becomes, using Equation (7.93) and the well-known expression for the sum of an infinite geometric series, in the limit:

$$\langle p^2(r)\rangle = \frac{W\rho c}{4\pi r^2}\left[1 + \frac{\beta_1 + \beta_2 + 2\beta_1\beta_2}{1 - \beta_1\beta_2}\right] \tag{7.94}$$

Equation (7.94) shows that the sound field, which includes both the direct field and the reverberant field, decays with the inverse square of the distance from the source. Equation (7.94) also shows that the reverberant field sound pressure may be greater than or less than the direct field at large distances from the source, depending upon the values of the energy reflection coefficients β_1 and β_2.

If the distance between the source and receiver is small and the energy reflection coefficients approach unity, then (Kuttruff, 1985):

$$\langle p^2(r)\rangle = \frac{W\rho c}{4\pi}\left[\frac{1}{r^2} + \frac{2\pi^2}{3a^2}\right] \tag{7.95}$$

In Equation (7.95), the first term on the right-hand side is the direct field term and the second term is the reverberant field term. Equation (7.95) shows that in the vicinity of the floor and ceiling and in the limiting case of a very reflective floor and ceiling, the direct field is dominant to a distance of $r = (2\sqrt{3}/\pi)a \approx 1.95\,a$ or about twice the distance from floor to ceiling. This distance at which the direct and reverberant fields are equal, is called the hall radius, r_h.

Equation (7.92) has been used to construct Figure 7.10 where the direct field and the reverberant field terms are plotted separately as a function of normalised distance, r/a, from the source for several values of the energy reflection coefficients $\beta_1 = \beta_2 = \beta$. The figure shows that at large distances, the reverberant field may exceed the direct field when the reflection coefficient is greater than one-third. This may readily be verified by setting the direct field term of Equation (7.92) equal to the far field reverberant field term of the same equation with $\beta_1 = \beta_2 = \beta$.

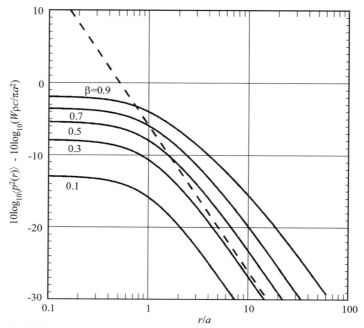

Figure 7.10 Direct and reverberant sound fields in a flat room of height a with specularly reflecting floor and ceiling as a function of the normalised distance from source to receiver. The reverberant field contribution is shown as a function of the energy reflection coefficient, β, assumed the same for floor and ceiling. The direct field is indicated by the dashed diagonal straight line.

7.9.2 Flat Room with Diffusely Reflecting Floor and Ceiling

The case of a flat room, which is furnished or occupied with machinery, or objects that tend to scatter incident sound, is considered in this section. A surface with scattering objects located upon it is replaced by an effective diffusely scattering surface for the purpose of this analysis. This approach, while obviously well suited for consideration of the floor of a furnished open plan office, may also be applied to the ceiling of a factory, which may be characterised by large open beams, conduits, pipes, corrugations, etc. This approach will simplify the analysis and provide a useful description of the sound field in a flat room. However, determination of effective surface reflection coefficients then becomes a problem, and generally an empirical approach will be necessary. For noise-control purposes, the proposed model is useful in spite of the mentioned limitation, as it provides the basis for consideration of the effectiveness of the introduction of measures designed to reduce the floor and/or ceiling reflection coefficients.

Based upon the proposed model, Kuttruff(1985) derived expressions for the sound intensities $I_1(r')$ and $I_2(r'')$, which characterise the contributions from the floor (surface 1) and ceiling (surface 2). The calculation of the mean square sound pressure at an observation point located by vector, r, requires that the quantity, $I_1(r')$, be integrated over surface 1 (floor) and $I_2(r'')$ be integrated over surface 2 (ceiling) to determine their respective contributions to the reverberant field. The total acoustic field is obtained by summing the reverberant field and the direct field contributions on an energy basis as the simple sum of mean square sound pressures. The direct field is given by Equation (5.14). For source and receiver at height h from surface 1 (floor), the calculation of the reverberant part of the acoustic field proceeds as follows:

$$\langle p^2(r) \rangle_R = \frac{\rho c}{\pi} \left[\beta_1 h \int_{S_1} \frac{I_1(r')}{R_1^3} \, dS + \beta_2 (a - h) \int_{S_2} \frac{I_2(r'')}{R_2^3} \, dS \right] \tag{7.96}$$

where

$$R_1 = \left[|r - r'|^2 + h^2 \right]^{1/2} \tag{7.97}$$

and

$$R_2 = \left[|r - r''|^2 + (a - h)^2 \right]^{1/2} \tag{7.98}$$

Here r' and r'' are, respectively, vector locations in surfaces S_1 and S_2, r is the vector from the source to the receiver and a is the distance from floor to ceiling. The origins to all three vectors project on to the same point on either bounding surface.

Kuttruff(1985) shows how expressions for I_1 and I_2 may be obtained, which when substituted in Equation (7.96) allow solution for several special cases of interest. Equation (7.96), and all of the special cases that follow, are to be compared with Equation (7.43) for the Sabine room. In the following analysis, the source will be located at the origin of the vector coordinate r and $r = |r|$.

For the case that the energy reflection coefficients of the bounding surfaces (floor and ceiling) are the same ($\beta_1 = \beta_2 = \beta$), Equation (7.96) may be simplified and the solution for the reverberant field contribution takes the following form (Kuttruff, 1985):

$$\langle p^2(r) \rangle_R = \frac{W \rho c \beta}{\pi a^2} \int_0^\infty \frac{e^{-z}}{1 - \beta z K_1(z)} J_0(rz/a) z \, dz \tag{7.99}$$

In Equation (7.99), $J_0(rz/a)$ is the zero order Bessel function and $K_1(z)$ is a modified Hankel function (Gradshteyn and Ryshik, 1965). In general, sufficient accuracy is achieved in evaluation of Equation (7.99) by use of the following approximation, otherwise the equation must be evaluated numerically. The following approximation holds, where the empirical constant Γ is evaluated according to Equation (7.100), using $K_1(1) = 0.6019$ so that the two sides of the expression are exactly equal for $z = 1$:

$$[1 - \beta z K_1(z)]^{-1} \approx 1 + \frac{\beta}{1 - \beta} e^{-\Gamma z} \tag{7.100}$$

where

$$\Gamma = \log_e\left[\frac{1 - 0.6019\beta}{(1 - \beta)\,0.6019}\right] \tag{7.101}$$

Substitution of Equation (7.100) into Equation (7.99) and integration gives the following closed form approximate solution:

$$\langle p^2(\mathbf{r})\rangle_R \approx \frac{W\rho c\beta}{\pi a^2}\left[\left[1 + \frac{r^2}{a^2}\right]^{-3/2} + \frac{\beta(\Gamma+1)}{1-\beta}\left[(\Gamma+1)^2 + \frac{r^2}{a^2}\right]^{-3/2}\right] \tag{7.102}$$

Equation (7.102) allows consideration of two limiting cases. For $r = 0$ the equation reduces to the following expression:

$$\langle p^2(\mathbf{r})\rangle_R = \frac{W\rho c\beta}{\pi a^2}\left[1 + \frac{\beta}{(\Gamma+1)^2(1-\beta)}\right] \tag{7.103}$$

Equation (7.103) shows that the reverberant field is bounded in the vicinity of the source and thus will be dominated by the source near (direct) field.

For $r \gg a$ Equation (7.102) takes the following approximate form:

$$\langle p^2(\mathbf{r})\rangle_R \approx \frac{W\rho c\beta}{\pi a^2}\left[1 + \frac{(\Gamma+1)\beta}{(1-\beta)}\right]\left[\frac{a}{r}\right]^3 \tag{7.104}$$

Equation (7.104) shows that the reverberant field decreases as the cube of the distance r from the source whereas the direct field decreases as the square of the distance from the source. Thus, the direct field will again be dominant at a large distance from the source.

In Figure 7.11 the direct field term calculated using Equation (5.14) and the reverberant field term calculated using Equation (7.104) are plotted separately as a function of normalised distance r/a from the source to illustrate the points made here. When the direct and reverberant fields are equal at large distance from the source, a second hall radius is defined. Setting Equation (7.104) equal to Equation (5.14) for the direct field, the second hall radius, r_{h2} may be calculated as follows:

$$r_{h2} \approx 4a\beta\left[1 + \frac{(\Gamma+1)\beta}{1-\beta}\right] \tag{7.105}$$

When the energy reflection coefficients of the bounding surfaces (floor and ceiling) of a flat room are unequal, Equation (7.99) must be replaced with a more complicated integral equation, which shows the dependency upon energy reflection coefficients β_1 and β_2 associated with bounding surfaces 1 (floor) and 2 (ceiling). In this case it will

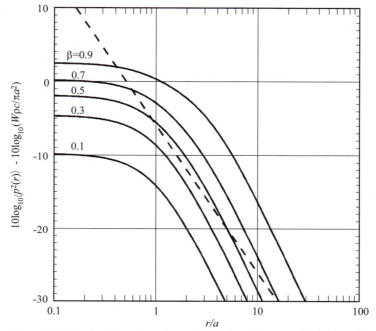

Figure 7.11 Direct and reverberant sound fields in a flat room of height a with diffusely reflecting floor and ceiling as a function of the normalised distance from source to receiver. The reverberant field contribution is shown as a function of the energy reflection coefficient, β, assumed the same for floor and ceiling. The direct field is indicated by the dashed diagonal straight line.

be convenient to introduce the geometric mean value, β_g, of the reflection coefficients of the floor and ceiling:

$$\beta_g = \sqrt{\beta_1 \beta_2} \tag{7.106}$$

As shown by Kuttruff (1985), the mean square sound pressure associated with the reverberant field in a flat room with unequal diffuse reflection coefficients is given by the following expression.

$$\langle p^2(\mathbf{r}) \rangle_R = \langle p^2(\mathbf{r}) \rangle_{R1}$$
$$+ \frac{W \rho c}{2 \pi a^2} \int_0^\infty \frac{\beta_1 e^{-2zh/a} + \beta_2 e^{-2z(1-h/a)} - 2\beta_g e^{-z}}{1 - (\beta_g z K_1(z))^2} J_0\left(\frac{r}{a} z\right) z \, dz \tag{7.107}$$

In Equation (7.107), the quantity $\langle p^2(\mathbf{r}) \rangle_{R1}$ is calculated using Equation (7.99) with $\beta = \beta_g$. The source is assumed to be located at the origin of coordinates and the receiver is located at the position given by the vector $\mathbf{r}$.

In general, Equation (7.107) will require numerical integration to obtain a solution. However, some special cases are of interest, which each allow a relatively simple closed form solution, and these will be considered. If the ceiling is removed so that $\beta_2 = 0$, and if the source and receiver are both at the same height, h, and separated by distance, r, then Equation (7.107) reduces to the following form (Kuttruff, 1985; Chien and Carroll, 1980):

$$\langle p^2(r)\rangle_R = \frac{W\rho c \beta_1 h}{\pi}\left[4h^2 + r^2\right]^{-3/2} \tag{7.108}$$

which describes the back scatter over a diffuse reflecting open plane.

If the source and receiver are located at distance r apart and both are midway between the two bounding planes so that $h = a/2$, then introducing the arithmetic mean value of the energy reflection coefficient $\beta_a = (\beta_1 + \beta_2)/2$, Equation (7.107) takes the following form:

$$\langle p^2(r)\rangle_R = \langle p^2(r)\rangle_{R1}$$
$$+ \frac{W\rho c}{\pi a^2}(\beta_a - \beta_g)\int_0^\infty \frac{e^{-z}}{1 - (\beta_g z K_1(z))^2}J_0\left[\frac{r}{a}z\right]z\,dz \tag{7.109}$$

In Equation (7.109), the quantity $\langle p^2(r)\rangle_{R1}$, is calculated using Equation (7.99) with $\beta = \beta_g$. A comparison between measured and predicted values using Equation (7.109) shows generally good agreement, with the theoretical prediction describing the mean of the experimental data (Kuttruff, 1985).

The second integral of Equation (7.109) may be evaluated using approximations similar to those used in deriving Equation (7.107) with the following result:

$$\langle p^2(r)\rangle_R = \frac{W\rho c}{\pi a^2}\left[\beta_a\left[1 + \frac{r^2}{a^2}\right]^{-3/2} + \beta_g^2\frac{(\Gamma + 1)}{1 - \beta_g}\left[\Gamma + 1 + \frac{r^2}{a^2}\right]^{-3/2}\right] \tag{7.110}$$

Consideration of Equation (7.110) shows that as Γ is of the order of unity (see Equation (7.101)) then for a large separation distance between the source and the receiver, so that $r \gg a$, the quantity $\langle p_2(r)\rangle_R$ decreases as the inverse cube of the separation distance or as $(a/r)^3$. Comparison of Equations (7.110) and (7.102) shows that they are similar. Indeed, it is found that for most values of the two energy reflection coefficients, the use of the mean energy reflection coefficient β_g in Equation (7.102) will give a sufficiently close approximation to the result obtained using β_1 and β_2 in Equations (7.107) and (7.110).

Some results of an investigation into the variation in height of the source and the receiver obtained using Equation (7.107), are shown in Figure 7.12. The energy reflection coefficient of the floor is $\beta_1 = 0.9$ and the energy reflection coefficient of the ceiling is $\beta_2 = 0.1$. Both surfaces are assumed to be diffuse reflectors. The results of

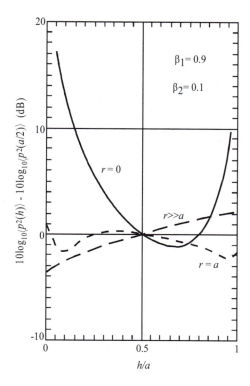

Figure 7.12 Reverberant sound field in a flat room of height a with diffusely reflecting floor and ceiling at three distances of the receiver from the source. The reflection coefficients of the floor and ceiling are respectively 0.9 and 0.1. The sound pressure is shown relative to that half way between floor and ceiling as a function of the distance from the surface of the greater reflection coefficient.

the calculations shown in the figure suggest that when the receiver and source are very close together (curve $r = 0$), the reverberant field will become very large, near either the floor or the ceiling. As the receiving point is moved away from the source, the reverberant field becomes fairly uniform from floor to ceiling when the distance from source to receiver is equal to the room height a (curve $r = a$). Finally, the reverberant field tends to a distribution at large distances in which the level is about 6 dB higher at the ceiling than at the floor (curve $r \gg a$).

7.9.3 Flat Room with Specularly and Diffusely Reflecting Boundaries

An open plan office might best be characterised as a flat room with a diffusely reflecting floor and a specularly reflecting ceiling. Indeed, a flat room characterised by one specularly reflecting boundary and one diffuse reflecting boundary may be the most common case. However, the cases already considered allow a simple extension to this case when it is observed that the room and sound source form an image room with image source when reflected through the specularly reflecting bounding surface. Thus a room of twice the height of the original room and with equally reflecting diffuse bounding surfaces is formed. The strength of the image source is exactly equal to the source strength and the double room is like a larger room with two identical sources symmetrically placed. The reflection loss at the specularly reflecting surface may be taken into account by imagining that the double room is divided by a curtain having a transmission loss equal to the energy reflection coefficient of the geometrically reflecting bounding surface. Thus, any ray that crosses the curtain is reduced by the magnitude of the energy reflection coefficient.

Using the analysis that has been outlined (Kuttruff, 1985), the following solution is obtained, which again requires numerical integration for the general case but which also has a useful closed form approximate solution for a source and receiver height, h, above the floor and floor to ceiling spacing of a:

$$\langle p^2(r) \rangle_R =$$

$$\frac{W\rho c}{4\pi a^2} \left[2\beta_1 \int_0^\infty \frac{(e^{-zh/a} + \beta_2 e^{-z(2-h/a)})^2}{1 - \beta_1\beta_2 2z K_1(2z)} J_0\left(\frac{r}{a}z\right) z\,dz + \beta_2 \left[4[1 - h/a]^2 + [r/a]^2\right]^{-1} \right] \quad (7.111)$$

Equation (7.112) which follows and Table 7.3 which provides calculated values of $\pi a^2 \langle p^2(0) \rangle_R / W\rho c$ for various values of β_1 and β_2, allows construction of approximate solutions.

Table 7.3 Calculated values of $\pi a^2 \langle p^2(0) \rangle_R / W\rho c$

β_2	β_1				
	0.1	0.3	0.5	0.7	0.9
0.1	0.078	0.184	0.291	0.398	0.507
0.3	0.133	0.253	0.376	0.502	0.634
0.5	0.190	0.324	0.467	0.619	0.784
0.7	0.247	0.399	0.566	0.754	0.973
0.9	0.304	0.476	0.675	0.916	1.244

Table 7.3 provides estimates of the local reverberant field, $r \approx 0$, while for large values $r \gg a$ and for $h \neq 0$ and $h \neq a$ the approximate solution for Equation (7.111) is as follows:

$$\langle p^2(r) \rangle_R$$

$$= \frac{W\rho c}{\pi a^2} \left[\frac{\beta_1(1 + \beta_2)}{1 - \beta_1\beta_2} \left[(1 - \beta_2)\frac{h}{a} + 2\beta_2 + \beta_1\beta_2(1 + \beta_2)\Gamma_1\right] \left(\frac{a}{r}\right)^3 + \frac{\beta_2}{4}\left(\frac{a}{r}\right)^2 \right] \quad (7.112)$$

In Figure 7.13, the normalised mean square sound pressure associated with the reverberant field for a sound source placed at height $h = a/2$ has been plotted as a function of the normalised distance r/a according to Equation (7.111) for two cases. In both cases, β_1 refers to the diffuse reflecting surface while β_2 refers to the specularly reflecting surface; for example, the furnished floor, surface 1, and smooth ceiling, surface 2, of an open plan office. In case (a), $\beta_1 = 0.9$ and $\beta_2 = 0.1$, while in case (b), $\beta_1 = 0.1$ and $\beta_2 = 0.9$. The figure shows that in case (a), the local reverberant field is about 2 dB higher, but decreases more rapidly with distance from the source than in case (b).

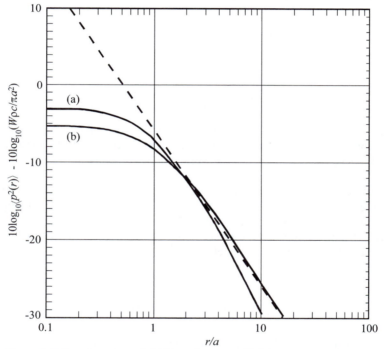

Figure 7.13 Reverberant sound field in a flat room of height a at $a/2$ with diffusely reflecting floor of energy reflection coefficient β_1 and specularly reflecting ceiling of energy reflection coefficient β_2. (a) $\beta_1 = 0.9$, $\beta_2 = 0.1$; (b) $\beta_1 = 0.1$, $\beta_2 = 0.9$. The direct field is indicated by the dashed diagonal straight line.

7.9.4 Long Room with Specularly Reflecting Walls

A long room with constant rectangular cross-section of height a and width b, and with specularly reflecting walls is considered here. It is convenient to introduce the coordinate r along the central axis of symmetry (long axis of the room). The reflection coefficients of all four walls are assumed to be the same and a point source placed at the origin is separated from the receiving point, also assumed to be located on the central axis of symmetry of the room, by distance r. More general cases of long rooms with geometrically reflecting walls have been discussed in the literature (Cremer *et al.*, 1982) and will not be discussed here. Rather, only this special case will be considered as a reference for the discussion of long rooms with diffusely reflecting walls.

Multiple reflections will produce two infinite series of image sources, which lie on vertical and horizontal axes through the source and which are normal to the long room central axis of symmetry. The source is assumed to emit a band of noise of power W, and the source and its images are assumed to be incoherent. Summing the contributions of the source and its images on a pressure squared basis leads to the following expression for the mean square sound pressure at the receiving point:

$$\langle p^2(r) \rangle = \frac{W \rho c}{4\pi} \left[\frac{1}{r^2} + \sum_{m=1}^{\infty} \sum_{n=1}^{\infty} \frac{4\beta^{m+n}}{(ma)^2 + (nb)^2 + r^2} \right.$$

$$\left. + \sum_{n=1}^{\infty} \frac{2\beta^n}{(nb)^2 + r^2} + \sum_{m=1}^{\infty} \frac{2\beta^m}{(ma)^2 + r^2} \right] \tag{7.113}$$

In Equation (7.113), the first term represents the direct field while the next three terms represent the reverberant field due to the contributions of the four lines of image sources extending away from each wall. Equation (7.113) has been used to construct Figure 7.14 for the case of a square cross-section long room of width b equal to height a, for some representative values of energy reflection coefficient, β. Note that a circular cross-section room of the same cross-sectional area as a square section room is also approximately described by Figure 7.14.

In the limit of very large distance, r, so that $a/r \approx b/r \approx 0$, the double sum of Equation (7.113) can be written in closed form as follows:

$$\langle p^2(r) \rangle \approx \frac{W \rho c}{4\pi r^2} \left[1 + \frac{4\beta}{(1-\beta)^2} \right] \tag{7.114}$$

If in the case considered of a long room of square cross-section of height and width a, the point source is replaced with an incoherent line source perpendicular to the axis of symmetry and parallel to two of the long walls of the room, then Equation (7.113) takes the following simpler form, where the power per unit length W' is defined so that $W'a = W$, the power of the original point source:

$$\langle p^2(r) \rangle = \frac{W' \rho c}{4} \left[\frac{1}{r} + \sum_{n=1}^{\infty} \frac{2\beta^n}{[(na)^2 + r^2]^{1/2}} \right] \tag{7.115}$$

Here it has been assumed that the line source emits a band of noise and thus the contributions of the source and its images add incoherently as the sum of squared

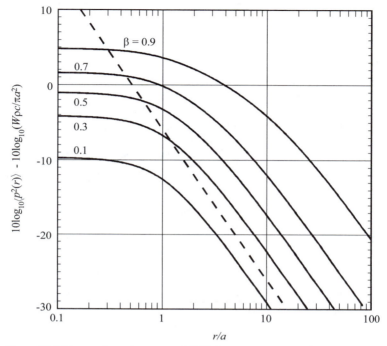

Figure 7.14 Direct and reverberant sound fields in a square cross-section, long room of height and width a with specularly reflecting floor, walls and ceiling as a function of the normalised distance from a point source to the receiver. The reverberant field is represented as solid lines for various values of the energy reflection coefficient, β, and the direct field is indicated by the dashed diagonal straight line.

pressures. Again in the limit of very large distance r, so that $a/r \approx 0$, the sum can be written in closed form as follows:

$$\langle p^2(r) \rangle = \frac{W'\rho c}{4r}\left[1 + \frac{2\beta}{1 - \beta}\right] \qquad (7.116)$$

Equation (7.116) shows that both the direct field (first term in brackets) and the reverberant field (second term in brackets) decay at the same rate with increasing distance r. The equation also shows by comparison of the two terms, that the direct field is equal to or greater than the reverberant field when the energy reflection coefficient $\beta < 1/3$.

Equation (7.115) has been used to construct Figure 7.15. In the figure, the direct field contribution given by the first term on the right-hand side of the equation, is plotted separately from the reverberant field contribution given by the second term so that their relative contributions may readily be assessed.

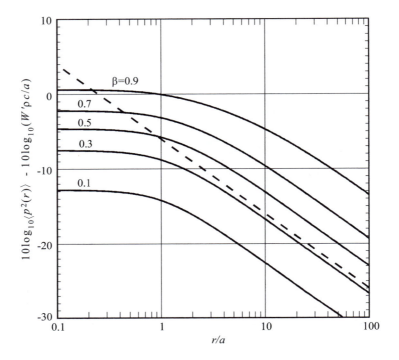

Figure 7.15 Direct and reverberant sound fields in a square cross-section long room of height and width a with specularly reflecting floor, walls and ceiling as a function of the normalised distance from a line source to the receiver. The reverberant field is represented as solid lines for various values of the energy reflection coefficient, β, and the direct field is indicated by the dashed diagonal straight line.

7.9.5 Long Room with Circular Cross-section and Diffusely Reflecting Wall

After some extensive mathematics, Kuttruff (1989) derives the following equation for the reverberant field in a long room of circular cross-section, radius a and diffusely reflecting wall:

$$\langle p^2(r)\rangle_R = \frac{2W\rho c\beta}{\pi^2 a^2}\int_0^\infty \frac{\left[\xi K_1(\xi)\right]^2}{1-\beta\lambda(\xi)}\cos\left(\frac{\xi r}{a}\right)d\xi \tag{7.117}$$

The function $\lambda(x)$ is defined in terms of the modified Bessel function $I_2(2x)$ and the Struve function $L_{-2}(2x)$ as follows (Abramowitz and Stegun, 1965):

$$\lambda(x) = \pi x\left[I_2(2x) - L_{-2}(2x)\right] \tag{7.118}$$

Equation (7.118) may be approximated as follows:

$$\lambda(x) \approx \left[1 + \frac{4}{3} x^2 \right]^{-1}$$

(7.119)

Numerical integration of Equation (7.117) gives the result shown in Figure 7.16. For comparison, the contribution of the direct field is also shown in the figure. Note that a square cross-section room of the same cross-sectional area as a circular section room is also approximately described by Figure 7.16. Comparison of Figure 7.16 with Figure 7.14 shows that the diffuse reflected reverberant field decreases more rapidly

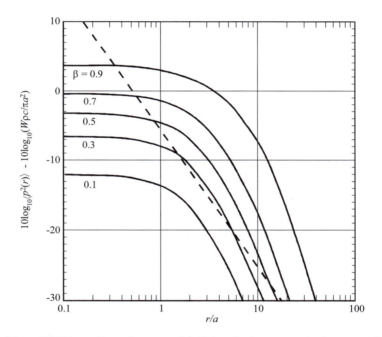

Figure 7.16 Direct and reverberant sound fields in a circular cross-section long room of diameter $2a$ with diffusely reflecting wall as a function of normalised distance from a point source to the receiver. The reverberant field is represented as solid lines for various values of the energy reflection coefficient, β, and the direct field is indicated by the dashed diagonal straight line.

than does the specularly reflected reverberant field. It may be shown that at large distances, the reverberant sound field decreases as $(r/a)^3$ or at the rate of 9 dB per doubling of distance. Reference to the figure shows that here again two hall radii (distance at which the direct and reverberant fields are equal) may be defined.

7.9.6 Long Room with Rectangular Cross-section

A rectangular cross-section long room of height, a, with a diffusely reflecting floor and ceiling and width, b, with specularly reflecting side walls is considered.

The energy reflection coefficients of the floor and ceiling are β, while the energy reflection coefficients of the side walls are unity; that is, they are assumed to reflect incident sound without loss. This case is typical of an industrial factory. A line source is assumed, which lies perpendicular to the room axis midway between the floor and ceiling and parallel to them. The source has a sound power of W' per unit length and is assumed to radiate incoherently. Kuttruff (1989) gives the following expression for the reverberant sound field (see Figure 7.17):

$$\langle p^2(\boldsymbol{r}) \rangle_R = \frac{2\,W'\rho c\beta}{\pi a} \int_0^\infty \frac{e^{-\xi}}{1 - \beta \xi K_1(\xi)} \cos\left(\frac{r}{a}\xi \right) d\xi \qquad (7.120)$$

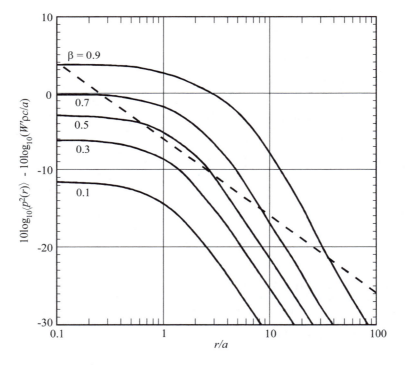

Figure 7.17 Direct and reverberant sound fields in a rectangular section long room of height a and width b with specularly reflecting walls and diffusely reflecting floor and ceiling as a function of the normalised distance from a line source (perpendicular to the room axis and parallel to the floor and ceiling) to the receiver. The reverberant field is represented as solid lines for various values of the energy reflection coefficient, β, of the floor and ceiling and the direct field is indicated by the dashed diagonal straight line. The energy reflection coefficient of the specularly reflecting walls is assumed to be unity.

Equation (7.120) has been used to construct Figure 7.17. In this case, the direct field diminishes as the inverse of the distance from the source, as a consequence of the assumption of a line source. However, because of the specularly reflecting walls, a point source and its images will look like a line source except in the immediate vicinity of the source ($r/a \le 1$), so that the expression also describes the sound field for a point source of the same total sound power output as the line source. Consideration of Figure 7.17 shows that for an energy reflection coefficient $\beta \ge 0.47$ there will exist a region in which the reverberant field will exceed the direct sound field so that in this case there will be two hall radii.

7.10 APPLICATIONS OF SOUND ABSORPTION

7.10.1 Relative Importance of the Reverberant Field

Consideration will now be given to determining when it is appropriate to treat surfaces in a room with acoustically absorbing material. The first part of the procedure is to determine whether the reverberant sound field dominates the direct sound field at the point where it is desired to reduce the overall sound pressure level, because treating reflecting surfaces with acoustically absorbing material can only affect the reverberant sound field. At locations close to the sound source (for example, a machine operator's position) it is likely that the direct field of the source will dominate, so there may be little point in treating a factory with sound absorbing material to protect operators from noise levels produced by their own machines. However, if an employee is affected by noise produced by other machines some distance away then treatment may be appropriate.

In the case of a Sabine room, the relative strength of the reverberant sound field may be compared with the direct field produced by a machine at a particular location by comparing the direct and reverberant field terms of the argument of Equation (7.43); that is, $4/R$ and $D_\theta/4\pi r^2$. For the case of flat rooms or long rooms see the discussion of Section 7.9. When the reverberant sound field dominates; for example, in the Sabine room when $4/R$ is much larger than $D_\theta/4\pi r^2$, then the introduction of additional absorption may be useful.

7.10.2 Reverberation Control

If the reverberant sound field dominates the direct field, then the sound pressure level will decrease if absorption is added to the room or factory. The decrease in reverberant sound pressure level, ΔL_p, to be expected in a Sabine room for a particular increase in sound absorption, expressed in terms of the room constant, R (see Equation (7.44)), may be calculated by using Equation (7.43) with the direct field term set equal to zero. The following equation is thus obtained, where R_o is the original room constant and R_f is the room constant after the addition of sound absorbing material.

$$\Delta L_p = 10 \log_{10} \left[\frac{R_f}{R_o} \right] \tag{7.121}$$

Referring to Equation (7.44) for the definition of the room constant R, it can be seen from Equation (7.121) that if the original room constant R_o is large then the amount of additional absorption to be added must be very large so that $R_f \gg R_o$ and ΔL_p is significant and worth the expense of the additional absorbent. Clearly, it is more beneficial to treat hard surfaces such as concrete floors, which have small Sabine absorption coefficients, because this will have greatest effect on the room constant.

To affect as many room modes as possible, it is better to distribute any sound-absorbing material throughout the room, rather than having it only on one surface. However, if the room is very large compared to the wavelength of sound considered, distribution of the sound-absorbing material is not so critical, because there will be many more oblique modes than axial or tangential modes in a particular frequency band. As each mode may be assumed to contain approximately the same amount of sound energy, then the larger percentage of sound energy will be contained in oblique modes, as there are more of them. Oblique modes consist of waves reflected from all bounding surfaces in the room and thus treatment anywhere will have a significant effect, although distribution of the sound absorbing material equally between the room walls, floor and ceiling will be more effective than having it only on one of those surfaces.

For flat rooms and long rooms (see Section 7.9) the discussion has been presented in terms of energy reflection coefficients β, which are related to absorption coefficients, α, by the relation, $\alpha + \beta = 1$.

7.11 AUDITORIUM DESIGN

The optimum design of auditoria for opera, music, drama or lectures is a challenging problem, especially if multi-use is expected. There are a number of acoustical parameters that are important in the design of auditoria, including reverberation time, early decay time, ratio of early to late energy (or clarity, C_{80}), envelopment (lateral energy fraction), background noise and total sound level (loudness). In practice, it is extremely difficult to measure the above quantities in an occupied auditorium, so methods have been developed to estimate the quantities for occupied auditoria from measurements made in unoccupied auditoria. Optimising the acoustical parameters for one type of use will result in a space unsuitable for other uses. Thus, for multi-use spaces, there must be some scope for altering the absorption and shape of the space, or introducing electroacoustics to enable the acoustical parameters to be varied within the range necessary.

7.11.1 Reverberation Time

As shown by Equation (7.52) or (7.57), the total sound absorption in a room is related directly to the room reverberation time. Various authors have suggested optimum reverberation times for occupied rooms for various purposes. One example (Stephens and Bate, 1950) is:

$$T_{60} = K\left[0.0118\, V^{1/3} + 0.1070\right] \tag{7.122}$$

In the equation, T_{60} is the 60 dB reverberation time (seconds), V is the room volume (m^3), and K equals 4 for speech, 4.5 for opera, 5 for orchestras, and 6 for choirs. An increase beyond that given by Equation (7.122) of about 10% at 250 Hz to 50% at 125 Hz and 100% at 63 Hz for T_{60} is advisable.

For classrooms, Hodgson and Nosal (2002) have shown that the optimum reverberation time for an occupied classroom is influenced by the level of background noise arising from sources such as air conditioning outlets and projectors. They conclude that in quiet classrooms, the optimum reverberation time would range from about 0.7 seconds for small class rooms (50 m^3) to 0.9 seconds for large classrooms (4000 m^3), for moderately noisy classrooms it would range from 0.4 to 0.5 seconds and for very noisy classrooms, the optimum range for maximum speech intelligibility would be between 0.2 and 0.1 seconds, this time decreasing with volume. Sound absorbing configurations that can be used to obtain the optimum reverberation times are discussed by Hodgson (1999, 2001). As a general rule, in order to achieve the correct reverberation time in a classroom, a volume of between 3 and 4 cubic metres is needed per seat.

Hodgson and Nosal (2002) provide the following expression relating the optimum reverberation time for an empty classroom to the previous optimum values for occupied classrooms:

$$\frac{1}{T_{60u}} = \left[\frac{1}{T_{60o}} - \frac{A_p N}{0.161\, V}\right] \tag{7.123}$$

where T_{60o} and T_{60u} are the reverberation times for the occupied and unoccupied classrooms respectively, A_p is the absorption (m^2) associated with each occupant (see Table 7.2), V is the room volume and N is the number of occupants.

Beranek (1996) suggests that for a concert hall to achieve a good rating, the bass ratio should be between 1.1 and 1.45 for halls with a T_{60} of 1.8 secs or less and between 1.1 and 1.25 for halls with higher values of T_{60}, where the bass ratio, BR, is defined as:

$$BR = \frac{T_{60}(125) + T_{60}(250)}{T_{60}(500) + T_{60}(1000)} \tag{7.124}$$

Meeting the reverberation time requirement as well as several other constraints (such as ensuring that the first reflected sound arrives at each listener less than 40 ms after the direct sound) effectively limits the seating capacity of a concert hall with good acoustics to 1600 (Siebein, 1994), although others claim that 3000 is possible (Barron, 1993).

For Sabine-type rooms, the optimum reverberation time in a particular room is achieved by the use of surface treatments having a suitable absorption coefficient in each frequency band. The total amount of absorption required can be calculated by using Equation (7.52), with the optimum reverberation time in place of T_{60}. The amount of absorption already in the room is measured as described earlier.

Often, better results for the calculation of reverberation time in a large enclosure such as a theatre, concert hall or studio, where the average absorption coefficient is 0.6 or less, are obtained by using calculated or measured statistical absorption coefficients and Equation (7.57) (Beranek, 1971) or Equation (7.61) (Neubauer, 2001). Given a desired value of T_{60}, Equation (7.57) may be used to find the required value of β_{st} ($= 1 - \bar{\alpha}_{st}$). Equation (7.59) is then used on a trial and error basis to determine the amount of absorptive material to add. Note that Equations (7.57) and (7.58) are approximate descriptions of the true behaviour of the sound field in an enclosure, and there will be situations where poor predictions may be obtained.

Recent papers (Beranek and Hidaka, 1998 and Nishihara et al., 2001) have expressed a preference for the use of the Sabine formula (Equation 7.52) for calculating reverberation times in auditoria. In using this formula, it is recommended that the seating area used in the calculation is the actual area of the flat surface over which the seats are mounted plus an edge correction, which is the area of a 0.5 m high edge around all parts of the seating area not adjacent to a wall. Thus the above authors recommend that auditoria reverberation times are calculated using:

$$
T_{60} = \frac{55.25\, V/c}{S_T \bar{\alpha}_T + S_R \bar{\alpha}_R + S_N \bar{\alpha}_N + S_R \alpha_a}
\tag{7.125}
$$

where the subscript, T refers to the audience area (excluding vertical surfaces but including the edge correction mentioned above), the subscript, R, refers to the bare auditorium without any special absorbing material, seats, audience or orchestra, the subscript, N, refers to any special sound absorbing material installed for sound control and the last term in the denominator is the air absorption given by Equation (7.38) with $S = S_R$. Representative values of $\bar{\alpha}_R$ and $S_T \bar{\alpha}_T$ (per seated person) for concert halls are provided in Table 7.2. Methods for measuring audience and seat absorption coefficients in a reverberation chamber are discussed by Nishihara (2001).

7.11.2 Early Decay Time (EDT)

The early decay time is the reverberation time based on a measure of the first 10 dB of decay of the sound field. It is expressed in the same way as the reverberation time. In a highly diffuse space where the decay rate is no different at the beginning of the decay to later on, the two quantities would be identical. However, in concert halls, the early decay time is often considered a more accurate measure of performance than the reverberation time. It is usually about 1.1 times greater than the reverberation time. The desired values are the same as those described in the previous section for reverberation time, which implies that the optimum values calculated for reverberation time using Equation (7.122) are about 10% too large. Note that the reverberation time and EDT are the most important parameters defining the acceptability of an auditorium. The other parameters that follow are of secondary importance but still need to be considered.

7.11.3 Clarity (C_{80})

Clarity is defined as (Barron, 1993):

$$C_{80} = 10\log_{10}\left[\frac{\text{energy arriving within 80 ms of direct sound}}{\text{energy arriving later than 80 ms of direct sound}}\right] \quad (7.126)$$

The quantity in the numerator includes the direct sound energy. The energy is measured by integrating the acoustic pressure squared over the measurement period.

For speech, the quantity, C_{50} is preferred (80 ms replaced with 50 ms in the above equation). The preferred value for C_{80} for a symphony concert is between -2 dB and +2 dB (Barron, 1993). For speech, C_{50} should be greater than -3 dB.

7.11.4 Envelopment

Envelopment or the early lateral energy fraction is defined as (Barron, 1993):

$$\text{envelopment} = \left[\frac{\text{energy arriving laterally within 80 ms of direct sound}}{\text{total energy arriving within 80 ms of direct sound}}\right] \quad (7.127)$$

The preferred value for envelopment for a symphony concert is between 0.1 and 0.35 (Barron, 1993).

7.11.5 Interaural Cross Correlation Coefficient, *IACC*

The *IACC* is a measure of the spaciousness of the sound or the apparent source width. In other words, it is a measure of the difference in the sounds arriving at the two ears at any instant. If the sounds were the same, then *IACC* = 0 (no spatial impression) and if they are completely different, *IACC* = 1. *IACC* is defined as:

$$IACC = \max\left[\frac{\int_{t_1}^{t_2}p_L(t)p_R(t+\tau)\,dt}{\int_{t_1}^{t_2}p_L^2(t)\,dt\ \int_{t_1}^{t_2}p_R^2(t)\,dt}\right]; \quad \text{for } -1 < \tau < +1 \text{ msec} \quad (7.128)$$

where time t_2 is 1000 milli-secs, t_1 is the arrival time of the direct sound and the integration includes the direct sound energy. The quantities, $p_L(t)$ and $p_R(t)$ represent the sound pressures arriving at the left and right ears respectively. For rating the apparent source width for concert halls, the value of t_2 is set equal to 80 milli-secs, which results in the quantity, $IACC_E$, where the subscript E refers to early energy. If only the three octave bands, 500 Hz, 1000 Hz and 2000 Hz are considered, the quantity is written as $IACC_{E3}$. Good concert halls have values of $IACC_{E3}$ between 0.6 and 0.72 (Beranek, 1996). For rating the envelopment of the listener, the value of t_1 is set to 80 milli-secs and the value of t_2 is set to 1000 milli-secs to give $IACC_L$ where the subscript L refers to late energy. If only the three octave bands, 500 Hz, 1000 Hz and 2000 Hz are considered, the quantity is written as $IACC_{L3}$.

7.11.6 Background Noise Level

The level of background noise due to air-conditioning systems and other external noise must be very low to avoid interference with opera and music performances and to avoid problems with speech intelligibility. Acceptable background noise levels range from NC15 for large concert halls to NC20 for drama theatres to NC25 for small auditoria of less than 500 seats. Achieving these levels is often quite a challenge for a noise control engineer.

7.11.7 Total Sound Level or Loudness, *G*

The total sound level, G (dB), is defined as (Barron, 1993):

> G = *total sound level at measurement position*
> *minus total sound level of direct sound* 10 m *from the source*

For symphony concerts, the total sound level in all seats (averaged over the octave bands from 125 Hz to 4000 Hz) should be greater than 0 dB and if just the 500 Hz and 1000 Hz octave bands are considered, it should be in the range 4 to 5.5 dB. For theatres and lecture halls, G should exceed 0 dB. A standard sound source with a reasonably uniform spectral distribution in the mid-frequency range (500 Hz to 2000 Hz) is used for the measurements.

7.11.8 Diffusion

For a concert hall, an adequate level of sound diffusion is important. This can be achieved using irregularities (both small and large scale) on the walls and ceiling. If a hall has inadequate diffusion, the quantity, $IACC_{L3}$, will be greater than about 0.16.

7.11.9 Speech Intelligibility

7.11.9.1 RASTI

The only measure of speech intelligibility to be incorporated into an IEC standard (IEC 60268 Part 16, 2003) is the Rapid Speech Transmission Index (RASTI). This is a simplification of the Speech Transmission Index (STI), which takes into account the influence of both the background noise and the room reverberation on the intelligibility of speech. The basic principle behind the measurement is that for good speech intelligibility, the envelope of the signal should not distort too much between the source and the receiver. To measure the distortion, a test signal is used that is modulated sinusoidally at frequencies between 0.4 and 20 Hz (corresponding to the modulation found in normal speech). The noise at the source is 100% modulated so that for a modulation frequency of 15 Hz (most common), there is a short time of silence every 0.067 seconds. The modulation depth (degraded by background noise and reverberation) of the received signal is measured over a range of 14 modulating frequencies at seven carrier frequencies corresponding to the centre frequencies of the octave bands from 125 Hz to 8000 Hz, a total of 98 separate measurements. The seven individual transmission index values are then weighted according to frequency and combined together to produce a single STI value. If this value is less than 0.3, speech intelligibility is bad. If it is greater than 0.75, speech intelligibility is excellent. As one can imagine, determining the STI for a space is a very complex and time-consuming procedure, so a new measure, Rapid Speech Transmission Index (RASTI), has been developed using the STI method as a basis. RASTI involves the use of nine modulation frequencies with measurements only being made at 500 Hz and 2000 Hz. The processing of the resulting data follows the same procedures as for the STI calculation, except that less data are used.

It is important to ensure that the sound source level used for the RASTI measurement is similar to an expected voice level and that all background noise sources are operating. In some implementations of RASTI, it is possible to numerically remove the background noise component contribution and obtain the result due to reverberation alone. It is also possible to numerically determine the influence of different background noises by entering specific levels into the calculation program.

7.11.9.2 Articulation Loss

Intelligibility can also be quantified in terms of articulation loss (Peutz, 1971) or in terms of the percentage articulation loss of consonants expressed as (Long, 2008):

$$\%ALcons = 200 \frac{T_{60}^2 r^2}{V} \tag{7.129}$$

where r is the distance between source and receiver and in a sound reinforcement system, the source would be considered to be the nearest loudspeaker. Values of the above quantity, for the 2000 Hz octave band that is usually used, which are less than $5 - 10\%$, are considered good (Long, 2008).

7.11.9.3 Signal to Noise Ratio

As discussed by Long (2008), Equation (7.129) is difficult to use when there are multiple sound sources such as in a typical sound reinforcement system in a theatre or concert hall. In this case a measure of intelligibility based on signal to noise ratio is used with different metrics using different definitions of the signal or the noise. One particularly useful metric according to Bistafa and Bradley (2000) that can be applied to multi-source systems is:

$$\%ALcons = 8.9 \left[\frac{T_{60} \times (\text{Reverberant noise energy})}{13.82 \times (\text{Direct signal energy})} \right] = 8.9 \left[\frac{T_{60} \times 10^{L_{pR}/10}}{13.82 \times 10^{L_{pD}/10}} \right] \quad (7.130)$$

where L_{pR} and L_{pD} represent the sound pressure levels of the reverberant and direct fields respectively. Again use of the above equation is usually restricted to the 2000 Hz octave band.

Another useful measure, which can be used with complex sound reinforcement systems, is the ratio of direct to reverberant energy level (Long, 2008):

$$\text{Direct-to-reverberant-level} = 10 \log_{10} \left[\frac{10^{L_{pD}/10}}{10^{L_{pR}/10}} \right] \quad (7.131)$$

and this is used for each of the three octave bands, 500 Hz, 1 kHz and 2 kHz. If the above expression has a value greater than -3 dB, then intelligibility excellent. If the value is less than -15 dB, then the speech intelligibility is very poor. In between these extremes, the speech intelligibility ranges from excellent to very good to good to fair to poor to very poor in 3 dB steps.

It becomes clear then that when a sound reinforcement system is used, intelligibility can be maximised by using high directivity loudspeakers or many distributed speakers close to intended receivers to raise the direct sound level energy. The reverberant field energy can be lowered by the above two strategies as well as by increasing absorption in the room (Long, 2008).

7.11.10 Sound Reinforcement

In many cases an auditorium has to be multi-purpose so it can be used for orchestras as well as for speeches and plays. For the latter two uses, speech intelligibility as discussed in the previous section is important and in many cases a sound reinforcement system becomes necessary to achieve intelligibility requirements. There are two things that need to be taken into account in the design of a sound reinforcement system in addition to the speech intelligibility aspect discussed above. These are: perception of direction; and feedback control, both of which will be discussed in the two sections to follow.

7.11.10.1 Direction Perception

It is important to maintain the illusion that the amplified sound is coming from the original source. This is done by controlling the time of the second arrival to be within 5 milliseconds of the first, which in turn is achieved by using speaker clusters above the sound source (capitalising on the fact that human perception of direction is less sensitive in the vertical plane than the horizontal plane). Speakers can also be added near the sound source and the signals fed to speakers amongst the audience can be delayed. This results in the sound being perceived as coming from the sound source even though the speakers in the audience area may be contributing more to the total energy arriving in the audience.

7.11.10.2 Feedback Control

Feedback occurs when the sound produced by a loudspeaker at the microphone that provides its signal is louder than the originating sound at the microphone. For no feedback to occur, the following equation must be satisfied (Long, 2008):

$$L_{DT} - L_{DS} - DI_M > 10 \text{ dB}$$

where L_{DT} is the direct field at the microphone produced by the original sound source, L_{DS} is the direct field produced by the loudspeaker system at the microphone and DI_M is the directivity of the microphone in the direction of the loudspeaker relative to the direction of the originating sound source (usually negative). To minimise feedback, the following should be done (Long, 2008):

- have the originating sound source as close to the microphone as possible so that system gains can be minimised;

- use a directional microphone that favours the originating sound source;

- use directional loudspeakers or a distributed system so that their influence at the microphone is minimised; and

- use equalisation, frequency shifting, compression and other electronic techniques.

7.11.11 Estimation of Parameters for Occupied Concert Halls

It is difficult to take measurements of the acoustical parameters described in the preceding sections for occupied concert halls due to the obvious annoyance caused to the audience. For this reason, Hidaka *et al.* (2001) developed empirical expressions to relate the occupied values to the unoccupied measurements. The quantity for the occupied space is obtained by multiplying the quantity for the unoccupied space by a constant, *a*, and then adding to the result another constant, *b*, where the constants, *a* and *b* are tabulated in Table 7.4 (Hidaka *et al.*, 2001).

Table 7.4 Values of constants *a* and *b* for calculating occupied room parameters from unoccupied values using, occupied parameter = (unoccupied parameter × *a*) + *b*

Frequency (Hz)	EDT (secs)		C_{80} (dB)		Total sound level (dB)	
	a	*b*	*a*	*b*	*a*	*b*
125	0.82	0.19	1.09	-0.07	0.52	1.31
250	0.74	0.25	0.75	0.71	0.80	-0.12
500	0.59	0.36	0.92	1.20	0.79	-0.59
1000	0.54	0.45	0.83	1.39	0.91	-1.19
2000	0.46	0.53	0.73	1.79	1.08	-2.67
4000	0.37	0.61	0.57	2.01	0.99	-1.89

7.11.12 Optimum Volumes for Auditoria

For concert halls, the volumes range from 8 to 12 m³ per occupant (recommended 10 m³), for opera houses the range is 4 to 6 m³ per occupant (recommended 5 m³), for theatres the range is 2.5 to 4 m³ per occupant (recommended 3 m³), for churches the range is 6 to 14 m³ per occupant (recommended 10 m³) and for rooms for lectures the recommended value is 3 m³ with a maximum of 6 m³ per occupant. Thus, the design of an auditorium must start with the desired size of the audience and the range of uses expected, which will then determine the required volume and reverberation time.

CHAPTER EIGHT

Partitions, Enclosures and Barriers

LEARNING OBJECTIVES

In this chapter the reader is introduced to:

- sound transmission through partitions and the importance of bending waves;
- transmission loss and its calculation for single (isotropic and orthotropic) and double panels;
- enclosures for keeping sound in and out;
- barriers for the control of sound out of doors and indoors; and
- pipe lagging.

8.1 INTRODUCTION

In many situations, for example where plant or equipment already exists, it may not be feasible to modify the characteristics of the noise source. In these cases, a possible solution to a noise problem is to modify the acoustic transmission path or paths between the source of the noise and the observer. In such a situation the first task for noise-control purposes, is to determine the transmission paths and order them in relative importance. For example, on close inspection it may transpire that, although the source of noise is readily identified, the important acoustic radiation originates elsewhere, from structures mechanically connected to the source. In this case structure-borne sound is more important than the airborne component. In considering enclosures for noise control one must always guard against such a possibility; if structure-borne sound is the problem, an enclosure to contain airborne sound can be completely useless.

In this chapter, the control of airborne sound is considered (the control of structure-borne sound will be considered in Chapter 10). Control of airborne sound takes the form of interposing a barrier to interrupt free transmission from the source to the observer; thus the properties of materials and structures, which make them useful for this purpose, will first be considered and the concept of transmission loss will be introduced. Complete enclosures will then be considered and means for estimating their effectiveness will be outlined. Finally, lagging for the containment of noise in conduits such as air ducts and pipes will be considered and means will be provided for estimating their effectiveness.

8.2 SOUND TRANSMISSION THROUGH PARTITIONS

8.2.1 Bending Waves

Solid materials are capable of supporting shear as well as compressional stresses, so that in solids shear and torsional waves as well as compressional (longitudinal) waves may propagate. In the audio-frequency range in thick structures, for example in the steel beams of large buildings, all three types of propagation may be important, but in the thin structures of which wall panels are generally constructed, purely compressional wave propagation is of negligible importance. Rather, audio-frequency sound propagation through panels and thus walls is primarily through the excitation of bending waves, which are a combination of shear and compressional waves.

In the discussion to follow, both isotropic and orthotropic panels will be considered. Isotropic panels are characterised by uniform stiffness and material properties, whereas orthotropic panels are usually characterised by a stiffness that varies with the direction of bending wave travel (for example, a corrugated or ribbed steel panel).

Bending waves in thin panels, as the name implies, take the form of waves of flexure propagating parallel to the surface, resulting in normal displacement of the surface. The speed of propagation of bending waves increases as the ratio of the bending wavelength to solid material thickness decreases. That is, a panel's stiffness to bending, B, increases with decreasing wavelength or increasing excitation frequency. The speed of bending wave propagation, c_B, for an isotropic panel is given by the following expression:

$$c_B = (B\omega^2/m)^{1/4} \quad \text{(m/s)} \tag{8.1}$$

The bending stiffness, B, is defined as:

$$B = EI'/(1 - v^2) = Eh^3/[12(1 - v^2)] \quad \text{kg m}^2 \text{ s}^{-2} \tag{8.2a,b}$$

In the preceding equations, ω is the angular frequency (rad/s), h is the panel thickness (m), ρ_m is the material density, $m = \rho_m h$ is the surface density (kg/m^2), E is Young's modulus (Pa), v is Poisson's ratio and $I' = h^3/12$ is the cross-sectional second moment of area per unit width (m^3), computed for the panel cross-section about the panel neutral axis.

As shown by Equation (8.1), the speed of propagation of bending waves increases with the square root of the excitation frequency; thus there exists, for any panel capable of sustaining shear stress, a critical frequency (sometimes called the coincidence frequency) at which the speed of bending wave propagation is equal to the speed of acoustic wave propagation in the surrounding medium. The frequency for which airborne and solid-borne wave speeds are equal, the critical frequency, is given by the following equation:

$$f_c = \frac{c^2}{2\pi} \sqrt{\frac{m}{B}} \quad \text{(Hz)} \tag{8.3}$$

where c is the speed of sound in air.

Substituting Equation (8.2b) into (8.3), and using Equations (1.1) and (1.4) for the longitudinal wave speed in a two-dimensional solid (or panel), the following equation is obtained for the critical frequency:

$$f_c = 0.55 c^2/(c_L h) \tag{8.4}$$

Here the longitudinal wave speed, c_L, for thin plates is given by:

$$c_L = \sqrt{E/[\rho_m(1 - v^2)]} \tag{8.5}$$

(see Section 1.3.4). Representative values of $\sqrt{E/\rho_m}$, the longitudinal wave speed in thin rods, are given in Appendix B. Using Equations (8.2b) and (8.5), the longitudinal wave speed may be written as:

$$c_L = \frac{\sqrt{12}}{h} \sqrt{\frac{B}{m}} \tag{8.6}$$

For a panel made of two layers of different materials bonded firmly together (such as the lead–aluminium jacket used in pipe lagging – see Section 8.6 and Figure 8.1), the bending stiffness and surface mass in the preceding equation must be replaced with an effective bending stiffness, B_{eff}, and surface mass, m_{eff}.

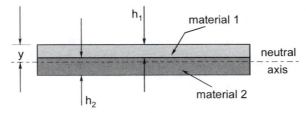

Figure 8.1 Composite material notation.

The effective bending stiffness may be calculated as follows:

$$
B_{eff} = \frac{E_1 h_1}{12(1 - v_1^2)}\left[h_1^2 + 12(y - h_1/2)^2\right]
$$
$$
+ \frac{E_2 h_2}{12(1 - v_2^2)}\left[h_2^2 + 12(y - (2h_1 + h_2)/2)^2\right]
\tag{8.7}
$$

where, the neutral axis location (see Figure 8.1) is given by:

$$y = \frac{E_1 h_1 + E_2 (2h_1 + h_2)}{2(E_1 + E_2)} \tag{8.8}$$

The surface mass used in Equation (8.6) for the double layer construction is simply the sum of the surface masses of the two layers making up the composite construction. That is, $m_{eff} = \rho_1 h_1 + \rho_2 h_2$, where ρ_1 and ρ_2 are the densities of the two panel materials. The critical frequency of this double layer construction is then:

$$f_c = \frac{c^2}{2\pi} \sqrt{\frac{m_{eff}}{B_{eff}}} \quad \text{(Hz)} \tag{8.9}$$

At the critical frequency, the panel bending wavelength corresponds to the trace wavelength of an acoustic wave at grazing incidence. A sound wave incident from any direction at grazing incidence, and of frequency equal to the critical frequency, will strongly drive a corresponding bending wave in the panel. Alternatively, a panel excited in flexure at the critical frequency will strongly radiate a corresponding acoustic wave.

As the angle of incidence between the direction of the acoustic wave and the normal to the panel becomes smaller, the trace wavelength of the acoustic wave on the panel surface becomes longer. Thus, for any given angle of incidence smaller than grazing incidence, there will exist a frequency (which will be higher than the critical frequency) at which the bending wavelength in the panel will match the acoustic trace wavelength on the panel surface. This frequency is referred to as a coincidence frequency and must be associated with a particular angle of incidence or radiation of the acoustic wave as illustrated in Figure 8.2(a). Thus, in a diffuse field, in the frequency range about and above the critical frequency, a panel will be strongly driven and will radiate sound well. However, the response is a resonance phenomenon, being strongest in the frequency range about the critical frequency and strongly dependent upon the damping in the system. This phenomenon is called coincidence, and it is of great importance in the consideration of transmission loss.

An important concept, which follows from the preceding discussion, is concerned with the difference in sound fields radiated by a panel excited by an incident acoustic wave and one excited by a mechanical localised force. In the former case, the structure will be forced to respond in modes that are characterised by bending waves having wavelengths equal to the trace wavelengths of the incident acoustic field. Thus, at excitation frequencies below the structure critical frequency, the modes that are excited will not be resonant, because the structural wavelength of the resonant modes will be smaller than the wavelength in the adjacent medium. Lower order modes will be excited by an acoustic field at frequencies above their resonance frequencies. As these lower order modes are more efficient than the unexcited higher order modes that would have been resonant at the excitation frequencies, the radiated sound will be higher than it would be for a resonantly excited structure having the same mean square velocity levels at the same excitation frequencies.

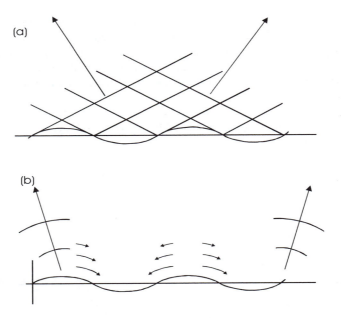

Figure 8.2 Coupling of the acoustic field and the panel flexural wave. (a) At and above the critical frequency, the panel radiates. (b) At frequencies less than the critical frequency, the disturbance is local – the panel does not radiate except at boundaries.

As excitation of a structure by a mechanical force results in resonant structural response, it can be concluded that sound radiation from an acoustically excited structure will be greater than that radiated by a structure excited mechanically to the same vibration level (McGary, 1988). A useful item of information that follows from this conclusion is that structural damping will only be effective for controlling mechanically excited structures because it is only the resonant structural response that is significantly influenced by damping.

At frequencies less than the critical frequency (lowest critical frequency for orthotropic panels), the structure-borne wavelength is shorter than the airborne acoustic wavelength and wave coupling is not possible (Cremer *et al.*, 1988). In this case, an infinite panel is essentially decoupled from an incident sound field. As illustrated in Figure 8.2(b), local disturbances are produced, which tend to cancel each other and decay very rapidly away from the panel. In finite panels, radiation coupling occurs at the edges and at stiffeners, where the disturbance is not matched by a compensating disturbance of opposite sign. At these places of coupling, the panel radiates sound or, alternatively, it is driven by an incident sound field.

As shown, the bending wave speed plays a very important role in the transmission of sound; thus, the difference between isotropic and orthotropic panels is of importance. Unless the panel is essentially isotropic in construction, the bending stiffness will be variable and dependent upon the direction of wave propagation. For example, ribbed or corrugated panels commonly found in industrial constructions are orthotropic, being stiffer along the direction of the ribs than across the ribs. The consequence is that orthotropic panels are characterised by a range of bending wave speeds (dependent upon the direction of wave propagation across the plate) due to the two different values of the cross-sectional second moment of area per unit width, I'. By contrast, isotropic panels are characterised by a single bending wave speed given by Equation (8.1).

For wave propagation along the direction of ribs or corrugations the bending stiffness per unit width may be calculated by referring to Figure 8.3, and using the following equation.

$$B = \frac{Eh}{(1 - v^2)\ell} \sum_{i=1}^{N} b_i \left(z_i^2 + \frac{h^2 + b_i^2}{24} + \frac{h^2 - b_i^2}{24} \cos 2\theta_i \right) \quad \text{(kg m}^2 \text{ s}^{-2}) \quad (8.10)$$

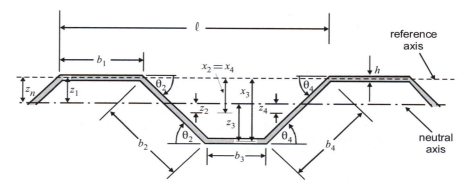

Figure 8.3 A typical cross-section of a ribbed panel.

The summation is taken over all sections in width ℓ (total, N) and the distances z_n are distances of the centre of the section from the neutral axis of the section. The location of the neutral axis is found by selecting any convenient reference axis, such as one that passes through the centre of the upper section (see Figure 8.3). Then the neutral axis location from the reference axis location is given by:

$$z_n = \frac{\displaystyle\sum_{i=1}^{N} x_i b_i h_i}{\displaystyle\sum_{i=1}^{N} b_i h_i} \quad (8.11)$$

For wave propagation in a stiffened panel, across the corrugations, the bending stiffness per unit width will be similar to that for an isotropic panel; that is:

$$B = \frac{Eh^3}{12(1 - v^2)\ell} \sum_{n} b_n \quad (8.12)$$

Note that Equation (8.12) follows from Equation (8.10) if in the latter equation z_n and θ_n are set equal to zero (see Figure 8.3).

The derivation of Equation (8.10) is predicated on the assumption that the wavelength of any flexural wave will be long compared to any panel dimension. Thus at high frequencies, where a flexural wavelength may be of the order of a characteristic dimension of the panel structure (for example b_n in Figure 8.3), the bending stiffness will approach that for an isotropic panel, as given by Equation (8.2b).

Although for an isotropic panel there exists just one critical frequency, for orthotropic panels the critical frequency is dependent upon the direction of the incident acoustic wave. However, as shown by Equation (8.3), the range of critical frequencies is bounded at the lower end by the critical frequency corresponding to a wave travelling in the panel stiffest direction (e.g. along the ribs for a corrugated panel) and at the upper end by the critical frequency corresponding to a wave propagating in the least stiff direction (e.g. across the ribs of a corrugated panel). For the case of an orthotropic panel, characterised by an upper and lower bound of the bending stiffness B per unit width, a range of critical frequencies will exist. The response will now be strong over this frequency range, which effectively results in a strong critical frequency response occurring over a much more extended frequency range than for the case of the isotropic panel.

As an interesting example, consider sound incident on one side of a floor or roof containing parallel rib stiffeners. At frequencies above the critical frequency, there will always be angles of incidence of the acoustic wave for which the projection of the acoustic wave on the structure will correspond to multiples of the rib spacing. If any one of these frequencies corresponds to a frequency at which the structural wavelength is equal to a multiple of the rib spacing, then a high level of sound transmission may be expected.

Another mechanism that reduces the transmission loss of ribbed or corrugated panels at some specific high frequencies is the resonance behaviour of the panel sections between the ribs. At the resonance frequencies of these panels, the transmission loss is markedly reduced.

8.2.2 Transmission Loss

When sound is incident upon a wall or partition some of it will be reflected and some will be transmitted through the wall. The fraction of incident energy that is transmitted is called the transmission coefficient τ. The transmission loss, TL (sometimes referred to as the sound reduction index, R_i), is in turn defined in terms of the transmission coefficient, as follows:

$$TL = -10 \log_{10} \tau \quad \text{(dB)} \tag{8.13}$$

In general, the transmission coefficient and thus the transmission loss will depend upon the angle of incidence of the incident sound. Normal incidence, diffuse field (random) incidence and field incidence transmission loss (denoted TL_N, TL_d and TL respectively) and corresponding transmission coefficients (denoted τ_N, τ_d and τ_F respectively) are terms commonly used; these terms and their meanings will be described in the following section. Field incidence transmission loss, TL, is the transmission loss commonly observed in testing laboratories and in the field, and reported in tables.

The transmission loss of a partition is usually measured in a laboratory by placing the partition in an opening between two adjacent reverberant rooms designed for such tests. Noise is introduced into one of the rooms, referred to as the source room, and part of the sound energy is transmitted through the test partition into the second room, referred to as the receiver room. The resulting mean space-average sound pressure levels (well away from the sound source) in the source and receiver rooms are measured and the difference in levels, called the noise reduction, *NR*, is determined. The receiver room constant is determined either by use of a standard sound power source or by measurements of the reverberation decay, as discussed in Section 7.6.2. The Sabine absorption in the room, including loss back through the test partition, is thus determined. An expression for the field incidence transmission loss in terms of these measured quantities can then be derived using the analysis of Section 7.4, as will now be shown.

The power transmitted through the wall is given by the effective intensity in a diffuse field (see Section 7.4.1) multiplied by the area, A, of the panel and the fraction of energy transmitted τ; thus, using Equation (7.34) one may write for the power transmitted:

$$W_t = \frac{\langle p_i^2 \rangle A \tau}{4\rho c} \tag{8.14}$$

The sound pressure level in the receiver room (from Equation (7.42)) is:

$$\langle p_r^2 \rangle = \frac{4 W_t \rho c (1 - \overline{\alpha})}{S\overline{\alpha}} = \frac{\langle p_i^2 \rangle A \tau (1 - \overline{\alpha})}{S\overline{\alpha}} \tag{8.15a,b}$$

and the noise reduction is thus given by:

$$NR = 10 \log_{10} \frac{\langle p_i^2 \rangle}{\langle p_r^2 \rangle} = TL - 10 \log_{10} \frac{A(1 - \overline{\alpha})}{S\overline{\alpha}} \tag{8.16a,b}$$

In reverberant test chambers used for transmission loss measurement, $\overline{\alpha}$ is always less than 0.1 and thus $S\overline{\alpha} / (1 - \overline{\alpha})$ may be approximated as $S\overline{\alpha}$. Equation (8.16) may then be rearranged to give the following expression, which is commonly used for the laboratory measurement of sound transmission loss:

$$TL = NR + 10 \log_{10}(A/S\overline{\alpha}) \quad \text{(dB)} \tag{8.17}$$

In the above equation, $S\overline{\alpha}$ is the Sabine absorption of the receiving room, including losses through the test partition, and A is the area of the test partition. S and $\overline{\alpha}$ are, respectively, the receiving room total surface area, including that of the test partition, and the mean Sabine absorption coefficient (including the test partition).

When conducting a transmission loss test, great care must be taken to ensure that all other acoustic transmission paths are negligible; that is, "flanking paths" must contribute an insignificant amount to the total energy transmitted. The test procedure is described in relevant standards publications (ISO 140, Parts 3 (1995) and 4 (1998), AS1191-2002, ASTM E336-07).

Care should be taken in the interpretation of measured TL data for which the TL is less than 10 dB (Bies and Davies, 1977), due to the absorption in the receiving room being influenced by coupling with the source room.

The sound transmission loss of a partition may also be determined using a single reverberant room as the source room and a not too reverberant space (preferably free field) as the receiving room. In this case, the power incident on the partition may be determined using Equation (8.14) with the quantity, τ, excluded and the transmitted power may be determined by measuring the average of the active sound intensity very close (500 to 100 mm) to the panel on the receiving room side. The transmitted power is then determined by multiplying the average sound intensity by the panel surface area, the transmission coefficient is determined as the ratio of the transmitted to incident power and the transmission loss is then determined using Equation (8.13). This latter method of transmission loss measurement is more accurate than the sound pressure measurement method and is becoming more accepted recently. It is described in detail in three ISO standards (ISO 15186-1, ISO 15186-2 and ISO 15186-3).

In practice, it is desirable to characterise the transmission loss of a partition with a single number descriptor to facilitate comparison of the performance of different partitions. For this reason, a single number rating scheme (see ASTM E413-04) called STC (or sound transmission class) has been introduced. To determine the STC for a particular partition, a curve fitting technique is used to fit the measured or calculated one third octave transmission loss (*TL*) data for the partition. Figure 8.4 shows a typical STC contour.

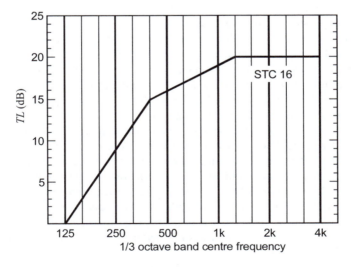

Figure 8.4 Example STC contour.

STC contours consist of a horizontal segment from 1250 to 4000 Hz, a middle segment increasing by 5 dB from 400 to 1250 Hz and a low frequency segment increasing by 15 dB from 125 to 400 Hz. The STC rating of a partition is determined by plotting the one-third octave band *TL* (rounded up or down to the nearest integer dB) of the partition and comparing it with the STC contours. The STC contour is shifted vertically downwards in 1 dB increments from a large value until the following criteria are met.

1.   The *TL* curve is never more than 8 dB below the STC contour in any one-third octave band.
2.   The sum of the deficiencies of the *TL* curve below the STC contour over the 16 one-third octave bands does not exceed 32 dB.

When the STC contour is shifted to meet these criteria, the STC rating is given by the integer value of the contour at 500 Hz.

The ISO method of determining a single number to describe the Sound Transmission Loss characteristics of a construction is outlined in ISO 717-1 (1996). Different terminology is used, otherwise the methods are very similar. The ISO standard uses Sound Reduction Index (R_i) instead of Sound Transmission Loss and Weighted Sound Reduction Index (R_w) instead of Sound Transmission Class (STC). The shape of the contour for 1/3 octave band data is identical to that shown in Figure 8.4, except that the straight line at the low frequency end continues down to 100 Hz and at the upper frequency end, the line terminates at 3150 Hz. In addition, there is no requirement to satisfy criterion number 1 listed above (the 8 dB criterion). However, measured *TL* values are rounded to the nearest 0.1 dB (rather than the 1 dB for STC) when calculating the deficiencies below the R_w curve. The ISO standard also allows for measurements to be made in octave bands between 125 Hz and 2 kHz inclusive. In this case the octave band contour is derived from the 1/3 octave band contour by connecting the values at the octave band centre frequencies. The value of 32 dB in the second criterion listed above is replaced with 10 dB for the octave band data.

The ISO method also provides a means of modifying (usually downgrading) the R_w value for different types of incident sound, by introducing correction factors, C and C_{tr} that are added to R_w. The correction factor, C, is used for incident sound consisting of living activities (talking, music, radio, TV), children playing, medium and high speed rail traffic, highway road traffic greater than 80 km hr^{-1}, jet aircraft at short distances and factories emitting mainly medium and high frequency sound. The correction factor, C_{tr}, is used for incident sound consisting of urban road traffic, low speed rail traffic, propeller driven aircraft, jet aircraft at long distances, disco music and factories emitting mainly low to medium frequency sound. For building elements, the Weighted Sound Reduction Index (which is a laboratory measurement) is written as $R_w (C; C_{tr})$, for example, 39 (-2; -6) dB. For stating requirements or performance of buildings a field measurement is used, called the Apparent Sound Reduction Index, R_w', and is written with a spectral correction term as a sum such as $R_w' + C_{tr} > 47$ (for example), where the measurements are conducted in the field according to ISO140-4 (1998) or ISO140-5 (1998).

The correction terms C and C_{tr} are calculated from values in Table 8.1 and the following equations.

$$C = -10\log_{10}\sum_{i=1}^{N} 10^{(L_{i,1} - R_i)/10} - R_w \qquad (8.18)$$

$$C_{tr} = -10\log_{10} \sum_{i=1}^{N} 10^{(L_{i,2} - R_i)/10} - R_w \qquad (8.19)$$

where $L_{i,1}$ and $L_{i,2}$ are listed for 1/3 octave or octave bands in Table 8.1, R_i is the transmission loss or sound reduction index for frequency band i and N is the number of bands used to calculate R_w (octave or 1/3 octave). Although the table shows values in the frequency range from 50 Hz to 5000 Hz, the standard frequency range usually used is 100 Hz to 3150 Hz. In this case (and for the case of the frequency range from 50 Hz to 3150 Hz), the octave and 1/3 octave band values in the table for $L_{i,1}$ (only) must be increased (made less negative) by 1 dB. When the expanded frequency range is used, the calculation of R_w is unchanged but the values of C and C_{tr} are different and indicated by an appropriate subscript; for example, $C_{50\text{-}3150}$ or $C_{50\text{-}5000}$ or $C_{100\text{-}5000}$.

A third rating scheme known as the Outdoor–Indoor Transmission Class (OITC) was described in ASTM E1332-90 (2003). However, recent work has shown that this rating method is not as useful as *STC* or R_w (Davy, 2000).

Table 8.1 Correction terms for Equations (8.18) and (8.19)

Band centre frequency	$L_{i,1}$		$L_{i,2}$	
	1/3 octave	octave	1/3 octave	octave
50	-41		-25	
63	-37	-32	-23	-18
80	-34		-21	
100	-30		-20	
125	-27	-22	-20	-14
160	-24		-18	
200	-22		-16	
250	-20	-15	-15	-10
315	-18		-14	
400	-16		-13	
500	-14	-9	-12	-7
630	-13		-11	
800	-12		-9	
1000	-11	-6	-8	-4
1250	-10		-9	
1600	-10		-10	
2000	-10	-5	-11	-6
2500	-10		-13	
3150	-10		-15	
4000	-10	-5	-16	-11
5000	-10		-18	

8.2.3 Impact Isolation

The ability of a construction such as a floor or ceiling to prevent transmission of impact noise such as footsteps is quantified in terms of its impact isolation, which is measured according to ASTM E989-06 (2006) using a standard tapping machine. A standard tapping machine can be purchased from suppliers of acoustic instrumentation and should conform to the specifications in ISO 140, part 6, 1998 and repeated in ASTM E1007-04. Basically such a machine consists of five standard hammers (weighing 0.5 kg each and consisting of steel cylinders with a radius of 500 mm on the end that strikes the floor) that bang on the floor sequentially from a height of 40 mm with 0.1 seconds between successive impacts. To measure the impact isolation of a floor, the standard tapping machine is placed on it and the resulting sound levels in the room on the opposite side of the floor or ceiling are measured in 1/3 or 1/1 octave bands (more commonly 1/3 octave bands) according to ASTM E1007-04. The sound pressure level must be averaged over 4 tapping machine locations and at least 4 microphone positions for each tapping machine location to give the average sound pressure level in the receiving room, L_p. As usual, this is an energy average, not a dB average. The normalised impact sound level for each measurement band is then calculated using:

$$L_n = L_p + 10\log_{10}\left(S\overline{\alpha}/10\right) \tag{8.20}$$

The quantity, $S\overline{\alpha}$, applies to the room in which the sound measurements are made and may be determined from the room reverberation time using Equation (7.52). From these measurements a single number Impact Insulation Class (IIC) may be determined. This is done in a similar way to determining the STC rating of a wall. The normalised sound pressure levels are rounded up or down to the nearest decibel and plotted on a set of axes similar to those used in Figure 8.5.

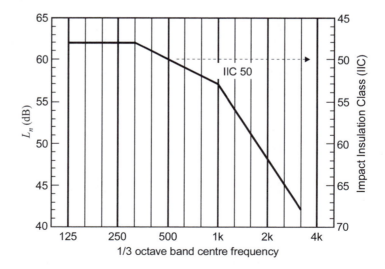

Figure 8.5 Example IIC contour.

The IIC curve shown in the figure is started at a low level and then shifted vertically upwards in 1 dB increments until the following conditions are met:

1. the L_n curve of normalised measured sound levels in the room on the opposite side of the floor or ceiling to the tapping machine is never more than 8 dB above the IIC contour in any one-third octave band; and
2. the sum of the deficiencies of the L_n curve above the IIC contour over the 16 one-third octave bands is as large as possible but does not exceed 32 dB. Note that the lower the IIC contour on the figure, the higher (and better) will be the IIC.

When the IIC contour has been adjusted to meet the above criteria, the IIC is the integer value of the contour at 500 Hz on the right of the figure or the value on the left ordinate (L_n) subtracted from 110.

It is possible to specify a similar quantity, called "Weighted Normalised Impact Sound Pressure Level, $L_{n,w}$" according to ISO 717-2 (1996). The calculation is similar to that for IIC except that the measured values of L_n are rounded to the nearest 0.1 dB (not 1 dB) and the first criterion above does not need to be satisfied. In addition, the ISO method can also be used for impact isolation between rooms as well as between building elements. The measurement of the required quantities is described in detail in ISO140-7 (1998). Again, the ISO requirement is for measurements in 1/3 octave bands from 100 Hz to 3150 Hz as opposed to the IIC requirement for measured data between 125 Hz and 4000 Hz. In ISO 140-7 (1998), a quantity, L_{nT}, called the "Standardised Impact Sound Pressure Level", is defined as:

$$L_{nT} = L_p + 10 \log_{10}\left(2 \times T_{60}\right) \qquad (8.21)$$

where T_{60} is the reverberation time the room in which the sound measurements are made. A corresponding Weighted Standardised Impact Sound Pressure Level, $L_{nT,w}$, is calculated using the results of each 1/3 octave band calculated using Equation (8.21) in the same way that the Weighted Normalised Impact Sound Pressure Level, $L_{n,w}$ is calculated with the results of Equation (8.20).

The standard, ISO717-2 (1996), also allows the use of octave band measurements of L_n (from 125 Hz to 2 kHz). In this case, the "modified" 1/3 octave band IIC contour is adjusted in 1 dB increments until the sum of the octave band deficiencies of the measured data above the curve in the 5 relevant octave bands is as large as possible but no more than 10 dB. The octave band IIC contours are identical to the 1/3 octave band contours except that they are truncated at 125 Hz and 2 kHz and the value of each contour at 2 kHz is increased by 1 dB over the 1/3 octave band value to account for the expected influence of the excluded 3150 Hz 1/3 octave band.

8.2.4 Panel Transmission Loss (or Sound Reduction Index) Behaviour

It will be instructive to consider the general behaviour of the field incidence transmission loss of a single uniform partition (isotropic panel) over the broad audio-frequency range. An illustration of typical behaviour is shown in Figure 8.6(a), in which various characteristic frequency ranges are indicated.

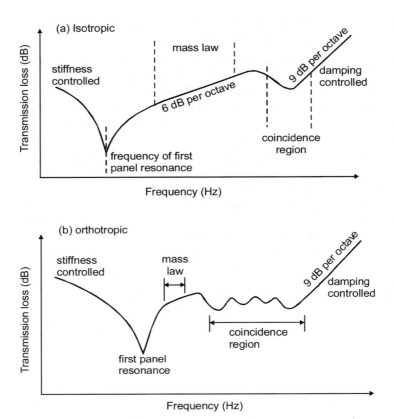

Figure 8.6 Typical single panel transmission loss as a function of frequency:
(a) isotropic panel characterised by a single critical frequency; (b) orthotropic panel
characterised by a critical frequency range.

At low frequencies, the transmission loss is controlled by the stiffness of the
panel. At the frequency of the first panel resonance, the transmission of sound is high
and consequently, the transmission loss passes through a minimum determined in part
by the damping in the system. Subsequently, at frequencies above the first panel
resonance, a generally broad frequency range is encountered, in which transmission
loss is controlled by the surface density of the panel. In this frequency range (referred
to as the mass law range, due to the approximately linear dependence of the
transmission loss on the mass of the panel) the transmission loss increases with
frequency at the rate of 6 dB per octave. Ultimately, however, at still higher
frequencies in the region of the critical frequency, coincidence is encountered. Finally,
at very high frequencies, the transmission loss again rises, being damping controlled,
and gradually approaches an extension of the original mass law portion of the curve.
The rise in this region is of the order of 9 dB per octave.

The transmission loss of orthotropic panels is strongly affected by the existence
of a critical frequency range. In this case the coincidence region may extend over two

decades for common corrugated or ribbed panels. Figure 8.6(b) shows a typical transmission loss characteristic of orthotropic panels. This type of panel should be avoided where noise control is important, although it can be shown that damping can improve the performance of the panel slightly, especially at high frequencies.

The resonance frequencies of a simply supported rectangular isotropic panel of width a, length b, and bending stiffness B per unit width may be calculated using the following equation:

$$f_{i,n} = \frac{\pi}{2}\sqrt{\frac{B}{m}}\left[\frac{i^2}{a^2} + \frac{n^2}{b^2}\right] \quad (\text{Hz}); \qquad i, n = 1, 2, 3, \ldots . \qquad (8.22)$$

The lowest order (or fundamental) frequency corresponds to $i = n = 1$. For an isotropic panel, Equations (8.2b) and (8.5) can be substituted into Equation (8.22) to give the following.

$$f_{i,n} = 0.453\, c_L h\left[\frac{i^2}{a^2} + \frac{n^2}{b^2}\right] \qquad (8.23)$$

The resonance frequencies of a simply supported rectangular orthotropic panel of width a and length b are (Hearmon, 1959):

$$f_{i,n} = \frac{\pi}{2m^{1/2}}\left(\frac{B_a\, i^4}{a^4} + \frac{B_b\, n^4}{b^4} + \frac{B_{ab}\, i^2 n^2}{a^2 b^2}\right)^{1/2} ; \qquad i, n = 1, 2, 3, \ldots . \qquad (8.24)$$

where,

$$B_{ab} = 0.5(B_a v + B_b v + Gh^3/3) \qquad (8.25)$$

In the preceding equations, $G = E/[2(1 + v)]$ is the material modulus of rigidity, E is Young's modulus, v is Poisson's ratio and B_a and B_b are the bending stiffnesses per unit width in directions a and b respectively, calculated according to Equations (8.10) or (8.12).

The following behaviour is especially to be noted. A very stiff construction tends to move the first resonance to higher frequencies but, at the same time, the frequency of coincidence tends to move to lower frequencies. Thus, the extent of the mass law region depends upon the stiffness of the panel. For example, steel-reinforced concrete walls of the order of 0.3 m thick, exhibit coincidence at about 60 Hz, and this severely limits the transmission loss of such massive walls. On the other hand, a lead curtain wall exhibits coincidence well into the ultrasonic frequency range, and its large internal damping greatly suppresses the first resonance, so that its behaviour is essentially mass-law controlled over the entire audio-frequency range.

The transmission coefficient for a wave incident on a panel surface is a function of the bending wave impedance, Z, which for an infinite isotropic panel is (Cremer, 1942):

$$Z = j2\pi fm\left[1 - \left(\frac{f}{f_c}\right)^2 (1 + j\eta)\sin^4\theta\right] \qquad (8.26)$$

where η is the panel loss factor (see Equation (7.23) and Appendix B) and m is the panel surface density (kg/m^2). For an infinite orthotropic panel the bending wave impedance is (Hansen, 1993):

$$Z = j2\pi fm \left[1 - \left[\frac{f}{f_{c_1}} \cos^2 \vartheta + \frac{f}{f_{c_2}} \sin^2 \vartheta \right]^2 (1 + j\eta) \sin^4 \theta \right] \tag{8.27}$$

where f_{c1} and f_{c2} are respectively, the lowest and highest critical frequencies of the panel and ϑ is the angle of incidence with respect to the axis about which the panel is least stiff (see Figure 8.7). For example, for a corrugated panel, it is with respect to the axis parallel to the corrugations.

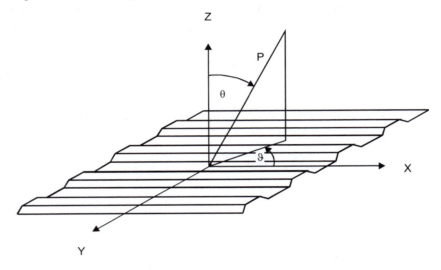

Figure 8.7 Geometry of a corrugated panel.

For a panel of infinite extent, the transmission coefficient at an angle (θ, ϑ) to the normal to the panel surface is given by Cremer (1942) as:

$$\tau(\theta, \vartheta) = \left| 1 + \frac{Z \cos \theta}{2\rho c} \right|^{-2} \tag{8.28}$$

The transmission coefficient for normal incidence, τ_N, is found by substituting $\theta = 0$ in Equation (8.28). The diffuse field transmission coefficient, τ_d, is found by determining a weighted average for $\tau(\theta, \vartheta)$ over all angles of incidence using the following relationship:

$$\tau_d = \frac{1}{\pi} \int_0^{2\pi} d\vartheta \int_0^{\pi/2} \tau(\theta, \vartheta) \cos \theta \sin \theta \, d\theta \tag{8.29}$$

The $\cos \theta$ term accounts for the projection of the cross-sectional area of a plane wave that is incident upon a unit area of wall at an angle, θ, to the wall normal. The $\sin \theta$ term is a coefficient that arises from the use of spherical coordinates.

For isotropic panels, Equation (8.29) can be simplified to:

$$\tau_d = \int_0^1 \tau(\theta) \, \mathrm{d}(\sin^2\theta) \tag{8.30}$$

and for orthotropic panels, Equation (8.29) becomes

$$\tau_d = \frac{2}{\pi} \int_0^{\pi/2} \mathrm{d}\vartheta \int_0^1 \tau(\theta, \vartheta) \mathrm{d}(\sin^2\theta) \tag{8.31}$$

as τ is a function of ϑ as well as θ.

In practice, panels are not of infinite extent and results obtained using the preceding equations do not agree well with results measured in the laboratory. However, it has been shown that good comparisons between prediction and measurement can be obtained if the upper limit of integration of Equation (8.30) is changed so that the integration does not include angles of θ between some limiting angle and 90°. Davy (1990) has shown that this limiting angle θ_L is dependent on the size of the panel as follows:

$$\theta_L = \cos^{-1} \sqrt{\frac{\lambda}{2\pi\sqrt{A}}} \tag{8.32}$$

where A is the area of the panel and λ is the wavelength of sound at the frequency of interest. Introducing the limiting angle, θ_L, allows the field incidence transmission coefficient, τ_F, of isotropic panels to be defined as follows:

$$\tau_F = \int_0^{\sin^2\theta_L} \tau(\theta) \, \mathrm{d}(\sin^2\theta) \tag{8.33}$$

Hansen (1993) has shown that the same reasoning is valid for orthotropic panels as well, giving:

$$\tau_F = \frac{2}{\pi} \int_0^{\pi/2} \mathrm{d}\vartheta \int_0^{\sin^2\theta_L} \tau(\theta, \vartheta) \, \mathrm{d}(\sin^2\theta) \tag{8.34}$$

Substituting Equation (8.26) or (8.27) into (8.28), then into (8.33) or (8.34) respectively and performing the numerical integration, allows the field incidence transmission coefficient to be calculated as a function of frequency for any isotropic or orthotropic panel, for frequencies above 1.5 times the first resonance frequency of the panel. At lower frequencies, the infinite panel model used to derive the equations is not valid and a different approach must be used as discussed in Section 8.2.4.1.

Third octave band results are obtained by averaging the τ_F results over a number of frequencies (at least 20) in each band. The field incidence transmission loss can then be calculated by substituting τ_F for τ in Equation (8.13). Results obtained by this procedure generally agree well with measurements made in practice.

To reduce the extent of the numerical calculations considerable effort has been made by various researchers to simplify the above equations by making various approximations. At frequencies below $f_c/2$ in Equation (8.26) or below $f_{c1}/2$ in Equation (8.27), the quantities in brackets in Equations (8.26) and (8.27) are in each case approximately equal to 1, giving for both isotropic and orthotropic panels:

$$Z = j 2\pi f m \tag{8.35}$$

Substituting Equation (8.35) into (8.28) and the result into Equation (8.13) gives the following expression for the mass law transmission loss of an infinite isotropic or orthotropic panel subject to an acoustic wave incident at angle θ to the normal to the panel surface:

$$TL_\theta = 10 \log_{10}\left[1 + \left(\frac{\pi f m}{\rho c} \cos\theta \right)^2 \right] \tag{8.36}$$

Normal incidence TL is obtained by substituting $\theta = 0$ in Equation (8.36).

8.2.4.1 Sharp's Prediction Scheme for Isotropic Panels

Sharp (1973) showed that good agreement between prediction and measurement in the mass law range is obtained for single panels by using a constant value for θ_L equal to about 78°. In this case, the field incidence transmission loss, TL, is related to the normal incidence transmission loss, TL_N, for predictions in 1/3 octave bands, for which $\Delta f / f = 0.236$, by:

$$TL = TL_N - 10 \log_{10}\left(1.5 + \log_e \frac{2f}{\Delta f} \right) = TL_N - 5.5 \quad \text{(dB)} \tag{8.37a,b}$$

In the preceding equation, if the predictions are required for octave bands of noise (rather than for 1/3 octave bands), for which $\Delta f / f = 0.707$, then the "5.5" is replaced with "4.0". Note that the mass law predictions assume that the panel is limp. As panels become thicker and stiffer, their mass law performance drops below the ideal prediction, so that in practice, very few constructions will perform as well as the mass law prediction.

Substituting Equation (8.36) with $\theta = 0$ into (8.37b) and rearranging gives the following for the field incidence transmission loss in the mass-law frequency range below $f_c/2$ for isotropic panels or $f_{c1}/2$ for orthotropic panels:

$$TL = 10 \log_{10}\left[1 + \left(\frac{\pi f m}{\rho c} \right)^2 \right] - 5.5 \quad \text{(dB)} \tag{8.38}$$

Equation (8.38) is not valid for frequencies below 1.5 times the first panel resonance frequency, but above this frequency, it agrees reasonably well with measurements taken in one-third octave bands. For octave band predictions, the 5.5

should be replaced with 4.0. Alternatively, better results are usually obtained for the octave band transmission loss, TL_o, by averaging logarithmically the predictions, TL_1, TL_2 and TL_3 for the three 1/3 octave bands included in each octave band as follows:

$$TL_o = -10 \log_{10} \frac{1}{3} \left[10^{-TL_1/10} + 10^{-TL_2/10} + 10^{-TL_3/10} \right] \quad \text{(dB)} \tag{8.39}$$

For frequencies equal to or higher than the critical frequency, Sharp gives the following equation for an isotropic panel:

$$TL = 10 \log_{10} \left[1 + \left(\frac{\pi f m}{\rho c} \right)^2 \right] + 10 \log_{10} \left[2\eta f / (\pi f_c) \right] \quad \text{(dB)} \tag{8.40}$$

Note that Equation (8.40) is only used until the frequency is reached at which the calculated TL is equal to that calculated using the mass law expression given by Equation (8.38) (see Figure 8.8a).

Values for the panel loss factor, η, which appears in the above equation, are listed in Appendix B. Note that the loss factors listed in Appendix B are not solely for the material but include the effects of typical support conditions found in wall structures.

The transmission loss between $0.5f_c$ and f_c is approximated by connecting with a straight line the points corresponding to $0.5f_c$ and f_c on a graph of TL versus $\log_{10}$ (frequency).

The preceding prediction scheme is summarised in Figure 8.8a, where a method for estimating the transmission loss for single isotropic panels is illustrated.

The lowest valid frequency for this scheme is 1.5 times the frequency of the first panel resonance. Occasionally, it may be of interest to be able to predict the TL at frequencies below this and for this purpose we adapt Fahy and Gardonio's (2007) analysis for a rigid panel on flexible supports. They define the resonance frequency of the rigid panel on flexible supports as:

$$f_0 = \sqrt{s/m} \quad \text{Hz} \tag{8.41}$$

where s and m are respectively the stiffness per unit area of the panel support and the mass per unit area of the panel. They then expresses the TL in the frequency range below the first panel resonance frequency in terms of the stiffness, s. Their model may be considered to be equivalent to that for a simply supported flexible panel vibrating in its first resonant mode (not necessarily at the resonance frequency). To re-write Fahy and Gardonio's expression in terms of the bending stiffness of a simply supported flexible panel, it is necessary to express the stiffness, s, in Equation (8.41) in terms of the panel bending stiffness, B. Comparing Equation (8.41) with Equation (8.22), evaluated for the first mode of vibration where $i = n = 1$, gives the following equivalence between B and s.

$$s = \pi^4 B \left[\frac{1}{a^2} + \frac{1}{b^2} \right]^2 \tag{8.42}$$

where a and b are the panel dimensions.

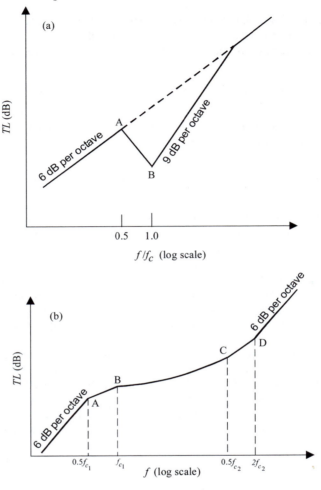

Figure 8.8 A design chart for estimating the transmission loss of a single panel. See Appendix B for values of f_c, $\rho\ (=mh)$ and η for typical materials. See the text for definitions of the quantities.

(a) A design chart for an isotropic panel. The points on the chart are calculated as follows:
point A: $TL = 20\log_{10}f_c m - 54$ (dB)

point B: $TL = 20\log_{10}f_c m + 10\log_{10}\eta - 44$ (dB)

(b) A design chart for an orthotropic (or ribbed) panel, with critical frequencies f_{c_1} and f_{c_2}, and small damping. For a well damped panel (see discussion following Equation (8.53)), Figure 8.8(a) is used where $f_c = f_{c2}$ and f_{c2} assumes a new value determined by the total panel surface density and the stiffness of a flat panel of identical thickness to the original material in the corrugated panel. The points in the chart are calculated as follows:

Point A: $TL = 20\log f_{c_1} m - 54$ (dB)

Between and including points B and C:
$$TL = 20\log_{10}f + 10\log_{10}m - 10\log_{10}f_{c_1} - 20\log_{10}\left[\log_e(4f/f_{c_1})\right] - 13.2 \quad \text{(dB)}$$
Point D: $TL = 10\log_{10}m + 15\log_{10}f_{c_2} - 5\log_{10}f_{c_1} - 17$ (dB)

Thus, Fahy and Gardonio's (2007) equation for the TL in the stiffness controlled region below half of the first resonance frequency of the panel can be written in terms of the panel bending stiffness and dimensions as:

$$
TL = 20\log_{10}\left[\pi^4 B\left(\frac{1}{a^2} + \frac{1}{b^2}\right)^2\right] - 20\log_{10}f - 20\log_{10}(4\pi\rho c)
$$

$$
= 20\log_{10}B - 20\log_{10}f + 20\log_{10}\left(\frac{1}{a^2} + \frac{1}{b^2}\right)^2 - 20\log_{10}(\rho c) + 17.8 \quad \text{(dB)}
$$

(8.43)

In the vicinity of the panel resonance frequency, Fahy and Gardonio (2007) state that provided the loss factor, $\eta \gg \rho c/2\pi f m$, the following expression may be used to calculate the *TL* of the panel:

$$
TL = 20\log_{10}f_{1,1} + 20\log_{10}m + 20\log_{10}\eta - 20\log_{10}(\rho c/\pi) \quad \text{(dB)}
$$

(8.44)

where $f_{1,1}$ is defined by Equation (8.22) with $i = n = 1$. Equation (8.44) can be used to estimate the panel TL over the frequency range from $0.5\,f_{1,1}$ to $1.5\,f_{1,1}$. If the loss factor, $\eta \ll \rho c/2\pi f m$, then the *TL* in this frequency range is set equal to 0.

8.2.4.2 Davy's Prediction Scheme for Isotropic Panels

A prediction scheme for the frequency range above $1.5\,f_{1,1}$, which is claimed to be more accurate and which allows variation of the limiting angle as a function of frequency to be taken into account according to Equation (8.32), has been proposed by Davy (1990).

In the frequency range below f_c:

$$
TL = 10\log_{10}\left[1 + \left(\frac{\pi f m}{\rho c}\right)^2\right] + 20\log_{10}\left[1 - (f/f_c)^2\right]
$$

$$
-10\log_{10}\left[\log_e\left(\frac{1 + a^2}{1 + a^2\cos^2\theta_L}\right)\right], \qquad f \leq 0.8 f_c
$$

(8.45)

where

$$
a = \left(\frac{\pi f m}{\rho c}\right)\left[1 - \left(\frac{f}{f_c}\right)^2\right]
$$

(8.46)

In the frequency range above f_c:

$$
TL = 10\log_{10}\left[1 + \left(\frac{\pi f m}{\rho c}\right)^2\right] + 10\log_{10}\left[\left(\frac{2\eta}{\pi}\right)\left(\frac{f}{f_c} - 1\right)\right], \qquad f \geq 1.7 f_c \quad (8.47)
$$

In the frequency range around the critical frequency:

$$TL = 10 \log_{10}\left[1 + \left(\frac{\pi f m}{\rho c}\right)^2\right] + 10 \log_{10}\left[\frac{2\eta \Delta_b}{\pi}\right], \qquad 0.95 f_c \geq f \leq 1.05 f_c \qquad (8.48)$$

where Δ_b is the ratio of the filter bandwidth to the filter centre frequency used in the measurements. For a 1/3 octave band, $\Delta_b = 0.236$ and for an octave band, $\Delta_b = 0.707$.

In the frequency range $1.05 f_c < f < 1.7 f_c$, the larger of the two values calculated using Equations (8.47) and (8.48) is used, while in the range $0.8 f_c > f > 0.95 f_c$, the larger of the two values calculated using Equations (8.45) and (8.48) is used.

Note that Equation (8.47) is the same as Equation (8.40) except for the "-1" in the argument of Equation (8.47). Also, Equation (8.48) is the same as (8.40) (with $f = f_c$) except for the Δ_b term in Equation (8.48).

It seems that Equation (8.48) agrees better with experiment when values for the panel loss factor, η, towards the high end of the expected range are used, whereas Equation (8.40) is in better agreement when small values of η are used. It is often difficult to decide which equation is more nearly correct because of the difficulty in determining a correct value for η. Ranges for η for some materials are given in Appendix B. The Davy method generally is more accurate at low frequencies while the Sharp method gives better results around the critical frequency of the panel.

8.2.4.3 Thickness Correction for Isotropic Panels

When the thickness of the panel exceeds about 1/6 of the bending wavelength, a correction is needed for the high frequency transmission loss. (Lgunggren, 1991). This is in the form of the maximum allowed transmission loss which the prediction result cannot exceed. This is given by:

$$TL_{max} = 20 \log_{10}\left(\frac{m}{c_L h}\right) + 10 \log_{10}\eta - 17 \quad \text{dB} \qquad (8.49)$$

and is implemented in the frequency range defined by:

$$f > \sqrt{\frac{B}{h^4 m}} \qquad (8.50)$$

8.2.4.4 Orthotropic Panels

Below half the first critical frequency, the Transmission Loss may be calculated using Equation (8.37). In the frequency range between the lowest critical frequency and half the highest critical frequency, the following relationship gives reasonably good agreement with experiment:

$$\tau_F = \frac{\rho c}{2\pi^2 fm} \frac{f_{c_1}}{f} \left(\log_e \frac{4f}{f_{c_1}} \right)^2 \tag{8.51}$$

This equation is an approximation to Equation (8.31) in which Equation (8.28) is substituted with $\eta = 0$ and has been derived by Heckl (1960). Equation (8.51) can be rewritten in terms of transmission loss using Equation (8.13) (with $\rho c = 414$) as follows:

$$TL = 20\log_{10} f + 10\log_{10} m - 10\log_{10} f_{c_1}$$
$$- 20\log_{10}\left[\log_e\left(\frac{4f}{f_{c_1}} \right) \right] - 13.2 \quad \text{(dB)} \quad f_{c_1} \le f < 0.5 f_{c_2} \tag{8.52}$$

Above $2f_{c_2}$, the *TL* is given by (Heckl, 1960):

$$TL = 20\log_{10} f + 10\log_{10} m - 5\log_{10} f_{c_1} - 5\log_{10} f_{c_2} - 23 \quad \text{(dB)} \tag{8.53}$$

Between $0.5f_{c2}$ and $2f_{c2}$, the *TL* is estimated by connecting the points $0.5f_{c2}$ and $2f_{c2}$, with a straight line on a graph of *TL* versus $\log_{10}$ (frequency). Between $f_{c1}/2$ and f_{c1}, the *TL* is also found in the same way. Note that although Equations (8.51) to (8.53) do not include the limiting angle as was done for isotropic panels, they provide reasonably accurate results and are satisfactory for most commonly used orthotropic building panels.

Nevertheless, there are two important points worth noting when using the above prediction schemes for orthotropic panels.

1. Particularly for small panels, the transmission loss below about $0.7f_{c1}$ is underestimated, the error becoming larger as the frequency becomes lower or the panel becomes smaller.
2. For common corrugated panels, there is nearly always a frequency between 2000 and 4000 Hz where there is a dip of up to 5 dB in the measured transmission loss curve, which is not predicted by theory. This corresponds either to an air resonance between the corrugations or one or more mechanical resonances of the panel. Work reported by Windle and Lam (1993) indicates that the air resonance phenomenon does not affect the *TL* of the panel and that the dips in the measured *TL* curve correspond to a few resonances in the panel which seem to be more easily excited than others by the incident sound field.

The transmission loss for a single orthotropic panel may be calculated using Figure 8.8(b). If the panel is heavily damped, then the transmission loss will be slightly greater (by about 1 to 4 dB) at higher frequencies, beginning with 1 dB at 500 Hz and increasing to 4 dB at 4000 Hz for a typical corrugated building panel.

8.2.5 Sandwich Panels

In the aerospace industry, sandwich panels are becoming more commonly used due to their high stiffness and light weight. Thus, it is of great interest to estimate the transmission loss of such structures. These structures consist of a core of paper honeycomb, aluminium honeycomb or foam. The core is sandwiched between two thin sheets of material commonly called the "laminate", which is usually aluminium on both sides or aluminium on one side and paper on the other. One interesting characteristic of these panels is that in the mid-frequency range it is common for the transmission loss of the aluminium laminate by itself to be greater than the honeycomb structure (Nilsson, 2001). Panels with thicker cores perform better than thinner panels at high frequencies but more poorly in the mid-frequency range. The bending stiffness of the panels is strongly frequency dependent. However, once a model enabling calculation of the stiffness as a function of frequency has been developed, the methods outlined in the preceding section may be used to calculate the transmission loss (Nilsson, 2001). Loss factors, η, for these panels when freely suspended are frequency dependent and are usually in the range 0.01 to 0.03. However, when included in a construction such as a ship's deck, the loss factors are much higher as a result of connection and support conditions and can range from 0.15 at low frequencies to 0.02 at high frequencies (Nilsson, 2001).

8.2.6 Double Wall Transmission Loss

When a high transmission loss structure is required, a double wall or triple wall is less heavy and more cost-effective than a single wall. Design procedures have been developed for both types of wall. However, the present discussion will be focussed mainly on double wall constructions. For a more thorough discussion of transmission loss, consideration of triple wall constructions and for some experimental data for wood stud walls, the reader is referred to the published literature (Sharp, 1973, 1978; Brekke, 1981; Davy, 1990, 1991; Bradley and Birta, 2001).

For best results, the two panels of the double wall construction must be both mechanically and acoustically isolated from one another as much as possible. Mechanical isolation may be accomplished by mounting the panels on separate staggered studs or by resiliently mounting the panels on common studs. Acoustic isolation is generally accomplished by providing as wide a gap between the panels as possible and by filling the gap with a sound-absorbing material, while ensuring that the material does not form a mechanical bridge between the panels. For best results, the panels should be isotropic.

8.2.6.1 Sharp Model for Double Wall TL

In the previous section it was shown that the transmission loss of a single isotropic panel is determined by two frequencies, namely the lowest order panel resonance $f_{1,1}$ and the coincidence frequency, f_c. The double wall construction introduces three new important frequencies. The first is the lowest order acoustic resonance, the second is the lowest order structural resonance, and the third is a limiting frequency related to the gap between the panels. The lowest order acoustic resonance, f_2 replaces the lowest order panel resonance of the single panel construction (below which the following procedure cannot be used) and may be calculated using the following equation:

$$f_2 = c/2L \qquad (8.54)$$

where c is the speed of sound in air and L is the longest cavity dimension.

The lowest order structural resonance may be approximated by assuming that the two panels are limp masses connected by a massless compliance, which is provided by the air in the gap between the panels.

Introducing the empirical constant 1.8, the following expression (Fahy, 1985) is obtained for the mass-air-mass resonance frequency, f_0, for panels that are large compared to the width of the gap between them:

$$f_0 = \frac{1}{2\pi}\left(\frac{1.8\rho c^2(m_1 + m_2)}{d m_1 m_2}\right)^{1/2} \quad \text{(Hz)} \qquad (8.55)$$

In Equation (8.55) m_1 and m_2 are, respectively, the surface densities (kg/m^2) of the two panels and d is the gap width (m). The empirical constant, "1.8" has been introduced by Sharp (1973) to account for the "effective mass" of the panels being less than their actual mass.

Finally, a limiting frequency f_ℓ, which is related to the gap width d (m) between the panels, is defined as follows:

$$f_\ell = c/2\pi d \approx 55/d \quad \text{(Hz)} \qquad (8.56)$$

The frequencies f_2, f_0 and f_ℓ, given by Equations (8.54)–(8.56) for the two-panel assembly, are important in determining the transmission behaviour of the double wall. Note that f_ℓ is equal to the lowest cavity resonance frequency, for wave propagation in the cavity normal to the plane of the panels, divided by π. The frequencies f_{c1} and f_{c2} calculated using Equation (8.3) for each panel are also important.

For double wall constructions, with the two panels completely isolated from one another both mechanically and acoustically, the expected transmission loss is given by the following equations (Sharp, 1978), where $k = 2\pi f/c$:

$$TL = \begin{cases} TL_M & f \le f_0 \\ TL_1 + TL_2 + 20\log_{10}(2kd) & f_0 < f < f_\ell \\ TL_1 + TL_2 + 6 & f \ge f_\ell \end{cases} \qquad (8.57)$$

In Equation (8.58), the quantities TL_1, TL_2 and TL_M are calculated by replacing m in Equations (8.39) and (8.40) with the values for the respective panel surface densities m_1 and m_2 and the total surface density, $M = m_1 + m_2$ respectively.

Equation (8.57) is formulated on the assumption that standing waves in the air gap between the panels are prevented, so that airborne coupling is negligible. To ensure such decoupling, the gap is usually filled with a sound-absorbing material. The density of material ought to be chosen high enough that the total flow resistance through it is of the order of $3\rho c$ or greater (see Appendix C). When installing a porous material, care should be taken that it does not form a mechanical coupling between the panels of the double wall; thus an upper bound on total flow resistance of $5\rho c$ is suggested or, alternatively, the material can be attached to just one wall without any contact with the other wall. Generally, the sound-absorbing material should be as thick as possible, with a minimum thickness of $15/f$ (m), where f is the lowest frequency of interest.

The transmission loss predicted by Equation (8.57) is difficult to realise in practice. The effect of connecting the panels to supporting studs at points (using spacers), or along lines, is to provide a mechanical bridge for the transmission of structure-borne sound from one panel to the other. Above a certain frequency, called the bridging frequency, such structure-borne conduction limits the transmission loss that can be achieved, to much less than that given by Equation (8.57). Above the bridging frequency, which lies above the structural resonance frequency, f_0, given by Equation (8.55), and below the limiting frequency, f_ℓ, given by Equation (8.56), the transmission loss increases at the rate of 6 dB per octave increase in frequency.

As the nature of the attachment of a panel to its supporting studs determines the efficiency of conduction of structure-borne sound from the panel to the stud and vice versa, it is necessary to distinguish between two possible means of attachment and, in the double panel wall under consideration, four possible combinations of such attachment. A panel attached directly to a supporting stud generally will make contact along the length of the stud. Such support is called line support and the spacing between studs, b, is assumed regular. Alternatively, the support of a panel on small spacers mounted on the studs is called point support; the spacing, e, between point supports is assumed to form a regular square grid. The dimensions b and e are important in determining transmission loss.

In the following discussion it is assumed that the two panels are numbered, so that the critical frequency of panel 1 is always less than or at most equal to the critical frequency of panel 2. With this understanding, four combinations of panel attachment are possible as follows: line–line, line–point, point–line and point–point. Of these four possible combinations of panel support, point–line will be excluded from further consideration, as the transmission loss associated with it is always inferior to that obtained with line–point support. In other words, for best results the panel with the higher critical frequency should be point supported if point support of one of the panels is considered.

In the frequency range above the bridging frequency and below about one-half of the critical frequency of panel 2 (the higher critical frequency), the expected transmission loss for the three cases (for adequate sound absorbing material in the cavity) is as follows (see Figure 8.9).

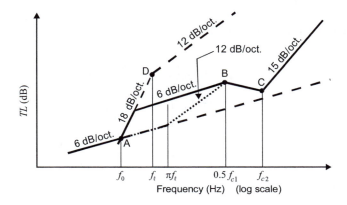

Figure 8.9 A design chart for estimating the transmission loss of a double panel wall, based on Sharpe's analysis (Sharp, 1973). In the following, the panels are assumed to be numbered, so that the critical frequency, f_{c1}, of panel 1 is always less than or equal to the critical frequency, f_{c2}, of panel 2, i.e., $f_{c1} \le f_{c2}$; m_1 and m_2 (kg m^{-2}) are the respective panel surface densities, and d (m) is the spacing between panels. b (m) is the spacing between line supports, while e (m) is the spacing of an assumed rectangular grid between point supports. c and c_L (m/s) are, respectively, the speed of sound in air and in the panel material, and h is the panel thickness. η_1 and η_2 are the loss factors respectively for panels 1 and 2. Calculate the points in the chart as follows:

Point A:

$$f_0 = 80\sqrt{(m_1 + m_2)/dm_1 m_2} \quad \text{(Hz)}; \qquad TL_A = 20 \log_{10}(m_1 + m_2) + 20 \log_{10} f_0 - 48 \qquad \text{(dB)}$$

Point B: $\qquad\qquad f_{c1} = 0.55 c^2/c_{L1} h_1 \quad$ (Hz)

The transmission loss, TL_B, at point B is equal to TL_{B1} if no sound absorptive material is placed in the cavity between the two panels; otherwise TL_B is equal to TL_{B2}, provided sufficient absorption is achieved:

$$TL_{B1} = TL_A + 20 \log_{10}(f_{c1}/f_0) + 20 \log_{10}(f_{c1}/f_\ell) - 22 \qquad \text{(dB)}$$

(a) Line–line support:

$$TL_{B2} = 20 \log_{10} m_1 + 10 \log_{10} b + 20 \log_{10} f_{c1} + 10 \log_{10} f_{c2} + 20 \log_{10}\left[1 + \frac{m_2 f_{c1}^{1/2}}{m_1 f_{c2}^{1/2}}\right] - 78 \quad \text{(dB)}$$

(b) Line–point support (f_{c2} is the critical frequency of the point supported panel):

$$TL_{B2} = 20 \log_{10} m_1 e + 20 \log_{10} f_{c1} + 20 \log_{10} f_{c2} - 99 \qquad \text{(dB)}$$

(c) Point–point support:

$$TL_{B2} = 20 \log_{10} m_1 e + 20 \log_{10} f_{c1} + 20 \log_{10}(f_{c2}) + 20 \log_{10}\left(1 + \frac{m_2 f_{c1}}{m_1 f_{c2}}\right) - 105 \qquad \text{(dB)}$$

Point C:

(a) $\quad f_{c2} \ne f_{c1}$, $\qquad\qquad TL_C = TL_B + 6 + 10 \log_{10} \eta_2 + 20 \log_{10} \dfrac{f_{c2}}{f_{c1}} \qquad$ (dB)

(b) $\quad f_{c2} = f_{c1}$, $\qquad\qquad TL_C = TL_B + 6 + 10 \log_{10} \eta_2 + 5 \log_{10} \eta_1 \qquad$ (dB)

Point D: $\qquad\qquad\qquad\qquad f_\ell = 55/d \quad$ (Hz)

The final TL curve for sound absorbing material in the cavity is the solid line in the figure. The dotted line deviation between f_0 and πf_ℓ is for no sound absorbing material in the cavity.

For line–line support (Sharp, 1973):

$$TL = 20 \log_{10} m_1 + 10 \log_{10}(f_{c2} b) + 20 \log_{10} f$$

$$+ 20 \log_{10} \left(1 + \frac{m_2 f_{c1}^{1/2}}{m_1 f_{c2}^{1/2}} \right) - 72 \quad (\text{dB}) \tag{8.58}$$

For point–point support:

$$TL = 20 \log_{10} m_1 + 20 \log_{10}(f_{c2} e) + 20 \log_{10} f$$

$$+ 20 \log_{10} \left(1 + \frac{m_2 f_{c1}}{m_1 f_{c2}} \right) - 99 \quad (\text{dB}) \tag{8.59}$$

For line–point support:

$$TL = 20 \log_{10} m_1 + 20 \log_{10}(f_{c2} e) + 20 \log_{10} f$$

$$+ 10 \log_{10}[1 + 2X + X^2] - 93 \quad (\text{dB}) \tag{8.60}$$

$$\text{where,} \quad X = \frac{77.7 m_2}{m_1 e \sqrt{f_{c1} f_{c2}}}$$

Based upon limited experimental data, Equation (8.58) seems to give very good comparison between prediction and measurement, whereas Equation (8.59) seems to give fair comparison. For line–point support the term X is generally quite small, so that the term in Equation (8.60) involving it may generally be neglected. Based upon limited experimental data, Equation (8.59) seems to predict greater transmission loss than observed. The observed transmission loss for point-point support seems to be about 2 dB greater than that predicted for line-point support.

If there is no absorption in the cavity, limited experimental data (Sharp, 1973) indicates that the double wall behaves as a single panel of mass equal to the sum of the masses of the individual panels up to a frequency of the first cavity resonance of πf_ℓ. Above this frequency, the TL increases at 12 dB/octave until it reaches $0.5 f_{c1}$.

A method for estimating transmission loss for a double panel wall is outlined in Figure 8.9. In the figure consideration has not been given explicitly to the lowest order acoustic resonance, f_2, of Equation (8.54). At this frequency it can be expected that somewhat less than the predicted mass-law transmission loss will be observed, dependent upon the cavity damping that has been provided. In addition, below the lowest order acoustic resonance, the transmission loss will again increase, as shown by the stiffness controlled portion of the curve in Figure 8.6. The procedure outlined in Figure 8.9 explicitly assumes that the inequality, $Mf > 2\rho c$, is satisfied. Two curves are shown; the solid curve corresponds to the assumption of sufficient acoustic absorbing material between the panels to suppress the acoustic resonances in the cavity and prevent acoustic coupling between the panels; the dotted (not dashed) line corresponds to no acoustic absorbing material in the cavity and in Figure 8.9, it is only different to the solid curve in the frequency range between f_0 and $0.5 f_{c1}$. Of course the TL at point B is different for the two cases but the curves for the two cases are constructed in the same way except for the frequency range between f_0 and $0.5 f_{c1}$. In some cases such as double glazed window constructions, it is only possible to put sound absorbing material in the cavity around the perimeter of the construction. Provided this material is at least 50 mm thick and it is fibreglass or rockwool of sufficient density, it will have almost as good an effect as if the material were placed in the cavity between the two panels. However, in these cases, the TL in the frequency range between f_0 and πf_ℓ will be slightly less than predicted.

8.2.6.2 Davy Model for Double Wall TL

The equations outlined in the previous section for a double wall are based on the assumption that the studs connecting the two leaves of the construction are infinitely stiff. This is an acceptable assumption if wooden studs are used but not if metal studs (typically thin-walled channel sections with the partition leaves attached to the two opposite flanges) are used (see Davy, 1990).

Davy (1990, 1991, 1993, 1998) presented a method for estimating the transmission loss of a double wall that takes into account the compliance, C_M (reciprocal of the stiffness) of the studs. Although this prediction procedure is more complicated than the one just discussed, it is worthwhile presenting the results here.

Below the mass-air-mass resonance frequency, f_0, the double wall behaves like a single wall of the same mass and the single wall procedures may be used to estimate the TL. Above f_0, the transmission from one leaf to the other consists of airborne energy through the cavity and structure-borne energy through the studs. The structure-borne sound transmission coefficient for all frequencies above f_0 is (Davy, 1993):

$$\tau_{F_c} = \frac{64\rho^2 c^3 D}{\left[g^2 + \left(4(2\pi f)^{3/2} m_1 m_2 c C_M - g \right)^2 \right] b (2\pi f)^2} \tag{8.61}$$

where

$$g = m_1 (2\pi f_{c_2})^{1/2} + m_2 (2\pi f_{c_1})^{1/2} \tag{8.62}$$

b is the spacing between the studs and $f_{c1} \le f_{c2}$.

For commonly used steel studs, the compliance (which is the reciprocal of the stiffness per unit length), $C_M = 10^{-6} \mathrm{m^2 N^{-1}}$ (Davy, 1990) and for wooden studs, $C_M = 0$.

However, Davy (1998) recommends that for steel studs, the compliance, C_M, is set equal to 0 as it is for wooden studs, and the transmission coefficient for structure-borne sound, τ_{Fc}, is decreased by a factor of 10 over that calculated using Equation (8.61) with $C_M = 0$. The quantity, D, is defined for line support on panel 2 as follows:

$$D = \begin{cases} \dfrac{2}{h} & \text{if } f < 0.9 \times f_{cl} \\[2em] \dfrac{\pi f_{cl}}{8 f \eta_1 \eta_2} \sqrt{\dfrac{f_{c2}}{f}} & \text{if } f \geq 0.9 \times f_{cl} \end{cases} \tag{8.63}$$

$$h = \left[1 - \left(\frac{f}{f_{cl}} \right)^2 \right]^2 \left[1 - \left(\frac{f}{f_{c2}} \right)^2 \right]^2 \tag{8.64}$$

where f_{cl} is the lower of the two critical frequencies corresponding to the two panels and η_1 and η_2 are the loss factors for panels 1 and 2 respectively.

The field incidence transmission coefficient for airborne sound transmission through a double panel (each leaf of area A), for frequencies between f_0 and $0.9f_{cl}$ (where f_{cl} is the lower of the two critical frequencies corresponding to the two panels), is:

$$\tau_{F_a} = \frac{1 - \cos^2 \theta_L}{\left[\dfrac{m_2^2 + m_1^2}{2m_1 m_2} + a_1 a_2 \bar{\alpha} \cos^2 \theta_L \right] \left[\dfrac{m_2^2 + m_1^2}{2m_1 m_2} + a_1 a_2 \bar{\alpha} \right]} \tag{8.65}$$

where

$$a_i = \left[\frac{\pi f m_i}{\rho c} \right] \left[1 - \left(\frac{f}{f_{c_i}} \right)^2 \right] \tag{8.66}$$

and the limiting angle, θ_L, is defined in Equation (8.32). Davy (1998) states that the limiting angle should not exceed $80°$.

In the above equations f_{ci} is the critical frequency of panel i ($i = 1, 2$), m_1, m_2 are the surface densities of panels 1 and 2 and $\bar{\alpha}$ is the cavity absorption coefficient, generally taken as 1.0 for a cavity filled with sound absorbing material, such as fibreglass or rockwool at least 50mm thick. At low frequencies, the maximum cavity absorption coefficient used in the above equation should not exceed kd, where d is the cavity width. For cavities containing no sound absorbing material, a value between 0.1 and 0.15 may be used for $\bar{\alpha}$ (Davy, 1998), but again it should not exceed kd.

At frequencies above $0.9f_{c1}$, the following equations may be used to estimate the field incidence transmission coefficient for airborne sound transmission:

$$\tau_{F_a} = \frac{\pi(\xi_1 + \xi_2)q_1}{4\,\overline{a}_1^2\,\overline{a}_2^2\,\eta_1\,\eta_2\,\xi_1\,\xi_2(q_1^2 + q_2^2)\overline{\alpha}^2} \tag{8.67}$$

$$\overline{a}_i = \frac{\pi f m_i}{\rho c}; \qquad i = 1, 2 \tag{8.68}$$

$$\xi_i = \left(\frac{f}{f_{c_i}}\right)^{\frac{1}{2}} \qquad i = 1, 2 \tag{8.69}$$

$$q_1 = \eta_1\xi_2 + \eta_2\xi_1 \tag{8.70}$$

$$q_2 = 4(\eta_1 - \eta_2) \tag{8.71}$$

The quantities η_1 and η_2 are the loss factors of the two panels and f is the one-third octave band centre frequency.

The overall transmission coefficient is:

$$\tau_F = \tau_{F_a} + \tau_{F_c} \qquad f > f_0 \tag{8.72}$$

The value of τ_F from Equation (8.72) is then used in Equation (8.13) to calculate the transmission loss (*TL*). The quantity, f_0, is defined by Equation (8.55).

Note that for frequencies between $2f_0/3$ and f_0, linear interpolation between the single panel TL result (for a wall of the same total mass) at $2f_0/3$ and the double panel result at f_0 should be used on a graph of TL vs log frequency.

Example 8.1

A double gypsum board wall is mounted at the perimeter in an opening of dimensions 3.0×2.44 m in a test facility. The spacing between the panels is 0.1 m. The surface densities and critical frequencies of each panel are, respectively, 12.16 kg/m² and 2500 Hz.

Calculate the expected transmission loss using Sharpe's theory. The space between the walls is well damped with a 50 mm thick layer of sound-absorbing material. However, the panels themselves have not been treated with damping material.

Solution

Reference is made to Figure 8.9. Calculate the coordinates of point A:

$$f_0 = 80.4\sqrt{2 \times 12.16/0.1 \times 12.16^2} = 103 \text{ Hz}$$

$$TL_A = 20\log_{10}(2 \times 12.16) + 20\log_{10}103 - 48 = 20 \text{ dB}$$

Calculate the coordinates of point B. Since the panel is supported at the edge, the area associated with each support is less than half of that assumed in the theory; and for this reason we empirically add 4 dB to the calculated transmission loss at point B. As thee is sound absorption in the cavity, $TL_B = TL_{B2}$

$$TL_{B2} = 20\log_{10}12.16 + 10\log_{10}2.44 + 30\log_{10}2500$$

$$+ 6 - 78 + 4 = 60 \text{ dB}; \quad \text{thus} \quad TL_B = 60\text{dB}$$

Calculate the coordinates of point C. In the absence of better information assume a loss factor $\eta = 0.1$ for each panel:

$$TL_c = 60 + 6 - 10 - 5 = 51 \text{ dB}$$

Construct the estimated transmission loss curve shown in the following figure (for comparison, experimentally determined points and the Davy method are also shown).

An important point regarding stud walls with gypsum board leaves is that a stud spacing of between 300 and 400 mm has been shown (Rindel and Hoffmeyer, 1991) to severely degrade the performance of the double wall in the 160 and 200 Hz one-third octave bands by up to 13 dB. Other stud spacings (even 100 and 200 mm) do not result in the same performance degradation, although smaller stud spacings improve the low frequency performance (below 200 Hz) at the expense of a few dB loss at all frequencies between 250 and 2000 Hz.

It is important not to use walls of the same thickness (and material) as this greatly accentuates the dip in the TL curve at the critical frequency. This is also important for double glazing constructions. As an aside, one problem with double glazing is that it can suffer from condensation, so if used, drainage holes are essential.

8.2.6.3 Staggered Studs

A staggered stud arrangement is commonly used to obtain high transmission loss. In this arrangement, studs of a common wall are alternately displaced. Panels fastened to them on opposite sides are then supported on alternate studs. The only common support between opposite panels is at the perimeter of the common wall, for example at the base and top.

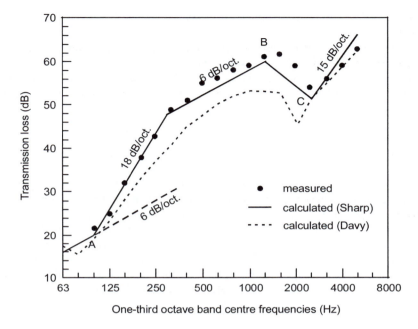

Example Figure 8.1

For the purpose of calculating expected transmission loss, the staggered stud construction could be modelled as a perimeter-supported double wall similar to that of Example 8.1 just considered. However, the introduction of studs improves the structure-borne coupling and degrades the transmission loss which is obtained. However, if care is taken to ensure that at least one of the panels is very well damped, even higher transmission loss may be obtained with staggered studs than shown by Example 8.1. Thus, if at least one panel is not very well damped, the expected transmission loss for a double wall on staggered studs will lie between that of perimeter mounting and line–line support given by Equation (8.58). Alternatively, if at least one panel is very well damped then the double wall may be modelled as perimeter supported and slightly higher transmission loss can be expected than predicted.

8.2.6.4 Panel Damping

A simple means for achieving the very high panel damping alluded to above is to construct a thick panel of two thin panels glued together at spots in a regular widely spaced grid. Slight motion of the panels results in shear between the panels in the spaces between attachments, resulting in very effective panel damping due to the shearing action, which dissipates energy in the form of heat. This mechanism can be considered to approximately double the loss factor of the base panels. Alternatively, the panels could be connected together with a layer of visco-elastic material to give a loss factor of about 0.2.

When glass is the material used for the wall or for a window, damping can be increased by using laminated glass, which is a sandwich of two layers of glass separated by a plastic sheet. Sound absorbing material may also be placed around the perimeter of the cavity between two glass walls to increase acoustic absorption without affecting the transparency of the glass.

8.2.6.5 Effect of the Flow Resistance of the Sound Absorbing Material in the Cavity

As the Sharp theory discussed in the preceding sections does not specify any properties of the sound absorbing material that is placed in the cavity of a double wall, experimental work has shown that the type of material used is important. Ideally the material should have a value of $R_1\ell/\rho c$ of between 2 and 5, where R_1 is the flow resistivity of the material, ℓ is the material thickness and ρ and c are, respectively, the density of air and the speed of sound in air. This should rule out the use of low density fibreglass (such as insulation batts used in house ceilings), as well as typical polyester blankets. In fact polyester blankets are likely to be completely ineffective.

8.2.7 Multi-leaf and Composite Panels

A multi-leaf panel, for the purposes of the following text, is a panel made up of two or more leaves of the same material, which are connected together in one of three ways: rigid, which is essentially glued very firmly; flexible, which is glued or nailed together at widely separated spots (0.3 to 0.6m apart); and visco-elastic, which is connected together with visco-elastic material such as silicone rubber (such as silastic). For the latter two constructions, the flexibility in the connections between the panels means that they essentially act separately in terms of bending waves propagating through them. Thus it makes sense to use the lowest critical frequency of the individual leaves for any TL calculations (as the thickness of each leaf does not have to be the same as any other). It is understood that this is an approximation only as one might expect differences in measured TL depending on where the thickest leaf is located amongst the various leaves. Thus, the TL for single and double walls is calculated following the procedures outlined previously, with the critical frequency calculated using the thickness and mass of the thickest leaf and the TL then calculated using this critical frequency together with the total mass per unit area of the entire panel including all the leaves that make it up. The loss factor used in the calculations is that described in Section 8.2.6.4.

When the leaves are connected rigidly together with glue covering the entirety of each leaf, the panel may be considered to act as a single leaf panel of thickness equal to the total thickness of all the leaves and mass equal to the total mass of all the leaves.

A composite panel for the purposes of this text is defined as a panel made up of two layers of different material, which are bonded rigidly together. The effective stiffness of the panel is calculated using Equation (8.7) and the critical frequency is calculated using Equation (8.9). Then the TL of the single or double panel may be calculated by following the procedures in the previous sections.

It is possible to have multi-leaf panel made up of composite leaves, where each leaf consists of two layers, each of a different material, bonded rigidly together. In this case, the effective bending stiffness and mass of each leaf are calculated first and the construction is then treated as a multi-leaf construction described above, except that each leaf will be a composite of two rigidly bonded layers.

8.2.8 Triple Wall Sound Transmission Loss

Very little work has been done in this area, but recent work reported by Tadeu and Mateus (2001) indicates that for double and triple glazed windows with the same total weight of glazing and total air gap, nothing is gained in using triple glazing over double glazing. However, this is because the cut-on frequency above which 3-D reflections occur in the cavity is above the frequency range of interest for typical panel separations used in windows (30 to 50mm). The cut-on frequency is given by the following equation:

$$f_{co} = c/2d \tag{8.73}$$

Note that the poorest performance is achieved with panes of glass separated by 10 to 30 mm (Tadeu and Mateus, 2001). Above the cut-off frequency, it is possible to achieve a marked improvement with a triple panel wall (Brekke, 1981).

Sharp (1973) reported that for constructions of the same total mass and total thickness, the double wall construction has better performance for frequencies below $4f_0$, whereas the triple wall construction performs better at frequencies above $4f_0$, where f_0 is the double panel resonance frequency defined by Equation (8.55), and using the total distance between the two outer panels as the air gap and the two outer panels as the masses m_1 and m_2. Below f_0, the two constructions will have the same transmission loss and this will be the same as for a single wall of the same total mass.

8.2.9 Common Building Materials

Results of transmission loss (field incidence) tests on conventional building materials and structures have been published both by manufacturers and testing laboratories. Some examples are listed in Table 8.2.

8.2.10 Sound-absorptive Linings

When an enclosure is to be constructed, some advantage will accrue by lining the walls with a porous material. The lining will prevent reverberant sound build-up, which would lessen the effectiveness of the enclosure for noise reduction, and at high frequencies it will increase the transmission loss of the walls. The transmission loss

Table 8.2 Representative values of airborne sound transmission loss for some common structures and materials

Panel construction	Thickness (mm)	Superficial weight (kg/m²)	Octave band centre frequency (Hz)							
			63	125	250	500	1000	2000	4000	8000
Panels of sheet materials										
1.5 mm lead sheet	1.5	17	22	28	32	33	32	32	33	36
3 mm lead sheet	3	34	24	30	31	27	38	44	33	38
20 g aluminum sheet, stiffened	0.9	2.5	8	11	10	10	18	23	25	30
6 mm steel plate	6	50	–	27	35	41	39	39	46	–
22 g galvanized steel sheet	0.55	6	3	8	14	20	23	26	27	35
20 g galvanized steel sheet	0.9	7	3	8	14	20	26	32	38	45
18 g galvanized steel sheet	1.2	10	8	13	20	24	29	33	39	44
16 g galvanized steel sheet	1.6	13	9	14	21	27	32	37	43	42
18 g fluted steel panels stiffened at edges, joints scaled	1.2	39	25	30	20	22	30	28	31	31
Corrugated asbestos sheet, stiffened and sealed	6	10	20	25	30	33	33	38	39	42
Chipboard sheets on wood framework	19	11	14	17	18	25	30	26	32	38
Fibreboard on wood framework	12	4	10	12	16	20	24	30	31	36
Plasterboard sheets on wood framework	9	7	9	15	20	24	29	32	35	38
2 layers 13 mm plaster board	26	22	–	24	29	31	32	30	35	–
Plywood sheets on wood framework	6	3.5	6	9	13	16	21	27	29	33
Plywood sheets on wood framework	12	7	–	10	15	17	19	20	26	–
Hardwood (mahogany) panels	50	25	15	19	23	25	30	37	42	46
Woodwork slabs, unplastered	25	19	0	0	2	6	6	8	8	10
Woodwork slabs, plastered (12 mm on each face)	50	75	18	23	27	30	32	36	39	43
Plywood	6	3.5	–	17	15	20	24	28	27	–
Plywood	9	5	–	7	13	19	25	19	22	–

Plywood	18	10	—	24	22	27	28	25	27	—
Lead vinyl curtains	3	7.3	—	22	23	25	31	35	42	—
Lead vinyl curtains	2	4.9	—	15	19	21	28	33	37	—

Panels of sandwich construction

Machine enclosure panels

16 g steel + damping with 100 mm of glass-fibre, covered by 22 g perforated steel	100	25	20	21	27	38	48	58	67	66
	100	31	25	27	31	41	51	60	65	66
As above, but 16 g steel replaced with 5 mm steel plate	100	50	31	34	35	44	54	63	62	68
1.5 mm lead between two sheets of 5 mm plywood	11.5	25	19	26	30	34	38	42	44	47
9 mm asbestos board between two sheets of 18 g steel	12	37	16	22	27	31	27	37	44	48
Compressed straw between two sheets of 3 mm hardboard	56	25	15	22	23	27	27	35	35	38

Single masonry walls

Single leaf brick, plastered on both sides	125	240	30	36	37	40	46	54	57	59
Single leaf brick, plastered on both sides	255	480	34	41	45	48	56	65	69	72
Single leaf brick, plastered on both sides	360	720	36	44	43	49	57	66	70	72
Solid breeze or clinker, plastered (12 mm both sides)	125	145	20	27	33	40	50	58	56	59
Solid breeze or clinker blocks, unplastered	75	85	12	17	18	20	24	30	38	41
Hollow cinder concrete blocks, painted (cement base paint)	100	75	22	30	34	40	50	50	52	53
Hollow cinder concrete blocks, unpainted	100	75	22	27	32	32	40	41	45	48
Thermalite blocks	100	125	20	27	31	39	45	53	38	62
Glass bricks	200	510	25	30	35	40	49	49	43	45
Plain brick	100	200	—	30	36	37	37	37	43	—
Aerated concrete blocks	100	50	—	34	35	30	37	45	50	—
Aerated concrete blocks	150	75	—	31	35	37	44	50	55	—

Table 8.2 (Continued)

Panel construction	Thickness (mm)	Superficial weight (kg/m²)	Octave band centre frequency (Hz)							
			63	125	250	500	1000	2000	4000	8000
Double masonry walls										
280 mm brick, 56 mm cavity, strip ties, outer faces plastered to thickness of 12 mm	300	380	28	34	34	40	56	73	76	78
280 mm brick, 56 mm cavity, expanded metal ties, outer faces plastered to thickness of 12 mm	300	380	27	27	43	55	66	77	85	85
Stud partitions										
50 mm × 100 mm studs, 12 mm insulating board both sides	125	19	12	16	22	28	38	50	52	55
50 mm × 100 mm studs, 9 mm plasterboard and 12 mm plaster coat both sides	142	60	20	25	28	34	47	39	50	56
Gypsum wall with steel studs and 16 mm-thick panels each side										
Empty cavity, 45 mm wide	75	26	–	20	28	36	41	40	47	–
Cavity, 45 mm wide, filled with fibreglass	75	30	–	27	39	46	43	47	52	–
Empty cavity, 86 mm wide	117	26	–	19	30	39	44	40	43	–
Cavity, 86 mm wide, filled with fibreglass	117	30	–	28	41	48	49	47	52	–
gypsum wall, 16 mm leaves, 200 mm cavity with no sound absorbing material and no studs	240	23	–	33	39	50	64	51	59	–
As above with 88 mm sound absorbing material	240	26	–	42	56	68	74	70	73	–
As above but staggered 4-inch studs	240	30	–	35	50	55	62	62	68	–

Gypsum wall, 16 mm leaves, 100 mm cavity, 56 mm thick sound absorbing material, single 4-inch studs with resilient metal channels on one side to attach the panel to the studs	140	28	–	25	40	48	52	47	52	–
Single glazed windows										
Single glass in heavy frame	4	10	–	20	22	28	34	29	28	–
Single glass in heavy frame	6	15	17	11	24	28	32	27	35	39
Single glass in heavy frame	8	20	18	18	25	31	32	28	36	39
Single glass in heavy frame	9	22.5	18	22	26	31	30	32	39	43
Single glass in heavy frame	16	40	20	25	28	33	30	38	45	48
Single glass in heavy frame	25	62.5	25	27	31	30	33	43	48	53
Laminated glass	13	32	–	23	31	38	40	47	52	57
Doubled glazed windows										
2.44 mm panes, 7 mm cavity	12	15	15	22	16	20	29	31	27	30
9 mm panes in separate frames, 50 mm cavity	62	34	18	25	29	34	41	45	53	50
6 mm glass panes in separate frames, 100 mm cavity	112	34	20	28	30	38	45	45	53	50
6 mm glass panes in separate frames, 188 mm cavity	200	34	25	30	35	41	48	50	56	56
6 mm glass panes in separate frames, 188 mm cavity with absorbent blanket in reveals	200	34	26	33	39	42	48	50	57	60
6 mm and 9 mm panes in separate frames, 200 mm cavity, absorbent blanket in reveals	215	42	27	36	45	58	59	55	66	70
3 mm plate glass, 55 mm cavity	63	25	–	13	25	35	44	49	43	–
6 mm plate glass, 55 mm cavity	70	35	–	27	32	36	43	38	51	–

Table 8.2 (Continued)

Panel construction	Thickness (mm)	Superficial weight (kg/m²)	Octave band centre frequency (Hz)							
			63	125	250	500	1000	2000	4000	8000
6 mm and 5 mm glass, 100 mm cavity	112	34	–	27	37	45	56	56	60	–
6 mm and 8 mm glass, 100 mm cavity	115	40	–	35	47	53	55	50	55	–
Doors										
Flush panel, hollow core, normal cracks as usually hung	43	9	1	12	13	14	16	18	24	26
Solid hardwood, normal cracks as usually hung	43	28	13	17	21	26	29	31	34	32
Typical proprietary "acoustic" door, double heavy sheet steel skin, absorbent in air space, and seals in heavy steel frame	100	–	37	36	39	44	49	54	57	60
2-skin metal door	35	16	–	26	26	28	32	32	40	–
Plastic laminated flush wood door	44	20	–	14	18	17	23	18	19	–
Veneered surface, flush wood door	44	25	–	22	26	29	26	26	32	–
Metal door; damped skins, absorbent core, gasketing	100	94	–	43	47	51	54	52	50	–
Metal door; damped skins, absorbent core, gasketing	180	140	–	46	51	59	62	65	62	–
Metal door; damped skins, absorbent core, gasketing	250	181	–	48	54	62	68	66	74	–
Two 16g steel doors with 25 mm sound-absorbing material on each, and separated by 180 mm air gap	270	86	–	50	56	59	67	60	70	–
Hardwood door	54	20	–	20	25	22	27	31	35	–
Hardwood door	66	44	–	24	26	33	38	41	46	–

Floors

Floors										
T & G boards, joints sealed	21	13	17	21	18	22	24	30	33	63
T & G boards, 12 mm plasterboard ceiling under, with 3 mm plaster skin coat	235	31	15	18	25	37	39	45	45	48
As above, with boards "floating" on glass-wool mat	240	35	20	25	33	38	45	56	61	64
Concrete, reinforced	100	230	32	37	36	45	52	59	62	63
Concrete, reinforced	200	460	36	42	41	50	57	60	65	70
Concrete, reinforced	300	690	37	40	45	52	59	63	67	72
126 mm reinforced concrete with "floating" screed	190	420	35	38	43	48	54	61	63	67
200 mm concrete slabs	200	280	–	34	39	46	53	59	64	65
As above, but oak surface	212	282	–	34	41	46	55	64	70	–
As above, but carpet + hair felt underlay, no of oak surface	200	281	–	34	36	46	55	66	72	–
Gypsum ceiling, mounted resiliently, and vinyl finished wood joist floor with glass-fibre insulation and 75 mm plywood	318	–	–	30	36	45	52	47	65	–

of a porous lining material is discussed in Appendix C. Calculated transmission loss values for a typical blanket of porous material are given in Table 8.3.

Table 8.3 Calculated transmission loss (TL) values (dB) for a typical blanket of porous acoustic material (medium density rockwool, 50 mm thick)

Frequency (Hz)	TL (dB)
1000	0.5
2000	1.5
4000	4
8000	12

8.3 NOISE REDUCTION vs TRANSMISSION LOSS

When a partition is placed between two rooms and one room contains a noise source which affects the other room (receiver room), the difference in sound level between the two rooms is related to the TL of the partition by Equation (8.16). When there are other paths, other than through the partition, for the sound to travel from one room to the other, the effective transmission loss of the panel in terms of the sound reduction from one room to the other will be affected. These alternative transmission paths could be through doors, windows or suspended ceilings. If the door or wall forms part of the partition, then the procedures for calculating the effective transmission loss are discussed in Section 8.3.1. If the transmission path is around the wall, then the effective transmission loss of this path needs to be calculated or measured in the laboratory according to such standards as ISO 140-10 (1991) or EN ISO 10848-1 (2006) and normalised to the area of the wall (which is done automatically if the procedures in the standards are followed). In this case, the effective transmission loss of the partition, including the flanking paths is calculated as described in Section 8.3.2.

8.3.1 Composite Transmission Loss

The wall of an enclosure may consist of several elements, each of which may be characterised by a different transmission loss coefficient. For example, the wall may be constructed of panels of different materials, it may include permanent openings for passing materials or cooling air in and out of the enclosure, and it may include windows for inspection and doors for access. Each such element must be considered in turn in the design of an enclosure wall, and the transmission loss of the wall determined as an overall area weighted average of all of the elements. For this calculation the following equation is used.

$$\tau = \frac{\displaystyle\sum_{i=1}^{q} S_i \tau_i}{\displaystyle\sum_{i=1}^{q} S_i} \tag{8.74}$$

In Equation (8.74), S_i is the surface area (one side only), and τ_i is the transmission coefficient of the ith element. The transmission coefficient of any element may be determined, given its transmission loss, TL, from the following equation:

$$\tau = 10^{(-TL/10)} \tag{8.75}$$

The overall transmission coefficient is then calculated using Equation (8.74) and, finally, the transmission loss is calculated using Equation (8.13).

If a wall or partition consists of just two elements, then Figure 8.10 is useful. The figure shows the transmission loss increment to be added to the lesser transmission loss of the two elements to obtain an estimate of the overall transmission loss of the two-element composite structure. The transmission loss increment, δTL, is plotted as a function of the ratio of the area of the lower transmission loss element divided by the area of the higher transmission loss element with the difference, ΔTL, between the transmission losses of the two elements as a parameter.

It can be seen from Figure 8.10 that low transmission loss elements within an otherwise very high transmission loss wall can seriously degrade the performance of the wall; the transmission loss of any penetrations must be commensurate with the transmission loss of the whole structure. In practice, this generally means that the transmission loss of such things as doors, windows, and access and ventilation openings, should be kept as high as possible, and their surface areas small.

A list of the transmission losses of various panels, doors and windows is included in Table 8.2. More comprehensive lists have been published in various handbooks, and manufacturers of composite panels generally supply data for their products. Where possible, manufacturer's data should be used; otherwise the methods outlined in Sections 8.2.3 and 8.2.4 or values in Table 8.2 may be used.

Example 8.2

Calculate the overall transmission loss at 125 Hz of a wall of total area 10 m² constructed from a material that has a transmission loss of 30 dB, if the wall contains a panel of area 3 m², constructed of a material having a transmission loss of 10 dB.

Solution

For the main wall, the transmission coefficient is:

$$\tau_1 = 1/[10^{(30/10)}] = 0.001$$

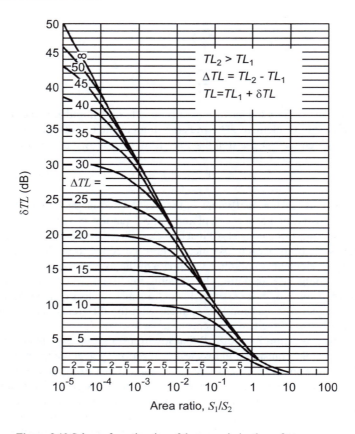

Figure 8.10 Scheme for estimation of the transmission loss of a two-element composite structure.

while for the panel:

$$\tau_2 = 1/[10^{(10/10)}] = 0.100$$

hence:

$$\tau = \frac{(0.001 \times 7) + (0.100 \times 3)}{10} = 0.0307$$

The overall transmission loss is therefore:

$$TL_{av} = 10 \log_{10}(1/0.0307) = 15 \text{ dB}$$

8.3.2 Flanking Transmission Loss

The effective Transmission Loss of a partition, including the effects of flanking transmission is given by (EN12354 - (2000)):

$$TL_{overall} = -10\log_{10}\left(10^{-TL_{flank}/10} + 10^{TL/10}\right) \quad \text{dB} \qquad (8.76)$$

where Tl_{flank} is the combined effective TL of all the flanking paths normalised to the area of the partition and TL is the transmission loss of the partition. When measuring the flanking path effects, the following quantity is reported:

$$D_{n,f} = L_1 - L_2 - 10\log_{10}\left(\frac{S\bar{\alpha}}{10}\right) \qquad (8.77)$$

where $S\bar{\alpha}$ is the absorption area of the receiving test chamber and L_1 and L_2 are the sound pressure levels in the source and receiving rooms respectively, with the receiver level due only to the flanking path or paths.

The flanking transmission loss, TL_{flank} or Flanking Sound Reduction Index, R_{flank} is calculated from the normalised sound pressure level difference quantity $D_{n,f}$ measured in the test facility as follows:

$$TL_{flank} = D_{n,f} - 10\log_{10}\left(\frac{10}{A}\right) \qquad (8.78)$$

where A is the area of the partition. If the field situation flanking condition exactly matches the laboratory situation, then the result of equation (8.78) may be used with Equation (8.76) to estimate an effective TL for use in a field installation. However, if the field situation is different to the laboratory configuration, then some adjustment needs to be made to Equation (8.78).

One example given in by EN12354 - (2000) is for a suspended ceiling where the ceiling dimensions for the installation are different to those used in the laboratory measurement. In this case, the quantity $D_{n,f}$ in Equation (8.78) is replaced by $D_{n,s}$, defined by:

$$D_{n,s} = D_{n,f} + 10\log_{10}\left(\frac{h_{pl}\ell_{ij}}{h_{lab}\ell_{lab}}\right) + 10\log_{10}\left(\frac{S_{cs,lab}S_{cr,lab}}{S_{cs}S_{cr}}\right) + C_\alpha \qquad (8.79)$$

where h_{pl} and h_{lab} are the heights of the space above the suspended ceiling in the actual installation and in the laboratory respectively, $\ell_{i,j}$ and ℓ_{lab} are the thicknesses of the partition where it connects to the suspended ceiling in the actual installation and in the laboratory respectively, S_{cs} and $S_{cs,lab}$ are the areas of the suspended ceiling in the source room in the actual installation and in the laboratory respectively, S_{cr} and $S_{cr,lab}$ are the areas of the suspended ceiling in the receiver room in the actual installation and in the laboratory respectively and C_α is defined as follows.

For no absorption in the space above the suspended ceiling, or if sound absorbing material exists and the condition, $f \le 0.015c/t_a$, (where t_a is the thickness of the absorbing material) is satisfied, then $C_\alpha = 0$. For absorption in this space, where the preceding condition is not satisfied:

$$C_\alpha = 10\log_{10}\left[\frac{h_{lab}}{h_{pl}}\sqrt{\frac{S_{cs}S_{cr}}{S_{cs,lab}S_{cr,lab}}}\right] \quad \text{for} \quad 0.015\frac{c}{t_a} < f < \frac{0.3c}{\min(h_{lab},h_{pl})} \quad (8.80)$$

$$C_\alpha = 10\log_{10}\left[\left(\frac{h_{lab}}{h_{pl}}\right)^2\sqrt{\frac{S_{cs}S_{cr}}{S_{cs,lab}S_{cr,lab}}}\right] \quad \text{for} \quad f \ge \frac{0.3c}{\min(h_{lab},h_{pl})} \quad (8.81)$$

Note that for an ISO test facility, $S_{cs,lab} = S_{cr,lab} = 20$ m² and $h_{lab} = 0.7$ m.

8.4 ENCLOSURES

8.4.1 Noise Inside Enclosures

The use of an enclosure for noise control will produce a reverberant sound field within it, in addition to the existing direct sound field of the source. Both the reverberant and direct fields will contribute to the sound radiated by the enclosure walls as well as to the sound field within the enclosure.

Equation (7.43) of Chapter 7 may be used to estimate the sound pressure level at any location within the enclosure, but with the restriction that the accuracy of the calculation will be impaired if the location considered is less than one half of a wavelength from the enclosure or machine surfaces.

8.4.2 Noise Outside Enclosures

The sound field immediately outside of an enclosure consists of two components. One component is due to the internal reverberant field and the other is due to the direct field of the source. Effectively, the assumption is that the transmission coefficients for the direct and reverberant fields are not the same but are equivalent to the normal incidence and field incidence transmission coefficients, τ_N and τ, respectively. The corresponding transmission losses TL_N and TL were discussed earlier. The expression for the total radiated sound power is:

$$W_t = S_E \langle p_1^2 \rangle / (\rho c) = W \tau_N + W(1 - \overline{\alpha}_i)[S_E/(S_i \overline{\alpha}_i)]\tau \qquad (8.82a,b)$$

In the above expression, subscript i indicates quantities interior to the enclosure. In writing Equation (8.82) the external radiated sound power, given by the integral over the external surface S_E of the sound intensity $\langle p^2 \rangle / \rho c$ (see Equation (1.74)), has been set equal to the fraction of the source sound power W, transmitted by the direct field, plus the fraction transmitted by the reverberant field. In the latter case, the power contributed by the reverberant field is determined by use of Equations (7.34) and (7.42). The quantity, S_i, is the enclosure internal surface area, including any machine surfaces. Note that $\overline{\alpha}_i$ is necessarily numerically greater than τ.

Consider the transmission of the direct sound field, and suppose that the transmission coefficient τ_N can be estimated by making use of Equation (8.36), where the angle of incidence $\theta = 0$. From Equation (8.37), $TL_N - TL = 5.5$ dB. Use of Equation (8.13) leads to the conclusion based on the above result that $\tau_N = 0.3\tau$. Substitution of the above expression in Equation (8.82) gives the following result.

$$S_E \langle p_1^2 \rangle / (\rho c) = W \tau_E \qquad (8.83)$$

where

$$\tau_E = \tau[0.3 + S_E(1 - \overline{\alpha}_i)/(S_i \overline{\alpha}_i)] \qquad (8.84)$$

Taking logarithms to the base ten of both sides of Equation (8.83) and assuming $\rho c = 400$ gives the following equation:

$$L_{p1} = L_w - TL - 10 \log_{10} S_E + C \qquad (8.85)$$

where:

$$C = 10 \log_{10}[0.3 + S_E(1 - \overline{\alpha}_i)/(S_i \overline{\alpha}_i)] \quad \text{(dB)} \qquad (8.86)$$

and TL is the field-incidence transmission loss.

The quantity L_{p1} is the average sound pressure level (dB re 20 μPa) immediately outside of the enclosure. Values of the constant C for various enclosure internal conditions are listed in Table 8.4. Alternatively, Equation (8.86) may be used. However, the calculations are very approximate and one cannot expect precision from the use of Equation (8.86); thus use of the table is recommended.

Example 8.3

A small pump has a sound power level of 80 dB re 10^{-12} W. It is to be contained in an enclosure of 2.2 m^2 surface area. Its sound power spectrum peaks in the 250 Hz octave band, and rolls off above and below at 3 dB per octave. Calculate the predicted sound pressure level in octave bands from 63 Hz to 8 kHz at the outside surface of the

Table 8.4 Values of constant C (dB) to account for enclosure internal acoustic conditions

Enclosure internal acoustic conditions[a]	Octave band centre frequency (Hz)							
	63	125	250	500	1000	2000	4000	8000
Live	18	16	15	14	12	12	12	12
Fairly live	13	12	11	12	12	12	12	12
Average	13	11	9	7	5	4	3	3
Dead	11	9	7	6	5	4	3	3

[a] Use the following criteria to determine the appropriate acoustic conditions inside the enclosure:
Live: All enclosure surfaces and machine surfaces hard and rigid
Fairly live: All surfaces generally hard but some panel construction (sheet metal or wood)
Average: Enclosure internal surfaces covered with sound-absorptive material, and machine surfaces hard and rigid
Dead: As for "Average", but machine surfaces mainly of panels.

enclosure, assuming average acoustical conditions within the enclosure and a wall transmission loss (TL), as shown in the Example 8.3 table for each of the octave bands.

Example 8.3 Table

	Octave band centre frequency (Hz)							
	63	125	250	500	1000	2000	4000	8000
Wall TL (dB)	8	11	12	15	18	23	25	30
Correction, C (dB)	13	11	9	7	5	4	3	3
Relative power spectrum (dB)	-6	-3	0	-3	-6	-9	-12	-15
L_w (dB) re 10^{-12} W)	69.6	72.6	75.6	72.6	69.6	66.6	63.6	60.6
L_p (dB re 20 µPa)	71	69	69	61	53	44	38	30

Solution

1. From Table 8.4 for average conditions, enter values for C in the table above.

2. The relative power spectrum as given is shown in row three of the example table. Logarithmic addition (see Section 1.11.4) of the values shown in row three gives the sum as 4.4 dB. The total is required to equal 80 dB; thus absolute levels in each band are determined by adding to the relative levels:

$$80 - 4.4 = 75.6 \text{ dB}$$

The resulting band sound power levels, L_w, are given in the table in row four.

3.   Use Equation (8.85) to calculate the required sound pressure levels

$$S_E = 2.2 \text{ m}^2, \qquad 10 \log_{10} S_E = 3.4 \text{ dB}$$

Therefore, assuming that $\rho c = 400$:

$$L_{p1} = L_w - 3.4 - TL + C$$

The estimates of L_{p1} based upon the above equation are indicated in the last row of the table.

If the enclosure is located outdoors, the following expression gives a reasonable approximation to the sound pressure level L_{p2} to be expected at a point some distance, r, from the enclosure:

$$L_{p2} = L_{p1} + 10 \log_{10} S_E + 10 \log_{10}(D_\theta/4\pi r^2) \qquad (\text{dB}) \qquad (8.87)$$

In Equation (8.85), the distance from the enclosure to the measurement position, r (m), is assumed to be large compared with the relevant enclosure face dimensions, and D_θ is the directivity factor for the enclosure. Normally, for an enclosure on a hard floor, $D_\theta = 2$ (see Table 5.1).

If the enclosure is located indoors, then the reverberant sound field due to the enclosing room must be considered. In this case, the sound pressure level, L_{p2}, at a position in the room is derived using the results of Chapters 6 and 7, as follows. The sound power W_t radiated by the enclosure can be written as:

$$W_t = \langle p_1^2 \rangle [S_E/(\rho c)] \qquad (\text{W}) \qquad (8.88)$$

or

$$L_{wt} = L_{p1} + 10 \log_{10} S_E \qquad (\text{dB}) \qquad (8.89)$$

where S_E is the total external surface area of the enclosure and the assumption has been made that $\rho c \approx 400$. In this case, use of Equation (7.43) gives:

$$L_{p2} = L_{wt} + 10 \log_{10} \left[\frac{D_\theta}{4\pi r^2} + \frac{4(1 - \bar{\alpha})}{S\bar{\alpha}} \right] \qquad (\text{dB}) \qquad (8.90)$$

where $S\bar{\alpha}$ is the total sound absorption in the room, and $\bar{\alpha}$ is the average sound-absorption coefficient.

Substituting Equation (8.89) into Equation (8.90) gives the following for L_{p2}:

$$L_{p2} = L_{p1} + 10 \log_{10} S_E + 10 \log_{10} \left[\frac{D_\theta}{4\pi r^2} + \frac{4(1 - \bar{\alpha})}{S\bar{\alpha}} \right] \qquad (\text{dB}) \qquad (8.91)$$

The noise reduction due to the enclosure may now be calculated. The sound pressure level at a position in the room with no enclosure is obtained using Equation (7.43) as follows:

$$L_{p2}' = L_w + 10 \log_{10} \left[\frac{D_\theta}{4\pi r^2} + \frac{4(1 - \bar{\alpha})}{S\bar{\alpha}} \right] \qquad (\text{dB}) \qquad (8.92)$$

Assuming that the enclosure has not altered the directivity characteristics of the machine or assuming that the direct field contribution is negligible at the measurement position, the noise reduction is given by $NR = L'_{p2} - L_{p2}$. Thus:

$$NR = L_w - L_{p1} - 10 \log_{10} S_E \quad \text{(dB)} \tag{8.93}$$

Substituting Equation (8.85) for L_{p1} into Equation (8.93) provides the following expression for the noise reduction:

$$NR = TL - C \quad \text{(dB)} \tag{8.94}$$

Note that the quantity C may be determined using Table 8.4 or, alternatively, using Equation (8.86). Note also that Equation (8.94) holds for an enclosure located outdoors or inside a building.

8.4.3 Personnel Enclosures

For personnel enclosures, the noise source is external and the purpose of the enclosure is to reduce levels within. Suppose that the enclosure is located within a space in which the reverberant field is dominant. The sound pressure level in the reverberant field, removed at least one half of a wavelength from the walls of the enclosure, is designated L_{p1}, (dB). Use will be made of Equation (7.40b), in which the mean absorption coefficient $\bar{\alpha}$ is replaced with the field-incidence wall transmission coefficient, τ. The power transmission W_i into the enclosure through the external walls of surface area, S_E, is:

$$W_i = S_E \langle p_1^2 \rangle \tau / (4\rho c) \tag{8.95}$$

Taking logarithms to the base ten of both sides of the equation, and noting that the numerical value of ρc is approximately 400, Equation (8.95) may be rewritten as follows:

$$L_{wi} = L_{p1} + 10 \log_{10} S_E - TL - 6 \quad \text{(dB)} \tag{8.96}$$

To estimate the sound field within the enclosure, use is made of Equation (7.43). In the latter equation, the direct field term is replaced with the reciprocal of the external surface area, S_E, and Equation (7.43) is rewritten as follows:

$$L_{pi} = L_{wi} + 10 \log_{10} \left[\frac{1}{S_E} + \frac{4(1 - \bar{\alpha}_i)}{S_i \bar{\alpha}_i} \right] \quad \text{(dB)} \tag{8.97}$$

Substitution of the above equation into Equation (8.96), and use of Equation (8.86) leads to the following result:

$$NR \approx L_{p1} - L_{pi} = TL - C \quad \text{(dB)} \tag{8.98}$$

The constant C may be calculated using Equation (8.86), or it may be estimated using Table 8.4.

The problem is not so simple when the direct field of the source is dominant at one or more walls of the enclosure. In this case, for the purposes of estimating the sound field incident on the exterior of each of the enclosure walls, it is necessary to treat the enclosure as a barrier. The method of approximate solution is illustrated in Example 8.8, and discussed in Section 8.5 on barriers.

Example 8.4

A small personnel enclosure of nominal dimensions 2 m wide, 3 m long and 2.5 m high is to be constructed of single leaf brick 125 mm thick, plastered on both sides. The floor will be of concrete but the ceiling will be of similar construction to the walls (not bricks but plastered, similar weight, etc.). Determine the expected noise reduction (*NR*) for the basic hard wall design. Assume that the external sound field is essentially reverberant and that any direct field from the source that is incident on the enclosure is negligible.

Solution

1. Use Tables 7.2 and 8.2 to determine values of wall and ceiling absorption coefficients and transmission loss. Enter the values in Example 8.4 table.

2. Calculate:

$$S_i \bar{\alpha}_i = [2(2 \times 2.5 + 3 \times 2.5) + 2 \times 3]\bar{\alpha}_w + [2 \times 3]\bar{\alpha}_f$$
$$= 31\bar{\alpha}_w + 6\bar{\alpha}_f$$

Enter values in the table below. Calculate $10\log_{10}(S_E/S_i\bar{\alpha}_i)$ and enter it in the table.

3. Calculate the external surface area excluding the floor: $S_E = 31$ m^2

4. Calculate noise reduction, *NR*, using Equation (8.17) rather than Equations (8.94) and (8.98) as $\bar{\alpha}$ is small. Note that in this case, the former test partition area A becomes the external area S_E exposed to the external sound field:

$$NR = TL - 10\log_{10}\left[\frac{S_E}{S_i\bar{\alpha}_i}\right]$$

The results are entered in the table on the next page.

Example 8.4 Table

	Octave band centre frequency (Hz)							
	63	125	250	500	1000	2000	4000	8000
TL from Table 8.2	30	36	37	40	46	54	57	59
$\bar{\alpha}_w$ from Table 7.2	0.013	0.013	0.015	0.02	0.03	0.04	0.05	0.06
$\bar{\alpha}_f$ from Table 7.2	0.01	0.01	0.01	0.01	0.02	0.02	0.02	0.03
$S_i\bar{\alpha}_i$ (m)	0.463	0.463	0.525	0.68	1.05	1.36	1.67	2.04
$S_E/S_i\bar{\alpha}_i$	67.0	67.0	59.0	45.6	29.5	22.8	18.6	15.2
$10\log_{10}(S_E/S_i\bar{\alpha}_i)$	18	18	18	17	15	14	13	12
NR (dB)	12	18	19	23	31	40	44	47

Example 8.5

Suppose that an opening of 0.5 m^2 is required for ventilation in the enclosure wall of Example 8.4. What transmission loss (*TL*) must any muffling provide for the ventilation opening if the noise reduction (*NR*) of the enclosure must not be less than 16 dB in the 250 Hz octave band?

Solution

1. Assume that $S_i\bar{\alpha}_i$ is essentially unchanged by the small penetration through the wall, then from the previous example at 250 Hz, $10\log_{10}S_i\bar{\alpha}_i = -2.8$ and $10\log_{10}S_E = 14.9$ dB.

2. In Equation (8.17), replace A with S_E; then $NR = TL + 10\log_{10} S_i\bar{\alpha}_i - 10\log_{10}S_E$. Let $NR = 16$. Then $TL = 16 + 2.8 + 14.9 = 33.7$ dB. Thus, the net *TL* required is 27.1 dB.

3. Calculate transmission coefficient τ of the ventilation opening.
    Equation (8.74) gives:

$$31\tau = (31 - 0.5)\tau_w + 0.5\tau_v$$

From the table of Example 8.4 and use of Equation (8.75):

$$\tau_w = 10^{-37/10} = 0.000200$$

Putting the above in Equation (8.74) gives:

$$TL = -10\log_{10}\tau = -10\log_{10}\left(\frac{30.5}{31} \times 0.0002 + \frac{0.5}{31}\tau_v\right) = 33.7 \text{ dB}$$

From the above, the following is obtained:

$$\frac{30.5}{31} \times 0.0002 + \frac{0.5}{31}\tau_v = 10^{-3.37} = 0.00043$$

or:

$$\tau_v = \frac{31}{0.5}\left(0.00043 - \frac{30.5}{31} \times 0.0002\right) = \frac{31}{0.5}(0.000233) = 0.0145$$

4. Calculate the required *TL* of muffling for the vent:

$$TL_v = -10\log_{10}\tau_v$$
$$= -10\log_{10}0.0145 = 18.0 \text{ dB}$$

8.4.4 Enclosure Windows

Inspection windows are usually double glazed. However, the use of double glazing may show no improvement over single glazing at low frequencies where the interaction of the mass of the glass panes and stiffness of the air trapped between them can produce a series of resonances. The lowest frequency corresponding to a resonance of this type and a corresponding poor value of transmission loss is given approximately by Equation (8.55) (see Appendix B for glass properties). There are other acoustic resonances of the cavity that are also important, but these usually have higher resonance frequencies (see Equations (7.17) and (8.54).

Reference to Figure 8.9 shows that the glass thickness and pane separation should be chosen so that f_0 (see Equation (8.55)) is well below the frequency range in which significant noise reduction is required. For example, a pane thickness of 6 mm and a separation of 150 mm gives $f_0 = 78$ Hz. Good transmission loss should not be expected at frequencies below about $1.15 f_0$ or 90 Hz. In the low-frequency range where these resonances occur, the transmission loss of a double-glazed window may be improved by placing a blanket of porous acoustic material in the reveals between the two frames supporting the glass (Quirt, 1982).

8.4.5 Enclosure Leakages

The effectiveness of an enclosure can be very much reduced by the presence of air gaps. Air gaps usually occur around removable panels or where services enter an enclosure. The effect of cracks around doors or around the base of a cover can be calculated with the help of Figure 8.11, which gives the transmission coefficient of a crack as a function of frequency and width. Note that if the crack is between one plane surface and another plane surface normal to it, the effective crack width must be doubled (because of reflection – for example, a crack under a door) before using Figure 8.11. However, note that the effective area of the crack is not doubled when overall *TL* values are calculated using Equation (8.74). Once the transmission

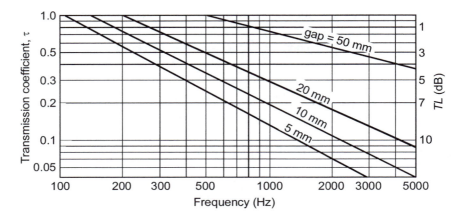

Figure 8.11 Transmission coefficients of long narrow cracks.

coefficient, τ, has been determined for a particular frequency, the procedure outlined in Section 8.3 is used for estimating the average value of τ for the enclosure wall or cover.

The importance of small air gaps is illustrated by reference to Figures 8.10 and 8.11. For example, consider a door with a 20 mm air gap beneath it, which for a typical door results in an area ratio S_1/S_2 of 0.01, where S_1 is the area of the gap at the bottom of the door and S_2 is the area of the door. The effective width of a 20 mm gap under a door is 40 mm due to reflection on the floor under the gap. Thus the transmission coefficient for a 20 mm gap under a door is 1.0 for frequencies below about 400 Hz. For $S_1/S_2 = 0.01$, Figure 8.10 shows that the best transmission loss that can be achieved at frequencies below 400 Hz is 20 dB.

Gomperts and Kihlman (1967) presented the following expression for calculating the transmission coefficient of a very thin slit (kw smaller than 0.5):

$$\tau = \frac{mkw}{2n^2\sin^2\left[kw\left(\dfrac{\ell}{w} + \dfrac{2}{\pi}\left(\log_e\dfrac{8}{kw} - 0.5772\right)\right)\right] + 2k^2w^2} \tag{8.99}$$

where k is the wavenumber, w is the height or width of the slit, ℓ is the depth of the slit and the length of the slit must exceed a wavelength at the frequency of interest. The coefficient, $m = 4$ for normally incident plane waves and $m = 8$ for a diffuse field incident on the slit. The coefficient, $n = 1$ if the slit is in the middle of a wall and $n = 0.5$ if the slit is at the edge of a wall, which takes into account the effect of the reflected path of sound travelling through the slit. With this expression, the slit transmission coefficient exceeds 1.0 most of the time, which indicates that the incident sound energy is concentrated in the vicinity of the slit as a result of its presence. As the slit becomes wider, this effect reduces.

For doors with no seals a leak height, w, of 1.2 mm around the full perimeter of the door is usually used with Equations (8.99) and (8.74) to calculate the effect on the overall TL of a wall with a door. For a door with weather stripping, a leak height of 0.5 mm is used and for a door with magnetic seals a leak height of 0.2 mm is used.

Example 8.6

Find the transmission loss at 1200 Hz of a solid door, 2.10×0.9m, with a 12.5 mm gap along its bottom edge. The door itself has a transmission loss of 30 dB ($\tau = 0.001$).

Solution

From Figure 8.11 a 25 mm gap has a transmission coefficient, $\tau = 0.3$ at 1200 Hz. (Note that, because the floor on either side of the door acts like a plane reflector, the effective width of the gap is doubled when Figure 8.11 is used but not when its area is calculated.):

$$\bar{\tau} = \frac{(0.001 \times 2.10 \times 0.90) + (0.3 \times 0.90 \times 0.0125)}{(2.10 \times 0.90) + (0.90 \times 0.0125)}$$

$$= \frac{0.00189 + 0.003375}{1.890 + 0.01125}$$

$$= 0.00277$$

$$TL = 10 \log_{10}(1/0.00277) = 25.6 \text{ dB}$$

8.4.6 Access and Ventilation

Most enclosures require some form of ventilation. They may also require access for passing materials in and out. Such necessary permanent openings must be treated with some form of silencing to avoid compromising the effectiveness of the enclosure. In a good design, the acoustic performance of access silencing will match the performance of the walls of the enclosure. Techniques developed for the control of sound propagation in ducts may be employed for the design of silencers (see Chapter 9).

If ventilation for heat removal is required but the heat load is not large, then natural ventilation with silenced air inlets at low levels close to the floor and silenced outlets at high levels, well above the floor, will be adequate. If forced ventilation is required to avoid excessive heating then the approximate amount of air flow needed can be determined using Equation (8.100).

$$\rho C_p V = H/\Delta T \tag{8.100}$$

where V is the volume (m³/s) of airflow required, H is the heat input (W) to the enclosure, ΔT is the temperature differential (°C) between the external ambient and the maximum allowble internal temperature of the enclosure, ρ is the gas (air) density (kg/m^3), and C_p is the specific heat of the gas (air) in SI units (1010 m^2 s^{-2} °C^{-1}).

If a fan is provided for forced ventilation, the silencer will usually be placed external to the fan so that noise generated by the fan will be attenuated as well. When high volumes of air flow are required, the noise of the fan should be considered very carefully, as this noise source is quite often a cause of complaint. As fan noise generally increases with the fifth power of the blade tip speed, large slowly rotating fans are always to be preferred over small high-speed fans.

8.4.7 Enclosure Vibration Isolation

Any rigid connection between the machine and enclosure must be avoided. If at all possible, all pipes and service ducts passing through the enclosure wall should have flexible sections to act as vibration breaks; otherwise, the pipe or duct must pass through a clearance hole surrounded by a packing of mineral wool closed by cover plates and mastic sealant.

It is usually advisable to mount the machine on vibration isolators (see Chapter 10), particularly if low-frequency noise is the main problem. This ensures that little energy is transmitted to the floor. If this is not done, there is a possibility that the floor surrounding the enclosure will re-radiate the energy into the surrounding space, or that the enclosure will be mechanically excited by the vibrating floor and act as a noise source.

Sometimes it is not possible to mount the machine on vibration isolators. In this case, excitation of the enclosure can be avoided by mounting the enclosure itself on vibration isolators. Note that great care is necessary, when designing machinery vibration isolators, to ensure that the machine will be stable and that its operation will not be affected adversely. For example, if a machine must pass through a system resonance when running up to speed, then "snubbers" can be used to prevent excessive motion of the machine on its isolation mounts. Isolator design is discussed in Chapter 10.

8.4.8 Enclosure Resonances

Two types of enclosure resonance will be considered. The first is mechanical resonance of the enclosure panels, while the second is acoustic resonance of the air space between an enclosed machine and the enclosure walls. At the frequencies of these resonances, the noise reduction due to the enclosure is markedly reduced from that calculated without regard to resonance effects.

The lowest order enclosure panel resonance is associated with a large loss in enclosure effectiveness at the resonance frequency. Thus the enclosure should be designed so that the resonance frequencies of its constituent panels are not in the frequency range in which appreciable sound attenuation is required. Only the lowest order, first few, panel resonances are of concern here. The panels may be designed in such a way that their resonance frequencies are higher than or lower than the frequency range in which appreciable sound attenuation is required. Additionally, the panels should be well damped.

If the sound source radiates predominantly high-frequency noise, then an enclosure with low resonance frequency panels is recommended, implying a massive enclosure. On the other hand, if the sound radiation is predominantly low frequency in nature then an enclosure with a high resonance frequency is desirable, implying a stiff but not massive enclosure. The discussion of Sections 8.2.4 and 8.2.6 may be used as a guide to resonance frequency and transmission loss calculations.

The resonance frequency of a panel may be increased by using stiffening ribs, but the increase that may be achieved is generally quite limited. For stiff enclosures with high resonance frequencies, materials with large values of Young's modulus to density ratio, E/ρ, are chosen for the wall construction to ensure large values of the longitudinal wave speed, c_L (see Section 1.3.4). For massive enclosures with small resonance frequencies, small values of E/ρ are chosen. Values of E and ρ for common panel materials are listed in Appendix B. In practice, stiff enclosures will generally be restricted to small enclosures.

If a machine is enclosed, reverberant build-up of the sound energy within the enclosure will occur unless adequate sound absorption is provided. The effect will be an increase of sound pressure at the inner walls of the enclosure over that which would result from the direct field of the source. A degradation of the noise reduction expected of the enclosure is implied.

In close-fitting enclosures, noise reduction may be degraded by yet another resonance effect. At frequencies where the average air spacing between a vibrating machine surface and the enclosure wall is an integral multiple of half wavelengths of sound, strong coupling will occur between the vibrating machine surface and the enclosure wall, resulting in a marked decrease in the enclosure wall transmission loss.

The effect of inadequate absorption in enclosures is very noticeable. Table 8.5 shows the reduction in performance of an ideal enclosure with varying degrees of internal sound absorption. The sound power of the source is assumed constant and unaffected by the enclosure. "Percent" refers to the fraction of internal surface area that is treated.

Table 8.5 Enclosure noise reduction as a function of percentage of internal surface covered with sound-absorptive material

Percent sound absorbent material	10	20	30	50	70
Noise reduction (dB)	-10	-7	-5	-3	-1.5

For best results, the internal surfaces of an enclosure are usually lined with glass or mineral fibre or open-cell polyurethane foam blanket. Typical values of absorption coefficients are given in Table 7.2.

Since the absorption coefficient of absorbent lining is generally highest at high frequencies, the high-frequency components of any noise will suffer the highest attenuation. Some improvement in low-frequency absorption can be achieved by using a thick layer of lining. See Section 7.7.3 for a discussion of suitable linings and their containment for protection from contamination with oil or water, which impairs their acoustical absorption properties.

8.4.9 Close-fitting Enclosures

The cost of acoustic enclosures of any type is proportional to size; therefore there is an economic incentive to keep enclosures as small as possible. Thus, because of cost or limitations of space, a close-fitting enclosure may be fitted directly to the machine which is to be quietened, or fixed independently of it but so that the enclosure internal surfaces are within, say, 0.5 m of major machine surfaces.

When an enclosure is close fitting, the panel resonance frequency predicted by Equation (8.23) may be too low; in fact, the resonance frequency will probably be somewhat increased due to the stiffening of the panel by the enclosed air volume. Thus, an enclosure designed to be massive with a low resonance frequency may not perform as well as expected when it is close fitting.

Furthermore, system resonances will occur at higher frequencies; some of these modes of vibration will be good radiators of sound, producing low noise reductions, and some will be poor radiators, little affecting the noise reduction. The magnitude of the decrease in noise reduction caused by these resonances may be controlled to some extent by increasing the mechanical damping of the wall. Thus, if low-frequency sound (less than 200 Hz) is to be attenuated, the close-fitting enclosure should be stiff and well damped, but if high-frequency sound is to be attenuated, the enclosure should be heavy and highly absorptive but not stiff. Doubling of the volume of a small enclosure will normally lead to an increase in noise reduction of 3 dB at low frequencies, so that it is not desirable to closely surround a source, such as a vibrating machine, if a greater volume is possible.

Generally, if sufficient space is left within the enclosure for normal maintenance on all sides of the machine, the enclosure need not be regarded as close fitting. If, however, such space cannot be made available, it is usually necessary to upgrade the transmission loss of an enclosure wall by up to 10 dB at low frequencies (less at high frequencies), to compensate for the expected degradation in performance of the enclosure due to resonances.

The analysis of close-fitting enclosures is treated in detail in the literature (Tweed and Tree, 1976; Byrne *et al.*, 1988).

8.4.10 Partial Enclosures

In many situations, where easy and continuous access to parts of a machine is necessary, a complete enclosure may not be possible, and a partial enclosure must be considered (Alfredson and Scow, 1976). However, the noise reductions that can be expected at specific locations from partial enclosures are difficult to estimate and will depend upon the particular geometry. Estimates of the sound power reduction to be expected from a partial enclosure, which is lined with a 50 mm thick layer of sound absorbing material, are presented in Figure 8.12.

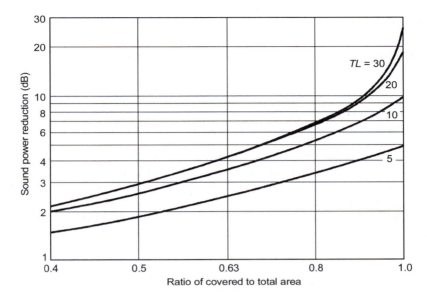

Figure 8.12 Sound attenuation due to a partial enclosure. The transmission loss (*TL*) of the enclosure wall is shown parametrically.

This figure shows the sound power reduction to be expected for various degrees of enclosure, with transmission loss through the enclosure as a parameter. The figure shows fairly clearly that the enclosure walls should have a transmission loss of about 20 dB, and the most sound power reduction that can be achieved for a realistic partial enclosure (90% closed) is about 10 dB.

However, sound pressure levels in some directions may be more greatly reduced, especially in areas immediately behind solid parts of the enclosure.

8.5 BARRIERS

Barriers are placed between a noise source and a receiver as a means of reducing the direct sound observed by the receiver. In rooms, barriers suitably treated with sound-absorbing material may also slightly attenuate reverberant sound field levels by increasing the overall room absorption.

Barriers are a form of partial enclosure, usually intended to reduce the direct sound field radiated in one direction only. For non-porous barriers having sufficient surface density, the sound reaching the receiver will be entirely due to diffraction around the barrier boundaries. Since diffraction sets the limit on the noise reduction that may be achieved, the barrier surface density is chosen to be just sufficient so that the noise reduction at the receiver is diffraction limited. For this purpose, the barrier surface density will usually exceed 20 kg/m². There are many proprietary designs for barriers; typical barriers are built of lightweight concrete blocks, but asbestos board, cement board, sheet metal, fibre-glass panels and high-density plastic sheeting have also been used.

The effect of barrier diffraction is considered together with ground effects and meteorological effects in outdoor sound propagation calculations as is discussed in Chapter 5.

8.5.1 Diffraction at the Edge of a Thin Sheet

In general, sound diffracted around a barrier may take many paths. To make the problem tractable, the idealised case of sound diffraction at the straight edge of a thin semi-infinite opaque plane barrier will first be considered, and the basis for all subsequent calculations will be established.

The discussion will make reference to Figure 8.13, in which a point source model is illustrated. The model may also be used for a line source that lies parallel to the barrier edge.

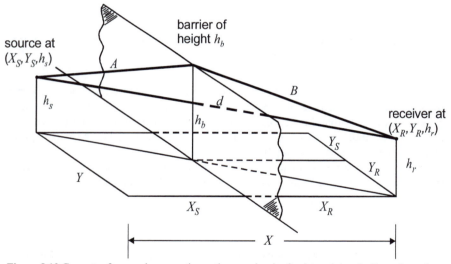

Figure 8.13 Geometry for sound propagation path over a barrier for determining the Fresnel number.

The lengths A and B are introduced as two segments of the shortest path over the barrier from source to receiver. The length d is the straight-line segment from source to receiver. In terms of these path lengths, the Fresnel number, N, is defined by the following equation:

$$N = \pm(2/\lambda)(A + B - d) \qquad (8.101)$$

where λ is the wavelength of the centre frequency of the narrow band of noise considered; for example, a one-third octave or octave band of noise.

The lengths, *d*, *A* and *B* shown in the figure are calculated using:

$$d = [X^2 + Y^2 + (h_r - h_s)^2]^{1/2}$$

$$A = [X_S^2 + Y_S^2 + (h_b - h_s)^2]^{1/2} \qquad (8.102)$$

$$B = [X_R^2 + Y_R^2 + (h_b - h_r)^2]^{1/2}$$

where

$$Y_R = YX_R/X \quad \text{and} \quad Y_S = \frac{X_S}{X_R} Y_R \qquad (8.103)$$

In Figure 8.14, an attenuation factor Δ_b associated with diffraction at an edge is plotted as a function of the Fresnel number (Maekawa, 1968, 1977, 1985). To enter the figure, the positive sign of Equation (8.100) is used when the receiver is in the shadow zone of the barrier, and the negative sign is used when the receiver is in the bright zone, in line of sight with the source. The horizontal scale in the figure is logarithmic for values of Fresnel number, *N*, greater than one, but it has been adjusted for values less than one to provide the straight-line representation shown.

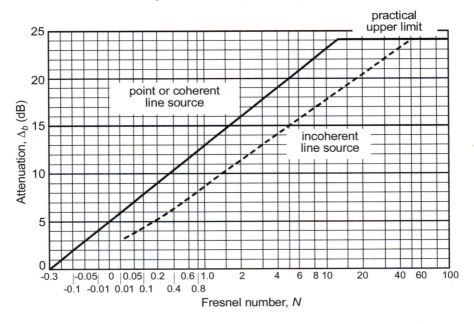

Figure 8.14 Sound attenuation by a semi-infinite screen in free space. If the receiver has direct line of sight to the source, then *N* is set negative.

The possibility that the source may be directional may be considered by introducing directivity factors in the direction from the source to the barrier edge $D_{\theta B}$

and in the direction from the source to the receiver $D_{\theta R}$. The attenuation, A_{bi}, of a single sound path, i, due to diffraction over the barrier is given by the following equation (Beranek, 1971, Chapter 7):

$$A_{bi} = \Delta_{bi} + 20\log_{10}[(A_i + B_i)/d_i] + D_{\theta R} - D_{\theta B} \qquad (8.104)$$

As an alternative to using Figure 8.14, Kurze and Anderson (1971) proposed the following expression, which is a very good approximation to the point source curve in Figure 8.14, for $N_i > 0.5$. Below $N_i = 0.5$, the amount by which Maekawa's curve exceeds the Kurze and Anderson formula gradually increases to a maximum of 1.5 dB at $N_i = 0.1$, and then gradually decreases again for smaller N_i.

$$\Delta_{bi} = 5 + 20\log_{10}\frac{\sqrt{2\pi N_i}}{\tanh\sqrt{2\pi N_i}} \qquad (8.105)$$

where N_i is the Fresnel number for path, i over the barrier. More recently, a correction to this expression was proposed by Menounou (2001) to make it more accurate for locations of the source or receiver close to the barrier or for the receiver close to the boundary of the bright and shadow zones. The more accurate equation for any particular path, i, over the barrier is:

$$A_{bi} = IL_s + IL_b + IL_{sb} + IL_{sp} + D_{\theta R} - D_{\theta B} \qquad (8.106)$$

where:

$$IL_s = 20\log_{10}\frac{\sqrt{2\pi N_i}}{\tanh\sqrt{2\pi N_i}} - 1 \qquad (8.107)$$

$$IL_b = 20\log_{10}\left[1 + \tanh\left(0.6\log_{10}\frac{N_2}{N_i}\right)\right] \qquad (8.109)$$

$$IL_{sb} = \left(6\tanh\sqrt{N_2} - 2 - IL_b\right)\left(1 - \tanh\sqrt{10N_i}\right) \qquad (8.110)$$

$$IL_{sp} = \begin{cases} 3\text{ dB} & \text{for plane waves} \\ 10\log_{10}(1 + (A + B)/d) & \text{for coherent line source} \\ 10\log_{10}((A + B)^2/d^2 + (A + B)/d) & \text{for point source} \end{cases} \qquad (8.108)$$

The term represented by Equation (8.110) should only be calculated when N_i is very small. The quantity, N_2, is the Fresnel number calculated for a wave travelling from the image source to the receiver where the image source is generated by reflection from the barrier (not the ground). Thus, the image source will be on the same side of the barrier as the receiver. The distance, d, used in Equation (8.101) to calculate the Fresnel number, N_2, is the straight line distance between the image source (due to reflection in the barrier) and receiver, and the distance $(A + B)$ is the same as used to calculate the Fresnel number for the actual source and receiver.

8.5.2 Outdoor Barriers

A common form of barrier is a wall, which may be very long so that only diffraction over the top is of importance, or it may be of finite length so that diffraction around the ends is also of importance. In either case, the problem is generally more complicated than that of simple diffraction at an edge, because of reflection in the ground plane (Gill, 1980b; Hutchins *et al.*, 1984). For example, the direct sound field of a tonal source and the reflected sound field from its virtual image in the ground plane may interfere to produce a relative minimum in the sound field at the position of the observer. The introduction of a barrier may effectively prevent such interference, with the result that the placement of a barrier may result in a net gain in level at the point of observation. However, this possibility will not be given further consideration here. It will be assumed that reflections are incoherent or that octave bands of noise are considered, in which case such frequency-dependent effects tend to wash out with the averaging process implied by octave band analysis and measurement.

Note that the use of narrowband analysis can result in large errors due to interference effects, which are ignored here but discussed in Chapter 5. For example, errors as large as 10 dB may be encountered when interference effects are ignored in conjunction with one-third octave band analysis, and the errors could be as large as 20 dB when tones are considered. Detailed analysis that takes account of coherence is discussed in Chapter 5 (see Section 5.11.10) and in the literature (Isei *et al.*, 1980; Embleton *et al.*, 1983; Nicolas *et al.*, 1983).

Diffraction over the top of a very long wall as illustrated in Figure 8.15(a) will be considered first. Due to possible reflection in the ground plane, a total of four possible diffraction paths must be considered, as shown in the figure. Referring to the figure, these paths are *SOR*, *SAOR*, *SAOBR* and *SOBR*. As indicated in the figure, placement of an image source and receiver in the diagram is helpful in determining the respective

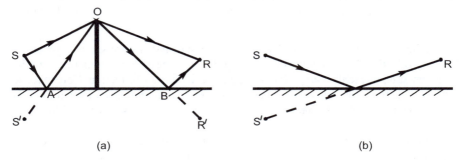

(a)   (b)

Figure 8.15 Diffraction paths for consideration of an infinite width barrier: (a) with barrier; (b) without barrier.

path lengths. These images are placed as far beneath the ground plane as the source and receiver are above it.

When paths involving one or more ground reflections are considered, the straight line distance, *d*, used in Equation (8.101) is the same as for non-reflected waves as the

Fresnel number and associated noise reduction are relative to the straight line propagation from S to R.

For the ground reflected wave on the side of the source, the quantity, A, corresponding to Equation (8.101) is:

$$A = [X_S^2 + Y_S^2 + (h_b + h_s)^2]^{1/2} \qquad (8.111)$$

and for the ground reflected wave on the side of the receiver the quantity, B, corresponding to Equation (8.101) is:

$$B = [X_R^2 + Y_R^2 + (h_b + h_r)^2]^{1/2} \qquad (8.112)$$

In the absence of a barrier, there are only two paths between the source and receiver, as shown in Figure 8.15(b). Thus, if the ground plane is hard and essentially totally reflective, and octave bands of noise are considered so that interference effects tend to wash out, the noise reduction due to the barrier is calculated as follows:

1. Making use of the image source and image receiver respectively, as indicated in Figure 8.15(a), the expected reduction in level is determined using Equation (8.103) for each of the four paths.

2. The four estimates are combined logarithmically as indicated in Equation (1.100).

3. The process is repeated for the two paths shown in Figure 8.15(b), again assuming total incoherent reflection, to produce a combined level at the receiver without a barrier. The assumption is implicit that the total power radiated by the source is constant; thus at large distances from the source this procedure is equivalent to the simple assumption that the source radiates into half-space.

4. The reduction due to the barrier is determined as the result of subtracting the level determined in step (2) from the level determined in step (3), as shown in Equation (1.101).

If the ground is not acoustically hard but is somewhat absorptive, as is generally the case, then the dB attenuation due to reflection must be added arithmetically to the dB attenuation due to the barrier for each path that includes a reflection. Note that one path over the top of the barrier includes two reflections, so for this path two reflection losses must be added to the barrier attenuation for this path. This process is illustrated in Example 8.7.

When a wall is of finite width, diffraction around the ends of the barrier may also require consideration. However, diffraction around the ends involves only one ground reflection and thus only consideration of two possible paths at each end, not four as in the case of diffraction over the top. The location of the effective point of reflection is found by assuming an image source, S', and image wall. Referring to Figure 8.16 the two paths SOR and $SO'O''R$ are, respectively, the shortest paths from source and image source to the receiver around one end of the wall and image wall. Again taking account of possible loss on reflection for one of the paths, the contributions over the two paths are determined using Figures 8.13 and 8.14 and Equation (8.104).

Referring to Figure 8.13, for definition of the symbols and defining h_b' as the height of point O above the ground and h_b'' as the height of point O$'$ above the ground, the following equations are easily derived using simple trigonometry:

$$h_b'' = \frac{X_S h_r - X_R h_s}{X_S + X_R}; \quad \text{and } h_b'' = \frac{X_R h_s - X_S h_r}{X_S + X_R} \qquad (8.113\text{a,b})$$

$$h_b' = \frac{X_R h_s + X_S h_r}{X_S + X_R} \qquad (8.114)$$

Equation (8.113a) applies if the ground reflection is on the source side of the barrier and Equation (8.113b) applies if the ground reflection is on the receiver side.

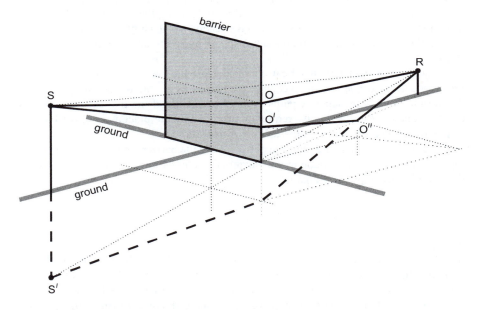

Figure 8.16 Geometry for diffraction around one edge of a finite width barrier. S and S$'$ are source and image source, R is the receiver, O is defined by the shortest path from S to R around the barrier edge, and O$'$ and O$''$ are defined by the shortest path from S$'$ to R around the image barrier edge.

Referring to the path lengths from the source to the barrier edge as A' and A'' for the direct and ground reflected paths respectively and similarly B' and B'' for the same two paths on the receiver side of the barrier, the following can be written:

$$A' = \sqrt{(h_s - h_b')^2 + X_S^2 + (Y_S - Y_B)^2} \qquad (8.115)$$

$$B' = \sqrt{(Y_R - Y_B)^2 + X_R^2 + (h_b' - h_r)^2} \tag{8.116}$$

$$A'' = \sqrt{(h_s - h_b'')^2 + X_S^2 + (Y_B - Y_S)^2} \tag{8.117}$$

$$B'' = \sqrt{(Y_R - Y_B)^2 + X_R^2 + (h_r + h_b'')^2} \tag{8.118}$$

where Y_B is the y-coordinate of the end edge of the barrier. The quantity d, is the same as it is for diffraction over the top of the barrier. If the source / receiver geometry is such that the point of reflection for the reflected wave is on the source side of the barrier, then expressions for A' and B' remain the same, but A'' and B'' are calculated by interchanging the subscripts r and s and interchanging the subscripts, R and S in Equations (8.113), (8.117) and (8.118). If the receiver is at a higher elevation than the source, the subscripts r and s and the subscripts, R and S are interchanged in Equation (8.114).

Equation (8.101) may then be used to calculate the Fresnel number and thus the noise reduction corresponding to each of the two paths around the end of the barrier. For a finite wall, eight separate paths should be considered and the results combined to determine the expected noise reduction provided by the placement of the barrier. In practice, however, not all paths will be of importance.

In summary, if there are multiple paths around the barrier (see next section), then the overall noise reduction due to the barrier is calculated using Equation (1.101) as illustrated in Example 8.7. That is:

$$A_b = 10 \log_{10}\left[1 + 10^{-(A_{Rw}/10)}\right] - 10 \log_{10}\sum_{i=1}^{n_A} 10^{-(A_{bi} + A_{Ri})/10} \tag{8.119}$$

where the reflection loss, A_R, due to the ground is added to each path that involves a ground reflection. The subscript i refers to the ith path around the barrier and the subscript w refers to the ground reflected path in the absence of the barrier.

For ground that is not uniform between the source and receiver, the reflection loss at the point of the ground corresponding to specular reflection is used. For plane wave reflection, A_R (equal to A_{Ri} or A_{Rw}) is given by:

$$A_R = -20 \log_{10}|R_p| \tag{8.120}$$

where R_p is defined in Equation (5.142) for extended reactive ground and (5.144) for locally reactive ground. If the more complex and more accurate spherical wave reflection model is used (see Sections 5.10.4 and 5.11.10.5):

$$A_R = -10 \log_{10}|\Gamma| \tag{8.121}$$

It is interesting to note that the first term in Equation (8.119) is equivalent to the excess attenuation due to ground effects in the absence of the barrier multiplied by -1.

For the case of source distributions other than those considered, the simple strategy of dividing the source into a number of equivalent line or point sources, which are then each treated separately may be used. Implicit in this approach is the assumption that the parts are incoherent, consistent with the analysis described here.

Barrier attenuation can often be increased by up to 8 dB by lining the source side with absorptive material. The attenuation due to the absorptive material increases as the source and receiver approach the barrier and the barrier height increases. A detailed treatment of absorptive barriers is given by Fujiwara *et al.* (1977b).

Note that in the procedure described in this section, the calculation of the barrier attenuation, A_b, involves subtracting the excess attenuation due to ground effects and thus A_g must be added to A_b in calculating the overall excess attenuation, as indicated by Equation (5.178).

8.5.2.1 Thick Barriers

Existing buildings may sometimes serve as barriers (Shimode and Ikawa, 1978). In this case it is possible that a higher attenuation than that calculated using Equation (8.103) may be obtained due to double diffraction at the two edges of the building. This has the same effect as using two thin barriers placed a distance apart equal to the building thickness (ISO 9613-2 (1996)). Double barriers are also discussed by Foss (1979). The effect of the double diffraction is to add an additional attenuation, ΔC, to the noise reduction achieved using a thin barrier (Fujiwara *et al.*, 1977a) located at the centre of the thick barrier:

$$\Delta C = K \log_{10}(2\pi b/\lambda) \tag{8.122}$$

where b is the barrier thickness, λ is the wavelength at the band centre frequency, and K is a coefficient that may be estimated using Figure 8.17. Note that to use Equation (8.122), the condition $b > \lambda/2$ must be satisfied. Otherwise the barrier may be assumed to be thin.

Similar results are obtained for earth mounds, with the effective barrier width being the width of the top of the earth mound. Any trees planted on top of the earth mound are not considered to contribute significantly to the barrier attenuation and can be ignored.

Example 8.7

A point source of low-frequency, broadband sound at 1 m above the ground introduces unwanted noise at a receiver, also 1 m above the ground at 4 m distance. The ground surface is grass. What is the effect in the 250 Hz octave band on the receiver, of a barrier centrally located, 2 m high and 6 m wide?

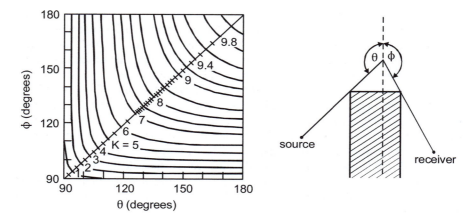

Figure 8.17 Finite width barrier correction factor, K (Equation (8.122)).

Solution

As octave bands are considered, waves reflected from the ground are combined incoherently with non-reflected waves. In fact, all sound waves arriving at the receiver by different paths are combined incoherently.

1. Calculate the reflection coefficient of the ground. The flow resistivity, R_1, for grass-covered ground is approximately 2×10^5 MKS rayls/m (see Table 5.2). This results in a value of flow resistance parameter, $\rho f / R_1 = 1.5 \times 10^{-3}$. For the geometry described, the angle of incidence from the horizontal for the ground reflection of the wave diffracted over the top of the barrier is 56°, and 15.5° for the waves diffracted around the ends. The reflection losses, A_R, corresponding to these angles of incidence (56° and 15.5°) and flow resistance parameter 1.5×10^{-3}, are 1.3 dB and 5.0 dB respectively (see Figure 5.20).

2. Next, calculate the noise level at the receiver due to each diffracted path. For each diffracted path calculate the required path lengths ($A + B$ and d, see Figure 8.13), Fresnel number, N (Equation (8.101)) attenuation A_b (Figure 8.14 and Equation (8.104)). Add the reflection loss A_R (where appropriate) to obtain the total attenuation $A_b + A_R$.

    (a) Reflected waves over the top (three paths): Image source–receiver path:

    $$d = 2\sqrt{5} = 4.5 \text{ m}; \quad A + B = \sqrt{13} + \sqrt{5} = 5.8 \text{ m}; \quad N = 1.9;$$

    $$A_b = 15.8 + 20\log_{10}[5.8/4.5] = 18.0 \text{ dB}; \quad A_R = 1.3 \text{ dB}; \quad A_b + A_R = 19.3 \text{ dB}$$

    Source–image receiver path (same as above):

    $$A_b + A_R = 19.3 \text{ dB}$$

Image source–image receiver path:

$$d = 4 \text{ m}; \quad A + B = 2\sqrt{13} = 7.2 \text{ m}; \quad N = 4.7;$$

$$A_b = 19.8 + 20\log_{10}[7.2/4] = 24.9 \text{ dB}; \quad A_R = 2.6 \text{ dB}; \quad A_b + A_R = 27.5 \text{ dB}$$

(b) Reflected waves around barrier ends (two paths):
Image source–receiver path (using Equations (8.113) to (8.118)):

$$d = 4.5 \text{ m}; \quad A'' + B'' = 2\sqrt{14} = 7.5 \text{ m}; \quad N = 4.3;$$

$$A_b = 19.5 + 20\log_{10}[7.5/4.5] = 23.9 \text{ dB}; \quad A_R = 5 \text{ dB}; \quad A_b + A_R = 28.9 \text{ dB}$$

(c) Non-reflected waves (three paths):
Source–receiver path over top of barrier:

$$d = 4 \text{ m}; \quad A + B = 2\sqrt{5} = 4.5 \text{ m}; \quad N = 0.7;$$

$$A_b = 12.0 + 20\log_{10}[4.5/4] = 13.0 \text{ dB}$$

Source–receiver path around barrier ends (two paths):
From Equation (8.115), $A' = B' = \sqrt{13}$. Thus,

$$d = 4 \text{ m}; \quad A' + B' = 2\sqrt{13} = 7.2 \text{ m}; \quad N = 4.6;$$

$$A_b = 19.8 + 20\log_{10}[7.2/4] = 24.9 \text{ dB}$$

3. Using Equation (1.101), combine all attenuations for each of the eight paths with the barrier present (19.3 dB, 19.3 dB, 27.5 dB, 28.9 dB, 28.9 dB, 13 dB, 24.9 dB and 24.9 dB) (NR_{Bi} in Equation (1.101)) and the attenuations of the two paths (NR_{Ai} in Equation (1.101)), when the barrier is absent (0 dB and 3 dB – see Figure 5.20 with $\beta = 27°$) to give an overall attenuation of approximately 12 dB. This is 3 dB less than the value that would have been obtained if all ground reflections were ignored, and only the diffraction over the top of the barrier were considered. This is a significant difference in this instance, although in many cases in which the width of the barrier is large in comparison with the height and source to receiver spacing, results of acceptable accuracy are often obtained by considering only diffraction over the top of the barrier and ignoring ground reflection. Note that the final result of the calculations is given to the nearest dB because this is the best accuracy which can be expected in practice and the accuracy with which Figure 8.14 may be read is in accord with this observation.

Example 8.8

Given an omnidirectional noise source with a sound power level of 127 dB re 10^{-12} W in the 250 Hz octave band, calculate the sound pressure level in the 250 Hz octave band inside a building situated 50 m away. One side of the building faces in the direction of the sound source and all building walls, including the roof (which is flat), have an average field incidence transmission loss of 20 dB in the 250 Hz octave band.

The building is rectangular in shape. Assume that the total Sabine absorption in the room (5 m × 5 m × 5 m) in the 250 Hz octave band is 15 m². Also assume that the ground between the source and building is acoustically hard (for example, concrete), and the excess attenuation due to ground effect is -3 dB.

Solution

First of all, calculate the sound pressure level incident, on average, on each of the walls and roof. Without the building, the sound pressure level 50 m from the source is determined by substitution of Equation (5.173), for a point source, into Equation (5.171) then calculating:

$$L_p = 127 - 20\log_{10}50 - 11 + 3 = 85 \text{ dB re } 20 \text{ μPa}$$

Next, calculate the incident sound power on each wall:

1. Wall facing the source:

$$L_w = 85 + 10\log_{10}25 = 99 \text{ dB re } 10^{-12} \text{ W}$$

2. Walls adjacent to the one facing the source, and the ceiling: the effective Fresnel number due to the barrier effect of the building is zero, resulting in a 5 dB reduction in noise level due to diffraction. Thus:

$$L_w = 99 - 5 = 94 \text{ dB re } 10^{-12} \text{ W}$$

3. Wall opposite the source: the effective Fresnel number due to the barrier effect of the building varies from $N = (2/\ell)(2.5)$ for sound diffracted to the centre of the wall ($A \approx d$, $B=0.5$) to zero ($A \approx d$, $B=0.0$) for sound diffracted to the edges of the wall. From Figure 8.14, this corresponds to a noise reduction ranging from 18 dB to 5 dB. The additional effect due to the finite thickness of the building is calculated using Equation (8.122) and is approximately 1 dB. Thus the area-weighted average noise reduction for the sound incident on this wall is approximately 12 dB. Thus $L_w = 99 - 12 = 87$ dB re 10^{-12} W.

Next, calculate the total power radiated into the enclosure:

1. Wall facing the source: as the sound field is normally incident, the normal incidence transmission loss must be used. For a 1/3 octave band of noise, $TL_N = TL + 5.5$ (Equation (8.37)). Thus the power radiated into the enclosure is 74 dB.

2. All other walls and ceiling: for these the transmission loss to be used is the field-incidence transmission loss. Thus the total sound power radiated into the room by these surfaces is the logarithmic sum (see Section 1.11.4) of 74 dB, 74 dB, 74 dB and 67 dB, and is equal to 79 dB. The total power radiated into the room is the logarithmic sum of the front wall contribution (74 dB) and all other walls

(79 dB), and is equal to 80 dB. The sound pressure level in the room is given by Equation (8.97) as:

$$L_p = 80 + 10 \log_{10}\left(\frac{1}{125} + \frac{4}{15} \right) \approx 74 \text{ dB re } 20\mu\text{Pa}$$

8.5.2.2 Shielding by Terrain

For outdoor sound propagation over undulating or mountainous terrain, the equivalent barrier effect due to the terrain is calculated using the geometry shown in Figure 8.18.

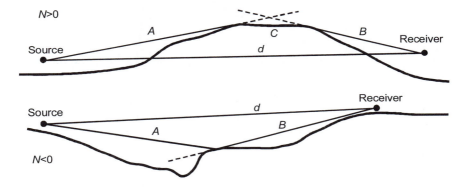

Figure 8.18 Geometry for calculation of barrier effect due to terrain.

The equivalent Fresnel number is:

$$N = \pm \frac{2}{\lambda}[A + B + C - d] \tag{8.123}$$

where the negative sign is used if there is direct line-of-sight between the source and receiver and the positive sign is used if there is no direct line of sight.

8.5.2.3 Effects of Wind and Temperature Gradients on Barrier Attenuation

The attenuation of a barrier outdoors is affected by vertical atmospheric wind and temperature gradients. This effect can be taken into account approximately as follows for diffraction over the top of a barrier (Tonin, 1985).

The effect of a positive vertical wind or temperature gradient (that is wind speed or temperature increasing with altitude) is to effectively move the source higher up and further from the barrier, as shown in Figure 8.19. The effective relative position of the receiver with respect to the wall is affected similarly.

The effect of negative gradients (that is wind speed or temperature decreasing with altitude) is to effectively move the source lower and closer to the barrier. Note that wind speed is measured in the direction from source to receiver.

The radius of curvature, R, of the curved sound ray shown in Figure 8.19 is given by Equation (5.203). If R in Equation (5.203) is positive, then rays are curved downwards as shown in Figure 8.19, resulting in less attenuation. If R is negative then the converse is true.

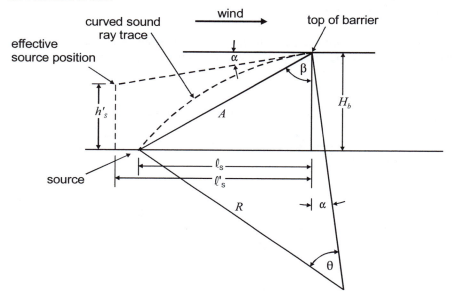

Figure 8.19 Geometry for locating the effective source position, assuming a positive vertical sonic gradient.

Referring to Figure 8.19, the effective source position given by ℓ'_s, which is the horizontal distance of the effective source from the barrier, and h'_s, which is the increase in height of the effective source above the actual source, may be calculated given the radius of curvature, R, using the following equations (Tonin, 1985):

$$\ell'_s = R\,\theta\cos\alpha$$

$$h'_s = H_b - R\,\theta\sin\alpha$$

$$\alpha = \frac{1}{2}(\pi - \theta) - \beta \qquad\qquad (8.124)$$

$$\beta = \cos^{-1}(H_b/A)$$

$$\theta = \pm\cos^{-1}[1 - (A^2/2R^2)], \quad |R| > A/2$$

where H_b is the actual difference between source and barrier heights, and A is the distance from the actual source to the barrier top. Note that θ must have the same sign as R.

The same analysis must be used when calculating the effective receiver location with respect to the barrier, but is not used for waves travelling around the sides of a barrier.

For the purposes of calculating the contributions of the reflected waves, for the case of positive wind and temperature gradients, the image source or receiver is as far below the ground as the new source or receiver position is above ground. The same applies for negative gradients except when the gradient results in the new source or receiver position being below ground. In this case the image source to be used for the reflected wave calculation would be twice the original source height below the new source. The same argument would apply to the location of the new image receiver.

8.5.2.4 ISO 9613-2 Approach to Barrier Insertion Loss Calculations

Following the International Standard procedure, the barrier noise reduction (or insertion loss) for path i around or over the barrier is calculated using the following equation instead of Figure 8.14:

$$A_{bi} = 10\log_{10}[3 + 10N_i C_3 K_{met}] - A_g \qquad \text{dB} \qquad (8.125)$$

where A_g is the excess attenuation due to the ground in the absence of the barrier and N_i is the Fresnel number (Equation (8.101)) for path i. The ground effect in the presence of the barrier is included in the term in square brackets in Equation (8.125). Equation (8.125) includes the effect of ground reflections either side of the barrier. If the ground reflected paths are included separately as described in Section 8.5.2, prior to 8.5.2.1, then the factor of "10" term in Equation (8.125) should be replaced with "20" and then Equation (8.125) is used for each path and the noise reductions of each path are combined using Equation (1.100).

The term, K_{met} (see Equation (8.132)) includes the effect of adverse wind and temperature gradients (such as temperature inversion and / or downwind propagation as discussed in the previous section). For diffraction around barrier ends, $K_{met} = 1$.

If it is decided to treat ground reflected paths separately, the overall excess attenuation due to the barrier with N propagation paths around and over it (including paths involving a ground reflection on one or both sides of the barrier) is given by:

$$A_b = -10\log_{10}\sum_{i=1}^{n_A} 10^{-(A_{bi} + A_{Ri})/10} \qquad (8.126)$$

where A_{Ri} is the ground reflection loss involved with path i over or around the barrier.

If ground reflected paths are not included separately, then equation (8.125) can be used to estimate the overall insertion loss of the barrier directly.

When calculating the total excess attenuation due to a barrier using the ISO model, the quantity, A_b, from the above equation is included in Equations (5.171) and (5.178) along with ALL other terms, including the ground term, A_g.

In Equation (8.125) C_3 is equal to 1 for a single edge diffraction and for double edge diffraction (such as for a thick barrier or two parallel barriers – see Figure 8.20) it is given by:

$$C_3 = [1 + (5\lambda/b)^2] / [(1/3) + (5\lambda/b)^2] \qquad (8.127)$$

where C_3 varies between 1 (for a single edge diffraction and 3 (for a well separated double edge diffraction $(b>>\lambda)$.

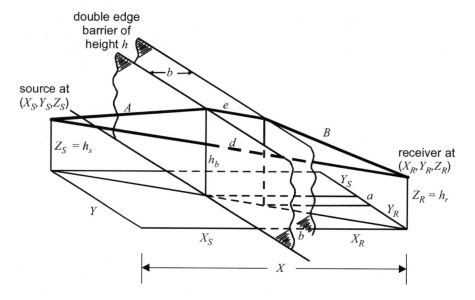

Figure 8.20 Geometry for double edge diffraction (single, thick barrier or two thin, parallel barriers).

For the double diffraction barrier, the Fresnel number is given by:

$$N = \pm(2/\lambda)(A + B + e - d) \qquad (8.128)$$

The quantities, A, B and d are calculated using the same equations as for the single diffraction barrier (Equations (8.101) and (8.102)) with the additional equation:

$$e = (a^2 + b^2)^{1/2} \qquad (8.129)$$

and where in this case, $X = X_R + X_S + b$ and $Y = Y_R + Y_S + a$.

If the two edges are at different heights, h_{b1} and h_{b2}, the quantity, e is given by:

$$e = \left(a^2 + b^2 + (h_{b1} - h_{b2})^2\right)^{1/2} \qquad (8.130)$$

An easier expression to evaluate for N is:

$$N = \pm\frac{2}{\lambda}\left\{\left[\left[\left(X_S^2 + (h_b - Z_S)^2\right)^{1/2} + \left(X_R^2 + (h_b - Z_R)^2\right)^{1/2} + b\right]^2 + Y^2\right]^{1/2} - d\right\} \qquad (8.131)$$

where for the case of different height barriers, the quantity, b, is replaced with $(b^2 + (h_{b1} - h_{b2})^2)^{1/2}$.

The quantity, K_{met}, of Equation (8.125) takes into account the effect of downwind or temperature inversion conditions and is given by:

$$K_{met} = \exp\left[-(1/2000)\sqrt{\frac{ABd}{2(A + B + e - d)}}\right] \qquad (8.132)$$

For the receiver in the bright zone or for diffraction around the ends of the barrier, or for distances between source and receiver of less than 100 m, $K_{met} = 1$. Note that the ISO procedure using Equation (8.125) calculated barrier attenuation for the worst case atmospheric conditions; that is, downwind conditions from source to receiver.

If there are more than two barriers between the source and receiver, ISO 9613 – 2 recommends that the two most effective barriers be included and that all other barriers be ignored.

The calculated total barrier attenuation in any octave band for the double diffraction case should not exceed 25 dB. For a single barrier it should not exceed 20 dB.

Barrier attenuation is adversely affected by reflections from vertical surfaces such as building facades. In this case the reflection must be considered as giving rise to an image source, which must be treated separately and its contribution added to the sound level at the receiver due to the non-reflected wave.

8.5.3 Indoor Barriers

Thus far, only the use of barriers outdoors has been considered or, more explicitly, the situation in which the contribution of the source direct field to the overall sound level is much larger than any reverberant field contribution. In this section, the effect of placing a barrier in a room where the reverberant sound field and reflections from other surfaces cannot be ignored will be considered.

The following assumptions are implicit in the calculation of the insertion loss for an indoor barrier:

1. The transmission loss of the barrier material is sufficiently large that transmission through the barrier can be ignored. A transmission loss of at least 20 dB in the frequency range of interest is recommended.

2. The sound power radiated by the source is not affected by insertion of the barrier.

3. The receiver is in the shadow zone of the barrier; that is, there is no direct line of sight between source and receiver.

4. Interference effects between waves diffracted around the side of the barrier, waves diffracted over the top of the barrier and reflected waves are negligible. This implies octave band analysis.

The barrier insertion loss, IL, in dB is (Moreland and Minto, 1976):

$$IL = 10 \log_{10}\left(\frac{D}{4\pi r^2} + \frac{4}{S_0 \overline{\alpha}_0} \right) - 10 \log_{10}\left(\frac{DF}{4\pi r^2} + \frac{4K_1 K_2}{S(1 - K_1 K_2)} \right) \quad (8.133)$$

In Equation (8.133), D is the source directivity factor in the direction of the receiver (for an omni-directional source on a hard floor, $D = 2$); r is the distance between source and receiver in the absence of the barrier; $S_0 \overline{\alpha}_0$ is the absorption of the original room before inserting the barrier, where S_0 is the total original room surface area and $\overline{\alpha}_0$ is the room mean Sabine absorption coefficient; S is the open area between the barrier perimeter and the room walls and ceiling; and F is the diffraction coefficient given by the following equation:

$$F = \sum_i \frac{1}{3 + 10N_i} \quad (8.134)$$

N_i is the Fresnel number for diffraction around the ith edge of the barrier, and is given by Equation (8.101). K_1 and K_2 are dimensionless numbers given by:

$$K_1 = \frac{S}{S + S_1 \overline{\alpha}_1}, \quad K_2 = \frac{S}{S + S_2 \overline{\alpha}_2} \quad (8.135a,b)$$

S_1 and S_2 are the total room surface areas on sides 1 and 2 respectively of the barrier i.e. $S_1 + S_2 = S_0 +$ (area of two sides of the barrier). The quantities $\overline{\alpha}_1$ and $\overline{\alpha}_2$ are the mean Sabine absorption coefficients associated with areas S_1 and S_2 respectively.

When multiple barriers exist, as in an open-plan office, experimental work (West and Parkin, 1978) has shown the following statements to be true in a general sense (test screens were 1.52 m high by 1.37 m wide):

1. No difference in attenuation is obtained when a 300 mm gap is permitted between the base of the screen and the floor.

2. When a number of screens interrupt the line of sight between source and receiver, an additional attenuation of up to 8 dB(A) over that for a single screen can be realised.

3. Large numbers of screens remove wall reflections and thus increase the attenuation of sound with distance from the source.

4. For a receiver immediately behind a screen, a local shadow effect results in large attenuation, even for a source a large distance away. This is in addition to the effect mentioned in (2) above.

5. For a screen less than 1 m from a source, floor treatment has no effect on the screen's attenuation.

6. A maximum improvement in attenuation of 4-7 dB as frequency is increased from 250 Hz to 2 kHz can be achieved by ceiling treatment. However, under most conditions, this greater attenuation can only be achieved at the higher frequencies.

7. Furnishing conditions are additive; that is, the attenuations measured under two
    different furnishing conditions are additive when the two conditions coexist.

8.6 PIPE LAGGING

Radiation from the walls of pipes or air conditioning ducts is a common source of
noise. The excitation usually arises from disturbed flow through valves or dampers,
in which case it is preferable to reduce the excitation by treatment or modification of
the source. However, as treatment of the source is not always possible, an alternative
is to acoustically treat the walls of the pipe or duct to reduce the transmitted noise. For
ventilation ducts the most effective solution is to line the duct internally with acoustic
absorbent, whereas with pipework an external treatment is normally used. The former
treatment is discussed in Section 9.8.3 and the latter treatment is discussed here.

8.6.1 Porous Material Lagging

The effect of wrapping a pipe with a layer of porous absorbent material may be
calculated by taking into account sound energy loss due to reflection at the porous
material surfaces and loss due to transmission through the material. Methods for
calculating these losses are outlined in Appendix C. The procedures outlined in
Appendix C only include the effects of the pipe fundamental "breathing" mode on the
sound radiation. Kanapathipillai and Byrne (1991) showed that pipe "bending" end
"ovaling" modes are also important and the sound radiation from these is influenced
in a different way by the lagging. Indeed, for a pipe vibrating in these latter modes, the
insertion loss of a porous lagging is negative at low frequencies, and a sound increase
is observed, partly as a result of the effective increase in sound radiating area. In
practice, it may be assumed that below 250 Hz, there is nothing to be gained in terms
of noise reduction by lagging pipework with a porous acoustic blanket.

8.6.2 Impermeable Jacket and Porous Blanket Lagging

Several theories have been advanced for the prediction of the noise reduction (or
insertion loss) resulting from wrapping a pipe first with glass-fibre and then with a
limp, massive jacket. Unfortunately, none of the current theories reliably predicts
measured results; in fact, some predicted results are so far removed from reality as to
be useless. Predictions based upon the theory presented in the following paragraphs
have been shown to be in best agreement with results of experiment, but even these
predictions can vary by up to 10 dB from the measured data. For this reason, the
analysis is followed with Figure 8.21, which shows the measured insertion loss for a
commonly used pipe lagging configuration; that is 50 mm of 70-90kg/m^3 glass-fibre
or rockwool covered with a lead–aluminium jacket of 6 kg/m^2 surface density. Before
proceeding with the analysis it should be noted that, where possible, manufacturer's
data rather than calculations should be used. In practice, it has been found that at

frequencies below about 300 Hz, the insertion loss resulting from this type of treatment is either negligible or negative. It has also been found that porous acoustic foam gives better results than fibre-glass or rockwool blanket because of it greater compliance. Note, however, that the following analysis applies to fibreglass or rockwool blankets only.

For the purposes of the analysis, the frequency spectrum is divided into three ranges (Hale, 1978) by two characteristic frequencies, which are the ring frequency, f_r, and the critical frequency, f_c, of the jacket ($f_r < f_c$) The jacket is assumed to be stiff and at the ring frequency the circumference of the jacket is one longitudinal wavelength long; thus $f_r = c_L / \pi d$. The critical frequency given by Equation (8.3) is discussed in Section 8.2.1. For jackets made of two layers (such as lead and aluminium), Equations (8.6) to (8.9) must be used to calculate f_c and c_L.

1. In the low-frequency range below the jacket ring frequency, f_r, the insertion loss is:

$$IL = 10 \log_{10}[1 - 0.012 X_r \sin 2C_r + (0.012 X_r \sin C_r)^2] \qquad (8.136)$$

where:

$$X_r = [1000(m/d)^{1/2}(\xi_r - \xi_r^2)^{1/4}] - [2d/\ell\xi_r)] \qquad (8.137)$$

and:

$$C_r = 30\xi_r \ell/d \qquad (8.138)$$

where $\xi_r = f/f_r$, d is the jacket diameter (m), m is the jacket surface density (kg m^{-2}), ℓ is the absorptive material thickness (m), f_r is the jacket ring frequency (Hz), c_L is the longitudinal wave speed (thin panel) in the jacket material, and f is the octave or one-third octave band centre frequency.

2. In the high-frequency range above the critical frequency, f_c, of the jacket (see Equation (8.3)), the insertion loss is:

$$IL = 10 \log_{10}[1 - 0.012 X_c \sin 2C_c + (0.012 X_c \sin C_c)^2] \qquad (8.139)$$

where:

$$X_c = [41.6(m/h)^{1/2}\xi_c(1 - 1/\xi_c)^{-1/4}] - [258h/\ell\xi_c)] \qquad (8.140)$$

and:

$$C_c = 0.232\xi_c \ell/h \qquad (8.141)$$

The quantity h is the jacket thickness (m), $\xi_c = f/f_c$, and f_c is the jacket critical frequency (Hz).

3.   In the mid-frequency range between f_r and f_c

$$IL = 10 \log_{10}[1 - 0.012X_m \sin 2C_c + (0.012X_m \sin C_c)^2]$$ (8.142)

where

$$X_m = [226(m/h)^{1/2}\xi_c(1 - \xi_c^2)] - [258h/(\ell\xi_c)]$$ (8.143)

As an alternative to the preceding prediction scheme, a simpler formula (Michelsen *et al.*, 1980) is offered, which provides an upper bound to the expected insertion loss at frequencies above 300 Hz. That is:

$$IL = \frac{40}{1 + 0.12/D} \log_{10}\frac{f\sqrt{m\ell}}{132}$$ (8.144)

where D is the pipe diameter (m) and $f \geq 120/(m\ell)^{1/2}$. The preceding equations are based on the assumption that there are no structural connections between the pipe and the jacket. If solid spacers are used to support the jacket, then the insertion loss will be substantially less.

In recent times, the use of acoustic foam in place of rockwool and fibreglass has become more popular. Manufacturers of the pre-formed foam shapes (available for a wide range of pipe diameters) claim superior performance over that achieved with the same thickness of rockwool. An advantage of the foam (over rockwool or fibreglass) is that it doesn't turn to powder when applied to a pipe suffering from relatively high vibration levels. However, acoustic foam is much more expensive than rockwool or fibreglass.

CHAPTER NINE

Muffling Devices

LEARNING OBJECTIVES

In this chapter the reader is introduced to:

- noise reduction, *NR*, and transmission loss, *TL*, of muffling devices;
- diffusers as muffling devices;
- classification of muffling devices as reactive and dissipative;
- acoustic impedance for analysis of reactive mufflers;
- acoustic impedance of orifices and expansion chambers;
- analysis of several reactive muffler types;
- pressure drop calculations for reactive mufflers;
- lined duct silencers as dissipative mufflers;
- design and analysis of lined ducts;
- duct break-out noise transmission calculations;
- lined plenum attenuators;
- water injection for noise control of exhausts; and
- directivity of exhaust stacks.

9.1 INTRODUCTION

Muffling devices are commonly used to reduce noise associated with internal combustion engine exhausts, high pressure gas or steam vents, compressors and fans. These examples lead to the conclusion that a muffling device allows the passage of fluid while at the same time restricting the free passage of sound. Muffling devices might also be used where direct access to the interior of a noise containing enclosure is required, but through which no steady flow of gas is necessarily to be maintained. For example, an acoustically treated entry way between a noisy and a quiet area in a building or factory might be considered as a muffling device.

Muffling devices may function in any one or any combination of three ways: they may suppress the generation of noise; they may attenuate noise already generated; and they may carry or redirect noise away from sensitive areas. Careful use of all three methods for achieving adequate noise reduction can be very important in the design of muffling devices, for example, for large volume exhausts.

9.2 MEASURES OF PERFORMANCE

Two terms, insertion loss, *IL*, and transmission loss, *TL*, are commonly used to describe the effectiveness of a muffling system. These terms are similar to the terms noise reduction, *NR*, and transmission loss, *TL*, introduced in Chapter 8 in connection with sound transmission through a partition.

The insertion loss of a muffler is defined as the reduction (in decibels) in sound power transmitted through a duct compared to that transmitted with no muffler in place. Provided that the duct outlet remains at a fixed point in space, the insertion loss will be equal to the noise reduction that would be expected at a reference point external to the duct outlet as a result of installing the muffler. The transmission loss of a muffler, on the other hand, is defined as the difference (in decibels) between the sound power incident at the entry to the muffler to that transmitted by the muffler.

Muffling devices make use of one or the other or a combination of two effects in their design. Either, sound propagation may be prevented (or strongly reduced) by reflection or suppression, or sound may be dissipated. Muffling devices based upon reflection or source sound power output suppression are called reactive devices and those based upon dissipation are called dissipative devices.

The performance of reactive devices is dependent upon the impedances of the source and termination (outlet). In general, a reactive device will strongly affect the generation of sound at the source. This has the effect that the transmission loss and insertion loss of reactive devices may be very different. As insertion loss is the quantity related to noise reduction, it will be used here to describe the performance of reactive muffling devices in preference to transmission loss (*TL*); however, *TL* will be also considered for some simple reactive devices.

The performance of dissipative devices, on the other hand, by the very nature of the mode of operation, tends to be independent of the effects of source and termination impedance. Provided that the transmission loss of a dissipative muffler is at least 5 dB it may be assumed that the insertion loss and the transmission loss are the same. This assertion is justified by the observation that any sound reflected back to the source through the muffler will be reduced by at least 10 dB and is thus small and generally negligible compared to the sound introduced. Consequently, the effect of the termination impedance upon the source must also be small and negligible.

9.3 DIFFUSERS AS MUFFLING DEVICES

A commonly used device, often associated with the design of dissipative mufflers for the reduction of high-pressure gas exhaust noise, is a gas diffuser. When properly designed, this device can very effectively suppress the generation of noise. Alternatively, if attention is not given to the design it can become a serious source of noise.

Aerodynamic sources were considered in Chapter 5, where it was shown that a fluctuating force or stress on a moving fluid can produce a dipole or quadrupole source of sound. As these sources were shown to generate sound power proportional to the sixth and eighth power respectively of the stream speed, it should be an aim of the diffuser design to minimise the generation of fluctuating forces and stresses. For this reason, a diffuser should have as its primary function the reduction of the pressure gradient associated with the exhaust. Thus, while the pressure drop may be fixed, the gradient may be reduced by extending the length over which the pressure drop is produced. This is quite often accomplished by providing a tortuous path for the exhausting gas.

In the mixing region between the exhausting gas and the ambient or slowly moving air, very large shear stresses can occur, giving rise to quadrupole noise generation. Such noise can become very serious at high discharge rates because of its eighth power dependence on stream speed. Thus a second function of the diffuser should be to reduce the shear in the mixing region between the exhausting gas and ambient or slowly moving air in the neighbourhood of the exhaust stream. This is quite often accomplished by breaking the exhaust stream up into many small streams. A common design, illustrated in Figure 9.1 is to force the exhaust to pass through the myriad holes in a perforated plate, perhaps incorporated in the muffler as a closed perforated cylinder. Such a design can also accomplish the task of reducing the pressure gradient.

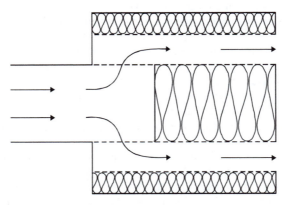

Figure 9.1 A common design for a high pressure gas exhaust muffler.

A third function of the diffuser should be to stabilise and reduce the magnitude of any shock waves developed in the exhaust. Stabilisation of shock waves is particularly important because an unstable, oscillating shock wave can be a very powerful generator of noise. In this connection, a diffuser that breaks the exhaust stream up into many smaller streams has been found experimentally to be quite effective. Such a device, illustrated in Figure 9.2, has been shown to accomplish by itself, without any additional muffling, a 10 dB insertion loss in broadband noise in a steam-generating plant blow-out operation.

9.4 CLASSIFICATION OF MUFFLING DEVICES

As an aid in the discussion that follows, it is convenient to classify commonly used muffling devices in some systematic way. A classification suitable for the present purpose is presented in Table 9.1, in which seven types of devices are identified in column one, and additionally classified according to the mechanism by which they

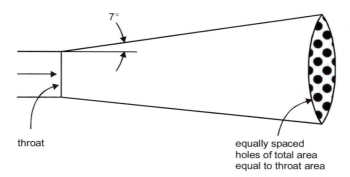

throat

equally spaced
holes of total area
equal to throat area

Figure 9.2 A noise suppression diffuser.

function in column two. Thus under the heading mechanisms, in column two, the devices are identified either as suppressive (reactive) or dissipative.

In the third, fourth and fifth columns of the table, the effective frequency range of the device and the critical dimensions are indicated. Both the effective frequency range and critical dimensions are given in dimensionless form, the latter being expressed in wavelengths of the sound to be attenuated. This choice of representation emphasises the fact that, in principle, the devices are not restricted to particular frequency ranges or sizes.

In practical designs, it will generally be found that reactive devices are favoured for the control of low-frequency noise, since they tend to be more compact than dissipative devices for the same attenuation. Conversely, dissipative devices are generally favoured for the attenuation of high-frequency noise, due to their simpler and cheaper construction and greater effectiveness over a broad frequency range.

The critical dimensions of columns four and five in Table 9.1 are arbitrary and the reader is warned of possible exceptions. Thus the width of the side branch resonator shown as not exceeding one-eighth could be increased to perhaps one-half for a special design, and similar comments are true of the other dimensions shown. However, in these cases, the lumped element analysis used in this chapter to determine the resonator dimensions for a particular design frequency may not be very accurate and some *in-situ* adjustment should be allowed for. Side branch resonators also exhibit noise attenuations at frequencies that are multiples of the design frequency.

The last column of the table shows the expected dependence of performance, of the devices listed, on the characteristics of the source and termination.

9.5 ACOUSTIC IMPEDANCE

Various types of impedance were discussed in Section 1.12. In the discussion of muffling devices and sound propagation in ducts, the most commonly used and useful type of impedance is acoustic impedance. This is an important parameter in the discussion of reactive muffling devices, where the assumption is implicit that the muffler cross-dimensions are small compared to a wavelength. In this case only plane waves propagate and a volume velocity, which is continuous at junctions in a ducted system, may be defined as the product of the particle velocity and muffler cross-section. The acoustic impedance relates the acoustic pressure to the associated volume velocity. At a junction between two ducts, the acoustic pressure will also be continuous (Kinsler *et al.*, 1982); thus the acoustic impedance has the useful property that it is continuous at junctions in a ducted system.

Table 9.1 Classification of muffling devices

Device	Mechanism	Effective frequency range	Critical dimensions $D = f\ell/c = \ell/\lambda$		Dependence of performance on end conditions
			Length	Width	
1. Lumped element	Suppressive	Band	$D < 1/8$	$D < 1/8$	Critical
2. Side branch resonator	Suppressive	Narrow band	$D \le 1/4$	$D < 1/8$	Critical
3. Transmission line	Suppressive	Multiple bands	$D > 1/8$	$D < 1/4$	Critical
4. Lined duct	Dissipative	Broadband	Unbounded[b]		Slightly dependent
5. Lined bend	Dissipative	Broadband	$D > \frac{1}{2}$	$D > \frac{1}{2}$	Not critical
6. Plenum chamber	Dissipative/ suppressive	Broadband	$D > 1$	$D > 1$	Not critical
7. Water injection	Dissipative	Broadband	Unbounded		Not critical

[a] f, c, λ and ℓ are respectively frequency, speed of sound, wavelength of sound, and critical dimension of the device.
[b] Theoretically, D is unbounded, but a practical lower bound for D is about 1/4.

As will be shown, a knowledge of the acoustic impedance of sources would be useful, but such information is generally not available. However, those devices characterised by fixed cyclic volume displacement, such as reciprocating pumps and compressors and internal combustion engines, are well described as constant acoustic volume–velocity sources of infinite internal acoustic impedance. On the other hand, other devices characterised by an induced pressure rise, such as centrifugal and axial fans and impeller-type compressors and pumps, are probably best described as constant acoustic-pressure sources of zero internal impedance. The case for the engine exhaust is a bit better, since the impedance for a simple exhaust pipe has been modelled as a vibrating piston at the end of a long pipe, analysed theoretically and verified by measurement (see Chapter 5, Figure 5.9), and such a termination is the most common case. A loudspeaker backed by a small air-tight cavity may be approximated as a constant volume velocity source at low frequencies where the wavelength is much greater than any cavity dimension.

9.6 LUMPED ELEMENT DEVICES

The discussion of lumped element devices and the side branch resonator will be carried out in terms of acoustical circuits, which are analogs of electrical circuits. For this purpose it will be necessary to develop expressions for the acoustical analogs of electrical inductances and capacitances. Using electrical analogies, voltage drop is the analog of acoustic pressure, and current is the analog of acoustic volume velocity. Thus the acoustic pressure in a volume or across an orifice is the analog of the voltage across a capacitance or an inductance respectively.

Note that, following the definition of acoustic volume velocity in Section 9.5 and acoustic intensity given by Equations (1.74) and (1.75), the acoustic power transmission in a duct is proportional to the mean square pressure or particle velocity.

9.6.1 Impedance of an Orifice or a Short Narrow Duct

Consideration of the acoustically induced cyclic flow through an orifice, as illustrated in Figure 9.3, shows that the acoustic energy will pass through the orifice largely in the form of kinetic energy. There will be a small volume of air somewhat larger than the

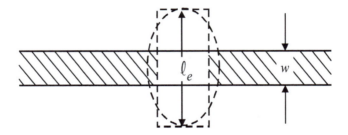

Figure 9.3 A schematic representation of acoustically induced flow through an orifice shown on edge. A cylinder of equivalent inertia of area A and effective length ℓ_e is indicated.

actual orifice that will participate in the induced motion. The bounds of this volume are obviously quite arbitrary, but the volume can be imagined as a little cylinder of air of cross-section equal to the area of the orifice and some effective length ℓ_e. Its motion will be governed by Newton's second law, written as:

$$- \frac{\partial p}{\partial x} = \rho \left[\frac{\partial u}{\partial t} + U \frac{\partial u}{\partial x} \right] \tag{9.1}$$

In Equation (9.1), p (Pa) is the acoustic pressure, x (m) is the displacement coordinate along the orifice axis, u (m/s) is the particle velocity, ρ (kg/m^3) is the density of the gas, t (s) is time, and U (m/s) is any convection velocity in the direction of sound propagation through the orifice of cross-sectional area, S.

To proceed, a solution is assumed for Equation (9.1) of the form:

$$u = u_0 e^{j(\omega t - kx)} \tag{9.2}$$

where ω (rad/s) is the angular frequency, $k = \omega/c = 2\pi/\lambda$ is the wave number (m⁻¹), c (m/s) is the speed of sound, λ (m) is the wavelength of sound and j is the imaginary number $\sqrt{-1}$. It will be assumed that sound propagation and direction of mean flow are the same. The case for mean flow opposite to the direction of sound propagation is not analysed in detail here, but would predict an increasing end correction $(1 + M)$, in the last term of Equation (9.6) below (instead of the $(1 - M)$ term) with a corresponding change in Equation (9.3) that follows. Substitution of Equation (9.2) into Equation (9.1) results in the following expression:

$$-\frac{\partial p}{\partial x} = \rho \left[j\omega u + U \left[\frac{1}{u_0} \frac{\partial u_0}{\partial x} - \frac{j\omega}{c} \right] u \right] \tag{9.3}$$

Equation (9.3) may be simplified by making use of the definition of the Mach number for the convection velocity $M = U/c$, and the following approximations:

$$-\frac{\partial p}{\partial x} \approx \frac{p}{\ell_e} \tag{9.4}$$

$$\frac{1}{u_0} \frac{\partial u_0}{\partial x} \approx \frac{1}{u_0} \frac{u_0}{\ell_e} \approx \frac{1}{\ell_e} \tag{9.5a,b}$$

Adding a resistive impedance, R_A, Equation (9.3) can be used to write:

$$Z_A = R_A + jX_A = R_A + \frac{p}{Su} = R_A + j\frac{\rho c}{S} k\ell_e (1 - M) \tag{9.6a,b,c}$$

Reference to Figure 9.3 shows that the effective length ℓ_e is made up of three parts; the length w of the orifice and an end correction at each end. The end corrections are also a function of the Mach number M of the flow through the orifice. Since the system under consideration is symmetric, the end correction for each end without flow is ℓ_0 and then the effective length is:

$$\ell_e \approx w + 2\ell_0 (1 - M)^2 \tag{9.7}$$

The "no flow" end correction, ℓ_0, will be considered in detail later in the next section.

For the case where the length, w, of the orifice is short compared to the effective length ℓ_e, substitution of Equation (9.7) into Equation (9.6) gives the following result:

$$Z_A \approx R_A + j\frac{\rho c}{S} k 2\ell_0 (1 - 3M + 3M^2 - M^3) \tag{9.8}$$

The first term (the real term) in Equations (9.6) and (9.8) can be evaluated using Equation (9.29).

For tubes of very small diameter (such that radius, $a < 0.002/\sqrt{f}$), Beranek (1954) indicates that the imaginary term in Equations (9.6) and (9.8) should be multiplied by 4/3 and the real part is given by Equation (9.32). For slits of cross-sectional area S, the imaginary term in Equations (9.6) and (9.8) should be multiplied by 6/5, with ℓ_e set equal to the depth of the slit, and the real term is given by Equation (9.33) (Beranek, 1954).

Equation (9.8) describes experimental data quite well. The results of experiments indicate that the convection velocity U could be either a steady superimposed flow through the hole or the particle velocity itself at high sound pressure levels. Equations (9.8) and (9.29) show that, for zero mean flow or low sound pressure levels, the acoustic impedance is essentially inductive and of the form:

$$Z_A \approx j\frac{\rho\omega\ell_e}{S} = jX_A \tag{9.9}$$

However, for high sound pressure levels, or in the presence of a significant mean flow, the resistive part of the impedance becomes important, as shown by Equation (9.29). From Equation (9.9), it can be seen that the acoustical inductance, analog of the electrical inductance, is $\rho\ell_e/S$, where S is the duct cross-sectional area.

The following alternative approach is presented as it shows how lumped element analysis is really a first approximation to the more general transmission-line analysis. Consider a plane acoustic wave that propagates to the left in a duct of cross-sectional area, A, and open end, as shown in Figure 9.4.

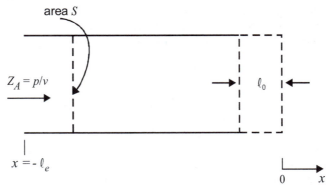

Figure 9.4 A schematic representation of an acoustic transmission line with an open end: Z_a is the acoustic impedance at cross-section of area S, and ℓ_0 is the end correction.

At the open end on the right, a wave will be reflected to the left, back along the duct toward the source. The approximation is now made that the acoustic pressure is essentially zero at a point distant from the duct termination an amount equal to the end correction ℓ_0, and for convenience, the origin of coordinates is taken at this point. Referring to equation (1.34) the equation for waves travelling in both directions in the duct is:

$$\varphi = B_1 e^{j(\omega t - kx)} + B_2 e^{j(\omega t + kx + \theta)} \tag{9.10}$$

Assuming that the wave is reflected at the open end on the right in Figure 9.4, with negligible loss of energy so that the amplitude of the reflected wave, B_2, is essentially equal to the amplitude of the incident wave, B_1, then $B_1 = B_2 = B$. At $x = 0$ the acoustic pressure and thus the velocity potential, φ, is required to be zero; thus, $\theta = \pi$. Use of Equation (1.11) gives for the pressure:

$$p = \rho \frac{\partial \varphi}{\partial t} = j\omega\rho B e^{j\omega t}\left[e^{-jkx} - e^{jkx}\right] \tag{9.11}$$

Similarly, use of Equation (1.10) gives for the particle velocity at any point x in the duct:

$$u = -\frac{\partial \varphi}{\partial x} = jkB e^{j\omega t}\left[e^{-jkx} + e^{jkx}\right] \tag{9.12}$$

The reactive part of the acoustic impedance at x, looking to the right, is:

$$jX_A = \frac{p}{v} = \frac{p}{Su} = -\frac{\rho c}{S}\frac{e^{jkx} - e^{-jkx}}{e^{jkx} + e^{-jkx}} \tag{9.13a-c}$$

or

$$jX_A = -j\frac{\rho c}{S}\tan kx \tag{9.14}$$

This expression can be used to estimate the impedance looking into a tube open at the opposite end, of cross-sectional area, A, and effective length, ℓ_e, by replacing x with $-\ell_e$. Then, for small ℓ_e, Equation (9.9) follows as a first approximation. If the mean flow speed through the tube is non-zero and has a Mach number, M, the following expression is obtained for the impedance looking into a tube open at the opposite end:

$$Z_A = j\frac{\rho c}{S}\tan(k\ell_e(1 - M)) + R_A \tag{9.15}$$

where R_A is the resistive part of the impedance, given by Equation (9.29).

Note that the term, "effective length" is used to describe the length of the tube. This is because the actual length is increased by an end effect at each end of the tube and the amount by which the effective length exceeds the actual length is called the "end correction". Generally, one end correction (ℓ_0) is added for each end of the tube so that the effective length, $\ell_e = w + 2\ell_0(1-M)^2$, where w is the physical length of the tube.

9.6.1.1 End Correction

Reference to Figure 9.3 shows that the end correction accounts for the mass reactance of the medium (air) just outside of the orifice or at the termination of an open-ended tube. The mass reactance, however, is just the reactive part of the radiation impedance presented to the orifice, treated here as a vibrating piston. As shown in Chapter 5, the radiation impedance of any source depends upon the environment into which the source radiates. Consequently, the end correction will depend upon the local geometry at the orifice termination. In general, for a circular orifice of radius a, centrally located in a baffle in a long tube of circular cross-section such that the ratio, ξ, of orifice diameter to tube diameter is less than 0.6, the end correction without either through or grazing flow for each side of the hole is (Bolt *et al.*, 1949):

$$\ell_0 = \frac{8a}{3\pi}(1 - 1.25\xi)$$ (9.16)

As ξ tends to zero, the value of the end correction tends to the value for a piston in an infinite baffle, as may be inferred from the discussion in Section 5.6. Note that when there is flow of Mach number, M, either through or past the hole, the end correction ℓ_0 must be multiplied by $(1 - M)^2$. End corrections for quarter wave tubes and Helmholtz resonators are discussed in Section 9.7.2.1

For orifices not circular in cross-section, an effective radius, a, can be defined, provided that the ratio of orthogonal dimensions of the orifice is not much different from unity. In such cases,

$$a \approx 2S/P_D$$ (9.17)

where S (m^2) is the cross-sectional area of the orifice and P_D (m) is its perimeter. Alternatively, if the orifice is of cross-sectional area, S, and aspect ratio (major dimension divided by minor orthogonal dimension) n, the effective radius, a, may be determined using the following equation:

$$a = K\sqrt{S/\pi}$$ (9.18)

The quantity, K, in the preceding equation is plotted in Figure 9.5 as a function of the aspect ratio n:

For tubes that are unflanged and look into free space, as will be of concern later in discussing engine exhaust tailpipes, the end correction without either through or grazing flow is (Beranek, 1954, p. 133):

$$\ell_0 = 0.61a$$ (9.19)

For holes separated by a distance q (centre to centre, where $q > 2a$) in a perforated sheet, the end correction for each side of a single hole, without either through or grazing flow is:

$$\ell_0 = \frac{8a}{3\pi}(1 - 0.43a/q)$$ (9.20)

The acoustic impedance corresponding to a single hole in the perforated sheet is then calculated using Equation (9.20) with Equations (9.15) and (9.7) where w is the thickness of the perforated sheet. The acoustic impedance for N holes is obtained by dividing the acoustic impedance for one hole by N, so that for a perforated sheet of area S_p, the acoustic impedance due to the holes is:

$$Z_{Ah} = \frac{100}{PS_p}\left(j\rho c\tan(k\ell_e(1 - M)) + R_A S\right)$$ (9.21)

where R_A is the acoustic resistance, defined in Equation (9.29), S is the area of a single hole and P is the % open area of the perforated panel, given by:

$$P = 100NS/S_p = 100N\pi a^2/S_p \qquad (9.22a,b)$$

where N is the total number of holes in area, S_p. The Mach number, M, refers to the speed of the flow either through the holes or across the face of the panel. The Mach number of the grazing flow across the panel is positive if it is on the same side of the holes as the side at which the impedance looking into the holes is required and negative if it is on the opposite side. The Mach number of the through flow is positive if the flow is in the same direction as the direction in which the impedance is required.

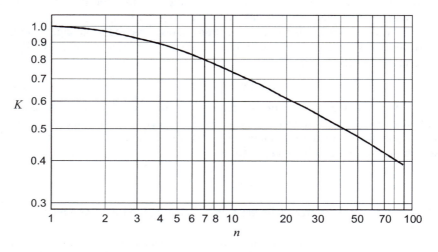

Figure 9.5 The ratio K of the effective radius for a non-circular orifice to the radius of a circular orifice of the same area. The non-circular orifice aspect ratio is n.

A more accurate expression for the impedance of a perforated plate involves including the mass of the panel between the holes and calculating the impedance assuming the panel mass impedance acts in parallel with the hole impedance as suggested by Bolt (1947). The acoustic impedance of the panel of mass per unit area, m, and area, S_p) is:

$$Z_{Ap} = j\omega m/S_p \qquad (9.23)$$

Thus, the acoustic impedance of the panel mass in parallel with the holes in the panel is:

$$Z_A = \frac{Z_{Ah}}{1 + Z_{Ah}/Z_{Ap}} = \frac{\dfrac{100}{PS_p}\left(j\rho c\tan(k\ell_e(1-M)) + R_A S\right)}{1 + \dfrac{100}{j\omega mP}\left(j\rho c\tan(k\ell_e(1-M)) + R_A S\right)} \qquad (9.24)$$

Note that for standard thickness (less than 3 mm) perforated sheets, the tan functions in the preceding equation may be replaced by their arguments.

The effective length, ℓ_e, of the holes is derived from Equations (9.7) and (9.20) as:

$$\ell_e = w + \left(\frac{16a}{3\pi}(1 - 0.43\,a/q) \right)(1 - M^2) \tag{9.25}$$

An alternative expression for the effective length, which may give slightly better results than Equation (9.25), for grazing flow across the holes, and which only applies for flow speeds such that $u_\tau/(\omega d) > 0.03$, is (Dickey and Selamet, 2001):

$$\ell_e = w + \left[\frac{16a}{3\pi}(1 - 0.43\,a/q) \right]\left[0.58 + 0.42\,e^{-23.6\left(u_\tau/(\omega d) - 0.03\right)} \right] \tag{9.26}$$

where d is the diameter of the holes in the perforated sheet and the friction velocity, u_τ, is:

$$u_\tau = 0.15U\,\mathrm{Re}^{-0.1} \tag{9.27}$$

Re is the Reynolds number of the grazing flow given by:

$$\mathrm{Re} = \frac{U d \rho}{\mu} \tag{9.28}$$

where μ is the dynamic gas viscosity (1.84×10^{-5} kg m^{-1} s^{-1} for air at 20°C), ρ is the gas density (1.206 for air at 20°C) and U is the mean flow speed.

9.6.1.2 Acoustic Resistance

The acoustic resistance R_A (kg m^{-4} s^{-1}) of an orifice or a tube of length w (m), cross-sectional area S (m^2) and internal duct cross-sectional perimeter D (m) may be calculated using the following equation:

$$R_A = \frac{\rho c}{S}\left[\frac{ktDw}{2S}\left[1 + (\gamma - 1)\sqrt{\frac{5}{3\gamma}} \right] + 0.288\,kt\log_{10}\left[\frac{4S}{\pi h^2} \right] + \varepsilon\frac{Sk^2}{2\pi} + M \right] \tag{9.29}$$

The derivation of Equation (9.29) was done with reference principally to Morse and Ingard (1968). In Equation (9.29), ρc is the characteristic impedance for air (415 MKS rayls for air at room temperature), γ is the ratio of specific heats (1.40 for air), $k\,(=\omega/c)$ is the wavenumber, ω is the angular frequency (rad/s), c is the speed of sound (m/s), M is the Mach number of any mean flow through the tube or orifice, the quantities h and ε are discussed in the following paragraphs, and t is the viscous boundary layer thickness given by:

$$t = \sqrt{2\mu/\rho\omega} \tag{9.30}$$

In the preceding equation, μ is the dynamic viscosity of the gas (1.84×10^{-5} kg m^{-1} s^{-1} for air at 20°C) and ρ is the gas density (1.2 kg/m^3 for air at STP).

The first term in Equation (9.29) accounts for attenuation along the length of the tube. This term is generally negligible except for small tubes or high frequencies. On the other hand, because the term depends upon the length w, it may become significant for very long tubes. It was derived using Equations (6.4.5), (6.4.31) and (9.1.12) from Morse and Ingard (1968).

The second term in Equation (9.29) accounts for viscous loss at the orifice or tube entry, and is a function of the quantity h. For orifices in thin plates of negligible thickness ($w \approx 0$), h is either the half plate thickness or the viscous boundary layer thickness t, given by Equation (9.30), whichever is bigger. Alternatively, if the orifice is the entry to a tube ($w \gg 0$), then h is the orifice edge radius or the viscous boundary layer thickness, whichever is bigger. This term was derived using Equations (6.4.31) and (9.1.23) from Morse and Ingard (1968).

The third term in Equation (9.29) accounts for radiation loss at the orifice or tube exit. For tubes that radiate into spaces of diameter much less than a wavelength of sound (for example, an expansion chamber in a duct) the parameter ε may be set equal to zero. Alternatively, for tubes that radiate into free space, but without a flange at their exit, ε should be set equal to 0.5. For tubes that terminate in a well-flanged exit or radiate into free space from a very large plane wall or baffle, ε should be set equal to 1. This term was derived using Equation (5.108) and is effectively the radiation efficiency of a piston radiator.

The fourth and last term of Equation (9.29) accounts for mean flow through the orifice and is usually the dominant term in the presence of a mean flow. It is valid only for Mach numbers less than about 0.2. This term is effectively the first term in Equation (11.3.37) of Morse and Ingard (1968) for the case of the orifice being small compared to the cross-section of the tube carrying the flow. Note that flow across an exit orifice will have a similar effect to flow through it. For high sound pressure levels in the absence of a mean flow, the mean flow Mach number M may be replaced with the Mach number corresponding to the particle velocity amplitude.

Alternatively, for grazing flow across a sheet with circular holes, each having a cross-sectional area, S, with a speed such that, $u_\tau /(\omega d) \geq 0.05$, Dickey and Selamet (2001) give the following expression for the acoustic resistance:

$$R_A = \frac{\rho c k d}{S}\left[-0.32 + \frac{9.57 u_\tau}{\omega d}\right] \tag{9.31}$$

where all variables have been defined previously.

For tubes that have a very small diameter (radius, $a < 0.002/\sqrt{f}$), Beranek (1954) gives the following expression for the acoustic resistance.

$$R_A = \frac{8\pi\mu w}{S^2} \tag{9.32}$$

where μ is the dynamic viscosity for air (= 1.84×10^{-5} N -s/m^2) which varies with absolute temperature, T in degrees Kelvin as $\mu \propto T^{0.7}$

For slits with an opening height, t (much smaller than the slit width), the acoustic resistance is given by (Beranek, 1954):

$$R_A = \frac{12\mu w}{t^2 S} \tag{9.33}$$

where w is the depth of the slit.

9.6.2 Impedance of a Volume

From the definition of capacitance as charge per unit induced voltage, it may be concluded that a volume should be the acoustical analog of the electrical capacitor. To begin, consider the adiabatic compression of a gas contained in a volume V. It has been shown both experimentally and theoretically that the heat conduction of air is so low that adiabatic compression is a very good approximation for compression in volumes of dimensions very much less than a wavelength of sound (Daniels, 1947; Golay, 1947). Thus:

$$p/P + \gamma(\Delta V/V) = 0 \tag{9.34}$$

In this equation, p is the acoustic pressure, P is the static (atmospheric) pressure, γ is the ratio of specific heats, V is the volume and ΔV is the incremental change in volume.

It will be assumed that the volume flow (or current) v into the volume is:

$$v = v_0 e^{j\omega t} \tag{9.35}$$

then, as a decrease in volume is achieved by a positive pressure and associated volume flow:

$$\Delta V = -\int v\,dt = -v/j\omega \tag{9.36}$$

Substitution of Equation (9.36) into Equation (9.34) and noting that the impedance is in the positive x-direction (see Figure 9.6) leads to the following

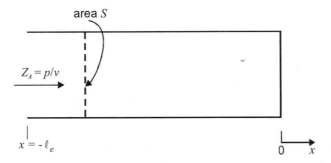

Figure 9.6 A schematic representation of an acoustic transmission line with a closed end: Z_A is the acoustic impedance at a cross-section of area S.

expressions for the impedance (assuming the acoustic resistance is negligible for this case):

$$Z_A \approx jX_A = \frac{p}{v} = -j\frac{\gamma P}{V\omega} = -j\frac{\rho c^2}{V\omega} \qquad (9.37a,b,c,d)$$

The last two alternate forms of Equation (9.37) follow directly from the relationship between the speed of sound, gas density and compressibility. Consideration of the alternate forms shows that the acoustical capacitance, the analog of electrical capacitance, is either $V/\gamma P$ or $V/\rho c^2$.

Equation (9.37) may be shown to be a first approximation to the more general transmission-line analysis represented by Equation (9.38). Referring to the closed end duct shown in Figure 9.6, and taking the coordinate origin at the closed end, the analysis is the same as was followed in deriving Equation (9.14), except that as the end is now closed, the sign of the reflected wave changes (phase θ changes from π to 0) so that at the reflecting end ($x = 0$) the pressure is doubled and the particle velocity is zero. The acoustic impedance looking right, into the tube, can be derived using the same procedure as for an open tube and including a resistance term, the acoustic impedance is written as:

$$Z_A = R_A + jX_A = R_A + j\frac{\rho c}{S}\cot(kx) \qquad (9.38a,b)$$

This expression can be used to estimate the impedance looking into a tube closed at the opposite end, and of cross-sectional area S and effective length ℓ_e, by replacing x with the effective length, $-\ell_e$, which includes an end correction so it is a bit larger than the physical length of the tube. For small ℓ_e, Equation (9.38) reduces to Equation (9.37) as a first approximation, where $V = S\ell_e$. For the situation of flow past the end of the tube, or through the tube, the impedance may be written as:

$$Z_A = R_A - j\frac{\rho c}{S}\cot(k\ell_e(1 - M)) \qquad (9.39)$$

Note that in most cases the resistive impedance associated with a volume is considered negligible, except in the case of a quarter wave tube.

9.7 REACTIVE DEVICES

Commercial mufflers for internal combustion engines are generally of the reactive type. These devices are designed, most often by cut and try, to present an essentially imaginary input impedance to the noise source in the frequency range of interest. The input power and thus the radiated sound power is then reduced to some acceptably low level.

The subject of reactive muffler design, particularly as it relates to automotive mufflers, is difficult, although it has received much attention in the literature (Jones, 1984; Davies, 1992a, 1992b, 1993). Consideration of such design will be limited to three special cases. Some attention will also be given in this section to the important matters of pressure drop. Flow-generated noise, which is often important in air conditioning silencers but is often neglected in muffler design, will also be mentioned.

9.7.1 Acoustical Analogs of Kirchhoff's Laws

The following analyses of reactive devices are based on acoustical analogies of the well known Kirchhoff laws of electrical circuit analysis. Referring to Section 9.5, it may be observed that the acoustic volume velocity is continuous at junctions and is thus the analog of electrical current. Similarly, the acoustic pressure is the analog of voltage. Thus, in acoustical terms, the Kirchhoff current and voltage laws may be stated respectively as follows:

1. The algebraic sum of acoustic volume velocities at any instant and at any location in the system must be equal to zero.
2. The algebraic sum of the acoustic-pressure drops around any closed loop in the system at any instant must be equal to zero.

In the remainder of the discussion on reactive silencers, the subscript, "A", which has been used to denote acoustic impedance as opposed to specific acoustic impedance, will be dropped to simplify the notation.

9.7.2 Side Branch Resonator

A particularly useful device for suppressing pure tones of constant frequency, such as might be associated with constant speed pumps or blowers is the side branch resonator. The side branch resonator functions by placing a very low impedance in parallel with the impedance of the remainder of the line at its point of insertion. It is most effective when its internal resistance is low, and it is placed at a point in the line where the impedance of the tone to be suppressed is real. This point will be considered further later.

The side branch resonator may take the form of a short length of pipe, for example a quarter wave stub, whose length (approximately a quarter of the wavelength of sound at the frequency of interest plus an end correction of approximately 0.3 × the pipe internal diameter) may be adjusted to tune the device to maximum effectiveness. The quarter wave tube diameter should be relatively constant along its length with no step changes as these will compromise the performance. The impedance of the quarter wave stub is:

$$Z_s = -\frac{j\rho c}{S}\cot k\ell_e + R_s \quad = 0 + R_s \quad \text{if} \quad \ell_e = \lambda/4 \tag{9.40}$$

An alternative type of side branch resonator is the Helmholtz resonator, which consists of a connecting orifice and a backing volume, as indicated schematically in Figure 9.7(a), with an equivalent acoustical circuit shown in Figure 9.7(b).

The side branch Helmholtz resonator appears in the equivalent acoustical circuit of Figure 9.7(b) as a series acoustic impedance, Z_s, in parallel with the downstream duct impedance, Z_d. The quantity, Z_u, is the acoustic impedance of the duct upstream

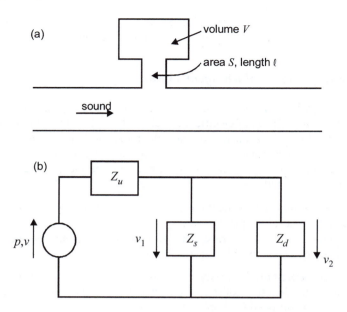

Figure 9.7 The side branch resonator: (a) acoustical system; (b) equivalent acoustical circuit.

of the resonator. The series acoustic impedance, Z_s, consists of a capacitance, X_C, inductance, X_L, and resistance, R_s, in series as shown by the following equation:

$$Z_s = \frac{j\rho\omega\ell_e}{S} - \frac{j\rho c^2}{V\omega} + R_s \qquad (9.41)$$

The capacitance is due to the resonator volume, and the inductance and resistance are due to the resonator neck. The side branch has a resonance or minimum impedance when the sum of the capacitive and inductive impedances is equal to zero. Using lumped element analysis and setting the inductive term given by Equation (9.9) equal to the negative of the capacitive term given by Equation (9.37d) gives the following expression for the frequency, ω_0 (rad/s) of the series resonance for the Helmholtz resonator.

$$\omega_0 = c\sqrt{S/\ell_e V} \qquad (9.42)$$

where S and ℓ_e are the cross-sectional area and effective length (actual length plus two end corrections) respectively of the connecting neck and V is the volume of the cavity.

Equation (9.42) has been shown to become inaccurate when the resonator dimensions exceed 1/16 of a wavelength or when the aspect ratio of the resonator cavity dimensions is much different to unity. Panton and Miller (1975) provided a more accurate expression for cylindrical Helmholtz resonators that takes into account axial wave propagation in the resonator chamber, but still requires that the neck dimensions are small compared to a wavelength of sound. Their expression is:

$$\omega_0 = c \sqrt{\frac{S}{\ell_e V + 0.33 \, \ell_V^2 S}} \qquad (9.43)$$

where ℓ_V is the cylindrical cavity length, which is concentric with the cylindrical neck.

Equation (9.43) is still not as accurate as really needed and the reader is referred to Equation (9.44) for a more accurate expression for which there are no restrictions on the cavity or neck lengths but the cavity diameter must be less than a wavelength at the resonance frequency (Li, 2003).

$$\omega_0 = c \sqrt{\frac{3\ell_e S_V + \ell_V S}{2\ell_e^3 S_V} + \sqrt{\left(\frac{3\ell_e S_V + \ell_V S}{2\ell_e^3 S_V}\right)^2 + \frac{3S}{\ell_e^3 S_V \ell_V}}} \qquad (9.44)$$

Note that if the orifice is off-set from the centre of the cavity, the above equations will significantly overestimate the resonator resonance frequency.

9.7.2.1 End Corrections for a Helmholtz Resonator Neck and Quarter Wave Tube

To calculate the effective length of a Helmholtz resonator neck or quarter wave tube, it is necessary to determine the end correction at each end of the neck. For a Helmholtz resonator, one end looks into the resonator volume and the other looks into the main duct on which the resonator is mounted. For the quarter wave tube, there is only the end correction for the end looking into the main duct.

For a cylindrical Helmholtz resonator where the cavity is concentric with the neck (of radius, a), the end correction for the end of the neck connected to the cavity is given by (Selamat and Ji, 2000), for configurations where $\xi < 0.4$, as:

$$\ell_0 = 0.82 a (1 - 1.33 \xi) \qquad (9.45)$$

which is very similar to Equation (9.16) for an orifice in an anechoically terminated duct, and where ξ is the ratio of the neck diameter to cavity diameter.

The end correction for the neck end attached to the main duct is difficult to determine analytically. However, Ji (2005) gives the following equations, based on a Boundary Element Analysis.

$$\ell_0 = a \begin{cases} 0.8216 - 0.0644\xi - 0.694\xi^2; & \xi \le 0.4 \\ 0.9326 - 0.6196\xi; & \xi > 0.4 \end{cases} \qquad (9.46)$$

where in this case, ξ is the ratio of neck diameter to main duct diameter. The above expression was only derived for the range $0 < \xi \le 1.0$. However, it is difficult to imagine a situation where ξ would be greater than 1.0.

Dang *et al.* (1998) undertook a series of measurements using a closed end tube attached to a duct and derived an empirical expression as follows, which applies to a quarter wave tube and also the neck duct interface for a Helmholtz resonator. This expression agrees remarkably well over the range $0 < \xi \leq 1.0$, with the expression derived by Ji (2005) and represented in Equation (9.46).

$$\ell_0 = a \left(\frac{1.27}{1 + 1.92\,\xi} - 0.086 \right); \qquad 0.2 < \xi < 5.0 \tag{9.47}$$

9.7.2.2 Quality Factor of a Helmholtz Resonator and Quarter Wave Tube

A quality factor, Q, may be defined for any resonance, as the ratio of energy stored divided by the energy dissipated during a cycle:

$$Q = 2\pi E_s / E_D \tag{9.48}$$

Energy stored, E_s, is proportional to the square root of the product of the capacitative and inductive impedances, while energy dissipated, E_D, is proportional to the resistive impedance. This leads to the following expression for the series circuit quality factor, Q, for a Helmholtz resonator:

$$Q = \frac{1}{R_s} \sqrt{X_C X_L} = \frac{\rho c}{R_s} \sqrt{\frac{\ell_e}{SV}} \tag{9.49a,b}$$

where ℓ_e, S and V are the effective length of the neck and cross-sectional area and volume of the resonator chamber, respectively.

In most instances, the acoustic resistance term, R_s, is dominated by the resistance of the neck of the resonator and may be calculated using Equation (9.29). However, the placement of acoustically absorptive material in the resonator cavity or, especially, in the neck will greatly increase the acoustic resistance. This will broaden the bandwidth over which the resonator will be effective at the expense of reducing the effectiveness in the region of resonance.

For a quarter wave tube of cross-sectional area, S, and effective length, ℓ_e, the resonance frequency and quality factor are given by (Beranek, 1954, pp. 129 and 138):

$$\omega_0 = \frac{\pi c}{2\ell_e}; \qquad Q = \frac{\pi \rho c}{6 R_s S} \tag{9.50a,b}$$

The preceding discussion for Helmholtz resonators and quarter wave tubes is applicable at low frequencies for which no dimension of the resonator exceeds a quarter of a wavelength. It is also only applicable for resonator shapes that are not too different to a sphere or a cube. Howard *et al.* (2000) showed, using finite element analysis, that the resonance frequencies of Helmholtz resonators are dependent on the resonator aspect ratio. Resonators (both 1/4 wave and Helmholtz) have many additional resonances at frequencies higher than the fundamental given by Equations (9.42) and (9.50a). At each of these frequencies, the resonator insertion loss is substantial. This multi-resonance nature is useful in the design of large industrial resonator mufflers that are considered later in this chapter.

The quality factor, Q, which is proportional to the resonator bandwidth, is a function of the ratio of the resonator (or quarter wave tube) neck cross-sectional area to the cross-sectional area of the duct on which it is installed. The larger this ratio, the smaller will be the quality factor. According to Equation 7.23(a), the smaller the quality factor, the larger will be the frequency bandwidth over which the resonator will provide significant noise reduction. However, even for area ratios approaching unity (ratios greater than unity are generally not practical), the quality factor is relatively high and the corresponding frequency bandwidth is small, resulting in noise attenuation that is dependent on the frequency stability of the tonal noise source. This, in turn, is dependent on the stability of the physical parameters on which the noise depends, such as duct temperature or rotational speed of the equipment generating the noise. Thus large changes in the effectiveness of the resonator can become apparent as these physical quantities vary. This problem could be overcome by using a control system to drive a moveable piston to change the volume of the resonator so that the tone is maximally suppressed. Alternatively, the limited bandwidth problem could be partly overcome by using two or more resonators tuned to slightly different frequencies, and separated by one wavelength of sound at the frequency of interest (Ihde, 1975).

9.7.2.3 Insertion Loss due to Side Branch

Referring to Figure 9.7(b), the following three equations may be written using the acoustical analogs of Kirchhoff's current and voltage laws (see Section 9.7.1):

$$p = v_2 Z_d + (v_1 + v_2) Z_u \tag{9.51}$$

$$v = v_1 + v_2 \tag{9.52}$$

$$v_1 Z_s = v_2 Z_d \tag{9.53}$$

The details of the analysis to follow depend upon the assumed acoustic characteristics of the sound source. To simplify the analysis, the source is modelled as either a constant acoustic volume–velocity (infinite internal impedance) source or a constant acoustic-pressure (zero internal impedance) source (see Section 9.5 for discussion).

For a constant volume–velocity source, consider the effect upon the power transmission to the right in the duct of Figure 9.7 when the side branch resonator is inserted into the system. Initially, before insertion, the power transmission is proportional to the supply volume velocity squared, v^2, whereas after insertion the power flow is proportional to the load volume velocity squared v^2_2. The insertion loss, *IL*, is defined as a measure of the decrease in transmitted power in decibels. A large positive *IL* corresponds to a large decrease.

$$IL = 20 \log_{10} \left| \frac{v}{v_2} \right| \tag{9.54}$$

Using the definition of insertion loss given by Equation (9.54), solution of Equations (9.52) and (9.53) gives the following expression for the insertion loss:

$$IL = 20 \log_{10} \left| 1 + \frac{Z_d}{Z_s} \right| \tag{9.55}$$

Alternatively, the source can be modelled as one of constant acoustic-pressure. In this case, by similar reasoning, the insertion loss IL may be defined as the ratio of the total acoustic pressure drop across Z_u and Z_d in the absence of the resonator to the acoustic pressure drop across Z_u and Z_d in the presence of the resonator as follows:

$$IL = 20 \log_{10} \left| \frac{p}{v_2 (Z_u + Z_d)} \right| \tag{9.56}$$

Solving Equations (9.51) and (9.55), and using Equation (9.56) gives the following expression for the insertion loss of a side branch resonator for a constant acoustic-pressure source:

$$IL = 20 \log_{10} \left| 1 + \frac{Z_u Z_d}{Z_s (Z_u + Z_d)} \right| \tag{9.57}$$

Comparison of Equations (9.55) and (9.57) shows that they are formally the same if, in the case of the constant acoustic-pressure source, the impedance Z_s of the side branch is replaced with the effective impedance $Z_s (1 + Z_d/Z_u)$.

To maximise the insertion loss for both types of sound source, the magnitude of Z_s must be made small while at the same time, the magnitude of Z_d (as well as Z_u for a constant pressure source) must be made large. Z_s is made small by making the side branch resonant (zero reactive impedance, such as a uniform tube one-quarter of a wavelength long), and the associated resistive impedance as small as possible (rounded edges, no sound absorptive material). The quantities Z_d and Z_u are made large by placing the side branch at a location on the duct where the internal acoustic pressure is a maximum. This can be accomplished by placing the side branch an odd multiple of quarter wavelengths from the duct exhaust, and as close as possible to the noise source (but no closer than about three duct diameters for fan noise sources to avoid undesirable turbulence effects). Because acoustic-pressure maxima in the duct are generally fairly broad, the resonator position need not be precise. On the other hand, the frequency of maximum attenuation is very sensitive to resonator physical dimensions, and it is wise to allow some means of fine-tuning the resonator on-site (such as a moveable piston to change the effective volume or length).

One should be prepared for loss in performance with an increase in mean flow through the duct, because the resistive part of the side branch resonator impedance will usually increase, as indicated by Equation (9.29). For example, a decrease in insertion loss from 30 dB to 10 dB at resonance, with flow speeds above about 40 m/s, as a result of the increased damping caused by the flow, has been reported in the literature (Gosele, 1965). Meyer *et al.* (1958) measured an increase of 20% in the resonance frequency in the presence of a flow rate of 70 m/s.

The value of the resistance, R_s, which appears in Equations (9.49) and (9.50), is difficult to calculate accurately and control by design, although Equation (9.29) provides an adequate approximation to the actual value in most cases. As a rule of thumb, the related quality factor, Q, for a side branch resonator may be expected to range between 10 and 100 with a value of about 30 being quite common.

Alternatively, as shown by Equation 7.23(a), if the sound pressure within the resonator (preferably at the closed end of a quarter wave tube or on the far wall of a Helmholtz resonator) can be measured while the resonator is driven by an external variable frequency source, the quality factor Q may readily be determined. This measurement will usually require mounting the resonator on the wall of a duct or large enclosure and introducing the sound into the duct or enclosure using a speaker backed by an additional small enclosure. Alternatively, the in-situ quality factor can be determined by mounting the resonator on the duct to be treated and then varying its volume around the design volume (Singh *et al.*, 2006).

It is interesting to note that the presence of a mean flow has the effect of increasing the resonator damping and increasing its resonance frequency.

9.7.2.4 Transmission Loss due to Side Branch

Sometimes it is useful to be able to compare the Transmission Loss of various mufflers even though it is not directly related to the noise reduction as a result of installing the muffler, as explained earlier in this chapter. The *TL* of a side branch may be calculated by referring to the harmonic solution of the wave equation (1.53) and Figure 9.7a. It will also be assumed that all duct and side branch dimensions are sufficiently small that only plane waves propagate and that the downstream duct diameter is equal to upstream duct diameter and each has a cross sectional area denoted S_d. The Transmission Loss is defined as the ratio of transmitted power to incident power and as we are only considering plane waves, this may be written as:

$$TL = 10\log_{10}\left|\frac{p_I}{p_T}\right|^2 = 10\log_{10}\left|\frac{A_I}{A_T}\right|^2 \qquad (9.58\text{a,b})$$

When a plane wave propagates down the duct from the left and encounters the side branch a transmitted wave and a reflected wave are generated. The total sound pressure in the duct to the left of the resonator is the sum of the incident pressure, p_I and the pressure, p_R reflected by the impedance mismatch in the duct caused by the side branch resonator. This pressure may be written as:

$$p_{inlet} = A_I e^{j(\omega t - kx)} + A_R e^{j(\omega t + kx + \beta_1)} \tag{9.59}$$

In the absence of any reflected wave from the downstream end of the duct, the transmitted pressure may be written as:

$$p_T = A_T e^{j(\omega t - kx + \beta_2)} \tag{9.60}$$

where β_1 and β_2 represent arbitrary phase angles which have no effect at all on the *TL*. Equations (1.10) and (1.11) may be used with the preceding two equations to write the following for the acoustical volume velocities (particle velocity multiplied by the duct cross sectional area) in the same locations:

$$v_{inlet} = \frac{S_d}{\rho c} \left[A_I e^{j(\omega t - kx)} - A_R e^{j(\omega t + kx + \beta_1)} \right] \tag{9.61}$$

$$v_T = \frac{S_d A_T}{\rho c} e^{j(\omega t - kx + \beta_2)} \tag{9.62}$$

If the duct axial coordinate, x is set equal to zero at the location in the duct corresponding to the centre of the side branch resonator, the acoustic pressure at the entrance to the side branch, p_s, is equal to the acoustic pressure in the duct at $x = 0$ so that at this location:

$$p_s = A_T e^{j(\omega t + \beta_2)} = A_I e^{j(\omega t)} + A_R e^{j(\omega t + \beta_1)} \tag{9.63}$$

At the junction of the side branch and duct, there must be continuity of volume velocity so that at $x = 0$, the incoming volume velocity from the left is equal to the sum of that moving in the duct to the right and that entering the side branch. The incoming and outgoing volume velocities in the duct are given by Equations (9.61) and (9.62) respectively, while the side branch volume velocity at $x = 0$ is:

$$v_s = \frac{p_s}{Z_s} = \frac{p_T}{Z_s} \tag{9.64}$$

where Z_s is the acoustic impedance of the side branch.

Thus, continuity of volume velocity at the side branch junction can be written as:

$$\frac{S_d}{\rho c} \left[A_I e^{j(\omega t - kx)} - A_R e^{j(\omega t + kx + \beta_1)} \right] = \frac{S_d A_T}{\rho c} e^{j(\omega t - kx + \beta_2)} + A_T e^{j(\omega t + \beta_2)} \tag{9.65a,b}$$

Simplifying and rearranging gives:

$$\frac{2 A_I S_d}{\rho c} = A_T e^{j\beta_2} \left(\frac{2 S_d}{\rho c} + \frac{1}{Z_s} \right) \tag{9.66}$$

Further rearranging gives:

$$\left|\frac{A_I}{A_T}\right| = \left|1 + \frac{\rho c}{2 S_d Z_s}\right| \qquad (9.67)$$

Thus the TL of the side branch resonator of impedance, Z_s, is:

$$TL = 20\log_{10}\left|1 + \frac{\rho c}{2 S_d Z_s}\right| \qquad (9.68)$$

The side branch acoustic impedance, Z_s, may be calculated using lumped analysis, and is found by adding together Equations (9.9) and (9.37d). However, more accurate results are obtained if the resonator is treated as a 1-D transmission line so that wave motion is allowed in the axial direction. In this case, the impedance is calculated using Equation (C.49) in Appendix C where all the specific acoustic impedances are replaced with acoustic impedances. The impedance Z_m corresponds to the air in the neck and is given by $\rho c/S$ where S is the cross sectional area of the neck, and the impedance Z_L is the load impedance, which is the impedance looking into the resonator chamber, given by the imaginary part of Equation (9.39) with $M = 0$ as only the zero flow condition is being considered and the acoustic resistance of the chamber is considered negligible. In this case, Equation (C.49) can be written in terms of the overall acoustic impedance of the side branch, Z_s, the acoustic impedance of the resonator cavity, Z_L, the length ℓ_V, of the cavity in the resonator axial direction, the volume, V, of the resonator cavity, the effective length, ℓ_e, and cross-sectional area, S, of the resonator neck, and the cross-sectional area, S_V of the resonator in the plane normal to the resonator axis, as:

$$Z_s = \frac{\rho c}{S}\frac{Z_L S/\rho c + j\tan(k\ell_e)}{1 + j(SZ_L/\rho c)\tan(k\ell_e)} \qquad (9.69)$$

Substituting Equation (9.39) for Z_L into Equation (9.69), with $R_A = M = 0$ gives:

$$Z_s = \frac{\rho c}{S}\frac{\dfrac{-jS}{S_V\tan(k\ell_V)} + j\tan(k\ell_e)}{1 + j\dfrac{-jS\tan(k\ell_e)}{S_V\tan(k\ell_V)}} = j\frac{\rho c}{S}\frac{(S_V/S)\tan(k\ell_e)\tan(k\ell_V) - 1}{(S_V/S)\tan(k\ell_V) + \tan(k\ell_e)} \qquad (9.70a,b)$$

Substituting Equation (9.70b) into Equation (9.68) gives:

$$TL = 10\log_{10}\left[1 + \left(\frac{S}{2S_d}\frac{(S_V/S)\tan(k\ell_V) + \tan(k\ell_e)}{(S_V/S)\tan(k\ell_e)\tan(k\ell_V) - 1}\right)^2\right] \qquad (9.71)$$

which is valid for frequencies above any axial resonances in the cavity and up to the resonance frequency of the first cross mode in the resonator cavity.

Equation (9.71) also applies to a 1/4 wave tube resonator by setting $S = S_V$ to give:

$$TL = 10\log_{10}\left[1 + \left(\frac{S}{2S_d}\tan(k(\ell_e + \ell_V))\right)^2\right] \qquad (9.72)$$

which is the *TL* of a quarter wave resonator of length $\ell_e + \ell_V$. Alternatively, if $S_V = 0$, then the side branch resonator has an effective length of ℓ_e and the *TL* is given by:

$$TL = 10\log_{10}\left[1 + \left(\frac{S}{2S_d}\tan(k\ell_e)\right)^2\right] \qquad (9.73)$$

If S_V approaches infinity, then there is effectively no cavity on the end and Equation (9.71) reduces to the *TL* for a side branch consisting of an open ended tube as:

$$TL = 10\log_{10}\left[1 + \left(\frac{S}{2S_d}\cot(k\ell_e)\right)^2\right] \qquad (9.74)$$

The axial resonance frequencies of the Helmholtz resonator can be determined by realising that they occur when the denominator in Equation (9.71) is zero and the *TL* is infinite. Thus, the resonance frequencies are those that satisfy:

$$(S_V/S)\tan(k\ell_e)\tan(k\ell_V) = 1 \qquad (9.75)$$

where k is the wavenumber defined as $k = 2\pi f/c$.

9.7.3 Resonator Mufflers

Resonator mufflers are used in industry where large low frequency noise reductions are needed and also in applications where it is not possible to use porous sound absorbing material in the muffler (due to possible contamination of the air flow or contamination of the sound absorbing material by particles or chemicals in the air flow). Resonator mufflers consist of a mix of 1/4 wave tubes and Helmholtz resonators, tuned to cover the frequency range of interest and attached to the walls of the duct through which the sound is propagating. Helmholtz resonators have a lower Q (and hence act over a broader frequency range) than 1/4 wave tubes, but their performance in terms of insertion loss is not as good. Thus, in practice, many mufflers are made up of a combination of these two types of resonator, with the 1/4 wave tubes tuned to tonal noise or frequency ranges where the greatest noise reduction is needed. To ensure good frequency coverage and to allow for possible variations in tonal frequencies and resonator manufacturing errors, Helmholtz resonators, covering a range of resonance frequencies about those of the 1/4 wave tubes, are used. It is important that adjacent resonators are not tuned to identical frequencies or even frequencies that are close together or they will interact (couple together) and substantially reduce the muffler performance at that frequency. In many cases, the duct cross-section through which the sound is travelling is so large that the resonator muffler is constructed using splitters to contain the resonators. This requires dividing the duct cross-section into a number of parallel sections using dividing walls and including resonators in each dividing wall as shown in Figure 9.8. Note that the resonators are usually angled towards the flow direction (see Figure 9.8(c)) to avoid filling up with particles from the air flow and to minimise the generation of tonal noise due to vortex shedding from the edge of the resonator inlet. To ensure that the pressure drop due to insertion of the muffler is not too great, it is usually sized so that the total open cross-sectional area between splitters is the same as the inlet duct cross-sectional area.

There are two types of resonator muffler – those that contain no sound absorbing material at all and those that include sound absorbing material in the resonator chambers. Those that contain no sound absorbing material at all have resonators with a relatively high Q, and many resonators are needed to cover a reasonable frequency range. A small amount of sound absorbing material in the resonator chambers will produce a muffler with much more uniform attenuation characteristics as a function of frequency than a muffler without sound absorbing material. However, the peak attenuation at some frequencies may not be as high as achieved by a resonator muffler with no sound absorbing material, for the case where an excitation frequency corresponds very closely to one of the side branch resonances.

In constructing resonator silencers, it is important that the walls of the resonator are made using sufficiently thick material so that their vibration does not significantly degrade the muffler performance.

In designing a resonator muffler, it is usually necessary to use a commercial finite element analysis package. The lumped element analysis described in this chapter is only valid for low frequencies and resonator dimensions less than one-quarter of a wavelength. Thus all resonances of the side branch resonators above the fundamental will not be taken into account with a simple lumped element analysis. In addition, the simple analysis does not have scope for taking account of the effect on the resonance frequency of resonator shapes that are significantly different to spheres or cubes. An example of a resonator muffler design process is given by Howard *et al.* (2000).

9.7.4 Expansion Chamber

9.7.4.1 Insertion Loss

A common device for muffling is a simple expansion chamber, such as that shown diagrammatically in Figure 9.9. Following the suggestion of Table 9.1, the expansion chamber will be assumed to be less than one-half wavelength long, so that wave propagation effects may be neglected. Under these circumstances, the expansion chamber may be treated as a lumped element device and the extension of the pipes shown in Figure 9.9(a) is of no importance; that is, the extension lengths shown as x

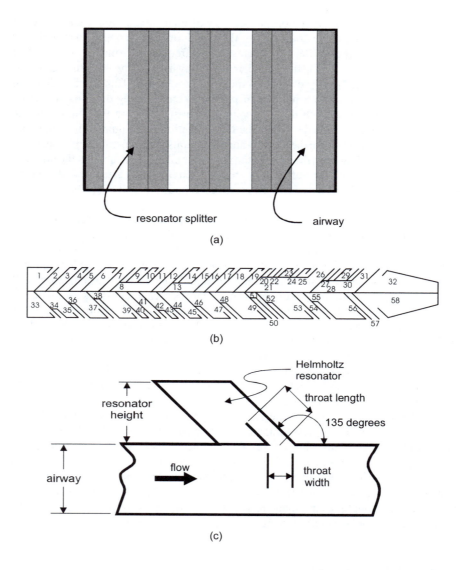

Figure 9.8 Resonator silencer details: (a) cross-section of splitter silencer (splitters are shaded); (b) detail of a splitter containing 58 resonators; (c) detail of a single resonator.

and y may be zero. The latter extensions become important at high frequencies, as the chamber length approaches and exceeds one-half wavelength. Similarly, the details of the location of the pipes around the perimeter of the expansion chamber only become important at high frequencies.

For the purpose of lumped element analysis, an equivalent acoustical circuit is shown in Figure 9.9(b). For convenience, the source has been assumed to be a constant

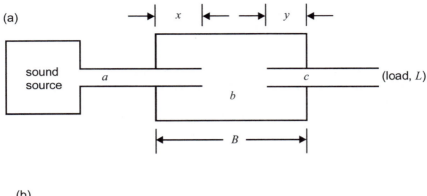

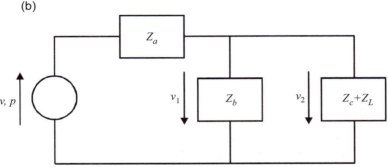

Figure 9.9 An expansion chamber muffler: (a) acoustical system; (b) equivalent acoustical circuit.

volume–velocity generator (see Section 9.5). The elements of the mechanical system labelled a, b, c and L are represented in the equivalent acoustical circuit as impedances Z_a, Z_b, Z_c and Z_L. A comparison of the equivalent acoustical circuits shown in Figures 9.7 and 9.9 shows that they are formally the same if Z_u, Z_s and Z_d of the former are replaced respectively with Z_a, Z_b and $(Z_c + Z_L)$ of the latter. For the constant volume–velocity source, the expression for insertion loss for the expansion chamber is formally identical to Equation (9.55). Thus, from Equation (9.55) the insertion loss of the expansion chamber is:

$$IL = 20\log_{10}\left|1 + \frac{Z_c + Z_L}{Z_b}\right| \tag{9.76}$$

Reference to Figure 9.9 shows that the impedance, Z_b, is capacitative, and may be calculated using Equation (9.37). The impedance sum $(Z_c + Z_L)$, is resistive and inductive and can be written as a total effective resistance $(R_c + R_L)$ and an effective inductance. In the latter case, Equation (9.14) provides the required expression, where the sum of the inductances is obtained by writing the expression in terms of an effective duct length, ℓ_e. These observations give the following result:

$$\frac{Z_c + Z_L}{Z_b} = j\frac{\omega V}{\rho c^2}\left[R_c + R_L + j\frac{\rho c}{S}\tan k\ell_e\right] \qquad (9.77)$$

Here V (m^3) is the volume of the expansion chamber, S (m^2) is the cross-sectional area of the tailpipe (c) and ℓ_e (m) is its effective length, ρ (kg/m^3) is the gas density and c (m/s) is the speed of sound.

For small values of $k\ell_e$, the preceding equation may be rewritten approximately as:

$$\frac{Z_c + Z_L}{Z_b} \approx -\left[\frac{\omega}{\omega_0}\right]^2 + j\frac{\omega}{\omega_0 Q} \qquad (9.78)$$

An approximate expression for insertion loss, for small values of $k\ell_e$, thus becomes:

$$IL \approx 10\log_{10}\left[\left(1 - (\omega/\omega_0)^2\right)^2 + Q^{-2}(\omega/\omega_0)^2\right] \qquad (9.79)$$

where the resonance frequency, ω_0 (rad/s), and the quality factor, Q, are given by Equations (9.42) and (9.49) respectively, where ℓ_e is the effective length of the tail pipe, S is its cross-sectional area, R_s is its acoustic resistance ($= R_c + R_L$) and V is the expansion chamber volume. If the tail pipe is not short compared to a wavelength (of the order of $\lambda/10$ or less) at the resonance frequency calculated using Equation (9.42), then the insertion loss must be calculated using Equations (9.76) and (9.77) rather than (9.78) and (9.79). Reference to Equation (9.79) shows that, at the frequency of resonance when $\omega = \omega_0$, the insertion loss becomes negative and a function of the quality factor. Equation (9.79) becomes:

$$IL\big|_{\omega = \omega_0} = -20\log_{10}Q \qquad (9.80)$$

In this case the expansion chamber amplifies the radiated noise!

However, well above resonance, appreciable attenuation may be expected. In this case:

$$IL \approx 40\log_{10}(\omega/\omega_0), \quad \omega \gg \omega_0 \qquad (9.81)$$

For the constant acoustic-pressure source, it is assumed that the pipes (a) and (c) of Figure 9.9(a) are part of the expansion chamber assembly. Thus, constant acoustic pressure at the outlet of the source or at the inlet to pipe (a) is assumed. In this case, the expression for the insertion loss of the expansion chamber assembly takes the following form:

$$IL = 20\log_{10}\left|\frac{p}{v_2 Z_L}\right| \qquad (9.82)$$

The formal similarity of the equivalent acoustical circuits of Figures 9.7 and 9.9 and use of Equations (9.51) and (9.53), having replaced Z_u with Z_a, Z_s with Z_b and Z_d with $(Z_c + Z_L)$, allows the preceding expression for the insertion loss to be rewritten as follows:

$$IL = 20 \log_{10}\left| 1 + \frac{Z_a(Z_c + Z_L + Z_b) + Z_b Z_c}{Z_b Z_L} \right| \tag{9.83}$$

Substituting the appropriate expressions for the various impedances, as was done for the constant volume–velocity source, it may be shown that well above resonance for a constant acoustic-pressure source:

$$IL \approx \log_{10}(\omega/\omega_0) + 20 \log_{10} Q ; \qquad \omega \gg \omega_0 \tag{9.84}$$

The results of experiment with such a device are shown in Figure 9.10, where the effect of flow is also shown. The experiment has been carried out to frequencies well

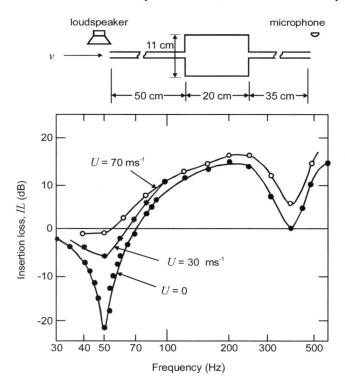

Figure 9.10 An experimental evaluation of an expansion chamber muffler. Near the resonance the insertion loss is increased by steady airflow; if the flow velocity is large enough, the resonance (and negative insertion loss) may be eliminated.

above the upper bound, about 220 Hz, for which the analysis is expected to apply. Thus the decrease in insertion loss at about 400 Hz must be accounted for by transmission-line theory which, however, is not considered here.

In Figure 9.10, the data below about 200 Hz generally confirm the predictions of Equation (9.79) if the resonance frequency is assumed to be about 50 Hz and the quality factor without flow to be about 10. The latter magnitude is quite reasonable. The data show further that the quality factor Q decreases with flow, in part because the inductances of Figure 9.9 decrease and in part because the corresponding resistances increase, as generally predicted by the considerations of Section 9.6.1.2 and shown by Equation (9.29). Note that the performance of the device improved with flow.

When wave propagation along the length of the expansion chamber is considered, the analysis becomes more complicated. It will be sufficient to state the result of the analysis for the case of zero mean flow. As before, the simplifying assumption is made that the chamber is driven by a constant volume–velocity source. A further simplification is made that the outlet pipe is much less than one half wavelength long, so that propagational effects may be neglected. The expansion chamber is assumed to have cross-sectional area, S and length B, as shown in Figure 9.9. The expression for the insertion loss is (Söderqvist, 1982):

$$IL = 10 \log_{10}\left[\frac{N}{(\rho c)^2 \cos^2 kx \ \cos^2 ky}\right] \qquad (9.85)$$

where $k = \omega/c$ and:

$$N = [\rho c \cos(kB - ky)\cos ky - S(X_c + X_L)\sin kB]^2 + [S(R_c + R_L)\sin kB]^2 \qquad (9.86)$$

The quantities X_c, X_L, R_c and R_L represent inductive and resistive impedances respectively of elements c and L (see Figure 9.9).

Referring to Figure 9.9 it is not difficult to show that when $x = y = 0$ and kB is small, Equation (9.85) reduces to Equation (9.79). However, Equation (9.85) has the advantage that it applies over a wider frequency range. For example, it accounts for the observed decrease in insertion loss shown in Figure 9.10 at 400 Hz.

9.7.4.2 Transmission Loss

Although the Transmission Loss (*TL*) of a reactive silencer is not necessarily directly translatable to the noise reduction that will be experienced when the silencer is installed, it is useful to compare the TL performance for various expansion chamber sizes as the same trends will be observable in their noise reduction performance. The simple expansion chamber is a convenient model to demonstrate the principles of Transmission Loss analysis.

The analysis is based on the model shown in Figure 9.9a and it is assumed that the inlet and discharge tubes do not extend into the expansion chamber. There will exist a right travelling wave in the inlet duct, which will be denoted p_I and a left travelling wave reflected from the expansion chamber inlet, denoted p_R. There will also be a left and a right travelling wave in the expansion chamber denoted p_A and p_B, respectively. In the exit pipe there will only be a right travelling wave, p_T, as an anechoic termination will be assumed. It will also be assumed that all muffler dimensions are sufficiently small that only plane waves will be propagating and that the exit pipe diameter is equal to the inlet pipe diameter and each has a cross-sectional area denoted S_1. The cross-sectional area of the expansion chamber is S_2. The Transmission Loss is defined as the ratio of transmitted power to incident power and as we are only considering plane waves, this may be written as:

$$TL = 10\log_{10}\left|\frac{p_I}{p_T}\right|^2 = 10\log_{10}\left|\frac{A_I}{A_T}\right|^2 \tag{9.87a,b}$$

The total sound pressure in the inlet pipe, which is the sum of the incident pressure, p_I and the pressure, p_R reflected from the expansion chamber entrance, may be written using the harmonic pressure solution to the wave equation (see Equation 1.53) as:

$$p_{inlet} = A_I e^{j(\omega t - kx)} + A_R e^{j(\omega t + kx + \beta_1)} \tag{9.88}$$

The total sound pressure in the expansion chamber may be written in terms of the right travelling wave and the reflected left travelling wave as:

$$p_{\exp} = A_A e^{j(\omega t - kx + \beta_2)} + A_B e^{j(\omega t + kx + \beta_3)} \tag{9.89}$$

The total sound pressure in the exit pipe may be written as

$$p_T = A_T e^{j(\omega t - kx + \beta_4)} \tag{9.90}$$

Equations (1.10) and (1.11) may be used with the preceding three equations to write the following for the acoustical particle velocities in the same locations:

$$u_{inlet} = \frac{1}{\rho c}\left[A_I e^{j(\omega t - kx)} - A_R e^{j(\omega t + kx + \beta_1)}\right] \tag{9.91}$$

$$u_{\exp} = \frac{1}{\rho c}\left[A_A e^{j(\omega t - kx + \beta_2)} - A_B e^{j(\omega t + kx + \beta_3)}\right] \tag{9.92}$$

$$u_{exit} = \frac{A_T}{\rho c} e^{j(\omega t - kx + \beta_4)} \tag{9.93}$$

Continuity of acoustic pressure and volume velocity at the junction of the inlet pipe and the expansion chamber where the coordinate system origin, $x = 0$ will be defined, gives:

$$A_I + A_R e^{j\beta_1} = A_A e^{j\beta_2} + A_B e^{j\beta_3} \tag{9.94}$$

and

$$S_1(A_I - A_R e^{j\beta_1}) = S_2(A_A e^{j\beta_2} - A_B e^{j\beta_3}) \tag{9.95}$$

At the junction of the expansion chamber and exit pipe, at $x = L$, continuity of acoustic pressure and volume velocity gives:

$$A_T e^{-jkL + j\beta_4} = A_A e^{-jkL + j\beta_2} + A_B e^{jkL + j\beta_3} \tag{9.96}$$

and

$$S_1 A_T e^{-jkL + j\beta_4} = S_2(A_A e^{-jkL + j\beta_2} - A_B e^{jkL + j\beta_3}) \tag{9.97}$$

Using Equations (9.94) to (9.97), the transmitted sound pressure amplitude can be written in terms of the incident sound pressure amplitude as:

$$A_T = \frac{2 A_I e^{jkL} e^{-j\beta_4}}{2 \cos kL + j(S_1/S_2 + S_2/S_1) \sin kL} \tag{9.98}$$

Substituting Equation (9.98) into (9.87b) gives:

$$TL = 10 \log_{10}\left[1 + \frac{1}{4}\left(\frac{S_1}{S_2} - \frac{S_2}{S_1} \right)^2 \sin^2 kL \right] \tag{9.99}$$

Equation (9.99) was derived using 1-D wave analysis rather than lumped analysis, so it takes into account the effect of axial modes but it is not valid if cross modes exist in the chamber. Equation (9.99) is plotted as a function of expansion ratio, $m = S_1/S_2$ vs wavenumber, k multiplied by the expansion chamber length, L in Figure 9.11. S_1 is always greater than S_2. Note that when $ka > 1.85$, higher order modes begin to propagate and notionally when $ka > 4$, the energy in the higher order modes exceeds the energy in the plane wave mode and experimental data will generally show smaller attenuation values than predicted in the figure. Also, Equation (9.99) has been derived without any consideration of resistive impedance so effectively damping has been excluded from the analysis which explains why the minima in TL are zero, instead of some positive number that would occur if damping were included.

9.7.5 Small Engine Exhaust

Small gasoline engines are commonly muffled using an expansion chamber and tailpipe. An important consideration in the design of such a muffling system is the

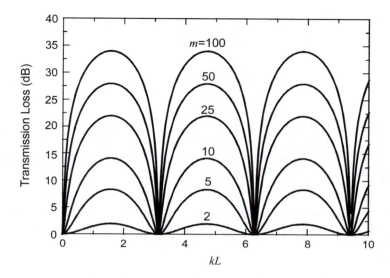

Figure 9.11 Transmission loss as a function of kL for an expansion chamber muffler of length, L and various area ratios, m.

effect of the back pressure imposed by the muffler on the performance of the engine. This matter has been both experimentally and analytically investigated, with the interesting result that an optimum tailpipe length has been shown to exist for any particular muffler configuration (Watters *et al.*, 1959).

A design procedure based upon the cited reference, which takes account of the observed optimum tailpipe length, is as follows:

1. Determine the exhaust volume flow rate, V_0 (m³/s), the speed of sound in the hot exhaust gas, c (m/s), and the effective ratio of specific heats of the exhaust gas γ.

2. Assume that the term involving the quality factor Q in Equation (9.79) may be neglected, and use the equation to determine the resonance frequency ω_0 (rad/s) to give the required insertion loss (or noise reduction, *NR*) at the expected engine speed (*RPM*). Note that ω_0 must always be smaller than ω (defined in 4. below).

3. Assume a volume V (m³) for the expansion chamber. Enter the required quantities and determine a pipe cross-sectional area S (m²) which satisfies the following equation:

$$\left[\frac{1}{\gamma} - \frac{18 V_0}{\omega V} - \frac{2 V_0^2}{S^2 c^2} \right] \frac{\omega_0^2 S^{1.5} V}{10 \sqrt{\pi} f_m V_0^2} > 1 \tag{9.100}$$

Note that $\omega = \pi \times RPM \times$(No. of cylinders)$\times$stroke/30, where stroke = 2 for a 2-stroke engine and 4 for a four stroke engine and for small steel pipes the friction factor $f_m \approx 8 \times 10^{-3}$.

4. Calculate the required effective length, ℓ_e, of tailpipe (including the end correction) using the following equation:

$$\ell_e = \frac{S}{V}\left[\frac{c}{\omega_0}\right]^2 \quad \text{(m)} \tag{9.101}$$

The physical tailpipe length, ℓ_t, may be calculated from the effective length in Equation (9.101) using Equations (9.16) and (9.19) and $\ell_t = \ell_e - 2\ell_0$. If ℓ_t is less than three times the diameter, then set it equal to ten times the diameter and use Equation (9.100) to re-calculate S.

5. Use Equation (9.79) to calculate the Insertion Loss over the frequency range of interest. The quality factor, Q, may be calculated as for an expansion chamber, using Equation (9.49). Note that the length and area terms and the acoustic resistance in Equation (9.49) refer to the tail pipe and V is the expansion chamber volume.

Experimental investigation has shown that the prediction of Equation (9.79) is fairly well confirmed for low frequencies. However, at higher frequencies for which the tailpipe length is an integer multiple of half wavelengths, a series of pass bands are encountered for which the insertion loss is less than predicted. Departure from prediction is dependent upon engine speed and can be expected to begin at about half the frequency for which the tailpipe is one half wavelength long. This latter frequency is called the first tailpipe resonance.

9.7.6 Lowpass Filter

A device commonly used for the suppression of pressure pulsations in a flowing gas is the lowpass filter. Such a device, sometimes referred to as a Helmholtz filter, may take various physical forms but, whatever the form, all have the same basic elements as the filter illustrated in Figure 9.12.

In Figure 9.12(a) the acoustical system is shown schematically as two expansion chambers b and d interconnected by a pipe c and in turn connected to a source and load by pipes a and e. In Figure 9.12(b) the equivalent acoustical circuit is shown as a system of interconnected impedances, identified by subscripts corresponding to the elements in Figure 9.12(a). Also shown in the figure are the acoustic volume flows, v (m³/s), through the various elements and the acoustic pressure, p (Pa), of the source. The impedances of pipes a and c are given by Equation (9.9), and the impedances of volumes b and d are given by Equation (9.37).

At low frequencies, where the elements of the acoustic device illustrated in Figure 9.12(a) are all much less than one half wavelength long, the details of construction are unimportant; in this frequency range the device may be analysed by reference to the equivalent acoustical circuit shown in Figure 9.12(b). The location of the inlet and discharge tubes in each of the chambers only take on importance above this frequency range.

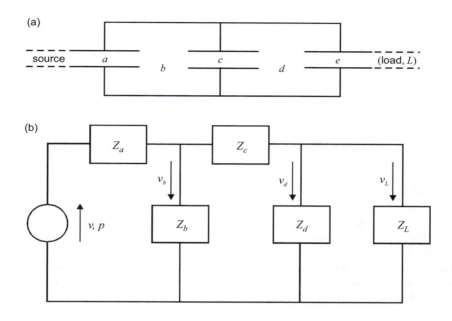

Figure 9.12 A low pass filter: (a) acoustical system; (b) equivalent acoustical circuit.

The impedances of the devices coupled to the filter affect its performance; thus, the lowpass filter cannot be analysed in isolation. However, only the details of the performance and not the basic function will be affected.

To simplify the analysis, certain assumptions will be made about the source and load attached to the filter. The source will be modelled either as a constant volume–velocity (infinite internal impedance) or a constant acoustic-pressure (zero internal impedance) source (see Section 9.5 for a discussion of internal source impedance). The further simplifying assumption will be made that the impedance, $(Z_e + Z_L)$, of the system with the filter is equal to the impedance, Z_L, of the original system; that is, the tailpipe length is unchanged by insertion of the filter.

The possibilities for the load impedance are myriad, but discussion will be restricted to just two, which are quite common. It will be assumed that either the filter is connected to a very long piping system, such that any pressure fluctuation entering the load is not reflected back toward the filter, or that the filter abruptly terminates in a short tailpipe (less than one-tenth of a wavelength long) looking into free space. In the former case, the impedance at the entry to pipe, e, of Figure 9.11(a) is simply the characteristic impedance of the pipe, so that where S_L is the cross-sectional area of the tailpipe of effective length, ℓ_e, the load impedance is:

$$Z_L = \rho c / S_L \tag{9.102}$$

For the case of the short tailpipe the following further simplifications are made:

1. the reactive part of the radiation impedance is accounted for as an end correction (see Section 9.6.1.1), which determines the effective length of pipe e; and

2. the real part of the radiation impedance is negligible compared to the pipe mass reactance in series with it. Thus, in this case the terminating impedance is:

$$Z_L = j\rho\omega\ell_e/S_L \qquad (9.103)$$

For lengths of pipe between the two extreme cases, the terminating impedance is the sum of the inductive impedance, calculated using Equation (9.14), and the resistive impedance, calculated using Equation (9.29). In all cases, the assumption is implicit that only plane waves propagate in the pipe; that is, the pipe diameter is less than 0.586 of a wavelength of sound. For larger diameter pipes in which cross modes propagate, the analysis is much more complex and will not be considered here.

The effect of introducing the lowpass filter will be described in terms of an insertion loss. The insertion loss provides a measure of the reduction in acoustical power delivered to the load by the source when the filter is interposed between the source and load. Referring to Figure 9.12(b), the insertion loss for the case of the constant volume–velocity source is:

$$IL = 20\log_{10}\left|\frac{v}{v_L}\right| \qquad (9.104)$$

while for the case of the constant acoustic-pressure source the insertion loss is:

$$IL = 20\log_{10}\left|\frac{p}{v_L Z_L}\right| \qquad (9.105)$$

Here, v (m³/s) is the assumed constant amplitude of the volume flow of the constant volume–velocity source, and for the constant acoustic-pressure source, p (Pa) is the constant amplitude of the acoustic pressure of the gas flowing from the source. v_L (m³/s) is the amplitude of the acoustic-volume flow through the load, and Z_L is the load impedance.

In the following analysis, acoustical resistances within the filter have been neglected for the purpose of simplifying the presentation, and because resistance terms only significantly affect the insertion loss at system resonance frequencies. It is also difficult to estimate acoustic resistances accurately, although Equation (9.29) may be used to obtain an approximate result, as discussed in Section 9.6.1.2.

Reference to Section 9.7.1 and to Figure 9.12 allows the following system of equations to be written for the constant volume–velocity source:

$$v = v_b + v_d + v_L \qquad (9.106)$$

$$0 = -v_b Z_b + v_d(Z_c + Z_d) + v_L Z_c \qquad (9.107)$$

$$0 = -v_d Z_d + v_L Z_L \qquad (9.108)$$

Alternatively, if the system is driven by a constant acoustic-pressure source, Equation (9.106) is replaced with the following equation:

$$p = Z_a(v_b + v_d + v_L) + v_b Z_b \tag{9.109}$$

Use of Equations (9.104), (9.106), (9.107) and (9.108) gives the following result for the constant volume–velocity source:

$$IL = 20 \log_{10} \left| 1 + \frac{Z_c}{Z_b} + \frac{Z_L}{Z_d} + \frac{Z_L}{Z_b} + \frac{Z_c Z_L}{Z_b Z_d} \right| \tag{9.110}$$

For the long pipe termination, use of Equations (9.9) and (9.37) allows Equation (9.110) to be written as:

$$IL = 10 \log_{10} \left[\left[1 - \frac{V_b \ell_c}{S_c} \left(\frac{\omega}{c} \right)^2 \right]^2 + \left[\frac{V_b + V_d}{S_L} \left(\frac{\omega}{c} \right) - \frac{V_b V_d \ell_c}{S_L S_c} \left(\frac{\omega}{c} \right)^3 \right]^2 \right] \tag{9.111}$$

Similarly, for the short tailpipe termination Equation (9.110) takes the following alternative form:

$$IL = 10 \log_{10} \left[1 - \left[\frac{V_b \ell_c}{S_c} + \frac{V_d \ell_L}{S_L} + \frac{V_b \ell_L}{S_L} \right] \left(\frac{\omega}{c} \right)^2 + \frac{V_b V_d \ell_L \ell_c}{S_L S_c} \left(\frac{\omega}{c} \right)^4 \right]^2 \tag{9.112}$$

Use of Equations (9.105), (9.107), (9.108) and (9.109) gives the following result for the constant acoustic-pressure source:

$$IL = 20 \log_{10} \left| \frac{Z_a}{Z_d} + \frac{Z_a}{Z_L} + \left(1 + \frac{Z_a}{Z_b} \right) \left(1 + \frac{Z_c}{Z_d} + \frac{Z_c}{Z_L} \right) \right| \tag{9.113}$$

For the long tailpipe termination, use of Equations (9.9) and (9.37) allows Equation (9.113) to be written as:

$$IL = 10 \log_{10} \left\{ \left[1 - \left(\frac{\ell_a V_d}{S_a} + \frac{\ell_a V_b}{S_a} + \frac{\ell_c V_d}{S_c} \right) \left(\frac{\omega}{c} \right)^2 + \frac{\ell_a \ell_c V_b V_d}{S_a S_c} \left(\frac{\omega}{c} \right)^4 \right]^2 \right. $$
$$+ \left. \left[\left(\frac{\ell_a S_L}{S_a} + \frac{\ell_c S_L}{S_c} \right) \left(\frac{\omega}{c} \right) - \frac{\ell_a \ell_c V_b S_L}{S_a S_c} \left(\frac{\omega}{c} \right)^3 \right]^2 \right\} \tag{9.114}$$

Similarly, for the short tailpipe termination, Equation (9.113) takes the following alternative form:

$$IL = 20 \log_{10} \left| 1 + \frac{\ell_a S_L}{\ell_L S_a} + \frac{\ell_c S_L}{\ell_L S_c} - \left[\frac{\ell_a V_d}{S_a} + \frac{\ell_a V_b}{S_a} + \frac{\ell_c V_d}{S_c} \right. \right.$$

$$\left. \left. + \frac{\ell_a \ell_c V_b S_L}{S_a S_c \ell_L} \right] \left(\frac{\omega}{c} \right)^2 + \frac{\ell_a \ell_c V_b V_d}{S_a S_c} \left(\frac{\omega}{c} \right)^4 \right| \tag{9.115}$$

Resonances may be expected at frequencies for which the arguments of Equations (9.111) to (9.115) are zero. Thus, at low frequencies, the lowpass filter of Figure 9.12 may be expected to amplify introduced noise. However, well above all such resonances, but still below the half wave resonances encountered at high frequencies, the system will behave as a lowpass filter, with the following limiting behaviour for each of the four cases considered:

1. Constant volume–velocity, long tailpipe termination:
$$IL = IL_0 - 20 \log_{10} S_L + 60 \log_{10}(\omega/c) \tag{9.116}$$

2. Constant volume–velocity, (short) tailpipe termination:
$$IL = IL_0 - 20 \log_{10} S_L + 20 \log_{10} \ell_L + 80 \log_{10}(\omega/c) \tag{9.117}$$

3. Constant acoustic-pressure, long tailpipe termination:
$$IL = IL_0 - 20 \log_{10} S_a + 20 \log_{10} \ell_a + 80 \log_{10}(\omega/c) \tag{9.118}$$

4. Constant acoustic-pressure, (short) tailpipe termination: same as (3) above.

In all cases:
$$IL_0 = 20 \log_{10} \left(V_b V_d \ell_c \right) - 20 \log_{10} S_c \tag{9.119}$$

The above equations hold at frequencies well above the resonances predicted by the preceding analysis. However, as the frequency range is extended upward, other resonances will be encountered when chamber or tube length dimensions approach integer multiples of one half wavelength. These resonances will introduce pass bands (suggesting small *IL*), which will depend upon dimensional detail and thus cannot be treated in an entirely general way; no such high-frequency analysis will be attempted here.

The possibility of high-frequency pass bands associated with half wavelength resonances suggests that their numbers can be minimised by choice of dimensions so that resonances tend to coincide. It is found, in practice, that when such care is taken, the loss in filter performance is minimal; that is, the filter continues to perform as a lowpass filter even in the frequency range of the predicted pass bands, although at reduced effectiveness. Presumably, inherent resistive losses and possibly undamping effects, particularly at high sound pressure levels such as may be encountered in a pumping system, account for the better than predicted performance. As stated earlier, these losses and undamping effects were neglected in the preceding analysis.

The following iterative procedure is recommended to obtain an approximate lowpass filter design which can then be modified slightly if necessary to achieve a required insertion loss at any desired frequency ω, by using Equations (9.111) to (9.115).

1.   Select the desired highest resonance frequency, f_0, for the filter to be approximately 0.5 times the fundamental frequency of the pressure pulsations.

2.   Select two equal chamber volumes, V_b and V_d, to be as large as practicable.

3.   Choose the length and diameter of the chambers so that the length is of similar magnitude to the diameter.

4.   If possible, configure the system so that the choke tube, c, connecting the two volume chambers exists completely inside the chambers, which are then essentially one large chamber divided into two by a baffle between them (see Figure 9.12(a)).

5.   Make the choke tube as long as possible, consistent with it being completely contained within the two chambers and such that its ends are at least one tube diameter from the end of each chamber.

6.   Choose the choke tube diameter to be as small as possible consistent with a pressure drop of less than 0.5% of the line pressure, but do not allow the choke tube diameter to be less than one half of nor greater than the inlet pipe diameter. Means for calculating pressure drops are discussed in Section 9.7.7.

7.   Calculate the highest resonance frequency of the system using the following approximate expression:

$$f_0 = \frac{c}{2\pi}\left[\frac{S_c}{\ell_c}\left[\frac{1}{V_b} + \frac{1}{V_d}\right]\right]^{1/2} \tag{9.120}$$

This expression is quite accurate for a constant volume–velocity source if the discharge pipework is much shorter than the choke tube (and is of similar or larger diameter), of if the discharge pipework is sufficiently long that pressure waves reflected from its termination may be ignored. For other situations, Equation (9.120) serves as an estimate only and can overestimate the resonance frequency by up to 50%.

8.   If f_0, calculated using Equation (9.120), is less than the frequency calculated in (1) above, then modify your design by: (a) reducing the chamber volumes, and/or (b) increasing the choke tube diameter.

9.   If f_0, calculated using Equation (9.120), is greater than the frequency calculated in (1) above, then modify your design by doing the reverse of what is described in (8) above.

10. Repeat steps (7), (8) and (9) until the quantity f_0, calculated using Equation (9.120), is within the range calculated in step (1).

It is also possible to solve Equation (9.120) directly by choosing all parameters except for one (e.g. choke tube cross-sectional area, S_c) and solving for the parameter not chosen. If the value of the calculated parameter is not satisfactory (e.g. too much pressure drop), then the other chosen parameters will need to be adjusted within acceptable bounds until a satisfactory solution is reached, following much the same iterative procedure as described in the preceding steps.

9.7.7 Pressure Drop Calculations for Reactive Muffling Devices

The introduction of reactive or dissipative muffling systems in a duct will impose a pressure drop. For example, an engine muffler will impose a back pressure on the engine, which can strongly affect the mechanical power generated. The total pressure drop of a muffling system is a combination of friction and dynamic losses through the system. The former friction losses, which are generally least important, will be proportional to the length of travel along tubes or ducts, while the latter dynamic losses will occur at duct discontinuities; for example, at contractions, expansions and bends. Reactive devices depend upon discontinuities in their design, whereas dissipative devices do not; thus, one can generally expect a greater pressure drop through reactive devices than through dissipative devices.

In this section, means will be provided for estimating expected pressure drops for the reactive devices discussed in this chapter. The analytical expressions provided here were derived by curve fitting empirical data (ASHRAE, 2005). Friction losses will be considered first.

For the case of laminar flow, friction losses depend upon the Reynold's number and are small. However, when the Reynold's number is greater than 2000 the flow will be turbulent, and the pressure drop will be independent of Reynold's number. Only the latter case is considered here, as it provides a useful upper bound on friction losses. The following equation may be used to estimate the expected pressure drop for flow through a duct:

$$\Delta P = f_m \left(\frac{L P_D}{4S} \right) \left(\frac{\rho U^2}{2} \right) \tag{9.121}$$

In the preceding expression, f_m is the friction factor, ΔP (Pa) is the pressure drop, U (m/s) is the mean flow speed through the duct, S (m^2) is the duct cross-sectional area, P_D (m) is the duct cross-sectional perimeter and L (m) is the length of the duct. The friction factor is defined as:

$$\begin{aligned} f_m &= f_m' && \text{if } f_m' \geq 0.018 \\ f_m &= 0.85 f_m' + 0.0028 && \text{if } f_m' < 0.018 \end{aligned} \tag{9.122}$$

where $f_m{}' = 0.11 \left(\dfrac{\varepsilon P_D}{4S} + \dfrac{68}{Re} \right)^{0.25}$

$$= 0.11 \left(\frac{\varepsilon P_D}{4S} + \frac{2.56 \times 10^{-4} P_D}{SU} \right)^{0.25} \quad \text{for standard air}$$

(9.123a,b)

As can be seen, the friction loss depends on the pipe or duct roughness, ε, which is usually taken as 1.5×10^{-4} m for galvanized steel ducts, pipes and tubes, such as considered in connection with engine mufflers, expansion chambers, and lowpass filters, and 9×10^{-4} m for fibreglass-lined ducts (ASHRAE, 2005). The quantity, Re is the Reynolds number, given by:

$$Re = \frac{4SU}{P_D \, \mu}$$

(9.124)

where μ is the dynamic viscosity of the gas flowing through the muffler.

Dynamic losses are calculated in terms of a constant K, dependent upon the geometry of the discontinuity, using the following equation:

$$\Delta P = \frac{1}{2} \rho U^2 K$$

(9.125)

Values of K may be determined by making reference to Figure 9.13, where various geometries and analytical expressions are summarised (ASHRAE, 2005).

9.7.8 Flow-generated Noise

Reactive muffling devices depend for their success upon the introduction of discontinuities in the conduits of the system. Some simple examples have been considered in previous sections. The introduction of discontinuities at the boundaries of a fluid conducting passage will produce disturbances in the fluid flow, which will result in noise generation. Regularly spaced holes in the facing of a perforated liner can result in fairly efficient "whistling", with the generation of associated tones. Such "whistling" can be avoided by choice of the shape or formation of the hole edge. For example, those holes that provide parallel edges crosswise to the mean flow will be more inclined to whistle than those that do not. Alternatively, arranging matters so that some small flow passes through the holes will inhibit "whistling".

Aside from the problem of "whistling", noise will be generated at bends and discontinuities in duct cross-sections. Fortunately, the associated noise-generating mechanisms are remarkably inefficient at low flow speeds and generally can be ignored. However, the efficiencies of the mechanisms commonly encountered increase with the cube and fifth power of the free stream local Mach number. An upper bound on flow speed for noise reduction for any muffling system is thus implied. At higher flow speeds "self-noise" generated in the device will override the noise reduction that it provides.

Figure 9.13 Dynamic pressure loss factors.

(a) Contracting bellmouth:

$$K = 0.03 + \frac{0.97}{1 + F_1(y)}$$

(b) Contracting bellmouth with wall:

$$K = 0.03 + \frac{0.47}{1 + F_1(y)}, \qquad y = 10r/D, \qquad F_1(y) = 6.82y^3 + 0.56y^2 + 1.25y$$

(c) Step contraction:

$$K = 0.5 - \frac{0.5}{1 + F_2(y)}, \qquad y = d_1/d_2 \geq 1, \qquad F_2(y) = 0.222y^2 + 1.892y - 2.114$$

(d) Gradual contraction:

$$K = 0.05$$

(e) Sharp edge, inward-projecting contractions:

$$K = 1.0$$

(f) Limited expansion:

$$K = \frac{[z^2 + 0.0047][1 - y^2]^2}{z^2 - 0.1682z + 0.0807}, \qquad z = (d_2 - d_1)/L, \qquad y = d_1/d_2 \leq 1$$

(g) Various unlimited expansions:

$$K = 1.0$$

(h) Mitered duct bends:

$$K = K_{MB} K_{RE}$$

$$K_{MB} = 0.34 \left[\frac{\theta}{45}\right]^{1.82} \left[\frac{1.11 + 2.03y^2 + 7.72y}{1.0 + 3.5y^2 + 6.36y}\right]$$

$$K_{RE} = 1.0 + \frac{0.613}{\mu} - \frac{0.213}{\mu^2}$$

where Reynold's number, $Re = 6.63 \times 10^4 UD$; $\mu = Re \times 10^{-4}$; circular duct, $y = 1$, $D = H = W$; rectangular duct, $y = H/W$, $D = 2HW/(W + H)$.

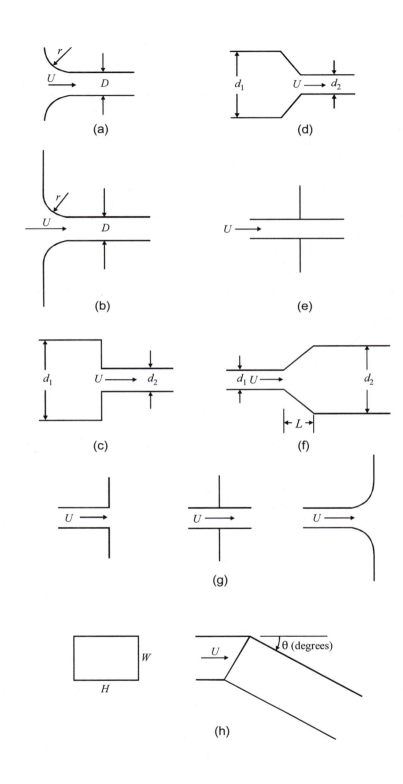

(a)

(b)

(c)

(d)

(e)

(f)

(g)

(h)

The problem of self-noise has long been recognised in the air-conditioning industry. The discussion of this section will depend heavily upon what information is available from the latter source. The analytical expressions provided in this section were derived by curve fitting of empirical data. The importance of self-noise generation in automotive muffling systems is well known. In the latter case, even without any mean flow, the high sound levels commonly encountered result in fluid movement of sufficiently large amplitude to generate enough noise to limit the effectiveness of an automotive muffling system for noise suppression purposes.

Flow noise generated at a mitred bend will be used here as a model. In the following discussion, reference should be made to Figure 9.14. At the inner (convex) corner, flow separation occurs at the sharp corner. Further downstream, flow re-attachment occurs. The point of re-attachment, however, is unsteady, resulting in an effective fluctuating drag force on the fluid. As shown in Section 5.3.2, such a fluctuating force acting on the stream can be interpreted as a dipole source. In the case considered here, the axis of the dipole is oriented parallel to the stream, and all frequencies propagate. Sound from this source increases with the sixth power of the stream speed. Alternatively, if the sound power is referenced to the stream power, then as shown in Section 5.3.3, the inner corner noise source will increase in efficiency with the cube of the local Mach number.

At the outer (concave) corner, flow separation also occurs, resulting in a fairly stable bubble in the corner. However, at the point of re-attachment downstream from the corner, very high unsteady shear stresses are induced in the fluid. As shown in Section 5.4, such a fluctuating shear stress acting on the stream can be interpreted as a quadrupole source. A longitudinal quadrupole may be postulated. Such a source, with its axis oriented parallel to the stream, radiates sound at all frequencies. The sound power produced by this type of source increases with the eighth power of the free-stream speed. Alternatively, if the sound power is again referenced to the stream power then, as shown in Section 5.4, the outer corner source efficiency will increase with the fifth power of the local Mach number.

Let the density of the fluid be ρ (kg/m^3), the cross-sectional area of the duct be S (m^2) and the free-stream speed be U (m/s), then the mechanical stream power level L_{ws} referenced to 10^{-12} W is:

$$L_{ws} = 30 \log_{10} U + 10 \log_{10} S + 10 \log_{10} \rho + 117 \quad \text{(dB)} \tag{9.126}$$

A dimensionless number, called the Strouhal number, is defined in terms of the octave band centre frequency f_C (Hz), the free-stream speed U (m/s) and the height of the elbow, H (see Figure 9.14), as follows:

$$N_s = f_C H / U \tag{9.127}$$

Experimental data for the sound power, L_{wB}, generated by a mitred bend without turning vanes is described by the following empirical equation.

$$L_{wB} - L_{ws} = -10 \log_{10}\left[1 + 0.165 N_s^2\right] + 30 \log_{10} U - 103 \quad \text{(dB)} \tag{9.128}$$

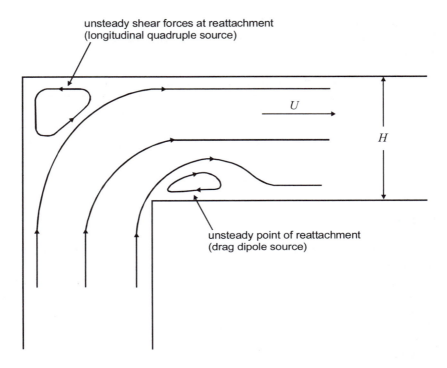

unsteady shear forces at reattachment
(longitudinal quadruple source)

U

H

unsteady point of reattachment
(drag dipole source)

Figure 9.14 A mitred bend as a source of flow noise.

The quantity, L_{wB}, is the octave band sound power level, and N_s is the Strouhal number corresponding to the octave band centre frequency.

The right-hand side of Equation (9.128) may be interpreted as a measure of efficiency of conversion of stream power into acoustic power or noise. Note that for small Strouhal numbers, the efficiency is proportional to the cube of the free-stream speed, but for large Strouhal numbers, the dependence of efficiency increases as the fifth power of the stream speed. Consideration of Equation (9.128) also shows that, for low flow speeds such as 1 m/s, the efficiency is very small and noise generation is negligible. However, for flow speeds of the order of 10 m/s or greater the efficiency of noise generation becomes significant; rather suddenly flow noise assumes importance.

Owing to the nature of "self-noise" sources, the behaviour of the duct bend is probably typical of duct discontinuities in general.

The self-noise of a range of commercial air conditioning dissipative-type silencers may be shown to obey the following relation (Iqbal *et al.*, 1977):

$$L_{wB} - L_{ws} = 50 \log_{10} U - 155 + C \quad \text{(dB)} \tag{9.129}$$

Here L_{wB} is the octave band sound power level of self-noise at the downstream end of the silencers, L_{ws} is again the power level of the free stream and U is the stream speed.

Dependence upon Strouhal number is implied by the frequency band corrections, C, which were empirically determined by measurement. Values of C are given in Table 9.2.

Table 9.2 Correction number, C, for Equation (9.129)

	\multicolumn{8}{c}{Octave band centre frequency (Hz)}
	63	125	250	500	1000	2000	4000	8000
Correction, C	0	0	0	-4	-13	-13	-19	-22

9.8 LINED DUCTS

Dissipative muffling devices are often used to muffle fans in air conditioning systems and other induced-draft systems. For example, a dissipative-type muffler may be used successfully to control the noise of a fan used in a wood-dust collection system. The dissipation of sound energy is generally accomplished by introducing a porous lining on one or more of the walls of the duct through which the induced draft and unwanted sound travel. In some cases (for example, the wood dust collection system), the porous lining material must be protected with a suitable facing.

A liner may also be designed to successfully make use of the undamping resistance of a simple hole through which, or across which, a modest flow is induced. This principle has been used in the design of a muffler for noise control of a large subsonic wind tunnel, in which a dissipative liner is used that consists of perforated metal sheets of about 4.8% open area, which are cavity backed.

When the dimensions of the lined duct are large compared to the free field wavelength of the sound that propagates, higher order mode propagation may be expected (see Section 9.8.3.2). Since each mode will be attenuated at its own characteristic rate that is dependent upon the design of the dissipative muffler, the insertion loss of a dissipative muffler will depend upon what modes propagate; for example, what modes are introduced at the entrance to the muffler. In general, what modes are introduced at the entrance and the energy distribution among them is unknown and consequently a definitive value for muffler insertion loss cannot be provided.

Two approaches have been taken to describe the attenuation provided by a lined duct when higher order mode propagation may be expected. Either the attenuation rate in decibels per unit length for the least attenuated mode may be provided, from which a lower bound for insertion loss for any length duct may be calculated, or the insertion loss for a given length of lined duct is calculated assuming introduction with equal energy distribution among all possible cut-on modes at the entrance. In the following discussion the rate of attenuation of the least attenuated mode is provided. For comparison, in Section 9.8.3 the insertion loss for rectangular lined ducts for the alternative case of equal energy distribution among all possible propagating modes is compared with the predicted insertion loss for the least attenuated mode.

9.8.1 Locally Reacting and Bulk Reacting Liners

When analysing sound propagation in a duct with an absorbent liner, one of two possible alternative assumptions is commonly made. The older, simpler, and more completely investigated assumption is that the liner may be treated as locally reacting and this assumption results in a great simplification of the analysis (Morse, 1939). In this case, the liner is treated as though it may be characterised by a local impedance that is independent of whatever occurs at any other part of the liner and the assumption is implicit that sound propagation does not occur in the material in any other direction than normal to the surface.

Alternatively, when sound propagation in the liner parallel to the surface is not prevented, analysis requires that the locally reacting assumption must be replaced with an alternative assumption that the liner is bulk reacting. In this case, sound propagation in the liner is taken into account.

The assumption that the liner is locally reacting has the practical consequence that for implementation, sound propagation in the liner must be restricted in some way to normal to the surface. This may be done by the placement of solid partitions in the liner to prevent propagation in the liner parallel to the surface. An example of a suitable solid partition is a thin impervious metal or wooden strip, or perhaps a sudden density change in the liner. Somewhat better performance may be achieved by such use of solid partitions but their use is expensive and it is common practice to omit any such devices in liner construction.

A generalised analysis, which has been published (Bies *et al.*, 1991), will be used here to discuss the design of dissipative liners. The latter analysis accounts for both locally reacting and bulk reacting liners as limiting cases. It allows for the possible effects of a facing, sometimes employed to protect the liner, and it also allows investigation of the effects of variation of design parameters. An error in the labelling of the ordinates of the figures in the reference has been corrected in the corresponding figures in this text.

9.8.2 Liner Specification

The dissipative devices listed as numbers 4 to 6 in Table 9.1 make use of wall-mounted porous liners, and thus it will be advantageous to discuss briefly, liners and their specifications before discussing their use in the listed devices. Generally, sound absorbing liners are constructed of some porous material such as fibreglass or rockwool, but any porous material may be used. However, it will be assumed in the following discussion that, whatever the liner material, it may be characterised by a specific acoustic impedance and flow resistivity. In general, for materials of generally homogeneous composition, such as fibrous porous materials, the specific acoustic impedance is directly related to the material flow resistivity and thickness.

Section 1.13 and Appendix C provide discussions of flow resistance and the related quantity flow resistivity. As shown in the latter reference, the flow resistivity of a uniform porous material is a function of the material density. Consequently, a liner consisting of a layer of porous material may be packed to a specified density, ρ_m, to give any required flow resistance.

Usually some form of protective facing is provided, but the facing may be omitted where, for example, erosion or contamination due to air flow or mechanical abuse is not expected. The protective facing may consist of a spray-on polyurethane coating, an impervious lightweight plastic sheet (20 μm thick polyester film is suitable), or a perforated heavy-gauge metal facing, or some equivalent construction or combination of constructions. Some possible protective facings are illustrated in Figure 9.15.

Referring to Figure 9.15, it is to be noted that the spacing between elements **A** and **C**, which is unspecified, is provided by a spacer, **B** and is essential for good performance. Spacer **B** can be implemented in practice using a coarse wire mesh, with holes of at least 12 mm × 12 mm in size and made of wire between 1 and 2 mm thick. If element **C** is either fiberglass cloth or a fine screen then the flow resistance of the

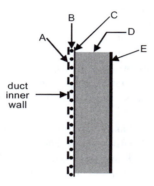

Figure 9.15 Protective facings for duct liners. The elements of the liner are: A, 20 gauge perforated facing, minimum 25% open area; B, spacer such as 2 mm thick wire mesh; C, light plastic sheet or fibreglass cloth or fine mesh metal screen; D, fibrous material of specified flow resistance; E, rigid wall or air cavity backing. Maximum flow speeds up to 8 m/s do not require A, B or C. Speeds up to 10 m/s require that the fibrous material of D be coated with a binder to prevent erosion of the fibres. Speeds up to 25 m/s require B and C, while speeds up to 90 m/s require A, B and C. Higher speeds are not recommended. In many installations, A is used regardless of flow speed, to provide mechanical protection for D.

element is negligible and the element plays no part in the predicted attenuation. On the other hand, if the element **C** is of the form of a limp impervious (plastic) membrane then it may play an important part in the predicted attenuation, as shown by Bies *et al.* (1991) and as discussed in Appendix C.

Care must be taken in the use of perforated facings for liner protection. Often, the result of flow across the regularly spaced holes in the perforated facing is the generation of "whistling" tones. Alternatively, a small induced flow through the perforations will prevent whistling. The perforated facing should have a minimum open area of 25% to ensure that its effect upon the performance of the liner is negligible.

Alternatively, where a perforated facing of percentage open area P less than 25 % is to be used (see element **A** of Figure 9.15), the effect of the facing may be taken into account by redefining element **C** of Figure 9.15. The effect is the same as either adding a limp membrane covering, if there is none, or of increasing the surface density of a limp membrane covering the porous material. The acoustic impedance of a single hole in a perforated sheet is given by Equation (9.15), which can be rewritten as:

$$Z_A = R_A + jX_A = R_A + j\omega M_A \qquad (9.130\text{a,b})$$

where R_A for a single hole is given by Equation (9.29) with A in that equation being the area of a single hole. The value of R_A for a perforated facing with N holes is $1/N$ what it is for one hole. The acoustic resistance for a perforated facing is then given by Equation (9.29) with the S in the denominator replaced with $P/100$ where P is the % open area of the perforated facing.

The mass reactance of Equation (9.130) for a single hole can be derived from Equation (9.15) and (9.130b) with $M = 0$ and $\tan k\ell_e \approx k\ell_e$ (for small $k\ell_e$). Thus:

$$M_A = \frac{\rho \ell_e}{\pi a^2} \qquad (9.131)$$

where a is the radius of a single hole, and ℓ_e is the effective length of the hole, defined in Equation (9.25). The effective mass per unit area (or specific mass reactance) of the air in the holes on the perforated sheet is equal to M_A multiplied by the area of a single hole divided by the fraction of the area of the perforated sheet that is open. Thus, if σ' is the surface density of the limp membrane, then the effective surface density, σ, to be used when entering the design charts to be discussed in Section 9.8.3, is obtained by adding to the limp membrane mass, the effective mass per unit area of the holes, which is their specific mass reactance, so that the effective surface density is:

$$\sigma \approx \sigma' + 100\rho\ell_e/P \qquad (9.132)$$

where ρ is the density of air, P is the percentage of open area of the perforated facing and ℓ_e is the effective length of the holes in the perforated facing, defined in Equation (9.25). The holes must be a distance apart of at least $2a$ (where a is the radius of the holes) for Equation (9.132) to hold. For a percentage open area P of greater than 25%, the effect of a perforated facing is generally negligible (Cummings, 1976).

For high frequencies and for significant flow of Mach number, M, past the perforated sheet, a more accurate version of Equation (9.132) may be derived from Equation (9.24) as:

$$\sigma \approx \sigma' + \frac{\dfrac{100\rho}{Pk}\tan(k\ell_e(1-M))}{1 + \dfrac{100\rho}{kmP}\tan(k\ell_e(1-M))} \qquad (9.133)$$

where m is the mass per unit area (surface mass) of the perforated facing.

9.8.3 Lined Duct Silencers

In the following sections, isotropic bulk reacting and locally reacting liners will be considered and design charts are provided in Figures 9.16 to 9.21, which allow determination of the rate of attenuation of the least attenuated mode of propagation for some special cases, which are not optimal, but which are not very sensitive to the accuracy in the value of flow resistance of the liner. However, the general design problem requires use of a computer program (Bies *et al.*, 1991). Alternative design procedures for determination of the insertion loss of cylindrical and rectangular lined ducts have been described in the literature (Ramakrishnan and Watson, 1991).

Procedures are available for the design of lined ducts for optimum sound attenuation but, generally, the higher the attenuation, the more sensitive is the liner flow resistance specification, and the narrower is the frequency range over which the liner is effective (Cremer, 1953, 1967).

The design charts shown in Figures 9.16 to 9.21 may be used for estimating the attenuation of the least attenuated propagating mode in lined ducts of both rectangular and circular cross-section. The design charts can be read directly for a rectangular duct lined on two opposite sides. For a lined circular duct, the value of $2h$ is the is the open width of a square cross-section duct of the same area as the circular section duct, and the values of attenuation given in the charts must be multiplied by two.

In Figure 9.16, attenuation data for a bulk reacting liner with zero mean flow in the duct are shown for various values of flow resistivity parameter, $R_1 \ell / \rho c$, and ratio of liner thickness to half duct height, ℓ / h, for a length of duct equal to half of the duct width. In this figure, the density parameter, $\sigma / \rho h$ is equal to zero, implying no plastic covering of the liner and no perforated sheet covering unless its open area exceeds 25%. Figures 9.17 and 9.18 are identical to Figure 9.16 except that the Mach number of flow through the lined duct is 0.1 and -0.1 respectively (positive Mach number implies flow in the same direction as sound propagation). In all three figures, 9.16–9.18, it is assumed that the flow resistance of the liner is the same in the direction normal to the duct axis as it is in the direction parallel to the duct axis.

In Figure 9.19, data are shown for the same cases as for Figure 9.16, except that the liner is assumed to be locally reacting. In practice, this is realised by placing rigid partitions in the liner normal to the duct axis so that sound propagation in the liner parallel to the duct axis is inhibited.

The data in Figures 9.20 and 9.21 are for various masses of limp membrane (usually plastic and including any perforated liner - see Equation (9.133)) covering of the liner. Densities of typical liners are given in Table C.3 in Appendix C. Figure 9.20 is for zero mean flow through the duct while Figure 9.21 is for a mean flow of Mach number = 0.1, with the figures on the left for flow in the same direction as sound propagation and the figures on the right representing flow in the opposite direction to sound propagation.

In Figures 9.16 to 9.21, the duct is assumed to be rectangular with two opposite walls lined, as shown in the insets. The open section (air way) of the duct is $2h$ wide, while the liner thickness on either side is ℓ. In determining the thickness, ℓ (see Figure 9.15), elements A and B and the spacing between them are generally neglected, so that ℓ refers to the thickness of element C. If the duct is lined on only one side then the

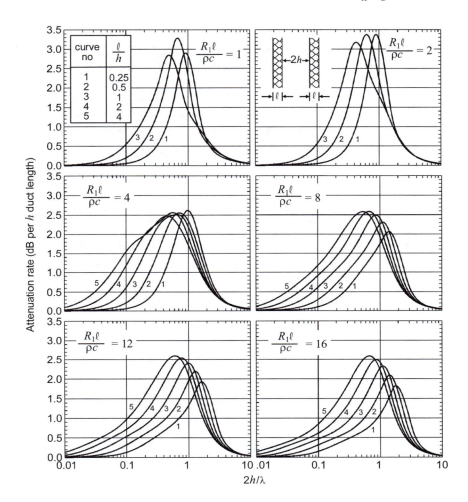

Figure 9.16 Predicted octave band attenuations for a rectangular duct lined on two opposite sides. Lined circular ducts or square ducts lined on all four sides give twice the attenuation shown here. The quantity ρ is the density of fluid flowing in the duct, c is the speed of sound in the duct, ℓ is the liner thickness, h is the half width of the airway, σ is the surface density of a limp membrane covering the liner, R_1 is the liner flow resistivity. Bulk reacting liner with no limp membrane covering (density ratio $\sigma/\rho h = 0$). Zero mean flow ($M = 0$).

attenuation would be the same as for a duct lined on two sides, which is twice as wide as the duct lined on one side.

For a duct lined on all four sides, the total attenuation may be approximated as the sum of the attenuations obtained by considering each pair of sides independently. This conclusion is based upon evaluation of a circular duct analysis (Bies *et al.*, 1991) and the observation that the attenuation for a lined circular duct is the same as that of a square duct of cross-sectional area equal to the circular duct area and lined on all four sides.

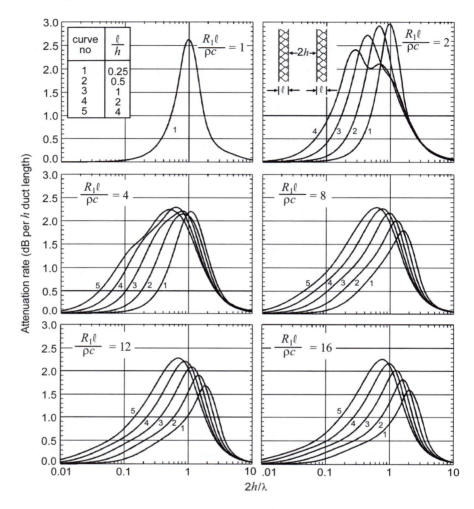

Figure 9.17 Predicted octave band attenuations for a rectangular duct lined on two opposite sides. Lined circular ducts or square ducts lined on all four sides give twice the attenuation shown here. The quantity ρ is the density of fluid flowing in the duct, c is the speed of sound in the duct, ℓ is the liner thickness, h is the half width of the airway, σ is the surface density of a limp membrane covering the liner, R_1 is the liner flow resistivity. Bulk reacting liner with no limp membrane covering (density ratio $\sigma/\rho h = 0$). Mean flow of Mach number, $M = 0.1$ (same direction as sound propagation).

The octave band attenuation predictions shown in Figures 9.16 to 9.21 can be used with reasonable reliability for estimating the attenuation of broadband noise in a duct, provided that calculated attenuations greater than 50 dB are set equal to 50 dB (as this seems to be a practical upper limit). On the other hand, the theory upon which the predictions are based gives values of attenuation at single frequencies that are difficult to achieve in practice. Use of octave band averaging has resulted in a significant smoothing and a reduction in peak values of the single-frequency

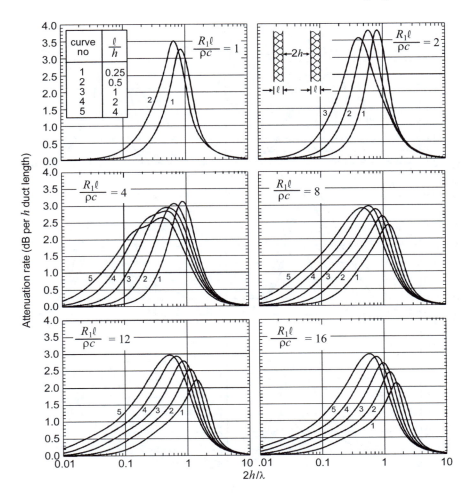

Figure 9.18 Predicted octave band attenuations for a rectangular duct lined on two opposite sides. Lined circular ducts or square ducts lined on all four sides give twice the attenuation shown here. The quantity ρ is the density of fluid flowing in the duct, c is the speed of sound in the duct, ℓ is the liner thickness, h is the half width of the airway, σ is the surface density of a limp membrane covering the liner, R_1 is the liner flow resistivity. Bulk reacting liner with no limp membrane covering (density ratio $\sigma/\rho h = 0$). Mean flow of Mach number, $M = -0.1$ (opposite direction to sound propagation).

predictions. Any point on a curve in the figures is the average response of 20 single-frequency predictions distributed within an octave band encompassing the point.

As explained above, the observed insertion loss for any lined duct will depend upon both the properties of the duct and the propagating sound field that is introduced at the lined duct entrance. In the figures, the assumption is implicit that only the least attenuated mode propagates and in general, the least attenuated mode corresponds to the plane wave mode in an unlined duct. In Table 9.3 experimentally determined values for insertion loss are listed for comparison with the predicted insertion loss for

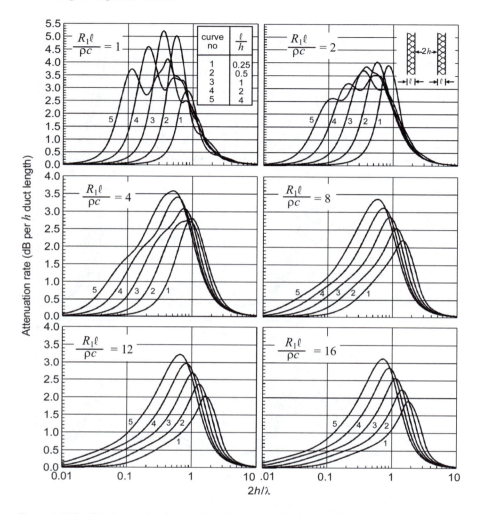

Figure 9.19 Predicted octave band attenuations for a rectangular duct lined on two opposite sides. Lined circular ducts or square ducts lined on all four sides give twice the attenuation shown here. The quantity ρ is the density of fluid flowing in the duct, c is the speed of sound in the duct, ℓ is the liner thickness, h is the half width of the airway, σ is the surface density of a limp membrane covering the liner, R_1 is the liner flow resistivity. Locally reacting liner with no limp membrane covering (density ratio $\sigma/\rho h = 0$). Zero mean flow ($M = 0$).

several lined rectangular ducts based: (1) upon the assumption of equal energy distribution among all propagating modes at the entrance (Ramakrishnan and Watson, 1991); and (2) the least attenuated mode only propagating.

As expected, the least attenuated mode approach underestimates the observed insertion loss at large duct dimension to wavelength ratios. On the other hand, the calculated insertion loss based upon the alternative assumption of equal energy distribution among all modes does not ensure accuracy of prediction either, because,

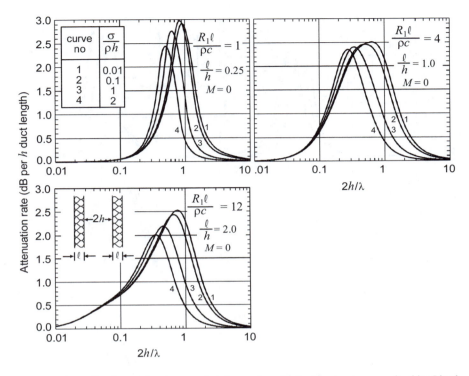

Figure 9.20 Predicted octave band attenuations for a rectangular duct lined on two opposite sides. Lined circular ducts or square ducts lined on all four sides give twice the attenuation shown here. The quantity ρ is the density of fluid flowing in the duct, c is the speed of sound in the duct, ℓ is the liner thickness, h is the half width of the airway, σ is the surface density of a limp membrane covering the liner, R_1 is the liner flow resistivity. Bulk reacting liner with various densities of limp membrane or equivalent perforated sheet ($\sigma/\rho h = 0.01$ to 2). Zero mean flow (Mach number, $M = 0$).

in practice, the assumption of equal energy distribution is unlikely to be satisfied even approximately.

9.8.3.1 Flow Effects

The assumption is implicit in the presentation of Figures 9.16 to 9.21 that, where flow is present, any velocity gradients are small and unimportant. Essentially, uniform flow in any duct cross-section (commonly referred to as plug flow) is assumed and this assumption generally will be adequate for the range of flow Mach numbers shown. The Mach number is the ratio of the stream speed in the duct to the local speed of sound, and is thus also dependent upon the local temperature. In the figures, a negative Mach number indicates sound propagation against or opposite to the flow, while a positive number indicates sound propagation in the direction of flow.

The primary effect of flow is to alter the effective phase speed of a propagating wave by convecting the sound wave with the flow (Bies and Zockel, 1976). Thus the

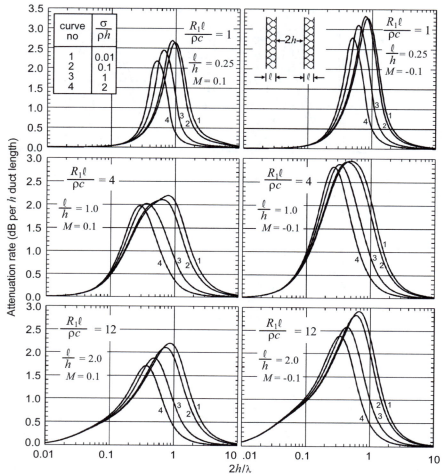

Figure 9.21 Predicted octave band attenuations for a rectangular duct lined on two opposite sides. Lined circular ducts or square ducts lined on all four sides give twice the attenuation shown here. The quantity ρ is the density of fluid flowing in the duct, c is the speed of sound in the duct, ℓ is the liner thickness, h is the half width of the airway, σ is the surface density of a limp membrane covering the liner, R_1 is the liner flow resistivity. Bulk reacting liner with various densities of limp membrane or equivalent perforated sheet ($\sigma/\rho h = 0.01$ to 2). Mean flow of Mach number, $M = 0.1$. For the figures on the left, the flow is in the same direction as sound propagation and for the figures on the right, the flow is in the opposite direction to sound propagation.

effective phase speed is $c(1 + M)$, where M, the Mach number, is less than 0.3. Consequently, the effective length of a duct lining is decreased (as measured in wavelengths) for downstream propagation, resulting in a decrease in attenuation, and increased for upstream propagation, resulting in an increase in attenuation, as may be verified by inspection of the curves in Figures 9.16 to 9.21.

In reality of course, the matter is not quite so simple as suggested above. The introduction of flow, in general, renders the problem of describing acoustic

Table 9.3 Comparison between measured and predicted insertion loss for rectangular splitter silencers (after Ramakrishnan and Watson, 1991)

Silencer unit size (mm)	w/h	Silencer length (mm)	See note below	Octave band centre frequency (Hz) insertion loss (dB)					
				125	250	500	1000	2000	4000
305	2	1525	e	8	20	38	47	51	34
			a	8	21	37	50	50	34
			b	8	21	37	50	50	34
305	1	1525	e	4	12	27	41	37	20
			a	4	12	26	44	36	13
			b	4	12	26	44	36	13
408	3.2	1525	e	17	27	38	48	50	31
			a	14	24	37	50	50	26
			b	14	24	37	50	50	26
408	1.13	1525	e	5	12	20	26	16	9
			a	6	14	21	29	17	8
			b	6	14	21	29	17	5
610	2	2135	e	17	24	36	49	33	18
			a	14	22	36	47	29	13
			b	14	22	36	47	27	9
610	1.4	1525	e	11	16	25	30	17	11
			a	8	13	21	26	12	7
			b	8	13	21	26	10	3
610	1.4	2775	e	18	25	37	50	30	16
			a	15	25	40	50	22	14
			b	15	25	40	50	22	9

a = equal energy among all possible significant modes at entrance;
b = least attenuated mode;
e = experimental data.

propagation and attenuation in a lined duct very complicated. For example, shear in the flow has the opposite effect to convection; sound propagating in the direction of flow is refracted into the lining, resulting in increased attenuation. Sound propagating opposite to the flow is refracted away from the lining, resulting in less attenuation than where shear is not present. At Mach numbers higher than those shown in Figures 9.16 to 9.21, where such effects as shear become important, the following empirical relation is suggested as a guide to expected behaviour:

$$D_M = D_0\left[1 - 1.5M + M^2\right] \tag{9.134}$$

where

$$-0.3 < M < 0.3 \tag{9.135}$$

In the equation, D_0 is the attenuation (in dB per unit length) predicted for a liner without flow, and D_M is the attenuation for the same liner with plug flow of Mach number M.

Flow can strongly affect the performance of a liner both beneficially and adversely, depending upon the liner design. In addition to the refraction effects mentioned earlier, the impedance matching of the wall to the propagating wave may be improved or degraded, resulting in more or less attenuation (Kurze and Allen, 1971; Mungar and Gladwell, 1968). The introduction of flow may also generate noise, for example, as discussed in Section 9.7.8.

9.8.3.2 Higher Order Mode Propagation

In the formulation of the curves in Figures 9.16 to 9.21 it has been explicitly assumed that only plane waves propagate and are attenuated. For example, inspection of the curves in the figures shows that they all tend to the same limit at high frequencies; none shows any sensible attenuation for values of the frequency parameter $2h/\ell$ greater than about three. As has been shown theoretically (Cremer, 1953), high-frequency plane waves tend to beam down the centre of a lined duct; any lining tends to be less and less effective, whatever its properties in attenuating plane waves, as the duct width to wavelength ratio grows large.

Waves which multiply reflect from the walls of a duct may also propagate. Such waves, called higher order modes or cross-modes, propagate at frequencies above a minimum frequency, called the cut-on frequency, f_{co}, which characterises the particular mode of propagation. For example, in a hard wall rectangular cross-section duct, only plane waves may propagate when the largest duct cross sectional dimension is less than 0.5 wavelengths, while for a circular duct, the required duct diameter is less than 0.5861 wavelengths.

Thus, for rectangular section ducts,

$$f_{co} = \frac{c}{2L_y} \tag{9.136}$$

where L_y is the largest duct cross-sectional dimension. For circular section ducts,

$$f_{co} = 0.586 \frac{c}{d} \tag{9.137}$$

where d is the duct diameter.

For ducts of greater dimensions than these, or for any ducts that are lined with acoustically absorptive material (i.e. soft-walled ducts), higher order modes may propagate as well as plane waves but, in general, the plane waves will be least rapidly attenuated. As plane waves are least rapidly attenuated, their behaviour will control the performance of a duct in the frequency range in which they are dominant; that is, in the range of wavelength parameter, $2h/\ell$, generally less than about one.

For explanation of the special properties of higher order mode propagation, it will be convenient to restrict attention to ducts of rectangular cross-section and to begin by generalising the discussion of modal response of a rectangular enclosure considered in Chapter 7. Referring to the discussion of Section 7.2.1, it may be concluded that a duct is simply a rectangular room for which one dimension is infinitely long. Letting $k_x = \omega/c_x$, and using Equation (7.16), Equation (7.17) can be rewritten as follows to give an expression for the phase speed c_x along the x-axis in an infinite rectangular section duct for any given frequency, $\omega = 2\pi f$:

$$c_x = \omega \left[\left(\frac{\omega}{c} \right)^2 - \left(\frac{\pi n_y}{L_y} \right)^2 - \left(\frac{\pi n_z}{L_z} \right)^2 \right]^{-1/2} \quad \text{(Hz)} \qquad (9.138)$$

For a circular section duct of radius, a, Morse and Ingard (1968) give the following for the phase speed along the x-axis.

$$c_x = \omega \left[\left(\frac{\omega}{c} \right)^2 - \left(\frac{\pi \alpha_{m,n}}{a} \right)^2 \right]^{-1/2} \quad \text{(Hz)} \qquad (9.139)$$

Values of $\alpha_{m,n}$ for the lowest order modes are given in Table 9.4.

Table 9.4 Values of the coeficient, $\alpha_{m,n}$ for circular section ducts (after Morse and Ingard, 1968)

$m \backslash n$	0	1	2	3	4
0	0	3.83	7.02	10.17	13.32
1	1.84	5.33	8.53	11.71	14.86
2	3.05	6.71	9.97	13.17	16.35
3	4.20	8.02	11.35	14.59	17.79
4	5.32	9.28	12.68	15.96	19.2
5	6.42	10.52	13.99	17.31	20.58

Consideration of Equations (9.138) and (9.139) shows that for rectangular section duct mode numbers n_y and n_z not both zero, and for circular section duct mode numbers, m, n, both not zero, there will be a frequency, ω, for which the phase speed, c_x, is infinite. This frequency is called either the cut-on or cut-off frequency but "cut-on" is the most commonly used term. Below cut-on, the phase speed is imaginary and no wave propagates; and any acoustic wave generated by a source in the duct or transmitted into the duct from outside will decay exponentially as it travels along the duct. Above cut-on, the wave will propagate at a phase speed, c_x, which depends upon frequency. With increasing frequency, the phase speed, c_x, measured as the trace along the duct, rapidly diminishes and tends to the free field wave speed, as illustrated in Figure 9.22. Evidently the speed of propagation of higher order modes is dispersive (frequency dependent) and for any given frequency, each mode travels at a speed different from that of all other modes.

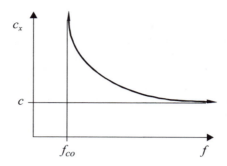

Figure 9.22 Phase speed of a higher order mode propagating in a duct as a function of frequency

Alternatively, letting the length, L_x, in Equation (7.17) tend to infinity gives the following equation for the cut-on frequencies for propagating higher order modes in a rectangular section duct characterised by mode numbers n_y and n_z:

$$f_{ny, nz} = \frac{c}{2}\sqrt{\left(\frac{n_y}{L_y}\right)^2 + \left(\frac{n_z}{L_z}\right)^2} \tag{9.140}$$

Referring to Equation (9.140), it can be observed that the result is the same as would be obtained by setting $n_x = 0$ in Equation (7.17) and in the latter case, sound propagation is between the opposite walls but not along the x-axis. Similarly, at cut-on, wave propagation is between opposite walls and consequently, the phase speed along the duct (x-axis) is infinite as the disturbance everywhere is in phase.

Equation (7.19) may be rewritten for the case of the infinitely long room for wave travel in the positive x-axis direction to give an expression for the propagating higher order mode, characterised by mode numbers n_y and n_z, as follows:

$$p = p_0 \cos\left[\frac{\pi n_y y}{L_y}\right] \cos\left[\frac{\pi n_z z}{L_z}\right] e^{j(\omega t - k_x x)} \tag{9.141}$$

Referring to Equation (9.141), it can be observed that a higher order mode is characterised by nodal planes parallel to the axis of the duct and wave fronts at any cross-section of the duct, which are of opposite phase on opposite sides of such nodes. For a circular section duct of radius, a, the cut-on frequencies are obtained by setting equal the two terms in square brackets in Equation (9.139) to give:

$$f_{m,n} = \frac{c\,\alpha_{m,n}}{2a} \tag{9.142}$$

At frequencies above the cut-on of the first higher order mode, the number of cut-on modes increases rapidly, being a quadratic function of frequency. Use of Equation (9.140) for representative values for rectangular cross-section ducts of width, L_y, and height, L_z, has allowed counting of cut-on modes and empirical determination of the following equation for the number of cut-on modes, N, in terms of the geometric mean of the duct cross-section dimensions, $L = \sqrt{L_y L_z}$ up to about the first 25 cut-on modes:

$$N = 2.57 (fL/c)^2 + 2.46 (fL/c) \tag{9.143}$$

A similar procedure has been used to determine the following empirical equation for the first fifteen cut-on modes of a circular cross-section duct of diameter D:

$$N = (fD/c)^2 + 1.5 (fD/c) \tag{9.144}$$

It should be noted that, in practice, one must always expect slight asymmetry in any duct of circular cross-section and consequently there will always be two modes of slightly different frequency where analysis predicts only one and they will be oriented normal to each other. The prediction of Equation (9.144) is based upon the assumption of a perfectly circular cross-section duct and should be multiplied by two to determine the expected number of propagating higher order modes in a practical duct.

The effect of a porous liner adds a further complexity to higher order mode propagation, since the phase speed in the liner may also be dispersive. For example, reference to Appendix C shows that the phase speed in a fibrous material tends to zero as the frequency tends to zero. Consequently, in fibrous, bulk reacting liners, where propagation of sound waves is not restricted to normal to the surface, cut-on of the first few higher order modes may occur at much lower frequencies than predicted based upon the dimensions of the airway.

The effect of mean flow in a duct is to decrease the frequency of cut-on and it is the same for either upstream or downstream propagation, since cut-on in either case is characterised by wave propagation back and forth between opposite walls of the duct. For the case of superimposed mean flow at cut-on, wave propagation upstream will be just sufficient to compensate wave convection downstream, whether upstream or downstream propagation is considered.

For ducts that are many wavelengths across, as are commonly encountered in air conditioning systems, one is concerned with cross-mode as well as plane-wave propagation. Unfortunately, attenuation in this case is very difficult to characterise, as it depends upon the energy distribution among the propagating modes, as well as the rates of attenuation of each of the modes. Thus, in general, attenuation in the frequency range for which the frequency parameter, $2h/\ell$, is greater than about unity, cannot be described in terms of attenuation per unit length as in Figures 9.16 to 9.21. Experimentally determined attenuation will depend upon the nature of the source and the manner of the test. Doubling or halving the test duct length will not give twice or half of the previously observed attenuation; that is, no unique attenuation per unit length can be ascribed to a dissipative duct for values of $2h/\ell$ greater than unity. However, at these higher frequencies, the attenuation achieved in practice will in all cases be greater than that predicted by Figures 9.16 to 9.21.

The consequence of the possibility of cross-mode propagation is that the performance of a lined duct is dependent upon the characteristics of the sound field introduced at the entrance to the attenuator. Since sound is absorbed by the lining, sound that repeatedly reflects at the wall will be more quickly attenuated than sound that passes at grazing incidence. Thus, sound at all angles of incidence at the entrance to a duct will very rapidly attenuate, until only the axial propagating portion remains. Empirically, it has been determined that such an effect may introduce an additional attenuation, as shown in Figure 9.23. The attenuation shown in the figure is to be treated as a correction to be added to the total expected attenuation of a lined duct. An example of an application of this correction is the use of a lined duct to vent an acoustic enclosure, in which case the sound entering the duct may be approximated as randomly incident.

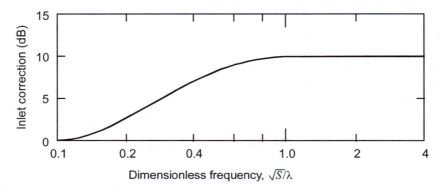

Figure 9.23 Duct inlet correction for random-incidence sound. The quantity λ is the sound wavelength, and S is the cross-sectional area of the open duct section. The data are empirical.

9.8.4 Cross-sectional Discontinuities

Generally, the open cross-section of a lined duct is made continuous with a primary duct, in which the sound to be attenuated is propagating. The result of the generally soft lining is to present to the sound an effective sudden expansion in the cross-section of the duct. The expansion affects the sound propagation in a similar way as the expansion chamber considered earlier. The effect of the expansion on the lined duct may be estimated using Figure 9.24. In using this figure the attenuation due to the lining alone is first estimated (using one of Figures 9.16 to 9.21), and the estimate is used to enter Figure 9.24 to find the corrected attenuation (see Section 9.2 for discussion of transmission loss).

Example 9.1

The open section of a duct lined on two opposite sides is 0.2 m wide. Flow is negligible. What must the length, thickness and liner flow resistance in the direction normal to the liner surface be to achieve 15 dB attenuation at 100 Hz, assuming that there is no protective covering on the liner?

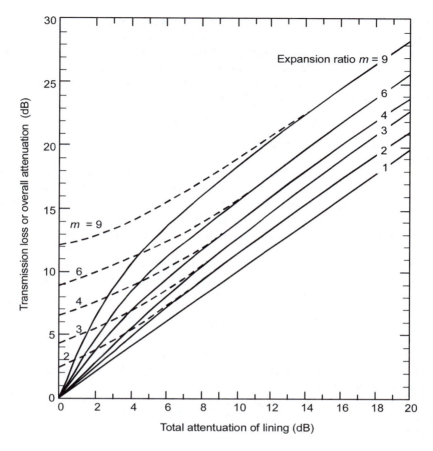

Figure 9.24 The transmission loss (*TL*) of a lined expansion chamber as a function of the area expansion ratio *m* and the total attenuation of the lining. The solid curves show *TL* for $kB = 0$, $\pi, \ldots, n\,\pi$, and the dashed curves show *TL* for $kB = \pi/2, 3\pi/2, \ldots, (2n + 1)\,\pi/2$. The quantity k is the wavenumber and B is the length of the expansion chamber.

Solution

There will be many possible solutions. In practice, the question becomes, "Are there any acceptable solutions?" For a liner with no solid partitions mounted normal to the liner surface to prevent axial wave propagation in the liner, the assumption of bulk reaction is appropriate. It will also be assumed that the liner flow resistivity in the axial direction is the same as in the normal direction. Thus Figure 9.16 is the appropriate figure to use for the liner attenuation calculations.

First calculate the frequency parameter taking the speed of sound as 343 m/s. That is, $2hf/c = 2 \times 0.1 \times 100/343 = 0.058$.

Referring to Figure 9.16, one notes that a thick liner is implied according to the table associated with the figure. An alternative choice, to be investigated, might be curve 4, this choice being a compromise promising a thinner liner. The best curve 5 of Figure 9.16 ($R_1\ell/\rho c = 4$) predicts an attenuation rate of 1.0 dB/h = 10.0 dB/m.

Next, Figure 9.24 is used to calculate the required liner attenuation for a total *TL* (or *IL*) of 10 dB. Calculate the area ratio and value of *kB* to enter the figure. Assume for now that the final liner length will be 1 m. If this guess is wrong, then it is necessary to iterate until a correct solution is obtained:

$$m = S_1/S_2 = 1 + \ell/h = 5; \quad \text{and} \quad kB = 2\pi fB/343 = 2\pi \times 100 \times 1.0/343 = 1.82$$

where S_1/S_2 is the ratio of total duct cross-sectional area to open duct cross-sectional area.

From the figure it can be seen that for an overall attenuation of 12 dB (transmission loss), the liner attenuation must be approximately 7 dB (as the value of *kB* = 1.82 is close to the dashed curve value). A duct 1 m long will give an attenuation of 10.0 dB which is a bit high. Iterating finally gives a required duct length of 0.6m corresponding to a liner attenuation of 6.0 dB and an overall attenuation (transmission loss) of 13 dB.

Alternatively, curve 4 of Figure 9.16, predicts an attenuation rate as follows: Attenuation rate = 0.6 dB/h = 6.0 dB/m. Assuming a liner length of 1.6 m calculate the area ratio and value of *kB* to enter Figure 9.24:

$$S_1/S_2 = 1 + \ell/h = 3; \quad \text{and} \quad kB = 2\pi \times 100 \times 1.6/343 = 2.92$$

From the figure it can be seen that for an overall attenuation (transmission loss) of 12 dB, the liner attenuation must be approximately 9.5 dB. A duct 1.6 m long will give an attenuation of 9.6 dB, which is the required amount.

In summary, use of curve 5 gives a duct 0.6 m long and 1.0 m wide, whereas curve 4 gives a duct 1.6 m long and 0.6 m wide. The required material flow resistance is such that $R_1\ell/\rho c = 4.0$.

9.8.5 Pressure Drop Calculations for Dissipative Mufflers

For lined dissipative ducts, an average absolute roughness, ε, of 9×10^{-4} m is appropriate and may be used with Equations (9.121) to (9.124) to calculate the pressure drop due to friction losses as a result of gas flowing through the muffler. Note that dynamic losses are additional to friction losses and must be calculated using Equation (9.125). The pressure drop due to a centre body in a circular cross-section muffler cannot be calculated using the expressions provided here.

9.9 DUCT BENDS OR ELBOWS

The lined bend was listed separately in Table 9.1, but such a device might readily be incorporated in the design of a lined duct. Figure 9.25 shows insertion loss data for lined and unlined bends with no turning vanes. The data shown are empirical and approximate (ASHRAE, 2007; Beranek, 1960). Data for lined bends are for bends

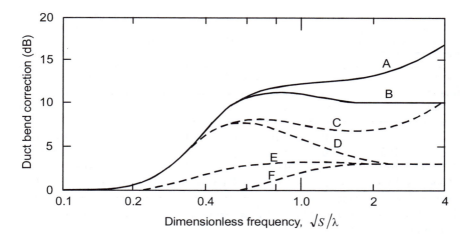

Figure 9.25 Correction for attenuation at a 90° duct bend with no turning vanes: A, rectangular lined duct and axial plane wave input; B, rectangular lined duct and diffuse input; C, rectangular unlined duct and axial plane wave input; D, rectangular unlined duct and diffuse input; E, circular unlined duct 0.2 m in diameter and diffuse input; F, circular unlined duct 1.5 m in diameter and diffuse input. The quantity S is the duct cross-sectional area, and λ is the sound wavelength. The circular duct curves are based on data from ASHRAE-2007 HVAC Applications.

lined both upstream and downstream for a distance of at least three duct diameters or three times the largest cross-sectional dimension.

9.10 UNLINED DUCTS

An unlined duct will exhibit sound attenuation properties. The amount of sound attenuation for an unlined rectangular duct may be estimated using Table 9.5 and for an unlined circular duct it may be estimated using Table 9.6.

Table 9.5 Approximate attenuation in unlined rectangular sheet metal ducts in dB/m[a]. Adapted from ASHRAE (2007)

Duct perimeter to area ratio (P/S) (m^{-1})[b]	Octave band centre frequency (Hz)			
	63	125	250	over 250
27	0.98	0.66	0.33	0.33
13	1.15	0.66	0.33	0.20
9.8	1.31	0.66	0.33	0.16
6.6	0.82	0.66	0.33	0.10
3.3	0.49	0.33	0.23	0.07
2.2	0.33	0.33	0.16	0.07

[a] If the duct is externally insulated then doubled these values.

[b] The quantity P is the perimeter and S is the area of the duct cross-section.

Table 9.6 Approximate attenuation in unlined circular sheet metal ducts in dB/m, inlet or outlet of the fan. If the total radiated sound power level is desired, add 3 dB to these values. Copyright 2007, American Society of Heating, Refrigerating and Air-Conditioning Engineers, Inc. (ASHRAE) www.ashrae.org. Used by permission

Duct diameter (mm)	Octave band centre frequency (Hz)						
	63	125	250	500	1000	2000	4000
D ≤ 180	0.10	0.10	0.16	0.16	0.33	0.33	0.33
180 < D ≤ 380	0.10	0.10	0.10	0.16	0.23	0.23	0.23
380 < D ≤ 760	0.07	0.07	0.07	0.10	0.16	0.16	0.16
760 < D ≤ 1520	0.03	0.03	0.03	0.07	0.07	0.07	0.07

The attenuation for an unlined circular section duct is unaffected by external insulation.

9.11 EFFECT OF DUCT END REFLECTIONS

The sudden change of cross-section at the end of a duct mounted flush with a wall or ceiling results in additional attenuation. This has been measured for circular and rectangular ducts and empirical results are listed in Table 9.7 (ASHRAE, 2007). Table 9.5 can also be used for rectangular section ducts, by calculating an equivalent diameter, D, using $D = \sqrt{4S/\pi}$, where S is the duct cross-sectional area.

Table 9.7 Duct reflection loss (dB)[a]. Adapted from ASHRAE (2007)

Duct diameter (mm)	Octave band centre frequency (Hz)					
	63	125	250	500	1000	2000
150	18(20)	13(14)	8(9)	4(5)	1(2)	0(1)
200	16(18)	11(12)	6(7)	2(3)	1(1)	0(0)
250	14(16)	9(11)	5(6)	2(2)	1(1)	0(0)
300	13(14)	8(9)	4(5)	1(2)	0(1)	0(0)
400	10(12)	6(7)	2(3)	1(1)	0(0)	0(0)
510	9(10)	5(6)	2(2)	1(1)	0(0)	0(0)
610	8(9)	4(5)	1(2)	0(1)	0(0)	0(0)
710	7(8)	3(4)	1(1)	0(0)	0(0)	0(0)
810	6(7)	2(3)	1(1)	0(0)	0(0)	0(0)
910	5(6)	2(3)	1(1)	0(0)	0(0)	0(0)
1220	4(5)	1(2)	0(1)	0(0)	0(0)	0(0)
1830	2(3)	1(1)	0(0)	0(0)	0(0)	0(0)

[a] Applies to ducts terminating flush with wall or ceiling and several duct diameters from other room surfaces. If closer to other surfaces use entry for next larger duct. Numbers in brackets are for ducts terminated in free space or at an acoustic suspended ceiling.

9.12 DUCT BREAK-OUT NOISE

9.12.1 Break-out Sound Transmission

In most modern office buildings, air conditioning ductwork takes much of the space between suspended ceilings and the floor above. Noise (particularly low-frequency rumble noise) radiated out of the ductwork walls is in many cases sufficient to cause annoyance to the occupants of the spaces below. In some cases, noise radiated into the ductwork from one space, propagated through the duct, and radiated out through the duct walls into another space, may cause speech privacy problems. Noise transmitted out through a duct wall is referred to as breakout transmission.

To predict in advance the extent of likely problems arising from noise "breaking-out" of duct walls, it is useful to calculate the noise level outside of the duct from a knowledge of the sound power introduced into the duct by the fan or by other external sources further along the duct. A prediction scheme, which is applicable in the frequency range between 1.5 times the fundamental duct wall resonance frequency and half the critical frequency of a flat panel equal in thickness to the duct wall, will be described. In most cases, the fundamental duct wall resonance frequency, f_0, is well below the frequency range of interest and can be ignored. If this is not the case, f_0 may be calculated (Cummings, 1980) and the transmission loss for the third octave frequency bands adjacent to and including f_0 should be reduced by 5 dB from that calculated using the following prediction scheme. Also, in most cases, the duct wall critical frequency is well above the frequency range of interest. If this is in doubt, the critical frequency may be calculated using Equation (8.3) and then the transmission loss predictions for a flat panel may be used at frequencies above half the duct wall critical frequency.

The sound power level, L_{wo}, radiated out of a rectangular section duct wall is given by (Ver, 1983):

$$L_{wo} = L_{wi} - TL_{out} + 10 \log_{10} \left[\frac{P_D L}{S} \right] + C \quad \text{(dB)} \qquad (9.145)$$

L_{wi} is the sound power level of the sound field propagating down the duct at the beginning of the duct section of concern (usually the fan output sound power level (dB) less any propagation losses from the fan to the beginning of the noise radiating duct section), TL_{out} is the transmission loss of the duct wall, S is the duct cross-sectional area, P_D is the cross-sectional perimeter, L is the duct length radiating the power and C is a correction factor to account for gradually decreasing values of L_{wi} as the distance from the noise source increases. For short, unlined ducts, C is usually small enough to ignore. For unlined ducts longer than 2 m or for any length of lined duct, C is calculated using:

$$C = 10 \log_{10} \left[\frac{1 - e^{-(\tau_a + \Delta/4.34)L}}{(\tau_a + \Delta/4.34)L} \right] \qquad (9.146)$$

Δ is the sound attenuation (dB/m) due to internal ductwork losses, which is 0.1 dB/m for unlined ducts (do not use tabulated values in ASHRAE, 2007 as these include losses due to breakout) and:

$$\tau_a = (P_D/S)10^{-TL_{out}/10} \tag{9.147}$$

The quantity TL_{out} may be calculated (ASHRAE, 2007) using the following procedure. First of all, the cross-over frequency from plane wave response to multi-modal response is calculated using:

$$f_{cr} = 612/(ab)^{\frac{1}{2}} \; ; \qquad \text{(Hz)} \tag{9.148}$$

where a is the larger and b the smaller duct cross-sectional dimension in metres.
    At frequencies below f_{cr}, the quantity TL_{out} may be calculated using:

$$TL_{out} = 10\log_{10}\left[\frac{fm^2}{(a+b)}\right] - 13 \quad \text{(dB)} \; ; \qquad f < f_{cr} \tag{9.149}$$

and at frequencies above f_{cr} and below $0.5f_c$:

$$TL_{out} = 20\log_{10}(fm) - 45 \quad \text{(dB)} \; ; \quad f_{cr} < f < f_c/2 \tag{9.150}$$

In the preceding equations, m (kg/m^2) is the mass/unit area of the duct walls and f (Hz) is the octave band centre frequency of the sound being considered.
    The minimum allowed value for TL_{out} is given by:

$$TL_{out} = 10\log_{10}\left[\frac{P_D L}{S}\right] \quad \text{(dB)} \tag{9.151}$$

The maximum allowed value for TL_{out} is 45 dB. For frequencies above half the critical frequency of a flat panel (see Chapter 8), TL predictions for a flat panel are used.
    The transmission loss for circular and oval ducts is difficult to predict accurately with an analytical model, although it is generally much higher than that for rectangular section ducts of the same cross-sectional area. It is recommended that the guidelines outlined by ASHRAE (2007) for the estimation of these quantities be followed closely.

9.12.2 Break-in Sound Transmission

Let L_{wo} be the sound power that is incident upon the exterior of an entire length of ductwork and assume that the incoming sound power is divided equally into each of the two opposing axial directions. Then the sound power entering into a rectangular section duct of cross-sectional dimensions a and b, and length, L, from a noisy area and propagating in one axial direction in the duct is:

$$L_{wi} = L_{wo} - TL_{in} - 3 \quad \text{(dB)} \tag{9.152}$$

For $a \geq b$, and $f < f_0$, where $f_0 = c / 2a$, the duct transmission loss, TL_{in}, for sound radiated into the duct is the larger of the following two quantities (ASHRAE, 2007):

$$TL_{in} = TL_{out} - 4 - 10\log_{10}(a/b) + 20\log_{10}(f/f_0) \qquad (9.153)$$

$$\text{or} \quad TL_{in} = 10\log_{10}(L/a + L/b) \qquad (9.154)$$

For $f > f_0$:

$$TL_{in} = TL_{out} - 3 \quad \text{(dB)} \qquad (9.155)$$

9.13 LINED PLENUM ATTENUATOR

A lined plenum chamber is often used in air conditioning systems as a device to smooth fluctuations in the air flow. It may also serve as a sound attenuation device. As shown in Table 9.1, such a device has dimensions that are large compared to a wavelength. The plenum thus acts like a small room and as such an absorptive liner, which provides a high random incidence sound absorption coefficient, is of great benefit. In general, the liner construction does not appear to be critical to the performance of a plenum, although some form of liner is essential. The geometry of such a device is shown in Figure 9.26(a) and, for later reference, Figure 9.26(b) shows the essential elements of the source field. The latter figure will be used in the following discussion.

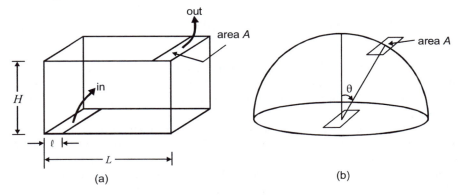

Figure 9.26 Lined plenum chamber: (a) physical acoustic system; (b) essentials of the source field.

In all cases discussed in this section, the values obtained for transmission loss will be the same as those for insertion loss provided that the plenum chamber is lined with sound absorbing material and providing that the transmission loss is greater than 5 dB.

9.13.1 Wells' Method

There are a number of analytical models that have been developed in the past by various authors for estimating the sound attenuation performance of a plenum chamber. The oldest known model is the Wells model (Wells, 1958) and this will be discussed first of all.

The acoustic power, W_o, which leaves the exit consists of two parts for the purpose of this analysis; a direct field, W_D, and a reverberant field, W_R. The acoustic power in the reverberant field is related to the input power W_i as:

$$W_R = W_i A/R \tag{9.156}$$

where

$$R = S\overline{\alpha}/(1 - \overline{\alpha}) \tag{9.157}$$

In the preceding equations, A (m^2) is the cross-sectional area of the plenum exit hole, R (m^2) is the plenum room constant, S (m^2) is the total wall area of the plenum, and $\overline{\alpha}$ is the mean Sabine wall absorption coefficient.

Referring to Figure 9.26(b) the power flow in the direct field is (Wells, 1958):

$$W_D = W_i (A/2\pi r^2)\cos\theta \tag{9.158}$$

where θ is the angular direction, and r is the line of sight distance from the plenum chamber entrance to the exit. It is recommended (ASHRAE, 2007) that for an inlet opening nearer to the edge than the centre of the plenum chamber wall, the factor of 2 in the preceding equation be deleted. If this is done, the transmission loss (see Section 9.2) of the plenum is:

$$TL = -10\log_{10}\frac{W_o}{W_i} = -10\log_{10}\left[\frac{A}{R} + \frac{A\cos\theta}{\pi r^2}\right] \tag{9.159a,b}$$

Equation (9.159) is only valid at frequencies for which the plenum chamber dimensions are large compared to a wavelength and also only for frequencies above the cut-on frequency for higher order mode propagation in the inlet duct (see Section 9.8.3.2).

If the room constant, R, is made large, then the effectiveness of the plenum may be further increased by preventing direct line of sight with the use of suitable internal baffles. When internal baffles are used, the second term in Equation (9.159b), which represents the direct field contribution, should be discarded. However, a better alternative is to estimate the insertion loss of the baffle by using the procedure outlined in Chapter 8, Section 8.5.3 and adding the result arithmetically to the IL of the plenum.

Equation (9.159b) agrees with measurement for high frequencies and for values of TL not large but predicts values lower than observed by 5 to 10 dB at low frequencies, which is attributed to neglect of reflection at the plenum entrance and exit and the modal behaviour of the sound field in the plenum.

9.13.2 ASHRAE Method

In 2004, Mouratidis and Becker published a modified version of Wells' method, and showed that their version approximated their measured data more accurately. Their analysis also includes equations describing the high-frequency performance. The Mouratidis and Becker approach has been adopted by ASHRAE (ASHRAE, 2007). Their expression for frequencies above the first mode cut on frequency, f_{co} in the inlet duct is:

$$TL = b\left[\frac{A}{\pi r^2} + \frac{A}{R}\right]^n \qquad (9.160)$$

which is a similar form to Equation (9.159). If the inlet is closer to the centre of the wall than the corner, then a factor of 2 is included in the denominator of the first term in the preceding equation. The constants, $b = 3.505$ and $n = -0.359$. The preceding equation only applies to the case where the plenum inlet is directly in line with the outlet. When the value of θ in Figure 9.26b is non-zero, corrections must be added to the *TL* calculated using the preceding equation and these are listed in Table 9.8 as the numbers not in brackets.

Table 9.8 Corrections (dB) to be added to the TL calculated using Equation (9.160) or Equation (9.161) for various angles θ defined by Figure 9.26b. The numbers not in brackets correspond to frequencies below the inlet duct cut on frequency and the numbers in brackets correspond to frequencies above the duct cut on frequency. The absence of numbers for some frequencies indicates that no data are available for these cases. Adapted from Mouratidis and Becker (2004)

1/3 Octave band centre frequency (Hz)	Angle, θ (degrees)				
	15	22.5	30	37.5	45
80	0	-1	-3	-4	-6
100	1	0	-2	-3	-6
125	1	0	-2	-4	-6
160	0	-1	-2	-3	-4
200	0 (1)	-1 (4)	-2 (9)	-3 (14)	-5 (20)
250	1 (2)	2 (4)	3 (8)	5 (13)	7 (19)
315	4 (1)	6 (2)	8 (3)	10 (4)	14 (5)
400	2 (1)	4 (2)	6 (3)	9 (4)	13 (6)
500	1 (0)	3 (1)	6 (2)	10 (4)	15 (5)
630	(1)	(2)	(3)	(5)	(7)
800	(1)	(2)	(2)	(3)	(3)
1000	(1)	(2)	(4)	(6)	(9)
1250	(0)	(2)	(4)	(6)	(9)
1600	(0)	(1)	(1)	(2)	(3)
2000	(1)	(2)	(4)	(7)	(10)
2500	(1)	(2)	(3)	(5)	(8)
3150	(0)	(2)	(4)	(6)	(9)
4000	(0)	(2)	(5)	(8)	(12)
5000	(0)	(3)	(6)	(10)	(15)

For frequencies below the duct cut on frequency, Mouratidis and Becker (2004) give the following expression for estimating the plenum transmission loss.

$$TL = A_f S + W_e \qquad (9.161)$$

where the constants A_f and W_e are given in Table 9.9. Again, when the value of θ in Figure 9.26b is non-zero, corrections must be added to the TL calculated using the preceding equation and these are listed in Table 9.8 as the numbers in brackets.

Table 9.9 Values of the constants in Equation (9.161). The numerical values given in the headings for W_e represent the thickness (ranging from 25 mm to 200 mm) of sound absorbing material between the facing and plenum wall. The material normally used is fibreglass or rockwool with an approximate density of 40 kg/m³. Adapted from ASHRAE (2007) and Mouratidis and Becker (2004)

1/3 Octave band centre frequency (Hz)	A_f Plenum Volume <1.4m³	>1.4m³	W_e 25mm fabric facing	50mm fabric facing	100mm perf. facing	200mm perf. facing	100mm solid metal facing
50	1.4	0.3	1	1	0	1	0
63	1.0	0.3	1	2	3	7	3
80	1.1	0.3	2	2	3	9	7
100	2.3	0.3	2	2	4	12	6
125	2.4	0.4	2	3	6	12	4
160	2.0	0.4	3	4	11	11	2
200	1.0	0.3	4	10	16	15	3
250	2.2	0.4	5	9	13	12	1
315	0.7	0.3	6	12	14	14	2
400	0.7	0.2	8	13	13	14	1
500	1.1	0.2	9	13	12	13	0

For an end in, side out plenum configuration (ASHRAE, 2007) corrections listed in Table 9.10 must be added to Equations (9.160) and (9.161) in addition to the corrections in Table 9.8.

9.13.3 More Complex Methods (Cummings and Ih)

Two other methods have been published for predicting the Transmission Loss of plenum chambers (Cummings, 1978; Ih, 1992). Cummings looked at high and low frequency range models for lined plenum chambers and Ih investigated the TL for unlined chambers. For the low frequency model, it was assumed that only plane waves existed in both the inlet and outlet ducts and higher order modes existed in the plenum chamber. This low frequency model is complicated to evaluate and the reader is referred to Cummings' original paper or the summary paper of Li and Hansen (2005).

Table 9.10 Corrections to be added to the TL calculated using Equations (9.160) and (9.161) for plenum configurations with an inlet on one end and an outlet on one side. The numbers not in brackets correspond to frequencies below the inlet duct cut on frequency and the numbers in brackets correspond to frequencies above the duct cut on frequency. The absence of numbers for some frequencies indicates that no data are available for these cases. Adapted from Mouratidis and Becker (2004)

1/3 Octave band centre frequency (Hz)	Elbow effect correction	1/3 Octave band centre frequency (Hz)	Elbow effect correction
50	2	630	(3)
63	3	800	(3)
80	6	1000	(2)
100	5	1250	(2)
125	3	1600	(2)
160	0	2000	(2)
200	-2 (3)	2500	(2)
250	-3 (6)	3150	(2)
315	-1 (3)	4000	(2)
400	0 (3)	5000	(1)
500	0 (2)		

For Cummings' high frequency model, it was assumed that higher order modes existed in the inlet and outlet ducts as well as the plenum chamber. After rather a complicated analysis, the end result is that the TL is given by:

$$TL = -10 \log_{10}\left[\frac{A}{R} + \frac{A \cos^2 \theta}{\pi r^2}\right] \tag{9.162}$$

where the chamber room constant, R, is calculated using Equation (9.157) but with the Sabine absorption coefficient, $\bar{\alpha}$, replaced by the statistical absorption coefficient, α_{st}. Equation (9.160) is very similar to Wells' corrected model of Equation (9.159).

Ih (1992) presented a model for calculating the Transmission Loss of unlined plenum chambers. As Ih's model assumes a piston driven rigid walled chamber, it is not valid above the inlet and outlet duct cut on frequencies. However, it is the only model available for unlined plenum chambers.

9.14 WATER INJECTION

Water injection has been investigated, both for the control of the noise of large rocket engines and for the control of steam venting noise. In both cases, the injection of large amounts of water has been found to be quite effective in decreasing high-frequency noise, at the expense of a large increase in low-frequency noise. This effect is illustrated in Figure 9.27 for the case of water injection to reduce steam venting noise. In both cases, a mass flow rate of water equal to the mass flow rate of exhaust gas was

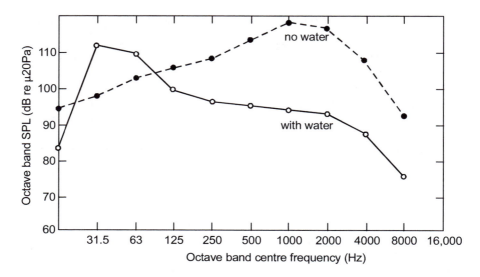

Figure 9.27 Effect of water injection on the reduction of steam venting noise.

injected directly into the flow to significantly cool the hot gases. Even more water would possibly prevent the low-frequency build-up, but the quantity of water required would be much greater.

Watters *et al.* (1955) investigated the effectiveness of water sprays used to cool hot exhaust gases in gas turbine exhausts. They provided the data shown here in Figure 9.28 for varying amounts of water, but cautioned against its universal use.

9.15 DIRECTIVITY OF EXHAUST DUCTS

The sound radiation from the exit of a duct may be quite directional, as shown in Figures 9.29 and 9.30. Figure 9.29 is based upon model studies (Sutton, 1990) in the laboratory and includes ducts of round, square and rectangular cross section. To get the rectangular section data to collapse on to the ka axis for circular ducts, where a is the duct radius, it was necessary to multiply the rectangular duct dimension ($2d$) in the direction of the observer by $4/\pi$. This factor takes into account the smaller radiating duct mouth in the direction of the observer for the circular duct. To be able to use the sets of curves for rectangular ducts it will be necessary to multiply the duct dimension in the direction of the observer by $4/\pi$ to get $2a$ and this is the value of a used to calculate ka prior to reading the *DI* from the figure. Also, as k is the wavenumber of sound at the duct exit, the temperature of the duct exhaust and exhaust gas properties must be used to calculate the speed of sound which is used to calculate k.

Figure 9.30 is based on extensive field measurements on circular ducts ranging in diameter from 305 mm to 1215 mm (Day and Bennett, 2008).

Figure 9.31 is based on a theoretical analysis of the problem (Davy, 2008a,b). The theory shows that the type and size of the sound source in the upstream end of the duct can significantly affect the results and for Figure 9.31, an effective source

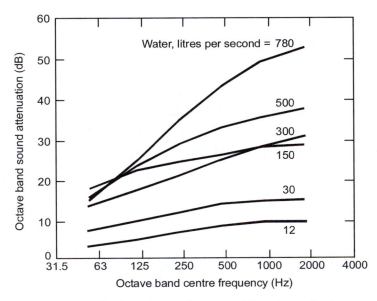

Figure 9.28 Octave band sound attenuation measured through several water spray systems (after Watters *et al.*, 1955).

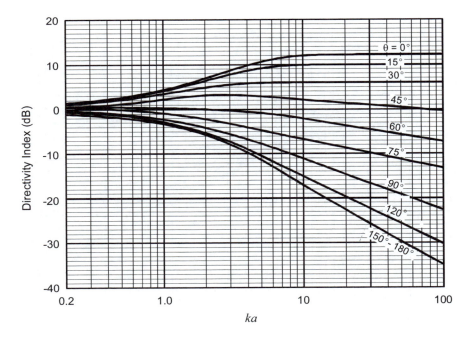

Figure 9.29 Exhaust stack directivity index measured in the laboratory vs *ka* where *a* is the stack inside radius. Curves fitted to data reported by Sutton (1990) and Dewhirst (2002).

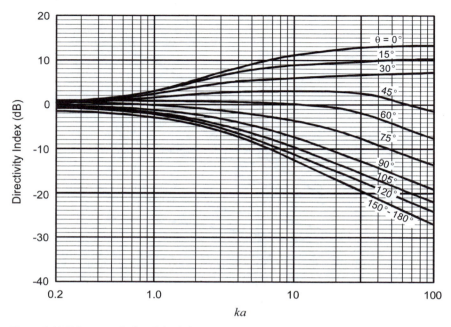

Figure 9.30 Exhaust stack directivity index measured in the field vs ka where a is the stack radius. Curves fitted to data reported by Day and Bennett (2008).

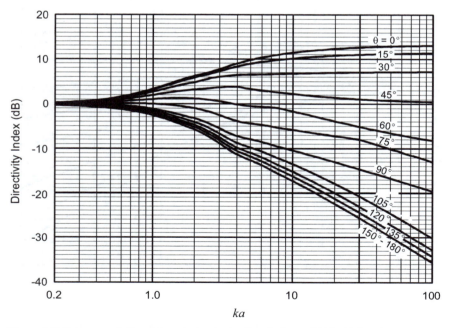

Figure 9.31 Exhaust stack directivity calculated using Davy's theory (2008a, b) for a $2a = 605$ mm diameter duct. The curves do not change significantly with duct radius.

dimension (or line source length), s_d, has been defined as $ks_d = 2\pi$. This seems to give the best agreement between theory and experiment. Although the duct length appears in the theoretical analysis, it is not an important parameter in terms of the final directivity result. The duct diameter, which was 610 mm for Figure 9.31, only has a slight effect on the results so Figure 9.31 can be used for duct diameters ranging from 100 mm to 2 m. The wavenumber, k, is the wavenumber in the gas at the exhaust duct exit and is dependent on the exhaust gas composition and temperature.

The final equations for calculating the duct directivity are given below but the reader is referred to Davy (2008a, 2008b) for a full derivation.

The directivity index is given by the intensity in any one direction divided by the intensity averaged over all possible directions which in this case is a sphere. Thus:

$$DI(\theta) = 10\log_{10}(I(\theta)/I_{av}) \tag{9.163}$$

As the preceding equation is a ratio, absolute values of the above quantities are not needed. To calculate the average intensity, I_{av}, a radial distance of unity is assumed and the intensity, $I(\theta)$, will be evaluated at the same location for an arbitrary source strength. Thus, the normalised average intensity is given by:

$$I_{av} = \frac{1}{4\pi}\int_0^{2\pi} d\varphi \int_0^{\pi} I(\theta)\sin(\theta)\,d\theta = \frac{1}{2}\int_0^{\pi} I(\theta)\sin(\theta)\,d\theta \tag{9.164a,b}$$

The directivity index $DI(\theta)$ and the intensity $I(\theta)$ also depend on the azimuthal angle for non-circular cross section ducts. This dependence will be ignored when calculating the directivity index in this section. The normalised sound intensity, $I(\theta)$, at angular

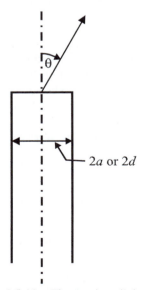

2a or 2d

Figure 9.32 Exhaust stack directivity definition. The quantity, a is the circular duct radius and $2d$ is the rectangular duct dimension in the direction of the observer.

location, θ, from the axis of an exhaust stack (see Figure 9.32) is given by (Davy, 2008a,b) (for $|\theta| \leq \pi/2$):

$$I(\theta) = \frac{1}{\rho c} p_1^2(\theta) p_2^2(\theta) \tag{9.165}$$

For $\pi/2 < |\theta| \leq \pi$, the normalised sound intensity is given by:

$$I(\theta) = \frac{I(\theta = \pi/2)}{1 - kz\cos(\theta)} \qquad \text{where} \qquad z = \frac{bd}{b + d} \tag{9.166a,b}$$

The first term in Equation (9.165) accounts for the expected sound pressure level at a particular location due to a point source and the second term effectively compensates for the size of the duct opening. The first term is given by:

$$p_1^2(\theta) = \int_{-\pi/2}^{\pi/2} \frac{w(\varphi)}{[2\rho c\,\sigma(\varphi)]^2} \left(\frac{\sin[kd(\sin\theta - \sin\varphi)]}{kd(\sin\theta - \sin\varphi)} \right)^2 d\varphi \tag{9.167}$$

which must be evaluated using numerical integration. The radiation efficiency, $\sigma(\varphi)$, of the duct exit is given by Davy (2008a,b) as:

$$\sigma(\varphi) = \begin{cases} \dfrac{1}{\pi/(2k^2 bd) + \cos(\varphi)} & \text{if } |\varphi| \leq \varphi_\ell \\[4mm] \dfrac{1}{\pi/(2k^2 bd) + 1.5\cos(\varphi_\ell) - 0.5\cos(\varphi)} & \text{if } \varphi_\ell < |\varphi| \leq (\pi/2) \end{cases} \tag{9.168}$$

where $2d$ is the duct cross-sectional dimension in the direction of the observer and $2b$ is the dimension at $90°$ to the direction of the observer, for a rectangular section duct. For a circular duct of radius, a, the relationship $d = b = \pi a/4$ in the preceding equations has been shown by Davy (2008a,b) to be appropriate.

As the graphs in figures 9.29 to 9.31 are in terms of ka for a circular section duct of radius, a, then for rectangular ducts, the scale on the x-axis (which should be kd) must be multiplied by $4/\pi$ The limiting angle, φ_ℓ, is defined as:

$$\varphi_\ell = \begin{cases} 0 & \text{if } \sqrt{\pi/(2kd)} \geq 1.0 \\[3mm] \arccos[\sqrt{\pi/(2kd)}] & \text{if } \sqrt{\pi/(2kd)} < 1.0 \end{cases} \tag{9.169}$$

The quantity, $w(\varphi)$ in Equation (9.167) is defined by (Davy, 2008a,b):

$$w(\varphi) = \left(\frac{\sin(ks_d \sin(\varphi))}{ks_d \sin(\varphi)} \right)^2 (1 - \alpha_{st})^{(L/2d)\tan|\varphi|} \tag{9.170}$$

where L is the length of the exhaust stack (from the noise source to the stack opening) and α_{st} is the statistical absorption coefficient of the duct walls, usually set equal to 0.05. The quantity, ks_d is a function of the source size and usually setting it equal to π seems to give results that agree with experimental measurements made using loudspeakers (Davy, 2008a,b). The quantity, $w(\varphi)$, is made up of two physical quantities. The first term in large brackets represents the directivity of the sound source at the end of the duct, which for the purposes of Equation (9.170), corresponds to a line source of length $2s_d$, where s_d is the radius of the loudspeaker sound source. Although this model works well for the loudspeaker sound sources used to obtain the experimental data in Figures 9.29 and 9.30, it may not be the best model for an industrial noise source such as a fan with the result that the directivity pattern for radiation from a stack driven by an industrial noise source could be slightly different to the directivities presented here.

The second term in Equation (9.170) accounts for the effect of reflections from the duct walls on the angular distribution of sound propagation in the duct.

The quantity, $p_2(\theta)$, in Equation (9.167) is defined as:

$$p_2(\theta) = \begin{cases} p_2(0) & \text{if } \cos(\varphi_\ell) \le \cos(\theta) \\ \dfrac{p_2(0)\cos(\theta) + \cos(\varphi_\ell) - \cos(\theta)}{\cos(\varphi_\ell)} & \text{if } 0 \le \cos(\theta) < \cos(\varphi_\ell) \end{cases} \tag{9.171}$$

where $p_2(0)$ is defined as:

$$p_2(0) = 1 + p_b p_d \tag{9.172}$$

$$p_b = \begin{cases} \sin(kb) & \text{if } kb \le \pi/2 \\ 1 & \text{if } kb > \pi/2 \end{cases} \quad \text{and} \quad p_d = \begin{cases} \sin(kd) & \text{if } kd \le \pi/2 \\ 1 & \text{if } kd > \pi/2 \end{cases} \tag{9.173a,b}$$

In comparing Figures 9.29, 9.30 and 9.31, it can be seen that the field measurements seem not to have as large a directivity as measured in the laboratory or predicted by the theory. One explanation for this may be scattering which adds to the noise levels calculated due to diffraction alone. For noise barriers, this limit has been set to $L_s = 24$ dB and there is no reason to assume anything different in practice for directivity from exhaust stacks. So at long distances, if we assume that the maximum sound pressure level is at $\theta = 0$, then the sound level at angle, θ, relative to that at $\theta = 0$ is:

$$L_p(\theta) - L_p(0) = 10 \log_{10} \left(\frac{I(\theta)}{I(0)} + 10^{-L_s/10} \right) \tag{9.174}$$

Note that the curves in figures 9.29 to 9.31 all show greater differences between the zero degree direction and the large angles than predicted by Equation (9.174). This is for different reasons in each case: the curves in Figure 9.29 were measured in still air in an anechoic room where scattering was far less than it would be outdoors; the curves in Figure 9.30 were measured within 1 or 2 metres of the duct exhaust where scattering is not as important as at larger distances; and the curves in Figure 9.31 were derived using a theoretical analysis that does not include scattering.

Curves showing the measured data at 0, 30, 45, 60, 90 and 120 degrees, are provided in Figure 9.33 for the field measured data, which gives an idea of the scatter that may be expected in practice. The data measured in the laboratory exhibited a similar amount of scatter.

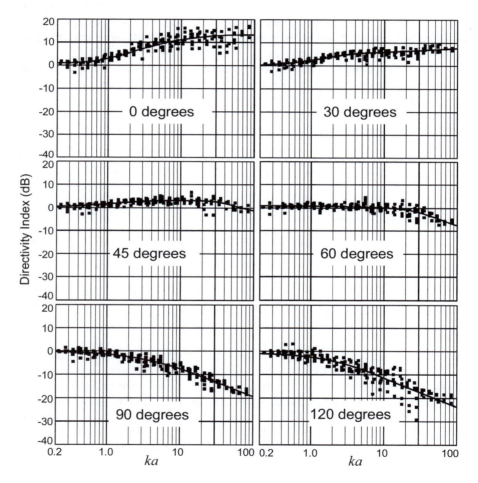

Figure 9.33 Scatter in the Directivity Index data measured in the field and reported by Day and Bennett(2008).

The results in Figures 9.29 to 9.31 only apply for unlined ducts. Ducts lined with sound absorbing material radiate more directionally so that higher on-axis sound levels are produced. As shown by Figures 9.29 to 9.31, a distinct advantage accrues from pointing an exhaust duct upwards. Alternatively, the figures show the importance of placing cooling towers such that line of sight to any nearby building is greater than about 30°.

Equation (9.170) is the best description that seems to be available to describe the energy propagation in a duct as a function of angle and this expression implies that cross-modes will contribute to the exit sound field, even for small values of *ka*. Thus it is difficult to obtain an accurate estimate of the radiated sound power from sound pressure level measurements at the duct exit as not all of the energy propagation is normal to the plane of the duct cross section at the exit.

The sound power can either be determined using the methods outlined in Chapter 6 or alternatively, if the sound pressure level, L_p, is measured at some distance, r, and some angle, θ, from the duct outlet, the sound power level, L_w, radiated by the duct outlet may be calculated with the help of Figure 9.31 and Equation (9.175):

$$L_w = L_p - DI_\theta + 10\log_{10} 4\pi r^2 + 10\log_{10}\frac{400}{\rho c} + A_E \qquad (9.175)$$

where the directivity index, DI_θ, may be obtained from Figure 9.31 and the excess attenuation, A_E, may be calculated as described in Section 5.11.6.

If it is desired to add an exhaust stack to the duct outlet, the resulting noise reduction may be calculated from a knowledge of the insertion loss, IL_s, of the stack due to sound propagation through it, and both the angular direction and distance from the stack axis to the receiver location (if different from the values for the original duct outlet). The excess attenuation, A_{Es}, must also be taken into account if it is different for propagation from the duct outlet without the stack. Thus:

$$NR = IL_s + A_{Es} - A_E + 20\log_{10}(r_s/r) + DI_\theta - DI_s \qquad (9.176)$$

where the subscript, *s*, refers to quantities with the stack in place.

The sound pressure level, L_{ps}, at location r_s with the stack in place is given by:

$$L_{ps} = L_w - IL_s + DI_s - 10\log_{10} 4\pi r_s^2 - 10\log_{10}\frac{400}{\rho c} - A_{Es} \qquad (9.177)$$

where the directivity, DI_s, of the exhaust stack in the direction of the receiver is obtained using Figure 9.31. Without the stack in place the sound pressure level may be determined using Equation (9.175).

CHAPTER TEN

Vibration Control

LEARNING OBJECTIVES

In this chapter the reader is introduced to:

- vibration isolation for single- and multi-degree-of-freedom systems;
- damping, stiffness and mass relationships;
- types of vibration isolators;
- vibration absorbers;
- vibration measurement;
- when damping of vibrating surfaces is and is not effective for noise control;
- damping of vibrating surfaces;
- measurement of damping.

10.1 INTRODUCTION

Many noise sources commonly encountered in practice are associated with vibrating surfaces, and with the exception of aerodynamic noise sources, the control of vibration is an important part of any noise-control program. Vibration is oscillatory motion of a body or surface about a mean position and occurs to some degree in all industrial machinery. It may be characterised in terms of acceleration, velocity, displacement, surface stress or surface strain amplitude, and associated frequency. On a particular structure, the vibration and relative phase will usually vary with location. Although high levels of vibration are sometimes useful (for example, vibrating conveyors and sieves), vibration is generally undesirable, as it often results in excessive noise, mechanical wear, structural fatigue and possible failure.

Any structure can vibrate and will generally do so when excited mechanically (for example, by forces generated by some mechanical equipment) or when excited acoustically (for example, by the acoustic field of noisy machinery). Any vibrating structure will have preferred modes in which it will vibrate and each mode of vibration will respond most strongly at its resonance frequency. A mode will be characterised by a particular spatial amplitude of response distribution, having nodes and anti nodes. Nodes are lines of nil or minimal response across which there will be abrupt phase changes from in-phase to opposite phase relative to a reference and anti nodes are regions of maximal response between nodes.

If an incident force field is coincident both in spatial distribution and frequency with a structural mode it will strongly drive that mode. The response will become stronger with better matching of the force field to the modal response of the structure. When driven at resonance, the structural mode response will only be limited by the damping of the mode. As will be discussed in Section 10.7, it is also possible to drive structural modes at frequencies other than their resonance frequencies. To avoid excessive vibration and associated problems, it is important in any mechanical system to ensure that coincidence of excitation frequency and structural resonance frequencies is avoided as much as possible.

With currently available analytical tools (e.g. statistical energy analysis, finite element analysis – see Chapter 12), it is often possible to predict at the design stage the dynamic behaviour of a machine and any possible vibration problems. However, vibration problems do appear regularly in new as well as old installations, and vibration control then becomes a remedial exercise instead of the more economic design exercise.

With the principal aim of noise control, five alternative forms of vibration control will be listed. These approaches, which may be used singly or in combination, are described in the following paragraphs.

The first form of vibration control is modification of the vibration generating mechanism. This may be accomplished most effectively at the design stage by choosing the process that minimises jerk, or the time rate of change of force. In a punch press, this may be done by reducing the peak level of tension in the press frame and releasing it over a longer period of time as, for example, by surface grinding the punch on a slight incline relative to the face of the die. Another way of achieving this in practice is to design tools that apply the load to the part being processed over as long a time period as possible, while at the same time minimising the peak load. This type of control is case specific and not amenable to generalisation; however, it is often the most cost effective approach and frequently leads to an inherently better process.

The second form of vibration control is modification of the dynamic characteristics (or mechanical input impedance) of a structure to reduce its ability to respond to the input energy, thus essentially suppressing the transfer of vibrational energy from the source to the noise-radiating structure. This may be achieved by stiffness or mass changes to the structure or by use of a vibration absorber. Alternatively, the radiating surface may be modified to minimise the radiation of sound to the environment. This may sometimes be done by choice of an open structure, for example, a perforated surface instead of a solid surface.

The third form of vibration control is isolation of the source of vibration from the body of the noise-radiating structure by means of flexible couplings or mounts.

The fourth form of vibration control is dissipation of vibrational energy in the structure by means of vibration damping, which converts mechanical energy into heat. This is usually achieved by use of some form of damping material.

The fifth form of vibration control is active control, which may be used either to modify the dynamic characteristics of a structure or to enhance the effectiveness of vibration isolators. Active control is discussed in Hansen (2001).

As has been mentioned, the first approach will not be discussed further and the fifth approach will be discussed elsewhere; the remaining three approaches, isolation, damping and alteration of the mechanical input impedance, will now be discussed with emphasis on noise control.

10.2 VIBRATION ISOLATION

Vibration isolation is considered on the basis that structure-borne vibration from a source to some structure, which then radiates noise, may indeed be as important or perhaps more important than direct radiation from the vibration source itself. Almost any stringed musical instrument provides a good example of this point. In every case, the vibrating string is the obvious energy source but the sound that is heard seldom originates at the string, which is a very poor radiator; rather, a sounding board, cavity or electrical system is used as a secondary and very much more efficient sound radiator.

When one approaches a noise-control problem, the source of the unwanted noise may be obvious, but the path by which it radiates sound may be obscure. Indeed, determining the propagation path may be the primary problem to be solved. Unfortunately, no general specification of simple steps to be taken to accomplish this task can be given. On the other hand, if an enclosure for a noisy machine is contemplated, then good vibration isolation between the machine and enclosure, between the machine and any pipework or other mechanical connections to the enclosure, and between the enclosure and any protrusions through it, should always be considered as a matter of course. Stated another way, the best enclosure can be rendered ineffective by structure-borne vibration. Thus, it is important to control all possible structural paths of vibration, as well as airborne sound, for the purpose of noise control.

The transmission of vibratory motions or forces from one structure to another may be reduced by interposing a relatively flexible isolating element between the two structures. This is called vibration isolation, and when properly designed, the vibration amplitude of the driven structure is largely controlled by its inertia. An important design consideration is the resonance frequency of the isolated structure on its vibration-isolation mount. At this frequency, the isolating element will amplify by a large amount the force transmission between the structure and its mount. Only at frequencies greater than 1.4 times the resonance frequency will the force transmission be reduced. Thus, the resonance frequency must be arranged to be well below the range of frequencies to be isolated. Furthermore, adding damping to the vibrating system, for the purpose of reducing the vibratory response at the resonance frequency, has the effect of decreasing the isolation that otherwise would be achieved at higher frequencies.

Two types of vibration-isolating applications will be considered; (1) those where the intention is to prevent transmission of vibratory forces from a machine to its foundation, and (2) those where the intention is to reduce the transmission of motion of a foundation to a device mounted on it. Rotating equipment such as motors, fans, turbines, etc. mounted on vibration isolators, are examples of the first type. An electron microscope, mounted resiliently in the basement of a hospital, is an example of the second type.

10.2.1 Single-degree-of-freedom Systems

To understand vibration isolation, it is useful to gain familiarity with the behaviour of single-degree-of-freedom systems, such as illustrated in Figure 10.1 (Church, 1963; Tse *et al.*, 1978; Rao, 1986). In the figure, the two cases considered here are illustrated with a spring, mass and dashpot. In the first case, the mass is driven by an externally applied force, $F(t)$, while in the second case, the base is assumed to move with some specified vibration displacement, $y_1(t)$ (Tse *et al.*, 1978).

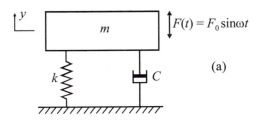

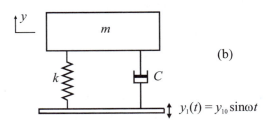

Figure 10.1 Single-degree-of-freedom system: (a) forced mass, rigid base; (b) vibrating base.

The equation of motion for the single-degree-of-freedom oscillator of mass, m (kg), damping constant, C (N-s/m), stiffness, k (N/m), displacement, y (m), and cyclic forcing function, $F(t)$ (N), shown in Figure 10.1(a) is:

$$m\ddot{y} + C\dot{y} + ky = F(t) \tag{10.1}$$

For sinusoidal motion $\ddot{y} = \omega^2 y$ and $\dot{y} = j\omega y$. In the absence of any excitation force, $F(t)$, or damping, C, the system, once disturbed, will vibrate sinusoidally at constant amplitude (dependent on the amplitude of the original disturbance) at its undamped resonance frequency, f_0. Solution of Equation (10.1) with $F(t) = C = 0$ gives for the undamped resonance frequency:

$$f_0 = \frac{1}{2\pi}\sqrt{\frac{k}{m}} \quad \text{(Hz)} \tag{10.2}$$

The static deflection, d, of the mass supported by the spring is given by $d = mg/k$ where g is the acceleration of gravity, so that Equation (10.2) may be written in the following alternative form:

$$f_0 = \frac{1}{2\pi}\sqrt{\frac{g}{d}} \qquad \text{(Hz)} \qquad (10.3)$$

Substitution of the value of g equal to 9.81 m/s gives the following useful equation (where d' is in metres):

$$f_0 = 0.5 / \sqrt{d'} \qquad \text{(Hz)} \qquad (10.4)$$

The preceding analysis is for an ideal system in which the spring has no mass, which does not reflect the actual situation. If the mass of the spring is denoted m_s, and it is uniformly distributed along its length, it is possible to get a first order approximation of its effect on the resonance frequency of the mass spring system by using Rayleigh's method and setting the maximum kinetic energy of the mass, m plus the spring mass, m_s, equal to the maximum potential energy of the spring. The velocity of the spring is zero at one end and a maximum of $\dot{y} = \omega y$ at the other end. Thus the kinetic energy in the spring may be written as:

$$KE_s = \frac{1}{2}\int_0^L u_m^2 \, dm_s \qquad (10.5)$$

where u_m is the velocity of the segment of spring of mass, dm_s and L is the length of the spring . The quantities, u_m and dm_s may be written as:

$$u_m = \frac{x\dot{y}}{L} \quad \text{and} \quad dm_s = \frac{m_s}{L}dx \qquad (10.6\text{a,b})$$

where x is the distance from the spring support to segment dm_s. Thus the KE in the spring may be written as:

$$KE_s = \frac{1}{2}\int_0^L \left(\frac{x\dot{y}}{L}\right)^2 \frac{m_s}{L}dx = \frac{m_s\dot{y}^2}{2L^3}\int_0^L x^2 dx = \frac{1}{2}\frac{m_s\dot{y}^2}{3} \qquad (10.7\text{a,b,c})$$

Equating the maximum KE in the mass, m and spring with the maximum PE in the spring gives:

$$\frac{1}{2}\frac{m_s}{3}\dot{y}^2 + \frac{1}{2}m\dot{y}^2 = \frac{1}{2}ky^2 \qquad (10.8)$$

Substituting $\dot{y} = \omega y$ in the above equation gives the resonance frequency as follows:

$$f_0 = \frac{1}{2\pi}\sqrt{\frac{k}{m + (m_s/3)}} \qquad (10.9)$$

Thus in the following analysis, more accurate results will be obtained if the suspended mass is increased by one-third of the spring mass.

The mass, m_s, of the spring is the mass of the active coils which for a spring with flattened ends is two less than the total number of coils. For a coil spring of overall diameter, D, and wire diameter, d, with n_C active coils of material density ρ_m, the mass is:

$$m_s = n_C \frac{\pi d^2}{4} \pi D \rho_m \tag{10.10}$$

For a coil spring, the stiffness (N/m) or the number of Newtons required to stretch it by 1 metre is given by:

$$k = \frac{G d^4}{8 n_C D^3} \tag{10.11}$$

where G is the modulus of rigidity (or shear modulus) of the spring material.

Of critical importance to the response of the systems shown in Figure 10.1 is the damping ratio, $\zeta = C/C_c$, where C_c is the critical damping coefficient defined as follows:

$$C_c = 2 \sqrt{km} \qquad \text{(kg/s)} \tag{10.12}$$

When the damping ratio is less than unity, the transient response is cyclic, but when the damping ratio is unity or greater, the system transient response ceases to be cyclic.

In the absence of any excitation force, $F(t)$, but including damping, $C < 1$, the system of Figure 10.1, once disturbed, will oscillate approximately sinusoidally at its damped resonance frequency, f_d. Solution of Equation (10.1) with $F(t) = 0$ and $C \neq 0$ gives for the damped resonance frequency:

$$f_d = f_0 \sqrt{1 - \zeta^2} \qquad \text{(Hz)} \tag{10.13}$$

When the excitation force $F(t) = F_0 e^{j\omega t}$ is sinusoidal, the system of Figure 10.1 will respond sinusoidally at the driving frequency $\omega = 2\pi f$. Let $f/f_0 = X$, then the solution of Equation (10.1) gives for the displacement amplitude $|y|$ at frequency, f:

$$\frac{|y|}{|F|} = \frac{1}{k} \left[(1 - X^2)^2 + 4\zeta^2 X^2 \right]^{-1/2} \tag{10.14}$$

The frequency of maximum displacement, which is obtained by differentiation of Equation (10.14) is:

$$f_{max\ dis} = f_0 \sqrt{1 - 2\zeta^2} \tag{10.15}$$

The amplitude of velocity $|\dot{y}| = 2\pi f |y|$ is obtained by differentiation of Equation (10.14), and is written as follows:

$$\frac{|\ddot{y}|}{|F|} = \frac{1}{\sqrt{km}}\left[\left(\frac{1}{X} - X\right)^2 + 4\zeta^2\right]^{-1/2}$$

(10.16)

Inspection of Equation (10.16) shows that the frequency of maximum velocity amplitude is the undamped resonance frequency:

$$f_{max\ vel} = f_0$$

(10.17)

Similarly, it may be shown that the frequency of maximum acceleration amplitude is:

$$f_{max\ acc} = f_0\left(1 - 2\zeta^2\right)^{-1/2}$$

(10.18)

Alternatively, if the structure represented by Figure 10.1 is hysteretically damped, which in practice is the more usual case, then the viscous damping model is inappropriate. This case may be investigated by setting $C = 0$ and replacing k in Equation (10.1) with complex $k(1 + j\eta)$ where η is the structural loss factor. Solution of Equation (10.1) with these modifications gives for the displacement amplitude of the hysteretically damped system, $y'(f)$, the following equation:

$$\frac{|y'|}{|F|} = \frac{1}{k}\left[\left(1 - X^2\right)^2 + \eta^2\right]^{-1/2}$$

(10.19)

For the case of hysteretic (or structural) damping the frequency of maximum displacement occurs at the undamped resonance frequency of the system as shown by inspection of Equation (10.19):

$$f'_{max\ dis} = f_0$$

(10.20)

Similarly the frequencies of maximum velocity and maximum acceleration for the case of hysteretic damping may be determined.

The preceding analysis shows clearly that maximum response depends upon what is measured and upon the nature of the damping in the system under investigation. Where the nature of the damping is known, the undamped resonance frequency and the damping constant may be determined using appropriate equations; however, in general where damping is significant, resonance frequencies can only be determined by curve fitting frequency response data (Ewins, 1984). Alternatively, for small damping the various frequencies of maximum response are essentially all equal to the undamped frequency of resonance.

Referring to Figure 10.1 (a) the fraction of the exciting force, F_0, acting on the mass, m, which is transmitted through the spring to the support is of interest. Alternatively, referring to Figure 10.1 (b), the fraction of the displacement of the base, which is transmitted to the mass, is often of greater interest. Either may be expressed in terms of the transmissibility, T_F, which in Figure 10.1(a) is the ratio of the force transmitted to the foundation to the force, F_0, acting on the machine, and in Figure 10.1(b) it is the ratio of the displacement of the machine to the displacement of the foundation. The transmissibility may be calculated as follows (Tse *et al.*, 1978):

$$T_F = \sqrt{\frac{1 + (2\zeta X)^2}{(1 - X^2)^2 + (2\zeta X)^2}}, \quad \text{where } X = f/f_0 \tag{10.21}$$

Figure 10.2 shows the fraction, expressed in terms of the transmissibility, T_F, of the exciting force (system (a) of Figure 10.1) transmitted from the vibrating body through the isolating spring to the support structure. The transmissibility is shown for various values of the damping ratio ζ, as a function of the ratio of the frequency of the vibratory force to the resonance frequency of the system.

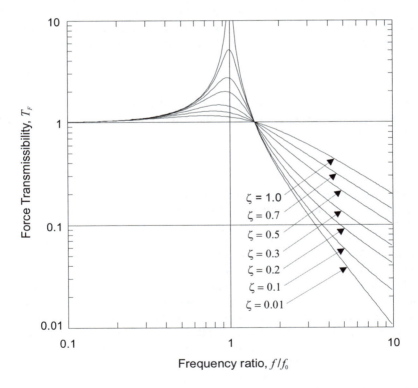

Figure 10.2 Force or displacement transmissibility of a viscously damped mass-spring system. The quantities f and f_0 are the excitation and undamped mass-spring resonance frequencies respectively, ζ is the system critical damping ratio, and T_F is the fraction of excitation force transmitted by the spring to the foundation. Note that for values of frequency ratio greater than $\sqrt{2}$, the force transmissibility increases with increasing damping ratio.

When the transmissibility, T, is identified with the displacement of the mass, m, of system (b), then Figure 10.2, shows the fraction of the exciting displacement amplitude transmitted from the base through the isolating spring to the supported mass, m. The figure allows one to determine the effectiveness of the isolation system for a single-degree-of-freedom system.

The vibration amplitude of a single-degree-of-freedom system is dependent upon its mass, stiffness and damping characteristics as well as the amplitude of the exciting force. This conclusion can be extended to apply to multi-degree of-freedom systems such as machines and structures. Consideration of Equation (10.21) shows that as X tends to zero, the force transmissibility, T_F, tends to one; the response is controlled by the stiffness k. When X is approximately one, the force transmissibility is approximately inversely proportional to the damping ratio; the response is controlled by the damping, C. As X tends to large values, the force transmissibility tends to zero as the square of X; the response is controlled by the mass, m.

The energy transmissibility, T_E, is related to the force transmissibility, T_F, and displacement transmissibility, T_D, by $T_E = T_F T_D$. As $T_F = T_D$, then $T_E = T_F^2$. The energy transmissibility, T_E, can be related to the expected increase or decrease, ΔL_w, in sound power radiated by the supported structure over that radiated when the vibrating mass is rigidly attached to the support structure as follows:

$$\Delta L_w = 10 \log_{10} T_E = 20 \log_{10} T_F \qquad (10.22a,b)$$

Differentiation of Equation (10.21) or use of Equation (10.15) gives for the frequency of maximum force transmissibility for a viscously damped system the following expression:

$$f_F = f_0 \sqrt{1 - 2\zeta^2} \qquad \text{(Hz)} \qquad (10.23)$$

The preceding equations and figures refer to viscous damping (where the damping force is proportional to the vibration velocity), as opposed to hysteretic or structural damping (where the damping force is proportional to the vibration displacement). Generally, the effects of hysteretic damping are similar to those of viscous damping up to frequencies of $f = 10 f_0$. Above this frequency, hysteretic damping results in larger transmission factors than shown in Figure 10.2.

The information contained in Figure 10.2 for the undamped case can be represented in a useful alternative way, as shown in Figure 10.3. However, it must be remembered that this figure only applies to undamped single-degree-of-freedom systems in which the exciting force acts in the direction of motion of the body.

Referring again to Figure 10.2, it can be seen that below resonance (ratio of unity on the horizontal axis) the force transmission is greater than unity and no isolation is achieved. In practice, the amplification obtained below a frequency ratio of 0.5 is rarely of significance so that, although no benefit is obtained from the isolation at these low frequencies, no significant detrimental effect is experienced either. However, in the frequency ratio range 0.5–1.4, the presence of isolators significantly increases the transmitted force and the amplitude of motion of the mounted body. In operation, this range is to be avoided. Above a frequency ratio of 1.4 the force transmitted by the

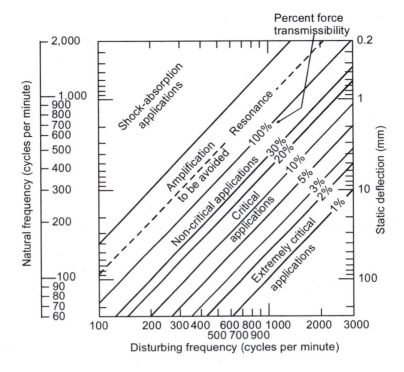

Figure 10.3 Force transmissibility as a function of frequency and static deflection for an undamped single-degree-of-freedom vibratory system.

isolators is less than that transmitted with no isolators, resulting in isolation of vibration; the higher the frequency the greater the isolation. Thus, for an isolator to be successful, its stiffness must be such that the mounted resonance frequency is less than 0.7 times the minimum forcing frequency.

All practical isolators have some damping, and Figure 10.2 shows the effect of damping; increasing the damping decreases the isolation achieved. For best isolation no damping would be desirable. On the other hand, damping is necessary in installations involving rotating equipment because the equipment rotational speed (and hence forcing frequency) will pass through the mounted resonance frequency on shutdown and start-up. In these cases, the amplitude of the transmitted force will exceed the exciting force, and indeed could build up to an alarming level.

Sometimes the machine can be accelerated or decelerated rapidly enough to pass through the region of resonance so quickly that the amplitude of the transmitted force does not have time to build up to the steady-state levels indicated by Figure 10.2. However, in some cases, the machine can only be accelerated slowly through the resonant range, resulting in a potentially disastrous situation if the isolator damping is inadequate. In this case the required amount of damping could be large; a damping ratio $\zeta = 0.5$ could be required.

An external damper can be installed to accomplish the necessary damping, but always at the expense of reduced isolation at higher frequencies. An alternative to using highly damped isolators is to use rubber snubbers to limit excessive motion of the machine at resonance. Snubbers can also be used to limit excessive motion. These have the advantage of not limiting high-frequency isolation. Active dampers, which are only effective below a preset speed, are also used in some cases. These also have no detrimental effect on high-frequency isolation and are only effective during machine shutdown and start-up. Air dampers can also be designed so that they are only effective at low frequencies (see Section 10.3.2).

10.2.1.1 Surging in Coil Springs

Surging in coil springs is a phenomenon where high frequency transmission occurs at frequencies corresponding to the resonance frequencies of wave motion in the coils. This limits the high frequency performance of such springs and in practical applications rubber inserts above or below the spring are used to minimise the effect. However it is of interest to derive an expression for these resonance frequencies so that in isolator design, one can make sure that any machine resonance frequencies do not correspond to surge frequencies.

The analysis proceeds by deriving an expression for the effective Young's modulus for the spring, which is then used to find an expression for the longitudinal wave speed in the spring. Finally, as in chapter 7 for rooms, the lowest order resonance is the one where the length of the spring is equal to a ½ wavelength. Higher order resonances are at multiples of half a wavelength..

Young's modulus is defined as stress over strain so that for a spring of length, L and extension, x due to an applied force:

$$E = \frac{\sigma}{\varepsilon} = \frac{kx/A}{x/L} = \frac{kL}{A} \qquad (10.24a,b)$$

The longitudinal wave speed in the spring is then:

$$c_L = \sqrt{\frac{E}{\rho}} = \sqrt{\frac{Lk/A}{m_s/LA}} = L\sqrt{\frac{k}{m_s}} \qquad (10.25a,b,c)$$

The surge frequency, f_s, occurs when the spring length, L, is equal to integer multiples of $\lambda/2$, so that:

$$L = n\frac{\lambda}{2} = n\frac{c_L}{2f_s} = \frac{nL}{2f_s}\sqrt{\frac{k}{m_s}} \,, \qquad n = 1, 2, 3,.......... \qquad (10.26a,b,c)$$

Rearranging gives for the surge frequencies:

$$f_s = \frac{n}{2}\sqrt{\frac{k}{m_s}} \qquad (10.27)$$

10.2.2 Four-isolator Systems

In most practical situations, more than one isolator is used to isolate a particular machine. This immediately introduces the problem of more than one system resonance frequency at which the force transmission will be large. If possible, it is desirable to design the isolators so that none of the resonance frequencies of the isolated system correspond to any of the forcing frequencies.

The most common example of a multi-degree-of-freedom system is a machine mounted symmetrically on four isolators (Crede, 1965). In general, a machine or body mounted on springs has six degrees of freedom. There will be one vertical translational mode of resonance frequency, f_0, one rotational mode about the vertical axis and two rocking modes in each vertical plane, as illustrated in Figure 10.4. The calculation of resonance frequencies for such a system in terms of the resonance frequency, f_0, will now be considered. The latter frequency may be calculated using either Equations (10.2) or (10.3) as for a single-degree-of-freedom system, with one spring having the combined stiffness of the four shown in Figure 10.4. Note that stiffnesses add linearly when springs are in parallel; that is, $k = k_1 + k_2$, etc.

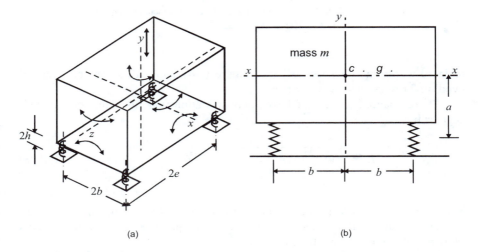

(a)   (b)

Figure 10.4 Vibration modes for a machine mounted on four isolators. The origin of the co-ordinates is coincident with the assumed centre of gravity (c.g.) at height $a + h$ above the mounting plane.

Rocking and horizontal mode resonance frequencies may be determined by reference to Figure 10.5. The resonance frequencies, f_a and f_b, for roll and horizontal motion are given, respectively, in parts (a) and (b) of the figure. The parameters in these figures are defined as follows: $W = (\delta/b)\sqrt{(k_x/k_y)}$, $M = a/\delta$ and $\Omega = (\delta/b)(f_i/f_0)$, where the subscript $i = a$ in Figure 10.5(a) and $i = b$ in Figure 10.5(b). k_x and k_y are the isolator stiffnesses in the x and y directions respectively and δ is the radius of gyration for rotation about the horizontal z axis through the centre of gravity (see Figure 10.4). The dimensions a and b are also defined in Figure 10.4. For motion in the orthogonal vertical plane, the same figures (10.5(a) and (b)) are used, with the quantities x and b

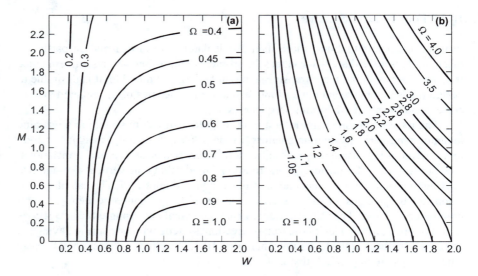

Figure 10.5 Charts for determining roots, Ω, of the characteristic equation:
$$\Omega^4 - \Omega^2(1 + W^2 + M^2W^2) + W^2 = 0.$$

replaced with z and e respectively (see Figure 10.4), and with δ now the radius of gyration for rotation about the x-axis.

The resonance frequency of the rotational vibration mode about the vertical y axis is given by:

$$f_y = \frac{1}{\pi}\sqrt{(b^2k_z + e^2k_x)/I_y} \qquad \text{(Hz)} \tag{10.28}$$

The quantities $2b$ and $2e$ are the distances between centre-lines of the support springs, k_z is the isolator stiffness in the z direction, usually equal to k_x, and I_y is the moment of inertia of the body about the y axis.

Values for the stiffnesses k_x, k_y and k_z, are usually available from the isolator manufacturer. Note that for rubber products, static and dynamic stiffnesses are often different. It is the dynamic stiffnesses that are required here. For a rectangular cross-section of dimensions $2d \times 2q$, the radius of gyration, δ, about an axis through the centre and perpendicular to the plane of the section is:

$$\delta = \sqrt{(d^2 + q^2)/3} \tag{10.29}$$

When placing vibration isolators beneath a machine, it is good practice to use identical isolators and to place them symmetrically with respect to the centre of gravity of the machine. This results in equal loading and deflection of the isolators.

The calculation of the force transmission for a multi-degree-of-freedom system is complex and not usually contemplated in conventional isolator design. However, the analysis of various multi-degree-of-freedom systems has been discussed in the literature (Mustin, 1968; Smollen, 1966). Generally, for a multi-degree-of-freedom system, good isolation is achieved if the frequencies of all the resonant modes are less than about two-fifths of the frequency of the exciting force. However, a force or torque may not excite all the normal modes, and then the natural frequencies of the modes that are not excited do not need to be considered, except to ensure that they do not actually coincide with the forcing frequency.

10.2.3 Two-stage Vibration Isolation

Two-stage vibration isolation is used when the performance of single stage isolation is inadequate and it is not practical to use a single stage system with a lower resonance frequency. As an example, two stage isolation systems have been used to isolate diesel engines from the hull of large submarines.

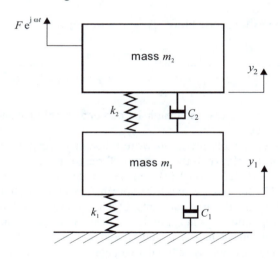

Figure 10.6 Two-stage vibration isolator.

A two-stage isolator is illustrated in Figure 10.6, where the machine to be isolated is represented as mass, m_2, and the intermediate mass is represented as mass, m_1. The intermediate mass should be as large as possible, but should be at least 70% of the machine mass being supported.

The equations of motion of the masses, m_1 and m_2, in Figure 10.6, may be written as:

$$m_1\ddot{y}_1 + C_1\dot{y}_1 + k_1 y_1 - C_2(\dot{y}_2 - \dot{y}_1) - k_2(y_2 - y_1) = 0 \tag{10.30}$$

and

$$m_2\ddot{y}_2 + C_2(\dot{y}_2 - \dot{y}_1) + k_2(y_2 - y_1) = Fe^{j\omega t} \tag{10.31}$$

These equations can be solved to give the complex displacements of each mass as:

$$\frac{y_1}{F} = \frac{k_2 + j\omega C_2}{(k_1 + k_2 - \omega^2 m_1 + j\omega C_1 + j\omega C_2)(k_2 - \omega^2 m_2 + j\omega C_2) - (k_2 + j\omega C_2)^2} \tag{10.32}$$

and:

$$\frac{y_2}{F} = \frac{k_1 + k_2 - \omega^2 m_1 + j\omega C_1 + j\omega C_2}{(k_1 + k_2 - \omega^2 m_1 + j\omega C_1 + j\omega C_2)(k_2 - \omega^2 m_2 + j\omega C_2) - (k_2 + j\omega C_2)^2} \tag{10.33}$$

The complex force transmitted to the foundation is:

$$F_T = y_1(k_1 + j\omega C_1) \tag{10.34}$$

and thus the complex transmissibility, $\overline{T}_F = F_T/F$, is given by:

$$T_F = \frac{(k_1 + j\omega C_1)(k_2 + j\omega C_2)}{(k_1 + k_2 - \omega^2 m_1 + j\omega C_1 + j\omega C_2)(k_2 - \omega^2 m_2 + j\omega C_2) - (k_2 + j\omega C_2)^2} \tag{10.35}$$

Complex variables have a real and imaginary part or an amplitude and phase relative to the excitation force.

It is a relatively simple matter to write a computer program to calculate the amplitude and phase (relative to the excitation force) of the complex displacements given by Equations (10.32) and (10.33), and the complex force transmissibility given by Equation (10.35), for given values of the parameters on the right hand side of the equations. Note that the damping constants, C_1 and C_2, are found by multiplying the the critical damping ratios, ζ_1 and ζ_2, by the critical damping, C_{c1} and C_{c2}, given by Equation (10.10), using stiffnesses, k_1 and k_2, and masses, m_1 and m_2, respectively.

As a two-stage isolation system has two degrees of freedom, it will have two resonance frequencies corresponding to high force transmissibility. The undamped resonance frequencies of the two-stage isolator may be calculated using (Beranek and Ver, 1992):

$$\left(\frac{f_a}{f_0}\right)^2 = Q - \sqrt{Q^2 - B^2} \quad \text{and} \quad \left(\frac{f_b}{f_0}\right)^2 = Q + \sqrt{Q^2 - B^2} \tag{10.36a,b}$$

where:

$$Q = 0.5\left(B^2 + 1 + \frac{k_1}{k_2}\right) \tag{10.37}$$

and:

$$B = \frac{f_1}{f_0} \qquad (10.38)$$

$$f_1 = \frac{1}{2\pi}\sqrt{\frac{k_1 + k_2}{m_1}} \qquad (10.39)$$

$$f_0 = \frac{1}{2\pi}\sqrt{\frac{k_1 k_2}{m_2(k_1 + k_2)}} \qquad (10.40)$$

The quantity, f_1, is the resonance frequency of mass, m_1, with mass, m_2, held fixed and f_0 is the resonance frequency of the single degree of freedom system with mass, m_1, removed. The upper resonance frequency, f_b, of the combined system is always greater than either f_1 or f_0, while the lower frequency is less than either f_1 or f_0.

At frequencies above twice the second resonance frequency, f_b, the force transmissibility for an undamped system will be approximately equal to $(f^2/(f_1 f_0))^2$, proportional to the fourth power of the excitation frequency, compared to a single-stage isolator, for which it is approximately $(f/f_0)^2$ above twice the resonance frequency, f_0.

In Figure 10.7, the force transmissibility for a two-stage isolator for a range of ratios of masses and stiffnesses is plotted for the special case where $\zeta_1 = \zeta_2$.

10.2.4 Practical Isolator Considerations

The analysis discussed thus far gives satisfactory results for force transmission at relatively low frequencies, if account is taken of the three-dimensional nature of the machine and the fact that several mounts are used. For large machines or structures this frequency range is generally infrasonic, where the concern is for prevention of physical damage or fatigue failure. Unfortunately, the analysis cannot be directly extrapolated into the audio-frequency range, where it is apt to predict attenuations very much higher than those achieved in practice. This is because the assumptions of a rigid machine and a rigid foundation are generally not true. In actuality, almost any foundation and almost any machine will have resonances in the audio-frequency range.

Results of both analytical and experimental studies of high-frequency performance of vibration isolators have been published (Ungar and Dietrich, 1966; Snowdon, 1965). This work shows that the effect of appreciable isolator mass and damping is to significantly increase, over simple classical theory predictions, the transmission of high-frequency forces or displacements. The effects begin to occur at forcing frequencies as low as 10 to 30 times the natural frequency of the mounted mass.

To minimise these effects, the ratio of isolated mass to isolator mass should be as large as possible (1000: 1 is desirable) and the damping in the isolated structure should be large. The effect of damping in the isolators is not as important, but nevertheless the isolator damping should be minimised.

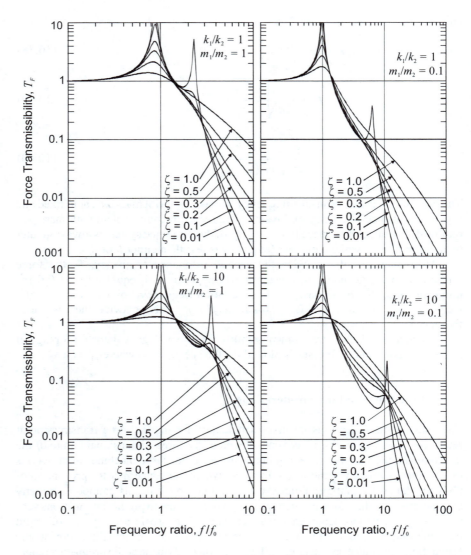

Figure 10.7 Force transmissibility for a two-stage vibration isolation system for various values of stiffness and mass ratio. In all figures, $\zeta_1 = \zeta_2 = \zeta$.

To continue, the concept of mobility will be introduced. The mobility of a system component is a complex, frequency dependent quantity, and is defined as the ratio of the velocity of response to a sinusoidal force input (reciprocal of mechanical impedance):

$$M = v/F \qquad (10.41)$$

The effectiveness of an isolator is related to the relative mobilities of the isolated mass, the isolators themselves and the foundation or attached structures. It may readily be shown using electrical circuit analysis that the relationship between the single isolator force transmissibility, T_F, and the mobilities of the components is as follows (Beranek, 1988):

$$T_F = \frac{M_m + M_f}{M_m + M_f + M_i} \qquad (10.42)$$

The quantity, M_m, is the mobility of the isolated mass, M_f is the mobility of the foundation and M_i is the mobility of the isolators. For a rigid isolated mass and a lightweight spring, the mobilities may be calculated using:

$$M_m = \frac{1}{j\omega m_m} \qquad (10.43)$$

$$M_i = \frac{j\omega}{k_i} \qquad (10.44)$$

$$M_f = j(k_f/\omega - \omega m_f)^{-1/2} \qquad (10.45)$$

In the preceding equations, m_m is the mass of the rigid mass supported on the spring, k_i is the stiffness of the "massless" isolator and k_f and m_f are the dynamic stiffness and dynamic mass of the support structure in the vicinity of the attachment of the isolating spring. The first two quantities are easy to calculate and are independent of the frequency of excitation. The latter two quantities are frequency dependent and difficult to estimate, so the foundation mobility usually has to be measured.

Equation (10.42) shows that an isolator is ineffective unless its mobility is large when compared with the sum of the mobilities of the machine and foundation.

The mobility of a simple structure may be calculated, and that of any structure may be measured (Plunkett, 1954, 1958). Some measured values of mobility for various structures have been published in the literature (Harris and Crede, 1976; Peterson and Plunt, 1982).

Attenuation of more than 20 dB ($T_F < 0.1$) is rare at acoustic frequencies with isolation mounts of reasonable stiffness, and no attenuation at all is common. For this reason, very soft mounts ($f_0 = 5$ to 6 Hz) are generally used where possible. As suggested by Equation (10.42), if a mount is effective at all, a softer mount (M_i larger) will be even more effective.

For a two-stage isolator, Equation (10.42) may be written as (Beranek and Ver, 1992):

$$\frac{1}{T_F} = \frac{M_{m2} + M_f + M_i}{M_{m2} + M_f} + \frac{(M_{i2} + M_{m2})(M_{i1} + M_f)}{M_{m1}(M_{m2} + M_f)} \qquad (10.46)$$

In Equation (10.46), the first term corresponds to a single isolator system, where the isolator moblity, M_i, is the same as the mobility of the two partial isolators in the two-stage system in series. The subscript, m_2, corresponds to the mobility of the machine being isolated, the subscript, m_1, corresponds to the mobility of the intermediate mass, the subscript, $i1$, corresponds to the mobility of the isolator between the intermediate mass and the foundation and the subscript, $i2$, corresponds to the mobility of the isolator between the intermediate mass and the machine being isolated.

Note that the second term in Equation (10.46) represents the improvement in performance as a result of using a two-stage isolator and that this improvement is inversely proportional to the mobility of the intermediate mass and thus directly proportional to the magnitude of the intermediate mass.

Once the total mobility, M_i, of the isolators has been selected, the optimum distribution between isolators 1 and 2 may be calculated using:

$$M_{i1} = r_i M_i \quad \text{and} \quad M_{i2} = (1 - r_i) M_i \tag{10.47}$$

and

$$\text{optimum } r_i = 0.5\left[1 + (M_f - M_{m2})/M_i\right] \tag{10.48}$$

10.2.4.1 Lack of Stiffness of Equipment Mounted on Isolators

If equipment is mounted on a non-rigid frame, which in turn is mounted on isolators, the mounted natural frequency of the assembly will be reduced as shown in Figure 10.8. In this case the mobility, M_m, of the isolated mass is large because of the non-rigid frame. According to Equation (10.42), the effectiveness of a large value of isolator mobility M_i in reducing the force transmissibility is thus reduced. Clearly a rigid frame is desirable.

10.2.4.2 Lack of Stiffness of Foundations

Excessive flexibility of the foundation is of significance when an oscillatory force generator is to be mounted on it. The force generator could be a fan, or an air conditioning unit, and the foundation could be the roof slab of a building. As a general requirement, if it is desired to isolate equipment from its support structure the mobility, M_i, of the equipment mounts must be large relative to the foundation mobility, M_f, according to Equation (10.42). A useful criterion is that the mounted resonance frequency should be much lower than the lowest resonance frequency of the support structure.

The equipment rotational frequency must be chosen so that it or its harmonics do not coincide with the resonance frequencies, which correspond to large values of the foundation (or support structure) mobility (see Equation (10.21)). If the support structure is flexible, the force generator should be placed on as stiff an area as possible, or supported on stiff beams, which can transfer the force to a stiff part of the

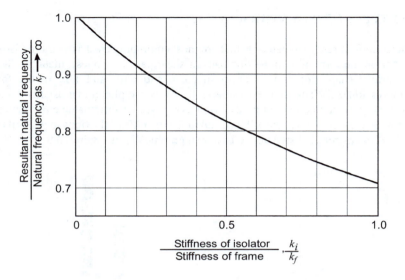

Figure 10.8 The natural frequency of equipment having a flexible frame and supported by an isolator, expressed as a fraction of the natural frequency obtained when the frame is infinitely rigid.

foundation. If the vibration mountings cannot be made sufficiently soft, their stiffnesses should be chosen so that the mounted resonance frequency does not coincide with support structure resonances, but lies in a frequency range in which the support structure has a small mobility.

When the support structure is non-rigid, substantially lower than normal mounting stiffnesses are required. As an example, for a machine speed of 1500 r.p.m. one manufacturer recommends the following static deflections for 95% vibration force isolation:

Installed in basement 8.6 mm
On a 10 m floor span 9.9 mm
On a 12 m floor span 10.7 mm
On a 15 m floor span 11.2 mm

10.2.4.3 Superimposed Loads on Isolators

If an external force (such as tension in a drive belt) is applied to a machine mounted on isolators, then the isolators must be designed to give the required stiffness under the combined action of the machine mass and external load. Any members transmitting an external force from the machine to the support structure must have a much lower stiffness than the isolators, or the mounted resonance frequency will increase and the isolators will become ineffective.

10.3 TYPES OF ISOLATORS

There are four resilient materials that are most commonly used as vibration isolators: rubber, in the form of compression pads or shear pads (or cones); metal, in the form of various shapes of springs or mesh pads; and cork and felt, in the form of compression pads. The choice of material for a given application is usually dependent upon the static deflection required as well as the type of anticipated environment (for example, oily, corrosive, etc.). The usual range of static deflections in general use for each of the materials listed above is shown graphically in Figure 10.9.

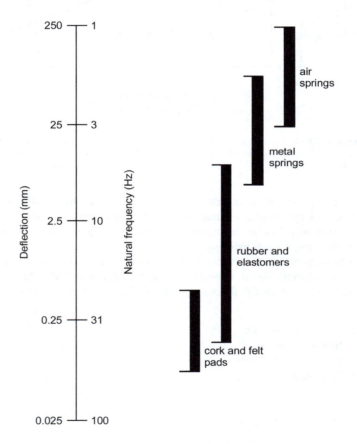

Figure 10.9 Ranges of application of different types of isolator.

10.3.1 Rubber

Isolators come in a variety of forms that use rubber in shear or compression, but rarely in tension, due to the short fatigue life experienced by rubber in tension. Isolator manufacturers normally provide the stiffness and damping characteristics of their products. As the dynamic stiffness of rubber is generally greater (by 1.3 to 1.8) than the static stiffness, dynamic data should be obtained whenever possible. Rubber can be used in compression or shear, but the latter use results in greater service life.

The amount of damping can be regulated by the rubber constituents, but the maximum energy that can be dissipated by damping tends to be limited by heat build-up in the rubber, which causes deterioration. Damping in rubber is usually vibration amplitude, frequency and temperature dependent.

Rubber in the form of compression pads is generally used for the support of large loads and for higher frequency applications (10 Hz resonance frequency upwards). The stiffness of a compressed rubber pad is generally dependent upon its size, and the end restraints against lateral bulging. Pads with raised ribs are usually used, resulting in a combination of shear and compression distortion of the rubber, and a static deflection virtually independent of pad size. However, the maximum loading on pads of this type is generally less than 550 kPa.

The most common use for rubber mounts is for the isolation of medium to lightweight machinery, where the rubber in the mounts acts in shear. The resonance frequencies of these mounts vary from about 5 Hz upwards, making them useful for isolation in the mid-frequency range.

10.3.2 Metal Springs

Next to rubber, metal springs are the most commonly used materials in the construction of vibration isolators. The load-carrying capacity of spring isolators is variable from the lightest of instruments to the heaviest of buildings. Springs can be produced industrially in large quantities, with only small variations in their individual characteristics. They can be used for low-frequency isolation (resonance frequencies from 1.3 Hz upwards), as it is possible to have large static deflections by suitable choice of material and dimensions.

Metal springs can be designed to provide isolation virtually at any frequency. However, when designed for low-frequency isolation, they have the practical disadvantage of readily transmitting high frequencies. Higher-frequency transmission can be minimised by inserting rubber or felt pads between the ends of the spring and the mounting points, and ensuring that there is no metal-to-metal contact between the spring and support structure.

Coil springs must be designed carefully to avoid lateral instability. For stable operation, the required ratio of unloaded spring length ℓ_0 to diameter D_0 for a given spring compression ratio, ξ (ratio of change in length when loaded to length unloaded) is shown in Figure 10.10.

Metal springs have little useful internal damping. However, this can be introduced in the form of viscous fluid damping, friction damping or, more cunningly, by viscous air damping. As an example of an air damper, at low frequencies in the region of the mounted resonance, air is pumped in and out of a dashpot by the motion of the spring, hence generating a damping force, but at higher frequencies the air movement and damping force are much reduced and the dashpot becomes an air spring in parallel with the steel spring. This configuration results in good damping at the mounted

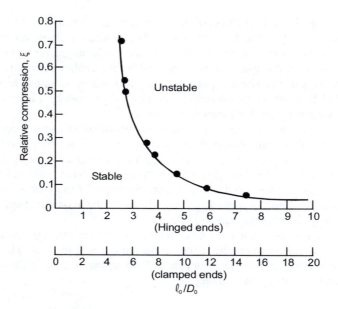

Figure 10.10 Stable and unstable values of relative compression for coil springs.

resonance frequency, and reduced damping at frequencies above resonance, thus giving better overall effectiveness. The accompanying increase in stiffness at higher frequencies normally has little effect on the isolation achieved.

Steel spring types include coil, torsion, coiled wire rope and mesh springs, as well as cantilever and beam springs of single or multi-leaf construction. Coiled wire rope springs are very undamping and must be carefully designed and tested, using similar vibration levels to those that will be experienced in practice, if they are to provide adequate isolation. In multi-leaf construction, interface friction between the leaves can provide friction damping, thus reducing higher-frequency transmission. By putting a suitable lubricant between the leaves, viscous or near viscous damping characteristics can be obtained.

Wire mesh springs consist of a pre-compressed block of wire mesh, which acts as a combined nonlinear spring and damper. Sometimes wire mesh springs are used in conjunction with a steel spring to carry part of the load. Damping is provided by friction within the mesh and between the mesh and the steel spring. This results in damping of the high-frequency vibration transmitted by the coil spring.

10.3.3 Cork

Cork is one of the oldest materials used for vibration isolation. It is generally used in compression and sometimes in a combination of compression and shear. The dynamic stiffness and damping of cork are very much dependent upon frequency. Also, the stiffness decreases with increasing load.

Generally, the machine or structure to be isolated is mounted on large concrete blocks, which are separated from the surrounding foundation by several layers of cork slabs, 2 to 15 cm thick. For optimum performance, the cork should be loaded to between 50 and 150 kPa. Increasing the cork thickness will lower the frequency above which isolation will be effective. However, large thicknesses, with associated stability problems, are required to achieve isolation at low frequencies. Although oil, water and moderate temperature have little effect upon its operating characteristics, cork does tend to compress with age under an applied load. At room temperature its effective life extends to decades; at 90°C it is reduced to less than a year.

10.3.4 Felt

To optimise the vibration isolation effectiveness of felt, the smallest possible area of the softest felt should be used, but in such a way that there is no loss of structural stability or excessive compression under static loading conditions. The felt thickness should be as great as possible. For general purposes, felt mountings of 1 to 2.5 cm thick are recommended, with an area of 5% of the total area of the machine base. In installations where vibration is not excessive it is not necessary to bond the felt to the machine. Felt has high internal damping ($\zeta \approx 0.13$), which is almost independent of load, and thus it is particularly suitable for reducing vibration at the mounted machine resonance frequency. In most cases felt is an effective vibration isolator only at frequencies above 40 Hz. Felt is particularly useful in reducing vibration transmission in the audio-frequency range, as its mechanical impedance is poorly matched with most engineering materials. Curves showing the resonance frequency of different grades of felt as a function of static load are illustrated in Figure 10.11.

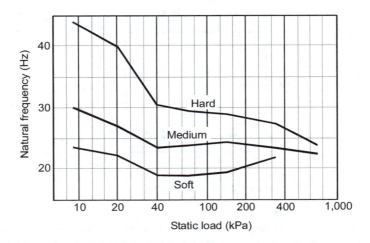

Figure 10.11 Natural frequency of 25 mm thick felt as a function of static load, expressed in units of pressure.

10.3.5 Air Springs

Although air springs can be used at very low frequencies, they become increasingly difficult and expensive to manufacture as the required resonance frequency is decreased; 0.7 Hz seems to be a practical lower limit and one that is achieved with difficulty. However, resonance frequencies of 1 Hz are relatively common.

Air springs consist of an enclosed volume of air, which is compressed behind a piston or diaphragm. Diaphragms are generally preferred to avoid the friction problems associated with pistons. The static stiffness of air springs is usually less than the dynamic stiffness, as a result of the thermodynamic properties of air. Machine height variations due to air volume changes, which are caused by ambient temperature changes, can be maintained by adding or removing air using a servo-controller.

One simple example of an air spring, which is very effective, is the inner tube from a car tyre, supported in a cutaway tire casing.

10.4 VIBRATION ABSORBERS

When vibration problems occur over a very narrow frequency range, a special-purpose device known as a dynamic vibration absorber may be useful. Such an absorber consists of a mass, m_2, attached via a spring of stiffness, k_2, to the vibrating structure or machine. The latter structure is idealised as a mass, m_1, suspended via a spring of stiffness, k_1, as illustrated in Figure 10.12.

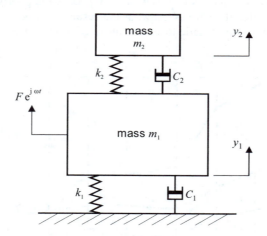

Figure 10.12 Vibration absorber system.

The equations describing the motion of this two-degree-of-freedom system, excited at frequency, ω (radians/sec), are similar to Equations (10.30) and (10.31) for a two-stage isolator and are as follows:

$$m_1 \ddot{y}_1 + C_1 \dot{y}_1 + k_1 y_1 - C_2 (\dot{y}_2 - \dot{y}_1) - k_2 (y_2 - y_1) = F e^{j\omega t} \tag{10.49}$$

and

$$m_2 \ddot{y}_2 + C_2 (\dot{y}_2 - \dot{y}_1) + k_2 (y_2 - y_1) = 0 \tag{10.50}$$

The steady state solutions for the motion of the two masses are (Soom and Lee, 1983):

$$y_1 = |y_1| \cos(\Omega t + \theta_1) \tag{10.51}$$

and

$$y_2 = |y_2| \cos(\Omega t + \theta_2) \tag{10.52}$$

The amplitudes, $|y_1|$ and $|y_2|$ are (Den Hartog, 1956; Soom and Lee, 1983):

$$\frac{|y_1|}{|F|} = \frac{1}{k_1} \sqrt{\frac{\left(\dfrac{2\zeta_2 \Omega m_1}{m_2}\right)^2 + \left(\Omega^2 - \dfrac{k_2 m_1}{k_1 m_2}\right)}{\left(\dfrac{2\zeta_2 \Omega m_1}{m_2}\right)^2 \left(\Omega^2 - 1 + \dfrac{m_2 \Omega^2}{m_1}\right)^2 + \left[\dfrac{k_2}{k_1}\Omega^2 - (\Omega^2 - 1)\left(\Omega^2 - \dfrac{k_2 m_1}{k_1 m_2}\right)\right]^2}} \tag{10.53}$$

and

$$|y_2| = |y_1| \left[(a/q)^2 + (b/q)^2 \right]^{1/2} \tag{10.54}$$

where, θ_1 and θ_2 are phase angles of the motion of the masses relative to the excitation force, $|F|$ is the amplitude of the excitation force and:

$$\Omega = \omega \sqrt{m_1/k_1} = f/f_0 \tag{10.55}$$

$$a = (k_2/k_1)^2 + 4\zeta_2^2 \Omega^2 - (m_2/m_1)(k_2/k_1)\Omega^2 \tag{10.56}$$

$$b = 2\zeta_2 (m_2/m_1)\Omega^3 \tag{10.57}$$

$$q = \left(\frac{k_2}{k_1} - \frac{m_2 \Omega^2}{m_1}\right)^2 + 4\zeta_2^2 \Omega^2 \tag{10.58}$$

The quantity, f_0 is the resonance frequency of the mass, m_1, with no absorber, the quantities, k_1, m_1, k_2 and m_2 are defined in Figure 10.12, and ζ_1 and ζ_2 are the critical damping ratios of the suspension of masses m_1 and m_2 respectively. These are defined as:

$$\zeta_1 = \frac{C_1}{C_{c1}} = \frac{C_1}{2\sqrt{k_1 m_1}} \tag{10.59}$$

and

$$\zeta_2 = \frac{C_2}{C_{c1}} = \frac{C_2}{2\sqrt{k_1 m_1}} \tag{10.60}$$

The resonance frequencies of each of the two masses without the other are given by:

$$2\pi f_j = \sqrt{k_j / m_i} \qquad j = 1, 2 \tag{10.61}$$

For a vibration absorber, the frequency, f_2, of the added mass-spring system is usually tuned to coincide with the resonance frequency, f_1, of the system with no absorber, thus causing the mass to vibrate out of phase with the structure, and hence to apply an inertial force opposing the excitation force. When the tuning frequency corresponds to the frequency of excitation and not the resonance frequency of the system without the absorber, the added spring mass system is referred to as a vibration neutraliser and is discussed in the next section.

The two natural frequencies, f_a and f_b, which result from the combination of absorber and machine, may be determined using Figures 10.5(a) and (b) with the following definition of parameters: $W = f_2/f_1$, $M = (m_2/m_1)^{1/2}$ and $\Omega = f_i/f_1$ where the subscript $i = a$ in Figure 10.5(a) and $i = b$ in Figure 10.5(b). Alternatively, Equations (10.36) to (10.38) may be used, noting the different definition of f_1 needed to use those equations.

The larger the mass ratio m_2/m_1, the greater will be the frequency separation of the natural frequencies, f_a and f_b, of the system with absorber from the natural frequency, f_1, of the system without absorber. The displacement amplitude of mass, m_2, is also proportional to the mass ratio, m_1/m_2. Thus, m_2 should be as large as possible.

If the frequency of troublesome vibration is constant, then the resonance frequency, f_2, of the absorber may be tuned to coincide with it and the displacement of mass, m_1, may be reduced to zero. However, as is more usual, if the frequency of troublesome vibration is variable and if no damping is added to the system, optimum design requires that the characteristic frequency, f_2, of the added system is made equal to the resonance frequency, f_1, of the original system to be treated, and a small displacement of m_1, is accepted. Alternatively, if damping, C_2, is added in parallel with the spring of the absorber of stiffness, k_2, then optimum tuning (for minimising the maximum displacement of the main mass, m_1, in the frequency domain) requires the following stiffness and damping ratios (Den Hartog, 1956):

$$\frac{k_2}{k_1} = \frac{m_1 m_2}{(m_1 + m_2)^2} = \frac{(m_2/m_1)}{(1 + m_2/m_1)^2} \tag{10.62}$$

$$\zeta_2^2 = \left(\frac{C_2}{C_{c1}}\right)^2 = \frac{3\,(m_2/m_1)^3}{8(1 + m_2/m_1)^3} \tag{10.63}$$

The predicted displacement amplitude of response $|y_1|$ of the mass, m_1 at the system resonance frequency which occurs in the absence of the absorber is:

$$\frac{|y_1|}{|F|} = \frac{1}{k_1}\sqrt{1 + 2(m_1/m_2)} \tag{10.64}$$

In the preceding equation, F is the excitation force and y_1 is the displacement of mass, m_1. A plot showing the effectiveness of an optimum absorber of varying mass is provided in Figure 10.13, where f_0 is the resonance frequency of mass, m_1, with no absorber.

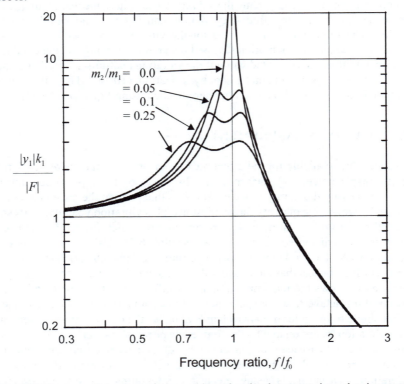

Figure 10.13 Response of absorber main mass with varying absorber mass ratios and optimum values for the stiffness and damping ratio. The main mass, m_1 has no damping. The optimum stiffness ratios, k_2/k_1, corresponding to mass ratios 0.05, 0.1 and 0.25 are 0.0454, 0.0826 and 0.16 respectively. The corresponding optimum critical damping ratios ζ_2 are 0.00636, 0.0168 and 0.0548 respectively.

Note that Den Hartog (1956) derived the above equations on the basis that there was no existing damping in the vibrating machine support system (that is, ζ_1 is assumed to be zero). Soom and Lee (1983) showed that as ζ_1 increases from zero to 0.1, the true optimum for ζ_2 increases from that given by Equation (10.63) by between

2% (for a mass ratio of 0.5) and 7% (for a mass ratio of 0.1) and the true optimum stiffness ratio (k_2/k_1) decreases from that given by Equation (10.57) by between 6% (for a mass ratio of 0.1) and 10% (for a mass ratio of 0.5). Soom and Lee (1983) also showed that there would be little benefit in adding a vibration absorber to a system that already had a critical damping ratio, ζ_1 greater than 0.2.

10.5 VIBRATION NEUTRALIZERS

A vibration neutraliser is of similar construction to a vibration absorber but differs from it in that a vibration neutraliser targets non-resonant vibration whereas a vibration absorber targets resonant vibration. The non-resonant vibration targeted by the vibration neutraliser occurs at a forcing frequency that is causing a structure to vibrate at a non resonant frequency. Thus the vibration neutraliser resonance frequency is made equal to the forcing frequency that is causing the undesirable structural vibration. The same equations as used for the vibration absorber may be used to calculate the reduction in vibration that will occur at the forcing frequency. First the vibration amplitude is calculated without the vibration neutraliser installed using the idealised SDOF system characterised by Equation (10.14). Then the vibration amplitude is calculated with the neutraliser attached using Equation (10.53).

10.6 VIBRATION MEASUREMENT

Transducers are available for the direct measurement of instantaneous acceleration, velocity, displacement and surface strain. In noise-control applications, the most commonly measured quantity is acceleration, as this is often the most convenient to measure. However, the quantity that is most useful is vibration velocity, as its square is related directly to the structural vibration energy, which in turn is often related directly to the radiated sound power (see Section 6.7). Also, most machines and radiating surfaces have a flatter velocity spectrum (see Appendix D) than acceleration spectrum, which means that the use of velocity signals is an advantage in frequency analysis as it allows the maximum amount of information to be obtained using an octave or third-octave filter, or spectrum analyser with a limited dynamic range.

For single frequencies or narrow bands of noise, the displacement, d, velocity, v, and acceleration, a, are related by the frequency, ω(rad/s), as $d\,\omega^2 = v\,\omega = a$. In terms of phase angle, velocity leads displacement by 90° and acceleration leads velocity by 90°. For narrow band or broadband signals, velocity can also be derived from acceleration measurements using electronic integrating circuits. Unfortunately, integration amplifies electronic noise at low frequencies and this can be a problem. On the other hand, deriving velocity and acceleration signals by differentiating displacement signals is generally not practical due primarily to the limited dynamic range of displacement transducers and secondarily to the cost of differentiating electronics.

One alternative, which is rarely used in noise control, is to bond strain gauges to the surface to measure vibration displacement levels. However, this technique will not be discussed further here.

10.6.1 Acceleration Transducers

Vibratory motion for noise-control purposes is most commonly measured with an accelerometer attached to the vibrating surface. The accelerometer most generally used consists of a small piezoelectric crystal, loaded with a small weight and designed to have a natural resonance frequency well above the anticipated excitation frequency range. Where this condition may not be satisfied and consequently a problem may exist involving excitation of the accelerometer resonance, mechanical filters are available which, when placed between the accelerometer base and the measurement surface, minimise the effect of the accelerometer resonance at the expense of the high-frequency response. This results in loss of accuracy at lower frequencies, effectively shifting the ±3 dB error point down in frequency by a factor of five. However, the transverse sensitivity (see below) at higher frequencies is also much reduced by use of a mechanical filter, which in some cases is a significant benefit. Sometimes it may also be possible to filter out the accelerometer resonance response using an electrical filter on the output of the amplifier, but this could effectively reduce the dynamic range of the measurements due to the limited dynamic range of the amplifier.

The mass-loaded piezoelectric crystal accelerometer may be thought of as a one-degree-of-freedom system driven at the base, such as that of case (b) of Figure 10.1. The crystal, which may be loaded in compression or shear, provides the stiffness and system damping as well as a small contribution to the inertial mass, while the load provides the major part of the system inertial mass. As may readily be shown (Tse *et al.*, 1978), the response of such a system driven well below resonance is controlled by the system (crystal) stiffness. Within the frequency design range, the difference $(y - y_1)$ (see Figure 10.1 (b)) between the displacement, y, of the mass mounted on the crystal and the displacement, y_1, of the base of the accelerometer , results in small stresses in the crystal. The latter stresses are detected as induced charge on the crystal by means of some very high-impedance voltage detection circuit, like that provided by an ordinary sound level meter or a charge amplifier. Although acceleration is the measured quantity, integrating circuitry is commercially available so that velocity and even displacement may also be measured.

Referring to Figure 10.1 (b), the difference in displacement, $y - y_1$, is (Tse *et al.*, 1978):

$$y - y_1 = y_1 X^2 / |Z| \tag{10.65}$$

where

$$|Z| = \left[(1 - X^2)^2 + (2X\zeta)^2\right]^{1/2}, \quad \text{and } X = f/f_0 \tag{10.66}$$

In the above equations, X is the ratio of the driving frequency to the resonance frequency of the accelerometer, ζ is the damping ratio of the accelerometer and $|Z|$ is the modulus of the impedance seen by the accelerometer mass, which represents the reciprocal of a magnification factor. The voltage generated by the accelerometer will be proportional to $(y - y_1)$ and, as shown by Equation (10.65), to the acceleration $y_1 X^2$ divided by the modulus of the impedance, $|Z|$.

If a vibratory motion is periodic it will generally have overtones. Alternatively, if it is not periodic, the response may be thought of as a continuum of overtones. In any case, if distortion in the measured acceleration is to be minimal, then it is necessary that the magnification factor be essentially constant over the frequency range of interest. In this case, the difference in displacement of the mass mounted on the crystal and the base of the accelerometer generates a voltage that is proportional to this difference and which, according to Equation (10.65) is also proportional to the acceleration of the accelerometer base. However, as the magnification factor, $1/|Z|$, in Equation (10.65) is a function of frequency ratio, X, it can only be approximately constant by design over some prescribed range and some distortion will always be present. The percent amplitude distortion is defined as:

$$\text{Amplitude distortion} = \left[(1/|Z|) - 1 \right] \times 100\% \tag{10.67}$$

To minimise distortion, the accelerometer should have a damping ratio of between 0.6 and 0.7, giving a useful frequency range of $0 < X < 0.6$.

Where voltage amplification is used, the sensitivity of an accelerometer is dependent upon the length of cable between the accelerometer and its amplifier. Any motion of the connecting cable can result in spurious signals. The voltage amplifier must have a very high input impedance to measure low frequency vibration and not significantly load the accelerometer electrically because the amplifier decreases the electrical time constant of the accelerometer and effectively reduces its sensitivity. Commercially available high impedance voltage amplifiers allow accurate measurement down to about 20 Hz, but are rarely used due to the above-mentioned problems.

Alternatively, charge amplifiers (which, unfortunately, are relatively expensive) are usually preferred, as they have a very high input impedance and thus do not load the accelerometer output; they allow measurement of acceleration down to frequencies of 0.2 Hz; they are insensitive to cable lengths up to 500 m and they are relatively insensitive to cable movement. Many charge amplifiers also have the capability of integrating acceleration signals to produce signals proportional to velocity or displacement. This facility should be used with care, particularly at low frequencies, as phase errors and high levels of electronic noise may be present, especially if double integration is used to obtain a displacement signal. Some accelerometers have in-built charge amplifiers and thus have a low impedance voltage output, which is easily amplified using standard low impedance voltage amplifiers.

The minimum vibration level that can be measured by an accelerometer is dependent upon its sensitivity and can be as low as 10^{-4} m/s^2. The maximum level is dependent upon size and can be as high as 10^6 m/s^2 for small shock accelerometers. Most commercially available accelerometers at least cover the range 10^{-2} to 5×10^4 m/s^2. This range is then extended at one end or the other, depending upon accelerometer type.

The transverse sensitivity of an accelerometer is its maximum sensitivity to motion in a direction at right angles to its main axis. The maximum value is usually quoted on calibration charts and should be less than 5% of the axial sensitivity. Clearly, readings can be significantly affected if the transverse vibration amplitude at the measurement location is an order of magnitude larger than the axial amplitude.

The frequency response of an accelerometer is regarded as essentially flat over the frequency range for which its electrical output is proportional to within ±5% of its mechanical input. The lower limit has been previously discussed. The upper limit is generally just less than one-third of the resonance frequency. The resonance frequency is dependent upon accelerometer size and may be as low as 2,500 Hz or as high as 180 kHz. In general, accelerometers with higher resonance frequencies are smaller in size and less sensitive.

When choosing an accelerometer, some compromise must always be made between its weight and sensitivity. Small accelerometers are more convenient to use; they can measure higher frequencies and are less likely to mass load a test structure and affect its vibration characteristics. However, they have low sensitivity, which puts a lower limit on the acceleration amplitude that can be measured. Accelerometers range in weight from miniature 0.65 grams for high-level vibration amplitude (up to a frequency of 18 kHz) on light weight structures, to 500 grams for low level ground vibration measurement (up to a frequency of 700 Hz). Thus, prior to choosing an accelerometer, it is necessary to know approximately the range of vibration amplitudes and frequencies to be expected as well as detailed accelerometer characteristics, including the effect of various types of amplifier (see manufacturer's data).

10.6.1.1 Sources of Measurement Error

Temperatures above 100°C can result in small reversible changes in accelerometer sensitivity up to 12% at 200°C. If the accelerometer base temperature is kept low using a heat sink and mica washer with forced air cooling, then the sensitivity will change by less than 12% when mounted on surfaces having temperatures up to 400°C. Accelerometers cannot generally be used on surfaces characterised by temperatures in excess of 400°C.

Strain variation in the base structure on which an accelerometer is mounted may generate spurious signals. Such effects are reduced using a shear-type accelerometer and are virtually negligible for piezo-resistive accelerometers.

Magnetic fields have a negligible effect on an accelerometer output, but intense **electric fields** can have a strong effect. The effect of intense electric fields can be minimised by using a differential pre-amplifier with two outputs from the same accelerometer (one from each side of the piezoelectric crystal with the accelerometer casing as a common earth) in such a way that voltages common to the two outputs are cancelled. This arrangement is generally necessary when using accelerometers near large generators or alternators.

If the test object is connected to ground, the accelerometer must be electrically isolated from it, otherwise an **earth loop** may result, producing a high level 50 Hz hum in the resulting acceleration signal.

10.6.1.2 Sources of Error in the Measurement of Transients

If the accelerometer charge amplifier lower limiting frequency is insufficiently low for a particular transient or very low frequency acceleration waveform, then the phenomenon of **leakage** will occur. This results in the waveform output by the charge amplifier not being the same as the acceleration waveform and errors in the peak measurement of the waveform will occur. To avoid this problem, the lower limiting frequency of the pre-amplifier should be less than $0.008/T$ for a square wave transient and less than $0.05/T$ for a half-sine transient, where T is the period of the transient in seconds. Thus, for a square wave type of pulse of duration 100 ms, the lower limiting frequency set on the charge amplifier should be 0.1 Hz.

Another phenomenon, called **zero shift**, that can occur when any type of pulse is measured is that the charge amplifier output at the end of the pulse could be negative or positive, but not zero and can take a considerable time longer (up to 1000 times longer than the pulse duration) to decay to zero. Thus, large errors can occur if integration networks are used in these cases. The problem is worst when the accelerometers are being used to measure transient acceleration levels close to their maximum capability. A mechanical filter placed between the accelerometer and the structure on which it is mounted can reduce the effects of zero shift.

The phenomenon of **ringing** can occur when the transient acceleration that is being measured contains frequencies above the useful measurement range of the accelerometer and its mounting configuration. The accelerometer mounted resonance frequency should not be less than $10/T$, where T is the length of the transient in seconds. The effect of ringing is to distort the charge amplifier output waveform and cause errors in the measurement. The effects of ringing can be minimised by using a mechanical filter between the accelerometer and the structure on which it is mounted.

10.6.1.3 Accelerometer Calibration

In normal use, accelerometers may be subjected to violent treatment, such as dropping, which can alter their characteristics. Thus, the sensitivity should be periodically checked by mounting the accelerometer on a shaker table which either produces a known value of acceleration at some reference frequency or on which a reference accelerometer of known calibration may be mounted for comparison.

10.6.1.4 Accelerometer Mounting

Generally, the measurement of acceleration at low to middle frequencies poses few mechanical attachment problems. For example, for measurements below 5 kHz, an accelerometer may be attached to the test surface simply by using double-sided adhesive tape. For the measurement of higher frequencies, an accelerometer may be attached with a hard epoxy, cyanoacrylate adhesive or by means of a stud or bolt. Use of a magnetic base usually limits the upper frequency bound to about 2 kHz. Beeswax may be used on surfaces that are cooler than 30°C, for frequencies below 10 kHz. Thus, for the successful measurement of acceleration at high frequencies, some care is required to ensure (1) that the accelerometer attachment is firm, and (2) that the mass loading provided by the accelerometer is negligible. With respect to the former it is suggested that the manufacturer's recommendation for attachment be carefully followed. With respect to the latter the following is offered as a guide.

Let the mass of the accelerometer be m_a grams. When the mass, m_a, satisfies the appropriate equation that follows, the measured vibration level will be at most 3 dB below the unloaded level due to the mass loading by the accelerometer. For thin plates:

$$m_a \leq 3.7 \times 10^{-4} \, (\rho c_L h^2 / f) \quad \text{(grams)} \tag{10.68}$$

and for massive structures:

$$m_a \leq 0.013 (\rho c_L^2 D_a / f^2) \quad \text{(grams)} \tag{10.69}$$

In the preceding equations ρ is the material density (kg/m³), h is the plate thickness (mm), D_a is the accelerometer diameter (mm), f is the frequency (Hz) and c_L is the longitudinal speed of sound (m/s). For the purposes of Equations (10.68) and (10.69) it will be sufficient to approximate c_L as $\sqrt{E/\rho}$ (see Appendix B).

As a general guide, the accelerometer mass should be less than 10% of the *dynamic* mass (or modal mass) of the vibrating structure to which it is attached. The effect of the accelerometer mass on any resonance frequency, f_s, of a structure is given by:

$$f_m = f_s \sqrt{\frac{m_s}{m_s + m_a}} \tag{10.70}$$

where f_m is the resonance frequency with the accelerometer attached, m_a is the mass of the accelerometer and m_s is the dynamic mass of the structure (often approximated as the mass in the vicinity of the accelerometer). One possible means of accurately determining a structural resonance frequency would be to measure it with a number of different weights placed between the accelerometer and the structure, plot measured resonance frequency *versus* added mass and extrapolate linearly to zero added mass.

If mass loading is a problem, an alternative to an accelerometer is to use a laser doppler velocimeter, (see Section 10.6.2).

10.6.1.5 Piezo-resistive Accelerometers

An alternative type of accelerometer is the piezo-resistive type, which relies upon the measurement of resistance change in a piezo-resistive element (such as a strain gauge) subjected to stress. Piezo-resistive accelerometers are less common than piezoelectric accelerometers and generally are less sensitive by an order of magnitude for the same size and frequency response. Piezo-resistive accelerometers are capable of measuring down to d.c. (or zero frequency), are easily calibrated (by turning upside down), and can be used effectively with low impedance voltage amplifiers. However, they require a stable d.c. power supply to excite the piezo-resistive element (or elements).

10.6.2 Velocity Transducers

Measurement of velocity provides an estimate of the energy associated with structural vibration; thus, a velocity measurement is often a useful parameter to quantify sound radiation.

Velocity transducers are generally of three types. The least common is the non-contacting magnetic type consisting of a cylindrical permanent magnetic on which is wound an insulated coil. As this type of transducer is only suitable for relative velocity measurements between two surfaces or structures, its applicability to noise control is limited; thus, it will not be discussed further.

The most common type of velocity transducer consists of a moving coil surrounding a permanent magnet. Inductive electromotive force (EMF) is induced in the coil when it is vibrated. This EMF (or voltage signal) is proportional to the velocity of the coil with respect to the permanent magnet. In the 10 Hz to 1 kHz frequency range, for which the transducers are suitable, the permanent magnet remains virtually stationary and the resulting voltage is directly proportional to the velocity of the surface on which it is mounted. Outside this frequency range the electrical output of the velocity transducer is not proportional to velocity. This type of velocity transducer is designed to have a low natural frequency (below its useful frequency range); thus it is generally quite heavy and can significantly mass load light structures. Some care is needed in mounting but this is not as critical as for accelerometers due to the relatively low upper frequency limit characterising the basic transducer.

The preceding two types of velocity transducer generally cover the dynamic range of 1 to 100 mm/s. Some extend down to 0.1 mm/s while others extend up to 250 mm/s. Sensitivities are generally high, of the order of 20 mV/ mm s^{-1}.

Low impedance, inexpensive voltage amplifiers are suitable for amplifying the signal. Temperatures during operation or storage should not exceed 120°C.

A third type of velocity transducer is the laser vibrometer (sometimes referred to as the laser doppler velocimeter), which is discussed in the next section.

Note that velocity signals can also be obtained by integrating accelerometer signals, although this often causes low-frequency electronic noise problems and signal phase errors.

10.6.3 Laser Vibrometers

The laser vibrometer is a specialised and expensive item of instrumentation that uses one or more laser beams to measure the vibration of a surface without any hardware having to contact the surface. They are much more expensive than other transducers but their application is much wider. They can be used to investigate the vibration of very hot surfaces on which it is not possible to mount any hardware and very lightweight structures for which the vibration is affected by any attached hardware.

Laser vibrometers operate on the principle of the detection of the Doppler shift in frequency of laser light that is scattered from a vibrating test object. The object scatters or reflects light from the laser beam, and the Doppler frequency shift of this scattered light is used to measure the component of velocity which lies along the axis of the laser beam. As the laser light has a very high frequency, its direct demodulation is not possible. An optical interferometer is therefore used to mix the scattered light with a reference beam of the same original frequency as the scattered beam before it encountered the vibrating object. A photo-detector is used to measure the intensity of the combined light, which has a beat frequency equal to the difference in frequency between the reference beam and the beam that has been reflected from the vibrating object. For a surface vibrating at many frequencies simultaneously, the beat frequency will contain all of these frequency components in the correct proportions, thus allowing broadband measurements to be made and then analysed in very narrow frequency bands.

Due to the non-contact nature of the laser vibrometer, it can be set up to scan surfaces, and three laser heads may be used simultaneously to scan a surface and evaluate the instantaneous vibration along three orthogonal axes over a wide frequency range, all within a matter of minutes. Sophisticated software provides maps of the surface vibration at any frequency specified by the user.

Currently available laser vibrometer instrumentation has a dynamic range typically of 80 dB or more. Instruments can usually be adjusted using different processing modules so that the minimum and maximum measurable levels can be varied, while maintaining the same dynamic range. Instruments are available that can measure velocities up to 20 m/s and down to 1 μm/sec (although not with the same processing electronics) over a frequency range from DC up to 20 MHz.

Laser vibrometers are also available for measuring torsional vibration and consist of a dual beam which is shone on to a rotating shaft. Each back-scattered laser beam is Doppler shifted in frequency by the shaft surface velocity vector in the beam direction. The velocity vector consists of both rotational and lateral vibration components. The processing software separates out the rotational component by taking the difference of the velocity components calculated by the doppler frequency shift of each of the two beams. The DC part of the signal is the shaft rpm and the AC part is the torsional vibration.

10.6.4 Instrumentation Systems

The instrumentation system that is used in conjunction with the transducers just described, depends upon the level of sophistication desired. Overall or octave band vibration levels can be recorded in the field using a simple vibration meter. If more detailed analysis is required, a portable spectrum analyser can be used. Alternatively, if it is preferable to do the data analysis in the laboratory, samples of the data can be recorded using a high quality DAC ot DAR recorder (see Section 3.11) and replayed through the spectrum analyser. This latter method has the advantage of enabling one to re-analyse data in different ways and with different frequency resolutions, which is useful when diagnosing a particular vibration problem.

10.6.5 Units of Vibration

It is often convenient to express vibration amplitudes in decibels. The International Standards Organisation has recommended that the following units and reference levels be used for acceleration and velocity. Velocity is measured as a root mean square (r.m.s.) quantity in millimetres per second and the level reference is one nanometre per second (10^{-6} mm/s) The velocity level expression, L_v, is:

$$L_v = 20 \log_{10}(v/v_{ref}); \qquad v_{ref} = 10^{-6} \, \text{mm s}^{-1} \qquad (10.71)$$

Acceleration is measured as an r.m.s. quantity in metres per second^2 (m/s^2) and the level reference is one micrometre per second squared (10^{-6} m/s^2). The acceleration level expression, L_a, is:

$$L_a = 20 \log_{10}(a/a_{ref}) \qquad a_{ref} = 10^{-6} \, \text{ms}^{-2} \qquad (10.72)$$

Although there is no standard for displacement, it is customary to measure it as a peak to peak quantity, d, in micrometres (μm) and use a level reference of one picometre(10^{-6} mm). The displacement level expression, L_d, is:

$$L_d = 20 \log_{10}(d/d_{ref}) \qquad d_{ref} = 10^{-6} \, \mu\text{m} \qquad (10.73)$$

When vibratory force is measured in dB, the standard reference quantity is 1 μN. The force level expression is then:

$$L_f = 20 \log_{10}(F/F_{ref}) \qquad F_{ref} = 1 \, \mu\text{N} \qquad (10.74)$$

10.7 DAMPING OF VIBRATING SURFACES

10.7.1 When Damping is Effective and Ineffective

In this section, the question of whether or not to apply some form of damping to a vibrating surface for the purpose of noise control is considered. Commercially available damping materials take many forms but generally, they are expensive and they may be completely ineffective if used improperly. Provided that the structure to be damped is vibrating resonantly, these materials generally will be very effective in damping relatively lightweight structures, and progressively less effective as the structure becomes heavier.

If the structure is driven mechanically by attachment to some other vibrating structure, or by impact of solid materials, or by turbulent impingement of a fluid, then the response will be dominated by resonant modes and the contribution due to forced modes, as will be explained, will be negligible. Damping will be effective in this case and the noise reduction will be equal to the reduction in surface vibration level.

Damping will be essentially ineffective in all other cases where the structure is vibrating in forced (or non-resonant) response.

Structures of any kind have preferred patterns known as modes of vibration to which their vibration conforms. A modal mass, stiffness and damping may be associated with each such mode, which has a corresponding resonance frequency at which only a small excitation force is required to make the structure vibrate strongly. Each such mode may conveniently be thought of as similar to the simple one-degree-of-freedom oscillator of Figure 10.1(a), in which the impedance of the base is infinite and its motion is nil. In general, many modes will be excited at once, in which case the response of a structure may be thought of as the collective responses of as many simple one-degree-of-freedom oscillators (Pinnington and White, 1981).

The acoustic load, like an additional small force applied to the mass of Figure 10.1(a), which is presented to a mechanically or acoustically driven panel or structural surface vibrating in air, is generally so small compared to the driving force that the surface displacement, to a very good approximation, is independent of the acoustic load. The consequence is that the modal displacement response of the surface determines its radiation coupling to the acoustic field (load). At frequencies well above (modal) resonance the displacement is independent of damping. In this high-frequency range the system response is said to be mass controlled.

In the consideration of the response of an extended system, such as a panel or structural surface subjected to distributed forces, a complication arises; it is possible to drive structural modes, when the forcing distribution matches the modal displacement distribution, at frequencies other than their resonance frequencies. The latter phenomenon is referred to as forced response. For example, in the mass-controlled frequency range of a panel (see Section 8.2.4), the modes of the panel are driven in forced response well above their resonance frequencies by an incident acoustic wave; their responses are controlled by their modal masses and are essentially independent of the damping.

If the acoustic radiation of a surface or structure is dominated by modes driven well above resonance in forced response, then the addition of damping will have very little or no effect upon the sound produced. For example, if a panel is excited by an incident sound field, forced modes will be strongly driven and will contribute most to the radiated sound, although resonant modes may dominate the apparent vibration response.

When a structure is excited with a single frequency or a band of noise, the resulting vibration pattern will be a superposition of many modes, some vibrating at or very near to their resonance frequencies and some vibrating at frequencies quite different from their resonance frequencies. The forced vibration modes will generally have much lower vibration amplitudes than the resonant modes. However, at frequencies below the surface-critical frequency (see Section 8.2.1) the sound-radiating efficiency of the forced vibration modes will be unity, and thus much greater than the efficiency of the large-amplitude resonant vibration modes. Thus a reduction of vibration levels, for example by the addition of damping, is not always associated with a reduction of radiated sound. In this case, damping as a means of noise control may be ineffective.

It is possible, and indeed is a common occurrence in the case of airborne sound excitation, that the vibration which is measured is controlled by modes which radiate sound poorly, while the sound radiation is controlled by modes which radiate efficiently and are of very much lower vibration level. In this case, the modes responsible for the sound radiation cannot be detected in the vibration because they are overwhelmed by the resonant modes, which radiate very much less efficiently, but vibrate more vigorously. For example, the addition of a damping material to a panel will generally have only a small effect upon transmission of sound through the panel, except in certain narrow ranges of resonant response where the effect will be large (see Section 8.2.4). These considerations suggest the exercise of caution in the use of surface damping as a means for noise control.

10.7.2 Damping Methods

Damping of sheet metal structures can be accomplished by the application of a damping material to the metal sheet, such as is used on car bodies. Many types of damping are available from various manufacturers for this purpose. They may take the form of tapes, sheets or sprays, which may be applied like paint. They all make use of some non-hardening, visco-elastic material. For optimum results, the weight of the layer of damping material should be at least equal to that of the base panel.

Damping materials can be applied more efficiently and effectively using a laminated construction (see Figure 10.14) of one or more thin sheet metal layers, each separated by a visco-elastic layer, the whole being bonded together. Very thin layers (approximately 0.4 mm) of visco-elastic material are satisfactory in these constrained-layer systems (see Cremer *et al.*, 1988, pp. 246-255). The greatest vibration reduction of the base structure occurs when the sheet metal constraining layer is equal in thickness to the base structure. For damping high frequency vibration, the visco-elastic damping layer should be stiffer than for damping low frequency vibration. A detailed design procedure for constrained layer damping is provided by Mead (1998).

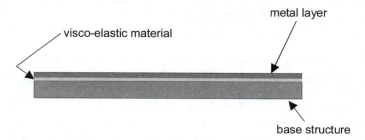

Figure 10.14 Constrained layer damping construction.

Rivetted metal constructions provide more damping than welded constructions. The damping mechanism is a combination of interfacial friction and air pumping through narrow gaps, although work on structures in a vacuum (Beranek, 1971) indicates that the latter mechanism dominates. Thus damped panels can be formed of layered materials that are rivetted, bolted or spot glued together.

10.8 MEASUREMENT OF DAMPING

Damping is associated with the modal response of structures or acoustic spaces; thus, the discussion of Section 7.3.2, in connection with modal damping in rooms, applies equally well to damping of modes in structures and need not be repeated here. In particular, the relationships between loss factor, η, quality factor, Q, and damping ratio, ζ, are the same for modes of rooms and structures. However, whereas the modal density of rooms increases rapidly with increasing frequency and investigation of individual modes is only possible at low frequencies, the modal density of structures such as panels is constant, independent of frequency, so that in the case of structures, investigation of individual modes is possible at all frequencies.

Damping takes many forms but viscous and hysteretic damping, described in Section 10.2.1, are the most common. As shown, they can be described relatively simply analytically, and consequently they have been well investigated. Viscous damping is proportional to the velocity of the structural motion and has the simplest analytical form. Viscous damping is implicit in the definition of the damping ratio, ζ, and is explicitly indicated in Figure 10.1 by the introduction of the dashpot. Damping of modes in rooms is well described by this type of damping (see Section 7.3.2).

Hysteretic (or structural) damping has also been recognised and investigated in the analysis of structures with the introduction of a complex elastic modulus (see Section 10.2.1). Hysteretic damping is represented as the imaginary part of a complex elastic modulus of elasticity of the material, introduced as a loss factor η, such that the elastic modulus E is replaced with $E(1 + j\eta)$. Hysteretic damping is thus proportional to displacement and is well suited to describe the damping of many, though not all, mechanical structures.

For the purpose of loss factor measurement, the excitation of modes in structures may be accomplished either by the direct attachment of a mechanical shaker or by shock excitation using a hammer. When direct attachment of a shaker is used, the coupling between the shaker and the driven structure is strong. In the case of strong coupling, the mass of the shaker armature and shaker damping become part of the oscillatory system and must be taken into account in the analysis. Alternatively, instrumented hammers are available, which allow direct measurement and recording of the hammer impulse applied to the structure. This information allows direct determination of the structural response and loss factor.

Hysteretic damping can be determined by a curve fitting technique using the experimentally determined frequency response function (Ewins, 1984); see Appendix D. For lightweight or lightly damped structures, this method is best suited to the use of instrumented hammer excitation, which avoids the shaker coupling problem mentioned above.

If a simpler, though less accurate, test method is sufficient, then one of the methods described in Section 7.3.2 may suffice. With reference to the latter section, if the reverberation decay method (making use of Equations (7.23) and (7.24)) is used, then it is important to avoid the problems associated with strong coupling mentioned above to ensure that measured damping is controlled by the structural damping which is to be measured and not by the damping of the excitation device.

The problems of strong coupling may be avoided by arranging to disconnect the driver from the structure when the excitation is shut off using a fuse arrangement. Alternatively, either a hammer or a non-contacting electromagnetic coil, which contains no permanent magnet, may be used to excite the structure. In the latter case the structure will be excited at twice the frequency of the driving source.

When frequency band filters are used to process the output from the transducer used to monitor the structural vibration, it is important to ensure that the filter decay rate is much faster than the decay rate that is to be measured so that the filter decay rate does not control the measured structural vibration decay rate. Typically this means that the following relation must be satisfied, where B is the filter bandwidth (Hz):

$$BT_{60} \geq 16 \tag{10.75}$$

If the steady-state determination of the modal bandwidth is used to determine the damping, making use of Equation (7.23), then it is necessary that the excitation force is constant over the frequency range of the modal bandwidth, Δf. The measurement requires that the frequency of resonance is determined and that the modal response at the 3 dB down points below and above resonance may be identified (using a sinusoidal excitation signal), as shown in Figure 10.15. In some cases, better results are obtained by using the bandwidth at the 7, 10 or 12.3 dB down points. In these cases the value of Δf used in Equation (7.23) is one-half, one-third or one-quarter, respectively, of the measured bandwidth, as illustrated for the 7 dB down point in Figure 10.15.

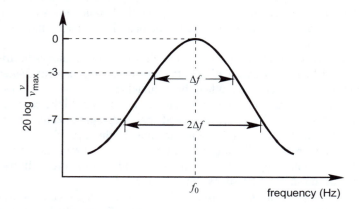

Figure 10.15 Determining system damping from frequency response function (FRF) bandwidth measurements.

The bandwidth may also be determined directly from the magnitude of the frequency response function in the vicinity of the resonance frequency, determined as the ratio of the structural acceleration response to the excitation force (Ewins, 1984).

In this latter case, it is not important that the excitation force be constant over the modal bandwidth and an impact hammer is often used as the excitation source.

Determination of the logarithmic decrement, δ (see Equation (7.23)) is one of the oldest methods of determining damping and depends upon determination of successive amplitudes of a vibrating system as the vibration decays after switching off the excitation source. If A_i is the amplitude of the ith cycle and A_{i+n} is the amplitude n cycles later, then the logarithmic decrement is:

$$\delta = \frac{1}{n} \log_e\left(\frac{A_i}{A_{i+n}}\right) = \frac{2\pi\zeta}{\sqrt{1-\zeta^2}} \approx \pi\eta \approx \frac{\pi}{Q} \qquad (10.76a,b,c,d)$$

where use has been made of Equation (7.23).

As a result of damping, the strain response of a structure lags behind the applied force by a phase angle, ε. Thus another measure of damping is the tan of this phase angle and this is related to the loss factor by:

$$\eta = \tan\varepsilon \qquad (10.77)$$

Another measure of damping is the SDC (specific damping capacity), which is calculated from the amplitudes of two successive cycles of a decaying vibration (following switching off of the excitation source). This cannot be directly related to the forms of damping measure mentioned above, and is defined as the percent energy dissipated in a single oscillation cycle as :

$$SDC = \frac{(A_n^2 - A_{n+1}^2)}{A_n^2} \times 100\% \qquad (10.78)$$

CHAPTER ELEVEN

Sound Power and Sound Pressure Level Estimation Procedures

LEARNING OBJECTIVES

In this chapter, procedures are outlined for estimating the noise radiated by:

- fans;
- compressors;
- cooling towers;
- pumps;
- fluid jets;
- control valves;
- fluid flow in pipes;
- boilers;

- turbines;
- internal combustion engines;
- furnaces;
- electric motors;
- generators;
- transformers;
- gears; and
- transportation vehicles.

11.1 INTRODUCTION

In this chapter, means are provided for estimating expected radiated sound power or sound pressure at 1 m, for a variety of types of mechanical equipment. Such information makes consideration of noise control at the design stage possible, a practice which experience has shown to be by far the most cost effective.

The prediction of the sound power level generated by equipment or machinery is generally very difficult, primarily because the possible noise-generating mechanisms are so many and varied in any but the simplest of devices, and the magnitude of the noise generated depends upon the environment presented to the noise source. It is possible to estimate the noise radiated by some aerodynamic noise sources in terms of an acoustic efficiency factor, as a fraction of the total stream power. However, this method cannot be applied in general to other noise-producing processes and mechanisms. Furthermore, it is not possible to make any sweeping simplifying assumptions on thermodynamic principles, because the power radiated as sound is generally only a very small part of the power balance of an operating machine. In short, greater efficiency does not necessarily mean less noise.

At the present time in most instances, quantitative laws for the generation of noise cannot be formulated. However, from the work that has been done and reported thus far, there does seem to emerge a general principle that may be stated as follows. The noise produced by any process seems to be proportional to the rate of change of acceleration of the parts taking part in that process. This has the consequence that of two or more methods of achieving a given end, one can expect the least noise to be produced by the method in which the rates of change of forces are least. Noise produced by the punch press has been shown to follow this rule (Richards 1979, 1981; Semple and Hall, 1981). Helical gears and mill cutters, which maintain continuous rather than interrupted or discontinuous contact, provide other examples (Richards, 1981).

Current noise-prediction schemes are largely empirical and, in a few cases, are augmented with established theoretical considerations. In this chapter a number of these schemes are reviewed. The noise sources considered are common industrial machines. The schemes presented should be useful for initial estimates of expected noise. It is stressed, however, that measured sound power level data are always to be preferred. Thus the procedures described in this chapter should be used as a guide, and only when measured data are not available or when the estimates are not critical.

In the following tables, all sound pressure levels L_p are in decibels relative to 20 μPa and all sound power levels L_w are in decibels relative to 10^{-12} W. Throughout the rest of this chapter, mechanical power in kilowatts will be represented as kW, or in megawatts as MW. Rotational speed in revolutions per minute (rev min^{-1}) will be represented as RPM.

The types of machines and machinery considered in this chapter are listed in Table 11.1, which provides an indication of what is estimated (sound power level or sound pressure level at 1 m) and means commonly used for noise control. The list of equipment considered here is by no means exhaustive. However, it is representative of much of the equipment commonly found in process and power generation plants. Wherever sound power estimation schemes are provided, sound pressure levels may be estimated using the methods of Chapters 7, 8 and 9. Where sound pressure levels at 1 m are given, then sound pressure levels at other distances may be estimated using the methods of Chapters 5 and 6. In general, the latter information is most useful for estimating sound pressure levels at the machinery operator's location, but sound power levels are more useful for estimating sound pressure levels at greater distances.

11.2 FAN NOISE

The fan is a common industrial noise source, which has been well documented (ASHRAE, 1987, 2007). The equations used in the past and reported in past editions of this text are no longer considered reliable (ASHRAE, 2007). The sound power of fans is dependent on the fan type and can vary considerably, depending on the operating point compared to the point of maximum efficiency. Higher air flow and lower static pressure than optimum increases noise levels. On the other hand, higher static pressure and lower air flow increase low frequency rumble noise.

All fans generate a tone at the blade pass frequency (*BPF*) given by:

$$BPF = N_b \times RPM/60 \qquad (11.1)$$

where N_b is the number of blades on the fan. The extent to which the tone is noticeable depends on the fan type and how close it is to its optimum efficiency. Fan types and some of their noise characteristics are listed in Table 11.2.

Table 11.1 Noise sources, prediction schemes and controls

Noise source	Quantity predicted L_p at 1m or L_w	Recommended control
Fans	L_p	Inlet and outlet duct fitted with commercial mufflers; blade passing frequency attenuated with a quarter wave tube.
Air compressors, small and large	L_p or L_w	As for fans, plus an enclosure built around the compressor
Reciprocating		
Centrifugal		
Axial		
Compressors for refrigeration units		As for fans, & enclosure
Reciprocating	L_w	
Rotary screw	L_w	
Centrifugal	L_w	
Cooling towers		Commercial mufflers
Propeller	L_p	
Centrifugal	L_p	
Pumps	L_p	Enclosure
Jet noise	L_p or L_w	Muffler
Gas vents	L_p	Muffler
Steam vents	L_p	Muffler
Control valves		Lagging, staging
Gases (including steam)	L_p or L_w	
Liquids	L_p or L_w	
Pipe flow		Mufflers, lagging
Gases	L_p	
Boilers	L_p	Enclosure
Steam and gas turbines	L_p	Enclosure
Reciprocating diesel and gas engines	L_p	Enclosure, muffler, barrier
Furnaces	L_p or L_w	Smaller fuel/air jet diameter (even if more jets are needed); fuel oil instead of gas
Electric motors	L_p	Enclosure
Generators	L_p	Enclosure
Transformers	L_w	Enclosure
Gears	L_w	Enclosure

Table 11.2 Some characteristics of various fan types (based on information in ASHRAE (2007))

Fan type	Description	Noise characteristics
Centrifugal - housed	Fan with housing in common use.	
Forward curved	Blades curve in same direction as rotation.	BPF less prominent and higher frequency than other fan types. low-frequency rumble from turbulence generated by blade tips if there is not 5 duct diameters of straight discharge duct or if fan is operating at lower volume flow than optimal efficiency.
Backward inclined	Blades curve in opposite direction to rotation.	Louder at BPF than FC fan for same duty. More energy efficient than FC. BPF increases in prominence with increasing fan speed. Quieter than FC at higher frequencies. Quieter below BPF as well.
Airfoil	Blades have an airfoil shape to increase efficiency.	Louder at BPF than FC fan for same duty. More energy efficient than FC. BPF increases in prominence with increasing fan speed. Quieter than FC at higher frequencies. Quieter below BPF as well.
Centrifugal - no housing	Discharges directly into plenum chamber. Inlet bell on chamber wall adjacent to fan inlet.	Substantially lower discharge noise but fan plenum must be correct size and acoustically treated. Discharge should not be in line with ductwork or BPF sound will be magnified.
Vaneaxial	Used in applications where the higher frequency noise can be managed with mufflers.	Lowest low-frequency sound levels. Noise is a function of blade tip speed and inlet airflow symmetry.
Propeller	Most commonly used on condensers and for power exhausts.	Noise dominated by low frequencies. BPF is usually low-frequency and level at BPF is sensitive to inlet obstructions. Shape of fan inlet also affects sound levels.

Manufacturer's data should always be used for fan sound power levels, These data are typically sound power levels of inlet, discharge and casing. The inlet and discharge levels are those that are inside the duct. To use these levels to calculate the sound power radiated externally through the fan casing and adjacent ductwork, the corrections listed in Table 11.3 may be used, or alternatively see Section 9.12. These corrections assume standard rectangular ductwork lined on the inside, beginning a short distance from the fan. To calculate the sound power emerging from the end of the duct, the attenuation due to duct linings, duct end reflections, duct bends and plenum chambers must be taken into account as discussed in Chapter 9.

Table 11.3 Octave band adjustments for sound power radiated by fan housings and adjacent ductwork

Octave band centre frequency (Hz)	Value to be subtracted from calculated in-duct sound power level, L_w (dB)
63	0
125	0
250	5
500	10
1000	15
2000	20
4000	22
8000	25

Data from Army, Air Force and Navy, USA (1983a).

In designing air handling systems, fan noise can be minimised by minimising air flow resistance and system pressure losses. In some cases the BPF tone or its harmonics may be amplified over the manufacturer's specifications due to a number of causes listed below:

- the BPF or its harmonics may correspond to an acoustic resonance in the ductwork;
- inlet flow distortions;
- unstable, turbulent or swirling inlet flow; and
- operation of an inlet volume control damper.

Fan noise in duct systems can be minimised by avoiding the conditions listed above by ensuring the following:

- sizing ductwork and duct elements for low air velocities
- avoiding abrupt changes in duct cross-sectional area or direction and providing smooth airflow through all duct elements;
- providing 5 to 10 duct diameters of straight ductwork between duct elements;
- using variable speed fans instead of dampers for flow control; and
- if dampers are used, locating them a minimum of 3 (preferably 5 to 10) duct diameters away from room air devices.

11.3 AIR COMPRESSORS

11.3.1 Small Compressors

Air compressors are a common source of noise. In this section several estimation procedures are presented for various types of compressor. If the compressors are of small to medium size then the data presented in Table 11.4 may be used in the power range shown, to estimate sound pressure levels at 1 m. In most cases the values will be conservative; that is, a little too high.

Table 11.4 Estimated sound pressure levels of small air compressors at 1 m distance in dB re 20 μPa

Octave band centre frequency (Hz)	Air compressor power (kW)		
	Up to 1.5	2–6	7–75
31.5	82	87	92
63	81	84	87
125	81	84	87
250	80	83	86
500	83	86	89
1000	86	89	92
2000	86	89	92
4000	84	87	90
8000	81	84	87

Data from Army, Air Force and Navy, USA (1983a).

11.3.2 Large Compressors (Noise Levels within the Inlet and Exit Piping)

The following equations may be used for estimating the sound power levels generated within the exit piping of large centrifugal, axial and reciprocating compressors (Heitner, 1968).

11.3.2.1 Centrifugal Compressors (Interior Noise Levels)

The overall sound power level measured at the exit piping inside the pipe is given by:

$$L_w = 20 \log_{10} kW + 50 \log_{10} U - 46 \quad \text{(dB re } 10^{-12} \text{ W)} \tag{11.2}$$

where U is the impeller tip speed (m/s) ($30 < U < 230$), and kW is the power of the driver motor (in kW). The frequency of maximum noise level is:

$$f_p = 4.1 U \quad \text{(Hz)} \tag{11.3}$$

The sound power level in the octave band containing f_p is taken as 4.5 dB less than the overall sound power level. The spectrum rolls off at the rate of 3 dB per octave above and below the octave of maximum noise level.

11.3.2.2 Rotary or Axial Compressors (Interior Noise Levels)

The following procedure may be used for estimating the overall sound power level at the exit piping within the pipe:

$$L_w = 68.5 + 20 \log_{10} kW \quad \text{(dB re } 10^{-12} \text{ W)} \tag{11.4}$$

The frequency of maximum noise output is the second harmonic, or:

$$f_p = B(RPM)/30 \quad \text{(Hz)} \tag{11.5}$$

where B is the number of blades on the compressor. The spectrum is obtained from the following equations:

For the 63 Hz octave:

$$L_w = 76.5 + 10 \log_{10} kW \quad \text{(dB re } 10^{-12} \text{ W)} \tag{11.6}$$

For the 500 Hz octave:

$$L_w = 72 + 13.5 \log_{10} kW \quad \text{(dB re } 10^{-12} \text{ W)} \tag{11.7}$$

For the octave containing f_p:

$$L_w = 66.5 + 20 \log_{10} kW \quad \text{(dB re } 10^{-12} \text{ W)} \tag{11.8}$$

For the octave containing f_h

$$L_w = 72 + 13.5 \log_{10} kW \quad \text{(dB re } 10^{-12} \text{ W)} \tag{11.9}$$

where

$$f_h = f_p^2/400 \tag{11.10}$$

To plot the total spectrum, a straight line is drawn between the L_w for the 63 Hz octave and the L_w for the 500 Hz octave on an L_w vs $\log_{10} f$ plot. A smooth curve is drawn from the L_w value for the 500 Hz octave through L_w values for the octave containing f_p and the octave containing f_h. The slope is continued as a straight line beyond the f_h octave.

Example 11.1

Estimate the interior sound power level spectrum for an axial compressor of 15 blades and 80 kW power turning at 3000 rev min⁻¹.

Solution

1. Use Equation (11.5) to calculate f_p :

$$f_p = 15(3000)/30 = 1500 \text{ Hz}$$

2. Use Equation (11.10) to calculate f_h:

$$f_h = 1500^2/400 = 5600 \text{ Hz}$$

3. Use Equations (11.6)–(11.9) to calculate power level in bands:

     For the 63 Hz octave band:

$$L_w = 76.5 + 10 \log_{10} 80 = 95.5 \text{ dB}$$

     For the 500 Hz octave band:

$$L_w = 72 + 13.5 \log_{10} 80 = 97.7 \text{ dB}$$

     f_p lies in the 2000 Hz octave band (see Table 1.2):

$$L_w = 66.5 + 20 \log_{10} 80 = 104.6 \text{ dB}$$

     f_h lies in the 4000 Hz octave band:

$$L_w = 72 + 13.5 \log_{10} 80 = 97.7 \text{ dB}$$

4. Using the values calculated above, follow the procedure outlined in the text to sketch the estimated spectrum (L_w vs $\log_{10} f$ plot).

11.3.2.3 Reciprocating Compressors (Interior Noise)

The overall sound power level within the exit piping of a reciprocating compressor can be calculated using:

$$L_w = 106.5 + 10 \log_{10} kW \quad (\text{dB re } 10^{-12} \text{ W}) \quad (11.11)$$

To determine the spectrum values, the octave band which contains the fundamental frequency, $f_p = B(RPM)/60$, where B is the number of cylinders of the machine, is determined. The power level in this band is taken as 4.5 dB less than the overall power level calculated using Equation (11.11). The levels in higher and lower octave bands decrease by 3 dB per octave. Implicit in the use of Equation (11.11) is the assumption that the radiated sound power is distributed over the first 15 m of downstream pipe.

11.3.3 Large Compressors (Exterior Noise Levels)

The sound power radiated by the compressor casing and exit pipe walls can be calculated using the equations for L_w inside the piping, which are included in Section 11.3.2, and subtracting the transmission loss of the casing and exit piping, which is calculated using the following equation (Heitner, 1968):

$$TL = 17 \log_{10}(mf) - 48 \quad \text{(dB)} \tag{11.12}$$

where m is the surface weight (kg/m^2) of the pipe wall, and f is the octave band centre frequency (Hz). This formula represents a simplification of a complex problem, and is based on the assumption of adequate structural rigidity. Thus, for large diameter, thin-walled, inadequately supported pipes, the transmission loss may be less than given above.

Alternatively, the following equations may be used to calculate the overall external sound power levels directly (Edison Electric Institute, 1978). Rotary and reciprocating compressors (including partially muffled inlets):

$$L_w = 90 + 10 \log_{10} kW \quad \text{(dB re } 10^{-12} \text{ W)} \tag{11.13}$$

Centrifugal compressors (casing noise excluding air inlet noise):

$$L_w = 79 + 10 \log_{10} kW \quad \text{(dB re } 10^{-12} \text{ W)} \tag{11.14}$$

Centrifugal compressors (un-muffled air inlet noise excluding casing noise):

$$L_w = 80 + 10 \log_{10} kW \quad \text{(dB re } 10^{-12} \text{ W)} \tag{11.15}$$

Octave band levels may be derived from the overall levels by subtracting the corrections listed in Table 11.5.

11.4 COMPRESSORS FOR CHILLERS AND REFRIGERATION UNITS

The compressor is usually the dominant noise source in a refrigeration unit, so that it is generally sufficient to consider noise generation from this source alone when considering a packaged chiller. Three types of compressor will be considered: centrifugal, rotary screw and reciprocating. Sound pressure levels measured at 1 m are listed in Table 11.6 for these machines; these levels will not be exceeded by 90–95% of commercially available machines. The machines are identified in the table both by type and by cooling capacity. Speed variations between the different commercially available units are insignificant.

Table 11.5 Octave band corrections for exterior noise levels radiated by compressors

Octave band centre frequency (Hz)	Correction (dB)		
	Rotary and reciprocating	Centrifugal casing	Centrifugal air inlet
31.5	11	10	18
63	15	10	16
125	10	11	14
250	11	13	10
500	13	13	8
1000	10	11	6
2000	5	7	5
4000	8	8	10
8000	15	12	16

Data from Edison Electric Institute (1978).

Table 11.6 Estimated sound pressure levels of packaged chillers at one metre (dB re 20 μPa). These levels are generally higher than observed (see manufacturer's data)

Type and capacity of machine	Octave band centre frequency (Hz)								
	31.5	63	125	250	500	1000	2000	4000	8000
Reciprocating compressors									
35–175 kW	79	83	84	85	86	84	82	78	72
175–615 kW	81	86	87	90	91	90	87	83	78
Rotary screw compressors									
350–1050 kW	70	76	80	92	89	85	80	75	73
Centrifugal compressors									
Under 1750 kW	92	93	94	95	91	91	87	80	–
≥1750 kW	92	93	94	95	93	98	98	93	87

Data from Army, Air Force and Navy, USA (1983a).

11.5 COOLING TOWERS

Various types of cooling towers are illustrated in Figure 11.1, and the corresponding estimated overall sound power levels are given by the following equations:

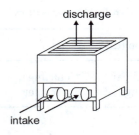

(a) Centrifugal fan,
    blow-through type

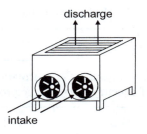

(b) axial flow

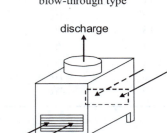

(c) Induced draft
    propeller-type

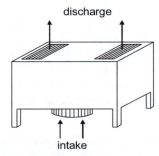

(d) Forced draft, propellor-type
    'underflow'

Figure 11.1 Principal types of cooling tower (Army, Air Force and Navy, USA 1983a).

1. Propeller-type cooling towers:

    Fan power up to 75 kW:

    $$L_w = 100 + 8 \log_{10} kW \quad \text{(dB re } 10^{-12} \text{ W)} \tag{11.16}$$

    Fan power greater than 75 kW:

    $$L_w = 96 + 10 \log_{10} kW \quad \text{(dB re } 10^{-12} \text{ W)} \tag{11.17}$$

    Subtract 8 dB if the fan is operated at half its rated speed.

2. Centrifugal type cooling towers:

    Fan power up to 60 kW:

    $$L_w = 85 + 11 \log_{10} kW \quad \text{(dB re } 10^{-12} \text{ W)} \tag{11.18}$$

    Fan power greater than 60 kW:

    $$L_w = 93 + 7 \log_{10} kW \quad \text{(dB re } 10^{-12} \text{ W)} \tag{11.19}$$

The octave band sound power levels may be calculated by subtracting the corrections listed in Table 11.7 from the appropriate overall values.

Table 11.7 Values (dB) to subtract from overall levels to obtain octave band sound power levels for cooling towers

Octave band centre frequency (Hz)	Propeller type	Centrifugal type
31.5	8	6
63	5	6
125	5	8
250	8	10
500	11	11
1000	15	13
2000	18	12
4000	21	18
8000	29	25

Data from Army, Air Force and Navy, USA (1983a).

The average sound pressure levels at various distances from the tower may be calculated using these numbers and the procedures outlined in Chapters 5 and 6. Table 11.8 gives the approximate corrections to add to the calculated average sound pressure levels to account for directivity effects at distances greater than 6 m from the tower (see also Section 9.15).

Table 11.9 gives estimates for the sound pressure levels close to the intake and discharge openings. Here and in many places in this chapter, the octave band sound power levels should add up logarithmically to the total sound power level and this should agree with the overall L_w (Equations (11.16 to 11.19 in this case). However, the corrections are just integer numbers and the total of the band levels can sometimes be a little different to the total calculated by the relevant equation. In this case the band levels should each be changed by the same number of dB which is equal to the difference between the total calculated by logarithmically summing octave band levels and that calculated by the relevant equation.

In some cases, where the overall level exceeds the sum of the octave band levels, the difference may be attributable to sound energy in octave bands not included in the summation, but this is more the exception than the rule.

11.6 PUMPS

Estimated sound pressure levels generated by a pump, at a distance of 1 m from its surface, as a function of pump power, are presented in Tables 11.10 and 11.11.

Table 11.8 Approximate corrections (dB) to average sound pressure level for directional effects of cooling towers[a]

Type of tower and location of measurement	Octave band centre frequency (Hz)								
	31.5	63	125	250	500	1000	2000	4000	8000
Centrifugal fan blow through type									
Front	+3	+3	+2	+3	+4	+3	+3	+4	+4
Side	0	0	0	-2	-3	-4	-5	-5	-5
Rear	0	0	-1	-2	-3	-4	-5	-6	-6
Top	-3	-3	-2	0	+1	+2	+3	+4	+5
Axial flow, blow through type									
Front	+2	+2	+4	+6	+6	+5	+5	+5	+5
Side	+1	+1	+1	-2	-5	-5	-5	-5	-4
Rear	-3	-3	-4	-7	-7	-7	-8	-11	-3
Top	-5	-5	-5	-5	-2	0	0	+2	+4
Induced draft, propeller type									
Front	0	0	0	+1	+2	+2	+2	+3	+3
Side	-2	-2	-2	-3	-4	-4	-5	-6	-6
Top	+3	+3	+3	+3	+2	+2	+2	+1	+1
"Underflow" forced draft propeller type.									
Any side	-1	-1	-1	-2	-2	-3	-3	-4	-4
Top	+2	+2	+2	+3	+3	+4	+4	+5	+5

[a] These corrections apply only when there are no reflecting or obstructing surfaces that would modify the normal radiation of sound from the tower. Add these corrections to the average sound pressure level calculated. Do not apply these corrections for close-in positions less than 6 m from the tower.

Data from Army, Air Force and Navy, USA (1983a).

11.7 JETS

11.7.1 General Estimation Procedures

Pneumatic devices quite often eject gas (air) in the form of high-pressure jets. Such jets can be very significant generators of noise. The acoustic power generated by a subsonic jet in free space is related to the mechanical stream power by an efficiency factor as follows:

$$W_a = \eta W_m \quad \text{(W)}$$

(11.20)

Table 11.9 Estimated close-in sound pressure levels (dB re 20 μPa) for the intake and discharge openings of various cooling towers

Type of tower and location of measurement	Octave band centre frequency (Hz)								
	31.5	63	125	250	500	1000	2000	4000	8000
Centrifugal fan blow through type									
Intake	85	85	85	83	81	79	76	73	68
Discharge	80	80	80	79	78	77	76	75	74
Axial blow through type									
Intake	97	100	98	95	91	86	81	76	71
Discharge	88	88	88	86	84	82	80	78	76
Propeller fan, induced draft type									
Intake	97	98	97	94	90	85	80	75	70
Discharge	102	107	103	98	93	88	83	78	73

Data from Army, Air Force and Navy, USA (1983a).

Table 11.10 Overall pump sound pressure levels (dB re 20 μPa) at 1 m from the pump

Speed range (rpm)	Drive motor nameplate power	
	Under 75 kW	Above 75 kW
3000–3600	$72 + 10 \log_{10} kW$	$86 + 3 \log_{10} kW$
1600–1800	$75 + 10 \log_{10} kW$	$89 + 3 \log_{10} kW$
1000–1500	$70 + 10 \log_{10} kW$	$84 + 3 \log_{10} kW$
450–900	$68 + 10 \log_{10} kW$	$82 + 3 \log_{10} kW$

Data from Army, Air Force and Navy, USA (1983a).

The stream mechanical power, W_m, in turn, is equal to the convected kinetic energy of the stream, which for a jet of circular cross-section is:

$$W_m = \rho U^3 \pi d^2 / 8 = U^2 \dot{m}/2 \quad \text{(W)} \qquad (11.21a,b)$$

In the above equations, U is the jet exit velocity (m/s), W_a is the radiated acoustic power (W), W_m is the mechanical stream power (W), $\dot{m}$ is the mass flow rate (kg/s), η is the acoustic efficiency of the jet, d is the jet diameter (m), and ρ is the density (kg/m³) of the flowing gas.

Table 11.11 Frequency adjustments for pump sound power levels[a]

Octave band centre frequency (Hz)	Value to be subtracted from overall sound pressure level (dB)
31.5	13
63	12
125	11
250	9
500	9
1000	6
2000	9
4000	13
8000	19

[a] Subtract these values from the overall sound pressure level to obtain octave band sound pressure levels. Data from Army, Air Force and Navy, USA (1983a).

The acoustical efficiency of the jet is approximately (Heitner, 1968):

$$\eta = (T/T_0)^2 (\rho/\rho_0) K_a M^5 \qquad (11.22)$$

for noise radiation due to turbulent flow in both subsonic and supersonic jets and for the range of the quantity $(T/T_0)^2(\rho/\rho_0)$ between 0.1 and 10. In Equation (11.22), ρ_0 is the density (kg/m^3) of the ambient gas, K_a is the acoustical power coefficient and is approximately 5×10^{-5}, M is the stream Mach number relative to the ambient gas, T is the jet absolute temperature ($^\circ$K), and T_0 is the absolute temperature ($^\circ$K) of the ambient gas. If Equation (11.22) gives a value of $\eta > 0.01$, then η is set equal to 0.01.

If the jet pressure ratio (ratio of pressure upstream of the jet to the ambient pressure) is less than 1.89, the jet is subsonic, and noise is only due to turbulent flow with the acoustical efficiency estimated using Equation (11.22). The radiated sound power is then found by substituting this acoustical efficiency value into Equation (11.20).

If the pressure ratio is greater than 1.89, the jet will be choked and the gas exit velocity will equal Mach 1. If the nozzle contraction is followed by an expansion, the gas exit velocity will exceed Mach 1. In both cases there will be shock wave generated noise in addition to turbulence noise. Equations (11.20) to (11.22) are used to calculate the turbulence part of the noise. The efficiency of the shock generated noise is calculated using Figure 11.2 and the associated sound power is calculated using Equation (11.20). The total sound power radiated by the choked jet is the sum of the sound power due to turbulence and that due to shock.

The overall sound power level of the jet (see Equation (1.80)) is:

$$L_w = 10 \log_{10} W_a + 120 \qquad \text{(dB re } 10^{-12} \text{ W)} \qquad (11.23)$$

This quantity can be used with the methods of Chapter 7 to estimate the sound pressure level in, for example, a room characterised by a given room constant. On the

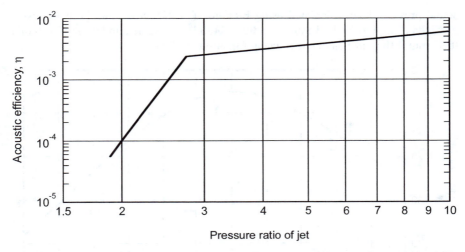

Figure 11.2 Acoustic efficiency of shock noise generated by choked jets (adapted from Heitner, 1968).

other hand, in a free field or close to the jet, the overall sound pressure level is:

$$L_p = L_w + DI - 10 \log_{10}(4\pi r^2) \quad \text{(dB re 20 } \mu\text{Pa)} \tag{11.24}$$

where DI is the jet directivity index and r is the distance (m) from the jet orifice to the observation point.

Values for the directivity index for a jet are given in Table 11.12 as a function of angle from the jet axis.

Table 11.12 Directional correction for jets (data from Heitner, 1968)

Angle from jet axis (°)	Directivity index, DI (dB)	
	Sub-sonic	Choked
0	0	-3
20	+1	+1
40	+8	+6
60	+2	+3
80	-4	-1
100	-8	-1
120	-11	-4
140	-13	-6
160	-15	-8
180	-17	-10

The spectrum shape for the jet is illustrated in Figure 11.3 (Ingard, 1959), where the sound pressure level in each of the octave bands is shown relative to the overall sound pressure level calculated using Equation (11.24). Note that the band levels obtained using Figure 11.3 must be adjusted (all by the same number of dB) so that their sum is the same as obtained using Equation (11.24).

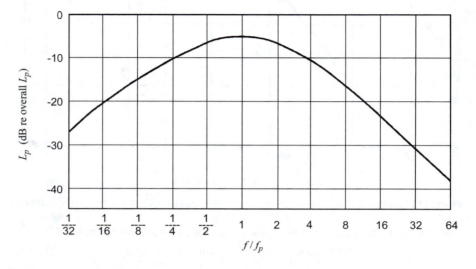

Figure 11.3 Noise spectrum for gas jets (adapted from Heitner, 1968).

In Figure 11.3, the frequency, f_p (Hz), is defined in terms of a Strouhal number, N_s, as:

$$N_s = f_p d / U \qquad (11.25)$$

N_s is generally about 0.2 for subsonic jets; d and U are, respectively, the jet diameter (m) and exit velocity (m/s).

Example 11.2

Gas is to be exhausted under pressure from a 100 mm diameter pipe to the atmosphere at a speed of 230 m/s. The density of the gas is 1.5 kg/m^3 and its temperature is essentially that of the ambient atmosphere into which it is to be exhausted. Determine the overall sound pressure levels at 10 m from the jet orifice and at various angles relative to the direction of the jet.

Solution

1. Determine the jet efficiency using Equation (11.22). First calculate:

$$\eta = (T/T_0)^2 (\rho/\rho_0) K_a M^5$$

The speed of sound has been taken as 343 m/s in the air surrounding the jet and $T = T_0$, $\rho_0 = 1.206$ kg/m^3. From Equation (11.22), using the above values, $\eta = 8.4 \times 10^{-6}$.

2. Calculate the mechanical stream power using Equation (11.21a):

$$W_m = \rho U^3 \pi d^2 / 8 = 71.7 \text{ (kW)}$$

3. Calculate the overall acoustic power using Equation (11.20):

$$W_a = \eta W_m = 0.60 \text{ (W)}$$

4. Calculate the overall sound power level using Equation (11.23):

$$L_w = 10 \log_{10} W_a + 120 = 117.8 \text{ (dB re } 10^{-12} \text{ W)}$$

5. Use Equation (11.24) to determine the overall sound pressure level at 10 m from the jet orifice:

$$L_p = L_w + DI - 10 \log_{10}(4 \pi r^2) = DI + 86.8 \quad \text{(dB re 20 } \mu\text{Pa)}$$

6. Use Table 11.12 and the equation of (5) above to construct the following table:

Example 11.2, Table 1

Angle from jet axis (°)	Overall sound pressure level (dB re 20 µPa)
0	87
20	88
40	95
60	89
80	83
100	79
120	76
140	74
160	72
180	70

7. Determine the octave band of maximum sound pressure level. First calculate the spectrum peak frequency, f_p, using the Strouhal number, $N_s = 0.2$ and Equation (11.25):

$$f_p = 0.2 U / d \quad \text{(Hz)} = 0.2 \times 230 / 0.1 = 460 \text{ Hz}$$

Reference to Table 1.2 shows that this frequency lies in the 500 Hz octave band.

8.   Use Figure 11.3 and the above information to construct the following table.

Example 11.2, Table 2

Angular position relative to direction of jet (°)	Octave band centre frequency (Hz)							
	63	125	250	500	1000	2000	4000	8000
	Sound pressure level (dB re 20 μPa)							
0	72	77	80	82	80	76	69	63
60	74	79	83	84	82	78	71	65
120	61	66	70	71	69	65	58	52
180	55	60	64	65	63	59	52	46

11.7.2 Gas and Steam Vents

Noise produced by gas and steam vents can be estimated by assuming they are free jets (unconstricted by pipe walls). The estimation procedure for free jets outlined above is then used to calculate sound power and sound pressure levels. However, for steam vents, if the calculated jet efficiency, η, is less than 0.005, more accurate estimates are obtained by adding 3 dB to the calculated noise levels.

11.7.3 General Jet Noise Control

Means that would be applicable for containment of noise from pneumatic jets were discussed in Chapter 8 and means for attenuation were discussed in Chapter 9. However, these means are generally less efficient than means that alter the noise-generating mechanism.

For example, quite large reductions can be obtained with aerodynamic noise sources when the basic noise-producing mechanism is altered. Some methods of reducing aerodynamic noise based on this principle are illustrated in Figure 11.4. The figure refers to silencers that discharge to the atmosphere, but some of these methods can result in appreciable noise reductions, even for discharge in the confinement of a pipe. For example, modified forms of the devices illustrated in the first two parts of Figure 11.4 have been found remarkably effective for the control of valve noise in piping systems. Alternatively, orifice plates (circular plates containing multiple holes) placed downstream from the jet or valve and as close as possible to it have been found to be very effective. The orifice plate should occupy the entire cross-section of the pipe.

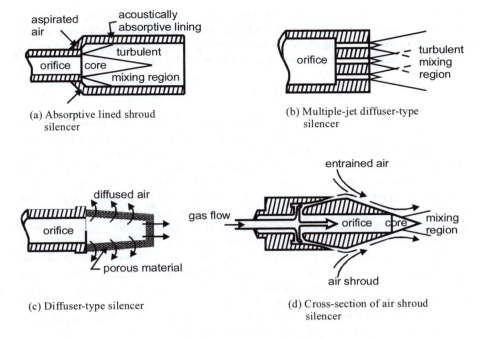

Figure 11.4 Discharge silencers for air.

11.8 CONTROL VALVES

A control valve allows the passage of a fluid from one piping system to another while at the same time controlling the pressure ratio between the two systems. For example, a fluid at static pressure P_1, in one system, may be transported to a fluid at a lower static pressure P_2, in the second system, by dissipating potential energy stored in the fluid at the higher pressure during the passage of the fluid through the control valve. The necessary energy dissipation is accomplished by conversion of the static pressure head to heat and a very small fraction to sound by intense turbulence and shock formation in the control valve.

Control valves are a common source of noise in many industries, especially when the operation of the valve is characterised by a large pressure drop. The primary noise-generating mechanism is the jet of fluid formed between the valve and its seat; thus valve noise is modelled as a confined jet. Consequently, the noise-generation mechanisms are turbulent mixing, turbulence boundary interaction, shock, shock/turbulence interaction and flow separation (Ng, 1980).

11.8.1 Internal Sound Power Generation

The observation that the radiated sound must be some fraction of the potential energy dissipated in the control valve suggests the existence of an energy conversion efficiency factor (or acoustical power coefficient), η, relating the mechanical stream power, W_m, entering the control valve, which is a function of the pressure drop across the valve, to the sound power, W_a, which is generated by the process. The proposed relation for the sound power radiated downstream of the valve is:

$$W_a = W_m \eta \, r_w \qquad (11.26)$$

The sound power, W_m, is defined in Equation (11.21) and the acoustical power coefficient, η, is defined later in Equations (11.34) to (11.38) for various valve flow conditions. The quantity, r_w, is the ratio of acoustic power propagated downstream of the valve to the total acoustic power generated by the valve and it varies between 0.25 and 1.0, depending on the valve type (see Table 11.13 and IEC 60534-8-3 (2000)). Note that the quantities on the right side of Equation (11.21) can refer to properties in any part of the gas flow. If properties in the valve vena contracta are used, then $d = d_j$ is the diameter of the jet in the vena contracta. If the flow is sonic through the vena contracta, then the stream power is calculated using the speed of sound as the flow speed. Note that the sound power radiated upstream of the valve is not considered significant. It is not clear from the standard what happens to the acoustic power that does not propagate downstream. Presumably it is dissipated in the valve, as, according to the standard, little escapes through the valve casing.

The control valve noise estimation procedure described here begins with consideration of possible regimes of operation. Consideration of the regimes of operation in turn provides the means for determining the energy conversion efficiency of the stream power entering the control valve.

To facilitate understanding and discussion of the noise-generation mechanisms and their efficiencies in a control valve, it will be convenient first to consider the possibility of pressure recovery within the valve, as illustrated in Figure 11.5. In the figure a valve opening and the corresponding static pressure distribution through the valve is schematically represented for one inlet pressure, P_1, and five possible outlet pressures, P_2.

Referring to Figure 11.5, when the fluid at higher pressure, P_1, enters the fluid at lower pressure, P_2, through the valve orifice, a confined jet is formed, which is characterised in turn by a vena contracta of diameter, d_j, and minimum static pressure, P_0, indicated by the three minima in the figure, after which the static pressure rises to P_2. As the figure shows, it is possible that the flow may be sonic in the vena contracta, although the pressure ratio across the valve is sub-critical. The extent of the pressure recovery is determined by the design of the valve and is determined empirically.

To quantify pressure recovery and the noise-generating behaviour of a valve, a quantity, F_L, with the unfortunate name "pressure recovery factor" has been defined. The definition, which forms the basis for its determination, is given in terms of the following relation involving the inlet static pressure, P_1, the outlet static pressure, P_2, and the static pressure, P_0, in the vena contracta:

$$F_L^2 = \frac{P_1 - P_2}{P_1 - P_0} \qquad (11.27)$$

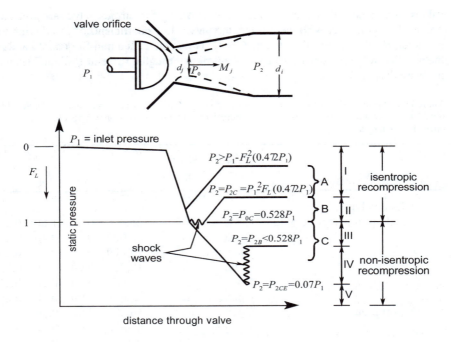

Figure 11.5 Schematic representation of the static pressure distribution through a control valve with inlet pressure P_1 and outlet pressure P_2. In pressure range A, there is subsonic flow and drag dipole turbulence noise. In pressure range B, there is sonic flow in the vena contracta, and dipole and quadrupole turbulence noise and shock noise. In pressure range C, there is sonic flow and shock noise is dominant. $F_L = 0$ when $P_2 = P_1$ and $F_L = 1$ when $P_2 = 0.528P_1$ (after Howe and Baumann, 1992).

The pressure in the vena contracta, which cannot be measured easily, is related to the upstream and downstream pressures by:

$$P_0 = P_1 - \left(\frac{P_1 - P_2}{F_L^2} \right) \tag{11.28}$$

This equation is used for calculating the vena contracta pressure for flow regime 1, but for other flow regimes, which are characterised by sonic flow in the vena contracta, Equation (11.29) to follow is used to calculate the pressure in the vena contracta.

Referring to Figure 11.5 or to Equation (11.27), it may be noted that F_L has the value 1 when $P_2 = P_0$ and there is no pressure recovery. However, in this case, sound generation must be maximal because it is always a fraction of the energy dissipated, which is now maximal. On the other hand, reference to either the figure or the equation shows that F_L has the value 0 when $P_2 = P_1$ and there is complete pressure recovery. In this case, no energy is dissipated and thus no energy has been converted to sound. A better name for F_L might have been "noise recovery factor". Nonetheless, manufacturers of control valves supply "pressure recovery factor" information for their valves, and values of the pressure recovery factor, F_L, for various valves are given in Table 11.13, together with the ratio of the valve flow coefficient, C_v, to the inside diameter, d_i (m), of the valve outlet, C_v/d_i^2. Note that C_v is a non-SI quantity and is the number of US gallons of water that will flow through a valve in 1 minute when a pressure drop of 1 psi exists.

Table 11.13 Typical values of valve noise prediction factors (after Howe and Baumann, 1992). See IEC 60534-8-3 for data for other valve types. The column labelled "flow to" indicates whether the flow acts to open or close the valve

Type of valve	Flow to	% of of capacity	C_v/d_i^2	F_L	F_d	r_w
1. Globe valves						
Single-port parabolic plug	Close	100	25,000	0.80	1.00	0.25
	Open	100	20,000	0.90	0.46	0.25
	Open	75	15,000	0.90	0.36	0.25
	Open	50	10,000	0.90	0.28	0.25
	Open	25	5,000	0.90	0.16	0.25
	Open	10	2,000	0.90	0.10	0.25
v-port plug	Open	100	16,000	0.92	0.50	0.25
		50	8,000	0.95	0.42	0.25
		30	5,000	0.95	0.41	0.25
Four-port cage	Open	100	25,000	0.90	0.43	0.25
	Open	50	13,000	0.90	0.36	0.25
Six-port cage	Open	100	25,000	0.90	0.32	0.25
	Open	50	13,000	0.90	0.25	0.25
		Angle of travel (°)				
2. Eccentric rotary plug	Open	50	20,000	0.85	0.42	0.25
		30	13,000	0.91	0.30	0.25
	Close	50	21,000	0.68	0.45	0.25
		30	13,000	0.88	0.30	0.25
3. Ball valve, segmented	Open	60	18,000	0.66	0.75	0.25
		30	5,000	0.82	0.63	0.25
4. Butterfly						
Swing-through vane		75	50,000	0.56	0.57	0.5
		60	30,000	0.67	0.50	0.5
		50	16,000	0.74	0.42	0.5
		40	10,000	0.78	0.34	0.5
		30	5,000	0.80	0.26	0.5
Fluted vane		75	40,000	0.70	0.30	0.5
		50	13,000	0.76	0.19	0.5
		30	7,000	0.82	0.08	0.5

The quantity F_d, also listed in the table, is the valve style modifier used later on in Equation (11.48). When there are fittings such as bends or T-pieces attached to the valve, the quantity F_L must be replaced by F_{LP}/F_P, where F_{LP} is the combined liquid pressure recovery and pipe geometry factor (supplied by the valve manufacturer) and F_P is the pipe geometry factor available from the International Standard, IEC 60534-2-1 (1998). The column titled "flow to" indicates whether the flow direction is acting to open or close the valve.

When considering noise generation in a control valve using this prediction procedure, three regimes of operation and associated energy dissipation and noise-generation mechanisms may be identified. These regimes may be understood by reference to Figure 11.5, where they are identified as A, B and C. In regimes A and B, hydrodynamic dipole and quadrupole sources are responsible for noise generation. Such noise generation is discussed in Chapter 5.

Referring to Figure 11.5, in regime A the flow through the valve is everywhere subsonic and all energy dissipation is accomplished by intense turbulence. In this case, noise generation is found to be proportional to the sixth power of the stream speed, which implies that it is dominated by fluctuating forces acting on the fluid and is thus dipole in nature (see Section 5.3.3). It is suggested here that a fluctuating re-attachment bubble in the region indicated by the dotted lines in Figure 11.5 (see also Figure 9.14) might account for a drag dipole with its axis oriented along the axis of the pipe and which radiates propagating sound at all frequencies. Referring to Figure 11.5, the bound between regimes A and B is seen to be determined by the pressure ratio P_2/P_1, at which the flow in the vena contracta becomes sonic. The flow speed in the vena contracta will be sonic when the pressure P_0 in the vena contracta is equal to the critical pressure, P_{0C}, which satisfies the following relation (Landau and Lifshitz, 1959):

$$\frac{P_{0C}}{P_1} = \left(\frac{2}{\gamma + 1} \right)^{\gamma/(\gamma - 1)} \tag{11.29}$$

For diatomic gases, where $\gamma = 1.4$:

$$P_{0C}/P_1 = 0.528 \tag{11.30}$$

Equations (11.29) and (11.30) are used for Regimes II to V to calculate the vena contracta pressure from the inlet pressure. For Regime I, Equation (11.28) is used. Values of the ratio of specific heats, γ, are given in Table 11.14 for various commonly used gases.

Solving Equation (11.27) for the pressure ratio P_2/P_1 and using Equation (11.29) gives the pressure ratio across the valve, $P_2/P_1 = P_{2C}/P_1$, at which the vena contracta flow is sonic, as follows:

$$\frac{P_{2C}}{P_1} = 1 - F_L^2 \left[1 - \left(\frac{2}{\gamma + 1} \right)^{\gamma/(\gamma - 1)} \right] \tag{11.31}$$

Table 11.14 Molecular weights and ratios of specific heats for some commonly used gases

Gas or vapour	Molecular weight, M kg/mole	Ratio of specific heats, γ
Acetylene	0.02604	1.30
Air	0.02897	1.40
Ammonia	0.01730	1.32
Argon	0.03995	1.67
Benzene	0.07811	1.12
Isobutane	0.05812	1.10
n-Butane	0.05812	1.11
Isobutylene	0.05611	1.11
Carbon dioxide	0.04401	1.30
Carbon monoxide	0.02801	1.40
Chlorine	0.07091	1.31
Ethane	0.03007	1.22
Ethylene	0.02805	1.22
Flourine	0.01900	1.36
Freon 11	0.13737	1.14
Freon 12	0.12091	1.13
Freon 13	0.10446	1.14
Freon 22	0.08047	1.18
Helium	0.00400	1.66
n-Heptane	0.10020	1.05
Hydrogen	0.00202	1.41
Hydrogen chloride	0.03646	1.41
Hydrogen flouride	0.02001	0.97
Methane	0.01604	1.32
Methyl chloride	0.05049	1.24
Natural gas (representative)	0.01774	1.27
Neon	0.02018	1.64
Nitric oxide	0.06301	1.40
Nitrogen	0.02801	1.40
Octane	0.11423	1.66
Oxygen	0.03200	1.40
Pentane	0.07215	1.06
Propane	0.04410	1.15
Propylene	0.04208	1.14
Saturated steam	0.01802	1.25-1.32
Sulphur dioxide	0.06406	1.26
Superheated steam	0.01802	1.315

Equations (11.30) and (11.31) have been used to establish the bounds between regimes B and C and between A and B in Figure 11.5.

In regime B the flow is sonic in the vena contracta and $P_2 > P_{0C}$. In this case, noise generation tends to be proportional to the eighth power of the stream speed. In this regime, the noise-generation efficiency of quadrupoles (see Section 5.4.3) associated with intense turbulence of the jet through the valve orifice will overtake the dipole sources that are still present, so that both are considered to contribute to noise generation. In addition, shocks will be formed, which will be rendered unstable by the passage of turbulent flow through them and these will provide a very efficient noise-generation mechanism.

The critical pressure ratio, when $P_2 = P_{0C}$ and flow through the valve is sonic (for diatomic gases), is shown in Figure 11.5 as the bound between regimes B and C. In regime C, the pressure ratio across the valve, P_2/P_1, is less than P_{0C}/P_1 given by Equation (11.30) and the flow is sonic everywhere in the valve. In this regime energy dissipation is dominated by shock formation and noise-generation efficiency is maximal.

The IEC standard (IEC 60534-8-3) has further divided regimes B and C into regimes II, III and IV, adding a regime V and renaming regime A as regime I (see Figure 11.5). To delineate regions IV and V, the IEC standard defines two additional pressures, P_{2B} and P_{2CE}:

$$P_{2B} = \frac{P_1}{\alpha}\left(\frac{1}{\gamma}\right)^{\gamma/(\gamma-1)}; \qquad P_{2CE} = \frac{P_1}{22\,\alpha} \qquad (11.32\text{a,b})$$

where the quantity, α, is defined as:

$$\alpha = \left[\left(\frac{\gamma+1}{2}\right)^{\gamma/(\gamma-1)} - F_L^2\left(\left(\frac{\gamma+1}{2}\right)^{\gamma/(\gamma-1)} - 1\right)\right]^{-1} \qquad (11.33)$$

Simplifying the equations used for calculating the efficiency in the standard, IEC 60534-8-3, the following expressions are obtained for the acoustical power coefficients for the regimes I to V defined in Figure 11.5. Note that the acoustical efficiency factors (η_1, η_2, etc.) used in the standard are related to the acoustical power coefficients, η, used here by $\eta = \eta_1 \times F_L^2$, $\eta = \eta_2 \times (P_1 - P_2)/(P_1 - P_{0C})$, $\eta = \eta_{3}, \eta_{4}$, and η_5.

In regime I, flow is sub-sonic everywhere in the valve and $P_2 \geq P_{2C}$:

$$\eta = 10^{-4}F_L^2\left[\frac{\dfrac{2}{\gamma-1}\left(1-\left(\dfrac{P_0}{P_1}\right)^{\frac{\gamma-1}{\gamma}}\right)}{\left(\dfrac{P_0}{P_1}\right)^{\frac{\gamma-1}{\gamma}}}\right]^{1.8} \qquad (11.34)$$

where the pressure, P_0 in the vena contracta may be calculated from the upstream and downstream pressures using Equation (11.28). When fittings are attached to the valve, the quantity, F_L, must be replaced with F_{LP}/F_P in Equations (11.33)–(11.38). However, this is generally only a small effect and a good approximation in the absence of the required data is to use F_L.

In regime II, flow is sonic in the vena contracta only and $P_{0C} \le P_2 < P_{2C}$:

$$\eta = 10^{-4} \left(\frac{P_1 - P_2}{P_1 - P_{0C}} \right) \left(\frac{2}{\gamma - 1} \left[\left(\frac{P_1}{\alpha P_2} \right)^{\frac{\gamma - 1}{\gamma}} - 1 \right] \right)^{3.3 F_L^2} \tag{11.35}$$

In regime III, flow is sonic everywhere in the valve and $P_{2B} \le P_2 < P_{0C}$:

$$\eta = 10^{-4} \left(\frac{2}{\gamma - 1} \left[\left(\frac{P_1}{\alpha P_2} \right)^{\frac{\gamma - 1}{\gamma}} - 1 \right] \right)^{3.3 F_L^2} \tag{11.36}$$

where α is defined in Equation (11.33).

In regime IV, flow is sonic everywhere in the valve and $P_{2CE} \le P_2 < P_{2B}$:

$$\eta = 10^{-4} \frac{1}{\gamma - 1} \left[\left(\frac{P_1}{\alpha P_2} \right)^{\frac{\gamma - 1}{\gamma}} - 1 \right] \times 2^{3.3 F_L^2} \tag{11.37}$$

In regime V, flow is sonic everywhere in the valve and $P_2 < P_{2CE}$:

$$\eta = 10^{-4} \frac{1}{\gamma - 1} \left[(22)^{(\gamma - 1)/\gamma} - 1 \right] \times 2^{3.3 F_L^2} \tag{11.38}$$

11.8.2 Internal Sound Pressure Level

The mean square sound pressure in the pipe downstream of the valve is approximated as follows:

$$\langle p_i^2 \rangle = \frac{4 W_a \rho_2 c_2}{\pi d_i^2} \tag{11.39}$$

where d_i is the internal diameter of the pipe and W_a is the sound power radiated downstream of the valve as defined in Equation (11.26).

It is assumed that the temperature upstream and downstream of the valve is essentially the same, so that:

$$\rho_2 \approx \rho_1 \frac{P_2}{P_1} \tag{11.40}$$

and

$$c_1 \approx c_2 = \sqrt{\frac{\gamma R T_1}{M}} \qquad (11.41)$$

Note that if the downstream temperature, T_2, is known, the downstream speed of sound, c_2, can be calculated more accurately by replacing T_1 with T_2 in Equation (11.41) and the downstream density, ρ_2 can be calculated more accurately using:

$$\rho_2 = \frac{P_2 M}{R T_2} \qquad (11.42)$$

The same Equation (11.42) can also be used to calculate the upstream density, given the upstream pressure, P_1 and the upstream temperature, T_1.

The speed of sound, c_0, in the vena contracta is related to the speed of sound, c_1, upstream of the valve as follows (Landau and Lifshitz, 1959):

$$c_0 = c_1 \sqrt{2/(\gamma + 1)} \qquad (11.43)$$

For sonic flow (regimes II to V), the flow speed, U_0, through the vena contracta is sonic and equal to c_0. For sub-sonic flow through the vena contracta (regime I), the flow speed, U_0, is calculated using:

$$U_0 = \sqrt{\left(\frac{2\gamma}{\gamma - 1}\right)\left[1 - \left(\frac{P_0}{P_1}\right)^{(\gamma - 1)/\gamma}\right]\frac{P_1}{\rho_I}} \qquad (11.44)$$

The preceding and following calculations require as input data the following properties of the gas flowing into the valve: ρ_1, T_1, P_1, γ, M, and $\dot{m}$ and the pressure, P_2 at the valve discharge. The mass flow rate, $\dot{m}$, may be calculated from a knowledge of the volume flow rate, $\dot{V}$, at any given pressure, P, and temperature, T, using the universal gas law:

$$\dot{m} = \frac{P \dot{V} M}{RT} \qquad (11.45)$$

where M is the molecular mass of the gas (kg/mole – see Table 11.14) and R ($= 8.314$ J/mole $^\circ$K) is the universal gas constant in consistent units. The mass flow rate is needed as an input in the valve noise calculations.

Taking logs of Equation (11.39) gives for the sound pressure level inside the pipe downstream of the valve:

$$L_{pi} = 10\log_{10}\left[\frac{3.18 \times 10^9 W_a \rho_2 c_2}{d_i^2}\right] \qquad (11.46)$$

The spectrum peak frequency, f_p, may be estimated using the following equations. (IEC 60534-8-3):

Regime I:

$$f_p = \frac{0.2 \, U_0}{d_j} \tag{11.47}$$

where U_0 is defined by Equation (11.44) and:

$$d_j \approx 4.6 \times 10^{-3} \, F_d \sqrt{C_v F_L} \quad \text{(m)} \tag{11.48}$$

When fittings are attached to the valve, the quantity, F_L, must be replaced with F_{LP}/F_P in Equation (11.48). The flow coefficient, C_v, may be calculated using the procedures outlined in IEC 60534-2-1 or it may be supplied by the valve manufacturer. Note that the actual C_v for the valve opening conditions (rather than the valve rated C_v) should be used. Values in Table 11.13 are based on the outlet diameter, d_i (m), of the valve and may be used in the absence of more accurate data. The quantity, F_d, called a valve style modifier, is empirically determined and listed in Table 11.13.

Regimes II and III:

$$f_p = \frac{0.2 \, M_j c_0}{d_j} \tag{11.49}$$

Regimes IV and V:

$$f_p = \frac{0.28 \, c_0}{d_j \sqrt{M_j^2 - 1}} \tag{11.50}$$

The stream Mach number is calculated as follows:

$$M_j = \left[\frac{2}{(\gamma - 1)} \left(\left(\frac{P_1}{\alpha P_2} \right)^{(\gamma - 1)/\gamma} - 1 \right) \right]^{1/2} \tag{11.51}$$

where α is defined in Equation (11.33).

The one-third octave band spectrum of the noise inside of the pipe may be estimated as follows, using L_{pi} calculated with Equation (11.46). The level, L_p, of the one-third octave band containing the spectrum peak frequency is $L_p = L_{pi} - L_x$. The quantity, L_x, is approximately 8 dB, but must be adjusted up or down in each case considered so that the total sound pressure level calculated by adding the one-third octave band levels as illustrated in Equation (1.98) is equal to L_{pi}. For frequencies greater than the peak frequency, the spectrum rolls off at the rate of 3.5 dB per octave. For frequencies less than the peak frequency, the spectrum rolls of at the rate of 5 dB per octave for the first two octaves and then at the rate of 3 dB per octave at lower frequencies.

11.8.3 External Sound Pressure Level

To calculate the transmission loss of the pipe wall, three frequencies need to be calculated. The first is the ring frequency, given by the following equation:

$$f_r = \frac{c_L}{\pi d_i} \tag{11.52}$$

In the above equation, d_i is the internal diameter of the pipe downstream of the valve and c_L is the compressional wave speed in the pipe wall. At the ring frequency, the mean circumference is just one compressional wavelength long. Below the ring frequency, the pipe wall bending wave response is controlled by the surface curvature, while above the ring frequency, the pipe wall bending wave response is essentially that of a flat plate and unaffected by the surface curvature.

A second important frequency is the internal coincidence frequency, f_0, of the gas in the pipe and this is (IEC 60534-8-3):

$$f_0 = \frac{f_r}{4} \left(\frac{c_2}{343} \right) \tag{11.53}$$

where c_2 is the speed of sound in the downstream fluid in the pipe. At this frequency, the transmission loss of the pipe wall is a minimum.

A third frequency is (IEC 60534-8-3):

$$f_g = \frac{6.49 \times 10^4}{t\, c_L} \tag{11.54}$$

where t is the pipe wall thickness (m).

There are two methods (Howe and Baumann, 1992 and IEC 60534-8-3) that have been described in the literature for calculating the pipe wall transmission loss (TL). For the Howe and Baumann method, the transmission loss is given by the following equation.

$$TL = 10 \log_{10} \left[\frac{t^2}{d_i^2} \left(\frac{P_2}{P_a} + 1 \right) \right] + 66.5 + \Delta_0 \tag{11.55}$$

which is applied only to the A-weighted internal level to get the A-weighted external level and is only applicable to control valve noise. In the above expression P_a is the external static pressure. Other quantities have been defined previously. The equation in the reference has been adjusted to convert the noise reduction at some distance, r, to transmission loss.

If the frequency of peak noise generation is below the internal coincidence frequency, f_0 of Equation (11.53), the correction term, Δ_0, has the following form:

$$\Delta_0 = 20 \log_{10} \frac{f_0}{f_p}; \qquad f_p \le f_0 \tag{11.56}$$

If the frequency of peak noise generation is between the first radial mode cut-on frequency and the ring frequency, the correction term Δ_0 has the following form:

$$\Delta_0 = 13 \log_{10} \frac{f_p}{f_r}; \quad f_0 < f_p \leq f_r \tag{11.57}$$

Finally, if the frequency of peak noise generation is above the ring frequency, the correction term Δ_0 has the following form:

$$\Delta_0 = 20 \log_{10} \frac{f_p}{f_r}; \quad f_p > f_r \tag{11.58}$$

The expression for the transmission loss given by IEC 60534-8-3 (inverted to match the definition of TL used in this book) is:

$$TL = 10 \log_{10} \left[\frac{t^2 f_p^2}{c_2^2} \left(\frac{\rho_2 c_2}{415 G_y} + 1 \right) \frac{1}{P_a G_x} \right] + 111.2 \tag{11.59}$$

which is applied only to the A-weighted internal level to get the A-weighted external level and is only applicable to control valve noise. The quantities G_x and G_y are defined as:

$$G_x = \begin{cases} \left(\dfrac{f_0}{f_r} \right)^{2/3} \left(\dfrac{f_p}{f_0} \right)^4; & f_p < f_0 \\[3ex] \left(\dfrac{f_p}{f_r} \right)^{2/3}; & f_0 \leq f_p < f_r \\[3ex] 1; & f_p \geq f_0 \text{ and } f_p \geq f_r \end{cases} \tag{11.60}$$

and

$$G_y = \begin{cases} \left(\dfrac{f_0}{f_g} \right); & f_p < f_0 < f_g \\[3ex] \left(\dfrac{f_p}{f_g} \right); & f_0 \leq f_p < f_g \\[3ex] 1; & f_0 \geq f_g \text{ and } f_0 > f_p \\[2ex] 1; & f_p \geq f_0 \text{ and } f_p \geq f_g \end{cases} \tag{11.61}$$

The A-weighted sound pressure level, L_{pAe}, external to the pipe at the outside diameter of the pipe is calculated as follows (IEC 60534-8-3):

$$L_{pAe} = L_{pi} - TL + 5 + L_g \tag{11.62}$$

The 5 dB correction term accounts for the many peaks in the internal noise spectrum.

The term L_g has been introduced in Equation (11.62) and is a correction term to account for the effect of gas flow within the pipe upon the sound energy transmitted through the pipe wall. The latter correction term is calculated as follows:

$$L_g = -16 \log_{10}\left(1 - \frac{4\dot{m}}{\pi d_i^2 \rho_2 c_2} \right) \tag{11.63}$$

If the second term in brackets of Equation (11.63) exceeds 0.3, it is set equal to 0.3.

The A-weighted sound pressure level, $L_{pAe,1m}$, external to the pipe at one metre from the pipe wall of diameter, d_i is calculated as follows (IEC 60534-8-3):

$$L_{pAe,1m} = L_{pAe} - 10\log_{10}\left[\frac{d_i + 2t + 2}{d_i + 2t} \right] \tag{11.64}$$

The overall, A-weighted sound power level radiated by the pipe is:

$$L_{wA} = L_{pAe} + 10\log_{10}\left(\frac{d_i + 2t}{2} \right) + 10\log_{10}\ell_p + 8 \tag{11.65}$$

where ℓ_p is the length of downstream pipe radiating sound.

The sound pressure level at any distance from the downstream pipe may be calculated by converting the sound power level to power in watts, using the analysis of Section 5.5 for a line source and then converting the mean square acoustic pressure to sound pressure level in dB.

Although the international standard, IEC 60534-8-3, 2000, does not address the spectral distribution of the sound level calculated using the preceding procedures, it has been discussed in a paper by Baumann and Hoffmann (1999). In their work, they divide the calculation of the spectrum shape into three regimes as follows, where the measurement band is 1/3 octave, f is the centre frequency of the 1/3 octave band of interest, f_p is the frequency of the peak noise level, f_0 is the internal coincidence frequency of the pipe (Equation (11.53)) and L_B is the 1/3 octave band sound pressure level (un-weighted) external to the pipe at 1 metre distance. Similar spectral distribution will apply ro sound power. Note that the final spectrum levels must all be adjusted by adding or subtracting a constant decibel number so that when A-weighted and added together, the result is identical to the A-weighted overall levels from Equations (11.64) and (11.65).

Regime 1, $f_p < f_0$

$$L_B = L_{pAe,1m} - 5 - 40\log_{10}(f_p/f); \quad f \le f_p$$

$$L_B = L_{pAe,1m} - 5; \quad f_p < f \le f_0$$

$$L_B = L_{pAe,1m} - 5 - 33\log_{10}(f/f_0); \quad f_0 < f < f_r$$

$$L_B = L_{pAe,1m} - 5 - 33\log_{10}(f_r/f_0) - 40\log_{10}(f/f_r); \quad f_r \le f$$

(11.66)

Regime 2, $f_0 \le f_p \le f_r$

$$L_B = L_{pAe,1m} - 5 - 40\log_{10}(f_0/f) - 7\log_{10}(f_p/f_0); \quad f \le f_0$$

$$L_B = L_{pAe,1m} - 5 - 7\log_{10}(f_p/f); \quad f_0 < f < f_p$$

$$L_B = L_{pAe,1m} - 5 - 33\log_{10}(f/f_p); \quad f_p < f < f_r$$

$$L_B = L_{pAe,1m} - 5 - 33\log_{10}(f_r/f_p) - 40\log_{10}(f/f_r); \quad f_r \le f$$

(11.67)

Regime 3, $f_r < f_p$

$$L_B = L_{pAe,1m} - 5 - 40\log_{10}(f_0/f) - 7\log_{10}(f_r/f_0); \quad f \le f_0$$

$$L_B = L_{pAe,1m} - 5 - 7\log_{10}(f_0/f) - 7\log_{10}(f_r/f_0); \quad f_0 < f \le f_r$$

$$L_B = L_{pAe,1m} - 5; \quad f_r < f < f_p$$

$$L_B = L_{pAe,1m} - 5 - 40\log_{10}(f/f_p); \quad f_p \le f$$

(11.68)

11.8.4 High Exit Velocities

The preceding calculation procedures are only valid for exit (downstream of the valve) Mach numbers of 0.3 or less. For higher exit velocities, calculation procedures are available in the standard, IEC 60534-8-3.

11.8.5 Control Valve Noise Reduction

As was explained in the previous section, the control valve necessarily functions as an energy dissipation device and the process of energy dissipation is accompanied by a small amount of energy conversion to noise. Consequently, noise control must take the form of reduction of energy conversion efficiency. In the previous sections it was shown that energy conversion efficiency increases with the Mach number and becomes even more efficient when shock waves are formed. Evidently, it is desirable from a noise reduction point of view to avoid critical flow and shock wave formation. This has been accomplished in practice by effectively providing a series of pressure drop devices across which the pressure drop is less than critical and intense turbulence is induced but shock formation is avoided. There are a number of control valves commercially available built upon this principle, but they are generally expensive.

11.8.6 Control Valves for Liquids

The following prediction procedure considers only noise generated by hydrodynamic processes and excludes noise that may be influenced by reflections, loose parts or resonances. The maximum downstream velocity for which the procedures are valid is 10 m/s. Calculating the noise generated by a cavitating valve is covered in the standard, IEC 60534-8-4 (1994) and will not be discussed here.

The stream mechanical power of a liquid flowing through a valve is given by Equation (11.21) (provided that no cavitation occurs). An alternative formulation that is easier to evaluate is (IEC 60534-8-4):

$$W_m = \frac{\dot{m}(P_1 - P_2)}{\rho_f} \tag{11.69}$$

where ρ_f is the density of the liquid and $\dot{m}$ is the mass flow rate. The acoustic power is obtained by multiplying W_m by an efficiency factor, η, which is taken as 10^{-8}. Thus, the sound power level inside the pipe downstream of the valve may be written as:

$$L_{wi} = 10\log_{10}\left(\frac{\eta\dot{m}(P_1 - P_2)}{\rho_f}\right) + 120 \quad (\text{dB re } 10^{-12} \text{ watts}) \tag{11.70}$$

The spectral distribution for the octave bands from 500Hz to 8kHz is given by:

$$L_{wi}(f_C) = L_{wi} - 10\log_{10}\left(\frac{f_C}{500}\right) - 2.9 \quad (\text{dB}) \tag{11.71}$$

where f_C is the octave band centre frequency. The un-weighted sound power radiated externally in each octave band is:

$$L_{we}(f_C) = L_{wi}(f_C) - 17.37\left(\frac{\ell_p}{2D}10^{-TL(f_C)/10}\right) - TL(f_C) + 10\log_{10}\left(\frac{4\ell_p}{D}\right) \tag{11.72}$$

where ℓ_p is the length of pipe radiating the noise (minimum of 3 m) and D is the pipe external diameter $= d_i + 2t$.

The transmission loss of the pipe wall is calculated using:

$$TL(f_C) = 10 + 10\log_{10}\left[\frac{c_L \rho_p t}{c_2 \rho_f D}\right] + 10\log_{10}\left[\frac{f_r}{f_C} + \left(\frac{f_C}{f_r}\right)^{1.5}\right]^2 \tag{11.73}$$

where c_L is the longitudinal wave speed in the pipe wall, ρ_p is the density of the pipe wall material, f_r is the pipe ring frequency $= c_L/[\pi(d_i + t)]$, c_2 is the speed of sound in the fluid downstream of the valve, ρ_f is the density of the downstream fluid and t is the pipe wall thickness.

The overall A-weighted sound power level external to the pipe is:

$$L_{wAe} = 10\log_{10}\sum_{n=1}^{5} 10^{L_{wAn}/10} \tag{11.74}$$

where L_{wAn} is the A-weighted sound power level of the nth octave band (calculated by applying the A-weighting correction to the un-weighted sound power level, $L_{we}(f_C)$ of Equation (11.72)), and the sum is over the five octave bands from 500 Hz to 8 kHz inclusive.

The overall A-weighted sound pressure level external to the valve, 1 m downstream of the outlet flange and 1 m from the pipe, is estimated from the A-weighted external sound power level using the following equation:

$$L_{pAe} = L_{wAe} - 10\log_{10}\left[\frac{\pi \ell_p}{\ell_0}\left(\frac{d_i}{D} + 1\right)\right] \tag{11.75}$$

where $\ell_0 = 3$ m and d_i is the pipe internal diameter.

The octave band external sound pressure levels may be calculated using Equations (11.72) and (11.75) with octave band sound power levels used in Equation (11.75) instead of overall sound power levels.

11.8.7 Control Valves for Steam

Steam valves are a common source of noise in industrial plants. Sound pressure levels generated by them can be estimated using the procedures outlined previously for gas valves. However, if the calculated jet efficiency is less than about 0.005, more accurate estimates are obtained by adding 3 dB to the calculated results.

Alternatively, the values listed in Table 11.15 may be used. These are conservative estimates based upon measurements of actual steam valves, and include noise radiation from the pipe connected to the valves.

Table 11.15 Estimated sound pressure levels (dB re 20 µPa at 1 m) for steam valves[a]

Octave band centre frequency (Hz)	Sound pressure level (dB)
31.5	70
63	70
125	70
250	70
500	75
1000	80
2000	85
4000	90
8000	90

[a] This table assumes simple, lightweight thermal wrapping of the pipe but no metal or heavy cover around the thermal wrapping. Both the valve and the connected piping radiate noise.

Data from Army, Air Force and Navy, USA (1983a).

11.9 PIPE FLOW

The calculation of flow noise from pipes will be restricted to consideration of flowing gases, as noise due to flowing liquids is generally insignificant. In straight pipes, frictional resistance to flow is considered to be the primary noise source. The mechanical power producing the noise is proportional to the frictional pressure drop in the pipe. Thus, the acoustic source power, W_a, may be written in terms of the pressure drop, ΔP (Pa), along the length of pipe, as follows (Heitner, 1968):

$$W_a = \eta(\Delta P)\dot{m}/\rho = \eta(\Delta P)AU \quad \text{(W)} \qquad (11.76a,b)$$

where A is the pipe cross-sectional area (m^2), U is the mean gas velocity (m/s) in the pipe, $\dot{m}$ is the mass flow rate (kg/s), of gas in the pipe, ρ is the density (kg/m^3), of gas in the pipe, and η is the acoustical efficiency of a free jet (see Equation (11.22)).

The sound power level, L_w, inside the pipe is calculated using Equations (11.76) and (1.83). The acoustic spectra inside the pipe and external to the pipe are calculated as for control valves (see Sections 11.8.2 and 11.8.3, Equation (11.46) onwards).

The method outlined above underestimates the noise level by 2-4 dB for gases having a density approximately equal to ambient air. On the other hand, for high-density gases (about 30 times the density of atmospheric air) the method overestimates the noise levels by 2-4 dB. For vacuum lines, the power calculated using Equation (11.76) for inside the pipe gives poor results for sound power level and average sound pressure level. However, the following expression gives results for sound power inside the pipe that are consistent with measured values for flow in vacuum lines (Heitner, 1968):

$$W_a = 0.5 \eta A U^3 \quad \text{(W)} \tag{11.77}$$

Again, acoustic spectra inside the pipe and external to the pipe are calculated as for control valves (see Sections 11.8.2 and 11.8.3, Equation (11.46) onwards).

11.10 BOILERS

For general-purpose boilers, the overall radiated sound power level is given by:

$$L_w = 95 + 4 \log_{10} kW \quad \text{(dB re } 10^{-12} \text{ W)} \tag{11.78}$$

For large power plant boilers, the overall sound power level is given by:

$$L_w = 84 + 15 \log_{10} MW \quad \text{(dB re } 10^{-12} \text{ W)} \tag{11.79}$$

The octave band levels for either type of boiler can be calculated by subtracting the appropriate corrections listed in Table 11.16.

Table 11.16 Values to be subtracted from overall power levels, L_w, to obtain band levels for boiler noise

Octave band centre frequency (Hz)	Octave band corrections (dB)	
	General-purpose boilers	Large power plant boilers
31.5	6	4
63	6	5
125	7	10
250	9	16
500	12	17
1000	15	19
2000	18	21
4000	21	21
8000	24	21

Data from Edison Electric Institute (1978).

11.11 TURBINES

The principal noise sources of gas turbines are the casing, inlet and exhaust. The overall sound power levels contributed by these components of gas turbine noise (with no noise control) may be calculated using the following equations (Army, Air Force and Navy, USA 1983a):

Casing:

$$L_w = 120 + 5 \log_{10} MW \quad \text{(dB re } 10^{-12} \text{ W)} \tag{11.80}$$

Inlet:

$$L_w = 127 + 15 \log_{10} MW \quad \text{(dB re } 10^{-12} \text{ W)} \tag{11.81}$$

Exhaust:

$$L_w = 133 + 10 \log_{10} MW \quad \text{(dB re } 10^{-12} \text{ W)} \tag{11.82}$$

For steam turbines, a good estimate for the overall sound power radiated is given by (Edison Electric Institute, 1978):

$$L_w = 93 + 4 \log_{10} kW \quad \text{(dB re } 10^{-12} \text{ W)} \tag{11.83}$$

The octave band levels for gas and steam turbines may be calculated by subtracting the corrections listed in Table 11.17. The approximate casing noise reductions due to various types of enclosure are listed in Table 11.18.

Table 11.17 Frequency adjustments (in dB) for gas turbine and steam turbine noise levels. Subtract these values from the overall sound power level, L_w, to obtain octave band and A-weighted sound power levels

Octave band centre frequency (Hz)	Value to be subtracted front overall L_w (dB)			
	Gas turbine			
	Casing	Inlet	Exhaust	Steam turbine
31.5	10	19	12	11
63	7	18	8	7
125	5	17	6	6
250	4	17	6	9
500	4	14	7	10
1000	4	8	9	10
2000	4	3	11	12
4000	4	3	15	13
8000	4	6	21	17
A-weighted (dB(A))	2	0	4	5

Data from Army, Air Force and Navy, USA (1983a).

Noise reductions due to inlet and exhaust mufflers can be calculated using the methods of Chapter 9, or preferably using manufacturer's data. Normally, the inlet and discharge cross-sectional areas of the muffler stacks are very large; thus additional noise reductions will occur due to the directivity of the stacks. This effect can be calculated using Figures 9.29–9.31.

Table 11.18 Approximate noise reduction of gas turbine casing enclosures

Octave band centre frequency (Hz)	Type 1[a]	Type 2[b]	Type 3[c]	Type 4[d]	Type 5[e]
31.5	2	4	1	3	6
63	2	5	1	4	7
125	2	5	1	4	8
250	3	6	2	5	9
500	3	6	2	6	10
1000	3	7	2	7	11
2000	4	8	2	8	12
4000	5	9	3	8	13
8000	6	10	3	8	14

[a] Glass fibre or mineral wool thermal insulation with lightweight foil cover over the insulation.
[b] Glass fibre or mineral wool thermal insulation covered with a minimum 20 gauge aluminum or 24 gauge steel.
[c] Enclosing metal cabinet for the entire packaged assembly, with open ventilation holes and with no acoustic absorptive lining inside the cabinet.
[d] Enclosing metal cabinet for the entire packaged assembly, with open ventilation holes and with acoustic absorptive lining inside the cabinet.
[e] Enclosing metal cabinet for the entire packaged assembly with all ventilation holes into the cabinet muffled and with acoustic absorptive lining inside the cabinet
Data from Army, Air Force and Navy, USA (1983a).

11.12 DIESEL AND GAS-DRIVEN ENGINES

The three important noise sources for this type of equipment are the engine exhaust, the engine casing and the air inlet. These sources will be considered in the following subsections.

11.12.1 Exhaust Noise

The overall sound power radiated by an unmuffled exhaust may be calculated using (Army, Air Force and Navy, USA, 1983a):

$$L_w = 120 + 10 \log_{10} kW - K - (\ell_{EX}/1.2) \quad \text{(dB re } 10^{-12} \text{ W)} \tag{11.84}$$

where $K = 0$ for an engine with no turbo-charger, $K = 6$ for an engine with a turbo-charger, and ℓ_{EX} is the length of the exhaust pipe (m).

The octave band sound power levels may be calculated by subtracting the corrections listed in Table 11.19 from the overall power level.

The approximate effects of various types of commercially available mufflers are shown in Table 11.20. Exhaust directivity effects may be calculated using Figures 9.29–9.31.

Table 11.19 Frequency adjustments for unmuffled engine exhaust noise (Equation (11.84))

Octave band centre frequency (Hz)	Value to be subtracted from overall sound power level (dB)
31.5	5
63	9
125	3
250	7
500	15
1000	19
2000	25
4000	35
8000	43
A-weighted (dB(A))	12

Data from Army, Air Force and Navy, USA (1983a).

Table 11.20 Approximate insertion loss (dB) of typical reactive mufflers used with reciprocating engines[a]

Octave band centre frequency (Hz)	Low pressure-drop muffler			High pressure-drop muffler		
	Small	Medium	Large	Small	Medium	Large
63	10	15	20	16	20	25
125	15	20	25	21	25	29
250	13	18	23	21	24	29
500	11	16	21	19	22	27
1000	10	15	20	17	20	25
2000	9	14	19	15	19	24
4000	8	13	18	14	18	23
8000	8	13	18	14	17	23

[a] Refer to manufacturers literature for more specific data.
Data from Army, Air Force and Navy, USA (1983a).

11.12.2 Casing Noise

The overall sound power radiated by the engine casing is given by (Army, Air Force and Navy, USA, 1983a):

$$L_w = 93 + 10 \log_{10} kW + A + B + C + D \quad \text{(dB re } 10^{-12} \text{ W)} \qquad (11.85)$$

The quantities A, B. C and D are listed in Table 11.21, and the corrections to be subtracted from the overall level to obtain the octave band levels are listed in Table 11.22.

Table 11.21 Correction terms to be applied to Equation (11.85) for estimating the overall sound power level (dB re 10^{-12} W) of the casing noise of a reciprocating engine

Speed correction term, A	
Under 600 rpm	-5
600-1500 rpm	-2
Above 1500 rpm	0
Fuel correction term, B	
Diesel only	0
Diesel and natural gas	0
Natural gas only (including a small amount of pilot oil)	-3
Cylinder arrangement term, C	
In-line	0
V-type	-1
Radial	-1
Air intake correction term, D	
Unducted air inlet to unmuffled roots blower	+3
Ducted air from outside the enclosure	0
Muffled roots blower	0
All other inlets (with or without turbo-charger)	0

Data from Army, Air Force and Navy, USA (1983a).

11.12.3 Inlet Noise

For engines with no turbo-charger, inlet noise is negligible in comparison with the casing and exhaust noise. However, for engines with a turbo-charger, the following equation may be used to calculate the overall sound power level of the inlet noise (Army, Air Force and Navy, USA, 1983a):

$$L_w = 95 + 5 \log_{10} kW - \ell/1.8 \quad \text{(dB re } 10^{-12} \text{ W)} \qquad (11.86)$$

where ℓ (m) is the length of the inlet ducting. The octave band levels may be calculated from the overall level by subtracting the corrections listed in Table 11.23.

Table 11.22 Frequency adjustments (dB) for casing noise of reciprocating engines: subtract these values from the overall sound power level (Equation (11.85)) to obtain octave band and A-weighted levels

Octave band centre frequency (Hz)	Engine speed under 600 rpm	Engine speed 600-1500 rev min^{-1}		Engine speed over 1500 rpm
		Without roots blower	With roots blower	
31.5	12	14	22	22
63	12	9	16	14
125	6	7	18	7
250	5	8	14	7
500	7	7	3	8
1000	9	7	4	6
2000	12	9	10	7
4000	18	13	15	13
8000	28	19	26	20
A-weighted (dB(A))	4	3	1	2

Data from Army, Air Force and Navy, USA (1983a).

Table 11.23 Frequency adjustments (dB) for turbo-charger air inlet. Subtract these values from the overall sound power level (Equation (11.86)) to obtain octave band and A-weighted levels

Octave band centre frequency (Hz)	Value to be subtracted from overall sound power level (dB)
31.5	4
63	11
125	13
250	13
500	12
1000	9
2000	8
4000	9
8000	17
A-weighted (dB(A))	3

Data from Army, Air Force and Navy, USA (1983a).

11.13 FURNACE NOISE

Furnace noise is due to a combination of three noise-producing mechanisms, which are the jet noise produced by the entering fuel gas and air, and the noise produced by the combustion process.

Fuel gas flow noise is calculated by using the procedure for estimating control valve noise (see Sections 11.8.2 and 11.8.3). This noise is dominant for burners having a high fuel gas pressure, whereas for burners using fuel oil, it is negligible. If the pressure drop associated with the fuel gas flowing into the furnace is low (less than 100 kPa), then the noise produced is calculated by assuming a free jet (see Section 11.7).

To calculate the overall noise 1 m from the burner and the radiated overall sound power produced by the flow of primary and secondary air, the following equations are recommended (Heitner, 1968):

$$L_p = 44 \log_{10} U + 17 \log_{10} \dot{m} + 44 + 10 \log_{10}\left(\frac{\rho c}{400} \right) \quad \text{(dB re 20 μPa)} \quad (11.87)$$

$$L_w = 44 \log_{10} U + 17 \log_{10} \dot{m} + 55 \quad \text{(dB re } 10^{-12} \text{ W)} \quad (11.88)$$

where U is the air velocity (m/s) through the register, $\dot{m}$ is the air mass flow rate (kg/s) and ρ, c are the density and speed of sound respectively in the gas surrounding the burner.

To estimate the octave band in which the maximum noise occurs, a Strouhal number of 1 is used (see Equation (11.25)). That is:

$$f_p d / U = 1 \quad (11.89)$$

where d is the smallest dimension of the air opening. The sound pressure level in this octave is 3 dB below the overall level. Above and below the octave of maximum noise, the level is reduced at a rate of 5 dB per octave. The total burner noise level is obtained by combining the levels in each octave band for fuel gas flow noise, combustion noise and air flow noise.

Combustion noise is usually important but sometimes not as significant as that produced by air and gas flow, and may be estimated using (Bragg, 1963):

$$W_a = \eta \dot{M} H \quad \text{(W)} \quad (11.90)$$

where W_a is the overall acoustical power (W), η is the acoustical efficiency (of the order of 10^{-6}), $\dot{M}$ is the flow rate of the fuel (kg/s), and H (Joules kg^{-1}) is the heating value of the fuel. Note that 1 calorie = 4.187 Joules. The maximum noise level occurs in the 500 Hz octave band and is 3 dB below the overall level. Above and below the 500 Hz band, the noise level is reduced at the rate of 6 dB per octave.

11.14 ELECTRIC MOTORS

11.14.1 Small Electric Motors (below 300 kW)

The overall sound pressure at 1 m generated by small electric motors can be estimated, for totally enclosed, fan cooled (TEFC) motors, using the following equations (Army, Air Force and Navy, USA, 1983a):

Under 40 kW:

$$L_p = 17 + 17 \log_{10} kW + 15 \log_{10} RPM \quad (\text{dB re } 20 \, \mu\text{Pa}) \tag{11.91}$$

Over 40 kW:

$$L_p = 28 + 10 \log_{10} kW + 15 \log_{10} RPM \quad (\text{dB re } 20 \, \mu\text{Pa}) \tag{11.92}$$

Drip-proof (DRPR) motors produce 5 dB less sound pressure level than TEFC motors. The octave band sound pressure levels may be obtained for both types of motor by subtracting the values in Table 11.24 from the overall levels. A TEFC motor with a quiet fan is likely to be 10 dB quieter than indicated by Equations (11.91) and (11.92).

Table 11.24 Octave band level adjustments (dB) for small electric motors

Octave band centre frequency (Hz)	Totally enclosed, fan cooled (TEFC) motor	Drip proof (DRPR) motor
31.5	14	9
63	14	9
125	11	7
250	9	7
500	6	6
1000	6	9
2000	7	12
4000	12	18
8000	20	27

Data from Army, Air Force and Navy, USA (1983a).

11.14.2 Large Electric Motors (above 300 kW)

Sound power levels radiated by electric motors with a power rating between 750 kW and 4000 kW are listed in Table 11.25. These levels can be reduced by 5 dB(A) for slow-speed motors and up to 15 dB(A) for high-speed motors, if specifically requested by the customer.

For motors rated above 4000 kW, add 3 dB to all levels in Table 11.25. For motors rated between 300 and 750 kW, subtract 3 dB from all levels.

Table 11.25 Sound power levels of large electric motors[a]

Octave band centre frequency (Hz)	1800 and 3600 rpm	1200 rpm	900 rpm	720 rpm and lower	250 and 400 rpm vertical
315	94	88	88	88	86
63	96	90	90	90	87
125	98	92	92	92	88
250	98	93	93	93	88
500	98	93	93	93	88
1000	98	93	96	98	98
2000	98	98	96	92	88
4000	95	88	88	83	78
8000	88	81	81	75	68

[a] Applies to induction motors rated between 750 and 4000 kW; includes drip-proof and P-1 and WP-2 enclosures (with no acoustical specification by the customer). Data from Edison Electric Institute (1978).

11.15 GENERATORS

The overall sound power levels radiated by generators (excluding the driver) can be calculated using the following equation:

$$L_w = 84 + 10 \log_{10} MW + 6.6 \log_{10} RPM \quad \text{(dB re } 10^{-12} \text{ W)} \tag{11.93}$$

To obtain the octave band levels, the values in Table 11.26 should be subtracted from the overall level calculated using Equation (11.93).

11.16 TRANSFORMERS

The sound power radiated in octave bands by a transformer is related to its NEMA (National Electrical Manufacturers Association, 1980) sound level rating and the total surface area of its four side walls by:

$$L_w = N_R + 10 \log_{10} S + C \quad \text{(dB re } 10^{-12} \text{ W)} \tag{11.94}$$

The quantity N_R is the NEMA sound level rating, which is the average sound pressure level measured around the transformer at a distance of 0.35 m, and is generally specified by the transformer manufacturer. The quantity, S is the surface area (m²) of the four transformer walls (excluding the top). Values of the correction term C are listed in Table 11.27.

Table 11.26 Octave band corrections for generator noise

Octave band centre frequency (Hz)	Value to be subtracted from overall sound power level L_w (dB)
31.5	11
63	8
125	7
250	7
500	7
1000	9
2000	11
4000	14
8000	19

Data from Army, Air Force and Navy, USA (1983a).

Table 11.27 Values of correction C of Equation (11.94) for transformer noise

Octave band centre frequency (Hz)	Octave band corrections (dB)		
	Location 1[a]	Location 2[b]	Location 3[c]
31.5	-1	-1	-1
63	5	8	8
125	7	13	13
250	2	8	12
500	2	8	12
1000	-4	-1	6
2000	-9	-9	1
4000	-14	-14	-4
8000	-21	-21	-11

[a] Outdoors, or indoors in a large mechanical room with a large amount of mechanical equipment.
[b] Indoors in small rooms, or large rooms with only a small amount of other equipment.
[c] Any critical location where a problem would result if the transformer should become noisy above its NEMA rating, following installation.
Data from Army, Air Force and Navy, USA (1983a).

Values of C are lower than actual for air-filled transformers and higher than actual for oil-filled transformers. Consult manufacturer's data for values of sound level rating. Typical values are listed in Table 11.28.

Table 11.28 Transformer sound level ratings, N_R for transformers built after 1995

Transformer power (kVA)	Sound level (dB)		
	Standard		Super quiet core
	No fan	Fan	
100-300	55	67	52
301-500	56	67	52
501-700	57	67	53
701-1000	58	67	56
1001-1500	60	67	58
1501-2000	61	67	60
2001-2500	62	67	62
2501-3000	63	67	63
3001-4000	64	67	63
4001-5000	65	67	64
5001-6000	66	68	65
6001-7500	67	69	66
7501-10000	68	70	67
10001-12000	69	71	68
12001-15000	70	72	69
15001-20000	71	73	70

11.17 GEARS

The following equation gives octave band sound pressure levels for gearboxes in all frequency bands at and above 125 Hz, at a distance of 1 m from the gearbox (Army, Air Force and Navy, USA, 1983a):

$$L_p = 78 + 4 \log_{10} kW + 3 \log_{10} RPM \quad \text{(dB re } 10^{-12} \text{ W)} \qquad (11.95)$$

where kW is the power transmitted by the gearbox, and RPM is the rotational speed of the slowest gear shaft. For the 63 Hz octave band, subtract 3 dB, and for the 31 Hz band subtract 6 dB from the value calculated using Equation (11.95).

These noise levels are applicable to spur gears, and may be reduced somewhat (by up to 30 dB) by replacing the spur gear with a quieter helical or herringbone design. The actual noise reduction (compared to a straight spur gear) is given very approximately by $13 + 20 \log_{10} Q_a$, where Q_a is the number of teeth that would be intersected by a straight line parallel to the gear shaft. For double helical or herringbone gears, the number of intersected teeth would only be for one helix, not both.

11.18 TRANSPORTATION NOISE

11.18.1 Road Traffic Noise

Traffic noise consists predominantly of engine/exhaust noise and tyre/road interaction noise. Engine/exhaust noise is dependent on the vehicle speed and the gear used, which in turn are dependent on vehicle technology, the grade of the road and driving behaviour. Tyre noise is dependent on the vehicle speed and the quality of the road surface. In automobiles, engine/exhaust noise generally predominates in first and second gear, engine/exhaust and tyre noise are equally loud in third, while tyre noise predominates in fourth gear.

There are a number of commercial models available for calculating the expected traffic noise for a particular vehicle number and speed and road surface. Two of these will be considered in detail here, the UK DOT (Department of Transport), CoRTN model and the USA Federal Highway Administration (FHWA) Traffic Noise Model, Version 1.1.

11.18.1.1 UK DoT model (CoRTN)

A relatively simple procedure to estimate the noise impact of a particular traffic flow, which is based on the model developed by the UK Dept. of Environment (referred to as CoRTN or Calculation of Road Traffic Noise) yields reasonable results, although it can result in significant errors in some cases (UK DoT, 1988; Delaney *et al.*, 1976).

For normal roads, the traffic flow in both directions is combined together to give the total traffic flow used for the sound pressure level calculations. However, if the two carriageways are separated by more than 5 metres, then the sound pressure level contribution at the receiver location due to each carriageway must be calculated separately and the results combined logarithmically using Equation (1.98) to give the total sound pressure level due to both carriageways. The nearside carriageway is treated as for a normal road. However, for the far side carriageway, the source line is assumed to be 3.5 m in from the far kerb and the effective edge of the carriageway is considered to be 7 m in from the far kerb.

The CoRTN model calculates A-weighted L_{10} (denoted here as L_{A10}) over 1-hour or 18-hour intervals. The A-weighted L_{10} (18hr) quantity is simply the arithmetic mean of the 18 separate one-hourly values of L_{10} covering the period 6:00 am to 12:00 pm on a normal working-day. The CoRTN model allows the sound level to be estimated at a distance, *d,* from the vehicle source using:

$$L_{A10}(18\text{hr}) = 29.1 + 10\log_{10}Q + C_{dist} + C_{use} + C_{grad} + C_{cond}$$
$$+ C_{ground} + C_{barrier} + C_{view} \quad \text{dB(A)} \tag{11.96}$$

Most recent regulations are expressed in terms of L_{Aeq} so there has been considerable interest in converting L_{10} estimates to L_{Aeq} estimates. The Transport Research Limited (TRL) (Abbott and Nelson, 2002) developed two conversion equations but these are not reproduced here due to some considerable disagreement with measured data (Kean, 2008). He showed that at 13.5 m from the roadway, the difference between L_{Aeq} and L_{A10} is very close to 3 dB(A) with L_{A10} being greater. This difference decreases with distance from the roadway and the reader is referred to Kean (2008) for more details.

A more accurate way of estimating L_{A10} (18hr) is to use estimates of L_{A10} (1hr) for each single hour in the 18-hour period:

$$L_{A10}(1\text{hr}) = 42.2 + 10 \log_{10}q + C_{dist} + C_{use} + C_{grad} + C_{cond}$$

$$+ C_{ground} + C_{barrier} + C_{view} \quad \text{dB(A)} \tag{11.97}$$

In Equations (11.96) and (11.97), Q is the total number of vehicles in the 18-hour period between 6:00 am and 12:00 pm, q is the number of vehicles per hour, C_{dist} is a correction factor to account for the distance of the observer from the road, C_{use} is a correction factor to account for the percentage of heavy vehicles, C_{grad} is a correction factor to account for the gradient of the road surface, C_{cond} is a correction factor to account for the type and condition of the road surface, C_{ground} is a correction factor to account for the effect of the ground surface and $C_{barrier}$ is a correction factor to account for the presence or otherwise of barriers. The distance correction is given by:

$$C_{dist} = -10 \log_{10}(r/13.5) \tag{11.98}$$

where r is the straight line distance from the source to the observer (dependent on source and observer height). The source line is assumed to be 3.5 m in from the near edge of the road and both carrriageways are treated together, except if they are more than 5 m apart (see previous page).

The use correction is given by:

$$C_{use} = 33 \log_{10}(v + 40 + (500/v)) + 10 \log_{10}(1 + (5P/v)) - 68.8 \quad \text{dB(A)} \tag{11.99}$$

where P is the percentage (0-100%) of heavy vehicles (weighing more than 1525 kg) and v is the average speed (km/h). Information on the average speed on most roads in the metropolitan area is available from the relevant government department responsible for road construction and maintenance. If it cannot be found or determined, the default values in Table 11.29 can be used for v.

The road gradient correction is $C_{grad} = 0.3G$ if the measured average speed is used and $C_{grad} = 0.2G$ if the design speed of the road is used, where G is the percentage gradient of the road. Note that no correction is used for vehicles travelling downhill.

The correction, C_{cond}, for the road surface is taken as zero for either sealed roads at speeds above 75 km/hr or gravel roads. For speeds below 75 km/hr on impervious sealed roads, the correction is -1 dB(A). For pervious road surfaces, the correction is -3.5 dB(A).

The correction, C_{ground} for $1.0 < h_r < (d/3 - 1.2)$ is:

$$C_{ground} = 5.2 P_d \log_{10} \left(\frac{3 h_r}{d + 3.5} \right) \quad \text{dB(A)} \tag{11.100}$$

where h_r is the observer height, d is the horizontal distance from the edge of the road to the observer (independent of source or receiver height and assumed greater than 4 m), P_d is the proportion (1.0 or less) of absorbent ground between the edge of the road and the observer and h_r is the height of the observer above the ground. For sound propagation over grass, P_d is set equal to 1.0 for L_{A10} calculations and 0.75 for L_{Aeq} calculations (Kean, 2008). Kean (2008) also points out that the ground correction factor was derived empirically and includes air absorption. Thus, the value of P_d should never be set less than 0.3 so that air absorption is included.

Table 11.29 Suggested average vehicle speeds for various road types and speed limits

Type of road	Speed limit	Value for v
Rural roads	110 km/h	108
Urban freeway	90 km/h	92
Urban highway	70 km/h	65
Urban street dual carriageway		60
Urban street single carriageway		55
Urban street single congested		50

If the observer height, h_r, is greater than $(d/3 - 1.2)$, then $C_{ground} = 0$. If the observer height is less than 1 m, then:

$$C_{ground} = 5.2 P_d \log_{10} \left(\frac{3}{d + 3.5} \right) \quad \text{dB(A)} \tag{11.101}$$

The relation between r (of Equation (11.98)) and d is (as the vehicle source is considered to be 0.5 m above the road and 3.5 m from the edge of the road), is:

$$r = \left[(d + 3.5)^2 + (h_r - 0.5)^2 \right]^{1/2} \tag{11.102}$$

Low barriers such as twin beam metal crash barriers can have less effect than soft ground. So if these are used with any proportion, P_d, of soft ground, their effect should be calculated by looking at the lower noise level (or the most negative correction) resulting from the following two calculations:

1. Soft ground correction ($0 < P_d < 1.0$), excluding the barrier correction; and
2. hard-ground correction ($P_d = 0$) plus the barrier correction.

If highway noise barriers exist, then their effect on the noise level at the observer may be calculated using:

$$C_{barrier} = \sum_{i=0}^{n} A_i X^i \quad dB(A) \tag{11.103}$$

where $X = 10\log_{10}\delta$ and δ is the difference (in metres) in the following two paths from the source line (3.5 m in from the edge of the road and at a height of 0.5 m) to the observer:

1. shortest path over the top of the barrier; and
2. shortest direct path in the absence of the barrier.

The constants, A_i, are listed in Table 11.30 and X^i means the quantity, X, raised to the ith power and $n = 5$ or 7 (see Table 11.30).

Table 11.30 Coefficients for barrier effect calculations for traffic noise

Coefficient	Shadow zone	Bright zone
A_0	-15.4	0.0
A_1	-8.26	+0.109
A_2	-2.787	-0.815
A_3	-0.831	+0.479
A_4	-0.198	+0.3284
A_5	+0.1539	+0.04385
A_6	+0.12248	–
A_7	+0.02175	–
Above valid for	$-3 \leq X \leq 1.2$	$-4 \leq X \leq 0$

For X outside the limits in the table, the following applies:
Shadow zone; for $X < -3$, $C_{barrier} = -5.0$ and for $X > 1.2$, $C_{barrier} = -30.0$
Bright zone; for $X < -4$, $C_{barrier} = -5.0$ and for $X > 0$, $C_{barrier} = 0.0$

Note that when multiple barriers of different heights screen the observer from the road, they should be evaluated separately and only the correction resulting in the lowest noise level should be used.

In some cases the angle of view of the road will include a range of different configurations such as bends in the road, intersections and short barriers. To accommodate this, the overall field of view must be divided into a number of segments, each of which is characterised by uniform propagation conditions. The overall sound level can be found by calculating the sound level due to each segment separately using Equations (11.96) and (11.97) and then adding the contributions (in dB) from each segment together logarithmically as for incoherent sources (see Section 1.11.4). In this case, the following correction is then applied to each segment.

$$C_{view} = 10 \log_{10}(\beta/180)$$ (11.104)

where β is the actual field of view in degrees. Note that in such segments, the road is always projected right along the field of view and the distance from the segment is measured perpendicular to the extended road, as illustrated in Figure 11.6.

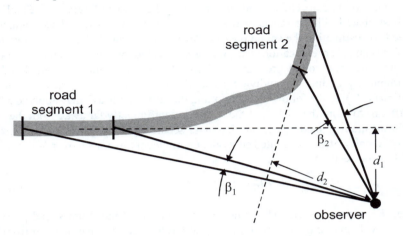

Figure 11.6 Arrangement defining β and d for two different road segments.

The segment method just outlined can be simplified if there are two propagation conditions that repeat. For example if there existed a barrier (or set of buildings) with regular or irregular gaps in it, then all the gaps could be added together to determine an effective β for the gaps and the same could be done for the barrier segments so that only two segments need be considered.

If there are dual carriageways (separated by more than 5 m) or multiple roads, then each road or carriageway is treated separately and the results added together logarithmically using Equation (1.98). However, for best accuracy (not proposed in CoRTN), it is best to treat each lane of traffic separately (as done in the FHWA model in the next section) and use the actual distance of the centre of the lane of traffic from the edge of the roadway, instead of the arbitrary 3.5 m as specified in CoRTN

Reflections from buildings and other hard surfaces increase the noise levels at the observer. The following general empirical corrections may be used to estimate the increase:

1.   If the observer is within 1m of a building facade, then the noise level is increased by 2.5 dB(A).

2.   Noise levels down side streets perpendicular to the road in question are 2.5 dB(A) higher due to reflections from adjacent houses.

3.   Reflective surfaces on the far side of the road increase the level by 1.5 dB(A).

11.18.1.2 United States FHWA Traffic Noise Model (TNM)

The United States FHWA Traffic Noise Model (TNM) is intended to be used for predicting noise impacts in the vicinity of highways. This is a reasonably complex model and is available as a software implementation from the FHWA. The FHWA (1995) also produces a general guide for analysing the impact of traffic noise and how it is best abated. The guide is available on the FHWA website. The calculation proceeds by estimating the sound pressure level due to a single lane of a single type of traffic at an observer location. The process is repeated for each lane and each traffic type, and the total sound pressure level for all lanes combined is calculated by logarithmically adding the individual sound pressure levels for each lane using Equation (1.98). In the following description, the term "roadway" may apply to a lane of traffic or to more than 1 lane if the receiver is sufficiently far away from the road that the accuracy gained by considering lanes separately is insignificant.

TNM calculates A-weighted equivalent noise levels, averaged over 1-hour using the following relation:

$$L_{Aeq,1h} = EL_i + A_{traff(i)} + A_d + A_s \qquad (11.105)$$

where, EL_i represents the vehicle noise emission level (maximum sound pressure level emitted by a vehicle pass-by at a reference distance of 15 m from the vehicle centre).

$A_{traff(i)}$ represents the adjustment for the vehicle volume and the speed for the vehicle of type, i.

A_d represents the adjustment for distance between the roadway and receiver and for the length of roadway.

A_s represents the adjustment for all shielding and ground effects between the roadway and receiver.

Note that the roadway section of interest must be divided into segments that subtend no more than 10 degrees at the receiver. Levels at the receiver due to each segment are added together logarithmically, using Equation (1.98).

When the hourly noise levels are combined together in the appropriate way, as discussed in Chapter 4, the average day–night sound level, L_{dn}, and the community noise equivalent level, L_{den}, can be calculated easily. After the noise levels corresponding to all of the different vehicle types and roadway segments have been calculated for a particular receiver location, they are added together logarithmically to give the total level.

The TNM data base for vehicle emission levels includes data for a number of different pavement conditions and vehicle types as well as for vehicles cruising, accelerating, idling and on grades. The data base includes 1/3 octave band spectra for cars (2 axles and 4 wheels), medium trucks (2 axles and 6 wheels), heavy trucks (3 or more axles), buses (2 or 3 axles and 6 or more wheels) and motorcycles (2 or 3 wheels). The data are further divided into two source locations; one at pavement height and one at 1.5 m in height (except for heavy trucks for which the upper height is 3.66 m). The data base is available in the FHWA Traffic Noise Model technical manual (Menge *et al.*, 1998). However, it is recommended that the FHWA Traffic

Noise Model software be purchased from the FHWA (USA) if accurate estimates of traffic noise impact are required. Alternatively, it is possible to take one's own measurements of particular vehicles and use those.

The TNM correction, $A_{traff(i)}$ for vehicle volume and speed is the same for all vehicle types and is given by the following equation.

$$A_{traff(i)} = 10 \log_{10}\left(\frac{V_i}{v_i}\right) - 13.2 \quad \text{(dB)} \tag{11.106}$$

where V_i is the vehicle volume in vehicles per hour of vehicle type, i and v_i is the vehicle speed in km/hr.

The adjustment, A_d, for distance from the elemental roadway segment to the receiver and for the length of the roadway segment for all vehicle types and source heights is given by:

$$A_d = 10 \log_{10}\left[\left(\frac{15}{d}\right)\left(\frac{\beta}{180}\right)\right] \quad \text{(dB)} \tag{11.107}$$

where d is the perpendicular distance in metres from the receiver to the line representing the centre of the roadway segment (or its extension) and β is the angle subtended at the receiver (in degrees) by the elemental roadway segment (that is, the field of view - see Figure 11.6). If $d < 0.3$ m AND $\beta < 20°$, the following equation should be used:

$$A_d = 10 \log_{10}\left[\frac{|d_2 - d_1|}{d_2 d_1}\right] + 12 \quad \text{(dB)} \tag{11.108}$$

where d_1 and d_2 are the distances from the receiver to each end of the roadway segment.

The calculation of the correction factor, A_s, for all shielding and ground effects between the roadway and receiver, is quite complicated and is explained in detail in the FHWA Traffic Noise Model technical manual (Menge *et al.*, 1998). Alternatively, the procedures outlined in Chapters 5 and 8 may be used.

The FHWA model is regarded as very accurate and more up to date than the CoRTN model.

11.18.1.3 Other Models

There are a number of other traffic noise models that are considerably more complex including the German Road Administration model (RLS-90, "guidelines for noise protection on streets"), the Acoustical Society of Japan model, which was later updated by Takagi and Yamamoto (1994), and the revised version of the joint Nordic prediction method for road traffic noise, published in 1989 and used mainly in Scandinavia. All of these models are similar in that they contain a source model for predicting the noise at the roadside (or close to it) and a propagation model that takes into account ground and atmospheric effects. The models have all been implemented in specialised software, which in most cases is available for a reasonable price. Useful reviews of the various models are available in the literature (Saunders *et al.*, 1983; Steele, 2001).

A comprehensive review of the effect of vehicle noise regulations on road traffic noise, changes in vehicle emissions over the past 30 years and recommendations for consideration in the drafting of future traffic noise regulations has been provided by Sandberg (2001).

11.18.2 Rail Traffic Noise

Train noise is usually dominated by wheel/rail interaction noise. As the train speed increases, the wheel/rail noise increases, but the locomotive engine noise decreases. Thus, it is often necessary to calculate the contribution from the two types of noise separately. The prediction of train noise is described in detail by Department of Transport – UK publications (U.K. DOT, 1995a, 1995b). The first publication describes the four stages involved in estimating noise from moving trains. These are:

1. **Divide the rail line into a number of segments such that the variation of noise over any segment is less than 2 dB(A).** This will be necessary in situations where there are bends in the track or if the gradient changes sufficiently or if the screening or ground cover changes to make a greater than 2 dB(A) difference. A long, straight track with constant gradient and noise propagation properties can usually be considered as a single segment. Clearly, crossings and train stations would also require segmentation of the track for the purpose of noise level calculations.

2. **For each segment determine the reference SEL (SEL_{ref}) at a given speed and at a distance of 25 m from the near-side of the track segment.** SEL is defined in Chapter 4. The first thing to do is obtain the single vehicle SEL value (SEL_v). This can be done by measuring the SEL at a distance of 25 m from the track for a train over a range of passing speeds. If a locomotive is involved, the SEL for that should be measured separately and then subtracted from the overall SEL to obtain the SEL for just the rolling stock. The SEL for a single vehicle is then calculated using:

$$SEL_v = SEL_T - 10 \log_{10} N \qquad (11.109)$$

where SEL_T is the overall SEL measured for N identical vehicles in the train. Next, a linear regression of SEL *vs* $\log_{10}$(speed) is then undertaken. Results provided by the Department of Transport – UK (U.K. DOT, 1995a) for rolling railway vehicles are:

$$SEL_v = 31.2 + 20 \log_{10} v + C_1 \qquad \text{dB(A)} \qquad (11.110)$$

and for locomotives under power or EuroStar fan noise:

$$SEL_v = 112.6 - 10 \log_{10} v + C_1 \quad \text{dB(A)} \tag{11.111}$$

where v is the vehicle speed in km/h and C_1 is a constant dependent on train type and listed in Table 11.31 for a few British trains (Department of Transport – UK (U.K. DOT, 1995a, 1995b)).

Table 11.31 Example SEL corrections, C_1 for various single railway vehicles, except for Eurostar for which the correction is for the entire train

Vehicle type	Correction, C_1 (dB(A))
Passenger coaches - tread braked	
Class 421 EMU or 422 EMU	10.8
British rail MK I or II	14.8
Passenger coaches - disc braked, 4 axles	
Class 319 EMU	11.3
Class 465 EMU and 466 EMU	8.4
Class 165 EMU and 166 EMU	7.0
British rail MK III or IV	6.0
Passenger coaches - disk braked, 6 axles	15.8
Passenger coaches - disk braked, 8 axles	14.9
Freight vehicles, tread braked, 2 axles	12.0
Freight vehicles, tread braked, 4 axles	15.0
Freight vehicles, disc braked, 2 axles	8.0
Freight vehicles, disc braked, 4 axles	7.5
Diesel locomotives (steady speed)	
Classes 20, 33	14.8
Classes 31, 37, 47, 56, 59, 60	16.6
Class 43	18.0
Diesel locomotives under full power	
Classes 20, 31, 33, 37, 43, 47, 56, 59	0.0
Class 60	-5.0
Electric locomotives	14.8
Eurostar rolling noise	
(2 powered cars separated by 14 or 18 coaches)	17.2
Eurostar fan noise	
(2 powered cars separated by 14 or 18 coaches)	-7.4

Note that different vehicle types must be considered as separate trains and the L_{Aeq} for each vehicle type (or train) is combined using Equation (1.98) to give the total L_{Aeq} for the entire train. For any specific train type consisting of N identical units, the quantity SEL_{ref} is calculated by adding $10\log_{10}N$ to SEL_v. In addition the track correction, C_2 from Table 11.32 must also be added so that:

$$SEL_{ref} = SEL_v + 10\log_{10}N + C_2 \tag{11.112}$$

The values of SEL_{ref} calculated using the procedures just described apply to continuously welded track with timber or concrete sleepers and ballast. For other track types, the corrections in Table 11.32 should be added to the SEL_{ref} values.

Table 11.32 Corrections C_2 to be added to SEL_{ref} to account for different track types

Description of rail	Correction C_2 (dB(A))	
Jointed track	2.5	
Points and crossings	2.5	
Slab track		2.0
Concrete bridges and viaducts		
(excludes shielding by parapet)	1.0	
Steel bridges		
(excludes shielding by parapet)	4.0	
Box girder with rails fitted directly to it	9.0	

3. **Determine the corrections for distance, ground effects, air absorption, barrier diffraction, angle of view of the observer to the track segment and reflections from buildings and barriers.** The SEL value at the observer is:

$$SEL = SEL_{ref} + C_{dist} + C_{abs} + C_{ground} + C_{barrier} + C_{view} \tag{11.113}$$

For this calculation, the source height used for rolling stock is the rail height. The source height used for the locomotive of EuroStar fan noise is 4 m above the track. In both cases, the distance to the observer is the shortest distance to the near side rail. The corrections listed in Table 11.32 are meant to be applied to overall A-weighted numbers, which is what SEL_{ref} is. Many of the corrections have been calculated based on typical noise spectra and would not apply to other types of sound source.

The **distance correction** is based on treating the noise source as a line source and for distances greater than 10 m, the correction to be added to SEL_{ref} is:

$$C_{dist} = -10\log_{10}(r/25) \tag{11.114}$$

where r is the straight line distance from the source to the observer and is defined for a diesel locomotive as:

$$r = \sqrt{d^2 + (h - 4.0)^2} \qquad (11.115)$$

and for everything else as:

$$r = \sqrt{d^2 + h^2} \qquad (11.116)$$

where h is the difference in height between the track and observer (observer height – track height) and d is the straight line normal distance from the track segment (or in many cases its extension – see Figure 11.6) to the observer.

The **air absorption correction** is:

$$C_{abs} = 0.2 - 0.008\,r \qquad (11.117)$$

The **ground correction** is:

$$
\begin{aligned}
C_{ground} &= 0 & \text{for } 10 < d \le 25 \text{ or } H > 6\,\text{m} \\
&= -0.6 P_d (6 - H) \log_{10}(d/25) & \text{for } 1.0 < H < 6\,\text{m} \\
&= -3 P_d \log_{10}(d/25) & \text{for } H \le 1.0\,\text{m}
\end{aligned} \qquad (11.118)
$$

where P_d is the fraction of absorbing ground between the source and receiver, d is the horizontal normal distance from the track segment (or in many cases its extension – see Figure 11.6) to the observer and H is the mean propagation height, $= 0.5 \times$ (source height + receiver height) for propagation over flat ground.

When **ballast** is used to support the railway sleepers, the SEL is reduced by a further 1.5 dB for all segments of track except for the one closest to the observer.

The **barrier correction** for the shadow zone (see Figure 11.7a) is:

$$
\begin{aligned}
C_{barrier} &= -21 \quad \text{dB(A)} & \text{for } \delta > 2.5\,\text{m} \\
&= -7.75 \log_{10}(5.2 + 203\,\delta) \quad \text{dB(A)} & \text{for } 0 < \delta < 2.5\,\text{m}
\end{aligned} \qquad (11.119)
$$

and for the bright zone (see Figure 11.7b) it is:

$$
\begin{aligned}
C_{barrier} &= 0 & \text{for } \delta > 0.4\,\text{m} \\
&= 0.88 + 2.14 \log_{10}(\delta + 10^{-3}) \quad \text{dB(A)} & \text{for } 0 < \delta < 0.4\,\text{m}
\end{aligned} \qquad (11.120)
$$

The quantity, δ, used in Equations (11.119) and (11.120), is the difference in the length of the two paths shown in each of the two parts of Figure 11.7. That is:

$$\delta = SB + BR - SR \qquad (11.121)$$

where S indicates the source location, R the receiver location and B the barrier edge location adjacent to where the line SR pierces the barrier. If several edges are involved, the reductions can be combined using Equation (1.101) and the procedures in the barrier section of Chapter 8.

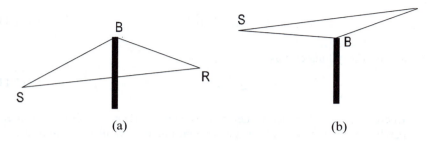

Figure 11.7 Geometry for calculating barrier correction: (a) shadow zone; (b) bright zone.

Note that when multiple barriers of different heights screen the observer from the road, they should be evaluated separately and only the correction resulting in the lowest noise level should be used. **Note that the barrier correction and ground correction are never used at the same time in Equation (11.113).** If the barrier correction is less than the ground correction, then the ground correction is used, otherwise the barrier correction is used.

The **view correction** (for $\alpha > \beta/2$) for all trains except diesel locomotives under full power is:

$$C_{view} = 10 \log_{10}[\beta - \cos(2\alpha)\sin\beta] - 5 \qquad (11.122)$$

and for diesel locomotives under full power it is:

$$C_{view} = -10 \log_{10}[\sin\alpha \ \sin(\beta/2)] \qquad (11.123)$$

where β is in radians and is defined along with α in Figure 11.8. Note that α is always less than 90° and is the acute angle between a line drawn through the observer, R, parallel to the track segment and the line bisecting the angle of view, β.

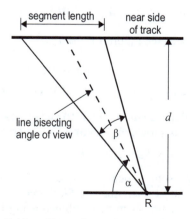

Figure 11.8 Definition of the angles used to calculate the view correction.

Reflections from buildings and other hard surfaces increase the noise levels at the observer. The following general empirical corrections may be used to estimate the increase:

(a)   If the observer is within 1 m of a building facade, then the noise level is increased by 2.5 dB(A).

(b)   Noise levels down side streets perpendicular to the railway in question are 2.5 dB(A) higher due to reflections from adjacent houses.

(c)   Reflective surfaces on the far side of the railway increase the level by 1.5 dB(A).

4.   **Convert values of SEL to L_{Aeq}.** This is done for each track segment and for each train using the following relationships:

$$L_{Aeq,6h} = SEL - 43.3 + 10\log_{10}Q_{NIGHT} \tag{11.124}$$

$$L_{Aeq,18h} = SEL - 48.1 + 10\log_{10}Q_{DAY} \tag{11.125}$$

where Q_{NIGHT} is the number of each train type passing the observer in the period, midnight to 6 am and Q_{DAY} is the number of each train type passing the observer during the period, 6 am to midnight.

Overall L_{Aeq} values for each period are obtained by logarithmically summing the component L_{Aeq} values for each train and each track segment using Equation (1.98). Note that the levels calculated using the preceding procedure can be up to 15 dB(A) higher if the track top surface is corrugated.

11.18.3 Aircraft Noise

The prediction of aircraft noise in the vicinity of airports is traditionally done using complex noise contour generation software. The calculation procedures are complex, as most airports are characterised by many flight paths, many aircraft types and many different engine power settings. Two noise-contour programmes that are widely used and are in the public domain are "NOISEMAP", which was developed over a number of years under sponsorship of the US Air Force and "INM" (Integrated Noise Model), which was developed under the sponsorship of the US Department of Transportation (DOT) and the Federal Aviation Administration (FAA). Others in use include FAA's helicopter noise model (HNM), the British AIRNOISE model for military aircraft (Berry and Harris, 1991), the US Air Force's BOOMMAP for supersonic aircraft, MOAMAP for modelling noise from flights in military operating areas and ROUTEMAP for modelling noise from operations along military training routes. Larson (1994) discusses the merits of the various models mentioned above. A detailed analysis of the differences between INM and NOISEMAP is undertaken by Chapkis (1980) and in greater detail by Chapkis *et al.* (1981). Another package, MITHRA (Jean, *et al.*, 2001) extends the accuracy and reliability of the INM model by using more sophisticated sound propagation modelling. However, the ground propagation modelling procedures in the INM package have more recently recently been updated

(Fleming *et al.*, 2000) to be comparable in reliability to the MITHRA model using the INM source data.

The outputs from the noise modelling software can be in the form of Sound Exposure Levels (*SEL* or L_{AE}), Effective Perceived Noise Level (*EPNL* or L_{PEN}) or Day/Night Equivalent Levels (*DNL* or L_{dn}). These quantities are defined in Chapter 4. In addition, alternative methods for calculating *EPNL* are discussed by Raney and Cawthorn (1998) and Zaporozhets and Tokarov (1998).

The noise models all contain extensive data bases of values of *SEL* as a function of distance of the closest approach of an aircraft to an observer for various engine power or thrust settings. Aircraft performance data are used to determine the height above ground and the engine thrust as a function of aircraft load and distance from brake release or landing threshold (Chapkis *et al.*, 1981).

For a specified flight path and track of a particular aircraft, the noise level at a specified ground location is determined by first calculating the distance to the ground point of the closest part of the flight path. Next the noise data base is used to find the SEL or EPNL corresponding to the distance and thrust setting. Next, adjustments to the level are applied to account for ground attenuation (mainly for shallow angles subtended from the horizontal at the aircraft by a line from the ground point), for fuselage shielding (again for shallow angles only) and for changes to the duration correction in SEL or EPNL as a result of curved flight paths and differences between actual and reference aircraft speeds. The noise contributions to the level at each ground point from all flights into and out of the airport in a given time period are summed on an energy basis to obtain the total noise exposure. This calculation is performed at a large number of ground points to allow contours of equal noise exposure to be generated.

As the calculation procedures are complex, it is recommended that one of the software packages mentioned above is used whenever a community noise exposure assessment as a result of aircraft operations is to be made.

CHAPTER TWELVE

Practical Numerical Acoustics
by Carl Howard

LEARNING OBJECTIVES

In this chapter the reader is introduced to:

- the basic theory underpinning various numerical analysis techniques;
- the use of commercially available software and free software for solving complex sound radiation and transmission problems;
- the difference in analysis approached between low frequency and high frequency problems;
- Rayleigh Integral Analysis;
- Boundary Element Method;
- Finite Element Analysis;
- Modal Coupling Analysis; and
- Statistical Energy Analysis.

12.1 INTRODUCTION

The determination of the sound power radiated by a machine or structure of complex shape at the design stage or the prediction of the distribution of sound in an enclosed space requires the application of numerical techniques. At low frequencies these include Boundary Element Analysis (BEM), Finite Element Analysis (FEA) and modal coupling analysis using MATLAB. At higher frequencies, Statistical Energy Analysis (SEA) must be used. There is insufficient space here to provide a complete description of the underlying theory for all of these techniques and a summary is all that will be presented along with some practical implementation examples.

If the expected forcing function can be determined (for example, by suitable measurements on a model in which the load impedance presented to the source is properly represented), then the sound power that will be radiated by the structure can be determined by using one of the above numerical or analytical techniques. Such an approach is particularly useful when the effects of modifications to existing structures, for the purpose of noise reduction, are to be investigated. For implementation of these analytical techniques three fundamental steps are necessary.

The first step is the determination and quantification of the force exciting the structure. A given exciting force is generally separated into a sum of sinusoidal components using Fourier analysis. The second step is the determination of the vibrational velocity distribution over the surface of the machine or structure in response to the excitation force. The final step is the calculation of the sound field, and hence the sound power generated by the vibrational response of the structure or machine surface. A similar approach is needed for the calculation of the distribution of a sound field in a room.

In the following paragraphs, numerical techniques, which have been used to determine radiated sound power, and the associated underlying theory, are described briefly. In-depth mathematical treatment is complicated and lengthy; and is covered adequately in the references.

The most fundamental point to understand is that the response of structures and machines can be described in terms of their normal modes of vibration. These normal modes can be excited at resonance or driven at frequencies different from resonance. The response of a machine or structure is always a combination of various vibration modes, most of which are driven off-resonance. In general, the heavier the machine or structure, the higher will be the frequency of the first modal resonance and the fewer will be the number of modes with resonance frequencies in any particular octave or one-third octave band. In the analysis of a structure, the first step is to divide the frequency range of interest into octave or one-third octave bands which, in turn, will lie either in a low-frequency or a high-frequency region. The low-frequency region is characterised by a paucity of modes in every frequency band, whereas the high-frequency region is defined as that region where there are consistently three or more vibration modes with resonance frequencies in the analysis bands.

12.2 LOW-FREQUENCY REGION

In the low-frequency region, the surface velocity distribution (or mode shape) is calculated for each vibration mode. For this purpose, a standard numerical analysis procedure such as finite element analysis may be used. For the analysis, the structure is divided into a finite number of surface elements. The element equilibrium and inter-connectivity requirements are satisfied using a system of differential equations. Many commercially available software packages exist, making this method relatively quick and straightforward to apply, even for a three-dimensional structure, once some basic fundamentals have been understood. However, it is only practical to use finite element analysis for the first few (up to 50) vibration modes of a structure. Beyond this, the required element size for accurate prediction becomes too small, the computational process becomes time consuming and prohibitively expensive, and the uncertainty in the accuracy of the model means that the results have a range of possible values. For these reasons Statistical Energy Analysis can be used for analyses at higher frequencies and is described later in this chapter.

If the overall velocity response of a structure is to be calculated using the finite element method, then knowledge of the damping of each vibration mode (or alternatively a global damping value) is needed for a given excitation force. Values of damping cannot be calculated and are generally estimated from measurements on, and experience gained with, similar structures or machines.

The vibrational velocity $v(\omega, r_0)$, at a point r_0 on a structural surface of area S, due to a sinusoidal excitation force of $F(\omega)$ at forcing frequency ω applied at r_F, is given by Ewins (1984) as:

$$v(\omega, \mathbf{r}_0) = j\omega F(\omega) \sum_{\ell=1}^{N_s} \frac{\psi_\ell(\mathbf{r}_0)\, \psi_\ell(\mathbf{r}_F)}{\Lambda_\ell Z_\ell} \tag{12.1}$$

where $\psi_\ell(\mathbf{r}_0)$ and $\psi_\ell(\mathbf{r}_F)$ are the normalised modal responses for mode, ℓ, at locations given by the vectors $\mathbf{r}_0$ and $\mathbf{r}_F$ respectively, and N_s is the number of modes that make a significant contribution to the response at frequency, ω. The time dependence, t, is implicit in the preceding equation.

The modal mass is given by:

$$\Lambda_\ell = \int_S m(\mathbf{r}_0)\, \psi_\ell^2(\mathbf{r}_0)\, \mathrm{d}S \tag{12.2}$$

where $m(\mathbf{r}_0)$ is the surface mass density ($\mathrm{kg/m^2}$) at location $\mathbf{r}_0$. The modal impedance, if hysteretic damping, characterised by loss factor, η_ℓ (the usual case for structures), is assumed, is given by:

$$Z_\ell = \omega_\ell^2 - \omega^2 + j\eta_\ell\,\omega_\ell^2 \tag{12.3}$$

Alternatively, if viscous damping, characterised by the critical damping ratio, ζ_ℓ, is assumed, then the modal impedance is given by the following equation.

$$Z_\ell = \omega_\ell^2 - \omega^2 + 2j\zeta_\ell\,\omega_\ell\,\omega \tag{12.4}$$

The modal impedance is complex; thus the quantity $v_0(\omega, \mathbf{r}_0)$ will be complex, having a magnitude and a phase relative to the forcing function $F_0(\omega)$. In the preceding equations, ω_ℓ is the resonance frequency, η_ℓ is the structural damping (loss factor) and ζ_ℓ is the viscous damping coefficient for mode, ℓ.

The space- and time-averaged mean square velocity over the structure is given by:

$$\langle v^2 \rangle_{S,t} = \langle F^2(\omega) \rangle_t \sum_{\ell=1}^{N_s} \left[\frac{1}{S} \iint_S \psi_\ell^2(\mathbf{r}_0)\, \mathrm{d}S \right] \frac{\psi_\ell^2(\mathbf{r}_F)}{M_\ell^2\, |Z_\ell|^2} \tag{12.5}$$

As the difference between ω and ω_ℓ becomes large, the contribution due to mode ℓ rapidly becomes small.

If a prototype machine or structure is available, mode shapes and modal damping can be determined from measured data using an experimental procedure known as modal analysis (Ewins, 1984). Modal analysis requires the measurement of the input force to a structure (generated by a shaker or instrumented hammer) and the structural response at a number of locations. Software packages, available from manufacturers of most spectrum analysers, will automatically calculate mode shapes, resonance frequencies and damping from these measurements. Again, this method is restricted to the first few structural resonances. From a knowledge of the surface velocity distribution for a given excitation force, the sound pressure field around the structure (and hence the radiated sound power) can be calculated by a number of methods that are described in the following sections. Each method has its advantages and disadvantages. The first method is generally referred to as the Helmholtz integral equation method (Section 12.2.1), which is implement in the Boundary Element Method using computational software (Section 12.2.2). The second method is referred to as the Rayleigh integral method (Section 12.2.3). The third method involves the use of Finite Element Analysis software (Section 12.2.4). The fourth method involves the use of the calculated mode shapes and resonance frequencies for the acoustic domain (Sections 12.2.5–12.2.6).

12.2.1 Helmholtz Method

In the Helmholtz method (Hodgson and Sadek, 1977, 1983; Koopmann and Benner, 1982), the acoustic pressure field generated by a closed vibrating body is described by the Helmholtz equation, which is just a re-organisation of the wave equation (Equation 1.15 in Chapter 1) such that the RHS is zero.

The acoustic pressure at a point r outside the vibrating surface at a frequency ω is given by Koopmann and Benner 1982, and Brod, 1984 as:

$$p(\omega, r) = -\int_S \left[p(\omega, r_0) \frac{\partial G(r, r_0)}{\partial n} + j\rho\omega\, v(\omega, r_0)\, G(r, r_0) \right] dS \qquad (12.6)$$

where again the time dependence is implicit and the Green's function, G, is defined by:

$$G(r, r_0) = \frac{e^{-jk|r - r_0|}}{4\pi |r - r_0|} \qquad (12.7)$$

This form of the Green's function applies to sound radiation by a vibrating surface into free space as well as sound radiation into a space enclosed by a vibrating surface. In the preceding equations, r is the vector distance from the origin of the coordinate system to the point at which the acoustic pressure $p(\omega, r)$ is to be calculated, r_0 is the vector distance from the surface element, dS, to the origin of the coordinate system, and $v(\omega, r_0)$ is the normal surface velocity for the surface element, dS, at frequency, ω.

The sound power radiated by the structure is found by integrating the product of the acoustic pressure and complex conjugate of the surface velocity over the surface.

The numerical implementation of the above method for a complex structure is known as the direct boundary element method.

12.2.2 Boundary Element Method (BEM)

The Boundary Element Method is a numerical implementation of the Helmholtz analysis method discussed in the previous section. The boundary element method, as its name implies, only involves discretising the boundary of an enclosed space or the boundary of a noise radiating structure. The method can be used to analyse acoustic problems such as the noise inside an enclosed volume, the noise radiated from a vibrating structure, and the acoustic field generated by the scattering of noise by objects in a free-field. On the other hand, Finite Element Analysis (FEA) involves discretising the enclosed volume for interior noise problems and a large space around a noise radiating structure for exterior noise problems. As a result, the boundary element formulation results in smaller computational models requiring less computer memory than FEA, but the downside is that more computer time may be needed to solve the matrix equations and produce the final result. The difference between the BEM and FEA formulation is illustrated in Figure 12.1, where a finite element model comprises nodes (the dots) and elements (the rectangles formed by the nodes for the FEM) and the BEM model comprises nodes (the dots) and lines.

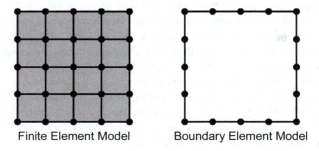

Finite Element Model   Boundary Element Model

Figure 12.1 Node locations for FEM and BEM.

Here the application of BEM will be illustrated by showing how publicly available BEM software may be used to solve a sound radiation problem inside a rigid walled room. There are two different boundary element methods that can be used to evaluate an acoustic field generated by a defined forcing function: the direct method and the indirect method. Both of these will be discussed in the following sections.

12.2.2.1 Direct Method

The direct method essentially involves solving equations based on the analysis of Section 12.1.1 with the additional consideration of the case where the pressure is evaluated on the surface, of area S.

The acoustic pressure at point r, within a volume due to vibration of an enclosing surface of velocity $v(\omega, r_0)$ at point r_0 on the surface is given by (Wu, 2000) as:

$$p(\omega, r) = -\frac{1}{C(r)} \int_S \left[p(\omega, r_0) \frac{\partial G(r, r_0)}{\partial n} + j\rho \omega v(\omega, r_0) G(r, r_0) \right] dS \qquad (12.8)$$

where dS is a differential area of the surface, and $C(r)$ is a constant depending on the location of the point r. $C(r) = 1$ when the point r is within the acoustic domain and $C(r_0) = \frac{1}{2}$ when the point r is on the surface (i.e. $r = r_0$). For mathematical completeness $C(r) = 0$ when the point r is not within the acoustic domain or on the enclosing surface; however this is rarely used as it applies to locations not in the acoustic medium, but inside the material of an enclosing structure or a structure that the within the sound field. These definitions for $C(r)$ apply for interior, exterior, and scattering acoustic problems. Note that the formulation for the boundary element method requires that the vector that is normal to the surface n points away from the acoustic domain. Hence for an interior acoustic problem, the normal vector of the surface must point outwards, away from the acoustic volume. For the analysis of an exterior acoustic field, the normal vectors point away from the acoustic domain, that is, inwards towards the enclosed surface. If both interior and exterior acoustic domains are of interest, then the indirect BEM must be used which is described in the following section. For this problem two surfaces are used: the surface that is in contact with the interior acoustic domain has surface normal vectors pointing outwards towards the exterior acoustic domain, and the surface that is in contact with the exterior acoustic domain has surface normal vectors that point inwards towards the interior acoustic domain.

In Equation (12.8), the time dependence may be ignored as it is the same on both sides of the equation, so the form of the acoustic pressure on the left-hand side will depend on the form of the acoustic pressure and particle velocity used on the right-hand side. So if amplitudes are used on the right-hand side, then the result on the left-hand side will be an acoustic pressure amplitude.

For acoustic scattering problems where an incident sound wave impinges on an object in an infinite acoustic domain, Equation (12.8) is modified slightly so that the total pressure p, which is the sum of the incident sound pressure p_i and the scattered sound pressure p_s, is given by:

$$p(\omega, r) = -\frac{1}{C(r)} \int_S \left[p(\omega, r_0) \frac{\partial G(r, r_0)}{\partial n} + j\rho\omega v(\omega, r_0) G(r, r_0) \right] dS + \frac{p_i(\omega, r)}{C(r)} \quad (12.9)$$

where the incident sound pressure amplitude $p_i(\omega, r)$ is without the presence of the object in the infinite domain and the sinusoidal time dependence term has been omitted.

The complex acoustic pressure amplitude (ignoring the sinusoidal time variation term, $e^{j\omega t}$) associated with an incident plane wave is:

$$p_i(\omega, r) = A e^{-j(k_x x + k_y y + k_z z)} \quad (12.10)$$

where k_x, k_y and k_z are the wave number components in the x, y and z directions respectively such that $k_x^2 + k_y^2 + k_z^2 = k^2 = (\omega/c)^2$ and A is the modulus of the acoustic pressure amplitude. The complex amplitude of the incident sound pressure associated with a spherical wave (or monopole) is (see Equation (5.86)):

$$p_i(\omega, r) = \frac{j\omega\rho Q_0}{4\pi r} e^{-jkr} \quad (12.11)$$

where Q_0 is the monopole source strength amplitude, r is the distance from the monopole to the surface on which the pressure is incident and ρ is the density of the fluid.

It should be noted that one problem with BEM for the analysis of exterior acoustic problems is that at certain frequencies the Helmholtz integral cannot be solved. This problem is overcome by using the Combined Helmholtz Integral Equation Formulation (CHIEF) method proposed by Schenck (1968), and is implemented in many BEM software packages. The details of this method are explained in Wu (2000) and von Estorff (2000).

12.2.2.2 Indirect Method

The indirect boundary integral formulation of the Helmholtz integral (Equation (12.6)) relies on boundary conditions involving the difference in the acoustic pressure and the difference in the pressure gradient. The indirect method can be used to calculate both interior and exterior acoustic fields as a result of a vibrating surface or acoustic sources, and can include openings that connect an enclosed region to a free-field region, or where free edges occur on a surface such as a stiffening rib attached perpendicular to a panel. The matrices resulting from this indirect method are fully populated and symmetric, which can result in faster solution times compared to solving unsymmetric matrices such as those associated with the direct method.

The formulation for the indirect boundary element method is given by the expression (Wu 2000):

$$\frac{\partial p(\boldsymbol{r})}{\partial x_i} = -j\rho\omega v_{x_i}(\boldsymbol{r}) = \int_S \left[\delta dp(\boldsymbol{r_S}) \frac{\partial G(\boldsymbol{r}, \boldsymbol{r_S})}{\partial x_i} - \delta p(\boldsymbol{r}) \frac{\partial^2 G(\boldsymbol{r}, \boldsymbol{r_S})}{\partial x_i \partial \boldsymbol{n}} \right] dS \quad (12.12)$$

where the time and frequency dependence of all acoustic pressure and particle velocity terms is implicit. The left hand side of the expression is the gradient of the pressure at any point $\boldsymbol{r}$ in the acoustic domain in the direction x_i, v_{xi} is the component of acoustic particle velocity in the direction x_i, and the right-hand side is an integral expression over the boundary surface. The term:

$$\delta p(\boldsymbol{r_S}) = p(\boldsymbol{r_{S1}}) - p(\boldsymbol{r_{S2}}) \quad (12.13)$$

is the difference in pressure across the surface of the boundary element model and is called the pressure jump or double layer potential, and $\boldsymbol{r_S}$ is a point on the boundary surface. The term:

$$\delta dp(\boldsymbol{r_S}) = \frac{\partial p(\boldsymbol{r_{S1}})}{\partial \boldsymbol{r_{S1}}} - \frac{\partial p(\boldsymbol{r_{S2}})}{\partial \boldsymbol{r_{S2}}} \quad (12.14)$$

is the difference in gradient of the pressure normal to the surface of the boundary element model at points $\boldsymbol{r_{S1}}$ and $\boldsymbol{r_{S2}}$ on opposite sides of the boundary surface and is called the single layer potential.

12.2.2.3 Meshing

Equation (12.8) must be evaluated numerically and this is achieved by discretising the surface with elements, similar to the process used in finite element analysis. If a system with a three-dimensional acoustic volume is to be analysed then surface patch elements are used, such as rectangular elements with four nodes, triangular elements with three nodes, or other elements with a greater number of nodes along the edges, which permits greater accuracy for modelling the pressure. For an acoustic system that can be modelled in two dimensions, where the acoustic domain is an area, the elements for the boundary element model will be line segments with two or more nodes.

Before solving Equation (12.12), it is necessary to define boundary conditions, in terms of the boundary variables. For the direct BEM, the acoustic pressure and acoustic particle velocity are used as the boundary variables. For the Indirect BEM, the boundary variables are the acoustic pressure and the difference of the pressure gradient across the boundary. Figure 12.2 shows the difference in boundary conditions between the two methods.

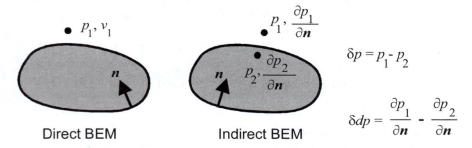

Figure 12.2 Direct vs Indirect BEM.

The direct BEM can be used to analyse interior or exterior problems, while the indirect BEM can be used to solve interior and exterior problems at the same time.

The boundary elements have nodes at the vertices or mid-span between two vertices, as shown in Figure 12.3. The pressure (or velocity) can be calculated anywhere inside the element using a mathematical expression of a weighted sum of the pressures (or velocities) at the nodes, which is the same method used in finite element analysis.

12.2.2.4 Problem Formulation

To solve a direct boundary element method problem, it is necessary to describe the problem using matrix equations that can solved using computer software.

The acoustic pressure and particle velocity within the boundary element can be written as (Wu (2000); von Estorff (2000)):

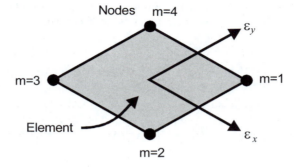

Figure 12.3 Boundary element coordinate system.

$$p(\omega, \varepsilon_x, \varepsilon_y) = \sum_{m=1}^{4} N_m(\omega, \varepsilon_x, \varepsilon_y) p_m \tag{12.15}$$

$$v_n(\omega, \varepsilon_x, \varepsilon_y) = \sum_{m=1}^{4} N_m(\omega, \varepsilon_x, \varepsilon_y) v_{nm} \tag{12.16}$$

where N_m is the shape function evaluated at the coordinates ε_x, ε_y at frequency ω, p_m is the pressure and $v_{n,m}$ is the normal particle velocity at node m. The pressures and particle velocities in the preceding equations may be instantaneous or amplitudes. In the following analysis, matrices and vectors will be denoted in bold font.

Equations (12.15) and (12.16) can be substituted into the Helmholtz integral equation (Equation (12.8)) resulting in the expression:

$$C(r) \left[\sum_{m=1}^{4} N_m p_m \right] = - \int_S \left[\sum_{m=1}^{4} N_m p_m \right] \frac{\partial G(r, r_0)}{\partial n} \, dS$$

$$- \int_S j\rho\omega \left[\sum_{m=1}^{4} N_m v_{nm} \right] G(r, r_0) \, dS \tag{12.17}$$

which can be rearranged into a matrix equation as:

$$[\mathbf{C}]\{\mathbf{p}\} + [\mathbf{H}]\{\mathbf{p}\} = [\mathbf{G}]\{\mathbf{v_n}\} \tag{12.18}$$

where $\{\mathbf{p}\}$ and $\{\mathbf{v_n}\}$ are vectors containing the nodal pressures and particle velocities normal to the boundary surface respectively. The acoustic pressures and particle velocities are frequency dependent. The elements of the matrices, **C**, **H** and **G** are:

$$c_i = -\int_{S_j} \frac{\partial G_L}{\partial \boldsymbol{n}} N_m \, dS = \begin{cases} 1 & \text{when the point is in the acoustic domain} \\ 1/2 & \text{when the point is on the surface} \end{cases} \quad (12.19)$$

$$h_i = \sum_{m=1}^{4} \int_{S_j} \frac{\partial G}{\partial \boldsymbol{n}} N_m \, dS \quad (12.20)$$

$$g_i = -j\rho\omega \sum_{m=1}^{4} \int_{S_j} G N_m \, dS \quad (12.21)$$

and these integral terms can be evaluated by using standard numerical integration methods such as Gaussian quadrature over each element S_j. The known pressures or normal acoustic particle velocities at the nodes, determined from the boundary conditions, can be substituted into Equation (12.18). Then the matrices on the left-hand side of Equation (12.18) can be combined and the expression re-arranged into the following form:

$$[\mathbf{A}]\{\mathbf{x}\} = \{\mathbf{b}\} \quad (12.22)$$

where the unknown boundary acoustic pressures or normal particle velocities are in the vector $\{\mathbf{x}\}$ and can be solved for by inverting the matrix $[\mathbf{A}]$. Once all the boundary pressures and velocities are solved for, the pressure at any point within the acoustic domain can be calculated using the Helmholtz integral equation.

Analysis of acoustic problems using Boundary Element Method involves the use of computational software. There are commercial and non-commercial software packages that are available. For demonstration purposes, the software called Helm3D that is available on a CD-ROM that accompanies the book by Wu (2000) is used here to calculate the sound pressure level distribution in a three-dimensional enclosure with rigid walls. A freely available graphical user interface (GUI) called GiD is used as a preprocessor to model the system. The model is then solved using the Helm3D software and the GiD software is used to review the results of the analysis.

How to install Helm3D and GiD to solve a BEM problem. Both the GiD and the Helm3D software are required to be installed on a computer with a Microsoft Windows operating system to solve this example problem. The GiD software can be obtained from http://gid.cimne.upc.es/. Once you have installed the GiD software you need to uncompress the file Helm3D.zip into the GiD "problem types" folder "C:\Program Files\GiD\GiD xxxx\problemtypes", where "xxxx" is the version number of the GiD software. A folder entitled Helm3d.gid containing the contents of the zip file should be created. The helm3D executable (helm3d.exe) obtained from the CD accompanying the book Wu (2000) also needs to be copied and placed within the helm3d.gid folder.

In the example described below, this software will be used to solve a problem using BEM.

Example 12.1

Consider a rigid walled box with dimensions 2.5m × 3m × 5m. In one corner of the box, a volume velocity acoustic source generates sound pressure in the room. The source has dimensions on the wall of 0.2m × 0.2m. Determine the sound pressure level distribution on the walls of the room at the first resonance frequency.

Solution

Start the GiD software and create a model of the room by firstly creating the floor plan. Select from the top menu "Utilities > Tools > Coordinates" Window, which will open a dialog box to aid in entering the coordinates for the model. Create the floor of the box by selecting from the top menu "Geometry > Create > Object > Rectangle". In the dialog box titled 'Coordinates window', ensure that the coordinates are $x = 0$, $y = 0$, $z = 0$, the coordinate system is Cartesian, and the Local Axes is "Global" as shown in Figure 12.4.

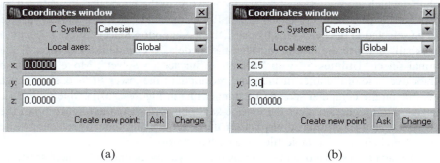

(a)   (b)

Figure 12.4 (a) The coordinates for the bottom left corner and (b) the top right corner of the rectangular box.

Click the Apply button at the bottom of the dialog box which will create a point at the origin. Change the entries Coordinates window to $x = 2.5$, $y = 3$ and click "Apply" and a box will be drawn which has blue outer lines and a smaller inner rectangle with pink lines. The inner rectangle with pink lines indicates that the rectangle is actually a surface, rather than just several lines. Change the view to Isometric by selecting "View > Rotate > Isometric".

Press "Control-C" and a dialog box will open to copy the surface. Change the entries in the Copy dialog box to: "Entities type: Surfaces"; "Transformation: Translation"; "Second Point z: 5.0"; "Do extrude: Surfaces"; as shown in Figure 12.5. Click the "Select" button and type ":" then click on "Enter" key in the command line at the bottom of the screen to select all the surfaces. You should notice the message in the window above "Added 1 new surfaces to the selection". Press the Escape key to end the selection and the box will be displayed. Select "View > Zoom > Frame" to rescale the model. Select "View > Normals > Surfaces > Normal" and then type ":" and press the "Enter" key in the command line. This will draw the normals on the

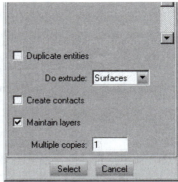

Figure 12.5 Settings for the Copy dialog box.

surfaces as shown in Figure 12.6. It can be seen that the arrow on the bottom face of the box is pointing inwards and should be pointing outwards, away from the acoustic volume, which is a requirement for the formulation of the boundary element method when considering the interior sound field. Rotate the model so that the lower surface can be selected in the next steps by using the menu option Select "View > Rotate > Trackball" and use the mouse to rotate the model. To change the direction of the normal select "Utilities > Swap Normals > Surfaces > Select". Move the mouse cursor over one of the lines of the inner rectangle of the lower surface, shown as a pink line on the computer screen (the thinner line in Figure 12.6), and left click the mouse button, and then press the Escape key. This will swap the orientation of the normal so that it is pointing outwards from the box. Save the model by selecting "File > Save" and choose a suitable filename such as "room".

The next step is to move the analysis problem to the boundary element software Helm3D. To do this select "Data > Problem Type > helm3d". A dialog window

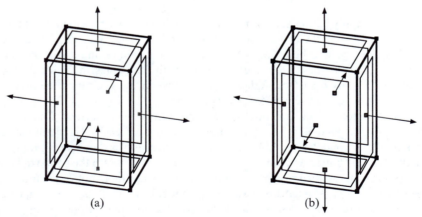

(a)   (b)

Figure 12.6 (a) The normal on the lower surface pointing inwards and (b) after correcting the normal to point outwards from the lower surface.

warning will appear warning that "all data information will be lost", which can be ignored by selecting "OK".

Next, create the acoustic source on one of the walls by drawing a small rectangle of dimensions 0.2 m × 0.2 m close to the corner. Select "Geometry > Create > Straight Line". Type 0,0.1,0.1 in the command line window at the bottom of the screen and press the Enter key. Next type 0,0.1,0.3 and press the Enter key, 0,0.3,0.3, press the Enter key, and 0,0.3,0.1, press the Enter key. To connect the last point to the original point type "join" and press the Enter key. Using the mouse, click on the point located at 0,0.1,0.1. You should now have a rectangle that will be used to define the source (see Figure 12.7a). Press the Escape key. The model should look similar that shown in Figure 12.7a.

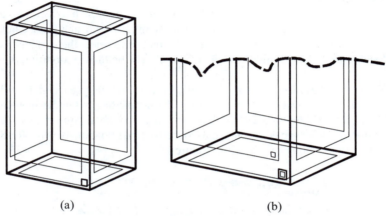

(a)   (b)

Figure 12.7 (a) The model of the box after the creation of the source region and (b) after dividing the surface into two.

At this stage, four lines have been created that define the size of the acoustic source. It is now necessary to divide the wall surface into two; the first is the entire surface minus the source rectangle and the second is the source rectangle by itself. To begin, rotate the model by selecting "View > Rotate > Trackball" and use the mouse and left mouse button to rotate the model so that the small square is facing the front. Then select "Geometry > Edit > Divide > Surfaces > Split". In the command window at the bottom of the screen will be written "Enter NurbSurface to split (ESC to leave)". Use the mouse and the left mouse button to select one of the lines of the inner rectangle along $x = 0$, which has pink lines on the computer screen and is indicated by the thinner lines in Figure 12.7a. The lines for the inner rectangle will change colour from pink to red to indicate it has been selected. In the command window will be the instruction "Enter division lines to split the surface". Using the mouse, select the four lines of the small rectangle for the source location. As each line is selected, they will change colour from blue to red. Once the four lines have been selected Press the escape key and the surface will be subdivided, with a separate source surface as shown in Figure 12.7b.

The next step is to ensure that all of the surface normals are facing in the correct direction. Follow the same procedure as previously described by selecting "Utilities > Swap Normals > Surfaces > Select". Click on all of the surfaces that have an inward pointing normal until all surface normals are oriented outwards, which indicates that the normal is pointing away from the acoustic volume. You might need to rotate the model to select an arrow head pointing in the incorrect direction. Press the "Escape" key to leave the "Swap Normals" mode.

The next step is to define the problem as a Helm3D BEM problem. To do this select "Data > Problem type > Helm3d". The only boundary conditions required for this problem are velocity boundary conditions. A velocity of 1+ 0j is needed to mimic the piston at the source location. All other surfaces use the default zero velocity. To do this select "Data > Conditions". Click on "Velocity". Set the real velocity component to 1. The imaginary part allows the relative phase between two or more sources to be taken into account when such problems are being analysed. Click on the "Assign" button in the dialog box and use the mouse to select the surface inside the small rectangle. Select "Finish" and then "Close". The boundary conditions have now been set.

The next tasks are to define the type of problem to be analysed, and the properties of the fluid. Select "Data > Problem data". Change the options under the "General" tab as follows. For the Project TITLE, select an appropriate name such as "Room". Ensure the Analysis type is selected to be "Internal", the Loading type to "Radiation", the symmetry to "None" as shown in Figure 12.8.

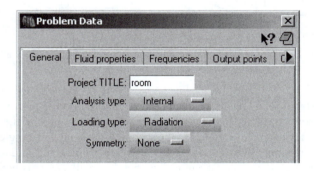

Figure 12.8 Parameters for the Problem Data dialog box.

Next select the Fluid Properties tab and change the following items. Change the density, speed of sound and reference pressure to 1.21, 343 and 20×10^{-6} respectively.

Next select the Frequency tab and change "Frequency start" and "frequency stop" to be 68.6, corresponding to the first resonance frequency of the room along the longest room dimension where the wavelength is 5 m, and then click "Accept" and then "Close".

The next step is to mesh the boundary of the room, using the default triangular meshing elements. Select "Mesh > Generate mesh". A dialog box will appear asking you to "Enter the size of elements to be generated", type in 0.4 and press the enter key. This will generate a triangular mesh where the nominal edge length along each side of the triangular elements is 0.4m. Note that the rule-of-thumb is that there should be at least 6 elements per wavelength. A dialog box will appear which states that a number of triangular elements have been created. Press "OK" and the mesh will appear, as shown in Figure 12.9 (a). To change the method of rendering the model, press the right button on the mouse, then choose "Render > Flat", as shown in Figure 12.9 (b).

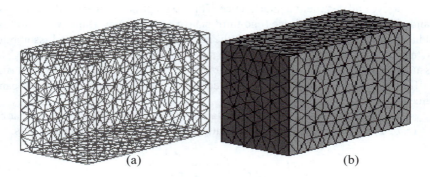

(a)   (b)

Figure 12.9 (a) Triangular mesh viewed as a wireframe; (b) triangular mesh viewed with flat shading.

A solution to the problem can now be calculated by selecting "Calculate > Calculate". A dialog box with the heading, "Process Info", will appear announcing that the calculations have finished. Click the button labelled "Postprocess", which will change the mode of the GiD software from the pre-processor to the post-processor. To display the pressure over the boundary of the room select "View results > Contour fill > SPL". It is possible to change the intervals for the contours by selecting "Options > Contour > Width Intervals" and change the value to 5. The result is shown in Figure 12.10, although it will be colour on your screen.

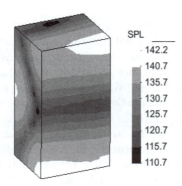

Figure 12.10 Contour plot of the sound pressure level inside the rectangular box at 68.6Hz.

12.2.3 Rayleigh Integral Method

The Rayleigh integral method involves evaluating the integral of the surface velocity of the object to calculate the radiated sound power. The Rayleigh method (Hayes and Quantz, 1982; Hansen and Mathews, 1983; Takatsubo *et al.*, 1983) is most effective when the machine or structure can be divided into a number of panels, which are approximately flat, but of any shape. Two broad assumptions must be made. First, it is assumed that the sound fields from adjacent panels do not interact to produce constructive and destructive interference. This assumption is satisfied if the analysis bandwidth is sufficiently large (one-third octave or octave), or if the wavelength of radiated sound is small compared with the panel dimensions. The second assumption is that the sound fields from the two sides of the panel do not interact. This assumption is satisfied if the panel is part of a closed surface or mounted in a rigid baffle, or if the wavelength of the radiated sound is small compared with the panel dimensions.

The Rayleigh integral for the calculation of the sound pressure, p, at location, $r = (r, \theta, \psi)$ (in the far field) from a radiating panel of area S, at frequency ω is:

$$p(\omega, r) = \frac{j\omega\rho}{2\pi} \iint_S v(\omega) \left[\frac{e^{-jkr}}{r} \right] dS \qquad (12.23)$$

where the time dependence is implicit as explained previously, $v(\omega)$ is the normal velocity on the surface element dS at frequency ω, and r is the distance from the surface element to the vector location, r in space.

The advantage of this method is that it can be used to calculate the panel radiated sound power, which can be used for the calculation of interior noise levels (for example, in a vehicle) as well as exterior noise levels.

The sound power radiated by each panel can be calculated by integrating the pressure amplitude over a hemispherical surface in the far field of the panel, using the plane of the panel as the base of the hemisphere. The total sound power radiated by the structure is calculated by adding logarithmically the power due to each panel making up the structure. Best results are obtained if data are averaged over one-third octave or octave frequency bands.

The Rayleigh Integral method is only valid when the sound radiating from the vibrating surface does not reflect or diffract around any parts of the vibrating object.

Figure 12.11 (a) shows a picture of a flat panel installed in a baffle that is vibrating and radiating sound into an infinite half-space. The panel has been discretised into four elements and five nodes. The Rayleigh integral of Equation (12.23) can be used to estimate the far-field sound pressure, which can then be used to estimate the sound power radiated by the structure. Although this system is represented in 2D, it also applies to 3D systems. The implicit assumption in the Rayleigh integral is that each element is a small independent source that produces a far field pressure given by Equation (5.86). The assumption of independent sources is a good engineering approximation only and should be used with caution when analysing vibrating structures that have surfaces which will cause the radiated sound to reflect or diffract, as shown in Figure 12.11 (b). Another complication occurs with

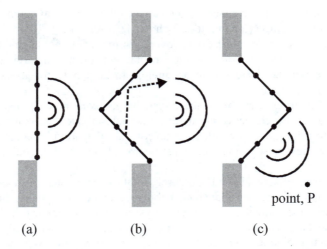

(a)   (b)   (c)

Figure 12.11 (a) Unobstructed surface; (b) concave surface with reflections; (c) convex surface with an obscured surface.

structures that have a convex shape, as shown in Figure 12.11 (c), where only part of the surface might be "visible" to a point in the far field, and other vibrating surfaces will be obscured by the structure. Herrin *et al.* (2003) describes a method that can be used to determine which surfaces are "visible" to a point in the acoustic field based on evaluating the product of the vector normal to the surface and the vector from a point on the surface to the point in the acoustic field. For structures with characteristics shown in Figures 12.11 (b) and (c) the use of the Rayleigh integral could give results that have large errors. In these cases, a numerical method such as the finite element or boundary element method should be used to estimate the radiated sound.

12.2.4 Finite Element Analysis (FEA)

Finite element analysis is a numerical method that can be used to calculate the response of a complex structure due to the application of forcing functions, which could be an acoustic field or a distribution of mechanical forces. FEA can also be used to estimate the sound power radiated by a structure or the distribution of the sound field in an enclosed space. Estimating the sound power radiated by a structure generally requires a large numerical model and the associated computer memory requirements are so large that it is not usually a practical alternative. Rather, it is better to use FEA to calculate the response of the noise radiating structure and then use a numerical evaluation of the Rayleigh Integral to calculate the radiated sound power. Alternatively, if the structure is excited by an external sound field, then FEA can be used to determine the structural response and also the resonance frequencies and mode shapes of the enclosed sound field. Then the actual sound pressure distribution in the enclosed space can be calculated using modal coupling analysis implemented with a programming tool such as MATLAB.

The underlying theory for FEA is covered in many text books and will not be repeated here. However, its practical implementation using a commercially available FEA package will be discussed in an attempt to help potential users apply the technique to acoustic analysis.

Finite element analysis of acoustic systems involves the discretisation of the acoustic volume into elements and nodes. An enclosed acoustic volume might be surrounded by rigid walls, a flexible structure, or walls that provide acoustic damping. Alternatively, the acoustic radiation of a structure into an anechoic field can also be examined.

The analysis of acoustic and structural vibration can be achieved using simple theoretical models for rectangular shaped objects. Any geometry more complex than a rectangular shaped object is onerous to analyse and vibro-acoustic practitioners opt for a numerical method such as finite element or boundary element analysis to solve their particular problem. Later in this chapter, Statistical Energy Analysis will be discussed for the analysis of "high" frequency problems that are characterised by sufficiently high modal density in the radiating structure or enclosed acoustic space. For "low" frequency problems, FEA and BEA as well offer a reasonably simple computational method for evaluating the vibro-acoustics of a system.

Finite element analysis of acoustic systems has numerous applications including the acoustic analysis of interior sound fields, sound radiation from structures, the transmission loss between of panels, the design of resonator type silences and diffraction around objects. The finite element method takes account of the bi-directional coupling between a structure and a fluid such as air or water.

In acoustic fluid-structure interaction problems, the structural dynamics equation needs to be considered along with the mathematical description of the acoustics of the system, given by the Navier-Stokes equations of fluid momentum and the flow continuity equation. The discretised structural dynamics equation can be formulated using structural finite elements. The fluid momentum (Navier-Stokes) and continuity equations are simplified to get the acoustic wave equation using the following assumptions:

- The acoustic pressure in the fluid medium is determined by the wave equation.
- The fluid is compressible where density changes are due to pressure variations.
- The fluid is inviscid with no dissipative effect due to viscosity.
- There is no mean flow of the fluid.
- The mean density and pressure are uniform throughout the fluid and the acoustic pressure is defined as the pressure in excess of the mean pressure.
- Finite element analyses are limited to relatively small acoustic pressures so that the changes in density are small compared with the mean density.

The acoustic wave equation (Equation (1.15)) is used to describe the acoustic response of the fluid. Because viscous dissipation of the fluid is neglected, the equation is referred to as the lossless wave equation. Suitable acoustic finite elements can be derived by discretising the lossless wave equation using the Galerkin method. For a derivation of the acoustic finite element the reader is referred to Craggs (1971).

There are two formulations of finite elements that are used to analyse acoustic problems: pressure and displacement. The most commonly used finite element to analyse acoustic problems is the pressure formulated element.

12.2.4.1 Pressure Formulated Acoustic Elements

The acoustic pressure p within a finite element can be written as:

$$p = \sum_{i=1}^{m} N_i p_i \tag{12.24}$$

where N_i a set of linear shape functions, p_i are acoustic nodal pressures, and m is the number of nodes forming the element. For pressure formulated acoustic elements the finite element equation for the fluid in matrix form is:

$$[\mathbf{M}_f]\{\ddot{\mathbf{p}}\} + [\mathbf{K}_f]\{\mathbf{p}\} = \{\mathbf{F_f}\} \tag{12.25}$$

where $[\mathbf{K}_f]$ is the equivalent fluid "stiffness" matrix, $[\mathbf{M}_f]$ is the equivalent fluid "mass" matrix, $\{\mathbf{F}_f\}$ is a vector of applied "fluid loads", $\{\mathbf{p}\}$ is a vector of unknown nodal acoustic pressures, and $\{\ddot{\mathbf{p}}\}$ is a vector of the second derivative of acoustic pressure with respect to time.

The equations of motion for the structure are:

$$[\mathbf{M}_s]\{\ddot{\mathbf{U}}\} + [\mathbf{K}_s]\{\mathbf{U}\} = \{\mathbf{F_s}\} \tag{12.26}$$

where $[\mathbf{K}_s]$ is the structural stiffness matrix, $[\mathbf{M}_s]$ is the structural mass matrix, $\{\mathbf{F}_s\}$ is a vector of applied structural loads", $\{\mathbf{U}\}$ is a vector of unknown nodal displacements and hence $\{\ddot{\mathbf{U}}\}$ is a vector of the second derivative of displacement with respect to time, equivalent to the acceleration of the node. The interaction of the fluid and structure occurs at the interface between the structure and the acoustic elements, where the acoustic pressure exerts a force on the structure and the motion of the structure produces a pressure. To account for the coupling between the structure and the acoustic fluid, additional terms are added to the equations of motion for the structure and fluid (of density, ρ) respectively, as:

$$[\mathbf{M}_s]\{\ddot{\mathbf{U}}\} + [\mathbf{K}_s]\{\mathbf{U}\} = \{\mathbf{F_s}\} + [\mathbf{R}]\{\mathbf{p}\} \tag{12.27}$$

$$[\mathbf{M}_f]\{\ddot{\mathbf{p}}\} + [\mathbf{K}_f]\{\mathbf{p}\} = \{\mathbf{F_f}\} - \rho[\mathbf{R}]^{\mathrm{T}}\{\ddot{\mathbf{U}}\} \tag{12.28}$$

where $[\mathbf{R}]$ is the *coupling* matrix that accounts for the effective surface area associated with each node on the fluid-structure interface. Equations (12.27) and (12.28) can be formed into a matrix equation including the affects of damping as:

$$\begin{bmatrix} \mathbf{M}_s & 0 \\ \rho\mathbf{R}^{\mathrm{T}} & \mathbf{M}_f \end{bmatrix}\begin{Bmatrix} \ddot{\mathbf{U}} \\ \ddot{\mathbf{p}} \end{Bmatrix} + \begin{bmatrix} \mathbf{C}_s & 0 \\ 0 & \mathbf{C}_f \end{bmatrix}\begin{Bmatrix} \dot{\mathbf{U}} \\ \dot{\mathbf{p}} \end{Bmatrix} + \begin{bmatrix} \mathbf{K}_s & -\mathbf{R} \\ 0 & \mathbf{K}_f \end{bmatrix}\begin{Bmatrix} \mathbf{U} \\ \mathbf{p} \end{Bmatrix} = \begin{Bmatrix} \mathbf{F_s} \\ \mathbf{F_f} \end{Bmatrix} \tag{12.29}$$

where $[\mathbf{C}_s]$ and $[\mathbf{C}_f]$ are the structural and acoustic damping matrices, respectively. This equation can be reduced to an expression without differentials as:

$$\begin{bmatrix} -\omega^2 \mathbf{M}_s + j\omega\mathbf{C}_s + \mathbf{K}_s & -\mathbf{R} \\ -\omega_2\rho\mathbf{R}^{\mathrm{T}} & -\omega^2\mathbf{M}_f + j\omega\mathbf{C}_f + \mathbf{K}_f \end{bmatrix} \begin{Bmatrix} \mathbf{U} \\ \mathbf{p} \end{Bmatrix} = \begin{Bmatrix} \mathbf{F}_s \\ \mathbf{F}_f \end{Bmatrix} \tag{12.30}$$

The important feature to notice about Equation (12.30) is that the matrix on the left-hand side is unsymmetric and solving for the nodal pressures and displacements requires the inversion of this unsymmetric matrix, which takes a significant amount of computer resources. The fluid-structure interaction method described above accounts for coupling between structures and fluids, and this is usually only significant if a structure is radiating into a heavier than air medium, such as water or if the structure is very lightweight, such as a car cabin.

A typical structural acoustic finite element model is shown in Figure 12.12, where the structural elements contain displacement DOFs, and most of the acoustic volume contains acoustic elements with only pressure degrees of freedom. At the interface between the acoustic fluid and the structure, a thin layer of elements, with pressure and displacement DOFs, one element wide is used to couple the structure to the fluid.

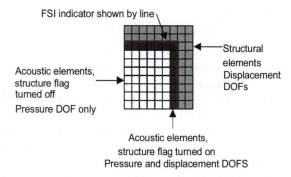

Figure 12.12 Structural acoustic finite element model with pressure formulated elements.

Modelling an acoustically rigid wall is achieved by not defining acoustic elements on an edge, shown in Figure 12.12 on the lower left corner of the model. Modelling a free surface is achieved by setting the pressure to be zero on the nodes of pure acoustic elements (i.e. only pressure DOF). Alternatively, if using acoustic elements with both pressure and displacement DOFs a free surface is modelled by not defining any loads, displacement constraints or structure. The motion of the surface can then be obtained by examining the response of the nodes on the surface.

12.2.4.2 Displacement Formulated Acoustic Elements

Another formulation of acoustic elements is based on nodal displacements and these elements are based on standard structural elements. A typical structural acoustic finite element model based on displacement formulated elements is shown in Figure 12.13. The difference between the solid structural elements and fluid elements is that the underlying material behaviour is altered to reflect the behaviour of a fluid, so that the stiffness terms associated with shearing stresses are set to near zero and Young's modulus is set equal to the bulk modulus of the fluid. What this means is that the element has no ability to resist shearing stresses and can lead to peculiar results. For example, a modal analysis of an acoustic space will produce a large number of zero energy modes that are associated with shearing mechanisms in the fluid, and these results have no relevance.

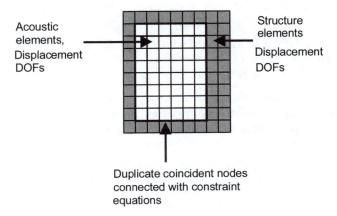

Acoustic
elements,
Displacement
DOFs

Structure
elements
Displacement
DOFs

Duplicate coincident nodes
connected with constraint
equations

Figure 12.13 Structural acoustic finite element model with displacement formulated elements.

A 3-dimensional (spatial) displacement formulated acoustic element has 3 displacement DOFs. In comparison, a 3-dimensional (spatial) pressure formulated acoustic element has 3 displacement DOFs and 1 pressure DOF. It is conceptually easier to understand the fluid-structure interaction mechanism if there are only displacement DOFs. However, only the nodal displacements of the acoustic elements that are normal to the structure should be coupled, as it is only the out-of-plane motion of the structure that generates acoustic pressure in the fluid, as the in-plane motion of the structure does not generate any acoustic pressure in the fluid. Hence, the fluid and the structure must retain independent displacement DOFs for motions that are tangential to the fluid-structure-interaction surface, which can be difficult to model. These systems are modelled by defining separate but coincident nodes for the fluid and the structure at the fluid-structure-interaction interface, and then coupling the appropriate nodal displacement DOFs, or defining mathematical relationships for the nodal displacement motion of the structure and fluid, to define the compatibility of displacements that are *normal* to the fluid-structure-interaction surface. In many cases it is advantageous to rotate the nodal coordinate systems of the structure and fluid meshes along the interface so that there is one axis that is normal, and two axes that are tangential to the interface surface.

One of the main advantages of using displacement formulated acoustic elements is that the matrix equation is symmetric and thus is quicker to solve than the unsymmetric matrix equation shown in Equation (12.30). Tables 12.1 and 12.2 list the advantages and disadvantages of pressure and displacement formulated acoustic elements.

Table 12.1 Advantages and disadvantages of pressure formulated elements

Advantages	Disadvantages
Only a single pressure DOF per node.	The set of equations to be solved in a general fluid-structure interaction analysis are unsymmetric, requiring more computational resources.
No zero-energy fluid modes are obtained in a modal analysis.	
The pressure DOF can be associated with either the total acoustic pressure (incident plus scattered) or only the scattered component of the acoustic pressure.	
Both displacement and pressure DOFs are available at a fluid structure interface. Hence, the definition of the fluid-structure coupling is relatively easy and does not require the use of duplicate nodes at the interface.	Modelling of free-field acoustic domains with 'infinite' elements could be improved.
Relatively easy to define a radiation boundary condition or a sound absorptive surface.	
Nodal acoustic pressures are output as solution quantities for direct use in post-processing.	

12.2.4.3 Practical Aspects of Modelling Acoustic Systems with FEA

The following paragraphs describe some practical considerations when modelling acoustic systems with Finite Element Analysis.

Mesh density. The use of the finite element method can be useful for low-frequency problems. However, as the excitation frequency increases, the number of nodes and elements required in a model increases exponentially, requiring greater computational resources and taking longer to solve. A general rule-of-thumb is that acoustic models should contain 6 elements-per-wavelength as a starting point. Accurate models can still be obtained for lower mesh densities; however, caution should be exercised. At regions in a model where there is a change in the acoustic impedance, for example where the diameter of a duct changes, at a junction of two or more ducts, or at the opening of the throat of a resonator into a duct, it is important to be wary of the mesh density in these regions to ensure that the complex acoustic field in these discontinuous areas are modelled accurately.

Mean flow. Many finite element software packages with acoustic finite elements require that there is no mean flow of the fluid, which is a significant limitation. When there is mean flow of fluid, a different formulation of the wave-equation is

Table 12.2 Advantages and disadvantages of displacement formulated elements

Advantages	Disadvantages
The set of equations to be solved in a general fluid-structure interaction analysis are symmetric.	There are 3 displacement DOFs per node which can result in a model with a large number of DOFs.
Displacement boundary conditions and applied loads have the same physical meaning as those used for standard structural elements.	The definition of the fluid-structure interface is complex requiring the use of duplicate nodes and expressions to couple the relevant DOFs.
Energy losses can be included in the displacement element via a fluid viscosity parameter as well as the standard techniques with solid elements.	Modal analyses can result in a large number of (near) zero frequencies associated with shearing of the fluid elements.
	The acoustic pressure at a point in the fluid cannot be expressed in terms of a known incident pressure and an unknown scattered pressure.
	The shape of the elements should be nearly square for good results.

required, which modifies the propagation of the acoustic disturbance (due to "convection"), depending on whether the flow is rotational or irrotational. However, it is still possible to conduct finite element modelling for low speed fluid flow, where the compressibility effects of the fluid are negligible, using "no flow" FEA software packages , but some assumptions that underpin the analysis will be violated. When there is mean flow in a duct, aero-acoustic phenomena might be important. For example, consider the situation of mean flow in a duct where the throat of a Helmholtz resonator attaches to the main duct, or over a sharp edge. It is possible that as air flows over the edge of the throat, noise will be generated, similar to blowing air over the top of a glass soda bottle. In some situations the flow over the structure might cause vortex shedding. Standard finite element models are not able to model these effects.

If the flow speed is significant or it is expected that there will be aero-acoustic phenomena, consider the use of Computational Fluid Dynamic (CFD) software to analyse the problem. However, this software also has limitations for the analysis of acoustic problems. Alternatively, some Boundary Element Analysis software packages are able to model acoustic systems with mean flow, but are not able to model noise generation from shedding type phenomena.

Rigid or flexible boundaries. Acoustic finite element models have rigid wall conditions at boundaries where no elements are defined. This assumption is valid where it is not expected that the motion of the boundary is likely to have any significant affect on the acoustics of the system. However, consider an automobile cabin comprising flexible sheet metal panels. Depending on the stiffness of these panels, acoustic excitation within the enclosure can cause the panels to vibrate, which in turn will affect the acoustic mode shapes and resonance frequencies of the enclosure. As highlighted above, modelling fluid-structure interaction can be computationally complex and can require substantial computer resources to solve. Hence careful consideration is required to decide whether the fluid-structure interaction should or must be modelled. A second subtle point is the consideration of re-radiation of structures in a different part of the acoustic model. Consider a duct with two Helmholtz resonators attached to a duct to reduce sound radiated from its exit as shown in Figure 12.14. A simple acoustic model could be constructed assuming rigid walls. However if parts of the system are in fact flexible, for example the wall dividing the two resonators, then high sound levels in the first resonator would vibrate the dividing wall, which would re-radiate sound into the second Helmholtz resonator. Alternatively, if the entire system were made from lightweight sheet metal, then vibrations could be transmitted along the duct work and result in the re-radiation of sound into the main duct.

The results from analyses are usually the acoustic pressure at discrete locations. Sometimes this level of detail is required but often it is not; instead an indicative

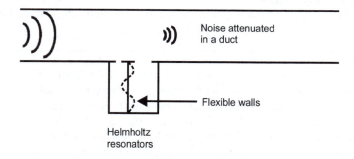

Figure 12.14 Re-radiation of sound from a structure into a different part of the acoustic model.

global noise level may be required for assessment, which will require post-processing of the results from the analysis. For higher frequency problems, statistical energy analysis methods may be more appropriate and significantly faster in obtaining a solution.

12.2.5 Numerical Modal Analysis

The resonance frequencies of an acoustic enclosure can be calculated theoretically for simple shapes such as a rectangular prism. For more complex geometries, a numerical method must be used. Two popular numerical methods that can be used to determine the resonance frequencies and mode shapes for enclosed spaces characterised by complex geometries are finite element and boundary element methods. For systems that involve a structure and an enclosed space, the resonance frequencies and mode shapes for each are usually calculated separately and then the interaction between them is evaluated using modal coupling analysis (see next section). The mode shapes of the enclosed space are calculated assuming rigid boundaries and the mode shapes of the structure are calculated assuming that it is vibrating in a vacuum. Of course it is possible to use BEA and FEA to evaluate the fully coupled system modes and thus avoid the need for modal coupling analysis. However, this uses an enormous amount of computing resources for most practical systems such as transportation vehicles.

The advantage of using boundary element methods compared to finite element methods to calculate the resonance frequencies and mode shapes of an acoustic or structural system is that the dimension of the problem can be reduced by one: the resonance frequencies inside a three-dimensional volume can be re-written as a problem involving a surface integral. This results in smaller matrices compared with a finite element formulation. The disadvantage is that these matrices are full (meaning that each entry in the matrix is occupied) and often a non-linear eigenvalue solver is required.

On the other hand, for a finite element formulation, although the matrices are larger in size, they are sparse matrices (meaning that there are few entries off the diagonal), so standard linear eigenvalue solvers can be used, and hence the matrices can be solved relatively easily to yield eigenvalues that correspond to system resonance frequencies. These eigenvalues are then used to calculate mode shapes.

For acoustic problems, the resonance frequencies of a volume are given by the eigenvalues of the Helmholtz equation, written in terms of the velocity potential, φ, as:

$$\nabla^2 \varphi + k^2 \varphi = 0 \tag{12.31}$$

The right-hand side of Equation (12.31) is set to zero, meaning that there is no acoustic source within the volume. To solve Equation (12.31), the boundary conditions must be specified, which are mathematical descriptions of the acoustic behaviour of the surface that surrounds the volume. The typical boundary condition considered for an acoustic modal analysis problem is that the boundaries are rigid such that the normal acoustic particle velocity at the boundary is zero. This is called the 'Neumann boundary condition' and is written as $v_n = 0$ or $\partial \varphi / \partial n = 0$ along the boundary.

The procedures for calculating resonance frequencies and mode shapes of an acoustic or structural system are implemented in most commercially available finite element and boundary software, and the underlying theory has been discussed in detail in a number of text books (Wu, 2000; von Estorff, 2000; Marburg and Nolte, 2008). Kirkup (2007) provides software on a CD-ROM that accompanies his text book, which can be used for modal analysis of a volume using boundary element analysis.

12.2.6 Modal Coupling using MATLAB

Fahy (1985) and Fahy and Gardonio (2007) describe equations for determining the coupled structural-acoustic displacement response of a system, $w(r_S)$, at some location, r_S, on the structure, in terms of the combination and summation of structural and acoustic mode shapes. The structural mode shapes are evaluated by assuming that the structure is vibrating in a vacuum and the acoustic mode shapes of the enclosure surrounded by the structure are evaluated by assuming that the surrounding structure is infinitely rigid. The structural displacement at frequency, ω, is described in terms of a summation over the *in vacuo* normal modes as:

$$w(\boldsymbol{r}_S, \omega) = \sum_{\ell=1}^{N_s} w_\ell(\omega)\varphi_\ell(\boldsymbol{r}_S) \qquad (12.32)$$

where the sinusoidal time dependency term, $e^{j\omega t}$ has been omitted from both sides of the equation as explained previously. The quantity, $\varphi_\ell(\boldsymbol{r}_S)$ is the mode shape of the ℓth structural mode at arbitrary location, $\boldsymbol{r}_S$ on the surface of the structure, and $w_\ell(\omega)$ is the modal participation factor of the ℓth mode at frequency, ω. Theoretically, the value of N_s should be infinity, but this is not possible to implement in practice so N_s is chosen such that the highest order mode considered has a resonance frequency between twice and four times that of the highest frequency of interest in the analysis, depending on the model being solved and the accuracy required. The N_s structural mode shapes and resonance frequencies can be evaluated using finite element analysis software, and the nodal displacements for a mode are described as a vector $\boldsymbol{\varphi}_\ell$ and then collated into a matrix $[\boldsymbol{\varphi}_1, \boldsymbol{\varphi}_2, ... \boldsymbol{\varphi}_{Ns}]$ for all the modes.

The acoustic pressure at frequency, ω, is described in terms of a summation of the acoustic modes of the fluid volume with rigid boundaries as:

$$p(\boldsymbol{r}, \omega) = \sum_{n=1}^{N_a} p_n(\omega)\psi_n(\boldsymbol{r}) \qquad (12.33)$$

where the time dependency term has been omitted as it is not used in the analysis. The quantity, $\psi_n(\boldsymbol{r})$ is the acoustic mode shape of the nth mode at arbitrary location, $\boldsymbol{r}$ within the volume of fluid, and p_n is the modal participation factor of the nth mode.

Theoretically, the value of N_a should be infinity, but this is not possible to implement in practice so N_a is chosen such that the highest order mode considered has a resonance frequency between twice and four times that of the highest frequency of interest in the analysis, depending on the model being solved and the accuracy required. Note that the $n = 0$ mode is the acoustic bulk compression mode of the cavity that must be included in the summation. When conducting a modal analysis using finite element analysis software, the bulk compression mode of the cavity is the pressure response at 0 Hz. The N_a acoustic mode shapes and resonance frequencies can be evaluated using finite element analysis software, where the nodal pressures for a mode, n, are described as a vector $\boldsymbol{\psi}_n$ and then collated into a matrix $[\boldsymbol{\psi}_1, \boldsymbol{\psi}_2, ... \boldsymbol{\psi}_{Na}]$ for all the modes from 1 to N_a.

The equation for the coupled response of the structure for structural mode, ℓ, is:

$$\ddot{w}_\ell + \omega_\ell^2 w_\ell = \frac{S}{\Lambda_\ell} \sum_{n=1}^{N_a} p_n C_{n\ell} + \frac{F_\ell}{\Lambda_\ell} \qquad (12.34)$$

where the frequency dependence of the pressures, forces and displacements is implicit; that is, these quantities all have a specific and usually different value for each frequency, ω. The quantity, ω_ℓ is the structural resonance frequency for the ℓth mode, Λ_ℓ is the modal mass (see Equation (12.2)), F_ℓ is the modal force applied to the structure for the ℓth mode, S is the surface area of the structure, and $C_{n\ell}$ is the dimensionless coupling coefficient between structural mode, ℓ, and acoustic mode, n, given by the integral of the product of the structural, φ_ℓ, and acoustic, ψ_n, mode shape functions over the surface of the structure, given by:

$$C_{n\ell} = \frac{1}{S} \int_S \psi_n(r_s)\,\varphi_\ell(r_s)\,\mathrm{d}S \tag{12.35}$$

The left hand side of Equation (12.34) is a standard expression to describe the response of a structure in terms of its modes. The right-hand side of Equation (12.34) describes the forces that are applied to the structure in terms of modal forces. The first term describes the modal force exerted on the structure due to the acoustic pressure acting on the structure. The second term describes the forces that act directly on the structure. As an example, consider a point force F_a acting normal to the structure at nodal location (x_a, y_a) for which the mode shapes and resonance frequencies have been evaluated using FEA. As the force acts on the structure at a point, the modal force, F_ℓ, at frequency, ω, for mode, ℓ is:

$$F_\ell(\omega) = \psi_\ell(x_a, y_a) F_a(\omega) \tag{12.36}$$

where $\psi_\ell(x_a, y_a)$ is the ℓth mode shape at the nodal location (x_a, y_a). Tangential forces and moment loadings on the structure can also be included in $F_\ell(\omega)$ and the reader is referred to Soedel (1993) and Howard (2007) for more information.

The dimensionless coupling coefficient $C_{n\ell}$ is calculated from finite element model results as:

$$C_{n\ell} = \frac{1}{S} \sum_{i=1}^{J_s} \psi_n(r_i)\,\varphi_\ell(r_i)\,S_i \tag{12.37}$$

where S is the total surface area of the structure in contact with the acoustic fluid, S_i is the nodal area of the ith node on the surface (and hence $S = \sum_{i=1}^{i=J_s} S_i$), J_s is the total number of nodes on the surface, $\psi_n(r_i)$ is the acoustic mode shape for the nth mode at node location r_i, and $\varphi_\ell(r_i)$ is the mode shape of the ℓth structural mode at node location r_i. The area associated with each node of a structural finite element is sometimes available and if so, can be readily extracted from the software. The nodal areas can also be calculated by using the nodal coordinates that form the elements.

The equation for the coupled response of the fluid (mode n) is given by:

$$\ddot{p}_n + \omega_n^2 p_n = -\left(\frac{\rho c^2 S}{\Lambda_n}\right) \sum_{\ell=1}^{N_s} \ddot{w}_\ell C_{n\ell} + \left(\frac{\rho c^2}{\Lambda_n}\right) \dot{Q}_n \tag{12.38}$$

where the frequency dependence of p, w and Q_n is implicit. The quantity, ω_n represents the resonance frequencies of the cavity, ρ is the density of the fluid, c is the speed of sound in the fluid, Λ_n is the modal volume defined as the volume integration of the square of the mode shape function:

$$\Lambda_n = \int \psi_n^2(r)\,dV \tag{12.39}$$

and $\dot{Q}_n$ is a modal volume acceleration defined as:

$$\dot{Q}_n(\omega) = \varphi_n(x_b, y_b)\dot{Q}_b(\omega) \tag{12.40}$$

where $\dot{Q}_b$ is the complex amplitude of the volume acceleration at nodal location (x_b, y_b), and $\varphi_n(x_b, y_b)$ is the nth mode shape at the nodal location (x_b, y_b). A common definition for an acoustic source has units of volume velocity, which in this case is Q_b, and hence the time derivative of this expression is the source volume acceleration $\dot{Q}_b$.

An important point to note is that because the acoustic mode shapes used in the structural-acoustic modal coupling method are for a rigid walled cavity, corresponding to a normal acoustic particle velocity at the wall surface equal to zero, the acoustic velocity at the surface resulting from the modal coupling method is incorrect (Jayachandran *et al.* 1998). However the acoustic pressure at the surface is correct, and this is all that is required for correctly coupling the acoustic and pressure modal equations of motion.

For simple systems such as rectangular, rigid-walled cavities and simple plates it is possible to write analytical solutions for the mode shapes and resonance frequencies. Anything more complicated than these simple structures nearly always involves the use of a discretised numerical model such as a finite element analysis, in which case, it is necessary to extract parameters from the finite element model to enable the calculation of the coupled response.

Cazzolato (1999) described a method to calculate the acoustic and structural modal masses from a finite element model. When using finite element analysis software to evaluate the acoustic pressure mode shapes, the vectors returned by the software can be normalised to either unity or to the mass matrix. By normalising the mode shapes to the mass matrix, the modal volume of the cavity can be obtained directly; that is:

$$\mathbf{\Psi}_n^T[\mathbf{M}_{fea}]\mathbf{\Psi}_n = 1 \tag{12.41}$$

where $\mathbf{\Psi}_n$ is the mass normalised mode shape function vector for the nth mode and $[\mathbf{M}_{fea}]$ is the fluid element mass matrix defined as:

$$[\mathbf{M}_{fea}] = \frac{1}{c^2}\int_{V_e}[\mathbf{N}][\mathbf{N}]^T\,dV_e \tag{12.42}$$

where $[\mathbf{N}]$ is the shape function for the acoustic element with a single pressure degree of freedom and V_e is the volume of the element. If the mode shape vectors are normalised to unity; that is, the maximum value in the vector is 1, then:

$$\hat{\mathbf{\Psi}}_n^T[\mathbf{M}_{fea}]\hat{\mathbf{\Psi}}_n = \frac{\Lambda_n}{c^2} \tag{12.43}$$

where $\hat{\boldsymbol{\Psi}}_n$ is the mode shape vector normalised to unity for the nth mode and Λ_n is the modal volume of the nth mode. It can be shown that the relationship between the mass normalised mode shape vector $\boldsymbol{\Psi}_n$ and the unity normalised mode shape vector $\hat{\boldsymbol{\Psi}}_n$ is (Ewins, 1995):

$$\boldsymbol{\Psi}_n = \frac{c}{\sqrt{\Lambda_n}} \hat{\boldsymbol{\Psi}}_n \tag{12.44}$$

Given that the maximum value of the unity normalised mode shape vector $\hat{\boldsymbol{\Psi}}_n = 1$, the maximum element of the mass-normalised mode shape vector is equal to the ratio of the speed of sound in air to the square root of the modal volume:

$$\Lambda_n = \frac{c^2}{\max(\boldsymbol{\Psi}_n{}^2)} \tag{12.45}$$

Hence, to extract the acoustic modal volume of a system using finite element analysis software, an acoustic modal analysis is conducted and the results are normalised to the mass matrix. Then Equation (12.45) is used to calculate the acoustic modal volumes for each mode. The unity normalised mode shapes can be calculated as:

$$\hat{\boldsymbol{\Psi}}_n = \frac{\boldsymbol{\Psi}_n}{\max(\boldsymbol{\Psi}_n)} \tag{12.46}$$

Equations (12.34) and (12.38) can form a matrix equation as:

$$\begin{bmatrix} \Lambda_\ell(\omega_\ell^2 - \omega^2) & -S[\mathbf{C}_{n\ell}] \\ S\omega^2[\mathbf{C}_{n\ell}]^T & \dfrac{\Lambda_n}{\rho c^2}(\omega_n^2 - \omega^2) \end{bmatrix} \begin{bmatrix} \mathbf{w}_\ell \\ \mathbf{p_n} \end{bmatrix} = \begin{bmatrix} \mathbf{F}_\ell \\ \dot{\mathbf{Q}}_n \end{bmatrix} \tag{12.47}$$

· where all the ℓ structural and n acoustic modes are included in the matrices, so that the square matrix on the left-hand side of Equation (12.47) has dimensions $(\ell + n) \times (\ell + n)$. The left-hand matrix in Equation (12.47) can be made symmetric by dividing all terms in the lower equation by $-\omega^2$. The structural, w_ℓ, and acoustic, p_n, modal participation factors, which are frequency dependent, can be calculated by pre-multiplying each side of the equation by the inverse of the square matrix on the left hand side. Once these factors are calculated, the vibration displacement of the structure can be calculated from Equation (12.32) and the acoustic pressure inside the enclosure can be calculated using Equation (12.33).

The equations described above do not have damping terms. In practice it is common to include damping in the structure by using a complex elastic modulus, and damping in the fluid by a complex bulk modulus.

The method described above can be used to make predictions of the vibro-acoustic response of an enclosed system, but it does have limitations. One mistake that is commonly made is to make numerical calculations with an insufficient number of structural and acoustic modes. This problem affects all numerical methods involving the summation of modes to predict the overall response and has been known since the early 1970s. Cazzolato *et al.* (2005) demonstrated the errors that can occur with modal truncation and how it can lead to erroneous conclusions. As a start, the analyst should consider including structural and acoustic modes that have resonance frequencies up to two octaves higher than the frequency range of interest. Methods have been proposed to reduce the number of modes required to be included in the analysis by including the affects of the higher-order modes in a residue or pseudo-static correction term (Tournour and Atalla 2000; Gu *et al.* 2001; Zhao *et al.* 2002).

The modal coupling method described above is applicable to vibro-acoustic systems where there is "light" coupling, such as between air and a structure. If the vibro-acoustic response of a system is to be calculated where there is "heavy" coupling due to the fluid loading, such as between water and a structure, then this method will generate erroneous results because it does not account for the cross-fluid-coupling terms; that is, the coupling between fluid modes.

One of the main advantages of using the modal coupling method is that the time taken to solve the system of equations is significantly less than conducting a full fluid-structure interaction analysis using finite element analysis. This is very important if optimisation studies are to be conducted that involve many FEA evaluations while converging to an optimum solution.

Example 12.2

Consider a rigid walled rectangular box containing air with dimensions $0.5 \times 0.3 \times 1.1$ m as shown in Figure 12.15. A simply supported panel is attached at one end of the box with dimensions 0.5×0.3 m. The panel is 3 mm thick aluminium with a Young's modulus of $E = 70.9$ GPa and a density of 2700 kg/m^3. A point force of $F = 1$ N acts on the panel at (0.10, 0.06, 0.0), which causes the panel to vibrate and radiate sound into the enclosure. Calculate the sound pressure level inside the enclosure at location (0.300, 0.105, 0.715).

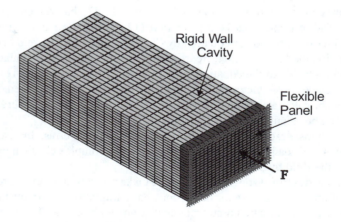

Figure 12.15 Rigid walled rectangular box model.

Solution

The problem was analysed using the numerical method described above and implemented in the software Matlab. An *in-vacuo* (meaning without the air inside the box) modal analysis of the structure was conducted and 21 mode shapes and resonance frequencies were calculated. Similarly, the modal response of the cavity assuming a rigid walled structure was calculated to determine 102 acoustic resonance frequencies and pressure mode shapes. The range of the resonance frequencies from the modal analyses covers two octaves higher than the frequency range of interest, and the distribution of resonance frequencies for the structure and the acoustic cavity are shown in Figure 12.16 below.

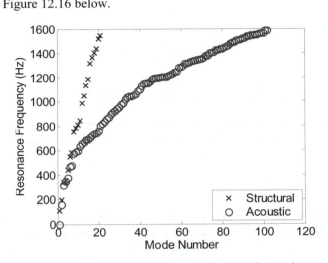

Figure 12.16 Structural and acoustic resonance frequencies.

The problem was also analysed using the ANSYS finite element analysis software. Pressure formulated acoustic elements (FLUID30) were used to model the acoustic cavity, and shell elements (SHELL63) were used to model the thin aluminium plate. For most of the cavity, the acoustic elements with only pressure degree of freedom were used. At the interface between the structure and the acoustic elements, the acoustic elements had pressure and structural displacement degrees of freedom, to enable the coupling between the structure and the acoustic fluid. The fluid-structure-interaction option was enabled for the structural elements in contact with the pressure and displacement formulated acoustic elements. The number of element divisions used along each side of the box was 20. At 1600Hz the wavelength is 343/1600 = 0.214 m, and the longest element length is 1.1/20 = 0.055m. Hence the ratio between the wavelength to largest element size is 0.214 / 0.055 = 3.9, which is less than the recommended 6 elements-per-wavelength. However in the frequency range of interest (400Hz) the ratio of elements-per-wavelength is ((343/400) / (1.1/20)) = 15.6.

Figure 12.17 shows a comparison of the sound pressure response calculated for a point (0.300, 0.105, 0.715) within the cavity using FEA with a full fluid-structure

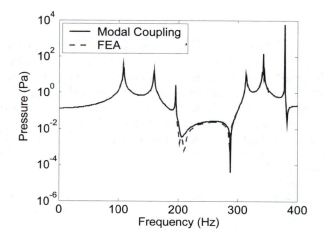

Figure 12.17 Pressure at a node (0.300, 0.105, 0.715) inside a box evaluated using modal coupling theory and FEA (ANSYS) using full fluid-structure-interaction coupling.

interaction formulation, and the modal coupling method implemented using Matlab. It can be seen that the results are almost identical over the frequency range of interest.

12.2.6.1 Acoustic Potential Energy

The acoustic potential energy $E_p(\omega)$ is a useful measure of the acoustic energy contained within a cavity at frequency, ω. This measure can be used to evaluate the effectiveness of noise control in an enclosure and is given by:

$$E_p(\omega) = \frac{1}{4\rho c^2} \int_V |p(r,\omega)|^2 \, dV \tag{12.48}$$

which can be implemented in a finite element formulation as:

$$E_p(\omega) = \frac{1}{4\rho c^2} \sum_{n=1}^{N_a} p_n^2(\omega) V_n \tag{12.49}$$

where p_n is the acoustic pressure at the nth node and V_n is the volume associated with the nth node. This equation can be re-arranged so that the acoustic potential energy is calculated in terms of the modal pressure amplitudes as:

$$E_p(\omega) = \sum_{n=1}^{N_a} \Lambda_n |p_n(\omega)|^2 = \mathbf{p}^H \mathbf{\Lambda} \mathbf{p} \tag{12.50}$$

where $\mathbf{\Lambda}$ is a $(N_a \times N_a)$ diagonal matrix for which the diagonal terms are:

$$\mathbf{\Lambda}(n,n) = \frac{\Lambda_n}{4\rho c^2} \tag{12.51}$$

12.3 HIGH-FREQUENCY REGION: STATISTICAL ENERGY ANALYSIS

In the high frequency region, a method generally known as statistical energy analysis (SEA) is used (Lyon, 1975; Sablik, 1985) to calculate the flow and storage of vibration and acoustic energy in a complex system. The total sound power radiated by a particular structure is calculated by summing that due to each of the individual panels or parts making up the structure.

It is necessary to consider frequency-averaged data, using at least one-third octave bandwidths, and preferably octave bandwidths. There should be at least three modes resonant in the frequency band being considered for each sub-system involved in the analysis and the modal overlap should be at least unity and even higher if possible.

For a part of the structure of area, S, the band-averaged radiated sound power, $W_{\Delta S}$, is calculated using:

$$W_{\Delta S} = S\rho c\sigma_{\Delta S}\langle v^2 \rangle_{St\Delta} \tag{12.52}$$

The subscript Δ denotes frequency band average, and $\sigma_{\Delta S}$ is the band-averaged radiation ratio (or radiation efficiency) for surface S. The quantity $\langle v^2 \rangle_{St\Delta}$ is the mean square surface velocity averaged over the frequency band, Δ, the surface S and time, t. It is calculated using an energy balance between the energy input to the structure and that dissipated within the structure.

Energy is input by the external excitation source and stored in the vibrational modes of the structure. Energy is dissipated by mechanical damping in the structure and transferred between the various parts of the structure, across interconnecting joints. The energy input to the structure by the exciting force is equal to the total energy in the structure plus that lost due to damping and sound radiation. The energy is assumed to be equally distributed among all vibration modes, so that each mode of a structure or system of connected structures has equal modal energy. For a given structural part or panel, the total vibratory power flowing into it is equal to the power dissipated by the panel plus the power flowing out of it. Energy is also dissipated at joints between adjacent panels. By setting up appropriate matrix equations, the modal energy in each part of the structure can be determined. The area, time and band averaged mean square velocity is determined by dividing the structural panel modal energy by the panel mass.

The radiation efficiencies used for this part of the analysis are the same as those referred to in Section 6.7. Calculated values for various other structures are available in the literature. Means for calculating the energy lost during transfer across various types of structural joints and connections are discussed in the literature (Lyon, 1975; Fahy, 1985; Fahy and Gardonio, 2007) and tables are also given later in this section.

One of the concepts behind Statistical Energy Analysis is that interconnected systems transfer vibro-acoustic energy between them and the total energy in the system must always be fully accounted for. Consider two generic vibro-acoustic systems as shown in Figure 12.18, which have a mechanism to transfer energy between the them. Examples of vibro-acoustic systems could be a vibrating panel, or a enclosed room.

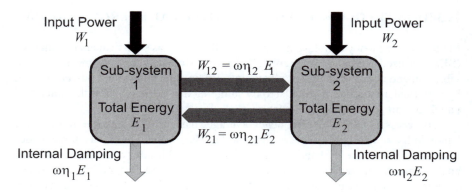

Figure 12.18 Two vibro-acoustic systems that are connected and have power flowing into (or out) them.

Imagine that a system boundary were drawn around sub-system 1 in Figure 12.18. An equation describing the energy balance of sub-system 1 is:

$$W_1 + W_{21} = \omega\eta_1 E_1 + W_{12} \tag{12.53}$$

where W_1 is the power entering sub-system 1, W_{21} is the power coming from sub-system 2 into system 1, $\omega\eta_1 E_1$ is the power dissipated in sub-system 1 by damping mechanisms, η_1 is the dissipative or damping loss factor (DLF) in sub-system 1, ω is the centre frequency of the band (in radians/sec), E_1 is the total energy in sub-system 1, W_{12} is the power that leaves sub-system 1 and goes into sub-system 2. The left-hand side of the energy balance Equation (12.53) is the power entering the sub-system, and the right-hand side is the power that is lost by the sub-system by damping and power transferred to a connected sub-system. Equation (12.53) can be re-arranged to give:

$$W_1 = [\omega\eta_1 E_1] + [\omega\eta_{12}E_1] - [\omega\eta_{21}E_2] \tag{12.54}$$

where $W_{12} = [\omega\eta_{12}E_1]$, $W_{21} = [\omega\eta_{21}E_2]$, η_{12} is the coupling loss factor from sub-system 1 to sub-system 2, and η_{21} is the coupling loss factor from sub-system 2 to sub-system 1. The coupling loss factor (CLF) is a ratio of the energy that is transferred from one sub-system to another, which varies with frequency, and can be determined theoretically, numerically using finite element analysis, or experimentally. The determination of CLFs is discussed in detail in the next section.

The estimates for DLFs can be obtained from text books or determined experimentally. The lower limits of the values in Appendix B are suitable as DLFs but the higher limits are not as these are for transmission loss calculations and have the damping associated with the panel mounting included. One of the purposes of using SEA is to determine the amplitude response of a system at various frequencies, and hence the damping in the system is often very important, but also hard to estimate. It is possible to obtain any response value by adjusting the values of the damping loss factor. However by doing so will also alter the energy levels in a system and hence the flow of energy between the interconnected systems.

A complex vibro-acoustic system modelled using the SEA framework can be thought of a network of sub-systems, where power flows in and out, and energy is stored within the systems. Altering the coupling or damping loss factors of the system has the effect of re-routing the power distribution throughout the network.

The relationship between the coupling loss factors is:

$$n_1 \eta_{12} = n_2 \eta_{21} \tag{12.55}$$

where n_i is the average density of vibro-acoustic modes defined as $n(\omega) = N / \Delta\omega$, N is the number of modes that are resonant in (one-, or one-third octave band) frequency band $\Delta\omega$, centred on frequency ω.

Consider the general case where there are k interconnected sub-systems. A system of energy balance equations from Equation (12.54) can be formed and put into a matrix equation as:

$$\omega \begin{bmatrix} (\eta_1 + \sum_{i \neq 1}^{k} \eta_{1i})\, n_1 & (-\eta_{12} n_1) & \cdots & (-\eta_{1k} n_1) \\ (-\eta_{21} n_2) & (\eta_2 + \sum_{i \neq 2}^{k} \eta_{2i})\, n_2 & \cdots & (-\eta_{2k} n_2) \\ \vdots & \vdots & \ddots & \vdots \\ (-\eta_{k1} n_k) & & \cdots & (\eta_k + \sum_{i \neq k}^{k} \eta_{ki})\, n_k \end{bmatrix} \begin{bmatrix} E_1/n_1 \\ E_2/n_2 \\ \vdots \\ E_k/n_k \end{bmatrix} = \begin{bmatrix} W_1 \\ W_2 \\ \vdots \\ W_k \end{bmatrix} \tag{12.56}$$

$$\omega\,[\mathbf{C}][\mathbf{E}] = [\mathbf{W}] \tag{12.57}$$

where $[\mathbf{C}]$ is the ($k \times k$) matrix of coupling loss factors, $[\mathbf{E}]$ is the ($k \times 1$) vector of the energies within each sub-system, and $[\mathbf{W}]$ is the ($k \times 1$) vector of input powers to each sub-system. The input power to each sub-system is known (from measurements or from the problem description) and hence the energy within each system $[\mathbf{E}]$ can be calculated by pre-multiplying each side of the equation by the inverse of $\omega\,[\mathbf{C}]$.

12.3.1 Coupling Loss Factors

The coupling loss factor for one-dimensional sub-systems connected at points is given by:

$$\eta_{12} = \frac{2}{\pi \omega n_1(\omega)} \frac{\mathrm{Re}(Z_1)\,\mathrm{Re}(Z_2)}{|\sum_i Z_i|^2} \tag{12.58}$$

where $n_1(\omega)$ is the modal density in the source sub-system, Z_1 and Z_2 are the impedances of the source and receiver, and the summation of impedances in the denominator is for all sub-systems directly connected to the source sub-system. Impedances and modal density expressions that can be used to calculate coupling loss factors for connections between various structures are given in Tables 12.3 and 12.4.

Table 12.3 Translational and rotational impedances of beams and plates

Axially excited semi-infinite beam
$$Z = \rho_m S c_L$$

Axially excited infinite beam
$$Z = 2\rho_m S c_L$$

Point force normal to thin semi-infinite beam
$$Z = (1 + j)\rho_m S c_B / 2$$

Point force normal to thin infinite beam
$$Z = 2(1 + j)\rho_m S c_B$$

Point moment at end of thin semi-infinite beam
$$Z_M = \frac{(1 - j)\rho_m S c_B^3}{2\omega^2}$$

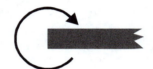

Point moment on infinite thin beam
$$Z_M = \frac{2(1 - j)\rho_m S c_B^3}{\omega^2}$$

Point force on thin infinite panel
$$Z = 8\sqrt{B\rho_m h} \approx 2.3\rho_m c_L h^2$$

Point moment on thin infinite panel
$$Z_M = \frac{16B}{\omega} \times \left[\frac{1}{1 - \frac{4j}{\pi} \log_e(0.9 k_B a)} \right]$$

radius a

where c_L is the longitudinal wave speed, c_B is the bending wave speed in a beam, given by $c_B = \left(EI\omega^2/\rho_m S \right)^{1/4}$, where E is Young's modulus, I is the second moment of area of the beam cross-section, ρ is the density, S is the cross sectional area. For a plate, the bending stiffness is given by Equation (8.2) as $B = Eh^3/[12(1 - \nu^2)]$, h is the thickness of the plate, ν is Poisson's ratio, and the bending wave number in a plate is $k_B = (\omega\rho_m h/B)^{1/4}$.

Table 12.4 Modal densities of sub-systems

Beam in compression

$$n = \frac{L}{\pi c_L}$$

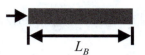
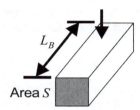

Beam in bending

$$n(\omega) = \frac{L}{2\pi}\left[\frac{\rho S}{EI\omega^2}\right]^{1/4}$$

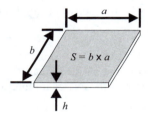

Finite plate in bending

$$n = \frac{\sqrt{3}S}{2\pi h c_L}$$

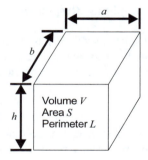

Acoustic cavity

$$n(\omega) = \frac{V\omega^2}{2\pi^2 c^3} + \frac{S\omega}{8\pi c^2} + \frac{L}{16\pi c}$$

The coupling loss factor for line connected sub-systems, such as between two plates, is given by:

$$\eta_{12} = \frac{c_g L}{\pi \omega A_s}\langle \tau(\theta)\cos\theta\rangle \qquad (12.59)$$

where c_g is the group velocity, L is the length of the line junction between the two sub-systems, A_s is the area of the source sub-system, θ is the angle of incidence, τ is the transmission coefficient and the brackets $\langle\ \rangle$ indicate the average over all angles.

The coupling loss factor for two acoustic spaces connected together through an open aperture is given by:

$$\eta_{12} = \frac{cA_w}{4\omega V_s} \langle \tau_{12} \rangle \tag{12.60}$$

where A_w is the area of the opening connecting the spaces, V_s is the volume of the source cavity, $\langle \tau_{12} \rangle$ is the average transmission coefficient.

The coupling loss factor for a panel connected to an acoustic cavity is given by:

$$\eta_{12} = \frac{\rho c A_p}{\omega m_p} \sigma \tag{12.61}$$

where A_p is the area of the panel, M_p is the total mass of the panel, and σ is the radiation efficiency (ratio) of the panel.

Another useful expression when dealing with acoustic cavities is the *damping* loss factor, which is derived using Equations (7.23), (7.24) and (7.52) is given by:

$$\eta = \frac{cS\bar{\alpha}}{25.1fV} \tag{12.62}$$

where S is the surface area of the cavity and $\bar{\alpha}$ is the average Sabine absorption coefficient of the enclosure interior surfaces.

12.3.2 Amplitude Responses

Once Equations (12.57) is calculated and the energies in each sub-system are calculated, the usual practice is to determine the amplitude response. For structural systems the average squared velocity of the structure is given by $_{St\Delta}$

$$\langle v^2 \rangle_{St\Delta} = E/M \tag{12.63}$$

where E is the energy in the sub-system, and M is the total mass of the sub-system. For acoustic systems the average square pressure is given by

$$\langle p^2 \rangle_{St\Delta} = E\rho c^2 / V \tag{12.64}$$

Analysis of vibro-acoustic problems using SEA can be conducted using commercial or free SEA software or they can be analysed using spreadsheet software package such as Microsoft Excel or OpenOffice Calc.

The implementation of SEA can be illustrated by application to the following example.

Example 12.3

Consider a system comprising a beam 2 m long, with a cross-section 20 mm × 20 mm that is welded to the centre of a large steel plate (3 m × 3 m) and has a thickness of

20 mm. The plate is mounted as a baffle in one wall of a room that has edge lengths of equal size of 5 m, hence a volume of 125 m³ʼ and the room has an average Sabine absorption coefficient of $\bar{\alpha} = 0.07$. A power of 1 W is injected into the beam, using a point moment at the end of the beam to generate flexural waves in the beam. The beam is point coupled to the plate, and in this case only generates bending waves in the plate. For the frequency range 63 Hz to 2 kHz, calculate the sound pressure level in the room. Assume that the damping loss factors for the beam and plate are 0.15.

Solution

The general procedure for solving SEA problems is:

1. Identify the sub-systems and their geometric and material parameters.
2. Sketch a network diagram of the system to identify the power flows between sub-systems and the method of power transmission at each junction. For example, consider if power transmission occurs due to translational and rotational coupling.
3. Calculate the input impedances for each sub-system.
4. Calculate the modal densities of the sub-systems.
5. Calculate the coupling loss factors between the sub-systems.
6. Calculate the damping loss factors for each sub-system.
7. Calculate the input powers to each sub-system.
8. Form the matrix equation that describes the power flows between sub-systems, and calculate the energy levels within each sub-system.
9. Calculate the amplitude responses of the sub-systems from the results of Step 8.

Step 1: Parameters
The parameters for the problem are contained in the problem description.

Step 2: Network Diagram
A system diagram can be drawn illustrating the power flow between sub-systems as shown in Figure 12.19.

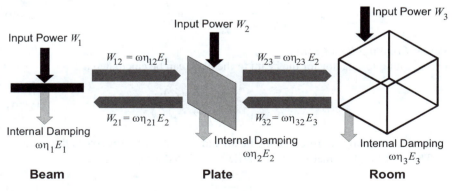

Figure 12.19 Power flow between sub-systems.

Step 3: Impedances of sub-systems
The impedances of the beam and plate can be calculated. From Table 12.3 the impedance of the beam for moment coupling is given by:

$$Z_{M\,\text{beam}}(\omega) = \frac{(1-j)\rho_m S c_B^3}{2\omega^2} \tag{12.65}$$

and the impedance of a thin plate due to an applied moment is given by:

$$Z_{M\,\text{plate}}(\omega) = \frac{16B}{\omega}\left[\frac{1}{1 - \dfrac{4j}{\pi}\log_e(0.9k_B a)}\right] \tag{12.66}$$

where a is the thickness of the beam.

Step 4: Modal Densities
The modal densities of the three sub-systems are calculated using Table 12.4 as:

Beam:

$$n_{\text{beam}}(\omega) = \frac{L_B}{2\pi}\left[\frac{\rho_m S}{EI\omega^2}\right]^{1/4} \tag{12.67}$$

Plate (independent of frequency):

$$n_{\text{plate}} = (\sqrt{3})S/(2\pi h c_L)$$

$$= (\sqrt{3})(3{\times}3)/(2\pi{\times}(20{\times}10^{-3}){\times}5050) \tag{12.68}$$

$$= 0.0245 \text{ modes per radian}$$

Room:
The modal density for a room varies with frequency and can be determined by converting the frequency, f in Hz in Equation (7. 21) to ω radians (by dividing by 2π) and then differentiating with respect to ω to give:

$$n_{\text{room}}(\omega) = \frac{V\omega^2}{2\pi^2 c^3} + \frac{S\omega}{8\pi c^2} + \frac{L}{16\pi c} \tag{12.69}$$

Step 5: Coupling Loss Factors
The coupling loss factor for the beam to the plate can be calculated as:

$$\eta_{12} = \frac{2}{\pi\omega n_1(\omega)}\frac{\text{Re}(Z_1)\ \text{Re}(Z_2)}{|Z_1 + Z_2|^2} \tag{12.70}$$

where the subscript "1" refers to the beam and the subscript 2 refers to the plate.
   The coupling loss factor for the baffled panel to the room is given by:

$$\eta_{23} = \frac{\rho c A_p}{\omega m_p}\sigma \tag{12.71}$$

where the radiation efficiency, σ, of the panel is determined by the frequency and the dimensions of the panel. The critical frequency of the panel is (from Equation (8.3)):

$$f_c = \left(\frac{c^2}{2\pi}\right)\sqrt{\frac{12\rho(1-v^2)}{Eh^2}} = \left(\frac{343^2}{2\pi}\right)\sqrt{\frac{12\times7850\times(1-0.3^2)}{210\times10^9\times(20\times10^{-3})^2}} = 598\,\text{Hz} \quad (12.72)$$

The radiation efficiencies for the baffled plate (from Figure 6.4) and the calculated coupling loss factors are given in the table below:

Octave band centre frequency (Hz)	σ (dB)	η_{12}	η_{23}
63	-15	0.00335	0.000213
125	-14	0.00301	0.000135
250	-10	0.00266	0.000170
500	-1	0.00232	0.000675
1000	+2	0.00198	0.000673
2000	0	0.00165	0.000212

Step 6: Damping Loss Factors
The damping loss factors for the beam and plate are $\eta_1 = \eta_2 = 0.15$. The damping loss factor for the room, η_3 is given by Equation (12.61).

Step 7: Input Powers
The input power to the system is 1W applied to the beam.

Step 8: System Equation and Solution
A matrix equation can be created as illustrated in Equations (12.56) and (12.57), and the energies in the sub-systems calculated accordingly.

Step 9: Amplitude Responses
Once the energy levels in each sub-system are created, the velocity of the beam and plate are calculated using Equation (12.63) and the sound pressure level in the room is calculated using Equation (12.64) giving the following results.

Octave band centre frequency (Hz)	63	125	250	500	1000	2000
SPL (dB re 20 µPa)	71	68	69	74	74	68
Velocity of beam (dB re 10^{-9} m/s)	152.5	149.5	146.5	143.5	140.5	137.5
Velocity of plate (dB re 10^{-9} m/s)	114.2	110.7	107.2	103.6	99.9	96.1

APPENDIX A

Wave Equation Derivation

The derivation of the acoustical wave equation is based on three fundamental fluid dynamical equations; the continuity (or conservation of mass) equation, Euler's equation (or the equation of motion), and the equation of state. Each of these equations are discussed separately in the following sections.

A.1 CONSERVATION OF MASS

Consider an arbitrary volume, V, as shown in Figure A.1. The total mass contained in this volume is $\int_V \rho_{tot} \, dV$. The law of conservation of mass states that the rate of mass leaving the volume, V, must equal the rate of change of mass in the volume. That is:

$$\int_A \rho_{tot} \, \boldsymbol{U_{tot}} \cdot \mathbf{n} \, dA \; = \; -\frac{d}{dt} \int_V \rho_{tot} \, dV \tag{A.1}$$

where A is the area of surface enclosing the volume, V, and $\mathbf{n}$ is the unit vector normal to the surface, A, at location, dA.

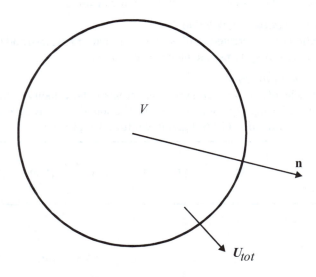

Figure A.1 Arbitrary volume for illustrating conservation of mass.

At this stage, it is convenient to transform the area integral on the left-hand side of Equation (A.2) to a volume integral by use of Gauss' integral theorem, which is written as follows:

$$\int_A \boldsymbol{\psi} \cdot \mathbf{n} \, dA = \int_V \nabla \cdot \boldsymbol{\psi} \, dV \tag{A.2}$$

where $\boldsymbol{\psi}$ is an arbitrary vector and the operator, ∇, is the scalar divergence of the vector $\boldsymbol{\psi}$. Thus, in cartesian coordinates:

$$\nabla \cdot \boldsymbol{\psi} = \frac{\partial \boldsymbol{\psi}}{\partial x} + \frac{\partial \boldsymbol{\psi}}{\partial y} + \frac{\partial \boldsymbol{\psi}}{\partial z} \tag{A.3}$$

and Equation (A.1) becomes:

$$\int_V \nabla \cdot (\rho_{tot} \, \boldsymbol{U}_{tot}) \, dV = -\frac{d}{dt} \int_V \rho_{tot} \, dV = -\int_V \frac{\partial \rho_{tot}}{\partial t} \, dV \tag{A.4a,b}$$

Rearranging gives:

$$\int_V \left[\nabla \cdot (\rho_{tot} \, \boldsymbol{U}_{tot}) + \frac{\partial \rho_{tot}}{\partial t} \right] dV = 0 \tag{A.5}$$

or:

$$\nabla \cdot (\rho_{tot} \, \boldsymbol{U}_{tot}) = -\frac{\partial \rho_{tot}}{\partial t} \tag{A.6}$$

Equation (A.6) is the continuity equation.

A.2 EULER'S EQUATION

In 1775 Euler derived his well-known equation of motion for a fluid, based on Newton's first law of motion. That is, the mass of a fluid particle multiplied by its acceleration is equal to the sum of the external forces acting upon it.

Consider the fluid particle of dimensions Δx, Δy and Δz shown in Figure A.2. The external forces F acting on this particle are equal to the sum of the pressure differentials across each of the three pairs of parallel forces. Thus:

$$F = \boldsymbol{i} \cdot \frac{\partial P_{tot}}{\partial x} + \boldsymbol{j} \cdot \frac{\partial P_{tot}}{\partial y} + \boldsymbol{k} \cdot \frac{\partial P_{tot}}{\partial z} = \nabla P_{tot} \tag{A.7}$$

where $\boldsymbol{i}, \boldsymbol{j}$ and $\boldsymbol{k}$ are the unit vectors in the x, y and z directions and where the operator ∇ is the grad operator and is the vector gradient of a scalar quantity.

The inertia force of the fluid particle is its mass multiplied by its acceleration and is equal to:

$$m\dot{\boldsymbol{U}}_{tot} = m\frac{d\boldsymbol{U}_{tot}}{dt} = \rho_{tot} V \frac{d\boldsymbol{U}_{tot}}{dt} \tag{A.8}$$

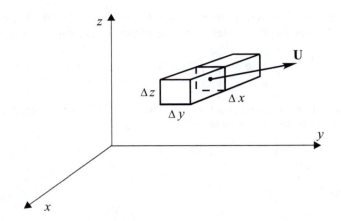

Figure A.2 Particle of fluid.

Assume that the fluid particle is accelerating in the positive x, y and z directions. Then the pressure across the particle must be decreasing as x, y and z increase, and the external force must be negative. Thus:

$$F = -\nabla P_{tot} = \rho_{tot} V \frac{dU_{tot}}{dt} \qquad (A.9a,b)$$

This is the Euler equation of motion for a fluid.

If we were interested in considering sound propagation through porous acoustic media, then it would be necessary to add the term, RU_{tot}, to the right-hand side of Equation (A.9), where R is a constant dependent upon the properties of the fluid.

The term dU_{tot}/dt on the right side of Equation (A.9) can be expressed in partial derivative form as follows:

$$\frac{dU_{tot}}{dt} = \frac{\partial U_{tot}}{\partial t} + (U_{tot} \cdot \nabla) U_{tot} \qquad (A.10)$$

where:

$$(U_{tot} \cdot \nabla) U_{tot} = \frac{\partial U_{tot}}{\partial x} \cdot \frac{\partial x}{\partial t} + \frac{\partial U_{tot}}{\partial y} \cdot \frac{\partial y}{\partial t} + \frac{\partial U_{tot}}{\partial z} \cdot \frac{\partial z}{\partial t} \qquad (A.11)$$

A.3 EQUATION OF STATE

As sound propagation is associated with only very small perturbations to the ambient state of a fluid, it may be regarded as adiabatic. Thus the total pressure P will be functionally related to the total density, ρ_{tot}, as follows:

$$P_{tot} = f(\rho_{tot}) \tag{A.12}$$

Since the acoustic perturbations are small, and P and ρ are constant, $dp = dP_{tot}$, $d\sigma = d\rho$ and Equation (A.12) can be expanded into a Taylor series as follows:

$$dp = \frac{\partial f}{\partial \rho}\, d\sigma + \frac{1}{2}\frac{\partial f}{\partial \rho}(d\sigma)^2 + \quad \text{higher order terms} \tag{A.13}$$

The equation of state is derived by using Equation (A.13) and ignoring all of the higher order terms on the right-hand side. This approximation is adequate for moderate sound pressure levels, but becomes less and less satisfactory as the sound pressure level exceeds 130 dB (60 Pa). Thus, for moderate sound pressure levels:

$$dp = c^2\, d\sigma \tag{A.14}$$

where $c^2 = \partial f/\partial \rho$ is assumed to be constant. Integrating Equation (A.14) gives:

$$p = c^2 \sigma + \text{const} \tag{A.15}$$

which is the linearised equation of state.

Thus the curve $f(\rho_{tot})$ of Equation (A.12) has been replaced by its tangent at P_{tot}, ρ_{tot}. The constant may be eliminated by differentiating Equation (A.15) with respect to time. Thus:

$$\frac{\partial p}{\partial t} = c^2 \frac{\partial \sigma}{\partial t} \tag{A.16}$$

Equation (A.16) will be used to eliminate $\partial \sigma / \partial t$ in the derivation of the wave equation to follow.

A.4 WAVE EQUATION (LINEARIZED)

The wave equation may be derived from Equations (A.6), (A.9) and (A.16) by making the linearising approximations listed below. These assume that the acoustic pressure, p, is small compared with the ambient pressure, P, and that P is constant over time and space. It is also assumed that the mean velocity, $U = 0$. Thus:

$$P_{tot} = P + p \approx P \tag{A.17a,b}$$

$$\rho_{tot} = \rho + \sigma \approx \rho \tag{A.18a,b}$$

$$U_{tot} = u \tag{A.19}$$

$$\frac{\partial P_{tot}}{\partial t} = \frac{\partial p}{\partial t} \tag{A.20}$$

$$\frac{\partial \rho_{tot}}{\partial t} = \frac{\partial \sigma}{\partial t} \tag{A.21}$$

$$\nabla P_{tot} = \nabla p \tag{A.22}$$

Using Equation (A.10), the Euler Equation (A.9) may be written as:

$$-\nabla P_{tot} = \rho_{tot}\left[\frac{\partial U_{tot}}{\partial t} + (U_{tot}\cdot\nabla)U_{tot}\right] \tag{A.23}$$

Using Equations (A.17) (A.18) and (A.19), Equation (A.23) may be written as:

$$-\nabla p = \rho\left[\frac{\partial u}{\partial t} + u\cdot\nabla u\right] \tag{A.24}$$

As u is small and ∇u is approximately the same order of magnitude as u, the quantity, $u \cdot \nabla u$ may be neglected and Equation (A.24) written as:

$$-\nabla p = \rho\frac{\partial u}{\partial t} \tag{A.25}$$

Using Equations (A.18), (A.19) and (A.21), the continuity Equation (A.6) may be written as:

$$\nabla\cdot(\rho u + \sigma u) = -\frac{\partial \sigma}{\partial t} \tag{A.26}$$

As σu is so much smaller than ρ, the equality in Equation (A.26) can be approximated as:

$$\nabla\cdot(\rho u) = -\frac{\partial \sigma}{\partial t} \tag{A.27}$$

Using Equation (A.16), Equation (A.27) may be written as:

$$\nabla\cdot(\rho u) = -\frac{1}{c^2}\frac{\partial p}{\partial t} \tag{A.28}$$

Taking the time derivative of Equation (A.28) gives:

$$\nabla\cdot\rho\frac{\partial u}{\partial t} = -\frac{1}{c^2}\frac{\partial^2 p}{\partial t^2} \tag{A.29}$$

Substituting Equation (A.25) into the left side of Equation (A.29) gives:

$$-\nabla\cdot\nabla p = -\frac{1}{c^2}\frac{\partial^2 p}{\partial t^2} \tag{A.30}$$

or:

$$\nabla^2 p = \frac{1}{c^2} \frac{\partial^2 p}{\partial t^2} \tag{A.31}$$

The operator, ∇^2 is the (div grad) or the Laplacian operator, and Equation (A.31) is known as the linearised wave equation or the Helmholtz equation.

The wave equation can be expressed in terms of the particle velocity by taking the gradient of the linearised continuity Equation (A.28). Thus:

$$\nabla(\nabla \cdot \rho \boldsymbol{u}) = -\nabla \left(\frac{1}{c^2} \frac{\partial p}{\partial t} \right) \tag{A.32}$$

Differentiating the Euler Equation (A.25) with respect to time gives:

$$-\nabla \frac{\partial p}{\partial t} = \rho \frac{\partial^2 \boldsymbol{u}}{\partial t^2} \tag{A.33}$$

Substituting Equation (A.33) into (A.32) gives:

$$\nabla(\nabla \cdot \boldsymbol{u}) = \frac{1}{c^2} \frac{\partial^2 \boldsymbol{u}}{\partial t^2} \tag{A.34}$$

However, it may be shown that grad div = div grad + curl curl, or:

$$\nabla(\nabla \cdot \boldsymbol{u}) = \nabla^2 \boldsymbol{u} + \nabla \times (\nabla \times \boldsymbol{u}) \tag{A.35}$$

Thus Equation (A.34) may be written as:

$$\nabla^2 \boldsymbol{u} + \nabla \times (\nabla \times \boldsymbol{u}) = \frac{1}{c^2} \frac{\partial^2 \boldsymbol{u}}{\partial t^2} \tag{A.36}$$

which is the wave equation for the acoustic particle velocity.

A convenient quantity in which to express the wave equation is the acoustic velocity potential, which is a scalar quantity with no particular physical meaning. The advantage of using velocity potential is that both the acoustic pressure and particle velocity can be derived from it mathematically, simply using differentiation and no integration (see Equations (1.10) and (1.11)). Thus is a convenient way to represent the solution of the wave equation for many applications.

It can be shown (Hansen and Snyder, 1997) that postulating a velocity potential solution to the wave equation causes some loss of generality and restricts the solutions to those that do not involve fluid rotation. Fortunately, acoustic motion in liquids and gases is nearly always rotationless.

Introducing Equation (1.11) for the velocity potential into Euler's Equation (A.25) gives the following expression.

$$-\nabla p = -\rho \frac{\partial \nabla \varphi}{\partial t} = -\rho \nabla \frac{\partial \varphi}{\partial t} \tag{A.37}$$

Integrating gives:

$$p = \rho \frac{\partial \varphi}{\partial t} + \text{const.} \tag{A.38}$$

Introducing Equation (A.38) into the wave Equation (A.31) for acoustic pressure, integrating with respect to time and dropping the integration constant gives:

$$\nabla^2 \varphi = \frac{1}{c^2} \frac{\partial^2 \varphi}{\partial t^2} \tag{A.39}$$

This is the preferred form of the Helmholtz equation as both acoustic pressure and particle velocity can be derived from the velocity potential solution by simple differentiation.

Properties of Materials

The properties of materials can vary considerably, especially for wood and plastic. The values in this table have been obtained from a variety of sources, including Simonds and Ellis (1943), Eldridge (1974), Levy (2001), Lyman (1961) and Green et al. (1999). The data vary significantly between different sources for plastics and wood and sometimes even for metals; however, the values listed below reflect those most commonly found.

Where values of Poisson's ratio were unavailable, they were calculated from data for the speed of sound in a 3-D solid using the equation at the end of this table. These data were unavailable for some plastics so for those cases values for similar materials were used. For wood products, the value for Poisson's ratio has been left blank where no data were available. Poisson's ratio is difficult to report for wood as there are 6 different ones, depending on the direction of stress and the direction of deformation. Here, only the value corresponding to strain in the longitudinal fibre direction coupled with deformation in the radial direction is listed.

The speed of sound values in column 4 were calculated from the values in columns 2 and 3. Where a range of values occurred in either or both of columns 2 and 3, a median value of the speed of sound was recorded in column 4.

The values in this table should be used with caution and should be considered as representative only. The values for the in-situ loss factor refer to the likely value of loss factor for a panel installed in a building and represents a combination of the material internal loss factor, the support loss factor and the sound radiation loss factor.

Table B.1 Properties of materials[a]

Material	Young's modulus, $E(10^9 \text{ N/m}^2)$	Density, $\rho(\text{kg/m}^3)$	$\sqrt{E/\rho}$ (m/s)	Internal– In-situ Loss factor, η	Poisson's ratio, v
Air (20°C)		1.206	343		
Fresh water (20°C)		998	1,497		0.5
Sea water (13°C)		1,025	1,530		0.5
Metals					
Aluminum sheet	70	2,700	5,150	0.0001– 0.01	0.35
Brass	95	8,500	3,340	0.001– 0.01	0.35
Brass (70%Zn 30%Cu)	101	8,600	3,480	0.001– 0.01	0.35
Carbon brick	8.2	1,630	2,240	0.001– 0.01	0.07
Carbon nanotubes	1000	1,330–1,400	27,000	0.001– 0.01	0.06
Graphite mouldings	9.0	1,700	2,300	0.001– 0.01	0.07
Chromium	279	7,200	6,240	0.001– 0.01	0.21
Copper (annealed)	128	8,900	3,790	0.002– 0.01	0.34
Copper (rolled)	126	8,930	3,760	0.001– 0.01	0.34

Material	Young's modulus, $E(10^9 \text{ N/m}^2)$	Density, $\rho(\text{kg/m}^3)$	$\sqrt{E/\rho}$ (m/s)	Loss factor, η	Poisson's ratio, v
Gold	79	19,300	2,020	0.001– 0.01	0.44
Iron	200	7,600	5,130	0.0005– 0.01	0.30
Iron (white)	180	7,700	4,830	0.0005– 0.01	0.30
Iron (nodular)	150	7,600	4,440	0.0005– 0.01	0.30
Iron (wrought)	195	7,900	4,970	0.0005– 0.01	0.30
Iron (gray (1))	83	7,000	3,440	0.0005– 0.02	0.30
Iron (gray (2))	117	7,200	4,030	0.0005– 0.03	0.30
Iron (malleable)	180	7,200	5,000	0.0005– 0.04	0.30
Lead (annealed)	16.0	11,400	1,180	0.015– 0.03	0.43
Lead (rolled)	16.7	11,400	1,210	0.015– 0.04	0.44
Lead sheet	13.8	11,340	1,100	0.015– 0.05	0.44
Magnesium	44.7	1,740	5,030	0.0001– 0.01	0.29
Molybdenum	280	10,100	5,260	0.0001– 0.01	0.32
Monel metal	180	8,850	4,510	0.0001– 0.02	0.33
Neodymium	390	7,000	7,460	0.0001– 0.03	0.31
Nickel	205	8,900	4,800	0.001– 0.01	0.31
Nickel-iron alloy (Invar)	143	8,000	4,230	0.001– 0.01	0.33
Platinum	168	21,400	2,880	0.001– 0.02	0.27
Silver	82.7	10,500	2,790	0.001– 0.03	0.36
Steel (mild)	207	7,850	5,130	0.0001–0.01	0.30
Steel (1% carbon)	210	7,840	5,170	0.0001–0.02	0.29
Stainless steel (302)	200	7,910	5,030	0.0001–0.01	0.30
Stainless steel (316)	200	7,950	5,020	0.0001–0.01	0.30
Stainless steel (347)	198	7,900	5,010	0.0001–0.02	0.30
Stainless steel (430)	230	7,710	5,460	0.0001–0.03	0.30
Tin	54	7,300	2,720	0.0001– 0.01	0.33
Titanium	116	4,500	5,080	0.0001– 0.02	0.32
Tungsten (drawn)	360	19,300	4,320	0.0001– 0.03	0.34
Tungsten (annealed)	412	19,300	4,620	0.0001– 0.04	0.28
Tungsten carbide	534	13,800	6,220	0.0001– 0.05	0.22
Zinc sheet	96.5	7,140	3,680	0.0003– 0.01	0.33
Building materials					
Brick	24	2,000	3,650	0.01– 0.05	0.12
Concrete (normal)	18-30	2,300	2,800	0.005–0.05	0.20
Concrete (aerated)	1.5-2	300-600	2,000	0.05	0.20
Concrete (high strength)	30	2,400	3,530	0.005–0.05	0.20
Masonry block	4.8	900	2,310	0.005–0.05	0.12
Cork	0.1	250	500	0.005–0.05	0.15
Fibre board	3.5–7	480–880	2,750	0.005–0.05	0.15

Table B.1 Properties of materials[a] (Continued)

Material	Young's modulus, $E(10^9$ N/m²)	Density, ρ(kg/m³)	$\sqrt{E/\rho}$ (m/s)	Loss factor, η	Poisson's ratio, v
Gypsum board	2.1	760	1,670	0.006–0.05	0.24
Glass	68	2,500	5,290	0.0006–0.02	0.23
Glass (pyrex)	62	2,320	5,170	0.0006–0.02	0.23
WOOD					
Ash (black)	11.0	450	4,940	0.04–0.05	0.37
Ash (white)	12.0	600	4,470	0.04–0.05	
Aspen (quaking)	8.1	380	4,620	0.04–0.05	0.49
Balsa wood	3.4	160	4,610	0.001–0.05	0.23
Baltic whitewood	10.0	400	5,000	0.04–0.05	
Baltic redwood	10.1	480	4,590	0.04–0.05	
Beech	11.9	640	4,310	0.04–0.05	
Birch (yellow)	13.9	620	4,740	0.04–0.05	0.43
Cedar (white-nthn)	5.5	320	4,150	0.04–0.05	0.34
Cedar (red-western)	7.6	320	4,870	0.04–0.05	0.38
Compressed hardboard composite	4.0	1,000	2,000	0.005–0.05	
Douglas fir	9.7–13.2	500	4,800	0.04–0.05	0.29
Douglas fir (coastal)	10.8	450	4,900	0.04–0.05	0.29
Douglas fir (interior)	8.0	430	4,310	0.04–0.05	0.29
Mahogany (African)	9.7	420	4,810	0.04–0.05	0.30
Mahogany (Honduras)	10.3	450	4,780	0.04–0.05	0.31
Maple	12.0	600	4,470	0.04–0.05	0.43
MDF	3.7	770	2190	0.005–0.05	
Meranti (light red)	10.5	340	5,560	0.04–0.05	
Meranti (dark red)	11.5	460	5,000	0.04–0.05	
Oak	12.0	630	4,360	0.04–0.05	0.35
Pine (radiata)	10.2	420	4,930	0.04–0.05	
Pine (other)	8.2–13.7	350–590	4,830	0.04–0.06	
Plywood (fir)	8.3	600	4,540	0.01–0.05	
Poplar	10.0	350–500	4,900	0.04–0.05	
Redwood (old)	9.6	390	4,960	0.04–0.05	0.36
Redwood (2nd growth)	6.6	340	4,410	0.04–0.05	0.36
Scots pine	10.1	500	4,490	0.04–0.05	
Spruce (sitka)	9.6	400	4,900	0.04–0.05	0.37
Spruce (engelmann)	8.9	350	5,040	0.04–0.05	0.42
Teak	14.6	550	5,150	0.02–0.05	
Walnut (black)	11.6	550	4,590	0.04–0.05	0.49
Wood chipboard (floor)	2.8	700	1,980	0.005–0.05	
Wood chipboard (std)	2.1	625	1,830	0.005–0.05	

Material	Young's modulus, $E(10^9$ N/m²)	Density, ρ(kg/m³)	$\sqrt{E/\rho}$ (m/s¹)	Loss factor, η	Poisson's ratio, v
Plastics and other					
Lucite	4.0	1,200	1,830	0.002–0.02	0.35
Plexiglass (acrylic)	3.5	1,190	1,710	0.002–0.02	0.35
Polycarbonate	2.3	1,200	1,380	0.003–0.1	0.35
Polyester (thermo)	2.3	1,310	1,320	0.003–0.1	0.40
Polyethylene					
(High density)	0.7–1.4	940–960	1,030	0.003–0.1	0.44
(Low density)	0.2–0.5	910–925	600	0.003–0.1	0.44
Polypropylene	1.4–2.1	905	1,380	0.003–0.1	0.40
Polystyrene					
(moulded)	3.2	1,050	1,750	0.003–0.1	0.34
(expanded foam)	0.0012–0.0035	16–32	300	0.0001–0.02	0.30
Polyurethane	1.6	900	1,330	0.003–0.1	0.35
PVC	2.8	1,400	1,410	0.003–0.1	0.40
PVDF	1.5	1760	920	0.003–0.1	0.35
Nylon 6	2.4	1,200	1,410	0.003–0.1	0.35
Nylon 66	2.7–3	1120–1150	1590	0.003–0.1	0.35
Nylon 12	1.2–1.6	1,010	1,170	0.003–0.1	0.35
Rubber–neoprene	0.01–0.1	1,100–1,200	190	0.05–0.1	0.49
Kevlar 49 cloth	31	1330	4830	0.008	-
Aluminum honeycomb					

Cell size (mm)	Foil thickness (mm)				
6.4	0.05	1.31	72	0.0001–0.01	
6.4	0.08	2.24	96		
9.5	0.05	0.76	48		
9.5	0.13	1.86	101		

[a] Loss factors of materials shown characterised by a very large range, are very sensitive to specimen mounting conditions. Use the upper limit for panels used in building construction. Speed of sound for a l-D solid, $= \sqrt{E/\rho}$; for a 2-D solid (plate), $c_L = \sqrt{E/[\rho(1-v^2)]}$; and for a 3-D solid, $c_L = \sqrt{E(1-v)/[\rho(1+v)(1-2v)]}$. For gases, replace E with γP, where γ is the ratio of specific heats (=1.40 for air) and P is the static pressure. For liquids, replace E with $V(\partial V/\partial p)^{-1}$, where V is the unit volume and $\partial V/\partial p$ is the compressibility. Note that Poisson's ratio, v, may be defined in terms of Young's Modulus, E, and the material Shear Modulus, G, as $v = E/(2G) - 1$ and it is effectively zero for liquids and gases.

APPENDIX C

Acoustical Properties of Porous Materials

C.1 FLOW RESISTANCE AND RESISTIVITY

If a constant differential pressure is imposed across a layer of bulk porous material of open cell structure, a steady flow of gas will be induced through the material. Experimental investigation has shown that, for a wide range of materials, the differential pressure, ΔP, and the induced normal velocity, U, of the gas at the surface of the material (volume velocity per unit surface area) are linearly related, provided that the normal velocity is small.

The ratio of differential pressure in Pascals to normal velocity in m/s is known as the flow resistance R_f (MKS rayls) of the material. It is generally assumed that the gas is air but a flow resistance may be determined for any gas.

If the material is generally of uniform composition, then the flow resistance is proportional to the material thickness. When the measured flow resistance is divided by the test sample thickness in metres, the flow resistivity, R_1, in MKS rayls per metre, is obtained, which is independent of the sample thickness and is characteristic of the material. Flow resistance and flow resistivity of porous materials are discussed in depth in the literature (Beranek, 1971; Bies and Hansen, 1979, 1980; Bies, 1981).

The flow resistance of a sample of porous material may be measured using an apparatus that meets the requirements of ASTM C522-03, such as illustrated in Figure C.1. Flow rates between 5×10^{-4} and 5×10^{-2} m/s are easily realisable, and yield good results. Higher flow rates should be avoided due to the possible introduction of undamping effects. To ensure that the pressure within the flow meter 7 is the same as that measured by the manometer 8, valve 6 must be adjusted so that flow through it is choked. The flow resistivity of the specimen shown in Figure C.1 is calculated from the measured quantities as follows:

$$R_1 = \rho \Delta P A / \dot{m} \ell = \Delta P A / V_0 \ell \qquad (C.1)$$

where ρ is the density of the gas (kg/m^3), ΔP is the differential pressure (N/m^2), A is the specimen cross-sectional area (m^2), $\dot{m}$ is the air mass flow rate (kg/s), V_0 is the volume flow rate through the sample (m^3/s) and ℓ is the specimen thickness (m).

Alternatively, acoustic flow resistance may be measured using a closed end tube, a sound source and any inexpensive microphones arranged as shown in Figure C.2 (Ingard and Dear, 1985). To make a measurement, the sound source is driven with a pure tone signal, preferably below 100 Hz, at a frequency chosen to produce an odd number of quarter wavelengths over the distance $w + \ell$ from the closed end to the sample under test. The first step is to satisfy the latter requirement by adjusting the

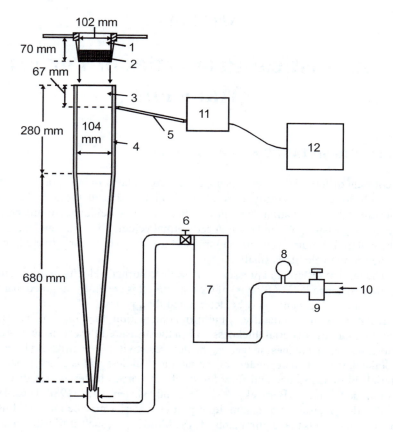

Figure C.1 Flow resistance measuring apparatus.
Key

1 sample holder and cutter	5 tube	9 pressure regulator
2 porous material	6 valve	10 air supply
3 O-ring seal	7 flow meter	11 barocell
4 conical tube to ensure uniform air flow through sample	8 manometer	12 electronic manometer

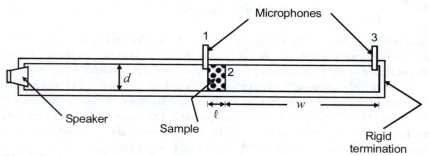

Figure C.2 An alternative arrangement for measuring flow resistance.

chosen frequency to achieve a minimum sound pressure level at microphone 1 in the absence of the sample. The sample is then inserted and the sound pressure level is measured at locations 1 and 3. The normalised flow impedance is a complex quantity made up of a real term (flow resistance) and an imaginary term (flow reactance) and is given by:

$$\frac{Z}{\rho c} = \frac{p_1 - p_2}{\rho c u_1} \tag{C.2}$$

As the tube is rigidly terminated and the losses along it are assumed small, the amplitude of the reflected wave will be the same as the incident wave and there will be zero phase shift between the incident and reflected waves at the rigid termination. If the coordinate system is chosen so that $x = 0$ corresponds to the rigid end and $x = -L$ (where $L = w + \ell$) at microphone location 1, then the particle velocity at location 1 can be shown to be given by:

$$u_1 = -\frac{jp_3}{\rho c} \sin(kL) \tag{C.3}$$

The acoustic pressure at location 2 is very similar to the pressure at location 1 in the absence of the sample and is given as:

$$p_2 = p_3 \cos(kL) \tag{C.4}$$

Thus Equation (C.2) can be written as:

$$\frac{Z}{\rho c} = \frac{jp_1}{p_3 \sin(kL)} + j\cot(kL) \tag{C.5}$$

If L is chosen to be an odd number of quarter wavelengths such that $L = (2n-1)\lambda/4$ where n is an integer (preferably $n=1$), the normalised flow impedance becomes:

$$\frac{Z}{\rho c} = j(-1)^n (p_1/p_3) \tag{C.6}$$

so the flow resistance is the imaginary part of p_1/p_3. Taking into account that the flow reactance is small at low frequencies, the flow resistance is given by the magnitude of the ratio of the acoustic pressure at location 1 to that at location 3. Thus the flow resistivity (flow resistance divided by sample thickness) is given by:

$$R_1 = \frac{\rho c}{\ell} 10^{(L_{p1} - L_{p3})/20} \tag{C.7}$$

As only the sound pressure level difference between locations 1 and 3 is required, then prior to taking the measurement, microphones 1 and 3 are placed together near the closed end in the absence of the sample so that they measure the same sound pressure and the gain of either one is adjusted so that they read the same level. The closed end location is chosen for this as here the sound pressure level varies only slowly with location. The final step is to place the calibrated microphones at positions 1 and 3 as shown in Figure C.1 and measure L_{p1} and $L_{p3,}$ then use Equation (C.7) to calculate the flow resistivity, R_1.

Measured values of flow resistivity for various commercially available sound-absorbing materials are available in the literature (Bies and Hansen, 1980, and manufacturers' data). For fibrous materials with a reasonably uniform fibre diameter and with only a small quantity of binder (such that the flow resistance is minimally affected), Figure C.3 may be used to obtain an estimate of flow resistivity. Figure C3 is derived from the empirical equation (Bies and Hansen, 1980):

$$R_1 = K_2 d^{-2} \rho_B^{K_I} \tag{C.8}$$

where ρ_B is the bulk density of material, d is the fibre diameter, $K_1 = 1.53$ and $K_2 = 3.18 \times 10^{-9}$.

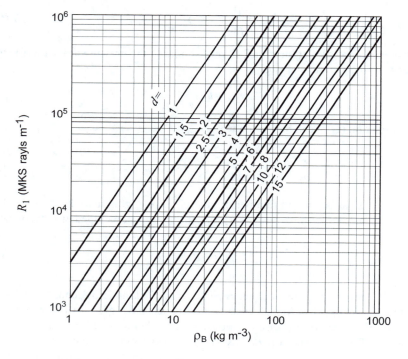

Figure C.3 Flow resistivity R_1 as a function of material bulk density ρ_B(kg/m³) and fibre diameter d (μm).

For fibres with diameters larger than those in Figure C.3, such as polyester, the flow resistivity may be estimated using Equation C.8 but with different values for K_1 and K_2. In this case Garai and Pompoli (2005) found that $K_1 = 1.404$ and $K_2 = 25.989 d^2$, with d the fibre diameter in metres.

C.2 SOUND PROPAGATION IN POROUS MEDIA

For the purpose of the analysis, the porous gas-filled medium is replaced by an effective medium, which is characterised in dimensionless variables by a complex density, ρ_m, and complex compressibility, κ. In terms of these quantities, a complex characteristic impedance and propagation constant are defined, analogous to the similar quantities for the contained gas in the medium. Implicit in the formulation of the following expressions is the assumption that time dependence has the form $e^{j\omega t}$, consistent with the practice adopted throughout the text.

The characteristic impedance of porous material may be written in terms of the gas density, ρ, the gas speed of sound c, the complex density, ρ_m, and complex compressibility, κ, as follows:

$$Z_m = \rho c \sqrt{\rho_m \kappa} \tag{C.9}$$

Similarly, a complex propagation constant, k_m, may be defined as:

$$k_m = \frac{2\pi}{\lambda}(1 - j\alpha_m) = (\omega/c)\sqrt{\rho_m/\kappa} \tag{C.10a,b}$$

where $\omega = 2\pi f$ is the angular frequency (rad/s) of the sound wave.

The quantities, ρ_m and κ may be calculated using the following procedure (Bies, 1981). This procedure gives results for fibrous porous materials within 4% of the mean of published data (Delaney and Bazley, 1969, 1970), and it tends to the correct limits at both high and low values of the dimensionless frequency $\rho f/R_1$. However this model and the Delaney and Bazley model have only been verified on fibreglass and rockwool materials with a small amount of binder and having short fibres smaller than 15 μm in diameter, which excludes such materials as polyester and acoustic foam.

$$\kappa = [1 + (1 - \gamma)\tau]^{-1} \tag{C.11}$$

$$\rho_m = [1 + \sigma]^{-1} \tag{C.12}$$

where γ is the ratio of specific heats for the gas ($=1.40$ for air), ρ is the density of gas ($=1.205$ kg/m³ for air at 20°C), f is the frequency (Hz) and R_1 is the flow resistivity of the porous material (MKS rayls/m) and:

$$\tau = 0.592\,a(X_1) + jb(X_1) \tag{C.13}$$

$$\sigma = a(X) + jb(X) \tag{C.14}$$

$$a(X) = \frac{T_3(T_1 - T_3)T_2^2 - T_4^2 T_1^2}{T_3^2 T_2^2 + T_4^2 T_1^2} \tag{C.15}$$

$$b(X) = \frac{T_1^2 T_2 T_4}{T_3^2 T_2^2 + T_4^2 T_1^2}$$ (C.16)

$$T_1 = 1 + 9.66X$$ (C.17)

$$T_2 = X(1 + 0.0966X)$$ (C.18)

$$T_3 = 2.537 + 9.66X$$ (C.19)

$$T_4 = 0.159(1 + 0.7024X)$$ (C.20)

$$X = \rho f / R_1$$ (C.21)

The quantities $a(X_1)$ and $b(X_1)$ are calculated by substituting $X_1 = 0.856X$ for the quantity X in Equations (C.9)–(C.14).

The relationships that have been generally accepted in the past (Delaney and Bazley, 1969, 1970), and which are accurate in the flow resistivity range $R_1 = 10^3$ to 5×10^4 MKS rayls/m are as follows:

$$Z_m = \rho c [1 + C_1 X^{-C_2} - j C_3 X^{-C_4}]$$ (C.22)

$$k_m = (\omega / c)[1 + C_5 X^{-C_6} - j C_7 X^{-C_8}]$$ (C.23)

The quantities X, Z_m, k_m, c and ρ have all been defined previously. Values of the constants $C_1 - C_8$ are given in Table C.1 below for various materials from various references.

Table C.1 Values of the constants $C_1 - C_8$ for various materials

Material type reference	C1	C2	C3	C4	C5	C6	C7	C8
Rockwool / fibreglass Delaney and Bazley (1970)	0.0571	0.754	0.087	0.732	0.0978	0.700	0.189	0.595
Polyester Garai and Pompoli (2005)	0.078	0.623	0.074	0.660	0.159	0.571	0.121	0.530
Polyurethane foam of low flow resistivity Dunn and Davern (1986)	0.114	0.369	0.0985	0.758	0.168	0.715	0.136	0.491
Porous plastic foams of medium flow resistivity Wu (1988)	0.212	0.455	0.105	0.607	0.163	0.592	0.188	0.544

C.3 SOUND REDUCTION DUE TO PROPAGATION THROUGH A POROUS MATERIAL

For the purpose of the calculation, three frequency ranges are defined; low, middle and high, as indicated in Figure C.4. The quantities in the parameters $\rho f/R_1$ and $f\ell/c$ are defined in Sections C.1 and C.2.

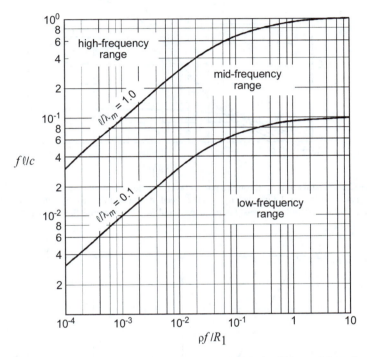

Figure C.4 Limits showing when low- and high-frequency models should be used for estimating the transmission loss through a porous layer. The low-frequency model should be used when the design point lies below the $\ell/\lambda_m = 0.1$ curve, and the high-frequency model should be used when the design point lies above the $l/\lambda_m = 1.0$ curve. The quantity λ_m is the wavelength of sound in the porous material, ρ is the gas density, c is the speed of sound in the gas, f is frequency, ℓ is the material thickness, and R_1 is the material flow resistivity.

In the low-frequency range, the inertia of the porous material is small enough for the material to move with the particle velocity associated with the sound wave passing through it. The transmission loss to be expected in this frequency range can be obtained from Figure C.5. If the material is used as a pipe wrapping, the noise reduction will be approximately equal to the transmission loss. In the high-frequency range, the porous material is many wavelengths thick and, in this case, reflection at both surfaces of the layer as well as propagation losses through the layer must be taken into account when estimating noise reduction. The reflection loss at an air/porous medium interface may be calculated using Figure C.6 and the transmission loss may be estimated using Figure C.7.

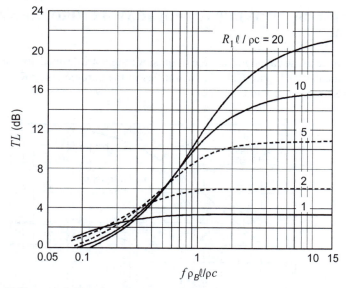

Figure C.5 Transmission loss through a porous layer for a design point lying in the low-frequency range. The quantity ρ is the gas density, c is the speed of sound in the gas, ρ_B is the material bulk density, ℓ is the material thickness, f is frequency, and R_1 is the material flow resistivity.

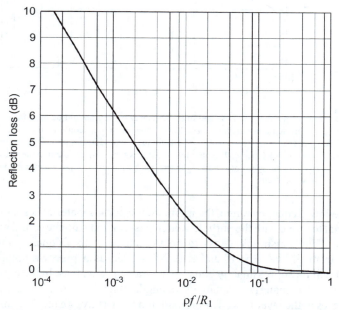

Figure C.6 Reflection loss (dB) at a porous material - air interface for a design point in the high-frequency range of Figure A3.4. The quantity ρ is the gas density, f is the frequency, and R_1 is the material flow resistivity.

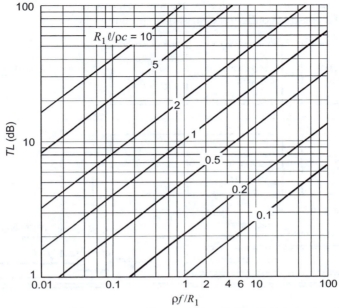

Figure C.7 The transmission loss of a porous lining of thickness ℓ and normalised flow resistance $R_1\ell/\rho c$ for a design point lying in the high frequency region. The quantity ρ is the gas density, c is the speed of sound in the gas, and R_1 is the material flow resistivity.

In the middle-frequency range it is generally sufficient to estimate the transmission loss graphically, with a faired, smooth curve connecting plotted estimates of the low- and high-frequency transmission loss versus log frequency.

C.4 MEASUREMENT AND CALCULATION OF ABSORPTION COEFFICIENTS

Absorption coefficients may be determined using impedance tube measurements of the normal incidence absorption coefficient, as an alternative to the measurement of the Sabine absorption coefficient using a reverberant test chamber. This same measuring apparatus may be used to obtain the normal impedance looking into the sample, which can be used to estimate the statistical absorption coefficient.

When a tonal sound field is set up in a tube terminated in an impedance Z, a pattern of regularly spaced maxima and minima along the tube will result, which is uniquely determined by the driving frequency and the terminating impedance. The absorption coefficients are related to the terminating impedance and the characteristic impedance ρc of air.

An impedance tube is relatively easily constructed and therein lies its appeal. Any heavy walled tube may be used for its construction. A source of sound should be placed at one end of the tube and the material to be tested should be mounted at the other end. Means must be provided for probing the standing wave within the tube. An example of a possible configuration is shown in Figure C.8.

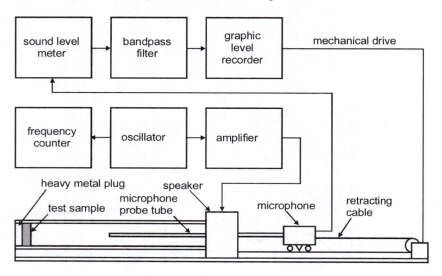

Figure C.8 Equipment for impedance tube measurement.

The older and simpler method by which the sound field in the impedance tube is explored, using a moveable microphone which traverses the length of the tube, will be described. In this case the impedance of the test sample is determined from measurements of the sound field in the tube. This method is slow but easily implemented. A much quicker method makes use of two fixed microphones and a digital frequency analysis system. However, both methods are described in ASTM standards listed in the references.

It is also possible to measure the specific acoustic surface impedance of materials and their absorption coefficient using two microphones close to the surface, in-situ (Allard and Champoux, 1989 and Dutilleax *et al.*, 2001). However, the procedure is quite complex.

Implicit in the use of the impedance tube is the assumption that only plane waves propagate back and forth in the tube. This assumption puts upper and lower frequency bounds on the use of the impedance tube. Let d be the tube diameter if it is circular in cross-section, or the larger dimension if it is rectangular in cross-section. Then the upper frequency limit (or cut-on frequency) is $0.586\ c/d$ for circular cross-section tubes and $0.5\ c/d$ for rectangular tubes. Here, c is the speed of sound and the frequency limit is given in Hz. The low frequency bound is given by the requirement that the length ℓ of the tube should be greater than $d + 3c/4f$ where f is the low frequency bound in Hz.

In general, the frequency response of the apparatus will be very much dependent upon the material under test. To reduce this dependence and to ensure a more uniform response, some sound absorptive material may be placed permanently in the tube at the source end. Furthermore, as energy losses due to sound propagation along the length of the tube are undesirable, the following equations may be used as a guide during design to estimate and minimise such losses:

$$L_{D1} - L_{D2} = a\lambda/2 \quad \text{(dB)} \tag{C.24}$$

$$a = 0.19137\sqrt{f}/cd \quad \text{(dB m}^{-1}) \tag{C.25}$$

In the above equations, a is the loss (in dB per metre of tube length) due to propagation of the sound wave down the tube, and L_{D1} and L_{D2} are the sound pressure levels at the first and second minima relative to the surface of the test sample. The other quantities are the frequency f (Hz), the corresponding wavelength λ (m) and the speed of sound c (m/s). For tubes of any cross-section, d (m) is defined as $d = 4S/P_D$ where S is the cross-sectional area and P_D is the cross-sectional perimeter.

The sound field within the impedance tube may be explored either with a small microphone placed at the end of a probe and immersed in the sound field or with a probe tube attached to a microphone placed externally to the field as illustrated in Figure C.8. Equation C.25 may be useful in selecting a suitable tube for exploring the field; the smaller the probe tube diameter, the greater will be the energy loss suffered by a sound wave travelling down it, but the smaller will be the disturbance to the acoustic field being sampled. An external linear scale should be provided for locating the probe. As the sound field will be distorted slightly by the presence of the probe, it is recommended that the actual location of the sound field point sampled be determined by replacing the specimen with a heavy solid metal face located at the specimen surface. The first minimum will then be $\lambda/4$ away from the solid metal face. Subsequent minima will always be spaced at intervals of $\lambda/2$. Thus, this experiment will allow the determination of how far in front of the end of the probe tube that the sound field is effectively being sampled.

It will be found that the minima will be subject to contamination by acoustic and electronic noise; thus, it is recommended that a narrow band filter, for example one which is an octave or one third octave wide, be used in the detection system.

When the material to be tested is in place and the resulting field within the impedance tube is explored, a series of maxima and minima will be observed. The maxima will be effectively constant in level but the minima will increase in level, according to Equation C.24, as one moves away from the surface of the test material. For best results, it is recommended that successive minima be determined and extrapolated back to the surface of the sample by drawing a straight line joining the first two minima to the location corresponding to the surface of the sample on the plot of sound pressure level in dB vs distance along the tube. The standing wave ratio L_0 is then determined as the difference between the maximum level L_{max}, and the extrapolated minimum level L_{min}.

It is of interest to derive an expression for the normal incidence absorption coefficient of a sample of acoustic material in an impedance tube as shown in Figure C.9.

For the following analysis, the impedance tube contains the material sample at the right end and the loudspeaker sound source at the left end as shown in Figure C.9. For simplicity, it is assumed that there are no losses due to dissipation in the tube; that

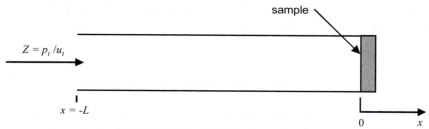

Figure C.9 Schematic of an impedance tube.

is, the quantity, a, in Equation (C.25) is assumed to be zero. The loudspeaker is not shown in the figure and the origin is set at the right end of the tube.

Reference should be made to Section 1.4.4, where it is shown that multiple waves travelling in a single direction may be summed together to give a single wave travelling in the same direction. For the case considered here, the multiple waves travelling in each direction are a result of multiple reflections from each end of the tube.

As the origin is at the right end of the tube, the resultant incident wave will be travelling in the positive x direction. Assuming a phase shift between the incident and reflected waves of β at $x = 0$, the incident wave and reflected wave pressures may be written as:

$$p_i = A\,e^{j(\omega t - kx)} \quad \text{and} \quad p_r = B\,e^{j(\omega t + kx + \beta)} \tag{C.26}$$

The total sound pressure is thus:

$$p_t = A\,e^{j(\omega t - kx)} + B\,e^{j(\omega t + kx + \beta)} \tag{C.27}$$

The first maximum pressure (closest to the sample) will occur when:

$$\beta = -2kx \tag{C.28}$$

and the first minimum will occur when:

$$\beta = -2kx + \pi \tag{C.29}$$

Thus:

$$p_{\max} = e^{-jkx}(A + B) \quad \text{and} \quad p_{\min} = e^{-jkx}(A - B) \tag{C.30a,b}$$

and the ratio of maximum to minimum pressures is $(A + B)/(A - B)$. The standing wave ratio, L_0, is the difference in decibels between the maximum and minimum sound pressures in the standing wave and is defined as:

$$10^{L_0/20} = \frac{A + B}{A - B} \tag{C.31}$$

Thus the ratio (B/A) is:

$$\frac{B}{A} = \left[\frac{10^{L_0/20} - 1}{10^{L_0/20} + 1} \right] \tag{C.32}$$

The amplitude of the pressure reflection coefficient squared is defined as: $|R_p|^2 = (B/A)^2$, which can be written in terms of L_0 as:

$$|R_p|^2 = \left[\frac{10^{L_0/20} - 1}{10^{L_0/20} + 1} \right]^2 \tag{C.33}$$

The normal incidence absorption coefficient is defined as:

$$\alpha_n = 1 - |R_p|^2 \tag{C.34}$$

and it can also be determined from Table C.2.

Table C.2 Normal incidence sound absorption coefficient, α_n versus standing wave ratio, L_0 (dB)

L_0	α_n	L_0	α_n	L_0	α_n	L_0	α_n
0	1.000	10	0.730	20	0.331	30	0.119
1	0.997	11	0.686	21	0.301	31	0.107
2	0.987	12	0.642	22	0.273	32	0.096
3	0.971	13	0.598	23	0.247	33	0.086
4	0.949	14	0.555	24	0.223	34	0.077
5	0.922	15	0.513	25	0.202	35	0.069
6	0.890	16	0.472	26	0.182	36	0.061
7	0.854	17	0.434	27	0.164	37	0.055
8	0.815	18	0.397	28	0.147	38	0.049
9	0.773	19	0.363	29	0.132	39	0.04

It is also of interest to continue the analysis to determine the normal impedance of the surface of the sample. This can then be used to determine the statistical absorption coefficient of the sample.

The total particle velocity can be calculated using Equations (1.10), (1.11) and (C.26) to give:

$$u_t = \frac{1}{\rho c}(p_i - p_r) \tag{C.35}$$

Thus:

$$u_t = \frac{1}{\rho c}\left(A e^{j(\omega t - kx)} - B e^{j(\omega t + kx + \beta)}\right) \tag{C.36}$$

The specific acoustic impedance (or characteristic impedance) at any point in the tube may be written as:

$$Z_s = \frac{p_t}{u_t} = \rho c \frac{A e^{-jkx} + B e^{jkx + j\beta}}{A e^{-jkx} - B e^{jkx + j\beta}} = \rho c \frac{A + B e^{j(2kx + \beta)}}{A - B e^{j(2kx + \beta)}} \tag{C.37}$$

At $x = 0$, the specific acoustic impedance is the normal impedance, Z_N, of the surface of the sample. Thus:

$$\frac{Z_N}{\rho c} = \frac{p_t}{\rho c u_t} = \frac{A + B e^{j\beta}}{A - B e^{j\beta}} \qquad (C.38)$$

The above impedance equation may be expanded to give:

$$\frac{Z_N}{\rho c} = \frac{A/B + \cos\beta + j\sin\beta}{A/B - \cos\beta - j\sin\beta} = \frac{(A/B)^2 - 1 + (2A/B)j\sin\beta}{(A/B)^2 + 1 - (2A/B)\cos\beta} \qquad (C.39)$$

In practice, the phase angle, β, is evaluated by measuring the distance, D_1, of the first sound pressure minimum in the impedance tube from the sample surface and the tube which corresponds to Figure C.9 is known as an impedance tube. Referring to Equation (C.29) and Figure C.9, the phase angle, β, may be expressed in terms of D_1 (which is a positive number) as:

$$\beta = 2kD_1 + \pi = 2\pi\left(\frac{2D_1}{\lambda} + \frac{1}{2}\right) \qquad (C.40a,b)$$

Equation (C.39) may be rewritten in terms of a real and imaginary components as:

$$\frac{Z_N}{\rho c} = R + jX = \frac{(A/B)^2 - 1}{(A/B)^2 + 1 - (2A/B)\cos\beta} + j\frac{(2A/B)\sin\beta}{(A/B)^2 + 1 - (2A/B)\cos\beta}$$

$$(C.41a,b)$$

where β is defined by Equation (C.40) and the ratio A/B is defined by the reciprocal of equation (C.32) where L_0 is the difference in dB between the maximum and minimum sound pressure levels in the tube.

In terms of an amplitude and phase, the normal impedance may also be written as:

$$Z_N/(\rho c) = \zeta e^{j\psi} \qquad (C.42)$$

where:

$$\zeta = \sqrt{R^2 + X^2} \qquad (C.43)$$

and:

$$\psi = \tan^{-1}(X/R) \qquad (C.44)$$

The statistical absorption coefficient, α_{st} is related to the specific acoustic normal impedance Z_N and may be calculated as follows:

$$\alpha_{st} = \frac{1}{\pi}\int_0^{2\pi}d\varphi\int_0^{\pi/2}\alpha(\theta)\cos\theta\sin\theta\,d\theta \qquad (C.45)$$

Rewriting the absorption coefficient in terms of the reflection coefficient using:

$$|R(\theta)|^2 = 1 - \alpha(\theta) \tag{C.46}$$

Equation (C.46) can be written as:

$$\alpha_{st} = 1 - 2 \int_0^{\pi/2} |R(\theta)|^2 \cos\theta \sin\theta \, d\theta \tag{C.47}$$

For bulk reacting materials, Equations (5.142) and (5.143) may be used to calculate the reflection coefficient, $R(\theta)$ and for locally reacting materials, Equation (5.144) may be used. Note that use of these equations requires a knowledge of the specific acoustic normal impedance, Z_N, of the material. This can be measured using an impedance tube in which a sample of the material, in the same configuration as to be used in practice in terms of how it is backed, is tested. Alternatively, the normal impedance for any configuration may be calculated using the methods described in the next section.

Using Equations (C.47) and (5.144), Morse and Bolt derive the following expression for the statistical absorption coefficient for a locally reactive surface of normal impedance, Z_N, given by Equation (C.42):

$$\alpha_{st} = \left\{\frac{8\cos\psi}{\xi}\right\} \left\{1 - \left[\frac{\cos\psi}{\xi}\right]\log_e(1 + 2\xi\cos\psi + \xi^2) \right. $$
$$\left. + \left[\frac{\cos(2\psi)}{\xi\sin\psi}\right] \tan^{-1}\left[\frac{\xi\sin\psi}{1 + \xi\cos\psi}\right]\right\} \tag{C.48}$$

Alternatively Figure C.10 may be used to determine the statistical absorption coefficient. Note that Equation (C.48) and Figure C.10 are based upon the explicit assumption that sound propagation within the sample is always normal to the surface. However, calculations indicate that the error in ignoring propagation in the porous material in other directions is negligible.

C.4.1 Porous Materials with a Backing Cavity

For porous acoustic materials, such as rockwool or fibreglass, the specific normal impedance of Equation (C.42) may also be calculated from the material characteristic impedance and propagation constant of Equations (C.9) and (C.10). For a material of infinite depth, the normal impedance is equal to the characteristic impedance of Equation (C.9). For a porous blanket of thickness, ℓ, backed by a cavity of any depth, L (including $L = 0$) with a rigid back, the normal impedance (in the absence of flow past the cavity) may be calculated using an electrical transmission line analogy (Magnusson, 1965) and is given by:

$$Z_N = Z_m \frac{Z_L + jZ_m \tan(k_m\ell)}{Z_m + jZ_L \tan(k_m\ell)} \tag{C.49}$$

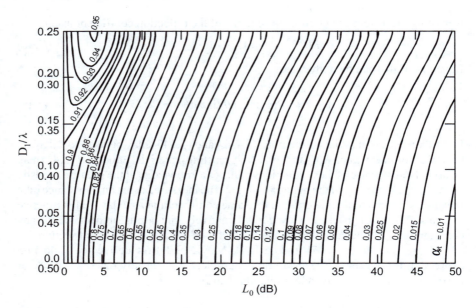

Figure C.10 A chart for determining the statistical absorption coefficient α_{st} from measurements in an impedance tube of the standing wave ratio L_0, and position D_1/λ of the first minimum sound pressure level. α_{st} is shown parametrically in the chart.

The quantities Z_m and k_m in Equation (C.49) are defined in Equations (C.9) and (C.10). The specific normal impedance, Z_L, of a rigidly terminated, partitioned backing cavity is given by:

$$Z_L = -j\rho c/\tan(2\pi fL/c) \qquad (C.50)$$

and for a rigidly terminated, non-partitioned backing cavity:

$$Z_L = -j\rho c \cos\theta/\tan(2\pi fL/c) \qquad (C.51)$$

where θ is the angle of incidence of the sound wave measured from the normal to the surface.

A partitioned cavity is one which is divided into compartments by partitions that permit propagation normal to the surface while inhibiting propagation parallel to the surface of the liner. The depth of each compartment is equal to the overall cavity depth.

If the porous material is rigidly backed so that $L = 0$ or, equivalently, L is an integer multiple of half wavelengths, Equation (C.38) reduces to:

$$Z_N = -jZ_m/\tan(k_m\ell) \qquad (C.52)$$

C.4.2 Multiple Layers of Porous Liner backed by an Impedance Z_L

Equation (C.49) can be easily extended to cover the case of multiple layers of porous material by applying it to each layer successively, beginning with the layer closest to the termination (rigid wall or cavity backed by a rigid wall) with impedance Z_L. The specific normal impedance looking into the ith layer surface that is furthest from the termination is:

$$Z_{N,i} = Z_{m,i} \frac{Z_{N,i-1} + jZ_{m,i}\tan(k_{m,i}\ell_i)}{Z_{m,i} + jZ_{N,i-1}\tan(k_{m,i}\ell_i)} \qquad (C.53)$$

The variables in the above equation have the same definitions as those in Equation (C.49), with the added subscript, i, which refers to the ith layer or the added subscript, i-1 that refers to the $(i-1)$th layer.

Equation (C.53) could also be used for materials whose density was smoothly varying by dividing the material into a number of very thin layers, where it was assumed that each layer had uniform properties.

C.4.3 Porous Liner Covered with a Limp Impervious Layer

If the porous material is protected by covering or enclosing it in an impervious blanket of thickness h and mass per unit area, σ', the effective specific normal impedance at the outer surface of the blanket, Z_{NB}, which can be used together with Equations (C.42) and (C.48) to find the statistical absorption coefficient of the construction, is as follows:

$$Z_{NB} = Z_N + j2\pi f\sigma' \qquad (C.54)$$

where f is the frequency of the incident tonal sound, tone or centre frequency of a narrow band of noise. Typical values for σ' and $\sqrt{E/\rho}$ are included in Table C.3 for commonly used covering materials.

Table C.3 Properties of commonly used limp impervious wrappings for environmental protection of porous materials

Material	Density (kg/m³)	Typical thickness (microns = 10^{-6} m)	σ' (kg/m²)[a]	c_L (approx.) (m/s)
Polyethylene (LD)	930	6-35	0.0055-0.033	460
Polyurethane	900	6-35	0.005-0.033	1330
Aluminium	2700	2-12	0.0055-0.033	5150
PVC	1400	4-28	0.005-0.033	1310
Melinex (polyester)	1390	15-30	0.021-0.042	1600
Metalised polyester	1400	12	0.017	1600

[a] σ' and c_L are, respectively, the surface density and speed of sound in the wrapping material.

Guidelines for the selection of suitable protective coverings are given by Andersson (1981).

C.4.4 Porous Liner Covered with a Perforated Sheet

If the porous liner were covered with a perforated sheet, the effective specific normal impedance at the outer surface of the perforated sheet is (Bolt, 1947):

$$Z_{NP} = Z_N + \cfrac{\dfrac{100}{P}\left(j\rho c\tan(k\ell_e(1-M)) + R_a A\right)}{1 + \dfrac{100}{j\omega mP}\left(j\rho c\tan(k\ell_e(1-M)) + R_a A\right)} \tag{C.55}$$

where Z_N is the normal specific acoustic impedance of the porous acoustic material with or without a cavity backing (and in the absence of flow), ℓ_e is the effective length of each of the holes in the perforated sheet as defined by Equation (9.22), ω is the radian frequency, P is the % open area of the holes, R_a is the acoustic resistance of each hole, A is the area of each hole, M is the mach number of the flow past the holes and m is the mass per unit area of the perforated sheet, all in consistent SI units.

C.4.5 Porous Liner Covered with a Limp Impervious Layer and a Perforated Sheet

In this case, the impedance of the perforated sheet and impervious layer are both added to the normal impedance of the porous acoustic material, so that:

$$Z_{NBP} = Z_N + \cfrac{\dfrac{100}{P}\left(j\rho c\tan(k\ell_e(1-M)) + R_a A\right)}{1 + \dfrac{100}{j\omega mP}\left(j\rho c\tan(k\ell_e(1-M)) + R_a A\right)} + j2\pi f\sigma' \tag{C.56}$$

It is important that the impervious layer and the perforated sheet are separated, using something like a mesh spacer, otherwise the performance of the construction as an absorber will be severely degraded.

APPENDIX D

Frequency Analysis

D.1 DIGITAL FILTERING

Spectral analysis is most commonly carried out in standardised octave, one-third octave, one-twelfth and one-twenty-fourth octave bands, and both analog and digital filters are available for this purpose. Such filters are referred to as constant percentage bandwidth filters meaning that the filter bandwidth is a constant percentage of the band centre frequency. For example, the octave bandwidth is always about 70.1% of the band centre frequency, the one-third octave bandwidth is 23.2% of the band centre frequency and the one-twelfth octave is 5.8% of the band centre frequency, where the band centre frequency is defined as the geometric mean of the upper and lower frequency bounds of the band.

The stated percentages are approximate, as a compromise has been adopted in defining the bands to simplify and to ensure repetition of the centre band frequencies. The compromise that has been adopted is that the logarithms to the base ten of the one-third octave centre band frequencies are tenth decade numbers (see discussion in Section 1.10.1).

Besides constant percentage bandwidth filters, instruments with constant frequency bandwidth filters are also available. However, these instruments have largely been replaced by fast Fourier transform (FFT) analysers which give similar results in a fraction of the time and generally at a lower cost. When a time-varying signal is filtered using either a constant percentage bandwidth or a constant frequency bandwidth filter, an r.m.s. amplitude signal is obtained, which is proportional to the sum of the total energy content of all frequencies included in the band.

When discussing digital filters and their use, an important consideration is the filter response time, $T_R(s)$, which is the minimum time required for the filter output to reach steady state. The minimum time generally required is the inverse of the filter bandwidth, B (Hz). That is

$$BT_R = \left(\frac{B}{f} \right) \cdot (fT_R) = bn_R \approx 1 \tag{D.1}$$

where the centre band frequency, f, the relative bandwidth, b, and the number of cycles, n_R, have been introduced. For example, for a one-third octave filter $b = 0.23$ and the number of cycles $n_R \approx 4.3$. A typical response of a one-third octave filter is illustrated in Figure D.1, where it will be noted that the actual response time is perhaps five cycles or more, depending upon the required accuracy.

Where the r.m.s. value of a filtered signal is required, it is necessary to determine the average value of the integrated squared output of the filtered signal over some prescribed period of time called the averaging time. The longer the averaging time, the more nearly constant will be the r.m.s. value of the filtered output.

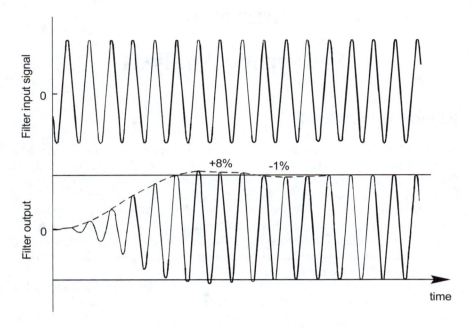

Figure D.1 Typical filter response of a one-third octave filter (after Randall, 1987).

For a sinusoidal input of frequency, f (Hz), or for several sinusoidal frequencies within the band, where f (Hz) is the minimum separation between components, the variation in the average value will be less than 1/4 dB for an averaging time, $T_A \geq 3/f$. For many sinusoidal components or for random noise and $BT_A \geq 1$, the error in the r.m.s. signal may be determined in terms of the statistical error, ε, calculated as follows:

$$\varepsilon = 0.5(BT_A)^{-1/2} \tag{D.2}$$

For random noise, the actual error has a 63.3% probability of being within the range $\pm \varepsilon$ and a 95.5% probability of being within the range, $\pm 2\varepsilon$,

The calculated statistical error may be expressed in decibels as follows:

$$20 \log_{10} e^{\varepsilon} = 4.34(BT_A)^{-1/2} \quad \text{dB} \tag{D.3}$$

D.2 DISCRETE FOURIER ANALYSIS

The process of decomposition of a time-varying signal into a spectrum of its component parts is sometimes referred to as transforming a signal from the time domain to the frequency domain. In the previous section, such a transformation process was described in which the frequency components of a time-varying signal were determined by the process of filtering, to produce a spectrum of the amplitudes of the frequency components.

A second method for transforming a signal from the time domain to the frequency domain is by Fourier analysis. Fourier analysis allows any time varying signal to be represented as the sum of a number of individual sinusoidal components, each characterised by a specific frequency, amplitude and relative phase. Whereas the filtering process described earlier provides information about the amplitudes of the frequency components, the information is insufficient to reconstruct the original time varying signal, and returning from the frequency domain to the time domain is not possible. By contrast, Fourier analysis provides sufficient information in the output to reconstruct the original time varying signal.

A general Fourier representation of a periodic time varying signal of period T and fundamental frequency $f_1 = 1/T$ such that $x(t) = x(t + nT)$ where $n = 1, 2, ...$ takes the following form:

$$x(t) = \sum_{n=1}^{\infty} [A_n \cos (2\pi n f_1 t) + B_n \sin (2\pi n f_1 t)] \qquad (D.4)$$

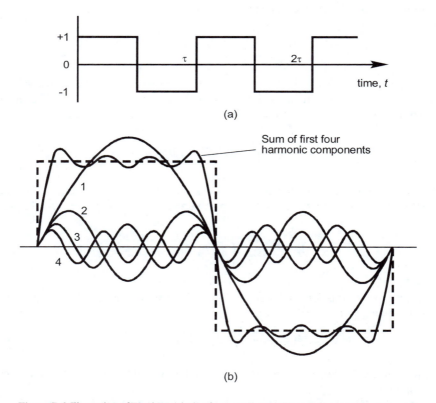

(a)

(b)

Figure D.2 Illustration of Fourier analysis of a square wave: (a) periodic square wave; (b) harmonic components of a square wave.

For example, the square wave illustrated in Figure D.2 is represented by Equation (D.4), where for n odd $A_n = 4/(\pi n)$, for n even $A_n = 0$ and $B_n = 0$. Note that the component characterised by frequency, nf_1, is usually referred to as the nth harmonic of the fundamental frequency, f_1, although some call it the $(n - 1)^{th}$ harmonic.

Use of Euler's well known equation (Abramowitz and Stegun, 1965) allows Equation (D.4) to be rewritten in the following alternative form:

$$x(t) = \frac{1}{2} \sum_{n=0}^{\infty} \left[(A - jB) \, e^{j2\pi nf_1 t} + (A + jB) \, e^{-j2\pi nf_1 t} \right]$$ (D.5)

A further reduction is possible by defining the complex spectral amplitude components $X_n = (A - jB)/2$ and $X_{-n} = (A + jB)/2$. Denoting by * the complex conjugate, the following relation may be written:

$$X_n = X_{-n}^*$$ (D.6)

Clearly Equation (D.6) is satisfied, thus ensuring that the right-hand side of Equation (D.5) is real. The introduction of Equation (D.6) in Equation (D.5) allows the following more compact expression to be written:

$$x(t) = \sum_{n=-\infty}^{\infty} X_n \, e^{j2\pi nf_1 t}$$ (D.7)

The spectrum of Equation (D.7) now includes negative as well as positive values of n, giving rise to components $-nf_1$. The spectrum is said to be two sided. The spectral amplitude components, X_n, may be calculated using the following expression:

$$X_n = \frac{1}{T} \int_{-T/2}^{T/2} x(t) \, e^{-j2\pi nf_1 t} \, dt$$ (D.8)

The spectrum of squared amplitudes is known as the power spectrum. The mean of the instantaneous power of the time-varying signal, $[x(t)]^2$, averaged over the period T is:

$$W_{mean} = \frac{1}{T} \int_0^T [x(t)]^2 \, dt$$ (D.9)

Substitution of Equation (D.4) in Equation (D.9) and carrying out the integration gives the following result:

$$W_{mean} = \frac{1}{2} \sum_{n=1}^{\infty} [A_n^2 + B_n^2]$$ (D.10)

Equation (D.10) shows that the total power is the sum of the powers of each spectral component.

Figure D.3 shows several possible interpretations of the Fourier transform in which is shown (1) the two sided power spectrum; (2) the one sided spectrum obtained by adding the corresponding negative and positive frequency components; (3) the r.m.s. amplitude (square root of the power spectrum); and (4) the logarithm to the base 10 of the power spectrum in decibels, which is the form usually used in the analysis of noise and vibration problems.

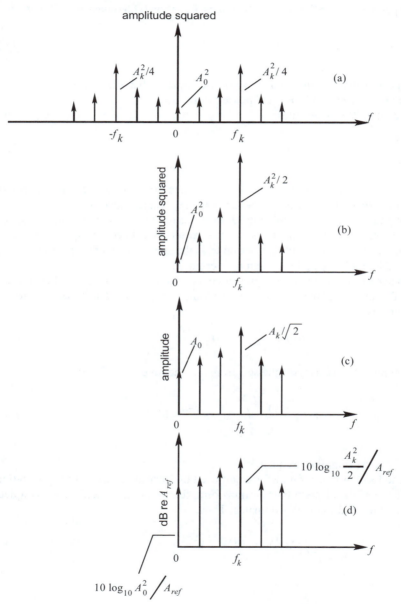

Figure D.3 Spectrum representations (after Randall, 1987): (a) two-sided power spectrum; (b) one sided power spectrum; (c) r.m.s. amplitude spectrum; (d) decibel spectrum.

The previous analysis may be extended to the more general case of non-periodic (or random noise) signals by allowing the period, T, to become indefinitely large. In this case, X_n becomes $X'(f)$, a continuous function of frequency, f. It is to be noted that whereas the units of X_n are the same as those of $x(t)$, the units of $X'(f)$ are those of $x(t)$ per hertz. With the proposed changes, Equation (D.8) takes the following form:

$$X'(f) = \int_{-\infty}^{\infty} x(t) e^{-j2\pi ft} \, dt \qquad (D.11)$$

The spectral density function, $X'(f)$, is complex, characterised by a real and an imaginary part (or amplitude and phase):

Equation (D.7) becomes:

$$x(t) = \int_{-\infty}^{\infty} X'(f) e^{j2\pi ft} \, df \qquad (D.12)$$

Equations (D.11) and (D.12) form a Fourier transform pair, with the former referred to as the forward transform and the latter as the inverse transform.

In practice, a finite sample time T always is used to acquire data and the spectral representation of Equation (D.8) is the result calculated by spectrum analysis equipment. This latter result is referred to as the spectrum and the spectral density is obtained by multiplying by the sample period, T.

Where a time function is represented as a sequence of samples taken at regular intervals, an alternative form of Fourier transform pair is as follows. The forward transform is:

$$X(f) = \sum_{k=-\infty}^{\infty} x(t_k) e^{-j2\pi ft_k} \qquad (D.13)$$

The quantity $X(f)$ represents the spectrum and the inverse transform is:

$$x(t_k) = \frac{1}{f_s} \int_{-f_s/2}^{f_s/2} X(f) e^{j2\pi ft_k} \, df \qquad (D.14)$$

where f_s is the sampling frequency.

The form of Fourier transform pair used in spectrum analysis instrumentation is referred to as the discrete Fourier transform, for which the functions are sampled in both the time and frequency domains. Thus:

$$x(t_k) = \sum_{n=0}^{N-1} X(f_n) e^{j2\pi nk/N} \qquad k = 1, 2, 3, \ldots N \qquad (D.15)$$

$$X(f_n) = \frac{1}{N} \sum_{k=0}^{N-1} x(t_k) e^{-j2\pi nk/N} \qquad n = 1, 2, 3, \ldots N \qquad (D.16)$$

where k and n represent discrete sample numbers in the time and frequency domains, respectively.

In Equation (D.15), the spacing between frequency components in hertz is dependent on the time, T, to acquire the N samples of data in the time domain and is equal to $1/T$ or f_s/N. Thus the effective filter bandwidth, B, is equal to $1/T$.

The four Fourier transform pairs are shown graphically in Figure D.4. In Equations (D.15), and (D.16), the functions have not been made symmetrical about the origin, but because of the periodicity of each, the second half of each sum also represents the negative half period to the left of the origin, as can be seen by inspection of the figure.

The frequency components above $f_s/2$ in Figure D.4 can be more easily visualised as negative frequency components and in practice, the frequency content of the final spectrum must be restricted to less than $f_s/2$. This is explained later when aliasing is discussed.

The discrete Fourier transform is well suited to the digital computations performed in instrumentation. Nevertheless, it can be seen by referring to Equation (D.15) that to obtain N frequency components from N time samples, N^2 complex multiplications are required. Fortunately this is reduced by the use of the fast Fourier transform (FFT) algorithm to $N \log_2 N$, which for a typical case of $N = 1024$, speeds up computations by a factor of 100. This algorithm is discussed in detail by Randall (1987).

D.2.1 Power Spectrum

The power spectrum is the most common form of spectral representation used in acoustics and vibration. For discussion, assume that the measured variable is randomly distributed about a mean value (also called expected value or average value). The mean value of $x(k)$ is obtained by an appropriate limiting operation in which each value, x, assumed by $x(k)$ is multiplied by its probability of occurrence, $p(x)$. This gives:

$$E[x(k)] = \int_{-\infty}^{\infty} x p(x) \, dx \tag{D.17}$$

where $E[\ ]$ represents the expected value over the index k of the term inside the brackets. Similarly, the expected value of any real, single-valued continuous function, $g(x)$, of the random variable, $x(k)$, is given by (Bendat and Piersol, 1971):

$$E[g(x(k))] = \int_{-\infty}^{\infty} g(x) p(x) \, dx \tag{D.18}$$

The two sided power spectrum is defined in terms of the amplitude spectrum $X(f) = X_n$ where $f = nf_1$ of Equation (D.13) in terms of the expected value of the average of a great many samples written as calculated using Equation (D.13) as follows:

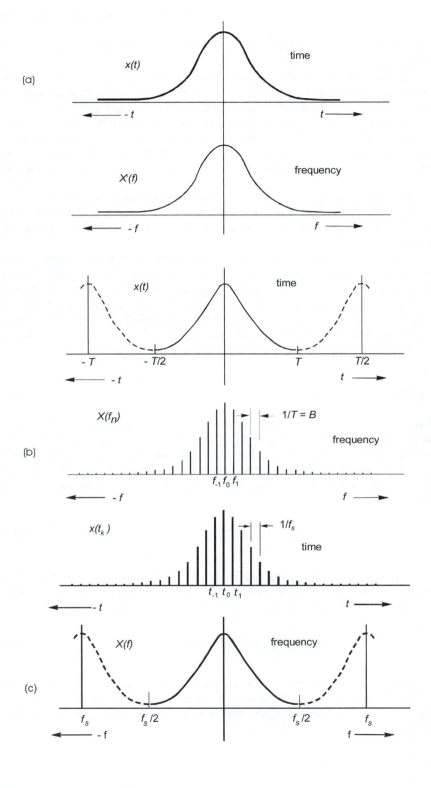

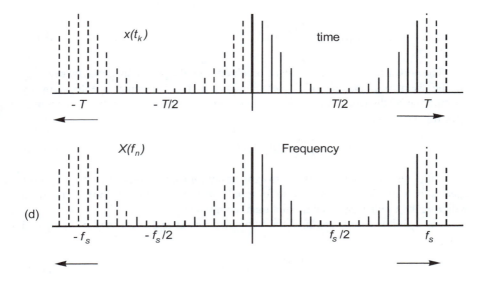

Figure D.4 Various Fourier transform pairs (after Randall, 1987).

(a) Integral transform; signal infinite and continuous in both the time and frequency domains

$$X(f_n) = \frac{1}{T} \int_{-T/2}^{T/2} g(t) e^{-j2\pi f_n t} \, dt \qquad \text{and} \qquad x(t) = \sum_{n=-\infty}^{\infty} X(f_n) e^{j2\pi f_n t}$$

(b) Fourier series; signal periodic in the time domain and discrete in the frequency domain

$$X(f) = \int_{-\infty}^{\infty} x(t) e^{-j2\pi ft} \, dt \qquad \text{and} \qquad x(t) = \int_{-\infty}^{\infty} X(f) e^{j2\pi ft} \, df$$

(c) Sampled function; signal discrete in the time domain and periodic in the frequency domain

$$X(f) = \sum_{k=-\infty}^{\infty} x(t_k) e^{-j2\pi f t_k} \qquad \text{and} \qquad x(t_k) = \frac{1}{f_s} \int_{-f_s/2}^{f_s/2} X(f) e^{j2\pi f t_k} \, df$$

(d) Discrete Fourier transform; signal discrete and periodic in both the time and frequency domains

$$X(f_n) = \frac{1}{N} \sum_{k=0}^{N-1} x(t_k) e^{-j2\pi nk/N} \qquad \text{and} \qquad x(t_k) = \sum_{n=0}^{N-1} X(f_n) e^{j2\pi nk/N}$$

$$S_{xx}(f) = \lim_{T \to \infty} E\left[X^*(f)X(f)\right] \tag{D.19}$$

The power spectral density can be obtained from the power spectrum by dividing by the frequency spacing of the components in the frequency spectrum or by multiplying by the time, T, to acquire one record of data. Thus, the two-sided power spectral density is:

$$S_{xx}'(f) = \lim_{T \to \infty} T E\left[X^*(f)X(f)\right] \tag{D.20}$$

Although it is often appropriate to express random noise spectra in terms of power spectral density, the same is not true for tonal components. Only the power spectrum will give the true energy content of a tonal component.

For a finite record length, T the two-sided power spectrum may be estimated using:

$$S_{xx}(f_n) \approx \frac{1}{q} \sum_{i=1}^{q} X_i^*(f_n) X_i(f_n) \tag{D.21}$$

where i is the spectrum number and q is the number of spectra over which the average is taken. The larger the value of q, the more closely will the estimate of $S_{xx}(f_n)$ approach its true value.

In practice, the two-sided power spectrum $S_{xx}(f_n)$ is expressed in terms of the one sided power spectrum $G_{xx}(f_n)$ where:

$$
\begin{aligned}
G_{xx}(f_n) &= 0 & f_n &< 0 \\
G_{xx}(f_n) &= S_{xx}(f_n) & f_n &= 0 \\
G_{xx}(f_n) &= 2S_{xx}(f_n) & f_n &> 0
\end{aligned}
\tag{D.22}
$$

Note that if successive spectra, $X_i(f_n)$, are averaged, the result will be zero, as the phases of each spectral component vary randomly from one record to the next. Thus in practice, power spectra are more commonly used, as they can be averaged together to give a more accurate result. This is because power spectra are only represented by an amplitude; phase information is lost when the spectra are calculated (see Equation (D.19)).

There remain a number of important concepts and possible pitfalls of frequency analysis, which will now be discussed.

D.2.2 Sampling Frequency and Aliasing

The sampling frequency is the frequency at which the input signal is digitally sampled. If the signal contains frequencies greater than half the sampling frequency, then these will be "folded back" and appear as frequencies less than half the sampling frequency.

For example if the sampling frequency is 20,000 Hz and the signal contains a frequency of 25,000 Hz, then the signal will appear as 5,000 Hz. Similarly, if the signal contains a frequency of 15,000 Hz, this signal also will appear as 5,000 Hz. This phenomenon is known as "aliasing" and in a spectrum analyser it is important to have analog filters that have a sharp roll off for frequencies above about 0.4 of the sampling frequency. Aliasing is illustrated in Figure D.5.

D.2.3 Uncertainty Principle

The uncertainty principle states that the frequency resolution (equal to the bandwidth, B) of a Fourier transformed signal is equal to the reciprocal of the time, T, to acquire the signal. Thus, for a single spectrum, $BT = 1$. An effectively higher BT product can be obtained by averaging several spectra together until an acceptable error is obtained according to Equation (D.2), where B is the filter bandwidth (Δf) or frequency resolution and T is the total sample time.

D.2.4 Real-time Frequency

Real-time frequency is often used in spectrum analyser specifications to characterise the analysis speed. It refers to the maximum frequency of an input signal, which can be continuously sampled by the analyser in such a way that the time taken to do an FFT analysis is equal to the time taken to acquire the necessary number of samples (usually 1024 or 2048). In this way, no data are lost while the analyser is processing data already obtained.

Although the total sampling time required for a spectrum analysis is fixed by the uncertainty principle, an analyser with a slower real time frequency will take longer to perform a frequency analysis if the upper analysis frequency is higher than the real-time frequency. Thus, some data will be lost between successive spectra.

Real-time frequency is also important when the analysis of transient or short duration time signals is undertaken. In this case, results for frequencies above the real time frequency will be inaccurate.

A more subtle aspect of the real-time frequency specification for a spectrum analyser is that it is related to the internal clock speed of the instrument. Thus an instrument with a lower real-time frequency will take longer to do any post-processing of data.

D.2.5 Weighting Functions

When sampling a signal in practice, a spectrum analyser must start and stop somewhere in time and this may cause a problem due to the effect of the discontinuity where the two ends of the record are effectively joined in a loop in the analyser. The solution is to apply a weighting function, called "windowing", which suppresses the

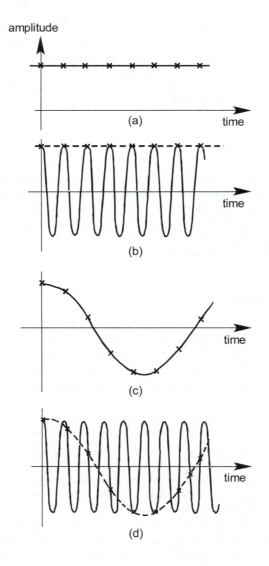

Figure D.5 Illustration of aliasing (after Randall, 1987): (a) zero frequency or DC component; (b) spectrum component at sampling frequency f_s interpreted as DC; (c) spectrum component at $(1/N)f_s$; (d) spectrum component at $[(N+1)/N]f_s$.

effect of the discontinuity. The discontinuity without weighting causes side lobes to appear in the spectrum for a single frequency, as shown by the solid curve in Figure D.6, which is effectively the same as applying a rectangular window weighting function. In this case, all signal samples before sampling begins and after it ends are multiplied by zero and all values in between are multiplied by one.

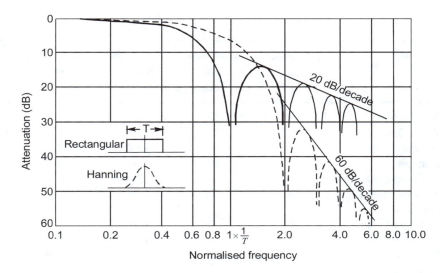

Figure D.6 Comparison of the filter characteristics of the RECTANGULAR and HANNING time weighting functions (after Randall, 1987).

A better choice of window is one that places less weight on the signal at either end of the window and maximum weight on the middle of the window. One such weighting, called a "Hanning window", is illustrated in Figure D.6. The result of weighting the input signal in this way is shown by the dashed curve in the figure. Even though the main lobe is wider, representing poorer frequency resolution, the side lobe amplitudes fall away more rapidly, resulting in less contamination of adjacent frequency bins. The properties of various weighting functions are summarised in Table D.1.

In the table, the "highest side lobe" is the number of dB that the signal corresponding to the highest side lobe will be attenuated compared to a signal at the filter centre frequency. The "side lobe fall off" is illustrated in Figure D.6, where the side lobes are the peaks to the right of the normalised frequency of 1.0. The "noise bandwidth" is the bandwidth of a rectangular filter that would let pass the same amount of broadband noise and the "maximum amplitude error" is the amount that a signal will be attenuated when it has a frequency that lies exactly mid-way between the centre frequencies of two adjacent filters (corresponding to normalised frequency = 0.5 in Figure D.6).

The best weighting function for amplitude accuracy in the frequency domain is the flat top (the name refers to the weighting in the frequency domain whereas the window shape refers to the weighting in the time domain). This is often used for calibration because of its uniform flat frequency response over the bandwidth, B ($=1/T$), which results in the measured amplitude of the spectral component being independent of small variations in signal frequency around the band centre frequency, thus making this window suitable for instrument calibration with a tonal signal. However, this window provides poor frequency resolution. Maximum frequency

Table D.1 Properties of the various time weighting functions

Window type	Highest sidelobe (dB)	Sidelobe fall off (dB decade)	Noise bandwidth	Maximum amplitude error (dB)
Rectangular	-13	-20	1.00	3.9
Triangular	-27	-40	1.33	1.8
Hanning	-32	-60	1.50	1.4
Hamming	-43	-20	1.36	1.8
Kaiser-Bessel (3.0)	-69	-20	1.80	1.0
Kaiser-Bessel (3.5)	-82	-20	1.93	0.9
Gaussian	-55	-20	1.64	1.3
Truncated Gaussian	-69	-20	1.90	0.9
Poisson (3.0)	-24	-20	1.65	1.5
Poisson (4.0)	-31	-20	2.08	1.0
Cauchy (4.0)	-35	-20	1.76	1.1
Cauchy (5.0)	-30	-20	2.06	1.3
Flat-top	-93	-	3.77	< 0.01

resolution (and minimum amplitude accuracy) is achieved with the rectangular window, so this is sometimes used to separate two spectral peaks which have a similar amplitude and a small frequency spacing. A good compromise commonly used is the Hanning window.

When a Fourier analysis is performed in practice using the FFT algorithm and a discrete Fourier transform, the resulting frequency spectrum is divided into a number of bands of finite width. Each band may be considered like a filter, the shape of which is dependent on the weighting function used. If the frequency of a signal falls in the middle of a band, its amplitude will be measured accurately. However, if it falls midway between two bands, the error in amplitude varies from 0.0 dB for the flat top window to 3.9 dB for the rectangular window. At the same time, the frequency bands obtained using the flat top window are 3.77 times wider so the frequency resolution is 3.77 times poorer than for the rectangular window. In addition, a signal at a particular frequency will also contribute to the energy in other nearby bands as can be seen by the shape of the filter curve in Figure D.6. This effect is known as spectral leakage and it is minimised by having a high value for the side lobe fall off in Table D.1.

D.2.6 Zoom Analysis

Zoom analysis is extremely important in a spectrum analyser as it allows part of a spectrum to be expanded and examined in greater detail. There are two types of zoom analysis commonly used. The first is real-time zoom in which data must be retaken over a longer period of time such that the uncertainty principle is satisfied. The second, referred to as "non-destructive zoom", uses the same data as was used to obtain the existing unexpanded spectrum. These data, stored in memory, are recycled a number of times through the analyser at a much lower sampling rate. Each time through, a different set of samples is acquired and the final result is averaged. Thus the extended time T required for a finer frequency resolution is achieved by using a lower sampling rate and recycling the data until all of it has been sampled once.

Whichever way it is done, if the frequency resolution is made finer by a factor of 10, then the time to acquire the necessary data will increase by a factor of 10, unless parallel processing is done on a stored spectrum.

D.3 IMPORTANT FUNCTIONS

The following functions, which involve the use of dual-channel spectrum analysers sampling two signals simultaneously, are used in noise analysis when it is desired to determine the relative phase and amplitude of two signals and whether both signals come from the same source. The cross-spectrum is also used in intensity measurement as discussed in Section 3.13.3.

D.3.1 Cross-spectrum

The two-sided cross-spectrum between two signals $x(t)$ and $y(t)$ in the frequency domain can be defined in a similar way as the power spectrum (or auto spectrum) of Equations (D.19) and (D.21) using Equation (D.18) as follows:

$$S_{xy}(f) = \lim_{T \to \infty} E\left[X^*(f)\, Y(f)\right] \tag{D.23}$$

or,

$$S_{xy}(f_n) \approx \frac{1}{q} \sum_{i=1}^{q} X_i^*(f_n)\, Y_i(f_n) \tag{D.24}$$

where the * represents the complex conjugate, $X_i(f)$ and $Y_i(f)$ are instantaneous spectra and $S_{xy}(f_n)$ is estimated by averaging over a number of instantaneous spectrum products obtained with finite time records of data. In contrast to the power spectrum which is real, the cross-spectrum is complex, characterised by an amplitude and a phase.

In practice, the amplitude of $S_{xy}(f_n)$ is the product of the two amplitudes $X(f_n)$ and $Y(f_n)$ and its phase is the difference in phase between $X(f_n)$ and $Y(f_n)$ ($= \theta_y - \theta_x$). This function can be averaged because for stationary signals, the relative phase between x and y is fixed and not random. The cross-spectrum S_{yx} has the same amplitude but opposite phase ($\theta_x - \theta_y$) to S_{xy}.

The cross-spectral density can also be defined in a similar way to the power spectral density by multiplying Equation (D.24) by the time taken to acquire one record of data (which is the reciprocal of the spacing between the components in the frequency spectrum, or the reciprocal of the spectrum bandwidth).

As for power spectra, the two-sided cross-spectrum can be expressed in single-sided form as:

$$G_{xy}(f_n) = 0 \qquad\qquad f_n < 0$$

$$G_{xy}(f_n) = S_{xy}(f_n) \qquad\qquad f_n = 0 \qquad\qquad\qquad (D.25)$$

$$G_{xy}(f_n) = 2S_{xy}(f_n) \qquad\qquad f_n > 0$$

Note that $G_{xy}(f)$ is complex, with real and imaginary parts referred to as the co-spectrum and quad-spectrum respectively. As for power spectra, the accuracy of the estimate of the cross-spectrum improves as the number of records over which the averages are taken increases. The statistical error for a stationary Gaussian random signal is given (Randall, 1987) as:

$$\varepsilon = \frac{1}{\sqrt{\gamma^2(f)q}} \qquad\qquad\qquad (D.26)$$

where $\gamma^2(f)$ is the coherence function (see next section) and q is the number of averages.

The amplitude of $G_{xy}(f)$ gives a measure of how well the two functions x and y correlate as a function of frequency and the phase angle of $G_{xy}(f)$ is a measure of the phase shift between the two signals as a function of frequency.

D.3.2 Coherence

The coherence function is a measure of the degree of linear dependence between two signals, as a function of frequency. It is calculated from the two auto-spectra (or power spectra) and the cross-spectrum as follows:

$$\gamma^2(f) = \frac{|G_{xy}(f)|^2}{G_{xx}(f) \cdot G_{yy}(f)} \qquad\qquad\qquad (D.27)$$

By definition, $\gamma^2(f)$ varies between 0 and 1, with 1 indicating a high degree of linear dependence between the two signals, $x(t)$ and $y(t)$. Thus in a physical system where $y(t)$ is the output and $x(t)$ is the input signal, the coherence is a measure of the degree to which $y(t)$ is linearly related to $x(t)$. If random noise is present in either $x(t)$ or $y(t)$ then the value of the coherence will diminish. Other causes of a diminished coherence are a undamping relationship between $x(t)$ and $y(t)$, insufficient frequency resolution in the frequency spectrum or poor choice of window function. A further cause of diminished coherence is a time delay, of the same order as the length of the record, between $x(t)$ and $y(t)$.

The main application of the coherence function is in checking the validity of frequency response measurements. Another more direct application is the calculation of the signal, S, to noise, N, ratio as a function of frequency:

$$S/N = \frac{\gamma^2(f)}{1 - \gamma^2(f)} \tag{D.28}$$

D.3.3 Frequency Response (or Transfer) Function

The frequency response function $H(f)$ is defined as:

$$H(f) = \frac{Y(f)}{X(f)} \tag{D.29}$$

Note that the frequency response function $H(f)$ is the Fourier transform of the system impulse response function, $h(t)$.

In practice, it is desirable to average $H(f)$ over a number of spectra, but as $Y(f)$ and $X(f)$ are both instantaneous spectra, it is not possible to average either of these. For this reason it is convenient to modify Equation (D.29). There are a number of possibilities, one of which is to multiply the numerator and denominator by the complex conjugate of the input spectrum. Thus:

$$H_1(f) = \frac{Y(f) \cdot X^*(f)}{X(f) \cdot X^*(f)} = \frac{G_{xy}(f)}{G_{xx}(f)} \tag{D.30a,b}$$

A second version is found by multiplying with $Y^*(f)$ instead of $X^*(f)$. Thus:

$$H_2(f) = \frac{Y(f) \cdot Y^*(f)}{X(f) \cdot Y^*(f)} = \frac{G_{yy}(f)}{G_{yx}(f)} \tag{D.31a,b}$$

Either of the above two forms of frequency response function are amenable to averaging, but $H_1(f)$ is the preferred version if the output signal $y(t)$ is more contaminated by noise than the input signal $x(t)$, whereas $H_2(f)$ is preferred if the input signal $x(t)$ is more contaminated by noise than the output (Randall, 1987).

References

Abbott, P. G. and Nelson, P. M. (2002). *Converting the UK traffic noise index, $L_{A10,18hr}$ to EU noise indices for noise mapping*. Transport Research Limited, Crowthorne, UK. Report PE/SE/4502/02.

Abramowitz, M. and Stegun, I. A. (1965). *Handbook of Mathematical Functions with Formulas, Graphs, and Mathematical Tables*. New York: Dover.

Agerkvist, F. T. and Jacobsen, F. (1993). Sound power determination in reverberation rooms at low frequencies. *Journal of Sound and Vibration*, **166**, 179–190.

Alfredson. R. J. and Scow, B. C. (1976). Performance of three sided enclosures. *Applied Acoustics*, **9**, 45–55.

Allard, J. F. and Champoux, Y. (1989). In situ two-microphone technique for the measurement of acoustic surface impedance of materials. *Noise Control Engineering Journal*, **32**, 15-23.

Andersson, P. (1981). Guidelines for film faced absorbers. *Noise and Vibration Control Worldwide*, January-February, 16–17.

Anderton, D. and Halliwell, N. A. (1980). *Noise from Vibration*. Warrendale, PA: Society of Automotive Engineers, Technical paper 800407.

Anonymous. *GiD Software*. International Centre for Numerical Methods in Engineering (CIMNE), Barcelona, Spain, available from http://gid.cimne.upc.es/.

Army, Air Force and Navy, USA (1983a). *Noise and Vibration Control for Mechanical Equipment*. Washington, DC: Technical manual TM 5-805-4 AFM 88–37 NAVFAC Dhl 3.10.

Army, Air Force and Navy, USA (1983b). *Power Plant Acoustics*. Technical manual TM 5–805–9 AFT 88–20 NAVFAC Dhf 3.14.

ASHRAE (American Society of Heating and Refrigeration Engineers) (2005). *Fundamentals Handbook*, Chapter 35. Atlanta, GA.

ASHRAE (American Society of Heating and Refrigeration Engineers) (1987). *Systems and Applications Handbook*, Chapter 52. Atlanta, GA.

ASHRAE (American Society of Heating and Refrigeration Engineers) (1991). *HVAC Applications Handbook*, Chapter 42. Atlanta, GA.

ASHRAE (American Society of Heating and Refrigeration Engineers) (2007). *HVAC Applications Handbook*, Chapter 47. Atlanta, GA.

Attenborough, K. (1994a). Prediction and exploitation of outdoor ground effect. In *Third International Congress on Air- and Structure-borne Sound and Vibration*. Montreal, Canada, 13–15 June.

Attenborough, K. (1994b). A note on short range ground characterization. *Journal of the Acoustical Society of America*, **95**, 3103–3108.

Attenborough, K., Sabik, I. H. and Lawther, J. M. (1980). Propagation of sound above a porous half-space. *Journal of the Acoustical Society of America*, **68**, 1493–1501.

Bacon, S.P. (2006). Auditory compression and hearing loss. *Acoustics Today*, **2**, 30–34.

Barron, M. (1993). *Auditorium Acoustics and Architectural Design*. London: E&FN Spon.

Baumann, H. D. (1987). *A Method for Predicting Aerodynamic Valve Noise Based on Modified Free Jet Noise Theories*. American Society of Mechanical Engineers 87-WA/NCA-7.

Baumann, H. D. (1984). Coefficients and factors relating to the aerodynamic sound level generated by throttling valves. *Noise Control Engineering Journal*, **22**, 1–6.

Baumann, H. D. and Hoffmann, H. (1999). Method for the estimation of frequency-dependent sound pressures at the pipe exterior of throttling valves. *Noise Control Engineering Journal*, **47**, 49–55.

Baumann, W. T., Ho, F. S. and Robertshaw, H. H. (1992). Active structural acoustic control of broadband disturbances. *Journal of the Acoustical Society of America*, **92**, 1998–2005.

Békésy, G von, Wever, E. G. (translator and ed.) (1960). *Experiments in Hearing*, American Institute of Physics by arrangement with McGraw Hill Book Company, New York.

Bendat, J. S. and Piersol, A. G. (1971). *Random Data: Analysis and Measurement Procedures*. New York: Wiley Interscience.

Beranek, L. L. (1954). *Acoustics*. New York: McGraw Hill, Chapter 5.

Beranek, L. L. (ed.) (1960). *Noise Reduction*. New York: McGraw Hill.

Beranek, L. L. (1962). *Music, Acoustics and Architecture*. New York: John Wiley.

Beranek, L. L. (ed.) (1971). *Noise and Vibration Control*. New York: McGraw Hill, Chapter 7.

Beranek, L. L. (1974). Protection of the public health and welfare from the effects of environmental noise. In *Noise Shock and Vibration Conference Proceedings*, J. D C. Crisp (ed.). Monash: Monash University, Clayton, Victoria, Australia.

Beranek, L. L. (1986). *Acoustics*. New York: American Institute of Physics.

Beranek, L. L. (ed.) (1988). *Noise and Vibration Control*, revised ed. Washington, DC: Institute of Noise Control Engineering.

Beranek, L.L. (1996). *Concert and Opera Halls. How They Sound*. Acoustical Society of America: New York.

Beranek, L. L. and Hidaka, T. (1998). Sound absorption in concert halls by seats, occupied and unoccupied and by the hall's interior surfaces. *Journal of the Acoustical Society of America*, **104**, 3169–3177.

Beranek, L. L. and Ver, I. L. (eds) (1992). *Noise and Vibration Control Engineering*. New York: McGraw Hill, Chapter 14, pp. 550–563.

Berglund, B., Lindvall, T. and Schwela, D.H. (1995). *Community Noise*. Stockholm: Stockholm University and Karolinska Institute.

Berglund, B., Lindvall, T. and Schwela, D. H. (eds) (1999). *Guidelines for Community Noise*. Geneva: World Health Organization.

Berry, B. F. and Harris, A. L. (1991). Military aircraft noise prediction and measurement. *Acoustics Bulletin*. November–December, 13–16.

Bies, D. A. (1961). Effect of a reflecting plane on an arbitrarily oriented multipole, *Journal of the Acoustical Society of America*, **33**, 286–288.

Bies, D. A. (1971). *Investigation of the Feasibility of Making Model Acoustic Measurements in the NASA Ames 40 × 80 ft Wind Tunnel*. BBN Report 2088 prepared for NASA under contract NAS2-6206, NASA Ames.

Bies, D. A. (1976). Uses of anechoic and reverberant rooms. *Noise Control Engineering Journal*, **7**, 154–163.

Bies, D. A. (1981). A unified theory of sound absorption in fibrous porous materials. In *Proceedings of the Australian Acoustical Society Annual Conference*. Cowes, Phillip Island.

Bies, D. A. (1984). Honours acoustics notes, Lecture 6. Department of Mechanical Engineering, University of Adelaide, Adelaide, South Australia 5005.

Bies, D. A. (1992). Circular saw aerodynamic noise. *Journal of Sound and Vibration*, **154**, 495–513.

Bies, D. A. (1994). An alternative model for combining noise and age-induced hearing loss. *Journal of the Acoustical Society of America*, **95**, 563–565

Bies, D. A. (1995). Notes on Sabine rooms. *Acoustics Australia*, **23**, 97–103.

Bies, D. A. (1996). The half-octave temporary threshold shift. *Journal of the Acoustical Society of America*, **100**, 2786 (A).

Bies, D. A. (1999). What the inner hair cells of the cochlea sense. *International Journal of Acoustics and Vibration*, **4**, 1–7.

Bies, D. A. (2000). The hydrodynamic forcing field of the cochlea. In *Proceedings of the 7th International Congress of Sound and Vibration*, 3–4 July, Garmisch-Partenkirchen, pp. 2259–2266.

Bies, D. A. and Bridges, G. E. (1993). Sound power determination in the geometric near field of a source by pressure measurements alone. In *Proceedings of the Australian Acoustical Society Annual Conference*. Glenelg, South Australia.

Bies, D. A. and Davies, J. M. (1977). An investigation of the measurement of transmission loss. *Journal of Sound and Vibration*, **53**, 203–221.

Bies, D. A. and Hansen, C. H. (1979). Notes on porous materials for sound absorption. *Bulletin of the Australian Acoustical Society*, **7**, 17–22

Bies, D. A. and Hansen, C. H. (1980). Flow resistance information for acoustical design. *Applied Acoustics*, **13**, 357–391.

Bies, D. A. and Hansen, C. H. (1990). An alternative mathematical description of the relationship between noise exposure and age-induced hearing loss. *Journal of the Acoustical Society of America*, **88**, 2743–2754

Bies, D. A. and Zockel, M. (1976). Noise attenuation in ducts with flow. In *Proceedings, Vibration and Noise Control Engineering Conference*, Institution of Engineers, Australia. 11–12 October. Publication no. 76/9.

Bies, D. A., Hansen, C. H. and Bridges, G. E. (1991). Sound attenuation in rectangular and circular cross-section ducts with flow and bulk-reacting liner. *Journal of Sound and Vibration*, **146**, 47–80.

Billone, M. C (1972). Mechanical stimulation of cochlear hair cells. PhD Thesis, Northwestern University, Evanston, IL.

Bistafa, S. R., and Bradley, J. S. (2000). Revisiting algorithms for predicting the articulation loss of consonants ALcons, *Journal of the Audio Engineering Society*, **48**, 531–544.

Blauert, J. (1983). *Spatial Hearing*. Boston, MA: MIT Press.

Bohne, B., Harding, G. W. and Lee, S. C. (2007). Death pathways in noise-damaged outer hair cells. *Hearing Research*, **223**, 61–70.

Bolt, R. H. (1947). On the design of perforated facings for acoustic materials. *Journal of the Acoustical Society of America*, **19**, 917–921.

Bolt, R. H., Labate, S. and Ingard, K. U. (1949). Acoustic reactance of small circular orifices. *Journal of the Acoustical Society of America*, **21**, 94.

Bork, I. (2000). A comparison of room simulation software: the 2nd round robin on room acoustical computer simulation. *Acustica - Acta acustica.*, **86**, 943-956.

Bradley, J. S. and Birta, J. A. (2001). On the sound insulation of wood stud exterior walls. *Journal of the Acoustical Society of America*, **110**, 3086–3096.

Bragg, S. L. (1963). Combustion noise. *Journal of the Institute of Fuel*, January, 12–16.

Brekke, A. (1981). Calculation methods for the transmission loss of single, double and triple partitions. *Applied Acoustics*, **14**, 225–240.

Brod, K. (1984). On the uniqueness of solution for all wavenumbers in acoustic radiation. *Journal of the Acoustical Society of America*, **76**, 1238–1243.

Broner, N. and Leventhall, H. G. (1983). A criterion for predicting the annoyance due to lower level low frequency noise. *Journal of Low Frequency Noise and Vibration*, **2**, 160–168.

Brooks, T. F. and Humphreys, W. M. (2006). A deconvolution approach for the mapping of acoustic sources (DAMAS) determined from phased microphone arrays. *Journal of Sound and Vibration*, **204**, 856–879.

Brownell, W. E., Bader, C. R., Bertrand, D. and De Ribaupierre, Y. (1985). Evoked mechanical responses of isolated cochlear hair cells. *Science*, **227**, 194–196.

Brüel, P.V. (1977). Do we measure damaging noise correctly? *Noise Control Engineering*, March–April, 2–10.

Brüel and Kjær (1973). Instructions and applications for 1/2-inch condenser microphones. Handbook. Copenhagen: Brüel and Kjær.

Brüel and Kjær (1986). *Master Catalog*. Copenhagen: Brüel and Kjær.

Brüel and Kjær (1996). *Microphone Handbook, Volume 1*. Copenhagen: Brüel and Kjær.

Bullmore, A. J., Nelson, P. A. and Elliott, S. J. (1986). *Active Minimisation of Acoustic Potential Enerqy in Harmonically Excited Cylindrical Enclosed Sound Fields*. AIAA paper AIAA86-1958.

Burns, W. and Robinson, D. W. (1970). *Hearing and Noise in Industry*. London: HMSO.

Byrne, K. P., Fischer, H. M. and Fuchs, H. V. (1988). Sealed, close-fitting machine-mounted, acoustic enclosures with predictable performance. *Noise Control Engineering Journal*, **31**, 7–15.

Carroll, M. M. and Miles, R. N. J. (1978). Steady-state sound in an enclosure with diffusely reflecting boundary. *Journal of the Acoustical Society of America*, **64**, 1424–1428.

Cazzolato, B.S. (1999). Sensing systems for active control of sound transmission into cavities. PhD thesis, Adelaide University, South Australia.

Cazzolato, B.S. and Hansen, C.H. (1999). Structural radiation mode sensing for active control of sound radiation into enclosed spaces. *Journal of the Acoustical Society of America*, **106**, 3732–3735.

Cazzolato, B. S., Petersen, C. D., Howard, C. Q. and Zander, A. C. (2005). Active control of energy density in a 1D waveguide: A cautionary note. *Journal of the Acoustical Society of America*, **117**, 3377-3380.

Chambers, C. (1972). *Specification Nos. NWG-1, NWG-2, NWG-3* (Revision 1). London: Oil Companies Materials Association.

Chapkis, R. L. (1980). Impact of technical differences between methods of INM and NOISEMAP. In *Proceedings of Internoise '80*, pp. 831–834.

Chapkis, R. L., Blankenship, G. L. and Marsh, A. H. (1981). Comparison of aircraft noise-contour prediction programs. *Journal of Aircraft.* **18**, 926 – 933.

Cheng, A. H. D. and Cheng, D. T. (2005). Heritage and early history of the boundary element method. *Engineering Analysis with Boundary Elements*, **29**, 268-302.

Chien, C. F. and Carroll, M. M. (1980). Sound source above a rough absorbent plane. *Journal of the Acoustical Society of America*, **67**, 827–829.

Christensen, J. J. and Hald, J. (2004). *Beamforming.* Technical Review 1–2004. Copenhagen: Brüel and Kjær.

Chung, Y. J. and Blaser, D. A. (1981). Recent developments in the measurement of acoustic intensity using the cross-spectral method. *Society of Automotive Engineers*, Technical paper 810396.

Church, A. (1963). *Mechanical Vibrations*, 2nd edn. New York: John Wiley.

Clifford, S. F. and Lataitis, R. J. (1983). Turbulence effects on acoustic wave propagation over a smooth surface. *Journal of the Acoustical Society of America*, **75**, 1545–1550.

CONCAWE (1981a). *The Propagation of Noise from Petroleum and Petrochemical Complexes to Neighbouring Communities*. The Hague: Report 4/81.

CONCAWE (1981b). *The Propagation of Noise from Petroleum and Petrochemical Complexes to Neighbouring Communities, Phase 1*. The Hague: AT/674 final report issued to Stichting.

Craggs, A. (1971). The transient response of a coupled plate-acoustic system using plate and acoustic finite elements. *Journal of Sound and Vibration*, **15**, 509-528.

Craven, N. J. and Kerry, G. (2001). *A Good Practice Guide on the Sources and magnitude of Uncertainty Arising in the Practical Measurement of Environmental Noise.* Salford, UK: University of Salford.

Crede, C. E. (1965). *Shock and Vibration Concepts in Engineering Design*. Englewood Cliffs, NJ: Prentice-Hall.

Cremer, L. (1942). Theorie der schalldämmung Dünner Wände bei schrägem einfall. *Akustische Zeitschrift*, **7**, 81–104.

Cremer, L. (1953). Theory regarding the attenuation of sound transmitted by air in a rectangular duct with an absorbing wall, and the maximum attenuation constant during this process. *Acustica*, **3**, 249–263 (in German).

Cremer, L. (1967). USA Patent 3 353 626.

Cremer, L. and Müller, H. A. (English translation by T. J. Schultz) (1982). *Principles and Applications of Room Acoustics*, Vol. I, London and New York: Applied Science Publishers, Chapter 2.1.

Cremer, L., Heckl, M. and Ungar, E. E. (1988). Structure-borne Sound, 2nd edn. New York: Springer-Verlag.

Crocker, M. J. and Kessler, F. M. (1982). *Noise and Noise Control*, Vol. II. Boca Raton: CRC Press.

Crocker, M. J. and Price, A. J. (1972–82). *Noise and Noise Control*. Cleveland, OH: CRC Press.

Cummings, A. (1976). Sound attenuation ducts lined on two opposite walls with porous material, with some applications to splitters. *Journal of Sound and Vibration*, **49**, 9–35.

Cummings, A. (1978). The attenuation of lined plenum chambers in a duct: I. Theoretical models. *Journal of Sound and Vibration*, **61**, 347–373.

Cummings, A. (1980). Low frequency acoustic radiation from duct walls. *Journal of Sound and Vibration*, **71**, 201–226.

Curle, N. (1955). The influence of solid boundaries on aerodynamic sound. *Proceedings of the Royal Society*, Series A, **231**, 505–514.

Daniels, F. B. (1947). Acoustical impedance of enclosures. *Journal of the Acoustical Society of America*, **19**, 569–571.

Dang, J., Shadle, C. H., Kawanishi, Y., Honda, K., Suzuki, H. (1998). An experimental study of the open end correction coefficient for side branch, *Journal of the Acoustical Society of America*, **104**, 1075–1084.

Davern, W. A. (1977). Perforated facings backed with porous materials as sound absorbers: an experimental study. *Applied Acoustics*, **10**, 85–112.

Davern, W. A. and Dubout, P. (1980). *First Report on Australasian Comparison Measurements of Sound Absorption Coefficients*. Commonwealth Scientific and Industrial Research Organization, Division of Building Research.

Davies, P. O. A. L. (1992a). *Intake and Exhaust Noise*.Southampton: ISVR Technical Report No. 207.

Davies, P. O. A. L. (1992b). *Practical Flow Duct Acoustic Modelling*. Southampton: ISVR Technical Report No. 213.

Davies, P. O. A. L. (1993). Realistic models for predicting sound propagation in flow duct systems. *Noise Control Engineering Journal*, **40**, 135–141.

Davy, J. L. (1990). Predicting the sound transmission loss of cavity walls. In *Proceedings of Interior Noise Climates, Australian Acoustical Society Annual Conference*, Adelaide, South Australia.

Davy, J. L. (1991). Predicting the sound insulation of stud walls. In *Proceedings of Internoise '91*, 251–54.

Davy, J. L. (1993). The sound transmission of cavity walls due to studs. In *Proceedings of Internoise '93,* 975–978.

Davy, J. L. (1998). Problems in the theoretical prediction of sound insulation. In *Proceedings of Internoise '98,* Paper No. 44.

Davy, J. L. (2000). The regulation of sound insulation in Australia. In *Proceedings of Acoustics 2000*. Australian Acoustical Society Conference, Western Australia, 15–17 November, 155-160.

Davy, J. L (2008a). The directivity of the forced radiation of sound from panels and openings including the shadow zone. In *Proceedings of Acoustics '08*, Paris, 29 June – 4 July, 3833-3838.

Davy, J. L. (2008b). Comparison of theoretical and experimental results for the directivity of panels and openings. In *Proceedings of Acoustics 2008*. Annual Conference of the Australian Acoustical Society, Geelong, Australia.

Day, A. and Bennett, B. (2008). Directivity of sound from an open ended duct. In *Proceedings of Acoustics 2008*. Annual Conference of the Australian Acoustical Society, Geelong, Australia.

de Bree, H. E., Leussink, R., Korthorst, T., Jansen, H. and Lammerink, M., (1996). The microflown: a novel device for measuring acoustical flows. *Sensors and Actuators A*, **54**, 552–557.

de Bree, H. E. and Druyvesteyn, W.F. (2005). A particle velocity sensor to measure the sound from a structure in the presence of background noise. In *Proceedings of Forum Acusticum*, Budapest.

DeJong, R. and Stusnik, E. (1976). Scale model studies of the effects of wind on acoustic barrier performance. *Noise Control Engineering*, **6**, 101–109.

de Vries, J. and de Bree, H. E. (2008). *Scan & Listen: A Simple and Fast Method to Find Sources*. SAE Paper 2008-36-0504.

Delany, M. E. (1972). *A Practical Scheme for Predicting Noise Levels (L_{10}) Arising from Road Traffic*. NPL Acoustics Report AC57.

Delany, M. E. and Bazley, E. N. (1969). *Acoustical Characteristics of Fibrous Absorbent Materials*. Aero Report Ac 37. National Physical Laboratory.

Delany, M. E. and Bazley, E. N. (1970). Acoustical properties of fibrous absorbent materials. *Applied Acoustics*, **3**, 105–106.

Delaney, M. E., Harland, D. G., Hood, R. A. and Scholes, W. E. (1976). The prediction of noise levels L_{10} due to road traffic. *Journal of Sound and Vibration*, **48**, 305-325.

Den Hartog, J. P. (1956). *Mechanical Vibrations*, 4th edn. New York: McGraw Hill.

Dewhirst, M. (2002). Exhaust stack directivity. Final year thesis for the Honours Degree of Bachelor of Engineering, University of Adelaide, South Australia.

Dickey, N. S. and Selamet, A. (2001). An experimental study of the impedance of perforated plates with grazing flow. *Journal of the Acoustical Society of America*, **110**, 2360–2370.

Diehl, G. M. (1977). Sound power measurements on large machinery installed indoors: two surface measurement. *Journal of the Acoustical Society of America*, **61**, 449–456.

Doelle, L. J. (1972). *Environmental Acoustics*. New York: McGraw Hill.

Dowling, A. and Ffowcs-Williams, J. (1982). *Sound and Sources of Sound*. Chichester: Halstead Press.

Driscoll, D. P. and Royster, L. H. (1984). Comparison between the median hearing threshold levels for an unscreened black nonindustrial noise exposed population (NINEP) and four presbycusis data bases. *American Industrial Association Journal*, **45**, 577–593.

Druyvesteyn, W. F. and de Bree, H. E. (2000). A new sound intensity probe; comparison to the Bruel and Kjaer p–p probe. *Journal of the Audio Engineering Society*, **48**, 10–20.

Dubout, P. and Davern, W. A. (1959). Calculation of the statistical absorption coefficient from acoustic impedance tube measurements. *Acustica*, **9**, 15–16.

Dunn, I. P. and Davern, W. A. (1986). Calculation of acoustic impedance of multi-layer absorbers. *Applied Acoustics*, **19**, 321–334.

Dutilleaux, G., Vigran, T. E. and Kristiansen, U. R. (2001). An in situ transfer function technique for the assessment of acoustic absorption of materials in buildings. *Applied Acoustics*, **62**, 555-572.

Ebbing, C. (1971). Experimental evaluation of moving sound diffusers in reverberant rooms. *Journal of Sound and Vibration*, **16**, 99–118.

Edge, P. M. Jr. and Cawthorn, J. M. (1976). *Selected methods for quantification of community exposure to aircraft noise*, NASA TN D-7977.

Edison Electric Institute (1978). *Electric Power Plant Environmental Noise Guide*. New York: Bolt Beranek and Newman Inc., Technical report.

Egan, M. D. (1987). *Architectural Acoustics*. New York: McGraw Hill.

Eldridge, H. J. (1974). *Properties of Building Materials*. Lancaster: Medical and Technical Publishing Co.

Elfert, G. (1988). Vergleich zwischen gemessenen und nach verschiedenen Verfahren berechneten Schulldruckpegeln in Flachräumen. In *Fortschritte der Akustik*, DAGA '88 DPG – Gmb H. Bad Honnef, 745.

Embleton, T. F. W. (1980). Sound propagation outdoors: improved prediction schemes for the '80s. *Internoise '80*, 17–30.

Embleton, T. F. W., Piercy, J. E. and Daigle, G. A. (1983). Effective flow resistivity of ground surfaces determined by acoustical measurements. *Journal of the Acoustical Society of America*, **74**, 1239–1244.

Ewins, D. J. (1984). *Modal Testing*. Letchworth, Somerset: Research Studies Press.

Ewins, D. J. (1995). *Modal Testing: Theory and Practice*. Letchworth, Somerset: Research Studies Press.

Fahy, F. J. (1977). A technique for measuring sound intensity with a sound level meter. *Noise Control Engineering*, **9**, 155–163.

Fahy, F. J. (1985). *Sound and Structural Vibration: Radiation, Transmission and Response*. London: Academic Press.

Fahy, F. J. (1995). *Sound Intensity, 2nd edn*. London: E&FN Spon.

Fahy, F. J. (2001). *Foundations of Engineering Acoustics*. London: Academic Press.

Fahy, F. J. and Walker, J. G. (1998). *Fundamentals of Noise and Vibration*. London: E&FN Spon.

Fahy, F. J. and Gardonio, P. (2007). Sound and Structural Vibration: Radiation, Transmission and Response, 2nd edn.. London: Academic Press.

Ffowcs Williams, J. E. and Hawkings, D. L. (1969). Sound generation by turbulence and surfaces in arbitrary motion. *Philosophical Transactions of the Royal Society*, **A264**, (1151), 321-342.

FHWA (1995). *Highway Traffic Noise Analysis and Abatement Guide*. US Dept of Transportation, Federal Highway Administration, Washington, DC.

Fitzroy, D. (1959). Reverberation formula which seems to be more accurate with nonuniform distribution of absorption. *Journal of the Acoustical Society of America*, **31**, 893-97.

Fleming, G. G., Burstein, J., Rapoza, A. S., Senzig, D. A. and Gulding, J. M. (2000). Ground effects in FAA's integrated noise model. *Noise Control Engineering Journal*, **48**, 16–24.

Foss, R. N. (1979). Double barrier noise attenuation and a predictive algorithm. *Noise Control Engineering*, **13**, 83–91.

Franken, P. A. (1955). *A Theoretical Analysis of the Field of a Random Noise Source Above an Infinite Plane*. NACA Technical Note 3557.

Frederiksen, E., Eirby, N. and Mathiasen, H. (1979). *Technical Review*, No. 4. Denmark: Bruel and Kjaer.

Fujiwara, K., Ando, Y. and Maekawa, Z. (1977a). Noise control by barriers. Part 1: noise reduction by a thick barrier. *Applied Acoustics*, **10**, 147–159.

Fujiwara, K., Ando, Y. and Maekawa, Z. (1977b). Noise control by barriers. Part 2: noise reduction by an absorptive barrier. *Applied Acoustics*, **10**, 167–170.

Garai, M. and Pompoli, F. (2005). A simple empirical model of polyester fibre materials for acoustical applications. *Applied Acoustics*, **66**, 1383–1398.

Gill, H. S. (1980a). *Measurement and Prediction of Construction Plant Noise.* ISVR Technical Report No. 112.

Gill, H. S. (1980b). *Effect of Barriers on the Propagation of Construction Noise.* ISVR Technical Report No. 113.

Golay, M. J. E. (1947). Theoretical consideration in heat and infra-red detection, with particular reference to the pneumatic detector. *Review of Scientific Instruments*, **18**, 347–356.

Gomperts, M. C. and Kihlman, T. (1967). The sound transmission loss of circular and slit-shaped apertures in walls. *Acustica*, **18**, 144–150.

Gosele. K. (1965). The damping behavior of reflection-mufflers with air flow. *VDI–Berickre*, **88**, 123–130 (in German).

Gradshteyn, I. S. and Ryshik, I. M. (1965). *Table of Integrals, Series and Products.* New York and London: Academic Press.

Green, D. W., Winandy, J. E., and Kretschmann, D. E. (1999). *Mechanical Properties of Wood.* Chapter 4 of *Wood handbook: Wood as an Engineering Material.* Gen. Tech. Rep. FPL–GTR–113. Madison, WI: US Department of Agriculture, Forest Service, Forest Products Laboratory.

Greenwood, D. D. (1990). A cochlear frequency-position function for several species – 29 years later. *Journal of the Acoustical of America*, **87**, 2592-2605.

Gu, J., Ma, Z. D. and Hulbert, G. M. (2001). Quasi-static data recovery for dynamic analyses of structural systems. *Finite Elements in Analysis and Design*, **37**, 825-841.

Hald, J., Morkholt, J. and Gomes, J. (2007). *Efficient Interior NSI Based on Various Beamforming Methods for Overview and Conformal Mapping Using SONAH Holography for Details on Selected Panels.* SAE Paper 2007-01-2276. Society of Automotive Engineers.

Hale, M. (1978). A comparison of several theoretical noise abatement prediction methods for pipe lagging systems. Paper presented at the 96th Acoustical Society of America meeting, Hawaii.

Hansen, C. H. (1993). Sound transmission loss of corrugated panels. *Noise Control Engineering Journal*, **40**, 187–197.

Hansen, C. H. (2001). *Understanding Active Noise Cancellation.* London: E&FN Spon.

Hansen, C. H. and Mathews, M. (1983). *Noise Reduction through Optimum Hull Design in Lightweight Tracked Vehicles.* Bolt, Beranek and Newman Inc., Report 5074.

Hansen, C. H. and Snyder, S. D. (1997). *Active Control of Sound and Vibration.* London: E&FN Spon.

Hardwood Plywood Manufacturers' Association (1962). *Design Procedure for the Sound Absorption of Resonant Plywood Panels.* Bolt, Beranek and Newman Inc., Report 925.

Harris, C. M. (ed.) (1979). *Handbook of Noise Control*, 2nd edn. New York: McGraw Hill.

Harris, C. M. and Crede, C. E. (1976). *Shock and Vibration Handbook*. NewYork: McGraw Hill.

Hayes, P. A. and Quantz, C. A. (1982). *Determining Vibration Radiation Efficiency and Noise Characteristics of Structural Designs Using Analytical Techniques*. SAE Technical Paper 820440.

Hearmon, R. F. S. (1959). The frequency of flexural vibration of rectangular orthotropic plates with clamped or supported edges. *Journal of Applied Mechanics*, **26**, 537–540.

Heilmann, G., Meyer, A. and Dobler, D. (2008). Beamforming in the time domain using 3D-microphone arrays. In *Proceedings of the Annual Conference of the Australian Acoustical Society*, Geelong.

Heckl, M. (1960). Untersuchungen an orthotropen platted *Acustica*, **10**, 109–115.

Heerema, H. and Hodgson, M. R. (1999). Empirical models for predicting noise levels, reverberation times, and fitting densities in industrial workshops. *Applied Acoustics*, **57**, 51–60.

Heitner, I. (1968). How to estimate plant noises. *Hydrocarbon Processing*, **47** (12), 67–74.

Herrin, D. W., Martinus, F.,Wu, T. W. , and Seybert A. F. (2003). *A New Look at the High Frequency Boundary Element and Rayleigh Integral Approximations*. SAE Paper 03NVC-114.

Hessler, G. F., Hessler, D. M., Brandstatte, P. and Bay, K. (2008). Experimental study to determine wind-induced noise and windscreen attenuation effects on microphone response for environmental wind turbine and other applications. *Noise Control Engineering Journal*, **56**, 300–309.

Hidaka, T., Nishihara, N. and Beranek, L.L. (2001). Relation of acoustical parameters with and without audiences in concert halls and a simple method for simulating the occupied state. *Journal of the Acoustical Society of America*, **109**, 1028–1041.

Hirsh, I. J. and Bilger, R. C. (1955). Auditory-threshold recovery after exposure to pure tones. *Journal of the Acoustical Society of America*, **27**, 1186–1194.

Hodge, G. C. and Garinther, G. R. (1970). Validation of the single-impulse correction factor of the CHABA impulse-noise damage risk criterion. *Journal of the Acoustical Society of America*, **48**, 1429–1430.

Hodgson, M. R. (1994a). When is diffuse-field theory accurate? In *Proceedings of the Wallace Clement Sabine Symposium*. Cambridge, MA, pp. 157–160.

Hodgson, M. R. (1994b). Are room surface absorption coefficients unique? In *Proc. of the Wallace Clement Sabine Symposium*. Cambridge, MA, 161–164.

Hodgson, M. R. (1999). Experimental investigation of the acoustical characteristics of university classrooms. *Journal of the Acoustical Society of America*, **106**, 1810–1819.

Hodgson, M. R. (2001). Empirical prediction of speech levels and reverberation in classrooms. *Journal of Building Acoustics*, **8**, 1–14.

Hodgson, M. R. (2003). Ray-tracing evaluation of empirical models for prediction noise in industrial workshops. *Applied Acoustics*, **64**, 1033–1048.

Hodgson, D. C. and Sadek, M. M. (1977). Sound power as a criterion for forging machine optimization. In *Proceedings of the International Conference on Machine Tool Dynamics and Research*. London: Imperial College, pp. 825–830.

Hodgson, D. C. and Sadek, M. M. (1983). A technique for the prediction of the noise field from an arbitrary vibrating machine. In *Proceedings of the Institution of Mechanical Engineers*, **179C**, 189–197.

Hodgson, M. R. and Nosal, E-M. (2002). Effect of noise and occupancy on optimal reverberation times for speech intelligibility in classrooms. *Journal of the Acoustical Society of America*, **111**, 931–939.

Hohenwarter, D. (1991). Noise radiation of (rectangular) plane sources. *Applied Acoustics*, **33**, 45–62.

Hoops, R. H. and Eriksson, L. J. (1991). *Rigid foraminous microphone probe for acoustic measurement in turbulent flow*. US Patent No. 4 903 249.

Hoover, R. M. (1961). *Tree Zones as Barriers for the Control of Noise Due to Aircraft Operations*. Bolt, Beranek and Newman, Report 844.

Howard, C. Q. (2007). Theoretical and experimental results of the transmission loss of a plate with discrete masses attached, *14th International Congress on Sound and Vibration*, Cairns, Australia, 9-12 July, Paper 14.

Howard, C. Q., Cazzolato, B. S. and Hansen, C. H. (2000). Exhaust stack silencer design using finite element analysis. *Noise Control Engineering Journal*, **48**, 113–120.

Howe, M. S. and Baumann, H. D. (1992). Noise of gas flows. In Noise and Vibration Control Engineering, L. L. Beranek and I. Ver (eds.). New York: McGraw Hill.

Hubner, G. (1981). Higher accuracy in sound power determination of machines under in-situ conditions by using a sound intensity meter. In *Proceedings of the International Congress on Recent Developments in Acoustic Intensity Measurements*. Senlis, France.

Huisman, W. H. T. and Attenborough, K. (1991). Reverberation and attenuation in a pine forest. *Journal of the Acoustical Society of America*, **90**, 2664–2677.

Hutchins, D. A., Jones, H. W. and Russell, L. T. (1984). Model studies of barrier performance in the presence of ground surfaces. *Journal of the Acoustical Society of America*, **75**, 1807–1826.

Ih, J-G. (1992). The reactive attenuation of reactive plenum chambers. *Journal of Sound and Vibration*, **157**, 93–122.

Ihde, W. M. (1975). Tuning stubs to silence large air handling systems. *Noise Control Engineering*, **5**, 131–135.

Ingard, K. U. (1959). Attenuation and regeneration of sound in ducts and jet diffusers. *Journal of the Acoustical Society of America*, **31**, 1206–1212.

Ingard, K. U. and Bolt, R. H. (1951). Absorption characteristics of acoustic material with perforated facings. *Journal of the Acoustical Society of America*, **23**, 533–540.

Ingard, K. U. and Dear, T. A. (1985). Measurement of acoustic flow resistance. *Journal of Sound and Vibration*, **103**, 567–572.

Ingard, K. U. and Ising, H. (1967). Acoustic nonlinearity of an orifice. *Journal of the Acoustical Society of America*, **42**, 6–17.

Iqbal, M. A., Wilson, T. K. and Thomas, R. J. (eds) (1977). *The Control of Noise in Ventilation Systems: A Designers Guide*. Atkins Research and Development. London: E&FN Spon.

Isakov, V. and Wu, S.F. (2002). On theory and applications of the HELS method in inverse acoustics. *Inverse Problems*, **18**, 1147–1159.

Isei, T., Embleton, T. F. W. and Piercy, J. E. (1980). A comparison of two different sound intensity measurement principles. *Journal of the Acoustical Society of America*, **67**, 46–58.

Jacobsen, F. and de Bree, H-E. (2005). A comparison of two different sound intensity measurement principles. *Journal of the Acoustical Society of America*, **118**, 1510–1517.

Jacobsen, F. and Liu, Y. (2005). Near field acoustic holography with particle velocity transducers. *Journal of the Acoustical Society of America*, **118**, 3139–3144.

Jacobsen, F. and de Bree, H-E. (2009). A comparison of two different sound intensity measurement principles. *Journal of the Acoustical Society of America*, in press.

Jansson, E. and Karlsson, K. (1983). Sound levels recorded within the symphony orchestra and risk criteria for hearing loss. *Scandinavian Audiology*. **12**, 215–221.

Jayachandran, V., Hirsch, S. and Sun, J. (1998). On the numerical modelling of interior sound fields by the modal expansion approach. *Journal of Sound and Vibration*, **210**, 243-254.

Jean, Ph., Rondeau, J.-F. and van Maercke, D. (2001). Numerical models for noise prediction near airports. In *Proceedings of the 8th International Congress on Sound and Vibration*, Hong Kong, 2-6 July, pp. 2929–2936.

Jenkins, R. H. and Johnson, J. B. (1976). The assessment and monitoring of the contribution from a large petrochemical complex to neighbourhood noise levels. *Noise Control Vibration and Insulation*, **7**, 328–335.

Jeyapalan, R. K. and Halliwell, N. A. (1981). Machinery noise predictions at the design stage using acoustic modelling. *Applied Acoustics*, **14**, 361–376.

Jeyapalan, R. K. and Richards, E. J. (1979). Radiation efficiencies of beams in flexural vibration. *Journal of Sound and Vibration*, **67**, 55–67.

Ji, Z. L. (2005). Acoustic length correction of a closed cylindrical side-branched tube. *Journal of Sound and Vibration*, **283**, 1180–1186.

Johnson, D. H. and Dudgeon, D. E. (1993). *Array Signal Processing: Concepts and Techniques*. New Jersey: Prentice Hall.

Jonasson, H. and Eslon, L. (1981). *Determination of Sound Power Levels of External Sources*. Report SP-RAPP 1981: 45, National Testing institute, Acoustics Laboratory, Borus, Sweden.

Jones, A. D. (1984). Modeling the exhaust noise radiated from reciprocating internal combustion engines: literature review. *Noise Control Engineering*, **23**, 12–31.

Josse, R. and Lamure, C. (1967). Transmission du son par unepario simple. *Acustica*, **14**, 267–280.

Kanapathipillai, S. and Byrne, K. P. (1991). Calculating the insertion loss produced by a porous blanket pipe lagging. *Proceedings of Internoise '91*, 283–286.

Karlsson, K., Lundquist, P. G. and Olausson, T. (1983). The hearing of symphonic musicians, *Scand. Audiol.*, **12**, 257–264.

Kean, S. (2008). Is CoRTN an Leq or L10 procedure? In *Proceedings of the 2008 Australian Acoustical Society Conference*, Geelong, November 24–26.

Keränen, J., Airo, E., Olkinuora, P., Hongisto, V. (2003). Validity of the ray-tracing method for the application of noise control in workplaces. *Acustica - acta acustica*, **89**, 863-874.

Kessel, R. G. and Kardon, R.H. (1979). *Tissues and Organs: A Text-Atlas of Scanning Electron Microscopy.* New York: W.H. Freeman and Co.

Kinsler, L. E., Frey, A. R., Coppens, A. B. and Sanders, J. V. (1982). *Fundamentals of Acoustics*, 3rd revised edn. New York: John Wiley.

Kirkup, S. M. (2007). *The Boundary Element Method in Acoustics*. Integrated Sound Software, published electronically at http://www.boundary-element-method.com.

Koopmann, G. H. and Benner, H. (1982). Method for computing the sound power of machines based on the Helmholtz integral. *Journal of the Acoustical Society of America*, **71**, 788.

Kosten, C. W. and Van Os, C. J. (1962). *Community Reaction Criteria for External Noises*. National Physical Laboratory Symposium No. 12. London: HMSO.

Kraak, W. (1981). Investigation on criteria for the risk of hearing loss due to noise. In *Hearing Research and Theory*, Vol. 1. San Diego: Academic Press.

Kraak, W., Kracht, L. and Fuder, G. (1977). Die Ausbildung von Gehörschäden als Folge der Akkomulation von Lärmeinwirkungen, *Acustica*, **38**, 102–117.

Krokstad, A., Strom, S. and Sorsdal, S. (1968). Calculation of the acoustical room response by the use of a ray tracing technique. *Journal of Sound and Vibration*, **8**, 118-125.

Kryter, K. D. (1959). Scaling human reactions to the sound from aircraft. *Journal of the Acoustical Society of America*, **31**, 1415–1429.

Kryter, K. D. (1970). *The Effects of Noise on Man*. New York: Academic Press.

Kurze, U. J. and Allen, C. H. (1971). Influence of flow and high sound level on the attenuation in a lined duct. *Journal of the Acoustical Society of America*, **49**, 1643.

Kurze, U. J. and Anderson, G. S. (1971). Sound attenuation by barriers. *Applied Acoustics*, **4**, 35–53.

Kuttruff, H. (1971). Simulierte nachhallkurven in rechteckräumen mit diffusem schallfeld. *Acustica*, **25**, 333–342.

Kuttruff, H. (1976). Nachhall und effektive absorption in räumen mit diffuser wandreflexion. *Acustica*, **35**, 141.

Kuttruff, H. (1979). *Room Acoustics*, 2nd edn. London: E&FN Spon.

Kuttruff, H. (1985). Stationare schallausbreitung in flachräumen. *Acustica*, **57**, 62–70.

Kuttruff, H. (1989). Stationare schallausbreitung in langräumen. *Acustica*, **69**, 53–62.

Kuttruff, H. (1994). Sound decay in enclosures with non-diffuse sound field. In *Proc. of the Wallace Clement Sabine Symposium*, Cambridge, MA, pp. 85–88.

Lam, Y. W. (1996). A comparison of three diffuse reflection modeling methods used in room acoustics computer models. *Journal of the Acoustical Society of America*, **100**, 2181-2192.

Landau, L. D. and Lifshitz, E. M. (translated by J. B. Sykes and W. H. Reid) (1959). *Fluid Mechanics*. Oxford: Pergamon Press.

Larson, H. (1978). *Reverberation at Low Frequencies*. Copenhagen: Brüel and Kjaer Technical Review No. 4.

Larson, K. M. S. (1994). The present and future of aircraft noise models: a user's perspective. In *Proceedings of Noise-Con '94*, 969 – 974.

Levine, H. and Schwinger, J. (1948). On the radiation of sound from an unflanged circular pipe. *Physics Review*, **73**, 383–406.

Levy, M. (ed.) (2001). *Handbook of Elastic Properties of Solids, Liquids and gases, Volumes II and III*. London: Academic Press.

Li, D. (2003). Vibroacoustic behaviour and noise control studies of advanced composite structures. PhD thesis, University of Pittsburgh.

Li, K. M. (1993). On the validity of the heuristic ray–trace–based modification to the Weyl–Van der Pol formula. *Journal of the Acoustical Society of America*, **93**, 1727–1735.

Li, K. M. (1994). A high frequency approximation of sound propagation in a stratified moving atmosphere above a porous ground surface. *Journal of the Acoustical Society of America*, **95**, 1840–1852.

Li, K. M., Taherzadeh, S. and Attenborough, K. (1998). An improved ray–tracing algorithm for predicting sound propagation outdoors. *Journal of the Acoustical Society of America*, **104**, 2077–2083.

Li, X. and Hansen, C. H. (2005). Comparison of models for predicting the transmission loss of plenum chambers. *Applied Acoustics*, **66**, 810-828.

Lighthill, J. M. (1952). On sound generated aerodynamically. In *Proceedings of the Royal Society (London)*, **A211**, 564.

Lighthill, J. M. (1978). *Waves in Fluids*. Cambridge: Cambridge University Press.

Lighthill, J. M. (1991). Biomechanics of hearing sensitivity. *Journal of Vibration and Acoustics*, **113**, 1–13.

Lighthill, J. M. (1996). Recent advances in interpreting hearing sensitivity. In *Proceedings of the Fourth ICSV*, St. Petersburg, Russia, 24–27 June, 5–11.

Ljunggren, S. (1991). Airborne sound insulation of thick walls. *Journal of the Acoustical Society of America*, **89**, 2238–2345.

Long, M. (2008). Sound system design. *Acoustics Today*, **4**, 23–30.

Lubman, D. (1969). Fluctuations of sound with position in a reverberation room. *Journal of the Acoustical Society of America*, **44**, 1491–1502.

Lubman, D. (1974). Precision of reverberant sound power measurement. *Journal of the Acoustical Society of America*, **56**, 523–533.

Lyman, T. (ed.) (1961). *Metals Handbook*. Metals Park, OH: American Society for Metals.

Lynch, C. T. (ed.) (1980). *Handbook of Materials Science*. Boca Raton, FL: CRC Press.

Lyon, R. H. (1975). *Statistical Energy Analysis of Dynamical Systems: Theory and Application*. Cambridge, MA: MIT Press.

Mackenzie. R. (ed.) (1979). *Auditorium Acoustics*. London: Elsevier Applied Science.

Macrae, J. H. (1991). Presbycusis and noise-induced permanent threshold shift. *Journal of the Acoustical Society of America*, **90**, 2513–2516.

Maekawa, Z. (1968). Noise reduction by screens. *Applied Acoustics*, **1**, 157–173.

Maekawa, Z. (1977). Shielding highway noise. *Noise Control Engineering*, **9**, 384.

Maekawa, Z. (1985). Simple estimation methods for noise reduction by various shaped barriers. *Conference on Noise Control Engineering*, Krakow, Poland, 24–27 September.

Magnusson, P. C. (1965). *Transmission Lines and Wave Propagation*. Boston, MA: Allyn and Bacon.

Maidanik, G. (1962). Response of ribbed panels to reverberant acoustic fields. *Journal of the Acoustical Society of America*, **34**, 809–826.

Mammano, F. and Nobili, R. (1993). Biophysics of the cochlea: linear approximation. *Journal of the Acoustical Society of America*, **93**, 3320–3332.

Manning, C. J. (1981). *The Propagation of Noise from Petroleum: and Petrochemical Complexes to Neighbouring Communities*. CONCAWE Report No. 4/81.

Marburg, S. and Nolte, B. (2008). *Computational Acoustics of Noise Propagation in Fluids: Finite and Boundary Element Methods*. Berlin: Springer Verlag.

Marsh, K. J. (1976). Specification and prediction of noise levels in oil refineries and petrochemical plants. *Applied Acoustics*, **9**, 1–15.

Marsh, K. J. (1982). The CONCAWE model for calculating the propagation of noise from open air industrial plants. *Applied Acoustics*, **15**, 411–428.

McGary, M. C. (1988). A new diagnostic method for separating airborne and structure-borne noise radiated by plates with application to propeller aircraft. *Journal of the Acoustical Society of America*, **84**, 830–840.

McLachlan, N. W. (1941). *Bessel Functions for Engineers*. London: Oxford University Press.

Mead, D.J. (1998). *Passive Vibration Control*. New York: John Wiley & Sons.

Meirovitch, L., Baruh, H. and Oz. H. (1983). A comparison of control techniques for large flexible systems. *Journal of Guidance*, **6**, 302–310.

Menge, C. W., Rossano, C. F., Anderson, G. S. and Bajdek, C. J. (1998). *FHWA Traffic Noise Model, Version 1.0, Technical Manual*. US Dept. Transportation, Washington, DC.

Menounou, P. (2001). A correction to Maekawa's curve for the insertion loss behind barriers. *Journal of the Acoustical Society of America*, **110**, 1828–1838.

Meyer, E. and Neumann, E. (1972). *Physical and Applied Acoustics* (translated by J. M. Taylor). New York: Academic Press, Chapter 5.

Meyer, E., Mechel, F. and Kurtez, G. (1958) Experiments on the influence of flow on sound attenuation in absorbing ducts. *Journal of the Acoustical Society of America*, **30**, 165–174.

Michelsen, R., Fritz, K. R. and Sazenhofan, C. V. (1980). Effectiveness of acoustic pipe wrappings (in German). In *Proceedings of DAGA '80*. Berlin: VDE-Verlag, 301–304.

Moore, B. C. J. (1982). *Introduction to the Psychology of Hearing*, 2nd edn. New York: Academic Press.

Moreland, J. and Minto, R. (1976). An example of in-plant noise reduction with an acoustical barrier. *Applied Acoustics*, **9**, 205–214.

Morse, P. M. (1939). The transmission of sound inside pipes. *Journal of the Acoustical Society of America*, **11**, 205–210.

Morse, P. M. (1948). *Vibration and Sound*, 2nd edn. New York: McGraw Hill.

Morse, P. M. and Bolt, R. H. (1944). Sound waves in rooms. *Reviews of Modern Physics*, **16**, 65–150.

Morse, P. M. and Ingard, K. U. (1968). *Theoretical Acoustics*. New York: McGraw Hill.

Mouratidis, E. and Becker, J. (2004). The acoustic properties of common HVAC plena. *ASHRAE Transactions*, **110**, Part 2, 597-606.

Mungar, P. and Gladwell, G. M. L. (1968). Wave propagation in a sheared fluid contained in a duct. *Journal of Sound and Vibration*, **9**, 28–48.

Mustin, G. S. (1968). *Theory and Practice of Cushion Design.* The Shock and Vibration Information Centre, US Department of Defense.

Nahin, P. J. (1990). Oliver Heaviside, *Scientific American*, **262**, June, 122–129.

National Electrical Manufacturer's Association (1980). *Transformer Regulators and Reactors,* NEMA TRI–1980.

Naylor, G. (1993). ODEON: Another hybrid room acoustical model. *Applied Acoustics,* **38**, 131–143.

Neise, W. (1975). Theoretical and experimental investigations of microphone probes for sound measurements in turbulent flow. *Journal of Sound and Vibration*, **39**, 371–400.

Neubauer, R. O. (2000). Estimation of reverberation times in non-rectangular rooms with non-uniformly distributed absorption using a modified Fitzroy equation. In *Proceedings of the 7th International Congress on Sound and Vibration*, Garmisch-Partenkirchen, Germany, July, pp. 1709–1716.

Neubauer, R. O. (2001). Existing reverberation time formulae - a comparison with computer simulated reverberation times. In *Proceedings of the 8th International Congress on Sound and Vibration*, Hong Kong, July, 805-812.

Ng, K. W. (1980). Control valve noise generation and prediction. In *Proceedings of Noisexpo, National Noise and Vibration Control Conference*, Chicago, April, 49–54.

Nicolas, J., Embleton, T. F. W. and Piercy, J. E. (1983). Precise model measurements vs. theoretical prediction of barrier insertion loss in the presence of the ground. *Journal of the Acoustical Society of America*, **73**, 44–54.

Nilsson, A. (2001). Wave propagation and sound transmission in sandwich composite plates. In *Proceedings of the 8th International Congress on Sound and Vibration*, Hong Kong, July, pp. 61–70.

Nishihara, N., Hidaka, T. and Beranek, L. L. (2001). Mechanism of sound absorption by seated audience in halls. *Journal of the Acoustical Society of America*, **110**, 2398–2411.

Olsen, E. S. (2005). Acoustical solutions in the design of a measurement microphone for surface mounting. *Technical Review No. 1, 2005.* Copenhagen: Brüel and Kjær.

Pan, J. and Bies, D. A. (1988). An experimental investigation into the interaction between a sound field and its boundaries. *Journal of the Acoustical Society of America*, **83**, 1436–1444.

Pan, J. and Bies, D. A. (1990a). The effect of fluid-structural coupling on sound waves in an enclosure: theoretical part. *Journal of the Acoustical Society of America*, **87**, 691–707.

Pan, J. and Bies, D. A. (1990b). The effect of fluid-structural coupling on sound waves in an enclosure: experimental part. *Journal of the Acoustical Society of America*, **87**, 708–717.

Pan, J. and Bies, D. A. (1990c). The effect of fluid-structural coupling on acoustical decays in a reverberation room in the high-frequency range. *Journal of the Acoustical Society of America*, **87**, 718–727.

Pan, J. and Bies, D. A. (1990d). The effect of a semi-circular diffuser on the sound field in a rectangular room. *Journal of the Acoustical Society of America*, **88**, 1454–1458.

Panton, R. L. and Miller, J. M. (1975). Resonant frequencies of cylindrical Helmholtz resonators. *Journal of the Acoustical Society of America*, **57**, 1533–1535.

Parkin, P. and Scholes, W. (1965). The horizontal propagation of sound from a jet engine close to the ground, at Hatfield. *Journal of Sound and Vibration*, **2**, 353.

Parkins, J. W. (1998). Active minimization of energy density in a three-dimensional enclosure. PhD thesis, Pennsylvania State University, USA.

Passchier-Vermeer, W. (1968). *Hearing Loss Due to Exposure to Steady State Broadband Noise*. Report No. 36. Institute for Public Health Eng., The Netherlands.

Passchier-Vermeer, W. (1977). *Hearing Levels of Non-Noise Exposed Subjects and of Subjects Exposed to Constant Noise During Working Hours*. Report B367, Research Institute for Environmental Hygiene, The Netherlands.

Pavic,G. (2006). Experimental identification of physical parameters of fluid-filled pipes using acoustical signal processing. *Applied Acoustics*, **67**, 864–881.

Peterson, B. and Plunt, J. (1982). On effective mobilities in the prediction of structure-borne sound transmission between a source structure and receiving structure: Parts 1 and 2. *Journal of Sound and Vibration*, **82**, 5171.

Peutz, V. M. A. (1971). Articulation loss of consonants as a criterion for speech transmission in a room. *Journal of the Audio Engineering Society*, **19**, 915–919.

Pickles, J. O. (1984). *An Introduction to the Physiology of Hearing*. New York: Academic Press.

Pierce, A. D. (1981). *Acoustics: An Introduction to its Physical Principles and Applications*. New York: McGraw Hill, Chapter 5.

Piercy, J., Embleton, T. and Sutherland, L. (1977). Review of noise propagation in the atmosphere. *Journal of the Acoustical Society of America*, **61**, 1403–1418.

Pinnington, R. J. and White, R. G. (1981). Power flow through machine isolators to resonant and non-resonant beams. *Journal of Sound and Vibration*, **75**, 179–197.

Pirinchieva, R. K. (1991). Model study of sound propagation over ground of finite impedance. *Journal of the Acoustical Society of America*, **90**, 2678–2682.

Plovsing, B. (1999). Outdoor sound propagation over complex ground. In *Proceedings of the Sixth International Congress on Sound and Vibration*, Copenhagen, Denmark, 685–694.

Plunkett, R. (1954). Experimental measurement of mechanical impedance or mobility. *Journal of Applied Mechanics*, **21**, 256.

Plunkett, R. (1958). Interaction between a vibrating machine and its foundation. *Noise Control*, **4**, 18–22.

Pobol, O. (1976). Method of measuring noise characteristics of textile machines. *Measurement Techniques, USSR*, **19**, 1736–1739.

Pope, J., Hickling, R., Feldmaier, D. A. and Blaser, D. A. (1981). *The Use of Acoustic Intensity Scans for Sound Power Measurements and for Noise Source Identification in Surface Transportation Vehicles*. Society of Automotive Engineers, Technical paper 810401.

Price, A. J. and Crocker, M. J. (1970). Sound transmission through double panels using Statistical energy analysis. *Journal of the Acoustical Society of America*, **47**, 154–158.

Quirt, J. D. (1982). Sound transmission through windows. I: single and double glazing. *Journal of the Acoustical Society of America*, **72**, 834–844.

Ramakrishnan, R. and Watson, W. R. (1991). Design curves for circular and annular duct silencers. *Noise Control Engineering Journal*, **36**, 107–120.

Rao, S. S. (1986). *Mechanical Vibrations*. Reading, MA: Addison-Wesley.

Randall, R. B. (1987). *Frequency Analysis*. Copenhagen: Brüel and Kjaer.

Raney, J. P. and Cawthorn, J. M. (1998). Aircraft noise, Chapter 47 in *Handbook of Acoustical Measurements and Noise Control*, 3rd edn reprint, edited by C. M. Harris, Acoustical Society of America, New York.

Raspet, R. and Wu, W. (1995). Calculation of average turbulence effects on sound propagation based on the fast field program formulation. *Journal of the Acoustical Society of America*, **97**, 147–153.

Raspet, R., L'Esperance, A. and Daigle, G.A. (1995). The effect of realistic ground impedance on the accuracy of ray tracing. *Journal of the Acoustical Society of America,* **97**, 683–693.

Rathe, E. J. (1969). Note on two common problems of sound propagation. *Journal of Sound and Vibration*, **10**, 472–479.

Reynolds, D. D. (1981). *Engineering Principles of Acoustics (Noise and Vibration Control)*. Boston, MA: Allyn and Bacon. See Chapters 5, 6 and 7.

Rice, C. G. (1974). Damage risk criteria for impulse and impact noise. In *Noise Shock and Vibration Conference Proceedings*, J. D. C. Crisp (ed.), 29–38. Monash University, Clayton, Victoria, Australia.

Rice, C. G. and Martin, A. M. (1973). Impulse noise damage risk criteria. *Journal of Sound and Vibration*, **28**, 359–367.

Richards, E. J. (1979). On the prediction of impact noise. Part 2: Ringing noise. *Journal of Sound and Vibration*, **65**, 419–451.

Richards, E. J. (1980). Vibration and noise relationships: some simple rules for the machinery engineer. *Journal of the Acoustical Society of America*, **68** (Supplement 1), S23 (abstract).

Richards, E. J. (1981). On the prediction of impact noise. Part 3: energy accountancy in industrial machines. *Journal of Sound and Vibration*, **76**, 187–232.

Rindel, J. H. and Hoffmeyer, D. (1991). Influence of stud distance on sound insulation of gypsum board walls. In *Proceedings of Internoise '91*, 279–282.

Royster, L. H., Royster, J. D. and Thomas, W. G. (1980). Representative hearing levels by race and sex in North Carolina industry. *Journal of the Acoustical Society of America*, **68**, 551–566.

Rudnick, I. (1947). The propagation of an acoustic wave along a boundary. *Journal of the Acoustical Society of America*, **19**, 348–356

Rudnick, I. (1957). Propagation of sound in the open air. *Handbook of Noise Control*, C. M. Harris (ed.). New York: McGraw Hill.

Sabine, W. C. (1993). *Collected Papers on Acoustics*. New York: Am. Inst. Physics.

Sablik, M. J. (1985). Statistical energy analysis, structural resonances and beam networks. *Journal of the Acoustical Society of America*, **77**, 1038–1045.

Sandberg, U. (2001). *Noise Emissions of Road Vehicles: Effect of Regulations*. Final Report 01-1 of the I-INCE Working Party on Noise Emissions of Road Vehicles. International Institute of Noise Control Engineering.

Saunders, R. E., Samuels, S. E., Leach, R. and Hall, A. (1983). *An Evaluation of the U.K. DoE Traffic Noise Prediction Method*. Research Report ARR No. 122. Australian Road Research Board, Vermont South, VIC., Australia.

Scharf, B. (1970). Critical bands. In *Foundations of Modern Auditory Theory*, Vol. 1, (ed. J. V. Tobias).

Schenck, H. (1968). Improved integral formulation for acoustic radiation problem. *Journal of the Acoustical Society of America*, **44**, 41-58.

Schomer, P. (2000). Proposed revisions to room noise criteria. *Noise Control Engineering Journal,* **48**, 85–96.

Schroeder, M. R. (1969). Effect of frequency and space averaging on the transmission responses of multimode media. *Journal of the Acoustical Society of America*, **46**, 277–283.

Selamet, A. and Ji, Z. L. (2000). Circular asymmetric Helmholtz resonators. *Journal of the Acoustical Society of America*, **107**, 2360–2369.

Semple, E. C. and Hall, R. E. I. (1981). Mechanisms of noise generation in punch presses and means by which this noise can be reduced. In *Proceedings of the Annual Conference of the Australian Acoustical Society*. Paper 1B3. Sydney: Australian Acoustical Society.

Sendra, J. J. (1999). *Computational Acoustics in Architecture*. Southampton: WIT.

Sharp, B. H. (1973). *A Study of Techniques to Increase the Sound Installation of Building Elements*. Wylie Laboratories Report WR 73-S, prepared for Department of Housing and Urban Development, Washington, DC, under contract H-1095.

Sharp, B. H. (1978). Prediction methods for the sound transmission of building elements. *Noise Control Engineering*, **11**, 5533.

Shimode, S. and Ikawa, K. (1978). Prediction method for noise reduction by plant buildings. In *Proceedings of Internoise '78*, 521.

Siebein, G. W. (1994). Architectural acoustics. Tutorial presented at the 127th meeting of the Acoustical Society of America, *Journal of the Acoustical Society of America*, **95**, 2930.

Simonds, H. R. and Ellis, C. (1943). *Handbook of Plastics*. New York: D. Van Nostrand Company Inc.

Singh, S., Hansen, C. H. and Howard, C, Q. (2006). The elusive cost function for tuning adaptive Helmholtz resonators. *Proceedings of First Australasian Acoustical Societies' Conference*, 20-22 November, Christchurch, New Zealand, 75–82.

Smith, B. J. (1971). *Acoustics*. London: Longman.

Smollen, L. E. (1966). Generalized matrix method for the design and analysis of vibration isolation systems. *Journal of the Acoustical Society of America*, **40**, 195–204.

Snowdon, J. C. (1965). Rubber-like materials, their internal damping and role in vibration isolation. *Journal of Sound and Vibration*, **2**, 175–193.

Society of Automotive Engineers (1962). Aerospace Shock and Vibration Committee G-5, *Design of Vibration Isolation Systems*.

Söderqvist, S. (1982). A quick and simple method for estimating the transmission or insertion loss of an acoustic filter. *Applied Acoustics*, **15**, 347–354.

Soedel, W. (1993) *Vibrations of plates and shells*, 2nd ed. New York: Marcel Dekker.

Soom, A. and Lee, M. (1983). Optimal design of linear and nonlinear vibration absorbers for damped systems. *Journal of Vibration, Acoustics, Stress and Reliability in Design*, **105**, 112–119.

Spoendlin, H. (1975). Neuroanatomical basis of cochlear coding mechanisms. *Audiology*, **14**, 383–407.

Steele, C. (2001). A critical review of some traffic noise prediction models. *Applied Acoustics*, **62**, 271-287.

Stephens, R. W. B. and Bate, A. E. (1950). *Wave Motion and Sound*. London: Edward Arnold.

Stevens, S. S. (1957). On the psychophysical law. *Psychological Review*, **64**, 153–181.

Stevens, S. S. (1961). Procedure for calculating loudness: Mark VI. *Journal of the Acoustical Society of America*, **33**, 1577–1585.

Stevens, S. S. (1972). Perceived level of noise by Mark VII and decibels (E). *Journal of the Acoustical Society of America*, **51**, 575–601.

Sutherland, L. (1979). *A Review of Experimental Data in Support of a Proposed New Method for Computing Atmospheric Absorption Losses*. DOT-TST-75-87.

Sutherland, L. and Bass, H. E. (1979). influence of atmospheric absorption on the propagation of bands of noise. *Journal of the Acoustical Society of America*, **66**, 885–894.

Sutherland, L. C., Piercy, J. F., Bass, H. E. and Evans, L. B. (1974). Method for calculating the absorption of sound by the atmosphere *Journal of the Acoustical Society of America*, **56**, Supplement I (abstract).

Sutton, M. (1990). Noise directivity of exhaust stacks. Final year thesis for the Honours Degree of Bachelor of Engineering, University of Adelaide, South Australia.

Sutton, P. (1976). Process noise: evaluation and control. *Applied Acoustics*, **9**, 17–38

Sutton, O. G. (1953). *Micrometeorology*. New York: MGraw-Hill.

Swanson, D. C. (1991). A stability robustness comparison of adaptive feed-forward and feedback control algorithms. In *Proceedings of Recent Advances in Active Control of Sound and Vibration*, 754–767.

Tadeu, A. J. B. and Mateus, D. M. R. (2001). Sound transmission through single, double and triple glazing. Experimental evaluation. *Applied Acoustics*, **62**, 307–325.

Takagi, K. and Yamamoto, K. (1994). Calculation methods for road traffic noise propagation proposed by ASJ. In *Proceedings of Internoise '94*. Yokohama, Japan, 289–294.

Takatsubo, J., Ohno, S and Suzuki, T. (1983). Calculation of the sound pressure produced by structural vibration using the results of vibration analysis. *Bulletin of the Japanese Society of Mechanical Engineers*, **26**, 1970–1976.

Templeton, D. (ed.) (1993). *Acoustics in the Built Environment*. Oxford: Butterworth – Heinemann.

Thomasson, S. I. (1978). Diffraction by a screen above an impedance boundary. *Journal of the Acoustical Society of America*, **63**, 1768–1781.

Tonin, R. (1985). Estimating noise levels from petrochemical plants, mines and industrial complexes. *Acoustics Australia*, **13**, 59–67.

Tournour, M. and Atalla, N. (2000). Pseudostatic corrections for the forced vibroacoustic response of structure-cavity system. *Journal of the Acoustical Society of America*, **107**, 2379-2386.

Tse, F. S., Morse, I. E. and Hinkle, R. T. (1979). *Mechanical Vibrations: Theory and Applications*, 2nd edn. Boston: Allyn and Bacon.

Tweed, L. W. and Tree, D. R. (1976). Three methods for predicting the insertion loss of close fitting acoustical enclosures. *Noise Control Engineering*, **10**, 74–79.

Tyzzer, F. G. and Hardy, H. C. (1947). Properties of felt in the reduction of noise and vibration. *Journal of the Acoustical Society of America*, **19**, 872–878.

U.K. DOT (1988). *Calculation of Road Traffic Noise*. Department of Transport. London: HMSO.

U.K. DOT (1995a). *Calculation of Railway Noise*. Department of Transport. London: HMSO.

U.K. DOT (1995b). *Calculation of Railway Noise. Supplement 1*. Department of Transport. London: HMSO.

Ungar, E. E. and Dietrich, C. W. (1966). High-frequency vibration isolation. *Journal of Sound and Vibration*, **4**, 224–241.

Van der Perre, G. (1991). Dynamic analysis of human bones. In *Functional Behaviour of Orthopedic Materials*, Vol. 1, Boca Raton, FL: CRC Press, Chapter 5.

Ver, I. L. (1983). *Prediction of Sound Transmission through Duct Walls: Breakout and Pickup*. Cambridge, MA: Bolt, Beranek and Newman, Report 5116.

Voldřich, L. (1978). Mechanical Properties of Basilar Membrane. *Acta Otolaryngologica*, **86**, 331–335.

von Estorff O. (ed.) (2000) *Boundary Elements in Acoustics*: *Advances and Applications*. Southampton: WIT Press.

von Gierke, H. E., Robinson, D. and Karmy, S. J. (1982). Results of the workshop on impulse noise and auditory hazard. *Journal of Sound and Vibration*, **83**, 579–584.

Wallace, C. E. (1972). Radiation resistance of a rectangular panel. *Journal of the Acoustical Society of America*, **51**, 946–952.

Wang, B. T., Dimitradis, E. K. and Fuller, C. R. (1990). Active control of structurally radiated noise using multiple piezoelectric actuators. In *Proceedings of AIAA SDM Conference*, AIAA paper 90-1172-CP, 2409–2416.

Ward, W. D. (1962). Damage risk criteria for line spectra. *Journal of the Acoustical Society of America*, **34**, 1610–1619.

Ward, W. D. (1974). The safe workday noise dose. In *Noise Shock and Vibration Conference Proceedings*, J. D. C. Crisp (ed.). Monash: Monash University, Clayton, Victoria, pp. 19–28.

Waterhouse, R. V. (1955). Interference patterns in reverberant sound fields. *Journal of the Acoustical Society of America*, **27**, 247–258.

Watters, B. G., Labate, S. and Beranek, L. L. (1955). Acoustical behavior of some engine test cell structures. *Journal of the Acoustical Society of America*, **27**, 449–456.

Watters, B. G., Hoover, R. M. and Franken, P. A. (1959). Designing a muffler for small engines. *Noise Control*, **5**, 18–22.

Wells, R. J. (1958). Acoustical plenum chambers. *Noise Control*, **4**, 9–15.

West, M. and Parkin, P. (1978). Effect of furniture and boundary conditions on the sound attenuation in a landscaped office. *Applied Acoustics*, **11**, 171–218.

Widrow, B. (2001). A Microphone Array for Hearing Aids. *Echoes, The newsletter of the Acoustical Society of America*, **11** (3) Summer 2001.

Wiener, F.M. and Keast, D.D. (1959). Experimental study of the propagation of sound over ground. *Journal of the Acoustical Society of America*, **31**, 724–733.

Williams, E. G. (1999). *Fourier Acoustics – Sound Radiation and Nearfield Acoustical Holography*. San Diego, CA: Academic Press.

Windle, R. M. and Lam, Y. W. (1993). Prediction of the sound reduction of profiled metal cladding. In *Proceedings of Internoise* '93, Leuven, Belgium.

Woebcken, W. (1995). *International Plastics Handbook*, 3rd edn. New York: Hanser Publishers.

Wu, Q. (1988). Empirical relations between acoustical properties and flow resistivity of porous plastic open-cell foam. *Applied Acoustics*, **25**, 141–148.

Wu, S. F. (2000). On reconstruction of acoustic pressure fields using the Helmholtz equation least squares method. *Journal of the Acoustical Society of America*, **107**, 2511–2522.

Wu, T. W. (ed.) (2000). *Boundary Element Acoustics: Fundamentals and Computer Codes*. Boston :WIT Press.

Xiangyang, Z., Kean, C., Jincai, S. (2003). On the accuracy of the ray-tracing algorithms based on various sound receiver models. *Applied Acoustics*, **64**, 433-441.

Yoshioka, H. (2000). Evaluation and prediction of airport noise in Japan. *Journal of the Acoustical Society of Japan* (E), **21**, 341–344.

Zander, A. C. and Hansen, C. H. (1992). Active control of higher order acoustic modes in ducts. *Journal of the Acoustical Society of America*, **92**, 244–257.

Zaporozhets, O. I. and Tokarev, V. I. (1998). Aircraft noise modelling for environmental assessment around airports. *Applied Acoustics*, **55**, 99–127.

Zhao, Y-Q., Chen, S-Q., Chai, S. and Qu., Q-W. (2002). An improved modal truncation method for responses to harmonic excitation. *Computers and Structures*, **80**, 99-103.

Zinoviev, A. and Bies, D. A. (2003). On acoustic radiation by a rigid object in a fluid flow. *Journal of Sound and Vibration*, **269**, 535-548.

Zwicker, E. (1958). Über psychologische und methodische Grundlagen der Lautheirt. *Acustica*, **8**, 237–258.

Zwicker, E. and Scharf B. (1965). A model of loudness summation. *Psychological Review*, **72**, 3–26.

List of Acoustical Standards

AMERICAN MILITARY STANDARDS

MIL-STD-1474D – 1991. Noise limits for military material.

AMERICAN NATIONAL STANDARDS INSTITUTE

A full list of ANSI standards relevant to acoustics is provided at
http://asa.aip.org/standards/NatCat.pdf

AMERICAN SOCIETY FOR TESTING OF MATERIALS

ASTM C384-04. Standard Test Method for Impedance and Absorption of Acoustical
Materials by Impedance Tube Method.

ASTM C423-08a. Standard Test Method for Sound Absorption and Sound Absorption
Coefficients by the Reverberation Room Method.

ASTM C522-03. Standard Test Method for Airflow Resistance of Acoustical Materials.

ASTM C634-08a. Standard Terminology Relating to Building and Environmental
Acoustics.

ASTM C919-08. Standard Practice for Use of Sealants in Acoustical Applications.

ASTM C1423-98(2003). Standard Guide for Selecting Jacketing Materials for Thermal
Insulation.

ASTM C1534-07. Standard Specification for Flexible Polymeric Foam Sheet Insulation
Used as a Thermal and Sound Absorbing Liner for Duct Systems.

ASTM E90-04. Standard Test Method for Laboratory Measurement of Airborne Sound
Transmission Loss of Building Partitions and Elements .

ASTM E336-07. Standard Test Method for Measurement of Airborne Sound
Attenuation between Rooms in Buildings.

ASTM E413-04. Classification for Rating Sound Insulation.

ASTM E477-06a. Standard Test Method for Measuring Acoustical and Airflow
Performance of Duct Liner Materials and Prefabricated Silencers.

ASTM E492-04. Standard Test Method for Laboratory Measurement of Impact Sound
Transmission Through Floor-Ceiling Assemblies Using the Tapping Machine.

ASTM E596-96(2002)e1. Standard Test Method for Laboratory Measurement of Noise
Reduction of Sound-Isolating Enclosures.

ASTM E756-05. Standard Test Method for Measuring Vibration-Damping Properties
of Materials.

ASTM E795-05. Standard Practices for Mounting Test Specimens During Sound
Absorption Tests.

ASTM E966-04. Standard Guide for Field Measurements of Airborne Sound Insulation
of Building Facades and Facade Elements.

ASTM E989-06. Standard Classification for Determination of Impact Insulation Class (IIC).

ASTM E1007-04e1. Standard Test Method for Field Measurement of Tapping Machine Impact Sound Transmission Through Floor-Ceiling Assemblies and Associated Support Structures.

ASTM E1014-08. Standard Guide for Measurement of Outdoor A-Weighted Sound Levels.

ASTM E1050-08. Standard Test Method for Impedance and Absorption of Acoustical Materials Using A Tube, Two Microphones and A Digital Frequency Analysis System.

ASTM E1110-06. Standard Classification for Determination of Articulation Class.

ASTM E1111-07. Standard Test Method for Measuring the Interzone Attenuation of Open Office Components

ASTM E1123-86(2003). Standard Practices for Mounting Test Specimens for Sound Transmission Loss Testing of Naval and Marine Ship Bulkhead Treatment Materials.

ASTM E1124-97(2004). Standard Test Method for Field Measurement of Sound Power Level by the Two-Surface Method.

ASTM E1130-08. Standard Test Method for Objective Measurement of Speech Privacy in Open Plan Spaces Using Articulation Index.

ASTME1179-07. Standard Specification for Sound Sources Used for Testing Open Office Components and Systems.

ASTM E1222-90(2002). Standard Test Method for Laboratory Measurement of the Insertion Loss of Pipe Lagging Systems.

ASTM E1264-08. Standard Classification for Acoustical Ceiling Products.

ASTM E1265-04. Standard Test Method for Measuring Insertion Loss of Pneumatic Exhaust Silencers.

ASTM E1332-90(2003). Standard Classification for Determination of Outdoor-Indoor Transmission Class.

ASTM E1374-06. Standard Guide for Open Office Acoustics and Applicable ASTM Standards.

ASTM E1408-91(2000). Standard Test Method for Laboratory Measurement of the Sound Transmission Loss of Door Panels and Door Systems.

ASTM E1414-06. Standard Test Method for Airborne Sound Attenuation Between Rooms Sharing a Common Ceiling Plenum.

ASTM E1425-07. Standard Practice for Determining the Acoustical Performance of Windows, Doors, Skylight, and Glazed Wall Systems.

ASTM E1433-04. Standard Guide for Selection of Standards on Environmental Acoustics.

ASTM E1503-06. Standard Test Method for Conducting Outdoor Sound Measurements Using a Digital Statistical Sound Analysis System.

ASTM E1573-02. Standard Test Method for Evaluating Masking Sound in Open Offices Using A-Weighted and One-Third Octave Band Sound Pressure Levels.

ASTM E1574-98(2006). Standard Test Method for Measurement of Sound in Residential Spaces.

ASTM E1686-03. Standard Guide for Selection of Environmental Noise Measurements and Criteria.

ASTM E1704-95(2002). Standard Guide for Specifying Acoustical Performance of Sound-Isolating Enclosures.

ASTM E1779-96a(2004). Standard Guide for Preparing a Measurement Plan for Conducting Outdoor Sound Measurements.

ASTM E1780-04. Standard Guide for Measuring Outdoor Sound Received from a Nearby Fixed Source.

ASTM E2179-03e2. Standard Test Method for Laboratory Measurement of the Effectiveness of Floor Coverings in Reducing Impact Sound Transmission Through Concrete Floors.

ASTM E2202-02. Standard Practice for Measurement of Equipment-Generated Continuous Noise for Assessment of Health Hazards.

ASTM E2235-04e1. Standard Test Method for Determination of Decay Rates for Use in Sound Insulation Test Methods.

ASTM E2249-02. Standard Test Method for Laboratory Measurement of Airborne Transmission Loss of Building Partitions and Elements Using Sound Intensity.

ASTM E2459-05. Standard Guide for Measurement of In-Duct Sound Pressure Levels from Large Industrial Gas Turbines and Fans.

ASTM F1334-08. Standard Test Method for Determining A-Weighted Sound Power Level of Vacuum Cleaners.

ASTM F2154-01(2007). Standard Specification for Sound-Absorbing Board, Fibrous Glass, Perforated Fibrous Glass Cloth Faced.

ASTM E2459-05. Standard Guide for Measurement of In-Duct Sound Pressure Levels from Large Industrial Gas Turbines and Fans.

ASTM F1334-08 Standard Test Method for Determining A-Weighted Sound Power Level of Vacuum Cleaners.

ASTM F2039-00(2006). Standard Guide for Basic Elements of Shipboard Occupational Health and Safety Program.

ASTM F2544-06e1. Standard Test Method for Determining A-Weighted Sound Power Level of Central Vacuum Power Units.

AUSTRALIAN STANDARDS

AS ISO 140.4-2006. Acoustics - Measurement of sound insulation in buildings and of building elements - Field measurements of airborne sound insulation between rooms (ISO 140-4:1998, MOD).

AS ISO 140.6-2006. Acoustics - Measurement of sound insulation in buildings and of building elements - Laboratory measurements of impact sound insulation of floors.

AS/NZS ISO 140.7:2006. Acoustics - Measurement of sound insulation in buildings and of building elements - Field measurements of impact sound insulation of floors (ISO 140-7:1998, MOD).

AS ISO 140.8-2006. Acoustics - Measurement of sound insulation in buildings and of building elements - Laboratory measurements of the reduction of transmitted impact noise by floor coverings on a heavyweight standard floor.

AS/NZS ISO 717.1:2004. Acoustics - Rating of sound insulation in buildings and of building elements - Airborne sound insulation.

AS ISO 717.2-2004. Acoustics - Rating of sound insulation in buildings and of building elements - Impact sound insulation.

AS 1055.1-1997. Acoustics - Description and measurement of environmental noise - General procedures.

AS 1055.2-1997. Acoustics - Description and measurement of environmental noise - Application to specific situations.

AS 1055.3-1997. Acoustics - Description and measurement of environmental noise - Acquisition of data pertinent to land use.

AS 1081.1-1990. Acoustics - Measurement of airborne noise emitted by rotating electrical machinery - Engineering method for free-field conditions over a reflective plane.

AS 1081.2-1990. Acoustics - Measurement of airborne noise emitted by rotating electrical machinery - Survey method.

AS 1191-2002. Acoustics - Method for laboratory measurement of airborne sound transmission insulation of building elements.

AS/NZS 1270:2002. Acoustics - Hearing protectors.

AS 1277-1983. Acoustics - Measurement procedures for ducted silencers.

AS/NZS 1591.1:1995. Acoustics - Instrumentation for audiometry - Reference zero for the calibration of pure-tone bone conduction audiometers.

AS/NZS 1591.4:1995. Acoustics - Instrumentation for audiometry - A mechanical coupler for calibration of bone vibrators.

AS/NZS 1935.1:1998. Acoustics - Determination of sound absorption coefficient and impedance in impedance tubes - Method using standing wave ratio.

AS ISO 1999-2003. Acoustics - Determination of occupational noise exposure and estimation of noise-induced hearing impairment.

AS 2012.1-1990. Acoustics - Measurement of airborne noise emitted by earth-moving machinery and agricultural tractors - Stationary test condition - Determination of compliance with limits for exterior noise.

AS 2012.2-1990. Acoustics - Measurement of airborne noise emitted by earth-moving machinery and agricultural tractors - Stationary test condition - Operator's position.

AS 2021-2000. Acoustics - Aircraft noise intrusion - Building siting and construction.

AS/NZS 2107:2000. Acoustics - Recommended design sound levels and reverberation times for building interiors.

AS 2254-1988. Acoustics - Recommended noise levels for various areas of occupancy in vessels and offshore mobile platforms.

AS 2363-1999. Acoustics - Measurement of noise from helicopter operations.

AS 2377-2002. Acoustics - Methods for the measurement of railbound vehicle noise.

AS/NZS 2399:1998. Acoustics - Specifications for personal sound exposure meters.

AS/NZS 2460:2002. Acoustics - Measurement of the reverberation time in rooms.

AS/NZS 2499:2000. Acoustics - Measurements of sound insulation in buildings and of buildings elements - Laboratory measurement of room-to-room airborne sound insulation of a suspended ceiling with a plenum above it.

AS 2533-2002. Acoustics - Preferred frequencies and band centre frequencies.

AS 2702-1984. Acoustics - Methods for the measurement of road traffic noise.

AS 2822-1985. Acoustics - Methods of assessing and predicting speech privacy and speech intelligibility.

AS 2900.7-2002. Quantities and units - Acoustics.

AS 2991.2-1991. Acoustics - Method for the determination of airborne noise emitted by household and similar electrical appliances - Particular requirements for dishwashers.

AS 3534-1988. Acoustics - Methods for measurement of airborne noise emitted by powered lawnmowers, edge and brush cutters and string trimmers.

AS ISO 354-2006. Acoustics - Measurement of sound absorption in a reverberation room.

AS 3657.2-1996. Acoustics - Expression of the subjective magnitude of sound or noise - Method for calculating loudness level.

AS 3671-1989. Acoustics - Road traffic noise intrusion - Building siting and construction.

AS 3755-1990. Acoustics - Measurement of airborne noise emitted by computer and business equipment.

AS 3756-1990. Acoustics - Measurement of high-frequency noise emitted by computer and business equipment.

AS 3757-1990. Acoustics - Declared noise emission values of computers and business equipment.

AS 3781-1990. Acoustics - Noise labelling of machinery and equipment.

AS 3782.1-1990. Acoustics - Statistical methods for determining and verifying stated noise emission values of machinery and equipment - General considerations and definitions.

AS 3782.2-1990. Acoustics - Statistical methods for determining and verifying stated noise emission values of machinery and equipment - Methods for stated values for individual machines.

AS 3782.3-1990. Acoustics - Statistical methods for determining and verifying stated noise emission values of machinery and equipment - Simple (transition) method for stated values for batches of machines.

AS 3782.4-1990. Acoustics - Statistical methods for determining and verifying stated noise emission values of machinery and equipment - Methods for stated values for batches of machines.

AS/NZS 3817:1998. Acoustics - Methods for the description and physical measurement of single impulses or series of impulses.

AS ISO 389.1-2007. Acoustics - Reference zero for the calibration of audiometric equipment - Reference equivalent threshold sound pressure levels for pure tones and supra-aural earphones.

AS ISO 389.2-2007. Acoustics - Reference zero for the calibration of audiometric equipment - Reference equivalent threshold sound pressure levels for pure tones and insert earphones.

AS ISO 389.3-2007. Acoustics - Reference zero for the calibration of audiometric equipment - Reference equivalent threshold force levels for pure tones and bone vibrators.

AS ISO 389.5-2003. Acoustics - Reference zero for the calibration of audiometric equipment - Reference equivalent threshold sound pressure levels for pure tones in frequency range 8 kHz to 16 kHz.

AS ISO 389.7-2003. Acoustics - Reference zero for the calibration of audiometric equipment - Reference threshold of hearing under free-field and diffuse-field listening conditions.

AS 4241-1994. Acoustics - Instruments for the measurement of sound intensity - Measurement with pairs of pressure sensing microphones.

AS/NZS 4476:1997. Acoustics - Octave-band and fractional-octave-band-filters.

AS ISO 7029-2003. Acoustics - Statistical distribution of hearing thresholds as a function of age.

AS ISO 11654-2002. Acoustics - Rating of sound absorption - Materials and systems.

AS ISO 12124-2003. Acoustics - Procedures for the measurement of real-ear acoustical characteristics of hearing aids.

DR 04039 CP. Acoustics - Measurement of sound insulation in buildings and of building elements - Part 7: Field measurements of impact sound insulation of floors.

DR 05299 CP. Acoustics - Test code for the measurement of airborne noise emitted by power lawn mowers, lawn tractors, lawn and garden tractors, professional mowers, and lawn and garden tractors with mowing attachments.

DR 07153 CP. Acoustics - Measurement, prediction and assessment of noise from wind turbine generators.

DR 08071. Acoustics - Audiometric test methods - Part 1: Basic pure tone air and bone conduction threshold audiometry.

DR 08072. Acoustics - Audiometric test methods - Part 3: Speech audiometry.

INTERNATIONAL ELECTROTECHNICAL COMMISSION

Some IEC standards relevant to audiology are listed at:
http://asa.aip.org/standards/IntnatCat.pdf. Other standards are listed below.

IEC 60034/9 – 1997. Rotating electrical machines. Part 9: Noise limits.

IEC 60050/801 – 1994. International Electrotechnical Vocabulary – Chapter 801: Acoustics and electroacoustics.

IEC 60268-16 – 1998. Sound system equipment – Part 16: Objective rating of speech intelligibility by speech transmission index.

IEC 60303 – 1970. IEC provisional reference coupler for the calibration of earphones used in audiometry.

IEC 60534/1 – 1987. Industrial-process control valves. Part 1: Control valve terminology and general considerations.

IEC 60534/2/1 – 1998. Industrial-process control valves. Part 2: Flow capacity – sizing equations for fluid flow under installed conditions.

IEC 60534/8/1 – 1986. Industrial-process control valves. Part 8: Noise considerations. Section 1: Laboratory measurement of noise generated by aerodynamic flow through control valves.

IEC 60534/8/2 – 1991. Industrial-process control valves. Part 8: Noise considerations. Section 2: Laboratory measurement of noise generated by hydrodynamic flow through control valves.

IEC 60534/8/3 – 2000. Industrial-process control valves. Part 8: Noise considerations. Section 3: Control valve aerodynamic noise prediction method.

IEC 60534/8/4 – 1994. Industrial-process control valves. Part 8: Noise considerations. Section 4: Prediction of noise generated by hydrodynamic flow.

IEC 60551 – 1987. Determination of transformer and reactor sound levels.

IEC 60651 – 1979. Sound level meters.

IEC 60704/1 – 1997. Household and similar electrical appliances. Test code for the determination of airborne acoustical noise. Part 1: General requirements.

IEC 60704/2/1 – 2000. Test code for the determination of airborne acoustical noise emitted by household and similar electrical appliances. Part 2: Particular requirements for vacuum cleaners.

IEC 60704/2/2 – 1985. Test code for the determination of airborne acoustical noise emitted by household and similar electrical appliances. Part 2: Particular requirements for forced draught convection heaters.

IEC 60704/2/3 – 2005. Test code for the determination of airborne acoustical noise emitted by household and similar electrical appliances. Part 2: Particular requirements for dishwashers.

IEC 60704/2/4 – 2001. Test code for the determination of airborne acoustical noise emitted by household and similar electrical appliances. Part 2: Particular requirements for washing machines and spin extractors.

IEC 60704/2/5 – 2005. Test code for the determination of airborne acoustical noise emitted by household and similar electrical appliances. Part 2: Particular requirements for room heaters of the storage type.

IEC 60704/2/6 – 2003. Test code for the determination of airborne acoustical noise emitted by household and similar electrical appliances Part 2: Particular requirements for tumble-dryers.

IEC 60704/2/7 – 1997. Household and similar electrical appliances. Test code for the determination of airborne acoustical noise. Part 2: Particular requirements for fans.

IEC 60704/2/8 – 1997. Household and similar electrical appliances. Test code for the determination of airborne acoustical noise. Part 2: Requirements for electric shavers.

IEC 60704/2/9 – 2003. Household and similar electrical appliances. Test code for the determination of airborne acoustical noise. Part 2-9: Particular requirements for electric hair care appliances.

IEC 60704/2/10 – 2004. Household and similar electrical appliances. Test code for the determination of airborne acoustical noise. Part 2-10: Particular requirements for electric cooking ranges, ovens, grills, microwave ovens and any combination of these.

IEC 60704/2/11 – 1998. Household and similar electrical appliances. Test code for the determination of airborne acoustical noise. Part 2–11: Particular requirements for electrically-operated food preparation.

IEC 60704/2/13 – 2008. Household and similar electrical appliances. Test code for the determination of airborne acoustical noise. Part 2–13: Particular requirements for range hoods.

IEC 60704/3 – 2006. Test code for the determination of airborne acoustical noise emitted by household and similar electrical appliances. Part 3: Procedure for determining and verifying declared noise emission values.

IEC 60942 – 2003. Electroacoustics: sound calibrators.

IEC 61027 – 1991. Instruments for the measurement of aural acoustic impedance/ admittance.

IEC 61043 – 1993. Electroacoustics: instruments for the measurement of sound intensity. Measurements with pairs of pressure sensing microphones.

IEC 61063 – 1991. Acoustics: measurement of airborne noise emitted by steam turbines and driven machinery.

IEC 61094/1 – 2000. Measurement microphones. Part 1: Specifications for laboratory standard microphones.

IEC 61094/4 – 1995. Measurement microphones. Part 4: Specifications for working standard microphones.

IEC 61183 – 1994. Electroacoustics: random-incidence and diffuse-field calibration of sound level meters.

IEC 61252 – 2002. Electroacoustics: specifications for personal sound exp. meters.

IEC 61260 – 1995. Electroacoustics: octave-band and fractional-octave-band filters.

IEC 61265 – 1995. Electroacoustics: instruments for measurement of aircraft noise. Performance requirements for systems to measure one-third-octave-band sound pressure levels in noise certification of transport-category aeroplanes.

IEC 61400/11 – 2006. Wind turbine generator systems. Part 11: Acoustic noise measurement techniques.

IEC 61669 – 2001. Electroacoustics: Equipment for the measurement of real-ear acoustical characteristics of hearing aids.

IEC 61672/1 – 2002 Electroacoustics: Sound level meters Part 1.

INTERNATIONAL STANDARDS ORGANIZATION (INCLUDING EUROPEAN STANDARDS)

A list of most of the ISO standards relevant to acoustics is provided at: http://asa.aip.org/standards/IntnatCat.pdf. Relevant Standards missing from that reference are listed below.

ISO 140-1 – 1997. Acoustics: measurement of sound insulation in buildings and of building elements. Part 1: Requirements for laboratories.

ISO 140-2 – 1991. Acoustics: measurement of sound insulation in buildings and of building elements. Part 2: Determination, verification and application of precision data.

ISO 140-3 – 1995. Acoustics: measurement of sound insulation in buildings and of building elements. Part 3: Laboratory measurements of airborne sound insulation of building elements.

ISO 140-4 – 1998. Acoustics: measurement of sound insulation in buildings and of building elements. Part 4: Field measurements of airborne sound insulation between rooms.

ISO 140-5 – 1998. Acoustics: measurement of sound insulation in building and of building elements. Part 5: Field measurements of airborne sound insulation of facade elements and facades.

ISO 140-6 – 1998. Acoustics: measurement of sound insulation in buildings and of building elements. Part 6: Laboratory measurements of impact sound insulation of floors.

ISO 140-7 – 1998. Acoustics: measurement of sound insulation in buildings and of building elements. Part 7: Field measurements of impact sound insulation of floors.

ISO 140-8 – 1997 Acoustics: measurement of sound insulation in buildings and of building elements. Part 8: Laboratory measurements of the reduction of transmitted impact noise by floor coverings on a standard floor.

ISO 140-9 – 1985. Acoustics: measurement of sound insulation in buildings and of building elements. Part 9: Laboratory measurement of room-to-room airborne sound insulation.

ISO 140-10 – 1991. Acoustics: measurement of sound insulation in buildings and of building elements. Part 10: Laboratory measurement of airborne sound insulation of small building elements.

ISO 140-11 – 2005. Acoustics: Measurement of sound insulation in buildings and of building elements. Part 11: Laboratory measurements of the reduction of transmitted impact sound by floor coverings on lightweight reference floors.

ISO 140-12 – 2000. Acoustics: measurement of sound insulation in buildings and of building elements. Part 12: Laboratory measurement of room-to-room airborne and impact sound insulation of an access floor.

ISO/TR 140-13 – 1997. Acoustics: measurement of sound insulation in buildings and of building elements. Part 13: Guidelines.

ISO 140-14 – 2004. Acoustics: Measurement of sound insulation in buildings and of building elements. Part 14: Guidelines for special situations in the field.

ISO 140-16 – 2006. Acoustics: Measurement of sound insulation in buildings and of building elements -- Part 16: Laboratory measurement of the sound reduction index improvement by additional lining.

ISO 140-18 – 2006. Acoustics: Measurement of sound insulation in buildings and of building elements. Part 18: Laboratory measurement of sound generated by rainfall on building elements.

ISO 354 – 2003. Acoustics: Measurement of sound absorption in a reverberation room.

ISO 717-1 – 1996. Acoustics: rating of sound insulation in buildings and of building elements. Part 1: Airborne sound insulation in buildings and of interior building elements.

ISO 717-2 – 1996. Acoustics: rating of sound insulation in buildings and of building elements. Part 2: Impact sound insulation.

ISO 717-3 – 1982. Acoustics: rating of sound insulation in buildings and of buildings elements. Part 3: Airborne sound insulation of facade elements and facades.

ISO 2151 – 2004. Measurement of airborne noise emitted by compressor/prime mover-units intended for outdoor use.

ISO 3382 – 1997. Acoustics: measurement of the reverberation time of rooms with reference to other acoustical parameters.

ISO 3382-2 – 2008. Acoustics: Measurement of room acoustic parameters. Part 2: Reverberation time in ordinary rooms.

ISO 3822-1 – 1999. Acoustics: laboratory tests on noise emission from appliances and equipment used in water supply installations. Part 1: Method of measurement.

ISO 3822-2 – 1995. Acoustics: Laboratory tests on noise emission from appliances and equipment used in water supply installations. Part 2: Mounting and operating conditions for draw-off taps and mixing valves.

ISO 3822-3 – 1997. Acoustics: laboratory tests on noise emission from appliances and equipment used in water supply installations. Part 3: mounting and operating conditions for in-line valves and appliances.

ISO 3822-4 – 1997. Acoustics: laboratory tests on noise emission from appliances and equipment used in water supply installations. Part 4: Mounting and operating conditions for special appliances.

ISO 5131 – 1996. Acoustics: tractors and machinery for agriculture and forestry. Measurement of noise at the operator's position. Survey method.

ISO 9052-1 – 1989. Acoustics: determination of dynamic stiffness. Part 1: Materials used under floating floors in dwellings.

ISO 9053 – 1991. Acoustics: materials for acoustical applications. Determination of airflow resistance.

ISO 10534-1 – 1996. Acoustics: determination of sound absorption coefficient and impedance in impedance tubes. Part 1: Method using standing wave ratio.

ISO 10534-2 – 1998. Acoustics: determination of sound absorption coefficient and impedance in impedance tubes. Part 2: Transfer-function method.

ISO 10848-1:2006. Acoustics: Laboratory measurement of the flanking transmission of airborne and impact sound between adjoining rooms. Part 1: Frame document.

ISO 10848-2 – 2006. Acoustics: Laboratory measurement of the flanking transmission of airborne and impact sound between adjoining rooms. Part 2: Application to light elements when the junction has a small influence.

ISO 10848-3 – 2006. Acoustics: Laboratory measurement of the flanking transmission of airborne and impact sound between adjoining rooms. Part 3: Application to light elements when the junction has a substantial influence.

ISO 11654 – 1997. Acoustics: sound absorbers for use in buildings. Rating of sound absorption.

ISO 15186-1 – 2000. Acoustics: measurement of sound insulation in buildings and of building elements using sound intensity: Part 1: Laboratory measurements.

ISO 15186-2 – 2003. Acoustics: Measurement of sound insulation in buildings and of building elements using sound intensity. Part 2: Field measurements.

ISO 15186-3 – 2002. Acoustics: Measurement of sound insulation in buildings and of building elements using sound intensity . Part 3: Laboratory measurements at low frequencies.

ISO 15712-1 – 2005. Building acoustics: Estimation of acoustic performance of buildings from the performance of elements. Part 1: Airborne sound insulation between rooms.

ISO 15712-2 – 2005. Building acoustics: Estimation of acoustic performance of buildings from the performance of elements. Part 2: Impact sound insulation between rooms.

ISO 15712-3 – 2005. Building acoustics: Estimation of acoustic performance of buildings from the performance of elements. Part 3: Airborne sound insulation against outdoor sound.

ISO 15712-4 – 2005. Building acoustics: Estimation of acoustic performance of buildings from the performance of elements. Part 4: Transmission of indoor sound to the outside.

ISO 16032 – 2004. Acoustics: Measurement of sound pressure level from service equipment in buildings. Engineering method.

ISO 18233 – 2006. Acoustics: Application of new measurement methods in building and room acoustics.

ISO/TR 25417 – 2007. Acoustics: Definitions of basic quantities and terms.

ISO 80000-8 – 2007. Quantities and units: Part 8: Acoustics.

Index